Essential Readings in

Evolutionary Biology

Essential Readings in Evolutionary Biology

Edited by

Francisco J. Ayala *and* John C. Avise

JOHNS HOPKINS UNIVERSITY PRESS
Baltimore

Printed in the United States of America on acid-free paper
9 8 7 6 5 4 3 2 1

Johns Hopkins University Press
2715 North Charles Street
Baltimore, Maryland 21218-4363
www.press.jhu.edu

Library of Congress Cataloging-in-Publication Data

Essential readings in evolutionary biology / edited by Francisco J. Ayala and John C. Avise.
pages cm
Includes index.
ISBN-13: 978-1-4214-1305-1 (hardcover : alk. paper)
ISBN-10: 1-4214-1305-1 (hardcover : alk. paper) 1. Evolution (Biology)—History. 2. Biology—History. I. Ayala, Francisco José, 1934– editor of compilation. II. Avise, John C., editor of compilation.
QH361.E87 2014
576.8—dc23 2013027718

A catalog record for this book is available from the British Library.

Special discounts are available for bulk purchases of this book. For more information, please contact Special Sales at 410-516-6936 or specialsales@press.jhu.edu.

Johns Hopkins University Press uses environmentally friendly book materials, including recycled text paper that is composed of at least 30 percent post-consumer waste, whenever possible.

Contents

Introduction ix
A Brief Timeline of Evolutionary Thought xi

On the origin of species by means of natural selection (excerpt) 1
C. Darwin (1859)

The descent of man and selection in relation to sex (excerpt) 16
C. Darwin (1871)

A new factor in evolution 32
J. M. Baldwin (1896)

The incidence of alkaptonuria: a study in chemical individuality 44
A. E. Garrod (1902)

Mendelian proportions in a mixed population 61
G. H. Hardy (1908)

Sex limited inheritance in *Drosophila* 64
T. H. Morgan (1910)

Artificial transmutation of the gene 68
H. J. Muller (1927)

The genetical theory of natural selection: a review 73
S. Wright (1930)

A critique of the species concept in biology 82
T. Dobzhansky (1935)

Genetic control of biochemical reactions in *Neurospora* 95
G. W. Beadle and E. L. Tatum (1941)

The species concept 104
G. G. Simpson (1951)

Molecular structure of nucleic acids: a structure for deoxyribose nucleic acid 119
J. D. Watson and F. H. C. Crick (1953)

Hybridization as an evolutionary stimulus 122
E. Anderson and G. L. Stebbins (1954)

An experimental study of interaction between genetic drift and natural selection. 134
T. Dobzhansky and O. Pavlovsky (1957)

An equilibrium theory of insular zoogeography 144
R. H. MacArthur and E. O. Wilson (1963)

The genetical evolution of social behavior: I 160
W. D. Hamilton (1964)

Butterflies and plants: a study in coevolution 184
P. R. Ehrlich and P. H. Raven (1964)

A molecular approach to the study of genic heterozygosity in natural populations: II. amount of variation and degree of heterozygosity in natural populations of *Drosophila pseudoobscura* 201
R. C. Lewontin and J. L. Hubby (1966)

Construction of phylogenetic trees 217
W. M. Fitch and E. Margoliash (1967)

Gene regulation for higher cells: a theory 224
R. J. Britten and E. H. Davidson (1969)

Protein polymorphism as a phase of molecular evolution 234
M. Kimura and T. Ohta (1971)

Punctuated equilibria: an alternative to phyletic gradualism 238
N. Eldredge and S. J. Gould (1972)

Nothing in biology makes sense except in the light of evolution 273
T. Dobzhansky (1973)

The logic of animal conflict 279
J. Maynard Smith and G. R. Price (1973)

Parent-offspring conflict 284
R. L. Trivers (1974)

Evolution at two levels in humans and chimpanzees 301
M.-C. King and A. C. Wilson (1975)

Evolution and tinkering 312
F. Jacob (1977)

Phylogenetic structure of the prokaryotic domain: the primary kingdoms 319
C. R. Woese and G. E. Fox (1977)

Mitochondrial DNA clones and matriarchal phylogeny within and among geographic populations of the pocket gopher, *Geomys pinetis* 323
J. C. Avise, C. Giblin-Davidson, J. Laerm, J. C. Patton, and R. A. Lansman (1979)

The spandrels of San Marco and the Panglossian paradigm: a critique of the adaptationist programme 329
S. J. Gould and R. C. Lewontin (1979)

Is a new evolutionary synthesis necessary? 348
G. L. Stebbins and F. J. Ayala (1981)

Biological classification: toward a synthesis of opposing methodologies 354
E. Mayr (1981)

Sexual selection, social competition, and speciation 362
M. J. West-Eberhard (1983)

The significance of responses of the genome to challenge 392
B. McClintock (1984)

On the virtues and pitfalls of the molecular evolutionary clock 403
F. J. Ayala (1986)

Intraspecific phylogeography: the mitochondrial DNA bridge between population genetics and systematics 414
J. C. Avise, J. Arnold, R. Martin Ball, E. Bermingham, T. Lamb, J. E. Neigel, C. A. Reeb, and N. C. Saunders (1987)

The evolution of sex and recombination 449
B. Charlesworth (1989)

Experimental phylogenetics: generation of a known phylogeny 454
D. M. Hillis, J. J. Bull, M. E. White, M. R. Badgett, I. J. Molineux (1992)

Gaia, nature worship and biocentric fallacies 459
G. C. Williams (1992)

Homeotic genes and the evolution of arthropods and chordates 468
S. B. Carroll (1995)

Transposable elements as sources of variation in animals and plants 476
M. G. Kidwell and D. Lisch (1997)

The chimeric eukaryote: origin of the nucleus from the karyomastigont in amitochondriate protists 485
L. Margulis, M. F. Dolan, and R. Guerrero (2000)

The evolutionary fate and consequences of duplicate genes 492
M. Lynch and J. S. Conery (2000)

Ecology and the origin of species 498
D. Schluter (2001)

Microbial minimalism: genome reduction in bacteria pathogens 508
N. Moran (2002)

Unpredictable evolution in a 30-year study of Darwin's finches 513
P. R. Grant and B. R. Grant (2002)

Genetics and the making of *Homo sapiens* 519
S. B. Carroll (2003)

Evolution experiments with microorganisms: the dynamics and genetic bases of adaptation 529
S. F. Elena and R. E. Lenski (2003)

Epilogue. Science and the Public 543

Index 545

Color illustrations follow the index

Introduction

A famous phrase in the biological sciences conveys a pithy truth: "Nothing in biology makes sense except in the light of evolution." When the evolutionary geneticist Theodosius Dobzhansky penned those words in 1973 (reproduced in this volume), he was wrestling with a striking contradiction: an overwhelming majority of scientists fully acknowledge evolution, whereas a thorough acceptance of evolution by the general populace remains notoriously low in many countries (including the United States). This disconnect has several causes, not the least being a widely perceived antagonism between objective science and faith-based religion. But, as Dobzhansky pointed out, it is possible to be an evolutionist and a theist: "Evolution is God's, or Nature's method of creation."

It remains true that modern science has not been particularly effective in conveying the many wonders of the evolutionary sciences to a broader audience. This book represents one attempt to ameliorate this educational shortfall. We provide an historical overview of evolutionary biology by reprinting nearly 50 classical papers from the field. These seminal articles helped to chart the course of scientific developments in evolution across the one-and-a-half centuries since Darwin.

When we conceived this project, we were surprised to discover that no such compilation of essential readings in evolutionary biology existed, despite the centrality of evolutionary thought in biology, and despite the availability of analogous collections of classical reprints in several biological fields such as ecology and wildlife management and conservation.

We begin this book with a short chapter that encapsulates the broad timeline of evolutionary thought across the past two millennia, beginning with the philosophers of ancient Greece and Rome, continuing to the dawn of modern evolutionary ideas in the late nineteenth century, and culminating with an entrance into the ongoing era of modern genomics that began in the waning years of the twentieth century. After this historical overview, the book then proceeds to its bulk: a collection of 48 papers reprinted from the primary scientific literature of evolutionary biology. Each classic paper is introduced by a commentary in which we explain why the article is important, place it into a broader conceptual context, and provide a short list of source papers that readers might wish to consult for further information.

The seminal reprints are arranged in chronological order, beginning with selected readings from Darwin's (1859) *On the Origin of Species* and *The Descent of Man and Selection in Relation to Sex* (1871). The readings then proceed through the dawn of the twentieth century with such accomplishments as the birth of biochemical genetics, the rediscovery of Mendelian principles, and an elucidation of the chromosomal basis of inheritance. During the first half of the twentieth century, notable achievements included an illumination of the nature of mutations, the conception and development of mathematical population genetics, the origin of biological species concepts, and an early account of biochemical operations in metabolic networks. Near midcentury, the so-called Modern Synthesis was ascendant, which integrated Darwinian thought with Mendelian principles, amalgamated mathematical population genetics with organismal biology, and incorporated ecological and speciation principles with all of the above.

The 1960s and 1970s witnessed the inception and early growth of the molecular revolution in evolutionary biology, which included such path-breaking developments as the rise of neutrality theory and the inauguration of the selection/neutrality controversies, the birth of molecular phylogenetics, and early ideas about possible contrasts between the evolution of biological molecules versus organismal phenotypes. The 1970s through the 1990s were also a time when many tenets of the Modern Synthesis were revisited in the light of new information from molecular biology, developmental biology, and possible levels at which natural selection operates with greatest efficacy.

Other noteworthy happenings included a controversy centered on punctuated equilibrium versus phyletic gradualism, the birth and gestation of phylogeography, which seeks to unify the study of microevolution (population genetics) with macroevolution (phylogenetics), the delineation of many branches in the tree of life, reconsideration of the evolutionary significance of sexual recombination, the emergence of "evo-devo" (the ontogenetic basis of evolutionary change), and the proliferation of experimental approaches to evolutionary studies. The 1900s concluded with the rise of genomic approaches to evolutionary analyses, thanks to vast improvements in DNA-sequencing techniques and genetic annotation methods. These efforts continue today as we enter the era of population genomics in which the management and interpretation of molecular data have become even more demanding than data acquisition itself. The book concludes with a brief epilogue on the challenge of conveying evolutionary reasoning to the broader public.

We hope that this collection of classic reprints will be a useful reference for students and practitioners of the evolutionary sciences and that it will also inform a broader audience of science historians and biologists in related disciplines such as genetics, ecology, molecular biology, and developmental biology. After all, "there is grandeur in this view of life" (Darwin 1859), and nothing in biology makes sense except in evolution's light.

RELATED READING

Darwin, C. 1859. On the origin of species by means of natural selection. Murray, London, England.

Real, L. A., and J. H. Brown, editors. 1991. Foundations of ecology: classic papers with commentaries. University of Chicago Press, Chicago, Illinois, USA.

Krausman, P. R., and B. D. Leopold. 2013. Essential readings in wildlife management and conservation. Johns Hopkins University Press, Baltimore, Maryland, USA.

A Brief Timeline of Evolutionary Thought

ca. 600 BC. The idea that different kinds of organisms can be transformed one into another, including humans into animals and dragons into human form, is a familiar theme in the mythology of many cultures. Among the philosophers of ancient Greece, Anaximander proposed that animals could metamorphose from one kind into another, and Empedocles speculated that organisms were made up of various combinations of preexisting parts. Some of these new organisms would be successful and thus become, by a kind of natural selection, those that continue to exist.

ca. 400 BC. The Greek philosopher Plato advanced that the objects, including organisms, that we perceive with our senses are imperfect representations of *forms*, which are perfect and timeless but transcend our perceptions.

ca. 350 BC. Plato's student Aristotle envisioned a *scala naturae,* or ladder of life, to establish the relationships of living things. Taoist philosophers in China speculated that species develop different features in response to their environments and that nature is always in flux.

ca. 70 BC. The Roman philosopher Titus Lucretius Carus asserted that the development of living things occurred by means of purely naturalistic (rather than preternatural) mechanisms.

ca. 400 AD. Augustine of Hippo, a Christian theologian, maintained that not all species of plants and animals were initially created by God; rather, some had evolved in historical times from God's creations.

9th century. The Muslim biologist and philosopher al-Jahiz described a struggle for existence among animals as being influenced by their environmental conditions.

12th and 13th centuries. Christian thinkers updated the ancient Greek concept of *scala naturae* to arrange God-made forms of life into a great chain or hierarchy from lowly worms to intermediate humans to higher angels and heaven. Albertus Magnus (1200–1280) and his student, Thomas Aquinas (1224–1274), asserted that Genesis and other Christian scriptures should not be read literally. Aquinas concluded, after consideration of the arguments, that the development of living creatures, such as maggots and flies, from nonliving matter, such as decaying meat, was not incompatible with Christian faith or philosophy, although he left it to others to decide whether this actually happened.

14th century. The Islamic writer Ibn Khaldun mused that humans may have developed from monkeys and also discussed broader notions of ascent and descent along a biological continuum.

17th and 18th centuries. Several theories with evolutionary overtones were proposed by various philosophers and biologists including René Descartes, Benoît de Maillet, Gottfried Leibniz, J. G. Herder, and Pierre Louis Maupertuis. Such theories generally were in opposition to the Christian wisdom of the time (natural theology and the argument from design) that species had been sculpted in their current forms by special Creation.

1762. The word *evolution* (from the Latin *evolutio*, "to unroll like a scroll") was introduced by Charles Bonnet, albeit strictly applied to the embryological development of individual organisms.

18th century. The French philosopher and naturalist Comte de Buffon referred to species as well-marked varieties that became modified through time through environmental factors; another French philosopher, Denis Diderot, somewhat anticipated natural selection when he wrote that species were constantly changing through trial and error (after having originally arisen via spontaneous generation).

18th and 19th centuries. Work by geologists—including James Hutton, William Smith, Alexandre Brongniart, and Georges Cuvier—helped to establish the Earth's great antiquity. Cuvier described fossils, making it evident that species could go extinct. In a published essay (*On the Principle of Population*) that was to greatly influence Charles Darwin, Thomas Malthus emphasized the capacity of organismal reproduction to outstrip available resources, leading to a "struggle for existence."

1796. Erasmus Darwin (Charles Darwin's grandfather) expressed evolutionary notions that involved connections among diverse animals in a temporal framework.

1802. William Paley published a highly influential book, *Natural Theology*, claiming that God had designed features of life for the functional purposes they serve.

1809. Jean-Baptiste Lamarck advanced the first complete theory of evolution, asserting a progression over time from simpler to more complex kinds of organisms by use and disuse in response to specific changes in the environment.

1831. The word *evolution* is used for the first time in the modern sense of species' change through time by Etienne Geoffroy Saint-Hilaire.

1830s. Charles Lyell challenged accepted scenarios of catastrophism by taking a uniformitarian stance that the Earth's geologic features register cumulative effects over time of the same gradual geologic forces that are observable today. Charles Darwin went on a five-year voyage aboard the *HMS Beagle* and began a series of private notebooks on the transmutation of species.

1844. The Scottish publisher Robert Chambers proposed an evolutionary scenario for origins of the solar system and life on Earth.

1858. Theories on natural selection by Charles Darwin and Alfred Russel Wallace are read jointly before the Linnean Society of London.

1859. Charles Darwin publishes the first edition of *On the Origin of Species.*

1871. Charles Darwin publishes *The Descent of Man, and Selection in Relation to Sex* in which he extends his evolutionary concepts to *Homo sapiens* and identifies sexual selection as a key driver (in addition to natural selection) of the evolutionary process.

2nd half of the 19th century. The British biologist Thomas H. Huxley defends and promotes Darwin's ideas to sometimes skeptical audiences. Evolution by natural selection becomes the mainstream scientific explanation for biological adaptations and the origin of species. Ernst Haeckel, a German biologist, launches a program to reconstruct the evolutionary history of life. Methodological naturalism or scientific materialism gradually supersedes divine intervention as a favored explanation for evolutionary processes. Neo-Lamarckism (the inheritance of acquired characteristics) continues to enjoy favor as a mechanism of heredity underlying evolution.

1900. Gregor Mendel's work on hereditary factors from the mid-1860s is rediscovered, demonstrating the laws of inheritance.

Early 20th century. Biologists split into two camps (Mendelians and Biometricians) regarding how Mendel's factors (genes) might underlie organismal phenotypes and evolution.

1920s–1940s. Theoretical work by population geneticists Ronald Fisher, J. B. S. Haldane, and Sewall Wright resolve the Mendelian/Biometrician controversy and pave the way for what would later become known as the modern evolutionary synthesis. Other major contributors to this synthesis are Theodosius Dobzhansky (genetics), Ernst Mayr (animals and systematics), George Gaylord Simpson (paleontology), and G. Ledyard Stebbins (plants).

2nd half of the 20th century. Technological advances allow molecular genetics and genomics to be added to the evolutionary synthesis.

Essential Readings in

Evolutionary Biology

Darwin, C. 1859. *On the origin of species by means of natural selection*. Murray, London, England. Chapter III, pp. 60–69; Chapter IV, pp. 80–81, 87–90.

Charles Darwin (1809–1882) occupies a prominent place in the history of Western thought, deservedly receiving credit for the theory of evolution. On December 27, 1831, a few months after graduating from the University of Cambridge, Darwin sailed as a naturalist on a round-the-world trip on the HMS *Beagle* that lasted until October 2, 1836. He often disembarked for extended trips ashore to collect natural specimens. The discovery of fossil bones from large extinct mammals in Argentina and the observation of giant tortoises and numerous species of finches in the Galápagos Islands were among the events that stimulated Darwin's interest in how species originate. The diversity of natural selection, Darwin's' awareness that it was a greatly significant discovery because it provided an explanation of the design of organisms, and Darwin's designation of natural selection as "my theory" can be traced in his "Red Notebook" and "Transmutation Notebooks B to E," which he started in March 1837, a few months after returning from his voyage on the *Beagle*, and completed in late 1839.

It was in the *Origin*, however, published in 1859, that he assembled the evidence demonstrating the evolution of organisms. In doing so, Darwin accomplished something much more important for intellectual history than demonstrating evolution. The *Origin* is, first and foremost, a sustained effort to account scientifically for the "design" of organisms, their complexity, diversity, and marvelous contrivances as the result of natural processes, which he accomplishes with his theory of natural selection. The evidence for evolution is brought in because it is a necessary consequence of natural selection as the explanation of design. Evolutionary change through time and evolutionary diversification ensue as byproducts of natural selection's fostering the adaptation of organisms to their milieu.

The introduction and chapters I through VIII of the *Origin* explain how natural selection accounts for the adaptations and behaviors of organisms. Natural selection implies that biological evolution occurs, which Darwin therefore seeks to demonstrate in most of the remainder of the book (chapters IX–XIII). In the concluding chapter, XIV, Darwin returns to the dominant theme of adaptation and design and concludes with an eloquent paragraph asserting that "there is grandeur in this view of life . . . , from so simple a beginning endless forms most beautiful and most wonderful have been, and are being evolved."

RELATED READING

Barlow, N., editor. 1946. Charles Darwin and the voyage of the *Beagle*. Philosophical Library, New York, New York, USA.

Berra, T. M. 2009. Charles Darwin: the concise story of an extraordinary man. Johns Hopkins University Press, Baltimore, Maryland, USA.

Desmond, A., and J. Moore. 1991. Darwin: the life of a tormented evolutionist. Warner Books, New York, New York, USA.

Eldredge, N. 2004. Darwin: discovering the tree of life. Norton, New York, New York, USA.

Jones, S. 1999. Darwin's ghost: the *Origin of Species* updated. Doubleday, London, England.

Reznick, D. N. 2010. The *Origin* then and now: an interpretive guide to the *Origin of Species*. Princeton University Press, Princeton, New Jersey, USA.

Weiner, J. 1994. The beak of the finch: a story of evolution in our time. Alfred Knopf, New York, New York, USA.

On the Origin of Species

BY

CHARLES DARWIN

Before entering on the subject of this chapter, I must make a few preliminary remarks, to show how the struggle for existence bears on Natural Selection. It has been seen in the last chapter that amongst organic beings in a state of nature there is some individual variability; indeed I am not aware that this has ever been disputed. It is immaterial for us whether a multitude of doubtful forms be called species or sub-species or varieties; what rank, for instance, the two or three hundred doubtful forms of British plants are entitled to hold, if the existence of any well-marked varieties be admitted. But the mere existence of individual variability and of some few well-marked varieties, though necessary as the foundation for the work, helps us but little in understanding how species arise in nature. How have all those exquisite adaptations of one part of the organisation to another part, and to the conditions of life, and of one distinct organic being to another being, been perfected? We see these beautiful co-adaptations most plainly in the woodpecker and missletoe; and only a little less plainly in the humblest parasite which clings to the hairs of a quadruped or feathers of a bird; in the structure of the beetle which dives through the water; in the plumed seed which is wafted by the gentlest breeze; in short, we see beautiful adaptations everywhere and in every part of the organic world.

Again, it may be asked, how is it that varieties, which I have called incipient species, become ultimately converted into good and distinct species, which in most cases obviously differ from each other far more than do the varieties of the same species? How do those groups of species, which constitute what are called distinct genera, and which differ from each other more than do the species of the same genus, arise? All these results, as we shall more fully see in the next chapter, follow inevitably from the struggle for life. Owing to this struggle for life, any variation, however slight and from whatever cause proceeding, if it be in any degree profitable to an individual of any species, in its infinitely complex relations to other organic beings and to external nature, will tend to the preservation of that individual, and will generally be inherited by its offspring. The offspring, also, will thus have a better chance of surviving, for, of the many individuals of any species which are periodically born, but a small number can survive. I have called this principle, by which each slight variation, if useful, is preserved, by the term of Natural Selection, in order to mark its relation to man's power of selection. We have seen that man by selection can certainly produce great results, and can adapt organic beings to his own uses, through the accumulation of slight but useful variations, given to him by the hand of Nature. But Natural Selection, as we shall hereafter see, is a power incessantly ready for action, and is as immeasurably superior to man's feeble efforts, as the works of Nature are to those of Art.

We will now discuss in a little more detail the struggle for existence. In my future work this subject shall be treated, as it well deserves, at much greater length. The elder De Candolle and Lyell have largely and philosophically shown that all organic beings are exposed to severe competition. In regard to plants, no one has treated this subject with more spirit and ability than W. Herbert, Dean of Manchester, evidently the result of his great horticultural knowledge. Nothing is easier than to admit in words the truth of the universal struggle for life, or more difficult—at least I have found it so—than constantly to bear this conclusion in mind. Yet unless it be thoroughly engrained in the mind, I am convinced that the whole economy of nature, with every fact on distribution, rarity, abundance, extinction, and variation, will be dimly seen or quite misunderstood. We behold the face of nature bright with gladness, we often see superabundance of food; we do not see, or we forget, that the birds which are idly singing round us mostly live on insects or seeds, and are thus constantly destroying life; or we forget how largely these songsters, or their eggs, or their nestlings, are destroyed by birds and beasts of prey; we do not always bear in mind, that though food may be now superabundant, it is not so at all seasons of each recurring year.

I should premise that I use the term Struggle for Existence in a large and metaphorical sense, including dependence of one being on another, and including (which is more important) not only the life of the individual, but success in leaving progeny. Two canine animals in a time of dearth, may be truly said to struggle with each other which shall get food and live. But a plant on the edge of a desert is said to struggle for life against the drought, though more properly it should be said to be dependent on the moisture. A plant which annually produces a thousand seeds, of which on an average only one comes to maturity, may be more truly said to struggle with the plants of the same and other kinds which already clothe the ground. The missletoe is dependent on the apple and a few other trees, but can only in a far-fetched sense be said to struggle with these trees, for if too many of these parasites grow on the same tree, it will languish and die. But several seedling missletoes, growing close together on the same branch, may more truly be said to struggle with each other. As the missletoe is disseminated by birds, its existence depends on birds; and it may metaphorically be said to struggle with other fruit-bearing plants, in order to tempt birds to devour and thus disseminate its seeds rather than those of other plants. In these several senses, which pass into each other, I use for convenience sake the general term of struggle for existence.

A struggle for existence inevitably follows from the high rate at which all organic beings tend to increase. Every being, which during its natural lifetime produces several eggs or seeds, must suffer destruction during some period of its life, and during some season or occasional year, otherwise, on the principle of geometrical increase, its numbers would quickly become so inordinately great that no country could support the product. Hence, as more individuals are produced than can possibly survive, there must in every case be a struggle for existence, either one individual with another of the same species, or with the individuals of distinct species, or with the physical conditions of life. It is the doctrine of Malthus applied with manifold force to the whole animal and vegetable kingdoms; for in this case there can be no artificial increase of food, and no prudential restraint from marriage. Although some species may be now increasing, more or less rapidly, in numbers, all cannot do so, for the world would not hold them.

There is no exception to the rule that every organic being naturally increases at so high a rate, that if not destroyed, the earth would soon be covered by the progeny of a single pair. Even slow-breeding man has doubled in twenty-five years, and at this rate, in a few thousand years, there would literally not be standing room for his progeny. Linnæus has calculated that if an annual plant produced only two seeds—and there is no plant so unproductive as this—and their seedlings next year produced two, and so on, then in twenty years there would be a million plants. The elephant is reckoned to be the slowest breeder of all known animals, and I have taken some pains to estimate its probable minimum rate of natural increase: it will be under the mark to assume that it breeds when thirty years old, and goes on breeding till ninety years old, bringing forth three pair of young in this interval; if this be so, at the end of the fifth century there would be alive fifteen million elephants, descended from the first pair.

But we have better evidence on this subject than mere theoretical calculations, namely, the numerous recorded cases of the astonishingly rapid increase of various animals in a state of nature, when circumstances have been favourable to them during two or three following seasons. Still more striking is the evidence from our domestic animals of many kinds which have run wild in several parts of the world: if the statements of the rate of increase of slow-breeding cattle and horses in South-America, and latterly in Australia, had not been well authenticated, they would have been quite incredible. So it is with plants: cases could be given of introduced plants which have become common throughout whole islands in a period of less than ten years. Several of the plants now most numerous over the wide plains of La Plata, clothing square leagues of surface almost to the exclusion of all other plants, have been introduced from Europe; and there are plants which now range in India, as I hear from Dr. Falconer, from Cape Comorin to the Himalaya, which have been imported from America since its discovery. In such cases, and endless instances could be given, no one supposes that the fertility of these animals or plants has been suddenly and temporarily increased in any sensible degree. The obvious explanation is that the conditions of life have been very favourable, and that there has consequently been less destruction of the old and young, and that nearly all the young have been enabled to breed. In such cases the geometrical ratio of increase, the result of which never fails to be surprising, simply explains the extraordinarily rapid increase and wide diffusion of naturalised productions in their new homes.

In a state of nature almost every plant produces seed, and amongst animals there are very few which do not annually pair. Hence we may confidently assert, that all plants and animals are tending to increase at a geometrical ratio, that all would most rapidly stock every station in which they could any how exist, and that the geometrical tendency to increase must be checked by destruction at some period of life. Our familiarity with the larger domestic animals tends, I think, to mislead us: we see no great destruction falling on them, and we forget that thousands are annually slaughtered for food, and that in a state of nature an equal number would have somehow to be disposed of.

The only difference between organisms which annually produce eggs or seeds by the thousand, and those which produce extremely few, is, that the slow-breeders would require a few more years to people, under favourable conditions, a whole district, let it be ever so large. The condor lays a couple of eggs and the ostrich a score, and yet in the same country the condor may be the more numerous of the two: the Fulmar petrel lays but one egg, yet it is believed to be the most numerous bird in the world. One fly deposits hundreds of eggs, and another, like the hippobosca, a single one; but this difference does not determine how many individuals of the two species can be supported in a district. A large number of eggs is of some importance to those species, which depend on a rapidly fluctuating amount of food, for it allows them rapidly to increase in number. But the real importance of a large number of eggs or seeds is to make up for much destruction at some period of life; and this period in the great majority of cases is an early one. If an animal can in any way protect its own eggs or young, a small number may be produced, and yet the average stock be fully kept up; but if many eggs or young are destroyed, many must be produced, or the species will become extinct. It would suffice to keep up the full number of a tree, which lived on an average for a thousand years, if a single seed were produced once in a thousand years, supposing that this seed were never destroyed, and could be ensured to germinate in a fitting place. So that in all cases, the average number of any animal or plant depends only indirectly on the number of its eggs or seeds.

In looking at Nature, it is most necessary to keep the foregoing considerations always in mind—never to forget that every single organic being around us may be said to be striving to the utmost to increase in numbers; that each lives by a struggle at some period of its life; that heavy destruction inevitably falls either on the young or old, during each generation or at recurrent intervals. Lighten any check, mitigate the destruction ever so little, and the number of the species will almost instantaneously increase to any amount. The face of Nature may be compared to a yielding surface, with ten thousand sharp wedges packed close together and driven inwards by incessant blows, sometimes one wedge being struck, and then another with greater force.

What checks the natural tendency of each species to increase in number is most obscure. Look at the most vigorous species; by as much as it swarms in numbers, by so much will its tendency to increase be still further increased. We know not exactly what the checks are in even one single instance. Nor will this surprise any one who reflects how ignorant we are on this head, even in regard to mankind, so incomparably better known than any other animal. This subject has been ably treated by several authors, and I shall, in my future work, discuss some of the checks at considerable length, more especially in regard to the feral animals of South America. Here I will make only a few remarks, just to recall to the reader's mind some of the chief points. Eggs or very young animals seem generally to suffer most, but this is not invariably the case. With plants there is a vast destruction of seeds, but, from some observations which I have made, I believe that it is the seedlings which suffer most from germinating in ground already thickly stocked with other plants. Seedlings, also, are destroyed in vast numbers by various enemies; for instance, on a piece of ground three feet long and two wide, dug and cleared, and where there could be no choking from other plants, I marked all the seedlings of our native weeds as they came up, and out of the 357 no less than 295 were destroyed, chiefly by slugs and insects. If turf which has long been mown, and the case would be the same with turf closely browsed by quadrupeds, be let to grow, the more vigorous plants gradually kill the less vigorous, though fully grown, plants: thus out of twenty species growing on a little plot of turf (three feet by four) nine species perished from the other species being allowed to grow up freely.

The amount of food for each species of course gives the extreme limit to which each can increase; but very frequently it is not the obtaining food, but the serving as prey to other animals, which determines the average numbers of a species. Thus, there seems to be little doubt that the stock of partridges, grouse, and hares on any large estate depends chiefly on the destruction of vermin. If not one head of game were shot during the next twenty years in England, and, at the same time, if no vermin were destroyed, there would, in all probability, be less game than at present, although hundreds of thousands of game animals are now annually killed. On the other hand, in some cases, as with the elephant and rhinoceros, none are destroyed by beasts of prey: even the tiger in India most rarely dares to attack a young elephant protected by its dam.

Climate plays an important part in determining the average numbers of a species, and periodical seasons of extreme cold or drought, I believe to be the most effective of all checks. I estimated that the winter of 1854–55 destroyed four-fifths of the birds in my own grounds; and this is a tremendous destruction, when we remember that ten per cent. is an extraordinarily severe mortality from epidemics with man. The action of climate seems at first sight to be quite independent of the struggle for existence; but in so far as climate chiefly acts in reducing food, it brings on the most severe struggle between the individuals, whether of the same or of distinct species, which subsist on the same kind of food. Even when climate, for instance extreme cold, acts directly, it will be the least vigorous, or those which have got least food through the advancing winter, which will suffer most. When we travel from south to north, or from a damp region to a dry, we invariably see some species gradually getting rarer and rarer, and finally disappearing; and the change of climate being conspicuous, we are tempted to attribute the whole effect to its direct action. But this is a very false view: we forget that each species, even where it most abounds, is constantly suffering enormous destruction at some period of its life, from enemies or from competitors for the same place and food; and if these enemies or competitors be in the least degree favoured by any slight change of climate, they will increase in numbers, and, as each area is already fully stocked with inhabitants, the other species will decrease. When we travel southward and see a species decreasing in numbers, we may feel sure that the cause lies quite as much in other species being favoured, as in this one being hurt. So it is when we travel northward, but in a somewhat lesser degree, for the number of species of all kinds, and therefore of competitors, decreases northwards; hence in going northward, or in ascending a mountain, we far oftener meet with stunted forms, due to the *directly* injurious action of climate, than we do in proceeding southwards or in descending a mountain. When we reach the Arctic regions, or snow-capped summits, or absolute deserts, the struggle for life is almost exclusively with the elements.

That climate acts in main part indirectly by favouring other species, we may clearly see in the prodigious number of plants in our gardens which can perfectly well endure our climate, but which never become naturalised, for they cannot compete with our native plants, nor resist destruction by our native animals.

How will the struggle for existence, discussed too briefly in the last chapter, act in regard to variation? Can the principle of selection, which we have seen is so potent in the hands of man, apply in nature? I think we shall see that it can act most effectually. Let it be borne in mind in what an endless number of strange peculiarities our domestic productions, and, in a lesser degree, those under nature, vary; and how strong the hereditary tendency is. Under domestication, it may be truly said that the whole organisation becomes in some degree plastic. Let it be borne in mind how infinitely complex and close-fitting are the mutual relations of all organic beings to each other and to their physical conditions of life. Can it, then, be thought improbable, seeing that variations useful to man have undoubtedly occurred, that other variations useful in some way to each being in the great and complex battle of life, should sometimes occur in the course of thousands of generations? If such do occur, can we doubt (remembering that many more individuals are born than can possibly survive) that individuals having any advantage, however slight, over others, would have the best chance of surviving and of procreating their kind? On the other hand, we may feel sure that any variation in the least degree injurious would be rigidly destroyed. This preservation of favourable variations and the rejection of injurious variations, I call Natural Selection. Variations neither useful nor injurious would not be affected by natural selection, and would be left a fluctuating element, as perhaps we see in the species called polymorphic.

Sexual Selection.—Inasmuch as peculiarities often appear under domestication in one sex and become hereditarily attached to that sex, the same fact probably occurs under nature, and if so, natural selection will be able to modify one sex in its functional relations to the other sex, or in relation to wholly different habits of life in the two sexes, as is sometimes the case with insects. And this leads me to say a few words on what I call Sexual Selection. This depends, not on a struggle for existence, but on a struggle between the males for possession of the females; the result is not death to the unsuccessful competitor, but few or no offspring. Sexual selection is, therefore, less rigorous than natural selection. Generally, the most vigorous males, those which are best fitted for their places in nature, will leave most progeny. But in many cases, victory will depend not on general vigour, but on having special weapons, confined to the male sex. A hornless stag or spurless cock would have a poor chance of leaving offspring. Sexual selection by always allowing the victor to breed might surely give indomitable courage, length to the spur, and strength to the wing to strike in the spurred leg, as well as the brutal cock-fighter, who knows well that he can improve his breed by careful selection of the best cocks. How low in the scale of nature this law of battle descends, I know not; male alligators have been described as fighting, bellowing, and whirling round, like Indians in a war-dance, for the possession of the females; male salmons have been seen fighting all day long; male stag-beetles often bear wounds from the huge mandibles of other males. The war is, perhaps, severest between the males of polygamous animals, and these seem oftenest provided with special weapons. The males of carnivorous animals are already well armed; though to them and to others, special means of defence may be given through means of sexual selection, as the mane to the lion, the shoulder-pad to the boar, and the hooked jaw to the male salmon; for the shield may be as important for victory, as the sword or spear.

Amongst birds, the contest is often of a more peaceful character. All those who have attended to the subject, believe that there is the severest rivalry between the males of many species to attract by singing the females. The rock-thrush of Guiana, birds of Paradise, and some others, congregate; and successive males display their gorgeous plumage and perform strange antics before the females, which standing by as spectators, at last choose the most attractive partner. Those who have closely attended to birds in confinement well know that they often take individual preferences and dislikes: thus Sir R. Heron has described how one pied peacock was eminently attractive to all his hen birds. It may appear childish to attribute any effect to such apparently weak means: I cannot here enter on the details necessary to support this view; but if man can in a short time give elegant carriage and beauty to his bantams, according to his standard of beauty, I can see no good reason to doubt that female birds, by selecting, during thousands of generations, the most melodious or beautiful males, according to their standard of beauty, might produce a marked effect. I strongly suspect that some well-known laws with respect to the plumage of male and female birds, in comparison with the plumage of the young, can be explained on the view of plumage having been chiefly modified by sexual selection, acting when the birds have come to the breeding age or during the breeding season; the modifications thus produced being inherited at corresponding ages or seasons, either by the males alone, or by the males and females; but I have not space here to enter on this subject.

Thus it is, as I believe, that when the males and females of any animal have the same general habits of life, but differ in structure, colour, or ornament, such differences have been mainly caused by sexual selection; that is, individual males have had, in successive generations, some slight advantage over other males, in their weapons, means of defence, or charms; and have transmitted these advantages to their male offspring. Yet, I would not wish to attribute all such sexual differences to this agency: for we see peculiarities arising and becoming attached to the male sex in our domestic animals (as the wattle in male carriers, horn-like protuberances in the cocks of certain fowls, &c.), which we cannot believe to be either useful to the males in battle, or attractive to the females. We see analogous cases under nature, for instance, the tuft of hair on the breast of the turkey-cock, which can hardly be either useful or ornamental to this bird;—indeed, had the tuft appeared under domestication, it would have been called a monstrosity.

Darwin, C. 1871. *The descent of man, and selection in relation to sex.* Murray, London, England. Chapter III, pp. 67–73, 81–85, 100–102.

Aristotle and other philosophers of classical Greece and Rome, as well as many other philosophers throughout the centuries, held that humans hold moral values by nature. A human is not only *Homo sapiens* but also *Homo moralis*. Darwin fully agrees: "Of all the differences between man and the lower animals, the moral sense or conscience is by far the most important" (*Descent*, p. 67). Biological evolution brings about two important issues: causation and timing. We do not attribute ethical behavior to animals. So, when did ethical behavior come about in human evolution? Did modern humans have an ethical sense from the beginning? Did Neanderthals hold moral values? What about the ancestral species *Homo erectus* and *Homo habilis*?

Darwin distinguishes the moral sense, by which some actions are judged to be moral, from the moral norms, which determine which actions are morally right or morally wrong. How did the moral sense evolve? Was it directly promoted by natural selection? Or did it come about as a byproduct of some other attribute (such as intelligence) that was the direct target of selection. Alternatively, is the moral sense an outcome of cultural evolution rather than of biological evolution? Darwin is quite explicit. The moral sense is a necessary consequence of the exalted intelligence with which humans are endowed: "any animal whatever . . . would inevitably acquire a moral sense or conscience, as soon as its intellectual powers had become as well developed, or nearly as well developed, as in man" (pp. 68–69). Darwin identifies four attributes of the advanced intelligence of human beings that connote a moral sense: sympathy, awareness of the motives and consequences of their actions, language, and habit.

The moral norms, according to Darwin, would not be determined by our biological makeup: "I do not wish to maintain that any strictly social animal, if its intellectual faculties were to become . . . as highly developed as in man, would acquire exactly the same moral sense as ours" (*Descent*, p.70). The norms by which an action is determined to be either right or wrong would be outcomes of cultural evolution. If social organization were as in the honeybees, the "unmarried females would . . . think a sacred duty to kill their brothers, and mothers would strive to kill their fertile daughters" (p. 70).

The Descent of Man was published in two volumes, the second one on sexual selection.

RELATED READING

Desmond, A., and J. Moore. 2009. Darwin's sacred cause: how a hatred of slavery shaped Darwin's views on human evolution. Houghton Mifflin, New York, New York, USA.

Huxley, Thomas H. 1863. Evidence as to man's place in nature. Williams and Norgate, London, England.

Richards, R. J. 1987. Darwin and the emergence of evolutionary theories of mind and behavior. University of Chicago Press, Chicago, Illinois, USA.

Richerson, P. J., and R. Boyd. 2005. Not by genes alone: how culture transformed human evolution. University of Chicago Press, Chicago, Illinois, USA.

Ruse, Michael, editor. 2013. The Cambridge encyclopedia of *Darwin* and evolutionary thought. Cambridge University Press, Cambridge, England.

Wilson, E. O. 2012. The social conquest of Earth. Norton, New York, New York, USA.

THE

DESCENT OF MAN,

AND

SELECTION IN RELATION TO SEX.

BY

CHARLES DARWIN, M.A., F.R.S., Etc.

I FULLY subscribe to the judgment of those writers[1] who maintain that, of all the differences between man and the lower animals, the moral sense or conscience is by far the most important. This sense, as Mackintosh[2] remarks, "has a rightful supremacy over every other principle of human action;" it is summed up in that short but imperious word *ought*, so full of high significance. It is the most noble of all the attributes of man, leading him without a moment's hesitation to risk his life for that of a fellow-creature; or after due deliberation, impelled simply by the deep feeling of right or duty, to sacrifice it in some great cause. Immanuel Kant exclaims, "Duty! Wondrous thought, that workest neither by fond insinuation,

[1] See, for instance, on this subject, Quatrefages, 'Unité de l'Espèce Humaine,' 1861, p. 21, etc.

[2] 'Dissertation on Ethical Philosophy,' 1837, p. 231, etc.

flattery, nor by any threat, but merely by holding up thy naked law in the soul, and so extorting for thyself always reverence, if not always obedience; before whom all appetites are dumb, however secretly they rebel; whence thy original?"[3]

This great question has been discussed by many writers[4] of consummate ability; and my sole excuse for touching on it is the impossibility of here passing it over, and because, as far as I know, no one has approached it exclusively from the side of natural history. The investigation possesses, also, some independent interest, as an attempt to see how far the study of the lower animals can throw light on one of the highest psychical faculties of man.

The following proposition seems to me in a high degree probable—namely, that any animal whatever, endowed with well-marked social instincts,[5] would inevitably acquire a moral sense or conscience, as soon as its intellect-

[3] 'Metaphysics of Ethics,' translated by J. W. Semple, Edinburgh, 1836, p. 136.

[4] Mr. Bain gives a list ('Mental and Moral Science,' 1868, pp. 543–725) of twenty-six British authors who have written on this subject, and whose names are familiar to every reader; to these, Mr. Bain's own name, and those of Mr. Lecky, Mr. Shadworth Hodgson, and Sir J. Lubbock, as well as of others, may be added.

[5] Sir B. Brodie, after observing that man is a social animal ('Psychological Inquiries,' 1854, p. 192), asks the pregnant question, "Ought not this to settle the disputed question as to the existence of a moral sense?" Similar ideas have probably occurred to many persons, as they did long ago to Marcus Aurelius. Mr. J. S. Mill speaks, in his celebrated work, 'Utilitarianism' (1864, p. 46), of the social feelings as a "powerful natural sentiment," and as "the natural basis of sentiment for utilitarian morality;" but, on the previous page, he says, "If, as is my own belief, the moral feelings are not innate, but acquired, they are not for that reason less natural." It is with hesitation that I venture to differ from so profound a thinker, but it can hardly be disputed that the social feelings are instinctive or innate in the lower animals; and why should they not be so in man? Mr. Bain (see, for instance, 'The Emotions and the Will," 1865,

ual powers had become as well developed, or nearly as well developed, as in man. For, *firstly*, the social instincts lead an animal to take pleasure in the society of its fellows, to feel a certain amount of sympathy with them, and to perform various services for them. The services may be of a definite and evidently instinctive nature; or there may be only a wish and readiness, as with most of the higher social animals, to aid their fellows in certain general ways. But these feelings and services are by no means extended to all the individuals of the same species, only to those of the same association. *Secondly*, as soon as the mental faculties had become highly developed, images of all past actions and motives would be incessantly passing through the brain of each individual; and that feeling of dissatisfaction which invariably results, as we shall hereafter see, from any unsatisfied instinct, would arise, as often as it was perceived that the enduring and always present social instinct had yielded to some other instinct, at the time stronger, but neither enduring in its nature, nor leaving behind it a very vivid impression. It is clear that many instinctive desires, such as that of hunger, are in their nature of short duration; and after being satisfied are not readily or vividly recalled. *Thirdly*, after the power of language had been acquired and the wishes of the members of the same community could be distinctly expressed, the common opinion how each member ought to act for the public good, would naturally become to a large extent the guide to action. But the social instincts would still give the impulse to act for the good of the community, this impulse being strengthened, directed, and sometimes even deflected, by public opinion, the power of which rests, as we shall presently see, on in-

p. 481) and others believe that the moral sense is acquired by each individual during his lifetime. On the general theory of evolution this is at least extremely improbable.

stinctive sympathy. *Lastly*, habit in the individual would ultimately play a very important part in guiding the conduct of each member; for the social instincts and impulses, like all other instincts, would be greatly strengthened by habit, as would obedience to the wishes and judgment of the community. These several subordinate propositions must now be discussed; and some of them at considerable length.

It may be well first to premise that I do not wish to maintain that any strictly social animal, if its intellectual faculties were to become as active and as highly developed as in man, would acquire exactly the same moral sense as ours. In the same manner as various animals have some sense of beauty, though they admire widely different objects, so they might have a sense of right and wrong, though led by it to follow widely different lines of conduct. If, for instance, to take an extreme case, men were reared under precisely the same conditions as hive-bees, there can hardly be a doubt that our unmarried females would, like the worker-bees, think it a sacred duty to kill their brothers, and mothers would strive to kill their fertile daughters; and no one would think of interfering. Nevertheless the bee, or any other social animal, would in our supposed case gain, as it appears to me, some feeling of right and wrong, or a conscience. For each individual would have an inward sense of possessing certain stronger or more enduring instincts, and others less strong or enduring; so that there would often be a struggle which impulse should be followed; and satisfaction or dissatisfaction would be felt, as past impressions were compared during their incessant passage through the mind. In this case an inward monitor would tell the animal that it would have been better to have followed the one impulse rather than the other. The one course ought to have been followed: the one would have been right and the other wrong; but to these terms I shall have to recur.

Sociability.—Animals of many kinds are social; we find even distinct species living together, as with some American monkeys, and with the united flocks of rooks, jackdaws, and starlings. Man shows the same feeling in his strong love for the dog, which the dog returns with interest. Every one must have noticed how miserable horses, dogs, sheep, etc., are when separated from their companions; and what affection at least the two former kinds show on their reunion. It is curious to speculate on the feelings of a dog, who will rest peacefully for hours in a room with his master or any of the family, without the least notice being taken of him; but, if left for a short time by himself, barks or howls dismally. We will confine our attention to the higher social animals, excluding insects, although these aid each other in many important ways. The most common service which the higher animals perform for each other, is the warning each other of danger by means of the united senses of all. Every sportsman knows, as Dr. Jaeger remarks,[6] how difficult it is to approach animals in a herd or troop. Wild horses and cattle do not, I believe, make any danger-signal; but the attitude of any one who first discovers an enemy, warns the others. Rabbits stamp loudly on the ground with their hind-feet as a signal: sheep and chamois do the same, but with their fore-feet, uttering likewise a whistle. Many birds and some mammals post sentinels, which in the case of seals are said[7] generally to be the females. The leader of a troop of monkeys acts as the sentinel, and utters cries expressive both of danger and of safety.[8] So-

[6] 'Die Darwin'sche Theorie,' s. 101.

[7] Mr. R. Browne in 'Proc. Zoolog. Soc.' 1868, p. 409.

[8] Brehm, 'Thierleben,' B. i. 1864, s. 52, 79. For the case of the monkeys extracting thorns from each other, see s. 54. With respect to the

cial animals perform many little services for each other: horses nibble, and cows lick each other, on any spot which itches: monkeys search for each other's external parasites; and Brehm states that, after a troop of the *Cercopithecus griseo-viridis* has rushed through a thorny brake, each monkey stretches itself on a branch, and another monkey sitting by "conscientiously" examines its fur and extracts every thorn or burr.

Animals also render more important services to each other: thus wolves and some other beasts of prey hunt in packs, and aid each other in attacking their victims. Pelicans fish in concert. The Hamadryas baboons turn over stones to find insects, etc.; and when they come to a large one, as many as can stand round, turn it over together and share the booty. Social animals mutually defend each other. The males of some ruminants come to the front when there is danger and defend the herd with their horns. I shall also in a future chapter give cases of two young wild-bulls attacking an old one in concert, and of two stallions together trying to drive away a third stallion from a troop of mares. Brehm encountered in Abyssinia a great troop of baboons which were crossing a valley: some had already ascended the opposite mountain, and some were still in the valley: the latter were attacked by the dogs, but the old males immediately hurried down from the rocks, and with mouths widely opened roared so fearfully, that the dogs precipitately retreated. They were again encouraged to the attack; but by this time all the baboons had reascended the heights, excepting a young one, about six months old, who, loudly calling for aid, climbed on a block of rock and was surrounded.

Hamadryas turning over stones, the fact is given (s. 76) on the evidence of Alvarez, whose observations Brehm thinks quite trustworthy. For the cases of the old male baboons attacking the dogs, see s. 79; and, with respect to the eagle, s. 56.

Now one of the largest males, a true hero, came down again from the mountain, slowly went to the young one, coaxed him, and triumphantly led him away—the dogs being too much astonished to make an attack. I cannot resist giving another scene which was witnessed by this same naturalist; an eagle seized a young Cercopithecus, which, by clinging to a branch, was not at once carried off; it cried loudly for assistance, upon which the other members of the troop with much uproar rushed to the rescue, surrounded the eagle, and pulled out so many feathers, that he no longer thought of his prey, but only how to escape. This eagle, as Brehm remarks, assuredly would never again attack a monkey in a troop.

Man a social animal.—Most persons admit that man is a social being. We see this in his dislike of solitude, and in his wish for society beyond that of his own family. Solitary confinement is one of the severest punishments which can be inflicted. Some authors suppose that man primevally lived in single families; but at the present day, though single families, or only two or three together, roam the solitudes of some savage lands, they are always, as far as I can discover, friendly with other families inhabiting the same district. Such families occasionally meet in council, and they unite for their common defence. It is no argument against savage man being a social animal, that the tribes inhabiting adjacent districts are almost always at war with each other; for the social instincts never extend to all the individuals of the same species. Judging from the analogy of the greater number of the Quadrumana, it is probable that the early ape-like progenitors of man were likewise social; but this is not of much importance for us. Although man, as he now exists, has few special instincts, having lost any which his early progenitors may have possessed, this is no reason why he should not have retained from an extremely remote period some degree of instinctive love and sympathy for his fellows. We are indeed all conscious that we do possess such sympathetic feelings;[19] but our con-

[19] Hume remarks ('An Enquiry Concerning the Principles of Morals,' edit. of 1751, p. 132), "there seems a necessity for confessing that the happiness and misery of others are not spectacles altogether indifferent to us, but that the view of the former . . . communicates a secret joy;

sciousness does not tell us whether they are instinctive, having originated long ago in the same manner as with the lower animals, or whether they have been acquired by each of us during our early years. As man is a social animal, it is also probable that he would inherit a tendency to be faithful to his comrades, for this quality is common to most social animals. He would in like manner possess some capacity for self-command, and perhaps of obedience to the leader of the community. He would from an inherited tendency still be willing to defend, in concert with others, his fellow-men, and would be ready to aid them in any way which did not too greatly interfere with his own welfare or his own strong desires.

The social animals which stand at the bottom of the scale are guided almost exclusively, and those which stand higher in the scale are largely guided, in the aid which they give to the members of the same community, by special instincts; but they are likewise in part impelled by mutual love and sympathy, assisted apparently by some amount of reason. Although man, as just remarked, has no special instincts to tell him how to aid his fellow-men, he still has the impulse, and with his improved intellectual faculties would naturally be much guided in this respect by reason and experience. Instinctive sympathy would, also, cause him to value highly the approbation of his fellow-men; for, as Mr. Bain has clearly shown,[20] the love of praise and the strong feeling of glory, and the still stronger horror of scorn and infamy, "are due to the workings of sympathy." Consequently man would be greatly influenced by the wishes, approbation, and blame of his fellow-men, as expressed by their gestures and language. Thus the social instincts, which must have been

the appearance of the latter . . . throws a melancholy damp over the imagination."

[20] 'Mental and Moral Science,' 1868, p. 254.

acquired by man in a very rude state, and probably even by his early ape-like progenitors, still give the impulse to many of his best actions; but his actions are largely determined by the expressed wishes and judgment of his fellow-men, and unfortunately still oftener by his own strong, selfish desires. But as the feelings of love and sympathy and the power of self-command become strengthened by habit, and as the power of reasoning becomes clearer so that man can appreciate the justice of the judgments of his fellow-men, he will feel himself impelled, independently of any pleasure or pain felt at the moment, to certain lines of conduct. He may then say, I am the supreme judge of my own conduct, and, in the words of Kant, I will not in my own person violate the dignity of humanity.

The more enduring Social Instincts conquer the less Persistent Instincts.—We have, however, not as yet considered the main point, on which the whole question of the moral sense hinges. Why should a man feel that he ought to obey one instinctive desire rather than another? Why does he bitterly regret if he has yielded to the strong sense of self-preservation, and has not risked his life to save that of a fellow-creature; or why does he regret having stolen food from severe hunger?

It is evident in the first place, that with mankind the instinctive impulses have different degrees of strength; a young and timid mother urged by the maternal instinct will, without a moment's hesitation, run the greatest danger for her infant, but not for a mere fellow-creature. Many a man, or even boy, who never before risked his life for another, but in whom courage and sympathy were well developed, has, disregarding the instinct of self-preservation, instantaneously plunged into a torrent to save a drowning fellow-creature. In this case man is impelled by the same instinctive motive, which caused the heroic little American monkey, formerly described, to attack the great and dreaded baboon, to save his keeper. Such actions as the above appear to be the simple result of the greater strength of the social or maternal instincts than of any other instinct or motive; for they are performed too instantaneously for reflection, or for the sensation of pleasure or pain; though if prevented distress would be caused.

I am aware that some persons maintain that actions performed impulsively, as in the above cases, do not come under the dominion of the moral sense, and cannot be called moral. They confine this term to actions done deliberately, after a victory over opposing desires, or to actions prompted by some lofty motive. But it appears scarcely possible to draw any clear line of distinction of this kind; though the distinction may be real. As far as exalted motives are concerned, many instances have been recorded of barbarians, destitute of any feeling of general benevolence toward mankind, and not guided by any religious motive, who have deliberately as prisoners sacrificed their lives,[21] rather than betray their comrades; and surely their conduct ought to be considered as moral. As far as deliberation and the victory over opposing motives are concerned, animals may be seen doubting between opposed instincts, as in rescuing their offspring or comrades from danger; yet their actions, though done for the good of others, are not called moral. Moreover, an action repeatedly performed by us, will at last be done without deliberation or hesitation, and can then hardly be distinguished from an instinct; yet surely no one will pretend that an action thus done ceases to be moral. On the con-

[21] I have given one such case, namely, of three Patagonian Indians who preferred being shot, one after the other, to betraying the plans their companions in war ('Journal of Researches,' 1845, p. 103).

trary, we all feel that an act cannot be considered as perfect, or as performed in the most noble manner, unless it be done impulsively, without deliberation or effort, in the same manner as by a man in whom the requisite qualities are innate. He who is forced to overcome his fear or want of sympathy before he acts, deserves, however, in one way higher credit than the man whose innate disposition leads him to a good act without effort. As we cannot distinguish between motives, we rank all actions of a certain class as moral, when they are performed by a moral being. A moral being is one who is capable of comparing his past and future actions or motives, and of approving or disapproving of them. We have no reason to suppose that any of the lower animals have this capacity; therefore when a monkey faces danger to rescue its comrade, or takes charge of an orphan-monkey, we do not call its conduct moral. But in the case of man, who alone can with certainty be ranked as a moral being, actions of a certain class are called moral, whether performed deliberately after a struggle with opposing motives, or from the effects of slowly-gained habit, or impulsively through instinct.

Summary of the last two Chapters.—There can be no doubt that the difference between the mind of the lowest man and that of the highest animal is immense. An anthropomorphous ape, if he could take a dispassionate view of his own case, would admit that though he could form an artful plan to plunder a garden—though he could use stones for fighting or for breaking open nuts, yet that the thought of fashioning a stone into a tool was quite beyond his scope. Still less, as he would admit, could he follow out a train of metaphysical reasoning, or solve a mathematical problem, or reflect on God, or admire a grand natural scene. Some apes, however, would probably declare that they could and did admire the beauty of the colored skin and fur of their partners in marriage. They would admit, that though they could make other apes understand by cries some of their perceptions and simpler wants, the notion of expressing definite ideas by definite sounds had never crossed their minds. They might insist that they were ready to aid their fellow-apes of the same troop in many ways, to risk their lives for them, and to take charge of their orphans; but they would be forced to acknowledge that disinterested love for all living creatures, the most noble attribute of man, was quite beyond their comprehension.

Nevertheless the difference in mind between man and the higher animals, great as it is, is certainly one of degree and not of kind. We have seen that the senses and intuitions, the various emotions and faculties, such as love, memory, attention, curiosity, imitation, reason, etc., of which man boasts, may be found in an incipient, or even sometimes in a well-developed condition, in the lower animals. They are also capable of some inherited improvement, as we see in the domestic dog compared with the wolf or jackal. If it be maintained that certain powers, such as self-consciousness, abstraction, etc., are peculiar to man, it may well be that these are the incidental results of other highly-advanced intellectual faculties; and these again are mainly the result of the continued use of a highly-developed language. At what age does the new-born infant possess the power of abstraction, or become self-conscious and reflect on its own existence? We cannot answer; nor can we answer in regard to the ascending organic scale. The half-art and half-instinct of language still bears the stamp of its gradual evolution. The ennobling belief in God is not universal with man; and the belief in active spiritual agencies naturally follows from his other mental powers. The moral sense perhaps affords the best and highest distinction between man and the lower animals; but I need not say any thing on this head, as I have so lately endeavored to show that the social instincts—the prime principle of man's moral constitution[39] —with the aid of active intellectual powers and the effects of habit, naturally lead to the golden rule, "As ye would that men should do to you, do ye to them likewise;" and this lies at the foundation of morality.

In a future chapter I shall make some few remarks on the probable steps and means by which the several mental and moral faculties of man have been gradually evolved. That this at least is possible ought not to be denied, when we daily see their development in every infant; and when we may trace a perfect gradation from the mind of an utter idiot, lower than that of the lowest animal, to the mind of a Newton.

[39] 'The Thoughts of Marcus Aurelius,' etc., p. 139.

Baldwin, J. M. 1896. A new factor in evolution. *American Naturalist* 30:441–451.

The interface of evolution and development, known more commonly by its nickname of evo-devo, is a hot research area in modern evolutionary biology, but its intellectual roots run deep. For example, Lamarck's (1809) theory regarding the inheritance of acquired characteristics was published in the year of Darwin's birth, preceding *On the Origin of Species* by a full half-century. Although many of the specifics of Lamarckian inheritance were widely discredited in the twentieth century, interest in the broader topic of the relationship of the evolution of species and the development of individual members of that species has surged again in recent years with the rise of a field known as *epigenetics,* which deals with the entire suite of regulatory mechanisms, developmental pathways, and social and other environmental influences by which complex genomes give rise to organismal phenotypes during the evolutionary process.

Further evidence for biologists' longstanding curiosity about ontogeny and its relationship to evolution can be found in this 1896 paper in which J. Mark Baldwin introduced what later became known as the "Baldwin effect": the idea that phenotypic accommodations to extreme conditions during individual ontogeny allow enhanced survival of appropriately responding individuals. The Baldwin effect remains controversial and rather difficult to grasp, with some authors (e.g., Robinson and Dukas 1999) seeing it as being of relatively minor or uncertain importance for evolution and other authors (e.g., West-Eberhard 2003) interpreting the notion as being extremely important and seeing the value of resurrecting it in modern discussions of evo-devo. This intriguing article was far ahead of its time and is well-worth reading even today.

RELATED READING

Baldwin, J. M. 1902. Development and evolution. MacMillan, New York, New York, USA.

Carey, N. 2012. The epigenetics revolution. Columbia University Press, New York, New York, USA.

Frank, S. A. 2011. Natural selection: II. developmental variability and evolutionary rate. Journal of Evolutionary Biology 24:2310–2320.

Lamarck, J.-B. 1809. Philosophie zoologique. Oxford University Press, Oxford, England.

Raff, R. A., and T. C. Kaufman. 1983. Embryos, genes, and evolution: the developmental-genetic basis of evolutionary change. Macmillan, New York, New York, USA.

Robinson, B. W., and R. Dukas. 1999. The influence of phenotypic modifications on evolution: the Baldwin effect and modern perspectives. Oikos 83:582–589.

West-Eberhard, M. J. 2003. Developmental plasticity and evolution. Oxford University Press, Oxford, England.

THE

AMERICAN NATURALIST

VOL. XXX. June, 1896.

A NEW FACTOR IN EVOLUTION.

BY J. MARK BALDWIN.

In several recent publications I have developed, from different points of view, some considerations which tend to bring out a certain influence at work in organic evolution which I venture to call "a new factor." I give below a list of references[1] to these publications and shall refer to them by number as this paper proceeds. The object of the present paper is to

[1] References:

(1). *Imitation: a Chapter in the Natural History of Consciousness*, *Mind*, (London), Jan., 1894. Citations from earlier papers will be found in this article and in the next reference.

(2). *Mental Development in the Child and the Race* (1st. ed., April, 1895; 2nd. ed., Oct., 1895; Macmillan & Co. The present paper expands an additional chapter (Chap. XVII) added in the German and French editions and to be incorporated in the third English edition.

(3). *Consciousness and Evolution*, *Science*, N. Y., August, 23, 1895; reprinted printed in the AMERICAN NATURALIST, April, 1896.

(4). *Heredity and Instinct* (I), *Science*, March 20, 1896. Discussion before N. Y. Acad. of Sci., Jan. 31, 1896.

(5). *Heredity and Instinct* (II), *Science*, April 10, 1896.

(6). *Physical and Social Heredity*, *Amer. Naturalist*, May, 1896.

(7). *Consciousness and Evolution*, *Psychol. Review*, May, 1896. Discussion before Amer. Psychol. Association, Dec. 28, 1895.

gather into one sketch an outline of the view of the process of development which these different publications have hinged upon.

The problems involved in a theory of organic development may be gathered up under three great heads: Ontogeny, Phylogeny, Heredity. The general consideration, the "factor" which I propose to bring out, is operative in the first instance, in the field of *Ontogeny*; I shall consequently speak first of the problem of Ontogeny, then of that of Phylogeny, in so far as the topic dealt with makes it necessary, then of that of Heredity, under the same limitation, and finally, give some definitions and conclusions.

I.

Ontogeny: "Organic Selection" (see ref. 2, chap. vii).—The series of facts which investigation in this field has to deal with are those of the individual creature's development; and two sorts of facts may be distinguished from the point of view of the *functions which an organism performs in the course of his life history.* There is, in the first place, the development of his heredity impulse, the unfolding of his heredity in the forms and functions which characterize his kind, together with the congenital variations which characterize the particular indiual—the phylogenetic variations, which are constitutional to him; and there is, in the second place, the series of functions, acts, etc., *which he learns to do himself in the course of his life.* All of these latter, the *special modifications which an organism undergoes during its ontogeny*, thrown together, have been called "acquired characters," and we may use that expression or adopt one recently suggested by Osborn,[2] "ontogenic variations" (except that I should prefer the form "ontogenetic variations"), if the word variations seems appropriate at all.

[2] Reported in *Science*, April 3rd.; also used by him before N. Y. Acad. of Sci., April 13th. There is some confusion between the two terminations "genic" and "genetic." I think the proper distinction is that which reserves the former, "genic," for application in cases in which the word to which it is affixed qualifies a term used *actively*, while the other, "genetic" conveys similarly a *passive* signification; thus agencies, causes, influences, etc., and "ontogenic phylogenic, etc.," while effects, consequences, etc, and "ontogenetic, phylogenetic, etc."

Assuming that there are such new or modified functions, in the first instance, and such "acquired characters," arising by the law of "use and disuse" from these new functions, our farther question is about them. And the question is this: How does an organism come to be modified during its life history?

In answer to this question we find that there are three different sorts of ontogenic agencies which should be distinguished—each of which works to produce ontogenetic modifications, adaptations, or variations. These are: first, the physical agencies and influences in the environment which work upon the organism to produce modifications of its form and functions. They include all chemical agents, strains, contacts, hindrances to growth, temperature changes, etc. As far as these forces work changes in the organism, the changes may be considered largely "fortuitous" or accidental. Considering the forces which produce them I propose to call them "physico-genetic." Spencer's theory of ontogenetic development rests largely upon the occurrence of lucky movements brought out by such accidental influences. Second, there is a class of modifications which arise from the spontaneous activities of the organism itself in the carrying out of its normal congenital functions. These variations and adaptations are seen in a remarkable way in plants, in unicellular creatures, in very young children. There seems to be a readiness and capacity on the part of the organism to "rise to the occasion," as it were, and make gain out of the circumstances of its life. The facts have been put in evidence (for plants) by Henslow, Pfeffer, Sachs; (for micro-organisms) by Binet, Bunge; (in human pathology) by Bernheim, Janet; (in children) by Baldwin (ref. 2, chap. vi.) (See citations in ref. 2, chap. ix, and in Orr, *Theory of Development*, chap. iv). These changes I propose to call "neuro-genetic," laying emphasis on what is called by Romanes, Morgan and others, the "selective property" of the nervous system, and of life generally. Third, there is the great series of adaptations secured by conscious agency, which we may throw together as "psycho-genetic." The processes involved here are all classed broadly under the term "intelligent," i. e., imitation, gregarious influences, maternal instruction, the lessons of pleasure and pain, and of experience generally, and reasoning from means to ends, etc.

We reach, therefore, the following scheme:

Ontogenetic Modifications.	*Ontogenic Agencies.*
1. Physico-genetic.	1. Mechanical.
2. Neuro-genetic.	2. Nervous.
3. Psycho-genetic.	3. Intelligent.
	Imitation.
	Pleasure and pain.
	Reasoning.

Now it is evident that there are two very distinct questions which come up as soon as we admit modifications of function and of structure in ontogenetic development: first, there is the question as to how these modifications can come to be adaptive in the life of the individual creature. Or in other words: What is the method of the individual's growth and adaptation as shown in the well known law of "use and disuse?" Looked at functionally, we see that the organism manages somehow to accommodate itself to conditions which are favorable, to repeat movements which are adaptive, and so to grow by the principle of use. This involves some sort of selection, from the actual ontogenetic variations, of certain ones—certain functions, etc. Certain other possible and actual functions and structures decay from disuse. Whatever the method of doing this may be, we may simply, at this point, claim the law of use and disuse, as applicable in ontogenetic development, and apply the phrase, "Organic Selection," to the organism's behavior in acquiring new modes or modifications of adaptive function with its influence of structure. The question of the method of "Organic Selection" is taken up below (IV); here, I may repeat, we simply assume what every one admits in some form, that such adaptations of function—"accommodations" the psychologist calls them, the processes of learning new movements, etc.—*do occur*. We then reach another question, second; what place these adaptations have in the general theory of development.

Effects of Organic Selection.—First, we may note the results of this principle in the creature's own private life.

1. *By securing adaptations, accommodations, in special circumstances the creature is kept alive* (ref. 2, 1st ed., pp. 172 ff.). This is true in all the three spheres of ontogenetic variation distinguished in the table above. The creatures which can stand the "storm and stress" of the physical influences of the environment, and of the changes which occur in the environment, *by undergoing modifications of their congenital functions or of the structures which they get congenitally—these creatures will live; while those which cannot, will not.* In the sphere of neurogenetic variations we find a superb series of adaptations by lower as well as higher organisms during the course of ontogenetic development (ref. 2, chap. ix). And in the highest sphere, that of intelligence (including the phenomena of consciousness of all kinds, experience of pleasure and pain, imitation, etc.), we find individual accommodations on the tremendous scale which culminates in the skilful performances of human volition, invention, etc. The progress of the child in all the learning processes which lead him on to be a man, just illustrates this higher form of ontogenetic adaptation (ref. 2, chap. x–xiii).

All these instances are associated in the higher organisms, and all of them unite to *keep the creature alive.*

2. By this means *those congenital or phylogenetic variations are kept in existence, which lend themselves to intelligent, imitative, adaptive, and mechanical modification during the lifetime of the creatures which have them.* Other congenital variations are not thus kept in existence. So there arises a more or less widespread series of *determinate variations in each generation's ontogenesis* (ref. 3, 4, 5).[3]

[3] "It is necessary to consider further how certain reactions of one single organism can be selected so as to adapt the organism better and give it a life history. Let us at the outset call this process "Organic Selection" in contrast with the Natural Selection of whole organisms. . . . If this (natural selection) worked alone, every change in the environment would weed out all life except those organisms, which by accidental variation reacted already in the way demanded by the changed conditions—in every case new organisms showing variations, not, in any case, new elements of life-history in the old organisms. In order to the latter we would have to conceive . . . some modification of the old reactions in an organism through the influence of new conditions. . . . We are, accordingly, left to the view that the new stimulations brought by changes in the environment

The further applications of the principle lead us over into the field of our second question, i. e., phylogeny.

II.

Phylogeny: Physical Heredity.—The question of phylogenetic development considered apart, in so far as may be, from that of heredity, is the question as to what the factors really are which show themselves in evolutionary progress from generation to generation. The most important series of facts recently brought to light are those which show what is called "determinate variation" from one generation to another. This has been insisted on by the paleontologists. Of the two current theories of heredity, only one, Neo-Lamarkism—by means of its principle of the inheritance of acquired characters—has been able to account for this fact of determinate phylogenetic change. Weismann admits the inadequacy of the principle of natural selection, as operative on rival organisms, to explain variations when they are wanted or, as he puts it, "the right variations in the right place" (*Monist,* Jan., '96).

I have argued, however, in detail that the assumption of determinate variations of function in ontogenesis, under the principle of neurogenetic and psychogenetic adaptation, does away with the need of appealing to the Lamarkian factor. In the case i. g., of instincts, "if we do not assume consciousness, then natural selection is inadequate; but if we do assume consciousness, then the inheritance of acquired characters is unnecessary" (ref. 5).

"The intelligence which is appealed to, to take the place of instinct and to give rise to it, uses just these partial variations which tend in the direction of the instinct; so the intelligence *supplements* such partial co-ordinations, makes them functional, and *so keeps the creature alive.* In the phrase of Prof.

themselves modify the reactions of an organism. . . . The facts show that individual organisms do acquire new adaptations in their lifetime, and that is our first problem. If in solving it we find a principle which may also serve as a principle of race-development, then we may possibly use it against the 'all sufficiency of natural selection' or in its support" (ref. 2, 1st. ed., pp. 175–6.)

Lloyd Morgan, this prevents the 'incidence of natural selection.' So the supposition that intelligence is operative turns out to be just the supposition which makes use-inheritance unnecessary. Thus kept alive, the species has all the time necessary to perfect the variations required by a complete instinct. And when we bear in mind that the variation required is not on the muscular side to any great extent, but in the central brain connections, and is a slight variation for functional purposes at the best, the hypothesis of use-inheritance becomes not only unnecessary, but to my mind quite superfluous" (ref. 4, p. 439). And for adaptations generally, "the most plastic individuals will be preserved to do the advantageous things for which their variations show them to be the most fit, and the next generation will show an emphasis of just this direction in its variations" (ref. 3, p. 221).

We get, therefore, from Organic Selection, certain results in the sphere of phylogeny:

1. *This principle secures by survival certain lines of determinate phylogenetic variation in the directions of the determinate ontogenetic adaptations of the earlier generation.* The variations which were utilized for ontogenetic adaptation in the earlier generation, being thus kept in existence, are utilized more widely in the subsequent generation (ref. 3, 4). "Congenital variations, on the one hand, are kept alive and made effective by their use for adaptations in the life of the individual; and, on the other hand, adaptations become congenital by further progress and refinement of variation in the same lines of function as those which their acquisition by the individual called into play. But there is no need in either case to assume the Lamarkian factor" (ref. 3). And in cases of conscious adaptation: "We reach a point of view which gives to organic evolution a sort of intelligent direction after all; for of all the variations tending in the direction of an adaptation, but inadequate to its complete performance, *only those will be supplemented and kept alive which the intelligence ratifies and uses.* The principle of 'selective value' applies to the others or to some of them. So natural selection kills off the others; and the *future*

development at each stage of a species' development must be in the directions thus ratified by intelligence. So also with imitation. Only those imitative actions of a creature which are useful to him will survive in the species, for in so far as he imitates actions which are injurious he will aid natural selection in killing himself off. So intelligence, and the imitation which copies it, will set the direction of the development of the complex instincts even on the Neo-Darwinian theory; and in this sense we may say that consciousness is a 'factor'" (ref. 4).

2. *The mean of phylogenetic variation being thus made more determinate, further phylogenetic variations follow about this mean, and these variations are again utilized by Organic Selection for ontogenetic adaptation.* So there is continual phylogenetic progress in the directions set by ontogenetic adaptation (ref. 3, 4, 5). "The intelligence supplements slight co-adaptations and so gives them selective value; but it does not keep them from getting farther selective value as instincts, reflexes, etc., by farther variation" (ref. 5). "The imitative function, by using muscular co-ordinations, supplements them, secures adaptations, keeps the creature alive, prevents the 'incidence of natural selection,' and so gives the species all the time necessary to get the variations required for the full instinctive performance of the function" (ref. 4). But, "Conscious imitation, while it prevents the incidence of natural selection, as has been seen, and so keeps alive the creatures which have no instincts for the performance of the actions required, nevertheless does not subserve the utilities which the special instincts do, nor prevent them from having the selective value of which Romanes speaks. Accordingly, on the more general definition of intelligence, which includes in it all conscious imitation, use of maternal instruction, and that sort of thing—no less than on the more special definition—we still find the principal of natural selection operative" (ref. 5).

3. *This completely disposes of the Lamarkian factor as far as two lines of evidence for it are concerned.* First, the evidence drawn from function, "use and disuse," is discredited; since by "organic selection," the reappearance, in subsequent generations, of the variations first secured in ontogenesis is ac-

counted for without the inheritance of acquired characters. So also the evidence drawn from paleontology which cites progressive variations resting on functional use and disuse. Second, the evidence drawn from the facts of "determinate variations;" since by this principle we have the preservation of such variations in phylogeny without the inheritance of acquired characters.

4. *But this is not Preformism in the old sense; since the adpatations made in ontogenetic development which "set" the direction of evolution are novelties of function in whole or part* (although they utilize congenital variations of structure). And it is only by the exercise of these novel functions that the creatures are kept alive to propagate and thus produce further variations of structure which may in time make the whole function, with its adequate structure, congenital. Romanes' argument from "partial co-adaptations" and "selective value," seem to hold in the case of reflex and instinctive functions (ref. 4, 5), as against the old preformist or Weismannist view, although the operation of Organic Selection, as now explained, renders them ineffective when urged in support of Lamarkism. "We may imagine creatures, whose hands were used for holding only with the thumb and fingers on the same side of the object held, to have first discovered, under stress of circumstances and with variations which permitted the further adaptation, how to make use of the thumb for grasping opposite to the fingers, as we now do. Then let us suppose that this proved of such utility that all the young that did not do it were killed off; the next generation following would be plastic, intelligent, or imitative, enough to do it also. They would use the same co-ordinations and prevent natural selection getting its operation on them; and so instinctive 'thumb-grasping' might be waited for indefinitely by the species and then be got as an instinct altogether apart from use-inheritance" (ref. 4). "I have cited 'thumb-grasping' because we can see in the child the anticipation, by intelligence and imitation, of the use of the thumb for the adaptation which the Simian probably gets entirely by instinct, and which I think an isolated and weak-minded child, say, would also come to do by instinct'" (ref. 4).

5. It seems to me also—though I hardly dare venture into a field belonging so strictly to the technical biologist—that *this principle might not only explain many cases of widespread "determinate variations" appearing suddenly, let us say, in fossil deposits, but the fact that variations seem often to be "discontinuous."* Suppose, for example, certain animals, varying, in respect to a certain quality, from *a* to *n* about a mean *x*. The mean *x* would be the case most likely to be preserved in fossil form (seeing that there are vastly more of them). Now suppose a sweeping change in the environment, in such a way that only the variations lying near the extreme *n* can accommodate to it and live to reproduce. The next generation would then show variations about the mean *n*. And the chances of fossils from this generation, and the subsequent ones, would be of creatures approximating *n*. Here would be a great discontinuity in the chain and also a widespread prevalence of these variations in a set direction. This seems especially evident when we consider that the paleontologist does not deal with successive generations, but with widely remote periods, and the smallest lapse of time which he can take cognizance of is long enough to give the new mean of variation, *n*, a lot of generations in which to multiply and deposit its representative fossils. Of course, this would be only the action of natural selection upon "preformed" variations in those cases which did not involve positive changes, in structure and function, *acquired in ontogenesis;* but in so far as such ontogenetic adaptations were actually there, the extent of difference of the *n* mean from the *x* mean would be greater, and hence the resources of explanation, both of the sudden prevalence of the new type and of its discontinuity from the earlier, would be much increased. This additional resource, then, is due to the "Organic Selection" factor.

We seem to be able also to utilize all the evidence usually cited for the functional origin of specific characters and groupings of characters. So far as the Lamarkians have a strong case here, it remains as strong if Organic Selection be substituted for the "inheritance of acquired characters." This is especially true where intelligent and imitative adaptations are involved, as in the case of instinct. This "may give the reason, e. g., that instincts are so often coterminous with the limits of species. Similar structures find the similar uses for their intelligence, and they also find the same imitative actions to be to their advantage. So the interaction of these conscious factors with natural selection brings it about that the structural definition which represents species, and the functional definition which represents instinct, largely keep to the same lines" (ref. 5).

6. It seems proper, therefore, to call the influence of Organic Selection "a new factor;" for it gives a method of deriving the determinate gains of phylogeny from the adaptations of ontogeny without holding to either of the two current theories. *The ontogenetic adaptations are really new, not performed; and they are really reproduced in succeeding generations, although not physically inherited.*

Garrod, A. E. 1902. The incidence of alkaptonuria: a study in chemical individuality. *Lancet* ii:1616–1620.

Sir Archibald Garrod, a prominent English physician, is often considered to be the "father of biochemical genetics." By the early 1900s, doctors had begun to appreciate that biochemical malfunctions inside the human body can produce physical ailments and abnormalities. This was a revolutionary concept, because, before that time, most biologists as well as philosophers had been "natural theologians" who interpreted nearly all biological traits to be near-perfect contrivances that offered pervasive evidence for skilled craftsmanship by a powerful Deity with human interests at heart. But Garrod's research on alkaptonuria and other metabolic disorders changed the tone of the discussion by demonstrating that human physiology is also characterized by endogenous biochemical imperfections with genetic etiologies. Today, modern extensions of Garrod's work have unearthed thousands of such cases in which endogenous biochemical-genetic flaws seriously compromise human health. Why would an intelligent designer have crafted the innermost machinery of humans to be so grossly error prone? This is just one subset of a broader "theodicy dilemma": why do biological flaws exist in a world supposedly run by an omnipotent and beneficent supernatural designer? In the natural sciences, the general answer now seems clear: evolutionary-genetic processes are directly responsible for biological flaws (as well as biological triumphs). This scientific revelation is but one of many reasons why the findings of evolutionary biology are highly relevant to philosophical discourse on numerous topics traditionally reserved for theologians and other religious practitioners.

RELATED READING

Avise, J. C. 1998. The genetic gods: evolution and belief in human affairs. Harvard University Press, Cambridge, Massachusetts, USA.

Avise, J. C. 2010. Inside the human genome: a case for non-intelligent design. Oxford University Press, Oxford, England.

Ayala, F. J. 2007. Darwin's gift to science and religion. Joseph Henry Press, Washington, D.C., USA.

Garrod, A. E. 1923. Inborn errors of metabolism, 2nd ed. Frowde, Hodder & Stoughton, London, England.

McKusick, V. A, editor. 1998. Mendelian inheritance in man, 12th ed. Johns Hopkins University Press, Baltimore, Maryland, USA.

Nyhan, W. L., B. A. Barstop, and A. I Al-Aqeel. 2012. Atlas of metabolic diseases, 3rd ed. Hodder Arnold, London.

Scriver, C. R., W. S. Sly, B. Childs, A. L. Beaudet, D. Valle, K. W. Kinzler, and B. Vogelstein. 2000. The metabolic and molecular basis of inherited disease, 8th ed. McGraw-Hill, New York, New York, USA.

Garrod, Archibald E. 1902. The Incidence of Alkaptonuria: A Study in Chemical Individuality. *Lancet*, vol. ii, pp. 1616-1620.

THE INCIDENCE OF ALKAPTONURIA: A STUDY IN CHEMICAL INDIVIDUALITY

ARCHIBALD E. GARROD

Physician to the Hospital for Sick Children, Great Ormondstreet,
Demonstrator of Chemical Pathology at St. Bartholemew's Hospital

ALL THE MORE RECENT WORK on alkaptonuria has tended to show that the constant feature of that condition is the excretion of homogentisic acid, to the presence of which substance the special properties of alkapton urine, the darkening with alkalies and on exposure to air, the power of staining fabrics deeply, and that of reducing metallic salts, are alike due. In every case which has been fully investigated since Wolkow and Baumann[1] first isolated and described this acid its presence has been demonstrated and re–examination of the material from some of the earlier cases also has led to its detection. The second allied alkapton acid, uroleucic, has hitherto only been found in the cases investigated by Kirk and in them in association with larger amounts of homogentisic acid.[2] By the kindness of Dr. R. Kirk I have recently been enabled to examine fresh specimens of the urines of his patients who have now reached manhood and was able to satisfy myself that at the present time even they are no longer excreting uroleucic acid. After as much of the homogentisic acid as possible had been allowed to separate out as the lead salt the small residue of alkapton acid was converted into the ethyl ester by a method recently described by Erich Meyer[3] and the

1 Wolkow and Baumann. 1891. *Z. Physiol. Chemie*. **XV:** 228.

2 R. Kirk. 1889. *Journal of Anatomy and Physiology*. **XXIII:** 69; Huppert. 1897. *Zeitschrift für Physiologische Chemie*. **XXIII:** 412.

3 E. Meyer. 1901. *Deutsches Archiv für Klinische Medicin*. **LXX:** 443.

crystalline product obtained had the melting point of ethyl homogentisate (120° C.). Further observations, and especially those of Mittelbach,[4] have also strengthened the belief that the homogentisic acid excreted is derived from tyrosin, but why alkaptonuric individuals pass the benzene ring of their tyrosin unbroken and how and where the peculiar chemical change from tyrosin to homogentisic acid is brought about, remain unsolved problems.

There are good reasons for thinking that alkaptonuria is not the manifestation of a disease but is rather of the nature of an alternative course of metabolism, harmless and usually congenital and lifelong. Witness is borne to its harmlessness by those who have manifested the peculiarity without any apparent detriment to health from infancy on into adult and even advanced life, as also by the observations of Erich Meyer who has shown that in the quantities ordinarily excreted by such persons homogentisic acid neither acts as an aromatic poison nor causes acid intoxication, for it is not excreted as an aromatic sulphate as aromatic poisons are, nor is its presence in the urine attended by any excessive output of ammonia. However, regarded as an alternative course of metabolism the alkaptonuric must be looked upon as somewhat inferior to the ordinary plan, inasmuch as the excretion of homogentisic acid in place of the ordinary end products involves a certain slight waste of potential energy. In this connexion it is also interesting to note that, as far as our knowledge goes, an individual is either frankly alkaptonuric or conforms to the normal type, that is to say, excretes several grammes of homogentisic acid per diem or none at all. Its appearance in traces, or in gradually increasing or diminishing quantities, has never yet been observed, even in the few recorded temporary or intermittent cases. In cases in which estimations have been carried out the daily output has been found to lie within limits which, considering the great influence of proteid food upon the excretion of homogentisic acid and allowing for differences of sex and age, may be described as narrow. This is well illustrated by Table I, in which the cases are arranged in order of age.

---

4 Mittelbach. 1901. *Ibid*. **LXXI:** 50.

TABLE I. SHOWING THE AVERAGE EXCRETION OF HOMOGENTISIC ACID

No.	Sex	Age	Average excretion of homogentisic acid per 24 hours on ordinary mixed diet	Name of observers
1	M	2½ years	3.2 grams	Erich Meyer
2	M	3½ years	2.6 grams	A. E. Garrod
3	M	8 years	2.7 grams	Ewald Stier
4	M	18 years	5.9 grams	P. Stange
5	M	44 years	4.6 grams	Mittelbach
6	M	45 years	4.7 grams	H. Ogden
7	M	60 years	5.3 grams	Hammarsten
8	F	60 years	3.2 grams	H. Emlslen
9	M	68 years	4.8 grams	Wolkow and Baumann

The information available as to the incidence of alkaptonuria is of great interest in connexion with the above view of its nature. That the peculiarity is in the great majority Of instances congenital cannot be doubted. The staining property of the urine allows of its being readily traced back to early infancy. This has been repeatedly done and in one of my cases[5] the staining of the napkins was conspicuous 57 hours after the birth of the child. The abnormality is apt to make its appearance in two or more brothers and sisters whose parents are normal and among whose forefathers there is no record of its having occurred, a peculiar mode of incidence which is well known in connexion with some other conditions. Thus of 32 known examples, which were presumably congenital, no less than 19 have occurred in seven families. One family contained four alkaptonurics, three others contained three, and the remaining three two each. The proportion of alkaptonuric to normal members is of some interest and Table II embodies such definite knowledge upon this point as is at present available regarding congenital cases.

5 A. E. Garrod. November 30, 1901. Lancet: 1481; 1902. *Transactions of the Royal Medical and Chirurgical Society*. **LXXXV:** 69.

TABLE II. SHOWING THE PROPORTION OF ALKAPTONURIC MEMBERS TO NORMAL MEMBERS IN 9 FAMILIES

No.	Total number of family (brothers and sisters)	Number of Alkaptonuric members	Number of normal members	Observers
1	14	4	10	Pavy
2	4	3	1	Kirk
3	7	3	4	Winternitz
4	2	1	1	Ewald Stier
5	2	2	0	Baumann, Embden
6	1	1	0	Erich Meyer
7	10	1	9	Noccioli and Domenici
8	5	2	3	A. E. Garrod
9	3	2	1	W. Smith, Garrod
Totals	48	19	29	—

The preponderance of males is very conspicuous. Thus, of the 40 subjects whose cases have hitherto been recorded 29 have been males and only 11 females.

In a paper read before the Royal Medical and Chirurgical Society in 1901 the present writer pointed out that of four British families in which 11 were congenitally alkaptonuric members no less than three were the offspring of marriages of first cousins who did not themselves exhibit this anomaly. This fact has such interesting bearings upon the etiology of alkaptonuria that it seemed desirable to obtain further information about as many as possible of the other recorded cases and especially of those which were presumably congenital. My inquiries of a number of investigators who have recorded such cases met with a most kindly response, and although the number of examples about which information could still be obtained proved to be very limited, some valuable facts previously unknown have been brought to light and indications are afforded of points which may be inquired into with advantage regarding cases which may come under observation in the future. In a number of instances the patients have been lost sight of, or for various reasons information can no longer be obtained concerning them. To those who have tried to help me with regard to such cases, and have in some instances been at great trouble in vain, my hearty thanks are no less due than to those who have been able to furnish fresh information.[6]

The following is a brief summary of the fresh information collected. Dr. Erich Meyer[7] who mentioned in his paper that the parents of his patient were related, informs me that as a matter of fact they are first cousins. Dr. H. Ogden[8] states that his patient is the seventh of a family of eight members and that his parents were first cousins. The three eldest children died in infancy; the fifth, a female, has three children, but neither is she nor are they alkaptonuric. There is no record of any other examples in the family. The patient, whose wife is not a blood relation, has three children none of whom are alkaptonuric. Professor Hammarsten[9] states that the parents of an alkaptonuric man, whose case he recently described, were first cousins. The patient, aged 61 years, has three brothers and the only brother whose urine has been seen is not alkaptonuric. I have learned from Professor Noccioli[10] that the parents of the woman whose case he investigated with Dr. Domenici were not blood relations. The patient, a twin, who is one of two survivors of a family of ten, states that none of her relations have exhibited the condition. Dr. Ewald Stier[11] informs me that the parents of his patient were not related and it is mentioned in his paper that they were not alkaptonuric. Professor Ebstein[12] states that the parents of the child with "pyrocatechinuria" whose case was investigated by him in conjunction with Dr. Willer in 1875 were not related, but I gather that he would not regard this as an ordinary case of alkaptonuria, the abnormal substance in the urine having been identified as pyrocatechin. Lastly, Professor Osler supplies the very interesting information that of two sons of the

---

6 To Hofrath Professor Huppert and to Professor Osler my very special thanks are due for invaluable aid in collecting information, and I would also express my most sincere gratitude to Professor Hammarsten, Geheimrath Professor Ebstein, Geheimrath Professor Fürbringer, Geheimrath Professor Erb, Professor Noccioli, and Professor Denigès, as also to Dr. F. W. Pavy, Dr. Kirk, Dr. Maguire, Dr. Futcher, Dr. Erich Meyer, Dr. H. Ogden, Dr. H. Embden, Dr. Mittelbach, Dr. Ewald Stier, Dr. Grassi, Dr. Carl Hirsch, and Dr. Winternitz, all of whom have been kind enough to help the inquiry in various ways.

7 E. Meyer. *Loc. cit.*

8 H. Ogden. 1895. *Z. Physiol. Chemie*. **XX:** 289.

9 Hammarsten. 1901. *Upsala Läkareförenings Förhandlingar*. **VII** 26.

10 Noccioli e Domenici. 1898. *Gazetta degli Ospedali*. **XIX:** 303.

11 Ewald Stier. 1898. *Berliner klinische Wochenschrift*. **XXXV:** 185.

12 Ebstein and Müller. 1875. *Virchow's Archiv*. **LXII:** 554.

alkaptonuric man previously described by Dr. Futcher[13] one is alkaptonuric. This is the first known instance of direct transmission of the peculiarity. The parents of the father, who has an alkaptonuric brother whose case was recorded by Marshall,[14] were not blood relations. The above particulars are embodied with those of the congenital British cases previously recorded in the following tabular epitome (Table III).

TABLE III. SHOWING THE LARGE PROPORTION OF ALKAPTONURICS WHO ARE THE OFFSPRING OF MARRIAGE OF FIRST COUSINS

A

Families the offspring of marriages of first cousins			
No.	Total number of family (brothers and sisters)	Number of known alkaptonuric members	Observers
1	14	4	Pavy
2	4	3	R. Kirk
3	5	2	A. E. Garrod
4	1	1	Erich Meyer
5	8	1	H. Ogden
6	4	1	Hammarsten
Totals	36	12	—

B

Families whose parents were not related and not alkaptonuric			
No.	Total number of family (brothers and sisters)	Number of known alkaptonuric members	Observers
1	3	2	Armstrong, Walter Smith, and Garrod
2	2	1	Ewald Stier
3	10	1	Noccioli and Domenici
4	?	2	Marshall and Futcher
Totals		6	—

13 Futcher. 1898. *New York Medical Journal*. **LXVII:** 69.

14 Marshall. 1887. *Medical News, Philadelphia*. **L:** 35.

C

Families in which alkaptonuria was directly inherited from a parent			
No.	Total number of family (brothers and sisters)	Number of known alkaptonuric members	Observers
1 *	?	1	Osler and Futcher
Totals		1	—

* B 4 and C 1 refer to two generations of one family. No information is forthcoming as to the absence of alkaptonuria in previous generations. Ebstein and Müller's case, which is not included in the table for reasons given above, would raise the number of families in list B to 5.

It will be seen that the results of further inquiries on the continent of Europe and in America confirm the impression derived from the British cases that of alkaptonuric individuals a very large proportion are children of first cousins. The above table includes 19 cases in all out of a total of 40 recorded examples of the condition, and there is little chance of obtaining any further information on the point until fresh cases shall come under observation. It will be noticed that among the families of parents who do not themselves exhibit the anomaly a proportion corresponding to 60 per cent are the offspring of marriages of first cousins. In order to appreciate how high this proportion is it is necessary to form some idea of the total proportion of the children of such unions to the community at large. Professor G. Darwin,[15] as the outcome of an elaborate statistical investigation, arrived at the conclusion that in England some 4 per cent of all marriages among the aristocracy and gentry are between first cousins; that in the country and smaller towns the proportion is between 2 and 3 per cent, whereas in London it is perhaps as low as 1.5 per cent. He suggests 3 per cent as a probable superior limit for the whole population. Assuming, although this is, perhaps, not the case, that the same proportion of these as of all marriages are fruitful, similar percentages will hold good for families, and assuming further that the average number of children results from such marriages they will hold good for individuals also. A very limited number of observations which I have made among hospital patients in London gave results which are quite compatible with the above figures. Thus, among 50 patients simultaneously inmates of St. Bartholomew's Hospital there

15 G. Darwin. 1875. *Journal of the Statistical Society*. **XXXVIII:** 153.

was one whose parents were first cousins. On another occasion one such was found among 100 patients, and there was one child of first cousins among 100 children admitted to my ward at the Hospital for Sick Children. It is evident, on the one hand, that the proportion of alkaptonuric families and individuals who are the offspring of first cousins is remarkably high, and, on the other hand, it is equally clear that only a minute proportion of the children of such unions are alkaptonuric. Even if such persons form only 1 per cent of the community their numbers in London alone should exceed 50,000, and of this multitude only six are known to be alkaptonuric. Doubtless there are others, but that the peculiarity is extremely rare is hardly open to question. A careful look–out maintained for several years at two large hospitals has convinced me of this, and although the subject has recently attracted much more attention than formerly the roll of recorded examples increases but slowly.

The question of the liability of children of consanguineous marriages to exhibit certain abnormalities or to develop certain diseases has been much discussed, but seldom in a strictly scientific spirit. Those who have written on the subject have too often aimed at demonstrating the deleterious results of such unions on the one hand, or their harmlessness on the other, questions which do not here concern us at all. There is no reason to suppose that mere consanguinity of parents can originate such a condition as alkaptonuria in their offspring, and we must rather seek an explanation in some peculiarity of the parents, which may remain latent for generations, but which has the best chance of asserting itself in the offspring of the union of two members of a family in which it is transmitted. This applies equally to other examples of that peculiar form of heredity which has long been a puzzle to investigators of such subjects, which results in the appearance in several collateral members of a family of a peculiarity which has not been manifested at least in recent preceding generations.

It has recently been pointed out by Bateson[16] that the law of heredity discovered by Mendel offers a reasonable account of such phenomena. It asserts that as regards two mutually exclusive characters, one of which tends to be dominant and the other recessive, cross–bred organisms will produce germinal cells (gametes) each of which, as regards the characters in question, conforms to one or other of the pure ancestral types and is therefore incapable of transmitting

16 W. Bateson. 1902. *Mendel's Principles of Heredity*, Cambridge.

the opposite character. When a recessive gamete meets one of the dominant type the resulting organism (the zygote) will usually exhibit the dominant character, whereas when two recessive gametes meet the recessive character will necessarily be manifested in the zygote. In the case of a rare recessive characteristic we may easily imagine that many generations may pass before the union of two recessive gametes takes place. The application of this to the case in question is further pointed out by Bateson, who, commenting upon the above observations on the incidence of alkaptonuria, writes as follows:[17] "Now there may be other accounts possible, but we note that the mating of first cousins gives exactly the conditions most likely to enable a rare, and usually recessive, character to show itself. If the bearers of such a gamete mate with individuals not bearing it the character will hardly ever be seen; but first cousins will frequently be the bearers of similar gametes, which may in such unions meet each other and thus lead to the manifestation of the peculiar recessive characters in the zygote." Such an explanation removes the question altogether out of the range of prejudice, for, if it be the true account of the matter, it is not the mating of first cousins in general but of those who come of particular stocks that tends to induce the development of alkaptonuria in the offspring. For example, if a man inherits the tendency on his father's side his union with one of his maternal first cousins will be no more liable to result in alkaptonuric offspring than his marriage with one who is in no way related to him by blood. On the other hand, if members of two families who both inherit the strain should intermarry the liability to alkaptonuria in the offspring will be as great as from the union of two members of either family, and it is only to be expected that the peculiarity will also manifest itself in the children of parents who are not related. Whether the Mendelian explanation be the true one or not, there seems to be little room for doubt that the peculiarities of the incidence of alkaptonuria and of conditions which appear in a similar way are best explained by supposing that, leaving aside exceptional cases in which the character, usually recessive, assumes dominance, a peculiarity of the gametes of *both* parents is necessary for its production.

Hitherto nothing has been recorded about the children of alkaptonuric parents, and the information supplied by Professor Osler and Dr. Ogden on this point has therefore a very special interest. Whereas Professor Osler's case shows that the condition may be

---

17 W. Bateson and Miss E. R. Saunders. 1902. *Report to the Evolution Committee of the Royal Society*. (1): 133*n*.

directly inherited from a parent Dr. Ogden's case demonstrates that none of the children of such a parent need share his peculiarity. As the matter now stands, of five children of two alkaptonuric fathers whose condition is known only one is himself alkaptonuric. It will be interesting to learn whether this low proportion is maintained when larger numbers of cases shall be available. That it will be so is rendered highly probable by the undoubted fact that a very small proportion of alkaptonurics are the offspring of parents either of whom exhibits the anomaly. It would also be extremely interesting to have further examples of second marriages of the parents of alkaptonurics. In the case of the family observed by Dr. Kirk the only child of the second marriage of the father, not consanguineous, is a girl who does not exhibit the abnormality. The only other available example is recorded by Embden. The two alkaptonurics studied by Professor Baumann and himself were a brother and sister born out of wedlock, and as far as could be ascertained the condition was not present in the children of the subsequent marriages which both parents contracted. The patient of Noccioli and Domenici was a twin, and I gather from Professor Noccioli's kind letter that the other twin was also a female, did not survive, and was not alkaptonuric. Further particulars are wanting, and the information was derived from the patient herself, who is described as a woman of limited intelligence but who was aware that in her own case the condition had existed from infancy. It is difficult to imagine that of twins developed from a single ovum one should be alkaptonuric and the other normal, but this does not necessarily apply to twins developed from separate ova.

It may be objected to the view that alkaptonuria is merely an alternative mode of metabolism and not a morbid condition, that in a few instances, not included in the above tables, it appears not to have been congenital and continuous but temporary or intermittent. In some of the cases referred to the evidence available is not altogether conclusive, and it is obvious that for the proof of a point of so much importance to the theory of alkaptonuria nothing can be regarded as wholly satisfactory which falls short of a complete demonstration of the presence of homogentisic acid in the urine at one time and its absence at another. The degree and rate of darkening of the urine vary at different periods apart from any conspicuous fluctuations in the quantity of homogentisic acid which it contains. The staining of linen in infancy is a much more reliable indication, especially if the mother of the child has had previous experience of alkaptonuric staining. In

Geyger's case[18] of a diabetic man the intermittent appearance in the urine of an acid which he identified with the glycosuric acid of Marshall was established beyond all doubt, and the melting point and proportion of lead in the lead salt render it almost certain that he was dealing with homogentisic acid. In Carl Hirsch's case[19] a girl, aged 17 years, with febrile gastrointestinal catarrh, passed dark urine which gave the indican reaction for three days. Professor Siegfried extracted by shaking with ether an acid which gave the reactions of homogentisic acid and formed a sparingly soluble lead salt. Neither the melting point of the acid nor any analytical figures are given. After three days the urine resumed its natural colour and reactions.

Von Moraczewski[20] also records a case of a woman, aged 43 years, who shortly before her death passed increasingly dark urine, rich in indican, from which he extracted an acid which had the melting point and reactions of homogentisic acid. Such increasing darkening of the urine as was here observed not infrequently occurs with urines rich in indoxyl–sulphate, as Baumann and Brieger first pointed out, and this was probably a contributory factor in the production of the colour which first called attention to the condition. Stange[21] has described a case in which the presence of homogentisic acid was very fully established, but he clearly does not regard the mother's evidence as to the intermittent character of the condition as conclusive. Zimnicki's[22] case of intermittent excretion of homogentisic acid by a man with hypertrophic biliary cirrhosis is published in a Russian journal which is inaccessible to me, and having only seen abstracts of his paper I am unacquainted with the details. Of hearsay evidence the most convincing is afforded in Winternitz's cases.[23] The mother of seven children, three of whom are alkaptonuric, was convinced that whereas two of her children had been alkaptonuric from the earliest days of life this had not been so with the youngest child in whom she had only noticed the peculiarity from the age of five years. This is specially interesting as supplying a link between the temporary and congenital cases. In a somewhat

18 A. Geyger. 1892. *Pharmazeutische Zeitung*: 488.

19 C. Hirsch. 1897. *Berliner klinische Wochenschrift*. **XXXIV:** 866.

20 W. von Moraczewski. 1896. *Zentralblatt für die Innere Medizin*. **XVII:** 177.

21 P. Stange. 1896. *Virchow's Archiv*. **CXLVI:** 86.

22 Zimnicki. 1900. Jeschenedelnik. Abstract, *Zentralblatt für Stoffwechsel und Verdauungs-Krankheiten*. **I**(4): 348.

23 Winternitz. 1899. *Münchener Medizinische Wochenschrift*. **XLVI:** 749. 11

similar case described by Maguire[24] the evidence of a late onset is not so conclusive. Slosse's case[25] in which, as in von Moraczewski's, the condition apparently developed in the last stages of a fatal illness, completes the list of those falling into the temporary class. Evidently we have still much to learn about temporary or intermittent alkaptonuria, but it appears reasonable to suppose that those who exhibit the phenomenon are in a state of unstable equilibrium in this respect, and that they excrete homogentisic acid under the influence of causes which do not bring about this result in normal individuals. There is reason to believe that a similar instability plays a not unimportant part in determining the incidence of certain forms of *disease* in which derangements of metabolism are the most conspicuous features. Thus von Noorden,[26] after mentioning that diabetes occasionally develops at an early age in brothers and sisters and comparatively seldom occurs in the children of diabetic parents, adds that in three instances he has met with this disease in the offspring of marriages of first cousins. In one such family two out of six children, in another two out of three, and in the third the only two children became diabetic at ages between one and four years.

The view that alkaptonuria is a "sport" or an alternative mode of metabolism will obviously gain considerably in weight if it can be shown that it is not an isolated example of such a chemical abnormality, but that there are other conditions which may reasonably be placed in the same category. In the phenomenon of albinism we have an abnormality which may be looked upon as chemical in its basis, being due rather to a failure to produce the pigments of the melanin group which play so conspicuous a part in animal colouration than to any defect of development of the parts in which in normal individuals such pigments are laid down. When we study the incidence of albinism in man we find that it shows a striking resemblance to that of alkaptonuria. It, too, is commoner in males than in females, and tends to occur in brothers and sisters of families in which it has not previously appeared, at least in recent generations. Moreover, there is reason to believe that an undue proportion of albinos are the offspring of marriages of first cousins. Albinism is mentioned by most authors who have discussed the effects of such

---

24 R. Maguire. 1884. *Brit. Med. J.* **II**: 808.

25 A. Slosse. 1895. *Annales de la Société Royale des Sciences Médicales et Naturelles, Bruxelles.* **IV:** 89.

26 Von Noorden. 1901. *Die Zuckerkrankheit*. 3, Aufgabe: 47.

marriages and Arcoleo,[27] who gives some statistics of albinism in Sicily, states that of 24 families in which there were 62 albino members five were the offspring of parents related to each other in the second canonical degree. On the other hand, Bemiss[28] found that of 191 children of 34 marriages of first or second cousins five were albinos. In a remarkable instance recorded by Devay[29] two brothers married two sisters, their first cousins. There were no known instances of albinism in their families, but the two children of the one marriage and the five children of the other were all albinos. After the death of his wife the father of the second family married again and none of the four children of his second marriage were albinos. Again, albinism is occasionally directly inherited from a parent, as in one instance quoted by Arcoleo, but this appears to be an exceptional occurrence. The resemblance between the modes of incidence of the two conditions is so striking that it is hardly possible to doubt that whatever laws control the incidence of the one control that of the other also.

A third condition which suggests itself as being probably another chemical "sport" is cystinuria. Our knowledge of its incidence is far more incomplete and at first sight direct inheritance appears to play here a more prominent part. However, when more information is forthcoming it may turn out that it is controlled by similar laws. In this connexion a most interesting family described by Pfeiffer[30] is very suggestive. Both parents were normal, but all their four children, two daughters and two sons, were cystinuric. The elder daughter had two children neither of whom was cystinuric. A number of other examples of cystinuria in brothers and sisters are recorded, but information about the parents is wanting, except in the cases of direct transmission. In some of the earlier cases such transmission through three generations was thought to be probable, but the presence of cystin in the urine of parent and child has only been actually demonstrated in two instances. In Joel's[31] often–quoted case it was only shown that the mother's urine contained excess of neutral

---

27 G. Arcoleo. 1871. *Sull' Albinismo in Sicilia*. See notice in Archivio per l'Anthropologia. **I**: 367,

28 Bemiss. 1857. *J. of Psychol. Med.* **X:** 368.

29 Devay. 1857. *Du Danger des Mariages Consanguins, &c.*, Paris.

30 E. Pfeiffer. 1894. *Zentralblatt für Krankheiten der Harn-und Sexual-Organe*. **V:** 187.

31 Joel. 1855. *Annalen der Chemie und Pharmacologie*: 247.

sulphur. E. Pfeiffer[32] found cystin in the urine of a father and son and in a family observed by Cohn[33] the mother and six of her children shared the peculiarity. As more than 100 cases are on record the proportion of cases of direct inheritance has not hitherto been shown to be at all high and Pfeiffer's first case shows that, as with alkaptonuria, the children of a cystinuric parent may escape. A large majority of the recorded cystinurics have been males. There is as yet no evidence of any influence of consanguinity of parents and in the only two cases about which I have information the parents were not related. Neither has it yet been shown that cystinuria is a congenital anomaly, although in one case, at any rate, it has been traced back to the first year of life. Observations upon children of cystinuric parents from their earliest infancy or upon newly–born brothers or sisters of cystinurics would be of great interest and should in time settle this question. Lastly, it seems certain that, like alkaptonuria, this peculiarity of metabolism is occasionally temporary or intermittent. The so frequent association with cystinuria of the excretion of cadaverine and putrescine adds to the difficulty of the problem of its nature and upon it is based the infective theory of its causation. However, it is possible that, as C. E. Simon[34] has suggested, these diamines may themselves be products of abnormal metabolism. Unlike alkaptonuria and albinism cystinuria is a distinctly harmful condition, but its ill effects are secondary to its deposition in crystalline form and the readiness with which it forms concretions. Its appearance in the urine is not associated with any primary morbid symptoms. All three conditions referred to above are extremely rare and all tend to advertise their presence in conspicuous manners. An albino cannot escape observation; the staining of clothing and the colour of the urine of alkaptonurics seldom fail to attract attention, and the calculous troubles and the cystitis to which cystinurics are so liable usually bring them under observation sooner or later. May it not well be that there are other such chemical abnormalities which are attended by no obvious peculiarities and which could only be revealed by chemical analysis? If such exist and are equally rare with the above they may well have wholly eluded notice up till now. A deliberate search for such, without some guiding indications, appears as hopeless an undertaking as the proverbial search for a needle in a haystack.

---

32 E. Pfeiffer. 1897. *Zentralblatt für Krankheiten der Harn-und Sexual-Organe*. **VIII:** 173.

33 J. Cohn. 1899. *Berliner klinische Wochenschrift*. **XXXVI**: 503.

34 C. E. Simon. 1890. *Am. J. Med. Sci*. **CXIX:** 39.

If it be, indeed, the case that in alkaptonuria and the other conditions mentioned we are dealing with individualities of metabolism and not with the results of morbid processes the thought naturally presents itself that these are merely extreme examples of variations of chemical behaviour which are probably everywhere present in minor degrees and that just as no two individuals of a species are absolutely identical in bodily structure neither are their chemical processes carried out on exactly the same lines. Such minor chemical differences will obviously be far more subtle than those of form, for whereas the latter are evident to any careful observer the former will only be revealed by elaborate chemical methods, including painstaking comparisons of the intake and output of the organism. This view that there is no rigid uniformity of chemical processes in the individual members of a species, probable as it is *a priori,* may also be arrived at by a wholly different line of argument. There can be no question that between the families, genera and species both of the animal and vegetable kingdoms, differences exist both of chemical composition and of metabolic processes. The evidences for this are admirably set forth in a most suggestive address delivered by Professor Huppert[35] in 1895. In it he points out that we find evidence of chemical specificity of important constituents of the body, such as the haemoglobins of different animals, as well as in their secretory and excretory products such as the bile acids and the cynuric acid of the urine of dogs. Again, in their behaviour to different drugs and infecting organisms the members of the various genera and species manifest peculiarities which presumably have a chemical basis, as the more recent researches of Ehrlich tend still further to show. To the above examples may be added the results of F. G. Hopkins's[36] well–known researches on the pigments of the pieridae and the recent observations of the precipitation of the blood proteids of one kind of animal by the serum of another. From the vegetable kingdom examples of such generic and specific chemical differences might be multiplied to an almost indefinite extent. Nor are instances wanting of the influence of natural selection upon chemical processes, as for example, in the production of such protective materials as the sepia of the cuttlefish and the odorous secretion of the skunk, not to mention the innumerable modifications of surface pigmentation. If,

35 Huppert. 1896. *Über die Erhaltung der Arteigenschaften*, Prague.

36 F. G. Hopkins. 1895. *Philosophical Transactions of the Royal Society*. **CLXXXVI** (B):661.

then, the several genera and species thus differ in their chemistry we can hardly imagine that within the species, when once it is established, a rigid chemical uniformity exists. Such a conception is at variance with all that is known of the origin of species. Nor are direct evidences wanting of such minor chemical diversities as we have supposed to exist within the species. Such slight peculiarities of metabolism will necessarily be hard to trace by methods of direct analysis and will readily be masked by the influences of diet and of disease, but the results of observations on metabolism reveal differences which are apparently independent of such causes, as for example, in the excretion of uric acid by different human individuals. The phenomena of obesity and the various tints of hair, skin, and eyes point in the same direction, and if we pass to differences presumably chemical in their basis idiosyncrasies as regards drugs and the various degrees of natural immunity against infections are only less marked in individual human beings and in the several races of mankind than in distinct genera and species of animals.

If it be a correct inference from the available facts that the individuals of a species do not conform to an absolutely rigid standard of metabolism, but differ slightly in their chemistry as they do in their structure, it is no more surprising that they should occasionally exhibit conspicuous deviations from the specific type of metabolism than that we should meet with such wide departures from the structural uniformity of the species as the presence of supernumerary digits or transposition of the viscera.

Hardy, G. H. 1908. Mendelian proportions in a mixed population. *Science* 28:49–50.

Shortly after the rediscovery of Mendel's laws of inheritance in 1900, some British scientists, the so-called Biometricians, following the work of Francis Galton and the leadership of Karl Pearson, argued that Mendelian heredity could not account for continuous variation, as would be the case for height, weight, fitness, predisposition to disease, and other "metric" traits.

A different sort of "objection" came from a confusion between the frequency of a trait in the progeny of a cross and its frequency in the population. The statistician G. Udney Yule (1871–1951) pointed out that, if Mendel's laws were correct, the frequency in the human population of a dominant trait, such as brachydactyly, would reach 75 percent. This, of course, would be the frequency of a dominant trait, such as brachydactyly, in the progeny of a cross where each one of the parents carried the dominant trait. When the Cambridge geneticist Reginald C. Punnett (1875–1967) told the mathematician Godfrey Harold Hardy (1877–1947) about Yule's assertion, Hardy was able to explain, with simple algebra, that such would not be the case but rather that, "in the absence of counteracting factors" (such as new mutations or natural selection), the frequency of brachydactyly, like that of any other trait, would remain constant in the population from generation to generation. Hardy considered the result mathematically trivial, but at the insistence of Punnett, published it in *Science*.

Consider two alleles, *a* and *b*, with frequencies p and q, such that $p + q = 1$. Assuming random mating, as it would be expected in a large population, the frequency of the genotypes would be given by the binomial expansion of the sum of gene frequencies: $(p+q)^2 = p^2 + 2pq + q^2$, corresponding to *aa*, *ab*, and *bb*. The geneticist Curt Stern noted in a 1943 *Science* article that the German physician Wilhem Weinberg (1862–1937) had published the same result in 1908 and had extended it to the case of three or more alleles. Consider three alleles *a, b, c*, with their respective frequencies p, q, and r adding to one. The equilibrium frequencies of the corresponding genotypes would be given by $(p+q+r)^2 = p^2$ (*aa*), q^2 (*bb*), r^2 (*cc*), $2pq$ (*ab*), $2pr$ (*ac*), and $2qr$ (*bc*).

This general result came to be known as the Hardy-Weinberg law, which is fundamental in population genetics, the study of genetic changes in populations. The Hardy-Weinberg law in genetics is analogous to Newton's first law in mechanics, which says that a body remains at rest or maintains a constant velocity when not acted upon by a net external force. Bodies are always acted upon by external forces, but the first law is the point of departure for applying other laws. The Hardy-Weinberg law says that, in the absence of disturbing processes, gene frequencies do not change over generations. Processes that change gene frequencies are always present in populations of organisms, and without them there would be no evolution. The Hardy-Weinberg law is the point of departure from which the effects of the processes of change are calculated.

RELATED READING

Dunn, L. C. 1965. A short history of genetics. McGraw-Hill, New York, New York, USA.

Castle, W. E. 1903. The laws of heredity of Galton and Mendel, and some laws governing race improvement by selection. Proceedings of the National Academy of Sciences, USA 39:223–242.

Punnett, R. C. 1950. Early days of genetics. Heredity 4:1–10.

Stern, C. 1943. The Hardy-Weinberg Law. Science 97:137–138.

Stern, C. 1962. Wilhelm Weinberg, 1862–1937. Genetics 47:1–5.

Weinberg, W. 1908. Über dem Nachweis der Vererbung beim Menschen. Ver. Vaterl. Naturk. Württemberg 64:369–382. English translation in S. H. Boyer. 1963. Papers on human genetics. Prentice-Hall, Englewood Cliffs, New Jersey, USA.

DISCUSSION AND CORRESPONDENCE

MENDELIAN PROPORTIONS IN A MIXED POPULATION

To the Editor of Science: I am reluctant to intrude in a discussion concerning matters of which I have no expert knowledge, and I should have expected the very simple point which I wish to make to have been familiar to biologists. However, some remarks of Mr. Udny Yule, to which Mr. R. C. Punnett has called my attention, suggest that it may still be worth making.

In the *Proceedings of the Royal Society of Medicine* (Vol. I., p. 165) Mr. Yule is reported to have suggested, as a criticism of the Mendelian position, that if brachydactyly is dominant "in the course of time one would expect, in the absence of counteracting factors, to get three brachydactylous persons to one normal."

It is not difficult to prove, however, that such an expectation would be quite groundless. Suppose that *Aa* is a pair of Mendelian characters, *A* being dominant, and that in any given generation the numbers of pure dominants (*AA*), heterozygotes (*Aa*), and pure recessives (*aa*) are as $p:2q:r$. Finally, suppose that the numbers are fairly large, so that the mating may be regarded as random, that the sexes are evenly distributed among the three varieties, and that all are equally fertile. A little mathematics of the multiplication-table type is enough to show that in the next generation the numbers will be as

$$(p+q)^2:2(p+q)(q+r):(q+r)^2,$$

or as $p_1:2q_1:r_1$, say.

The interesting question is—in what circumstances will this distribution be the same as that in the generation before? It is easy to see that the condition for this is $q^2=pr$. And since $q_1^2=p_1r_1$, whatever the values of p, q and r may be, the distribution will in any case continue unchanged after the second generation.

Suppose, to take a definite instance, that *A* is brachydactyly, and that we start from a population of pure brachydactylous and pure normal persons, say in the ratio of 1:10,000. Then $p=1$, $q=0$, $r=10{,}000$ and $p_1=1$, $q_1=10{,}000$, $r_1=100{,}000{,}000$. If brachydactyly is dominant, the proportion of brachydactylous persons in the second generation is 20,001:100,020,001, or practically 2:10,000, twice that in the first generation; and this proportion will afterwards have no tendency whatever to increase. If, on the other hand, brachydactyly were recessive, the proportion in the second generation would be 1:100,020,001, or practically 1:100,000,000, and this proportion would afterwards have no tendency to decrease.

In a word, there is not the slightest foundation for the idea that a dominant character should show a tendency to spread over a whole population, or that a recessive should tend to die out.

I ought perhaps to add a few words on the effect of the small deviations from the theoretical proportions which will, of course, occur in every generation. Such a distribution as $p_1:2q_1:r_1$, which satisfies the condition $q_1^2=p_1r_1$, we may call a *stable* distribution. In actual fact we shall obtain in the second generation not $p_1:2q_1:r_1$ but a slightly different distribution $p_1':2q_1':r_1'$, which is not "stable." This should, according to theory, give us in the third generation a "stable" distribution $p_2:2q_2:r_2$, also differing slightly from $p_1:2q_1:r_1$; and so on. The sense in which the distribution $p_1:2q_1:r_1$ is "stable" is this, that if we allow for the effect of casual deviations in any subsequent generation, we should, according to theory, obtain at the next generation a new "stable" distribution differing but slightly from the original distribution.

I have, of course, considered only the very simplest hypotheses possible. Hypotheses other that that of purely random mating will give different results, and, of course, if, as appears to be the case sometimes, the character is not independent of that of sex, or has an influence on fertility, the whole question may be greatly complicated. But such complications seem to be irrelevant to the simple issue raised by Mr. Yule's remarks.

G. H. Hardy

Trinity College, Cambridge,
April 5, 1908

P. S. I understand from Mr. Punnett that he has submitted the substance of what I have said above to Mr. Yule, and that the latter would accept it as a satisfactory answer to the difficulty that he raised. The "stability" of the particular ratio 1:2:1 is recognized by Professor Karl Pearson (*Phil. Trans. Roy. Soc.* (A), vol. 203, p. 60).

Morgan, T. H. 1910. Sex limited inheritance in Drosophila. *Science* 32:120–122.

Thomas Hunt Morgan (1886–1945) received in 1933 the Nobel Prize in Physiology or Medicine for his discovery of "heredity transmission mechanisms in Drosophila." Morgan's name remains forever recognized for his chromosome theory of heredity, asserting that genes are responsible for hereditary traits, sequentially organized along chromosomes.

Morgan's early research was in embryology. His research in genetics with *Drosophila* flies started in 1908 at Columbia University (New York), where his laboratory became one of the most important centers of genetics research for decades to come. Several of Morgan's Ph.D. students and postdoctorals also worked there, such as A. H. Sturtevant, H. J. Muller, C. B. Bridges, and the eminent evolutionist Theodosius Dobzhansky (1900–1975), who joined Morgan's lab at Columbia University in 1927 and moved with him to California Institute of Technology in 1928. In 1940, Dobzhansky returned to Columbia as a faculty member.

In 1909 Morgan discovered one white-eyed male in a *Drosophila* culture. Following a series of crosses between this male and normal red-eyed females, Morgan discovered sex-linked inheritance. The gene for eye color is on the X chromosome. The females carry two X chromosomes and the males, only one. Morgan's paper is a model example of the scientific (hypothetico-deductive) method. The results of some early crosses inspired Morgan to suggest an explanatory hypothesis (sex-linked inheritance), which he then proceeded to test with two series of experiments. The male/female *Drosophila* chromosome composition, denoted as XY/XX, is also characteristic of humans, mammals, and many sorts of organisms. In birds, the females are the heterogametic sex (ZW) and the males are homogametic (WW). In grasshoppers and some other insects, the sex chromosome is diploid (two sets of chromosomes) in females (XX) but haploid (one set of chromosomes) in males (XO). Other chromosome modes of sex determination occur in honeybees and other social insects, where the females are diploid and the males are haploid. Temperature and other environmental conditions are determinants of sex in some reptiles and fishes.

For more than a decade after Morgan's work, the Y chromosome was assumed to be genetically empty. The first demonstration of a gene located on the Y chromosome was made by Antonio de Zulueta in 1925. The beetle *Phytodecta variabilis* has four phenotypes: lined, yellow, red, and black, distinguished by the color and pattern of the elytra. These phenotypes are determined by four alleles of a single gene located in the Y chromosome. In humans the Y chromosome carries genes, many in multiple copies, with mostly testis-specific functions.

RELATED READING

Bridges, C. B. 1925. Sex in relation to chromosomes and genes. American Naturalist 59:127–137.

Charlesworth, B. 1991. The evolution of sex chromosomes. Science 251:1030–1033.

Rice, W. R. 1996. Evolution of the Y sex chromosome in animals. BioScience 46:331–343.

Steinman, M., S. Steinman, and F. Lottspeich. 1993. How Y chromosomes become inert. Proceedings of the National Academy of Sciences, USA 9:5737–5741.

Zulueta, A., de. 1925. La herencia ligada al sexo en el coleóptero *Phytodecta variabilis* (OP.). EOS, Revista Española de Entomología 1:203–209.

SPECIAL ARTICLES

SEX LIMITED INHERITANCE IN DROSOPHILA

In a pedigree culture of *Drosophila* which had been running for nearly a year through a considerable number of generations, a male appeared with white eyes. The normal flies have brilliant red eyes.

The white-eyed male, bred to his red-eyed sisters, produced 1,237 red-eyed offspring, (F_1), and 3 white-eyed males. The occurrence of these three white-eyed males (F_1) (due evidently to further sporting) will, in the present communication, be ignored.

The F_1 hybrids, inbred, produced:

2,459 red-eyed females,
1,011 red-eyed males,
782 white-eyed males.

No white-eyed females appeared. The new character showed itself therefore to be sex limited in the sense that it was transmitted only to the grandsons. But that the character is not incompatible with femaleness is shown by the following experiment.

The white-eyed male (mutant) was later crossed with some of his daughters (F_1), and produced:

129 red-eyed females,
132 red-eyed males,
88 white-eyed females,
86 white-eyed males.

The results show that the new character, white eyes, can be carried over to the females by a suitable cross, and is in consequence in this sense not limited to one sex. It will be noted that the four classes of individuals occur in approximately equal numbers (25 per cent.).

An Hypothesis to Account for the Results.—The results just described can be accounted for by the following hypothesis. Assume that all of the spermatozoa of the white-eyed male carry the "factor" for white eyes "W"; that half of the spermatozoa carry a sex factor "X" the other half lack it, *i. e.*, the male is heterozygous for sex. Thus the symbol for the male is "WWX," and for his two kinds of spermatozoa WX—W.

Assume that all of the eggs of the red-eyed female carry the red-eyed "factor" R; and that all of the eggs (after reduction) carry one X, each, the symbol for the red-eyed female will be therefore RRXX and that for her eggs will be RX—RX.

When the white-eyed male (sport) is crossed with his red-eyed sisters, the following combinations result:

WX — W (male)
RX — RX (female)

RWXX (50 %) — RWX (50 %)
Red female — Red male

When these F_1 individuals are mated, the following table shows the expected combinations that result:

RX — WX (F_1 female)
RX — W (F_1 male)

RRXX	RWXX	RWX	WWX
(25 %)	(25 %)	(25 %)	(25 %)
Red female	Red female	Red male	White male

It will be seen from the last formulæ that the outcome is Mendelian in the sense that there are three reds to one white. But it is also apparent that all of the whites are confined to the male sex.

It will also be noted that there are two classes of red females—one pure RRXX and one hybrid RWXX—but only one class of red males (RWX). This point will be taken up later. In order to obtain these results it is necessary to assume, as in the last scheme, that, when the two classes of the spermatozoa are formed in the F_1 red male (RWX), R and X go together—otherwise the results will not follow (with the symbolism here used). This all-important point can not be fully discussed in this communication.

The hypothesis just utilized to explain these results first obtained can be tested in several ways.

Verification of Hypothesis

First Verification.—If the symbol for the white male is WWX, and for the white female WWXX, the germ cells will be WX—W (male) and WX—WX (female), respectively. Mated, these individuals should give

WX — W (male)
WX — WX (female)

WWXX (50 %) — WWX (50 %)
White female — White male

All of the offspring should be white, and male and female in equal numbers; this in fact is the case.

Second Verification.—As stated, there should be two classes of females in the F_2 generation, namely, RRXX and RWXX. This can be tested by pairing individual females with white males. In the one instance (RRXX) all the offspring should be red—

RX — RX (female)
WX — W (male)

RWXX — RWX

and in the other instance (RWXX) there should be four classes of individuals in equal numbers, thus:

RX — WX (female)
WX — W (male)

RWXX — WWXX — RWX — WWX

Tests of the F_2 red females show in fact that these two classes exist.

Third Verification.—The red F_1 females should all be RWXX, and should give with any white male the four combinations last described. Such in fact is found to be the case.

Fourth Verification.—The red F_1 males (RWX) should also be heterozygous. Crossed with white females (WWXX) all the female offspring should be red-eyed, and all the male offspring white-eyed, thus:

RX — W (red male)
WX — WX (white female)

RWXX — WWX

Here again the anticipation was verified, for all of the females were red-eyed and all of the males were white-eyed.

Crossing the New Type with Wild Males and Females

A most surprising fact appeared when a white-eyed female was paired to a wild, red-eyed male, *i. e.*, to an individual of an unrelated stock. The anticipation was that wild males and females alike carry the factor for red eyes, but the experiments showed that all wild males are heterozygous for red eyes, and that all the wild females are homozygous. Thus when the white-eyed female is crossed with a wild red-eyed male, all of the female offspring are red-eyed, and all of the male offspring white-eyed. The results can be accounted for on the assumption that the wild male is RWX. Thus:

RX — W (red male)
WX — WX (white female)

RWXX (50 %) — WWX (50 %)

The converse cross between a white-eyed male RWX and a wild, red-eyed female shows that the wild female is homozygous both for X and for red eyes. Thus:

WX — W (white male)
RX — RX (red female)

RWXX (50 %) — RWX (50 %)

The results give, in fact, only red males and females in equal numbers.

General Conclusions

The most important consideration from these results is that in every point they furnish the converse evidence from that given by Abraxas as worked out by Punnett and Raynor. The two cases supplement each other in every way, and it is significant to note in this connection that in nature only females of the sport *Abraxas lacticolor* occur, while in *Drosophila* I have obtained only the male sport. Significant, too, is the fact that analysis of the result shows that the wild female *Abraxas grossulariata* is heterozygous for color and sex, while in *Drosophila* it is the male that is heterozygous for these two characters.

Since the wild males (RWX) are heterozygous for red eyes, and the female (RXRX) homozygous, it seems probable that the sport arose from a change in a single egg of such a sort that instead of being RX (after reduction) the red factor dropped out, so that RX became WX or simply OX. If this view is correct it follows that the mutation took place in the egg of a female from which a male was produced by combination with the sperm carrying no X, no R (or W in our formulæ). In other words, if the formula for the eggs of the normal female is RX—RX, then the formula for the particular egg that sported will be WX; *i. e.*, one R dropped out of the egg leaving it WX (or no R and one X), which may be written OX. This egg we assume was fertilized by a male-producing sperm. The formula for the two classes of spermatozoa is RX—O. The latter, O, is the male-producing sperm, which combining with the egg OX (see above) gives OOX (or WWX), which is the formula for the white-eyed male mutant.

The transfer of the new character (white eyes) to the female (by crossing a white-eyed male, OOX to a heterozygous female (F_1)) can therefore be expressed as follows:

OX — O (white male)
RX — OX (F_1 female)

RXOX	RXO	OOXX	OOX
Red female	Red male	White female	White male

It now becomes evident why we found it necessary to assume a coupling of R and X in one of the spermatozoa of the red-eyed F_1 hybrid (RXO). The fact is that this R and X are combined, and have never existed apart.

It has been assumed that the white-eyed mutant arose by a male-producing sperm (O) fertilizing an egg (OX) that had mutated. It may be asked what would have been the result if a female-producing sperm (RX) had fertilized this egg (OX)? Evidently a heterozygous female RXOX would arise, which, fertilized later by any normal male (RX—O) would produce in the next generation pure red females RRXX, red heterozygous females RXOX, red males RXO, and white males OOX (25 per cent.). As yet I have found no evidence that white-eyed sports occur in such numbers. Selective fertilization may be involved in the answer to this question.

T. H. Morgan

Woods Hole, Mass.,
July 7, 1910

Muller, H. J. 1927. Artificial transmutation of the gene. *Science* 66:84–87.

Following his discovery in 1909–1910 of sex-linked Mendelian heredity, T. H. Morgan (see the previous article) established a *Drosophila* genetics laboratory at Columbia University in New York City. This lab famously became known as the "fly room." Alfred H. Sturtevant (1891–1970) and Calvin B. Bridges (1889–1867) became incorporated in 1910–1911, and shortly thereafter Hermann J. Muller (1890–1967) joined the team. All three coauthored with Morgan *The Mechanism of Mendelian Heredity* (1915; revised edition, 1922), which together with Morgan's *The Theory of the Gene* (1926) became leading genetics textbooks in the early decades of the twentieth century. Muller's greatest interest at the time was the characterization of mutations as the source of genetic variation and of evolution. He determined that spontaneous gene mutations are mostly recessive and deleterious. Most significant was the discovery that mutations can be induced by the ionizing radiation of X-rays, for which he received the Nobel Prize for Physiology or Medicine in 1946.

In the 1930s, Muller worked first for a year in Germany, and from 1933 to 1937 in Moscow at the Institute of Genetics, where he had been invited by the great Russian plant geneticist Nicolai I. Vavilov, an invitation to which Muller was favorably predisposed by his leftist political preferences, which would eventually change. In collaboration with Newton E. Morton and James F. Crow, Muller developed a theory about the accumulation of deleterious mutations (the *mutational load*) in humans, as a consequence of exposure to environmental mutagens and of medicine's cure of carriers of severe hereditary diseases, which then would transmit the deleterious genes to their descendants. After the atomic bombings of Hiroshima and Nagasaki in Japan, Muller became much involved in political campaigns about the dangers of radiation. His efforts eventually led to the elimination of atomic testing in the atmosphere.

RELATED READING

Carlson, E. A. 1966. The gene: a critical history. Saunders, Philadelphia, Pennsylvania, USA.

Carlson, E. A. 1981. Genes, radiation, and society: the life and work of H. J. Muller. Cornell University Press, Ithaca, New York, USA.

Muller, H. J. 1950. Our load of mutations. American Journal of Human Genetics 2:111–176.

Muller, H. J. 1962. Studies in genetics. Indiana University Press, Bloomington, Indiana, USA.

Sturtevant, A. H. 1965. A history of genetics. Harper and Row, New York, New York, USA.

ARTIFICIAL TRANSMUTATION OF THE GENE

Most modern geneticists will agree that gene mutations form the chief basis of organic evolution, and therefore of most of the complexities of living things. Unfortunately for the geneticists, however, the study of these mutations, and, through them, of the genes themselves, has heretofore been very seriously hampered by the extreme infrequency of their occurrence under ordinary conditions, and by the general unsuccessfulness of attempts to modify decidedly, and in a sure and detectable way, this sluggish "natural" mutation rate. Modification of the innate nature of organisms, for more directly utilitarian purposes, has of course been subject to these same restrictions, and the practical breeder has hence been compelled to remain content with the mere making of recombinations of the material already at hand, providentially supplemented, on rare and isolated occasions, by an unexpected mutational windfall. To these circumstances are due the wide-spread desire on the part of biologists to gain some measure of control over the hereditary changes within the genes.

It has been repeatedly reported that germinal changes, presumably mutational, could be induced by X or radium rays, but, as in the case of the similar published claims involving other agents (alcohol, lead, antibodies, etc.), the work has been done in such a way that the meaning of the data, as analyzed from a modern genetic standpoint, has been highly disputatious at best; moreover, what were apparently the clearest cases have given negative or contrary results on repetition. Nevertheless, on theoretical grounds, it has appeared to the present writer that radiations of short wave length should be especially promising for the production of mutational changes, and for this and other reasons a series of experiments concerned with this problem has been undertaken during the past year on the fruit fly, *Drosophila melanogaster*, in an attempt to provide critical data. The well-known favorableness of this species for genetic study, and the special methods evolved during the writer's eight years' intensive work on its mutation rate (including the work on temperature, to be referred to later), have finally made possible the finding of some decisive effects, consequent upon the application of X-rays. The effects here referred to are truly mutational, and not to be confused with the well-known effects of X-rays upon the distribution of the chromatin, expressed by non-disjunction, non-inherited crossover modifications, etc. In the present condensed digest of the work, only the broad facts and conclusions therefrom, and some of the problems raised, can be presented, without any details of the genetic methods employed, or of the individual results obtained.

It has been found quite conclusively that treatment of the sperm with relatively heavy doses of X-rays induces the occurrence of true "gene mutations" in a high proportion of the treated germ cells. Several hundred mutants have been obtained in this way in a short time and considerably more than a hundred of the mutant genes have been followed through three, four or more generations. They are (nearly all of them, at any rate) stable in their inheritance, and most of them behave in the manner typical of the Mendelian chromosomal mutant genes found in organisms generally. The nature of the crosses was such as to be much more favorable for the detection of mutations in the X-chromosomes than in the other chromosomes, so that most of the mutant genes dealt with were sex-linked; there was, however, ample proof that mutations were occurring similarly throughout the chromatin. When the heaviest treatment was given to the sperm, about a seventh of the offspring that hatched from them and bred contained individually detectable mutations in their treated X-chromosome. Since the X forms about one fourth of the haploid chromatin, then, if we assume an equal rate of mutation in all the chromosomes (per unit of their length), it follows that almost "every other one" of the sperm cells capable of producing a fertile adult contained an "individually detectable" mutation in some chromosome or other. Thousands of untreated parent flies were bred as controls in the same way as the treated ones. Comparison of the mutation rates under the two sets of conditions showed that the heavy treatment had caused a rise of about fifteen thousand per cent. in the mutation rate over that in the untreated germ cells.

Regarding the types of mutations produced, it was found that, as was to have been expected both on theoretical grounds and on the basis of the previous mutation studies of Altenburg and the writer, the lethals (recessive for the lethal effect, though some were dominant for visible effects) greatly outnumbered the non-lethals producing a visible morphological abnormality. There were some "semi-lethals" also (defining these as mutants having a viability ordinarily between about 0.5 per cent. and 10 per cent. of the normal), but, fortunately for the use of lethals as an index of mutation rate, these were not nearly so numerous as the lethals. The elusive class of "invisible" mutations that caused an even lesser reduction of viability, not readily confusable with lethals, appeared larger than that of the semi-lethals, but they were not subjected to study. In addition, it was also possible to obtain evidence in these experiments for the first time, of the occurrence of dominant lethal genetic changes, both in the X and in the other chromosomes. Since the zygotes receiving these never developed to maturity, such lethals could not be detected individually, but their number was so great that through egg counts and effects on the sex ratio evidence could be obtained of them *en masse*. It was found that their numbers are of the same order of magnitude as those of the recessive lethals. The "partial sterility" of treated males is, to an appreciable extent at least, caused by these dominant lethals. Another abundant class of mutations not previously recognized was found to be those which, when heterozygons, cause sterility but produce no detectable change in appearance; these too occur in numbers rather similar to those of the recessive lethals, and they may hereafter afford one of the readiest indices of the general mutation rate, when this is high. The sterility thus caused, occurring as it does in the offspring of the treated individuals, is of course a separate phenomenon from the "partial sterility" of the treated individuals themselves, caused by the dominant lethals.

In the statement that the proportion of "individually detectable mutations" was about one seventh for the X, and therefore nearly one half for all the chromatin, only the recessive lethals and semi-lethals and the "visible" mutants were referred to. If the dominant lethals, the dominant and recessive sterility genes and the "invisible" genes that merely reduce (or otherwise affect) viability or fertility had been taken into account, the percentage of mutants given would have been far higher, and it is accordingly evident that in reality the great majority of the treated sperm cells contained mutations of some kind or other. It appears that the rate of gene mutation after X-ray treatment is high enough, in proportion to the total number of genes, so that it will be practicable to study it even in the case of individual loci, in an attack on problems of allelomorphism, etc.

Returning to a consideration of the induced mutations that produced visible effects, it is to be noted that the conditions of the present experiment allowed the detection of many which approached or overlapped the normal type to such an extent that ordinarily they would have escaped observation, and definite evidence was thus obtained of the relatively high frequency of such changes here, as compared with the more conspicuous ones. The belief has several times been expressed in the *Drosophila* literature that this holds true in the case of "natural" mutations in this organism, but it has been founded only on "general impressions"; Baur, however, has demonstrated the truth of it in *Antirrhinum*. On the whole, the visible mutations caused by raying were found to be similar, in their general characteristics, to those previously detected in non-rayed material in the extensive observations on visible mutations in *Drosophila* carried out by Bridges and others. A considerable proportion of the induced visible mutations were, it is true, in loci in which mutation apparently had never been observed before, and some of these involved morphological effects of a sort not exactly like any seen previously (*e.g.*, "splotched wing," "sex-combless," etc.), but, on the other hand, there were also numerous repetitions of mutations previously known. In fact, the majority of the well-known mutations in the X-chromosome of *Drosophila melanogaster*, such as "white eye," "miniature wing," "forked bristles," etc., were reobtained, some of them several times. Among the visible mutations found, the great majority were recessive, yet there was a considerable "sprinkling" of dominants, just as in other work. All in all, then, there can be no doubt that many, at least, of the changes produced by X-rays are of just the same kind as the "gene mutations" which are obtained, with so much greater rarity, without such treatment, and which we believe furnish the building blocks of evolution.

In addition to the gene mutations, it was found that there is also caused by X-ray treatment a high proportion of rearrangements in the linear order of the genes. This was evidenced in general by the frequent inherited disturbances in crossover frequency (at least 3 per cent. were detected in the X-chromosome alone, many accompanied but some unaccompanied by lethal effects), and evidenced specifically by various cases that were proved in other ways to involve inversions, "deficiencies," fragmentations, translocations, etc., of portions of a chromosome. These cases are making possible attacks on a number of genetic problems otherwise difficult of approach.

The transmuting action of X-rays on the genes is not confined to the sperm cells, for treatment of the unfertilized females causes mutations about as readily as treatment of the males. The effect is produced both on oöcytes and early oögonia. It should be noted especially that, as in mammals, X-rays (in the doses used) cause a period of extreme infertility, which commences soon after treatment and later is partially recovered from. It can be stated positively that the return of fertility does not mean that the new crop of eggs is unaffected, for these, like those mature eggs that managed to survive, were found in the present experiments to contain a high proportion of mutant genes (chiefly lethals, as usual). The practice, common in current X-ray therapy, of giving treatments that do not certainly result in permanent sterilization, has been defended chiefly on the ground of a purely theoretical conception that eggs produced after the return of fertility must necessarily represent "uninjured" tissue. As this presumption is hereby demonstrated to be faulty it would seem incumbent for medical practice to be modified accordingly, at least until genetically sound experimentation upon mammals can be shown to yield results of a decisively negative character. Such work upon mammals would involve a highly elaborate undertaking, as compared with the above experiments on flies.

From the standpoint of biological theory, the chief interest of the present experiments lies in their bearing on the problems of the composition and behavior of chromosomes and genes. Through special genetic methods it has been possible to obtain some information concerning the manner of distribution of the transmuted genes amongst the cells of the first and later zygote generations following treatment. It is found that the mutation does not usually involve a permanent alteration of all of the gene substance present at a given chromosome locus at the time of treatment, but either affects in this way only a portion of that substance, or else occurs subsequently, as an after-effect, in only one of two or more descendant genes derived from the treated gene. An extensive series of experiments, now in project, will be necessary for deciding conclusively between these two possibilities, but such evidence as is already at hand speaks rather in favor of the former. This would imply a somewhat compound structure for the gene (or chromosome as a whole) in the sperm cell. On the other hand, the mutated tissue is distributed in a manner that seems inconsistent with a general applicability of the theory of "gene elements" first suggested by Anderson in connection with variegated pericarp in maize, then taken up by Eyster, and recently reenforced by Demerec in *Drosophila virilis.*

A precociously doubled (or further multiplied) condition of the chromosomes (in "preparation" for later mitoses) is all that is necessary to account for the above-mentioned *fractional effect* of X-rays on a given locus; but the theory of a divided condition of each gene, into a number of (originally identical) "elements" that can become separated somewhat indeterminately at mitosis, would lead to expectations different from the results that have been obtained in the present work. It should, on that theory, often have been found here, as in the variegated corn and the eversporting races of *D. virilis,* that mutated tissue gives rise to normal by frequent "reverse mutation"; moreover, treated tissues not at first showing a mutation might frequently give rise to one, through a "sorting out" of diverse elements, several generations after treatment. Neither of these effects was found. As has been mentioned, the mutants were found to be stable through several generations, in the great majority of cases at least. Hundreds of non-mutated descendants of treated germ cells, also, were carried through several generations, without evidence appearing of the production of mutations in generations subsequent to the first. Larger numbers will be desirable here, however, and further experiments of a different type have also been planned in the attack on this problem of gene structure, which probably can be answered definitely.

Certain of the above points which have already been determined, especially that of the fractional effect of X-rays, taken in conjunction with that of the production of dominant lethals, seem to give a clue to the especially destructive action of X-rays on tissues in which, as in cancer, embryonic and epidermal tissues, the cells undergo repeated divisions (though the operation of additional factors, *e.g.,* abnormal mitoses, tending towards the same result, is not thereby precluded); moreover, the converse effect of X-rays, in occasionally producing cancer, may also be associated with their action in producing mutations. It would be premature, however, at this time to consider in detail the various X-ray effects previously considered as "physiological," which may now receive a possible interpretation in terms of the gene-transmuting property of X-rays; we may more appropriately confine ourselves here to matters which can more strictly be demonstrated to be genetic.

Further facts concerning the nature of the gene may emerge from a study of the comparative effects of varied dosages of X-rays, and of X-rays administered at different points in the life cycle and under varied conditions. In the experiments herein reported, several different dosages were made use of, and while the figures are not yet quite conclusive they make it probable that, within the limits used, the number of recessive lethals does not vary directly with the X-ray energy absorbed, but more nearly with the square root of the latter. Should this lack of exact proportionality be confirmed, then, as Dr. Irving Langmuir has pointed out to me, we should have to conclude that these mutations are not caused directly by single quanta of X-ray energy that happen to be absorbed at some critical spot. If the transmuting effect were thus relatively indirect there would be a greater likelihood of its being influenceable by other physico-chemical agencies as well, but our problems would tend to become more complicated. There is, however, some danger in using the total of lethal mutations produced by X-rays as an index of gene mutations occurring in single loci, for some lethals, involving changes in crossover frequency, are probably associated with rearrangements of chromosome regions, and such changes would be much less likely than "point mutations" to depend on single quanta. A reexamination of the effect of different dosages must therefore be carried out, in which the different types of mutations are clearly distinguished from one another. When this question is settled, for a wide range of dosages and developmental stages, we shall also be in a position to decide whether or not the minute amounts of gamma radiation present in nature cause the ordinary mutations which occur in wild and in cultivated organisms in the absence of artificially administered X-ray treatment.

As a beginning in the study of the effect of varying other conditions, upon the frequency of the mutations produced by X-rays, a comparison has been made between the mutation frequencies following the raying of sperm in the male and in the female receptacles, and from germ cells that were in different portions of the male genital system at the time of raying. No decisive differences have been observed. It is found, in addition, that aging the sperm after treatment, before fertilization, causes no noticeable alteration in the frequency of detectable mutations. Therefore the death rate of the mutant sperm is no higher than that of the unaffected ones; moreover, the mutations can not be regarded as secondary effects of any semi-lethal physiological changes which might be supposed to have occurred more intensely in some ("more highly susceptible") spermatozoa than in others.

Despite the "negative results" just mentioned, however, it is already certain that differences in X-ray influences, by themselves, are not sufficient to account for all variations in mutation frequency, for the present X-ray work comes on the heels of the determination of mutation rate being dependent upon temperature (work as yet unpublished). This relation had first been made probable by work of Altenburg and the writer in 1918, but was not finally established until the completion of some experiments in 1926. These gave the first definite evidence that gene mutation may be to any extent controllable, but the magnitude of the heat effect, being similar to that found for chemical reactions in general, is too small, in connection with the almost imperceptible "natural" mutation rate, for it, by itself, to provide a powerful tool in the mutation study. The result, however, is enough to indicate that various factors besides X-rays probably do affect the composition of the gene, and that the measurement of their effects, at least when in combination with X-rays, will be practicable. Thus we may hope that problems of the composition and behavior of the gene can shortly be approached from various new angles, and new handles found for their investigation, so that it will be legitimate to speak of the subject of "gene physiology," at least, if not of gene physics and chemistry.

In conclusion, the attention of those working along classical genetic lines may be drawn to the opportunity, afforded them by the use of X-rays, of creating in their chosen organisms a series of artificial races for use in the study of genetic and "phaenogenetic" phenomena. If, as seems likely on general considerations, the effect is common to most organisms, it should be possible to produce, "to order," enough mutations to furnish respectable genetic maps, in their selected species, and, by the use of the mapped genes, to analyze the aberrant chromosome phenomena simultaneously obtained. Similarly, for the practical breeder, it is hoped that the method will ultimately prove useful. The time is not ripe to discuss here such possibilities with reference to the human species.

The writer takes pleasure in acknowledging his sincere appreciation of the cooperation of Dr. Dalton Richardson, Roentgenologist, of Austin, Texas, in the work of administering the X-ray treatments.

H. J. Muller

University of Texas

Wright, S. 1930. The genetical theory of natural selection: a review. *Journal of Heredity* 21:349–356.

The understanding of natural selection as the main guiding process in evolution began in earnest with the birth of population genetics in the 1920s, largely emerging from the work of Ronald A. Fisher (1890–1962), John B. S. Haldane (1892–1964), and Sewall Wright (1889–1988). Fisher's book *The Genetical Theory of Natural Selection* (1930), reviewed here by Wright, was the first systematic attempt in English to harmonize Darwin's observations on natural variation with Mendelian genetics. Haldane's main contribution was *The Causes of Evolution* (1932). In the early 1920s, Wright published several papers investigating the dynamics and equilibrium frequencies of gene variants, but his most significant early contribution was a 60-page article published in 1931, "Evolution in Mendelian populations." He published numerous papers over the following decades, consummating his contributions with his magnum opus, *Evolution and the Genetics of Populations*, in four volumes, published in 1968, 1969, 1977 and 1978, late in his long life, when he was between 78 and 88 years old. Volume 3, *Experimental Results and Evolutionary Deductions*, and volume 4, *Variability within and among Natural Populations,* are primarily dedicated, respectively, to the analysis of laboratory experiments and of studies of variability in natural populations, while the first two volumes are mostly theoretical.

In his review of Fisher's book, Wright indicates similarities and differences between their two views concerning significant genetic components of the evolutionary processes, pointing out that Fisher overlooks the contributions of chromosome aberrations and hybridization to speciation, as well as the role of inbreeding leading to the differentiation of local populations. Most of Wright's review, however, contrasts their different mathematical treatments of the population dynamics and equilibrium conditions of genetic factors. Wright also points out their "differences in interpretation" of the theoretical results. Wright notices in particular Fisher's formulation of the "fundamental theorem of natural selection," which would play a preeminent role in the future development of evolutionary genetics: "The rate of increase in fitness of any organism at any time is equal to its genetic variation in fitness at that time."

RELATED READING

Chetverikov, S. S. 1959. On certain aspects of the evolutionary process from the standpoint of genetics. Proceedings of the American Philosophical Society 105:167–195. (English translation of Chetverikov's original paper, published in Russian in 1929, often considered the first theoretical treatment of the population dynamics of genetic factors.)

Crow, J. F. 1994. Sewall Wright. December 21, 1889–March 3, 1988: a biographical memoir. The National Academies Press, Washington, D.C., USA.

Haldane, J. B. S. 1933. Science and human life. Harper, New York, New York, USA.

Provine, W. B. 1986. Sewall Wright and evolutionary biology. University of Chicago Press, Chicago, Illinois, USA.

Wright, S. 1931. Evolution in Mendelian populations. Genetics 16:97–159.

THE GENETICAL THEORY OF NATURAL SELECTION

A Review

SEWALL WRIGHT

Department of Zoology, University of Chicago

DURING the latter part of the nineteenth century, increasing difficulty was felt in accepting Darwin's conception of the evolutionary process as one in which variation merely plays the subordinate (though necessary) rôle of providing a field of potentialities, through which the actual direction of advance is determined by natural selection. Theories were developed according to which the "origin of species" was to be sought more directly in the "origin of variation." Most of these were Lamarckian, others, of which de Vries' theory was most important, were not. The rediscovery of Mendelian heredity was a direct consequence of the mutation theory of the origin of species and was naturally seized upon as supporting this view. Only gradually has it become apparent that the real implications of Mendelian heredity are exactly the opposite and that in fact, it supplies the answer to some of the main difficulties felt with Darwin's theory. Dr. Fisher has played a leading part in developing the statistical consequences of Mendelian heredity and here brings together his views in a unified form.* It is a book which is certain to take rank as one of the major contributions to the theory of evolution.

The first chapter is concerned with a comparison of the consequences of blending and particulate heredity. A consequence of blending heredity, which Dr. Fisher shows was well understood by Darwin, and which was felt by him, and others, as a major difficulty with his theory, is the fact that under such heredity, the variability of a population tends to be greatly reduced in each successive generation. The portion of the variance lost per generation is one-half (if there is no assortative mating) and after ten generations only one-tenth of one per cent is left. Thus Darwin felt constrained to believe that an enormous amount of new variation appears in each generation, the differences among brothers being of this sort. This variability must be seized upon at once by natural selection or it will be lost. With even a slight departure from randomness in its occurrence, direction of mutation, rather than natural selection becomes the guiding principle of evolution.

All of this was changed with the demonstration of particulate inheritance and orderly segregation. The frequencies of zygotes of the types *aa, Aa* and *AA* tend to remain indefinitely in the proportions of a binomial square $p^2+2pq+q^2$ where p and q are the proportion in which alternative genes are represented in the population. In a population of limited size, to be sure, there is some variability of gene frequency, due to the accidents of sampling from generation to generation, but this brings about only a very low rate of reduction of variance. As to the actual rate of reduction of variance (and of heterozygosis), Fisher here confirms the figure which I had obtained by the method of path coefficients, viz. $\frac{1}{2n}$ per generation, where n is the effective size of the breeding population. The modern geneticist may get an appreciation of the difficulties which confronted

*The Genetical Theory of Natural Selection, by R. A. Fisher Sc. D., F. R. S., Price $6.00. Oxford University Press, New York. 1930.

Darwin, in attempting to account for natural variability and to apply selection as a guiding principle, by considering the case of a self-fertilized line, in which the loss of variance actually is 50% per generation, the same as with a random breeding population under blending heredity. The difference that with any initial variability, the inbred line tends to split up into many diverse lines, while the population under blending heredity becomes fixed as of one type, further emphasizes the difficulty. That pure lines actually show very little genetic variability, Fisher points out, is convincing evidence that substantially all inheritance is Mendelian. A quotation will bring out his conclusions with regard to mutation and selection:

> For mutations to dominate the trend of evolution it is thus necessary to postulate mutation rates immensely greater than those which are known to occur and of an order of magnitude which in general would be incompatible with particulate inheritance. * * * The whole group of theories which ascribe to hypothetical physiological mechanisms, controlling the occurrence of mutations, a power of directing the course of evolution, must be set aside once the blending theory of inheritance is abandoned. The sole surviving theory is that of natural selection and it would appear impossible to avoid the conclusion that if any evolutionary phenomenon appears to be inexplicable on this theory it must be accepted at present merely as one of the facts which in the present state of knowledge seems inexplicable.

I may state at this point that I am in accord with Dr. Fisher on the rôle of mutation, except that I would perhaps allow occasional significance to chromosome aberration, and to hybridization, as direct species forming agencies. It appears to me, however, that in this statement and throughout the book, he overlooks the rôle of inbreeding as a factor leading to nonadaptive differentiation of local strains, through selection of which, adaptive evolution of the species as a whole may be brought about more effectively than through mass selection of individuals.

Distribution of Gene Frequencies

The central problem in the analysis of the statistical consequences of Mendelian heredity is that of determining the distribution of gene frequencies under the pressures of mutation, selection, migration, etc., and not least important, as affected by size of population. Under given conditions, what proportion of the genes will be fixed? How many will have frequencies in the neighborhood of 50%? How many 99%? How rapidly will new mutations attain fixation under favorable selection? Two of the chapters (IV, V), are devoted to a mathematical investigation of such questions. As I have recently presented certain results in this field,* it may be of interest to bring out the points of agreement and disagreement.

My approach to the subject was from a different angle than Dr. Fisher's in being through the problem of inbreeding. I found that the decrease in heterozygosis, to be expected under inbreeding (but ignoring new mutations and selection) could be obtained by an application of the method of path coefficients.[4] The method could be applied to complex pedigrees encountered in livestock, and studies of the history of the Shorthorn breed of cattle have been made by means of it by Dr. McPhee and myself[5] of Clydesdale horses by Calder,[1] and of Jersey cattle by Buchanan Smith.[3] In the case of random mating in a population of N_m males and N_f females, it gave as a close approximation $\frac{1}{8N_m}+\frac{1}{8N_f}$ as the rate of loss of heterozygosis (and hence of variance) per generation. With an equal number of males and females in a total breeding population of n this reduces to the $\frac{1}{2n}$ referred to above. Fisher, studying the problem of evolution of large popula-

*These results were presented at the 1929 meeting of the A. A. A. S. An abstract appeared in the *Anatomical Record* (Vol. 44, p. 287, 1929). The full paper is to appear in *Genetics*.

tions, made the first attempt to find the actual distribution of gene frequencies under various conditions.[2] He reached a solution for the case of unselected genes not replenished by mutation, which indicated loss of variance at the rate of $\frac{1}{4n}$ per generation, just half of the rate indicated by my method. His formula for the distribution of gene frequencies was expressed on a scale of the logarithms of the ratio of alternative gene frequencies $\log\frac{q}{1-q}$, a scale which has the advantage of stretching the important regions close to 0% and 100% and also of making the effect of simple selection uniform at all points. It is interesting to note, however, that on transforming his formula to the simple scale of percentage frequencies it indicates an equal number of genes at all frequencies (y = 1). He also obtained a solution for the case in which decrease in heterozygosis is just balanced by mutation.

On noting the discrepancy between his result and mine for decrease in the rate of heterozygosis, I was not able to correct a questionable point in his derivation, but was able to reach a formula for the distribution of gene frequencies in a different way. The result agreed with his solution in form (y = 1), but with the rate of decline as $\frac{1}{2n}$ per generation. In the case of loss of variance, balanced by mutation, the distribution differed considerably in form, being $y = \frac{1}{2n[.577+\log(2n)]q(1-q)}$ instead of his $\frac{1}{\pi\sqrt{q(1-q)}}$. It appeared further from this method that a selective advantage such that genes A and a reproduce in the ratio $1:1-S$ introduced an exponential term e^{4nsq} into the formula. This is valid, however, only for irreversible mutation and then only for extremely small values of the selection coefficient. It now appears that for reversible mutation and in any case for values considerably larger than $\frac{1}{2n}$ it should be e^{4nsq}. Appreciable rates of recurrence of mutation (u) and reverse mutation (v) were stated* to throw the formula into the form $y = Cq^{4nv-1}(1-q)^{4nu-1}$ a curve which for high mutation rates (relative to $\frac{1}{2n}$) approaches the form of a probability curve and indicates a random drifting of gene frequency about an equilibrium point. The case which has seemed most important to me is that of the effects of migration in a population which is a sub-group of a large one. The formula is similar mathematically to that for mutation. It is given below in a revised form.

These results were communicated to Dr. Fisher, who now finds on reexamination of his method, that the addition of a term which had seemed unimportant gives a confirmation of my formulae in the first two cases. He obtains on the other hand, a somewhat different form for the effect of selection, viz., $y = \frac{2}{aq}\left(\frac{1-e^{-4na(1-q)}}{1-e^{-4na}}\right)$ where his a is my s, except for a change of sign. The exact case which he deals with, is not one which I had considered, a fact which reflects our differences in viewpoint on the general problem. His formula refers to flux equilibrium with respect to an inexhaustible supply of irreversible mutations. On solving for it by my method I get results substantially identical with his as long as the selection coefficient is less than $\frac{1}{4n}$. Above this there is rapid divergence. His formula is undoubtedly a better approximation, and in fact, I may say that on reexamination, I find that I, in turn, have here neglected terms which should be taken account of. The general formula for a partially isolated population by my method, as now revised, is as follows: $y = Ce^{4nsq}q^{4n(mq_m+v)-1}(1-q)^{4n[m(1-q_m)+u]-1}$ where m is the rate of population exchange with the species as a whole, q_m is the gene frequency in the latter and s measures the differential selection of the group as compared with the species as a whole. If v is actually zero (completely irreversible mutation from an inexhaustible supply of genes), and no im-

*Presented without proof in *American Nat.* 63:556-561 (1929).

migration is assumed, the formula takes a somewhat different form and in fact reduces to Dr. Fisher's result, identically. Summing up, our mathematical results on the distribution of gene frequencies are now in complete agreement as far as comparable, although based on very different methods of attack. He has not yet checked my conclusions as to the effects of recurrent and reversible mutation and of immigration by his method.

Differences in Interpretation

There are, however, important differences in interpretation. Dr. Fisher is interested in the figure $\frac{1}{2n}$, measuring decrease in variance, only because of its extreme smallness, from which he argues that the effects of random sampling are negligible in evolution (except as bearing on the chances of loss of a recently originated gene). I, on the contrary, have attributed to the inbreeding effect, measured by this coefficient, an essential rôle in the theory of evolution, arguing that the effective breeding population, represented by n of the formula may after all be relatively small compared with the actual size of the population. In this view I have been encouraged by the rather high coefficients of inbreeding found even in entire breeds of livestock. Calder, for example, finds a rate of increase of the inbreeding coefficient in Scotch Clydesdale horses of nearly 1% per generation which let it be emphasized again is a direct determination of the value of $\frac{1}{2n}$ for this large breed, assuming as seems to be justified, that there is no important subdivision into local strains.

The core of Dr. Fisher's theory of selection is given in Chapter II. He reaches a formula on which he lays great emphasis as "the fundamental theorem of natural selection." "The rate of increase in fitness of any organism at any time is equal to its genetic variance in fitness at that time." This is given as exact for idealized populations in which fortuitous fluctuations in genetic composition have been excluded *i. e.,* in indefinitely large populations. He calculates the standard error of the rate of advance in fitness, due to such fluctuations, and concludes that this is negligibly small; even over a single generation, in populations of the order of size of natural species. This means that the small random fluctuations in the frequencies of individual genes balance each other in their effect on a selected character to such an extent that irregularities in evolutionary advance are of the second order with respect to the rate of advance. He compares this principle to the regular increase of entropy in a physical system. The only effective offset to undeviating increase in fitness, which he recognizes, is change of environment, living or non-living, which he points out must usually be for the worse. The net effect of natural selection, and change of environment is registered in an increase or decrease in numbers and a somewhat winding course of evolution.

The splitting of species, he attributes to differences in the direction of selection in different parts of the range. The process may be facilitated by geographic (or other) isolation, but he holds that it may also be brought about wholly by selection, the primary selection tending to set up secondary processes (including especially preferential mating *i. e.,* sexual selection), which in the end may lead to complete fission of the species.

It will be seen that Dr. Fisher's conception of evolution is pure Darwinian selection. The extent to which he carries the principle is well illustrated in his theory of dominance (chapter III) in which he attempts to account for the prevalent dominance of type genes, over mutant genes by the natural selection of modifiers of dominance.* I

*This theory was first elaborated by Fisher in papers which appeared in *The American Naturalist* (62:115-126, 571-574, 1928). In a criticism of it (*American Naturalist* 63:274-

have pointed out elsewhere and he has agreed, that the selection pressure on the modifiers is here of the second order, compared with the rate of mutation of the primary gene. It seemed probable to me that such a minute selection pressure would ordinarily be of the second order compared with other selection pressures acting on the same gene, and therefore negligible. Dr. Fisher on the other hand, adheres to the effectiveness of selection in this case.

In order to bring out the point at which we part company with respect to the efficacy of selection, it will be necessary to return to Dr. Fisher's fundamental theorem: "The rate of increase in fitness of any organism at any time is equal to its genetic variance in fitness at that time." One's first impression is that the genetic variance in fitness must in general be large and that hence if the theorem is correct the rate of advance must be rapid. As Dr. Fisher insists, however, the statement must be considered in connection with the precise definition which he gives of the terms. He uses "genetic variance" in a special sense. It does not include all variability due to differences in genetic constitution of individuals. He assumes that each gene is assigned a constant value, measuring its contribution to the character of the individual (here fitness) in such a way that the sums of the contributions of all genes will equal as closely as possible the actual measures of the character in the individuals of the population. Obviously there could be exact agreement in all cases only if dominance and epistatic relationships were completely lacking. Actually, dominance is very common and with respect to such a character as fitness, it may safely be assumed that there are always important epistatic effects. Genes favorable in one combination, are, for example, extremely likely to be unfavorable in another. Thus allelomorphs which are held in equilibrium by a balance of opposing selection tendencies (possibilities of which are discussed in Chapter V) may contribute a great deal to the total genetically determined variance but not at all to the genetic variance in Fisher's special sense, since at equilibrium there is no difference in their contributions. The formula itself seems to need revision in the case of another important class of genes, ones slightly deleterious in effect but maintained at a certain equilibrium in frequency by recurrent mutation (or migration). These contribute to the genetic variance of the species, but not to the increase in fitness. Terms involving mutation (and migration) rates seem to be omitted in the formula as given.

Mutational Flux as a Factor

Consider now the case of a population so large that fortuitous variation of gene frequency is negligible. According to my view, such a population is one in which all mutations which can occur will recur at measurable rates. All genes which are not fixed will be held in equilibrium by opposing selections, or by selection opposed by mutation, the cases just discussed. Thus while there may be a great deal of genetically determined variance, there will be no movement of gene frequencies and hence no evolution as long as external conditions remain constant. This state of equilibrium may be upset by change of external conditions, bringing changes in the direction and intensity of selection. All gene frequencies may then be expected to shift in an orderly fashion until the equilibrium consistent with the new conditions is attained. On return to the old conditions, all gene frequencies should shift back to the old positions. It may be granted that an irregular sequence

297, 1929). I proposed an alternative directly physiological interpretation of the phenomenon. Further discussion may be found in Fisher's reply to this criticism (*American Naturalist* 63:553-556, a counter-reply *ibid* 556-561 and a paper by J. B. S. Haldane, *American Naturalist* 64:87-90 1930).

of environmental condition would result occasionally in irreversible changes (because of epistatic relationships) thus giving a real, if very slow, evolutionary process; but this is not Dr. Fisher's scheme under which evolution should proceed under constant external conditions. He would have the system of equilibria of gene frequencies kept in motion by a steady flux of novel mutations. These to be effective must be advantageous practically from the first, since non-recurrent, unfavorable mutations would be lost (in an indefinitely large population) before they could reach such a frequency as to have any appreciable effect on the situation. Even those advantageous at once would also usually be lost within a few generations of their appearance. They would, however, as Dr. Fisher shows, have a finite chance of reaching high frequencies and ultimately fixation. In their progress, they may be expected to unsettle the equilibria of other genes by creating new favorable (or unfavorable) combinations. Thus the entire system of gene frequencies is thrown into motion and may yield the steady adaptive advance of the theory.

As noted above, this scheme appears to depend on an inexhaustible flow of new favorable mutations. Dr. Fisher does not go into this matter of inexhaustibility but presumably it may be obtained by supposing that each locus is capable of an indefinitely extended series of multiple allelomorphs, each new gene becoming a potential source of genes which could not have appeared previously. The greatest difficulty, seems to be in the posited favorable character of the mutations. Dr. Fisher, elsewhere, presents cogent reasons as to why the great majority of all mutations should be deleterious. He shows that all mutations affecting a metrical character "unless they possess countervailing advantages in other respects will be initially disadvantageous." He shows that in any case the greater the effect, the less the chance of being adaptive. Add to this the point that mutations as a rule probably have multiple effects, and that the sign of the net selection pressure is determined by the greater effects, and it will be seen that the chances of occurrence of new mutations, advantageous from the first are small indeed.

Partial Isolation as a Factor

I would not deny the possibility of very slow evolutionary advance through this mechanism but it has seemed to me that there is another mechanism which would be much more effective in preventing the system of gene frequencies from settling into a state of equilibrium, than the occurrence of new immediately favorable mutations. If the population is not too large, the effects of random sampling of gametes in each generation brings about a random drifting of the gene frequencies about their mean positions of equilibrium. In such a population we can not speak of single equilibrium values but of probability arrays for each gene, even under constant external conditions. If the population is too small, this random drifting about leads inevitably to fixation of one or the other allelomorph, loss of variance, and degeneration. At a certain intermediate size of population, however (relative to prevailing mutation and selection rates), there will be a continuous kaleidescopic shifting of the prevailing gene combinations, not adaptive itself, but providing an opportunity for the occasional appearance of new adaptive combinations of types which would never be reached by a direct selection process. There would follow thorough-going changes in the system of selection coefficients, changes in the probability arrays themselves of the various genes and in the long run an essentially irreversible adaptive advance of the species. It has seemed to me that the conditions for evolution would be more favorable here than in the indefinitely large population of Dr. Fisher's scheme. It would, however, be very slow, even in terms of geologic time, since it can be shown to be limited by mutation rate. A much more favorable condition would be that of a large population, broken up into imperfectly isolated local strains. The probability array for genes within such a local strain has been given on a previous page. The rate of evolutionary change depends primarily on the balance between the effective size of population in the local strain (n) and the amount of interchange of individuals with the species as a whole (m) and is therefore not limited by mutation rates. The consequence would seem to be a rapid differentiation of local strains, in itself non-adaptive, but permitting selective increase or decrease of the numbers in different strains and thus leading to relatively rapid adaptive advance of the species as a whole. Thus I would hold that a condition of subdivision of the species is important in evolution not merely as an occasional precursor of fission, but also as an essential factor in its evolution as a single group. Between the primary gene mutations, gradually carrying each locus through an endless succession of allelomorphs, and the control of the major trends of evolution by natural selection, I would interpolate a process of largely random differentiation of local strains. As to the existence of such strain differences, the situations described in the herring by Heinke, in Zoarces and Lebistes by J. Schmidt, and in deer mice by Sumner, as well as the situation in man, may be called to mind.

Sexual Selection and Mimicry

To the general biological reader, the later chapters of the book dealing with concrete applications of the selection principle may prove most attractive. A well sustained attempt to rehabilitate Darwin's theory of sexual selection has already been noted. Another chapter deals with mimicry. The validity of both Batesian and Müllerian mimicry is accepted and the possibility of accounting for the origin of each sort by natural selection is developed after careful analysis of opposing arguments which have been widely accepted, especially in the case of Müllerian mimicry. The author naturally applies his theory of direct progress through mass selection. It appears to me, however, that these cases fall at least equally well under the viewpoint which I have developed, which does not require such a minutely continuous path of selective advantage between the original pattern of the species and that ultimately reached.

Evolution In Man

More than one-third of the book is devoted to discussion of the trend of evolution in man. This portion deserves the most careful consideration by all interested in problems of Eugenics. The course of the argument may be summarized briefly as follows: One might expect to find that civilization once started on the earth would give such an advantage that its history would be an uninterrupted succession of triumphs. Instead of this, we find that every civilization, after a period of prosperity, has fallen into decay, and succumbed to the onslaughts of numerically weak, barbarous peoples. The cause of this decay, he finds reason to believe, is genetical rather than social. Evidence indicates that differences in fertility are in part hereditary, whether dependent on physical or mental qualities. The bulk of the evidence from civilized communities, ancient and modern, indicates that fertility is lowest in the upper classes of the population, where qualities which make for individual ability and leadership are most frequent. The reason for this inversion of the normal relation is seen in the tendency (first pointed out by Galton) for infertility as well as ability to rise in the social scale. The result is a tendency to extinction of ability, applying to all classes in society. Examination of conditions in more primitive societies organized on the clan basis, lead to the conclusion that the play of natural selection is here exactly the opposite. The evolution of individual qualities he believes reaches its climax just before civilization begins.

The final chapter deals with the conditions necessary for a permanent civilization. The author holds that only a wage system definitely designed to remove the present severe social penalty on fertility and indeed tending to promote fertility would adequately oppose the present tendency toward racial deterioration.

Literature Cited

1. Calder, A., 1927. *Proc. Roy. Soc. Edinburgh.* 47:118-140.

2. Fisher, R. A., 1922. *Proc Roy. Soc. Edinburgh.* 42:321-341.

3. Smith, A. D. B., 1926. *Eugenics Review.* 14:189-204. 1928. *Report of Brit. Ass. Adv. Science,* 649-655.

4. Wright, S., 1921. *Genetics.* 6:111-178.

5. Wright, S., 1922. *Amer. Nat.* 61:330-338. 1923, *Jour. Her.* 14:339-348, 405-422. McPhee, H. C., and S. Wright. 1925-26. *Jour. Hered.* 16:205-215, 17:397-401.

Dobzhansky, T. 1935. A critique of the species concept in biology. *Philosophy of Science* 2:344–355.

The biological species concept (BSC)—the notion that species are groups of actually or potentially interbreeding populations—is perhaps most often associated with the famous evolutionary biologist Ernst Mayr (1963), but the idea has deeper roots, as this early paper by Theodosius Dobzhansky indicates. Indeed, whether (and, if so, why) organic discontinuities characterize the biological world has been a major topic of discussion since before Darwin's time.

The debates have continued even into the modern era (e.g., Wheeler and Meier 2000). Members of one school of thought would like to abandon the BSC altogether and replace it with a phylogenetic species concept that focuses on identifying small monophyletic units to be equated with species (Cracraft 1983). More generally, recent decades have witnessed a great proliferation of alternative species concepts (few of which are mutually exclusive). Given evolutionary biology's longstanding obsession with species concepts, we have included Dobzhansky's early critique in this volume, not only because of its historical significance but also because, given all the brouhaha over competing species notions, we find that Dobzhansky's cardinal ideas and masterful prose continue to reassure us that the BSC is probably on the correct track.

RELATED READING

Cracraft, J. L. 1983. Species concepts and speciation analysis. Pages 159–187 *in* Current ornithology, Richard F. Johnston, editor. Plenum Press, New York, New York, USA.

Hey, J. 2001. Genes, categories, and species. Oxford University Press, New York, New York, USA.

Mayr, E. 1963. Animal species and evolution. Harvard University Press, Cambridge, Massachusetts, USA.

Wheeler, Q. D., and R. Meier, editors. 2000. Species concepts and phylogenetic theory. Columbia University Press, New York, New York, USA.

A Critique of the Species Concept in Biology

BY

TH. DOBZHANSKY

". . . though we cannot strictly define species, they yet have properties which varieties have not, and . . the distinction is not merely a matter of degree."—W. BATESON.

THE PROBLEM

THE species concept is one of the oldest and most fundamental in biology. And yet it is almost universally conceded that no satisfactory definition of what constitutes a species has ever been proposed. The present article is devoted to an attempt to review the status of the problem from a methodological point of view. Since the species is one of the many taxonomic categories, the question of the nature of these categories in general needs to be entered into.

The only biological category possessing an undisputable ontological significance is that of a living individual. Albeit in certain cases, as in those of the symbiotic and colonial forms, the delimitation of an individual meets with difficulties, the category of individual is saved by the admission of the existence of individualities of different orders. However, this category alone is plainly insufficient. We are confronted with (a) a stupendous number of existing individuals, and with (b) their immense diversity. Hence, a coherent knowledge of the living world is possible only with the aid of a hierarchic classification. Regardless of what principle is chosen as a foundation of such a classification, introduction of categories of different orders is indispensable. The category of individual is the lowest in the classification now accepted, that of the species is one of the higher ones. As such, the category of species has a clear pragmatic value.

It remains to be ascertained whether the species is a purely artificial device employed for making the bewildering diversity of living beings intelligible, or corresponds to something tangible in the outside world. Having evolved a classification based on arbitrary categories the investigator may be only too prone to mistake the product of his own mentality for some preëxisting "order of nature." Is, then, the species a part of the "order of nature," or a part of the order-loving mind?

NATURAL AND ARTIFICIAL CLASSIFICATION

It is customary to make a distinction between natural and artificial classifications. A critical evaluation shows that the "naturalness" of a classification may vary quantitatively; more and less natural classifications are possible. A natural classification may be defined as one reflecting empirically existing discontinuities in the materials to be classified. Any apparent discontinuity may serve as a basis for construction of a natural classification. Organisms vary in color, size, in external and internal structures, in their physiologies, in descent, etc. A classification is the more natural the larger is the number of discontinuities it subsumes in each division. An ideal classification would include all the discontinuities, and the knowledge of the position of an organism in such a classification would permit the formation of a sufficient number of deductive propositions for a complete description of this organism.[1]

[1] Since the post-Darwin period a "natural classification" has meant in biology a classification based on the hypothetical common descent of the organisms. This restriction of the meaning of the term is unjustified. The actual mode of descent has been ascertained, and can be ascertained, only for very few groups. But even granting the possibility of establishing the complete phylogenetic history of every organism, it has never been adequately proven that the degree of similarity between the organisms is always proportional to the closeness of their blood relationships. Some palaeontological data cast a grave doubt on this point.

According to the above definition, in a continuously varying living world only a purely artificial classification would be possible. In such a world all the individuals would be arrangeable into a single linear or dichotomic series, in which the extremes would be connected by an infinite number of transition forms (other forms of continuous variability can also be conceived). Applied to a continuous series of forms, a species concept would by no means be meaningless (the opposite opinion has been repeatedly expressed in literature). In fact, a hierarchic classification would still remain the only instrument with the aid of which the diversity of forms might be made describable and intelligible. A perfectly continuous series of forms may be cut at such points, and into as many sections, as deemed desirable for the purposes of the investigator. The segregation of a group of forms into two, three, or more "species" would be determined merely by expediency.

Cases of continuous variability are actually encountered. The species of the fossils are readily separable only as long as the known representatives of a given group from the older strata are scarce. New findings fill up the gaps between the species, until a continuous series of forms finally emerges. A similar condition is observed among some of the now living organisms, principally in the asexually or parthenogenetically reproducing groups. The Bacterium coli—Bacterium typhi complex consists of a large number of strains that differ from each other in their behavior in cultures, in their metabolism, and in pathogenic properties. Some strains differ widely from each other and are readily distinguishable, but the numerous intermediates make the splitting of the group into two, three, or more species equally arbitrary.

And yet, bacteriologists find it more useful to speak in this case about a group of species rather than about a single greatly variable one. Some genera of plants (Rubus, Hieracium, Rosa) approximate this condition rather closely, with a result that the "species" in these genera becomes an arbitrary unit.

DISCONTINUOUS VARIABILITY

Continuous variability is, however, an exception rather than the rule among the now existing living things. A superficial as well as a most searching investigation reveals not continuums but discrete groups of forms, every member of each group being more similar to every other member of the same group than to any member of any other group. The degree of discontinuity is, of course, highly variable. The intermediates connecting the different groups may be either wholly wanting, or they may be merely less frequent that the modal members of each group. Expressed quantitatively, the organisms do not form a single probability distribution, but rather a large number of separate probability distributions, or of more or less sharply multimodal curves.

The manifest tendency of life toward formation of discrete arrays is not deducible from any a priori considerations. It is simply a fact to be reckoned with. Formation of discrete groups is observed among animals as well as among plants, among the lowest as well as the most complex organisms. It is a universal rule, and the existence of some cases of continuous variability is far from sufficient to invalidate it. A causal analysis of this phenomenon is one of the major tasks of the biological sciences. Strange as it may seem, the predominance of the evolutionary trend of thought in some branches of biology during the last half century diverted the attention of the investigators away from this task rather than toward it, for an evolutionist is frequently more interested in the "bridging the gaps" between groups of organisms than in the nature of the gaps themselves. Only with the advent of genetics has the situation changed somewhat (cf. W. Bateson, "Problems of Genetics").

Discontinuous variability constitutes a foundation of the biological classification. The segregation of a given array of forms into two or more separate groups (be they classes, genera, or species) implies that a corresponding number of discontinuities have been observed among the constituents of this array. As long as such a correspondence is retained, the classification not only serves the purpose of a well-ordered catalogue, but is firmly rooted in phenomenology. The very existence of the classification becomes a symbol of the predominance of discontinuous variability among organisms.

The above reasoning is evidently applicable to the categories above the species (genera, families, orders), below the species (subspecies, aberrations), and to the species itself. It should be emphasized, however, that the usage of the different categories remains arbitrary, in the sense that a given discrete group of forms may be classed as a genus, subfamily, or a family as may be deemed desirable by the investigator. Only in the case of species this freedom of usage may be restricted by the introduction of a separate criterion, thus making the category of species methodologically more valuable than the rest.

SPECIES AS "SYNGAMEONS"

The rôle of the sexual method of reproduction as an obstacle in the way of the formation of discontinuous groups of individuals has been clearly recognized for a long time. Darwin devoted no little attention to this point, and all the more recent evolutionists were fully aware of it. The development of genetics brought a clarification of the understanding of the mechanism involved. Every discrete group of individuals represents a definite constellation of genes. If the different groups interbreed freely with each other, a new equilibrium is established in which the different genic constellations become fused into a single one. It necessarily follows that no discontinuous variation can exist in a perfectly panmictic population.[2] Mutatis mutandis, the existence of two or more discrete groups of individuals is a proof that free interbreeding between them is prevented by some factor or factors.

It is a matter of observation that closely similar individuals habitually interbreed, and those less similar do not. Hence, a stage must exist in the process of evolutionary divergence, at which an originally panmictic population becomes split into two or more populations that interbreed with each other no longer. A fundamental importance of this stage has been especially emphasized by Bateson, and more recently by Lotsy. Lotsy

[2] This statement applies, of course, only to the discontinuities caused by gene complexes (related species, not to speak about higher categories, differ from each other usually in many genes). Changes in single genes may produce variants so far removed from the original form that their characteristics do not overlap. Discontinuities of this nature may be preserved indefinitely in a panmictic population.

coined the word "syngameon" to designate "an habitually interbreeding community" of individuals. According to Lotsy, a species is a syngameon.

Here, then, we have a definition of species which is especially attractive because of its simplicity. Unfortunately, in its original form this definition meets with difficulties, since by no means all the groups of individuals that are potentially capable of interbreeding actually do so. Colonies of individuals living on islands or in isolated mountain valleys may not interbreed for centuries, and nevertheless their gene structures may be so similar that it were a paradox to classify them as separate species. Similarly, in most of the widely distributed species the communities of individuals inhabiting the remote parts of the specific area never interbreed, though they may be connected by a chain of communities interbreeding to a certain extent with their neighbors.

In spite of the above objections, Lotsy's attempt to clarify the species concept is sound in principle. The emphasis should be placed however not on the absence of actual interbreeding between the different form complexes, but rather on the presence of physiological mechanisms making interbreeding difficult or impossible.

ISOLATING MECHANISMS

Although the mechanisms preventing free and unlimited interbreeding of related forms are as yet little understood, it is already clear enough that a large number of very different mechanisms of this kind are functioning in nature. This diversity of the isolating mechanisms is in itself remarkable and difficult to explain. It is unclear how such mechanisms can be created at all by natural selection, that is what use the organism derives directly from their development. We are almost forced to conjecture that the isolating mechanisms are merely by-products of some other differences between the organisms in question, these latter differences having some adaptive value and consequently being subject to natural selection. The experimental evidence in favor of this conjecture is however scanty in the extreme.

The physiological mechanisms producing isolation may be

divided for descriptive purposes into two classes, namely mechanisms preventing the appearance of hybrid offspring (incompatibility of the parental forms), and mechanisms making the hybrids sterile and, consequently, incapable of propagating further (hybrid sterility). Some examples of each kind are given below.

Sexual isolation (in animals often termed psychological isolation) is very commonly encountered. Males of one species fail to mate with females of the other, or females avoid the alien males. In plants the pollination of the flowers of one species with the pollen of another may be prevented by the structure of the flowers themselves, or a failure of the foreign pollen to germinate properly on the stigma. Failure of spermatozoa of one species to be attracted by the eggs of another, and death or malformation of the developing embryos, are observed in some animals, although these methods of isolation are encountered mostly between rather remote forms.

Since the times of Ormancey and Dufour it is generally believed that unlike matings, especially among insects, may be prevented by dissimilarities in the structure of the genitalia that make copulation impossible. This is termed mechanical isolation. There is no doubt that the mechanical isolation actually exists in some cases (cf. Standfuss's and Federley's experiments on hybridisation in different moths), but in general its importance has been greatly exaggerated. In most insects female and male genitalia are not at all related to each other as a positive and a negative (lock and key according to Dufour), and experimental evidence shows that normal copula is frequently possible between forms having rather differently built genitalia.

Under laboratory conditions some closely related forms cross readily and produce fertile offspring, and nevertheless hybrids between them are rare or absent in nature. This may be due to the parental forms occupying different ecological stations (ecological isolation), or having different breeding seasons (temporal isolation), and consequently never meeting each other except in captivity. A very clear case of this type has recently been described by Cuenot in Sepia officinalis, where two rather similar forms live at different depths in the sea, and each have a rather short breeding season that does not coincide with that of the other. The ecological and temporal isolations are less dependable in their effects than other isolating mechanisms, and probably many if not most of the naturally occuring interspecific hybrids (especially among plants) come from species that are isolated from each other only in this way.

Hybrid sterility proves to be a common name far rather diverse mechanisms preventing the formation of functional germ cells by a hybrid organism. Probably the best understood kind of sterility is that observed in hybrids between species that are members of polyploid series (i.e. one having a chromosome number which is a simple multiple of that in the other). Such hybrids carry one set of chromosomes from one of the parents, and two or more sets from the other. As shown first by Rosenberg in Drosera, the result is the formation at meiosis of a number of bivalents approximately equal to the number of chromosomes in the parent having fewer chromosomes, while the remaining chromosomes are left as univalents. At the meiotic divisions the univalents are distributed irregularly, causing unbalanced chromosome complements in the resulting cells, which are therefore abnormal and non-functional. Since polyploid series of chromosome numbers are encountered in some plants, and very seldom in animals, this type of sterility is on the whole a rather special case.

In hybrids whose parents have similar chromosome numbers, or numbers that are not multiples of each other, the mechanism of sterility is different from the above. I have pointed out in other publications that at least two kinds of sterility are observed, namely chromosomal and genic sterility. Chromosomal sterility is caused by dissimilarities in the gross structure of the chromosomes of the parental species. The gene arrangement in the chromosomes of related species is often different. Hence, the hybrids become what is known as structural heterozygotes carrying inversions, translocations, and perhaps also duplications and deficiencies for chromosome sections. Since chromosome pairing at meiosis is due to a mutual attraction between homologous loci in the chromosomes, and not between chromosomes as wholes, an accumulation of structural differences between the chromosomes leads to irregularities or to failure of chromosome pairing in the hybrids. This, in turn, causes abnormalities in the functioning of the cell division mechanisms, and a consequent degeneration of the gametes. There exist some evidence that the sterility of many interspecific hybrids in plants, and also in some animals, is due to these causes.

Some species are so similar in the gross structure of their chromosomes that the sterility of the hybrids between them can not be accounted for by mechanical difficulties in chromosome pairing. Good examples of such species are Drosophila melanogaster and Drosophila simulans (Sturtevant), and the A and B "races" of Drosophila pseudoobscura (Dobzhansky). The sterility of their hybrids is apparently due to the action of complementary genetic factors contributed by each of the parental species. Certain combinations of these factors produce disturbances in the processes of the formation of germ cells in the hybrids, making the latter sterile. It may be conjectured that this type of sterility (the genic type) is most commonly encountered in interspecific hybrids, at least among animals. The classical case of interspecific sterility, that of the mule, probably belongs here.

Other mechanisms of hybrid sterility are probably also encountered (sterility due to an intersexul condition in the hybrids observed by Goldschmidt in Lymantria, the plasmatic type suggested by Michaelis's studies on Epilobium). Moreover, in some cases two or more of these mechanisms combine to cause the isolation of even closely related species. Remarkably enough, an accumulation of genic differences between the parental forms does not necessarily produce either hybrid incompatibility or sterility. The genetic factors responsible for the production of these isolating mechanisms appear to constitute rather a class by themselves. Thus, mechanisms preventing a free interbreeding may apparently develop in forms that are rather similar genotypically, and, vice versa, genotypically more different forms may remain potentially interfertile.

THE STATIC AND THE DYNAMIC CONCEPTIONS OF THE SPECIES

Lotsy's definition of what constitutes a species should be modified thus: a species is a group of individuals fully fertile inter se, but barred from interbreeding with other similar groups by its physiological properties (producing either incompatibility of parents, or sterility of the hybrids, or both). Since the organisms propagating sexually can be experimentally shown to be segregated in nature into discrete non-interbreeding groups, this definition rests on a solid factual basis. However, an attempt to apply this definition in practice will soon show that in some cases classification of two groups of individuals as separate species or as varieties of the same species remains arbitrary. The reason for this is exceedingly simple: neither the mechanisms producing incompatibility, nor those producing sterility, function on an all-or-none principle. For instance, sexual isolation may be incomplete, and individuals belonging to different groups may sometimes, though seldom, copulate. Similarly, some hybrids are only semi-sterile, or sterile in one sex only.

An analysis of the situation shows where the difficulty lies. The concepts of the taxonomic categories, as all the taxonomic concepts, are essentially static. Taxonomy attempts to isolate from the actual experience a sort of a photographic image of the living world, and to give a comprehensible description of the regularities observed in this image. The concepts of taxonomy are derived from reality by a process of abstraction, which leaves out of consideration the dynamism which is one of the most essential, if not the most essential attribute of life. By this method the flowing and changing patterns of life are made to conform more or less to the static limits of the taxonomic categories. But actually the taxonomic categories in general, and species in particular, are not static but dynamic units. The discontinuity of the living world is constantly emerging from a continuity, while continuity tends to increase in extent as a result of the extinction of some of the discrete groups previously formed. The unstable equilibrium between these two opposing trends, and the factors tending to displace it in either direction constitute one of the main problems of any theory of evolution. Considered dynamically, the species represents that stage of evolutionary divergence, at which the once actually or potentially interbreeding array of forms becomes segregated into two or more separate arrays which are physiologically incapable of interbreeding.

The fundamental importance of this stage is due to the fact that it is only the development of the isolating mechanisms that makes possible the coexistence in the same geographic area of different discrete groups of organisms. As pointed out above, every discrete group owes its distinctness from all other groups to a definite and unique system of genes it carries. As long as such groups are virtually fully interbreedable, they can remain separate only if they inhabit different and sufficiently remote geographic regions. The boundary lines between such groups become rapidly obliterated soon after they come in contact in the same locality, and free interbreeding sets in. On the other hand, development of isolating mechanisms renders the differences between groups relatively fixed and irreversible, and permits them to dwell side by side without loosing their differentiating characteristics. This, in turn, opens the possibility for the organisms dwelling together to become adapted to different places in the general economy of nature. The usage of the term "species" can and should be made to reflect the attainment by a group of organisms of this evolutionary stage.

SPECIES FROM THE GENETIC AND THE TAXONOMIC STANDPOINTS

In cross-fertilizing organisms species may be separated from each other on the basis of the presence or absence of physiological hindrances to interbreeding. An accurate determination of the degree of isolation attained by a given pair of presumed species is possible, of course, only by means of an experimental analysis of their relationship. The question immediately presents itself: to what extent do the species thus delimited coincide with the species of taxonomists? Indeed, in a tremendous majority of cases taxonomists have no direct evidence bearing on the crossability of forms which they describe as separate species. Although a detailed discussion of this problem lies outside of the scope of the present article, it should be pointed out that the two classifications are of necessity not too dissimilar. The criteria used by taxonomists for separating species (especially clearly formulated by A. P. Semenov-Tian-Shansky) give a sort of circumstantial evidence for distinguishing non-interbreeding groups of forms from those that regularly interbreed. Taxonomists assume that different species exhibit different cycles of variability separated from each other by more or less clearly pronounced gaps, while varieties of the same species belong to the same cycle of variability and are not separated by such gaps. Similarly, discrete groups coexisting in the same locality without producing intermediates are considered separate species, while forms replacing each other in different geographic regions (so that a single form occurs in every region), and grading into each other in the intervening regions, are classed as subspecies or varieties. Discrepancies between the "genetic" and the "taxonomic" species are to be expected mainly in those relatively rare cases where different groups do not interbreed despite the scarcity or absence of morphological differences between them (a good example of this sort are the "races" of Trichogramma minutum described by Harland and Atteck), or where geographically isolated races, without losing the ability to interbreed, have diverged so widely in their morphological characters that taxonomists feel compelled to consider them separate species (for instance, some "species" of pheasants).

Among organisms reproducing exclusively by parthenogenesis or asexually, species in our sense do not exist at all. The classification of these organisms must be based solely on the observable discontinuities in their morphological structures and physiologies. Methodologically this classification is similar to the combining of species into higher taxonomic groups.

California Institute of Technology
Pasadena, California

Beadle, G. W., and E. L. Tatum. 1941. Genetic control of biochemical reactions in *Neurospora*. *Proceedings of the National Academy of Sciences USA* 27:499–506.

In the first half of the twentieth century, a widespread assumption was that proteins would prove to be the genetic material of life. The complexity of organisms seemed to demand a comparable complexity for hereditary factors, and proteins—being composed of lengthy stretches of about 20 different types of amino acid—seemed ideal candidates for comprising genes. DNA seemed an unlikely candidate for genetic material, because nucleic acids are composed of only four types of chemical building blocks, thus supposedly making them too simple to underlie life's richness. It came as quite a surprise in the 1940s when key experiments proved that nucleic acids were the molecules of heredity (see Watson and Crick 1953 in this volume).

At about that same time, biochemists began recording details that implied some special relationship between genes and proteins. This paper by George Beadle and Edward Tatum is a prime example. These researchers demonstrated, based on their research on the bread mold *Neurospora crassa*, that one gene seemed to equate somehow to one protein or polypeptide in a metabolic pathway. This finding set the stage for the eventual cracking of the genetic code and the elaboration of the sequential processes of RNA transcription and protein translation that are the hallmarks of gene expression. Gene expression, in turn, has proved to be key to understanding the biochemistry underlying many of the biological adaptations that are of special interest to evolutionary biologists. In 1958, the two men shared a Nobel Prize for their foundational work in biochemical genetics.

RELATED READING

Hochachka, P. W., and G. N. Somero. 1973. Strategies of biochemical adaptation. Saunders, Philadelphia, Pennsylvania, USA.

*GENETIC CONTROL OF BIOCHEMICAL REACTIONS IN NEUROSPORA**

BY G. W. BEADLE AND E. L. TATUM

BIOLOGICAL DEPARTMENT, STANFORD UNIVERSITY

Communicated October 8, 1941

From the standpoint of physiological genetics the development and functioning of an organism consist essentially of an integrated system of chemical reactions controlled in some manner by genes. It is entirely tenable to suppose that these genes which are themselves a part of the system, control or regulate specific reactions in the system either by acting directly as enzymes or by determining the specificities of enzymes.[1] Since the components of such a system are likely to be interrelated in complex ways, and since the synthesis of the parts of individual genes are presumably dependent on the functioning of other genes, it would appear that there must exist orders of directness of gene control ranging from simple one-to-one relations to relations of great complexity. In investigating the rôles of genes, the physiological geneticist usually attempts to determine the physiological and biochemical bases of already known hereditary traits. This approach, as made in the study of anthocyanin pigments in plants,[2] the fermentation of sugars by yeasts[3] and a number of other instances,[4] has established that many biochemical reactions are in fact controlled in specific ways by specific genes. Furthermore, investigations of this type tend to support the assumption that gene and enzyme specificities are of the same order.[5] There are, however, a number of limitations inherent in this approach. Perhaps the most serious of these is that the investigator must in general confine himself to a study of non-lethal heritable characters. Such characters are likely to involve more or less non-essential so-called "terminal" reactions.[5] The selection of these for genetic study was perhaps responsible for the now rapidly disappearing belief that genes are concerned only with the control of "superficial" characters. A second difficulty, not unrelated to the first, is that the standard approach to the problem implies the use of characters with visible manifestations. Many such characters involve morphological variations, and these are likely to be based on systems of biochemical reactions so complex as to make analysis exceedingly difficult.

Considerations such as those just outlined have led us to investigate

TABLE 1

GROWTH OF PYRIDOXINLESS STRAIN OF *N. sitophila* ON LIQUID MEDIUM CONTAINING INORGANIC SALTS,[9] 1% SUCROSE, AND 0.004 MICROGRAM BIOTIN PER CC. TEMPERATURE 25°C. GROWTH PERIOD, 6 DAYS FROM INOCULATION WITH CONIDIA

MICROGRAMS B_6 PER 25 CC. MEDIUM	STRAIN	DRY WEIGHT MYCELIA, MG.
0	Normal	76.7
0	Pyridoxinless	1.0
0.01	"	4.2
0.03	"	5.7
0.1	"	13.7
0.3	"	25.5
1.0	"	81.1
3.0	"	81.1
10.0	"	65.4
30.0	"	82.4

the general problem of the genetic control of developmental and metabolic reactions by reversing the ordinary procedure and, instead of attempting to work out the chemical bases of known genetic characters, to set out to determine if and how genes control known biochemical reactions. The ascomycete *Neurospora* offers many advantages for such an approach and is well suited to genetic studies.[6] Accordingly, our program has been built around this organism. The procedure is based on the assumption that x-ray treatment will induce mutations in genes concerned with the control of known specific chemical reactions. If the organism must be able to carry out a certain chemical reaction to survive on a given medium, a mutant unable to do this will obviously be lethal on this medium. Such a mutant can be maintained and studied, however, if it will grow on a medium to which has been added the essential product of the genetically blocked reaction. The experimental procedure based on this reasoning can best be illustrated by considering a hypothetical example. Normal strains of *Neurospora crassa* are able to use sucrose as a carbon source, and are therefore able to carry out the specific and enzymatically controlled

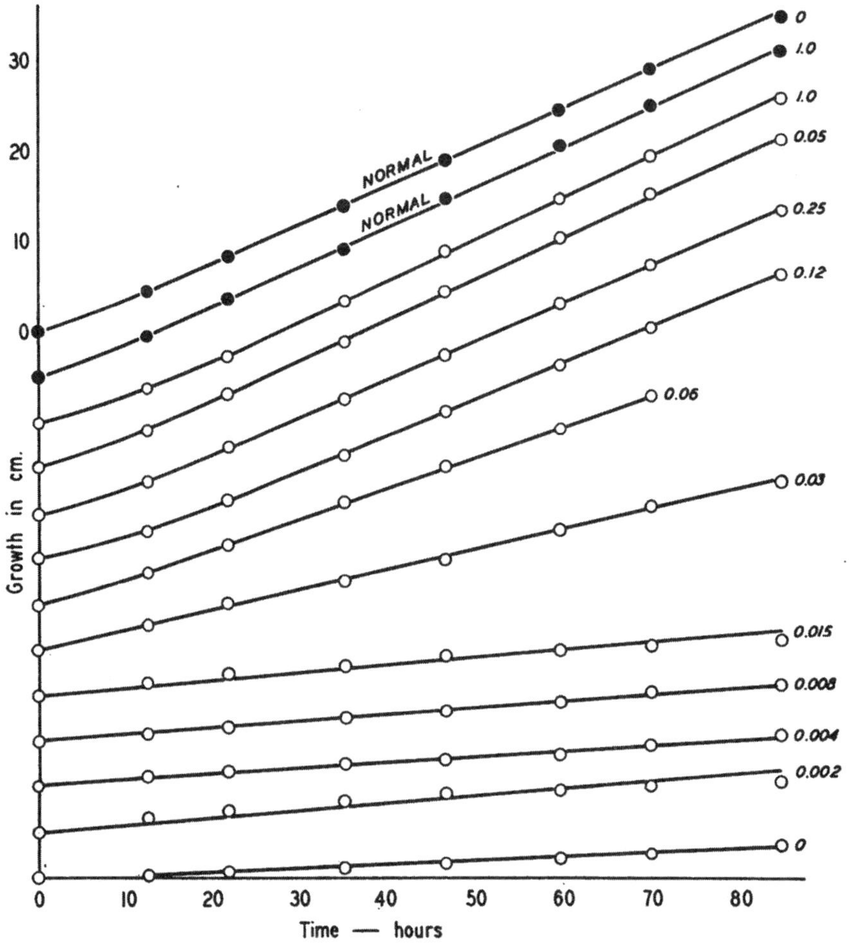

FIGURE 1

Growth of normal (top two curves) and pyridoxinless (remaining curves) strains of *Neurospora sitophila* in horizontal tubes. The scale on the ordinate is shifted a fixed amount for each successive curve in the series. The figures at the right of each curve indicate concentration of pyridoxine (B_6) in micrograms per 25 cc. medium.

reaction involved in the hydrolysis of this sugar. Assuming this reaction to be genetically controlled, it should be possible to induce a gene to mutate to a condition such that the organism could no longer carry out sucrose hydrolysis. A strain carrying this mutant would then be unable to grow on a medium containing sucrose as a sole carbon source but should be able to grow on a medium containing some other normally utilizable carbon source. In other words, it should be possible to establish and maintain such a mutant strain on a medium containing glucose and detect its inability to utilize sucrose by transferring it to a sucrose medium.

Essentially similar procedures can be developed for a great many metabolic processes. For example, ability to synthesize growth factors (vitamins), amino acids and other essential substances should be lost through gene mutation if our assumptions are correct. Theoretically, any such metabolic deficiency can be "by-passed" if the substance lacking can be supplied in the medium and can pass cell walls and protoplasmic membranes.

In terms of specific experimental practice, we have devised a procedure in which x-rayed single-spore cultures are established on a so-called "complete" medium, i.e., one containing as many of the normally synthesized constituents of the organism as is practicable. Subsequently these are tested by transferring them to a "minimal" medium, i.e., one requiring the organism to carry on all the essential syntheses of which it is capable. In practice the complete medium is made up of agar, inorganic salts, malt extract, yeast extract and glucose. The minimal medium contains agar (optional), inorganic salts and biotin, and a disaccharide, fat or more complex carbon source. Biotin, the one growth factor that wild type *Neurospora* strains cannot synthesize,[7] is supplied in the form of a commercial concentrate containing 100 micrograms of biotin per cc.[8] Any loss of ability to synthesize an essential substance present in the complete medium and absent in the minimal medium is indicated by a strain growing on the first and failing to grow on the second medium. Such strains are then tested in a systematic manner to determine what substance or substances they are unable to synthesize. These subsequent tests include attempts to grow mutant strains on the minimal medium with (1) known vitamins added, (2) amino acids added or (3) glucose substituted for the more complex carbon source of the minimal medium.

Single ascospore strains are individually derived from perithecia of *N. crassa* and *N. sitophila* x-rayed prior to meiosis. Among approximately 2000 such strains, three mutants have been found that grow essentially normally on the complete medium and scarcely at all on the minimal medium with sucrose as the carbon source. One of these strains (*N. sitophila*) proved to be unable to synthesize vitamin B_6 (pyridoxine). A second strain (*N. sitophila*) turned out to be unable to synthesize vitamin B_1 (thiamine). Additional tests show that this strain is able to synthesize the pyrimidine half of the B_1 molecule but not the thiazole half. If thiazole alone is added to the minimal medium, the strain grows essentially normally. A third strain (*N. crassa*) has been found to be unable to synthesize para-aminobenzoic acid. This mutant strain appears to be entirely normal when grown on the minimal medium to which *p*-aminobenzoic acid has been added. Only in the case of the "pyridoxinless" strain has an analysis of the inheritance of the induced metabolic defect been investigated. For this reason detailed accounts of the thiamine-deficient and *p*-aminobenzoic acid-deficient strains will be deferred.

Qualitative studies indicate clearly that the pyridoxinless mutant, grown on a medium containing one microgram or more of synthetic vitamin B_6 hydrochloride per 25 cc. of medium, closely approaches in rate and characteristics of growth normal strains grown on a similar medium with

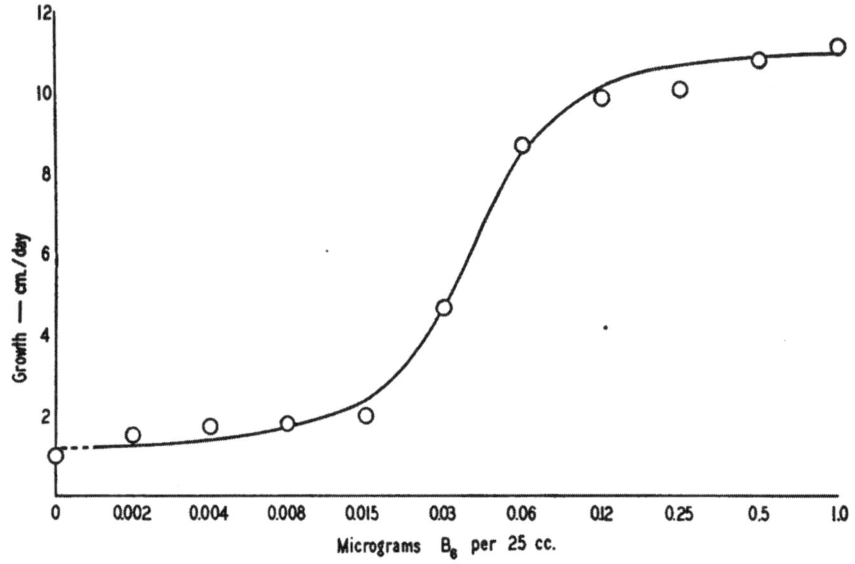

FIGURE 2

The relation between growth rate (cm./day) and vitamin B_6 concentration.

no B_6. Lower concentrations of B_6 give intermediate growth rates. A preliminary investigation of the quantitative dependence of growth of the mutant on vitamin B_6 in the medium gave the results summarized in table 1. Additional experiments have given results essentially similar but in only approximate quantitative agreement with those of table 1. It is clear that additional study of the details of culture conditions is necessary before rate of weight increase of this mutant can be used as an accurate assay for vitamin B_6.

It has been found that the progression of the frontier of mycelia of *Neurospora* along a horizontal glass culture tube half filled with an agar medium provides a convenient method of investigating the quantitative effects of growth factors. Tubes of about 13 mm. inside diameter and about 40 cm. in length are used. Segments of about 5 cm. at the two ends are turned up at an angle of about 45°. Agar medium is poured in so as to fill the tube about half full and is allowed to set with the main segment of the tube in a horizontal position. The turned up ends of the tube are stoppered with cotton plugs. Inoculations are made at one end of the agar surface and the position of the advancing front recorded at convenient intervals. The frontier formed by the advancing mycelia is remarkably well defined, and there is no difficulty in determining its position to within a millimeter or less. Progression along such tubes is strictly linear with time and the rate is independent of tube length (up to 1.5 meters). The rate is not changed by reducing the inside tube diameter to 9 mm., or by

TABLE 2

RESULTS OF CLASSIFYING SINGLE ASCOSPORE CULTURES FROM THE CROSS OF PYRIDOXINLESS AND NORMAL *N. sitophila*

ASCUS NUMBER	1	2	3	4	5	6	7	8
17	—	*pdx*	*pdx*	*pdx*	*N*	*N*	*N*	—
18	—	—	*N*	*N*	—	—	*pdx*	*pdx*
19	—	*pdx*	—	—	—	—	—	*N*
20	—	—	*N*	—	—	—	—	*pdx*
22	—	—	*N*	—	—	—	—	—
23	—	*	*	*	*N*	*N*	*pdx*	*pdx*
24	*N*	*N*	*N*	*N*	*pdx*	*pdx*	*pdx*	*pdx*

N, normal growth on B_6-free medium. *pdx*, slight growth on B_6-free medium. Failure of ascospore germination indicated by dash.

* Spores 2, 3 and 4 isolated but positions confused. Of these, two germinated and both proved to be mutants.

sealing one or both ends. It therefore appears that gas diffusion is in no way limiting in such tubes.

The results of growing the pyridoxinless strain in horizontal tubes in which the agar medium contained varying amounts of B_6 are shown graphically in figures 1 and 2. Rate of progression is clearly a function of vitamin B_6 concentration in the medium.[10] It is likewise evident that there is no significant difference in rate between the mutant supplied with B_6 and the normal strain growing on a medium without this vitamin. These results are consistent with the assumption that the primary physiological difference between pyridoxinless and normal strains is the inability of the former to carry out the synthesis of vitamin B_6. There is certainly more than one step in this synthesis and accordingly the gene differential involved is presumably concerned with only one specific step in the biosynthesis of vitamin B_6.

In order to ascertain the inheritance of the pyridoxinless character, crosses between normal and mutant strains were made. The techniques for hybridization and ascospore isolation have been worked out and described by Dodge, and by Lindegren.[6] The ascospores from 24 asci of the cross were isolated and their positions in the asci recorded. For some unknown reason, most of these failed to germinate. From seven asci, however, one or more spores germinated. These were grown on a medium containing glucose, malt-extract and yeast extract, and in this they all grew normally. The normal and mutant cultures were differentiated by growing them on a B_6 deficient medium. On this medium the mutant cultures grew very little, while the non-mutant ones grew normally. The results are summarized in table 2. It is clear from these rather limited data that this inability to synthesize vitamins B_6 is transmitted as it should be if it were differentiated from normal by a single gene.

The preliminary results summarized above appear to us to indicate that the approach outlined may offer considerable promise as a method of learning more about how genes regulate development and function. For example, it should be possible, by finding a number of mutants unable to carry out a particular step in a given synthesis, to determine whether only one gene is ordinarily concerned with the immediate regulation of a given specific chemical reaction.

It is evident, from the standpoints of biochemistry and physiology, that the method outlined is of value as a technique for discovering additional substances of physiological significance. Since the complete medium used can be made up with yeast extract or with an extract of normal *Neurospora*, it is evident that if, through mutation, there is lost the ability to synthesize an essential substance, a test strain is thereby made available for use in isolating the substance. It may, of course, be a substance not previously known to be essential for the growth of any organism. Thus we may expect to discover new vitamins, and in the same way, it should be possible to discover additional essential amino acids if such exist. We have, in fact, found a mutant strain that is able to grow on a medium containing Difco yeast extract but unable to grow on any of the synthetic media we have so far tested. Evidently some growth factor present in yeast and as yet unknown to us is essential for *Neurospora*.

Summary.—A procedure is outlined by which, using *Neurospora*, one can discover and maintain x-ray induced mutant strains which are characterized by their inability to carry out specific biochemical processes.

Following this method, three mutant strains have been established. In one of these the ability to synthesize vitamin B_6 has been wholly or largely lost. In a second the ability to synthesize the thiazole half of the vitamin B_1 molecule is absent, and in the third para-aminobenzoic acid is not synthesized. It is therefore clear that all of these substances are essential growth factors for *Neurospora*.[11]

Growth of the pyridoxinless mutant (a mutant unable to synthesize vitamin B_6) is a function of the B_6 content of the medium on which it is grown. A method is described for measuring the growth by following linear progression of the mycelia along a horizontal tube half filled with an agar medium.

Inability to synthesize vitamin B_6 is apparently differentiated by a single gene from the ability of the organism to elaborate this essential growth substance.

NOTE: Since the manuscript of this paper was sent to press it has been established that inability to synthesize both thiazole and *p*-aminobenzoic acid are also inherited as though differentiated from normal by single genes.

* Work supported in part by a grant from the Rockefeller Foundation. The authors are indebted to Doctors B. O. Dodge, C. C. Lindegren and W. S. Malloch for stocks and for advice on techniques, and to Miss Caryl Parker for technical assistance.

[1] The possibility that genes may act through the mediation of enzymes has been suggested by several authors. See Troland, L. T., *Amer. Nat.*, **51**, 321–350 (1917); Wright, S., *Genetics*, **12**, 530–569 (1927); and Haldane, J. B. S., in *Perspectives in Biochemistry*, Cambridge Univ. Press, pp. 1–10 (1937), for discussions and references.

[2] Onslow, Scott-Moncrieff and others, see review by Lawrence, W. J. C., and Price, J. R., *Biol. Rev.*, **15**, 35–58 (1940).

[3] Winge, O., and Laustsen, O., *Compt. rend. Lab. Carlsberg, Serie physiol.*, **22**, 337–352 (1939).

[4] See Goldschmidt, R., *Physiological Genetics*, McGraw-Hill, pp. 1–375 (1939), and Beadle, G. W., and Tatum, E. L., *Amer. Nat.*, **75**, 107–116 (1941) for discussion and references.

[5] See Sturtevant, A. H., and Beadle, G. W., *An Introduction to Genetics*, Saunders, pp. 1–391 (1931), and Beadle, G. W., and Tatum, E. L., loc. cit., footnote 4.

[6] Dodge, B. O., *Jour. Agric. Res.*, **35**, 289–305 (1927), and Lindegren, C. C., *Bull. Torrey Bot. Club*, **59**, 85–102 (1932).

[7] In so far as we have carried them, our investigations on the vitamin requirements of *Neurospora* corroborate those of Butler, E. T., Robbins, W. J., and Dodge, B. O., *Science*, **94**, 262–263 (1941).

[8] The biotin concentrate used was obtained from the S. M. A. Corporation, Chagrin Falls, Ohio.

[9] Throughout our work with *Neurospora*, we have used as a salt mixture the one designated number 3 by Fries, N., *Symbolae Bot. Upsalienses*, Vol. 3, No. 2, 1–188 (1938). This has the following composition: NH_4 tartrate, 5 g.; NH_4NO_3, 1 g.; KH_2PO_4, 1 g.; $MgSO_4 \cdot 7H_2O$, 0.5 g.; NaCl, 0.1 g.; $CaCl_2$, 0.1 g.; $FeCl_3$, 10 drops 1% solution; H_2O, 1 l. The tartrate cannot be used as a carbon source by *Neurospora*.

[10] It is planned to investigate further the possibility of using the growth of *Neurospora* strains in the described tubes as a basis of vitamin assay, but it should be emphasized that such additional investigation is essential in order to determine the reproducibility and reliability of the method.

[11] The inference that the three vitamins mentioned are essential for the growth of normal strains is supported by the fact that an extract of the normal strain will serve as a source of vitamin for each of the mutant strains.

Simpson, G. G. 1951. The species concept. *Evolution* 5:285–298.

The general acceptance of natural selection as the fundamental process that accounts for the adaptations of organisms and their diversity may be traced to the theoretical work of R. A. Fisher, J. B. S. Haldane, and Sewall Wright (see Wright 1930 in this volume). These theoretical works had limited impact on the biology of the time for various reasons: they were formulated for the most part in difficult mathematical language; they were almost exclusively theoretical without empirical corroboration—that is, they showed how evolution *could* occur but not that it *did* occur according to the theory—and they were limited in scope. The full integration of genetics and the theory of natural selection was accomplished in the ensuing decades through the work of numerous biologists, notably Theodosius Dobzhansky (*Genetics and the Origin of Species*, 1937), Julian S. Huxley (*Evolution: The Modern Synthesis*, 1942), Ernst Mayr (*Systematics and the Origin of Species*, 1942), George Gaylord Simpson (*Tempo and Mode in Evolution*, 1944), and G. Ledyard Stebbins (*Variation and Evolution in Plants*, 1950). This integration of genetics and natural selection became known as the Synthetic Theory of Evolution or the Modern Synthesis.

George Gaylord Simpson (1902–1984) was one of the most influential paleontologists of the twentieth century. In addition to his prolific research on the taxonomy of fossil and living mammals, he contributed to evolutionary theory with important books, such as *The Meaning of Evolution* (1949), *The Major Features of Evolution* (1953), *This View of Life* (1954) and others. He wrote monographs on horses (*Horses*, 1951) and penguins (*Penguins*, 1976), a taxonomy treatise (*Principles of Animal Taxonomy*, 1961), a successful textbook (*Life: An Introduction to Biology*, 1967), and several other books, including his autobiography, *Concessions to the Improbable* (1978). In "The Species Concept," he seeks to extend the biological concept of species to paleontology, where organisms classified as different species may come from fossils of animals that lived at different times and often had undetermined geographic distribution. Among the alternative considerations that he explores in order to define (paleontological) species, some (panel D in figure 2) are similar to those later advanced by the proponents of cladistics taxonomy.

RELATED READING

Avise, J. C. 2000. Phylogeography: the history and formation of species. Harvard University Press, Cambridge, Massachusetts, USA.

Dobzhansky, T., F. J. Ayala, G. L. Stebbins, and J. W. Valentine. 1977. Evolution. Freeman, San Francisco, California, USA.

Henning, W. 1966. Phylogenetic systematics. University of Illinois Press, Urbana, Illinois, USA.

Laporte, L. F. 2000. George Gaylord Simpson, paleontologist and evolutionist. Columbia University Press, New York, New York, USA.

Olson, E. C. 1991. George Gaylord Simpson 1902–1984: Biographical memoirs. The National Academies Press, Washington, D.C., USA.

Wiley, E. O. 1981. Phylogenetics: the theory and practice of phylogenetic systematics, 1st ed. Wiley, New York, New York, USA.

EVOLUTION

INTERNATIONAL JOURNAL OF ORGANIC EVOLUTION

PUBLISHED BY

THE SOCIETY FOR THE STUDY OF EVOLUTION

Vol. V DECEMBER, 1951 No. 4

THE SPECIES CONCEPT

George Gaylord Simpson

The American Museum of Natural History and Columbia University, New York, New York

Received April 24, 1951

Introduction

The species concept is focal in evolutionary studies and, indeed, in all biological thought. Its endless discussion is sometimes boring and seemingly fruitless, but is not wholly futile. In the course of such discussion the concept has been clarified, comprehension and a consensus have tended to develop, and the concept has changed in a significant way. There have recently been two more flurries of attention to this perennial topic. One, mostly in Evolution, by Burma (1949a, 1949b), Mayr (1949), Dunbar (1950), Elias (1950), and Gregg (1950), was originally concerned with whether the species is a "fiction" or is "objective," but also treated such matters as the relationships of neontological and paleontological species concepts. The other, in the *Journal of Paleontology,* by Weller (1949), Jeletzky (1950), Bell (1950), and Wright (1950), was concerned with the bases and practices of paleontological systematics and also with "morphological" versus "phylogenetic" or "natural" versus "unnatural" classification. These papers, among others, and an attempt to grapple with the whole problem for a class in systematics have inspired the following remarks.

I agree with most of what all the authors just cited have said. I believe, however, that it is possible to say much of this in a somewhat different and therefore possibly clarifying way, to combine some of their apparently but not really conflicting views into one consistent statement, and to add a few significant considerations not explicit, at least, in any of their papers.

Parts of the discussions cited and some of the apparent conflicts are primarily semantic. By a ponderous application of symbolic logic, Gregg sought to show that the issue raised by Burma and Mayr is not a genuine taxonomic problem or, at least, that if it does relate to a taxonomic problem it does so in the wrong words. It is, of course, important that words be used as accurately as possible and that they do not obscure properly taxonomic questions. Nevertheless, Burma and Mayr (as well as subsequent discussants) *were* considering a genuine taxonomic problem, in words perhaps not logically impeccable but, taken in context, adequately performing their main semantic function, that of communicating understandably among colleagues.

The semantics of the systematists' vocabulary is a fascinating subject, which surely does have its own importance but which has the danger of merely diverting attention from the systematists' proper business, systematics. I believe that most of the purely semantic confusion on the present subject can be avoided if such terms as "real," "natural," or "objective," and opposite or contrasting terms, are not applied to taxonomic categories or methods of classification, and if the two terms "arbitrary" and "non-arbitrary" are used in specially defined senses.

Definitions of taxonomic categories, such as a species, specify the sort of data or of inferences from data that are to be used in assignment of organisms to a group ranked in that category. For instance, the category definition of a species as a group of "actually or potentially interbreeding natural populations which are reproductively isolated from other such groups" (Mayr) specifies that data and inferences as to interbreeding and its absence are to be used. In some cases the data or inferences used will indicate essential continuity among the organisms to be grouped, and in other cases they will indicate essential discontinuity. Under the preceding genetic definition, actual or potential interbreeding is continuity and reproductive isolation is discontinuity. With a morphological-associational definition, continuity would be overlap in variation between compared populations and discontinuity the absence of overlap. Essential continuity or discontinuity in geographic, ecological, or temporal distribution has obvious meaning.

I propose to call taxonomic procedure arbitrary when organisms are placed in separate groups although the information about them indicates essential continuity in respects pertinent to the definition being discussed, or when they are placed in a single group although essential discontinuity is indicated. Conversely, procedure is non-arbitrary when organisms are grouped together on the basis of pertinent, essential continuity and separated on the basis of pertinent, essential discontinuity. Of course no stigma is meant to attach to taxonomic procedure thus defined as arbitrary. A completely non-arbitrary classification is impossible. It would be possible to extend discussion to such points as the precise definition of "essential continuity" or other parts of these definitions, but I think their meaning will be clear to all taxonomists, now or as discussion proceeds, and that semantics may be dropped at this point.

Typology, Morphology, and Genetical Groups

The typological concept of a taxonomic group is that the group corresponds with an abstract or ideal morphological pattern. Variation may be dealt with by a fixed or intuitive standard as to allowable deviation from the pattern, in which case the grouping is arbitrary. (It may either include discontinuities or draw a line across continuity.) Or, somewhat less naïvely, at a given level, usually that of species, the criterion of continuity in variation around the pattern may be used, a non-arbitrary procedure for that category.

The typological concept is pre-evolutionary and non-evolutionary. It still underlies a great deal of taxonomic practice but is now seldom favored in theory. Arkell (1950), an experienced paleontological taxonomist, seems to be accepting it when he says that, "Theoretically, at least, the number of species named reflects the number of forms, and so is more or less an objective matter," but he was mainly concerned with the highly laudable desire to keep super-specific categories conveniently manageable. The only serious modern theoretical support for frankly typological taxonomy comes from those few students who believe that species arise by abrupt morphological change from one "morphotype" to another, notably Schindewolf (1950).

Most of the data actually used in the practice of taxonomy are morphological. It is therefore not surprising that practical taxonomists suggest from time to time that classification should be morphological, in principle, as Weller (1950) has recently done. But a purely morphological classification would be based strictly on degrees of morphological difference between organisms, and this is really so impractical that no one, not even Weller, really tries to do it consistently. It is a commonplace that the degree of morphological difference within what everyone, morphologist, geneticist, or other, calls a single species is frequently greater than that between what all call separate, related species. It is also quite impractical to obtain a valid, over-all measure of total morphological difference between two organisms. Characters are always selected, weighted, and interpreted. As Wright (1950) pointed out in criticism of Weller, the usual and meaningful basis for selection, weighting, and interpretation is phylogenetic. Even typological classification, more strictly morphological than others, requires definition of the morphotype from characters in a group already set up on grounds not, in practice, purely morphological. Typological or not, practical morphological classification starts with some sort of grouping and in most practice this is usually an attempt to recognize what is (whether so called or not) a genetically defined population. Thus Bell (1950) cogently argues the value of stratigraphic evidence in paleontological taxonomy because it bears on pertinent and useful biological taxonomic criteria that are not morphological.

The fundamental point here for taxonomy is the modern idea that it is populations, not specimens, that are being classified. Newell (1948) has stressed this point for invertebrate paleontology in criticism of the practice of naming variants which are not populations. Jeletzky (1950) also emphasizes this point of view and its usefulness (one might say, necessity) for phylogenetic classification, although his argument is greatly weakened by his statement that variants are "natural groups" within the population and by his contrasting of phylogenetic with statistical methods, as if statistical methods were not a means of reaching conclusions about populations and hence about phylogeny. I have also insistently recommended population concepts in taxonomy, and so have many others. A few paleontologists have been mentioned first because, on the whole, paleontologists have been rather slower to grasp or accept the population concept of taxonomic groups. Despite some conservatives and reactionaries, the concept is widely accepted among neontologists (in botany, e.g., Camp, 1951; or in zoology, e.g., Mayr, 1942).

If classification is to start with populations, category definitions at and below the species level should refer to populations which, further, should be meaningful biologically. It seems to me, and few systematists are likely now to question this, that such groups should likewise have evolutionary significance. Here is the most serious fault of typological or of purely morphological definitions. Unless by chance or unless a hidden genetical criterion is actually used, they do not define biological populations or have clear evolutionary significance.

Attention to biologically significant populations is the basis and justification for the now usual neontological definitions of the species category in terms of interbreeding and reproductive isolation, i.e. of genetical factors, like the definition already quoted from Mayr. As Wright (1950) has mentioned, the fact that a species, as a group, is actually diagnosed in morphological terms, does not conflict with definition of the species, as a category, in genetical terms. The basis for definition of a category is quite different from the evidence available for decision as to whether a particular group of organisms meets that definition. And although the evidence used is mainly morphological in practice, it also almost always includes other sorts of data as well: distribution or association, at least, and preferably also other information.

The genetical definition of a species as

a group of actually or potentially interbreeding organisms reproductively isolated from other such groups is non-arbitrary both in its inclusion and its exclusion. Its criteria are reproductive continuity and discontinuity. The group defined is co-extensive with the continuity and bounded by the discontinuity. A species under this definition is the largest group with non-arbitrary exclusion and the smallest group with non-arbitrary inclusion. By the criteria of this definition and in cases to which it applies, infraspecific groups are non-arbitrary as to what they include (being reproductively continuous, by definition), but more or less arbitrary as to what they exclude (having boundaries without full reproductive discontinuity, by definition). Under the same criteria and circumstances, supra-specific groups are arbitrary as to inclusion, because by definition they do or may include two or more groups between which there is discontinuity, but non-arbitrary as to exclusion, because their boundaries coincide with the non-arbitrary boundaries of included species.

Thus under this particular concept and in the particular cases to which it applies, the species is defined as the one taxonomic category that is non-arbitrary both in exclusion and in inclusion. This is another way of expressing what is clearly intended by statements that the species (so defined) is the "objective" or the "real" taxonomic unit. If my usage of "non-arbitrary" is accepted and discussion of the meaning of "objective" or "real" is avoided, it should not be seriously questioned that the statement of the first sentence of this paragraph is valid. Objections, which may also be entirely valid in their own terms, are of five principal sorts:

1. The genetical concept of species is not the only one possible, and for certain groups and in particular circumstances it may be less desirable than some other.

2. Application of the genetical definition to actual cases, even those to which it could theoretically apply, sometimes turns out to be vague or impractical.

3. There are many groups of organisms, or circumstances involved in their taxonomic grouping, to which the stated genetical definition does not apply even in theory.

4. The genetical definition implies but does not adequately state or overtly take into consideration more definitely evolutionary criteria on which it does or should depend, criteria as to the evolutionary role of a lineage, to be discussed below.

5. Application of this or of related evolutionary concepts of the species does not correspond with past and current usage in certain groups and by certain taxonomists.

It seems to me that all these objections have considerable force, more force than is granted them by some students whose taxonomic work is in the circumscribed fields where the genetical definition is in fact most practical or those whose interests are not primarily taxonomic. Yet I do not think that the objections invalidate the genetical concept or remove it from a central and basic position in taxonomic theory. They merely require that it be modified in certain applications and that it be supplemented by other concepts to meet situations to which it does not properly or practically apply. The rest of this paper is devoted mainly to discussion of some desirable or necessary modifications and supplemental concepts. One pertinent subject not directly discussed, because of limitations of space and ability, is the taxonomy of asexual groups, clones, apomicts, agamic complexes, and the like. Some of what is said below bears inferentially on this subject, and it has been briefly but ably considered by Stebbins (1950) with references to older literature.

Genetical and Evolutionary Species

As Mayr (1946, 1950) has emphasized, the usefulness of the genetical concept in taxonomy and the non-arbitrary definition of the genetical species (its "objective reality") are most evident in what he calls "non-dimensional species," those established in biotas living in one place at one time. Under such conditions, discontinuities in morphological and associated physiological variation are usually evident. In sexually reproducing groups it is almost always easy under these circumstances to establish by observation, experimentation, and inference which morphological discontinuities reflect reproductive discontinuities and to designate these as species boundaries.

But, as Mayr has also recognized, the fact that genetical species are usually rather obvious under these special limitations does not mean that they are equally clear and the genetical definition equally adequate under other and perhaps more important conditions. Populations do have extension in time and space and a non-dimensional taxonomy cannot cope with many essentials of life and of its evolution. With extension in space, the criteria of genetical continuity and discontinuity, of actual or potential interbreeding or its absence, cease in many cases to be absolute and clearly non-arbitrary and become merely relative. The similar and related local populations may not in fact interbreed over a period of years and yet may reasonably be considered as still having that potentiality. On the other hand, quite extensive interbreeding may occur between adjacent populations which nevertheless retain their own individualities, morphologically and genetically, so clearly that any consensus of modern systematists would call them different species. In some groups of plants, even though species are defined and considered as genetical entities, occurrence of some hybridization between adjacent species may be the rule rather than the exception. In such cases the species are in part arbitrarily bounded even though the gene flow is less between than within species, and the genus may become the most fully non-arbitrary unit (a thought expressed in other words by Camp, 1951). A rigidly genetical zoologist might then insist that such botanical genera equal zoological species, but evidently most botanists feel that in some way their species *are* analogous with zoological species and they can make out a good, even though not an absolutely clear-cut, case. (See Stebbins, 1950.)

In practice, even by zoologists who adhere strictly to genetical concepts of taxonomic units and who work on groups to which the concepts are applicable, it is often clear that the criterion of interbreeding or its absence is not taken as wholly decisive. Species may be distinguished even though they interbreed (hybridize) to some extent, and populations may be referred to a single species even though there is evidence that they are not in fact interbreeding. Other criteria are given weight additional to that of their evidence on interbreeding, e.g., morphological divergence, partial or full intersterility, and especially occurrence with discontinuity in the same area.

The genetical definition is meaningful because it is related to the evolutionary processes that give rise to the groups being classified. Yet the genetical criteria are not related to evolutionary change directly but only, as a rule, by implication. The following seems to be the strictly *evolutionary* criterion implied: a phyletic lineage (ancestral-descendent sequence of interbreeding populations) evolving independently of others, with its own separate and unitary evolutionary role and tendencies, is a basic unit in evolution. The genetical definition tends to equate the species with such an evolutionary unit. Most of the vagueness and differences of opinion involved in use of the genetical definition are clarified, at least, if not wholly resolved by taking the genetical criterion, or interbreeding, not as definitive in itself but as evidence on whether the evolutionary definition is fulfilled. Thus the species as actually used by many progressive systematists in both animals and plants does tend to approximate a unitary phyletic lineage of separate evolutionary role even though in both cases outbreeding, hybridization, may occur and in some groups of plants this is widespread and usual. Emphasis on unitary evolutionary role may even resolve the theoretical difficulty of defining species in asexually reproducing groups.

This redefinition, or shift of emphasis, or revealing of the implicit basis of much modern evolutionary taxonomy, introduces the element of time into the concept of species, even in the so-called non-dimensional situation. It designates the species, including the "non-dimensional" species, as a unit which has been evolving separately, or which will do so, or, as a rule, both. Decision that populations will evolve separately involves prediction. Such points as wide geographical discontinuity (especially with a strong intervening barrier), morphological divergence, sympatric occurrence without interbreeding, and intersterility are clearly items of evidence for this sort of prediction. Their bearing seems to me more meaningful in evolutionary terms than in the definition of actually or potentially interbreeding populations, although of course the evolutionary species usually is also such a group. The special importance of intersterility, even though no modern taxonomist makes it an absolute requirement for specific separation, is, for instance, evident in this context: intersterility makes the prediction of separate evolutionary roles certain.

The Species in Paleontology: Data and Discontinuities

Part of the endless discussion on species concepts is concerned with the relationship between neontological and paleontological species. Opinions vary from the view that the two usually are quite different (e.g. Elias, 1950) to the view that they are usually essentially the same or that one is only an extension of the other concept (e.g. Mayr, 1950). Both views are correct in the sense that species just like those (by any definition) of neontology do occur in paleontology, but that actual practice regarding them may be more difficult or, at least, necessarily somewhat different in paleontology and that there also occur in paleontology taxonomic groups to which no strictly neontological species concept can properly be applied.

That paleontological data and materials are different from neontological is well known and sometimes overstressed. Direct genetical methods are unavailable in paleontology, but they are very rarely used in neontological taxonomy. The paleontologist usually has parts, only, of the organisms concerned, but the neontologist commonly uses parts, only, of recent organisms. Different parts may be available or used in the two cases, but inferences from them regarding populations may nevertheless be closely analogous or actually identical. Nearly or exactly the same general sorts of data, morphological, distributional, and associational, are frequently used in the practice of paleontological and neontological taxonomy. (Fuller discussion of these points was given in Simpson, 1943.)

The "non-dimensional" species is encountered more frequently in paleontology than in neontology, in spite of the fact that paleontology is inherently more multidimensional than neontology. The neontologist is seldom forced to confine himself to collections from one locality, and is never justified in doing so unless forced. Much paleontological taxonomy is necessarily and properly based on quarry collections or mass collections from one local stratum, associations without appreciable dispersion in space or time and ideally non-dimensional. In such cases neontological concepts and definitions of genetical and evolutionary species apply without modification. (Even Elias, 1950, outspoken opponent of the current rapprochement of neontological with paleontological systematics and of both with genetics, admits that in such cases paleontologists "may be obliged to resort to neontological concepts of taxonomic terms.")

Discontinuities are more frequent and of more varied sorts in paleontology than in neontology. This has certain disadvantages for paleontological theory and interpretation, but it also has some practical advantages. Discontinuities of observation, only, due to inadequate sampling of local populations or inadequate distribution of sampling stations, occur in both fields but are generally harder to fill in when paleontological. Discontinuities of record, that is, in the organisms actually present and available for sampling in the field, are a particular paleontological problem and may concern both time and space. When samples have been obtained from different localities or horizons, rocks and fossils intermediate between them may not exist. Such discontinuities are, as of now, facts in nature. Their use to delimit taxonomic groups is non-arbitrary, by definition. Yet they do not necessarily coincide with any particular sort of discontinuity that existed when the organisms were alive. Hence their relationship to the sorts of units defined in neontology may be and remain ambiguous.

The special questions involved in succession or sequence will be discussed separately, but it should be noted here that paleontological samples discontinuous in space are often also discontinuous in time and that the possibility can seldom be discarded. With such samples of similar organisms it is always difficult and it may be quite impossible to determine whether:

(a) They represent local populations that were genetically continuous, and hence infraspecific groups by genetical and evolutionary definition.

(b) They represent separate phyletic lineages, and hence distinct genetical and evolutionary species.

(c) They represent ancestral and descendent populations, and hence a special and peculiarly paleontological situation discussed below.

In such a case, the preferred practical procedure is:

1. To consider the two (or more) lots of associated specimens as samples of different local populations and to derive from them estimates of morphological variation in those populations.
2. If the population estimates indicate no significant mean difference, to consider the samples as representative of essentially a single population and hence taxonomic group.
3. If the population estimates indicate significant mean difference but overlap in range of variation, to consider the samples as drawn from different subspecies of one species.
4. If the population estimates indicate no overlap in range of variation (for at least one well-defined character), to consider the samples as drawn from different species.

Species recognized in this way are non-arbitrary in exclusion and inclusion by combined morphological and distributional criteria. They are morphologically similar to most genetical and evolutionary species. In many cases they will in fact be genetical or evolutionary species, but under the stated conditions it is virtually impossible to determine this equivalence with any high degree of probability.

The Species in Paleontology: Succession

Succession on a small scale and involving short periods of time occurs in some neontological data and there involves some special taxonomic problems, but on the whole succession is distinctively paleontological.

Discontinuities of observation and of record are frequent in paleontological study of successive populations. They frequently correspond with discontinuities of time, already mentioned. Diastems, geologically brief intervals of non-deposition (with or without erosion), are abundant in most stratigraphic sequences. They represent local discontinuities in time, but may be considered taxonomically unimportant if there was no significant change in the populations being studied or if intervening fossils of the same or closely similar populations are available from other localities. Larger and regional stratigraphic unconformities are also, although less, common and they usually represent taxonomically significant discontinuities in time.

Discontinuities in succession may also be caused by migration, by change of (biotic, and commonly of correlated stratigraphic) facies, or, frequently, by a combination of both (see especially Bell, 1950, also Newell, 1948). Such discontinuities often coincide with discontinuities in time but, as Bell has stressed, they need not do so. For instance figure 1 shows diagrammatically a situation in which there is a discontinuity of facies, with fossil populations as limited by facies successive wherever found, but really contemporaneous and without discontinuity (or, indeed, regional succession) in time.

When discontinuities in succession are present in the data, they may be dealt with in practice as outlined above for paleontological discontinuities in general. They may similarly permit non-arbitrary delimitation of morphological-distributional species which may approximate, but cannot usually be clearly equated with, genetical-evolutionary species. This greatly simplifies paleontological procedures and in many particular cases it averts the special taxonomic problems inherent in continuity of succession.

Essential continuity in sequences long enough to involve significant progression or diversification of populations is far from universal in paleontology. It is, however, frequent and becomes steadily more so as collecting becomes more extensive. The special problems involved therefore do have great and increasing practical importance. They are of supreme importance for paleontological taxonomic theory. No one seriously doubts that the whole of life has factually been a continuum of populations when the whole sequence is considered, in spite of the innumerable discontinuities in the record.

The genetical-evolutionary concept of species is applicable as between different phyletic branches, evolving lineages, especially if they are contemporary but also if they are not. Thus in figure 2A, *a, b,* and *c* are three different species, by genetical or, more clearly, by evolutionary definitions, although *a* and *b* are contem-

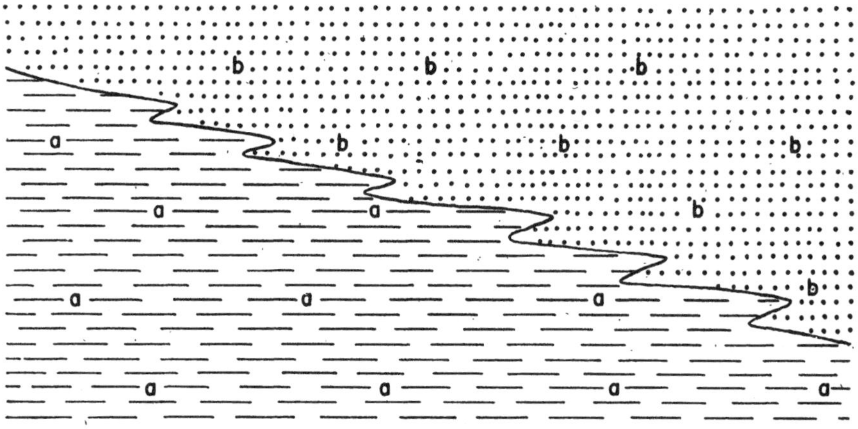

FIG. 1. Diagram of change of facies and fossil succession. Broken lines and dots represent two different rock facies each with a characteristic fossil, *a* and *b*. Although *b* everywhere occurs above *a* in any given local section, the two are, in fact, contemporaneous.

porary with each other but not with *c*. Serious problems in theory, and usually also in practice, arise rather regarding the parts of such a pattern that cannot be distinguished as separate branches.

One possible solution, diagrammed in figure 2B, is to recognize central lines as species and to distinguish branches as other species. This procedure is "correct" from an evolutionary point of view, or, better, the species so designated do fulfill the proposed evolutionary definition even though their delimitation is genetically arbitrary at the points of branching. For rather small groups under exceptionally favorable circumstances the procedure is also practicable and is actually used. Its practicability depends, however, on recognition of an essentially unchanging central line, *a,* and main branches, *b* and *c*. It is, however, more usual even within rather small groups and universal within really large groups and long sequences for all lines to evolve materially. Then it is not practical taxonomy to designate the whole of any one line as a single species, and there is no meaningful criterion for designating "main" or "central" lines and branches. Thus the four alternatives of figure 2C are all equally acceptable interpretations of the same phyletic facts as in figure 2A, in terms of main lines and branches, if all lines are undergoing progressive change. The only reasonable criterion of choice would be designation of certain terminal branches as more important, or somehow definitive, than others. A logical extreme would be, for instance, to take *Homo sapiens* as the supreme species and to consider its ancestry, from the beginning of life (or even before) as the main line, not specifically separable from *H. sapiens*. This arrangement has in fact been seriously proposed by a philosopher (Miller, 1949). Taxonomists will surely agree that this result and the whole procedure involved are impractical if not absurd.

Another possible approach is to recognize each evolutionary lineage as a unitary species until it divides and then to consider the descendent branches as species distinct from each other and from the single ancestral line, as diagrammed in figure 2D. This grouping meets an evolutionary definition of species, although delimitations between adjacent species are arbitrary by genetical criteria. It is, however, both undesirable and impractical. It frequently happens that a population undergoes no essential change even though a branch, a separate species, has arisen from a part of it. E.g. in figure 2D, *d* and *e* may be genetically and morphologically identical in all essentials. It is then not meaningful taxonomy to designate them as separate species. An even more serious objection is practical: the pattern of branching in a paleontological sequence is gradually discovered, perhaps never fully known, and generally depends as much on opinion as on unequivocal data. The taxonomy of long-known species would be changed every time a new branch was discovered or inferred and would be excessively and unnecessarily subject to personal disagreement. Moreover, a phyletic line may change radically between branches (say within *e* of figure 2D), and it is then not useful taxonomy to classify it as the same thing throughout.

The difficulties involved here are merely obscured by the presence of phyletic branching. They arise, regardless of whether or not branching occurs, from the problem of classifying ancestral and descendent stages in a continuously evolving population. Such a population may be diagrammatically represented, as in figure 3A, by a curve of variation (both genetical and morphological), moving through time and also being displaced as its genetical and morphological characters change. A cross-section represents the population at a particular instant in time, as it would be represented by a fossil sample from a single horizon. Such a cross-section is a genetical non-dimensional species, as seen both in neontology and in paleontology.

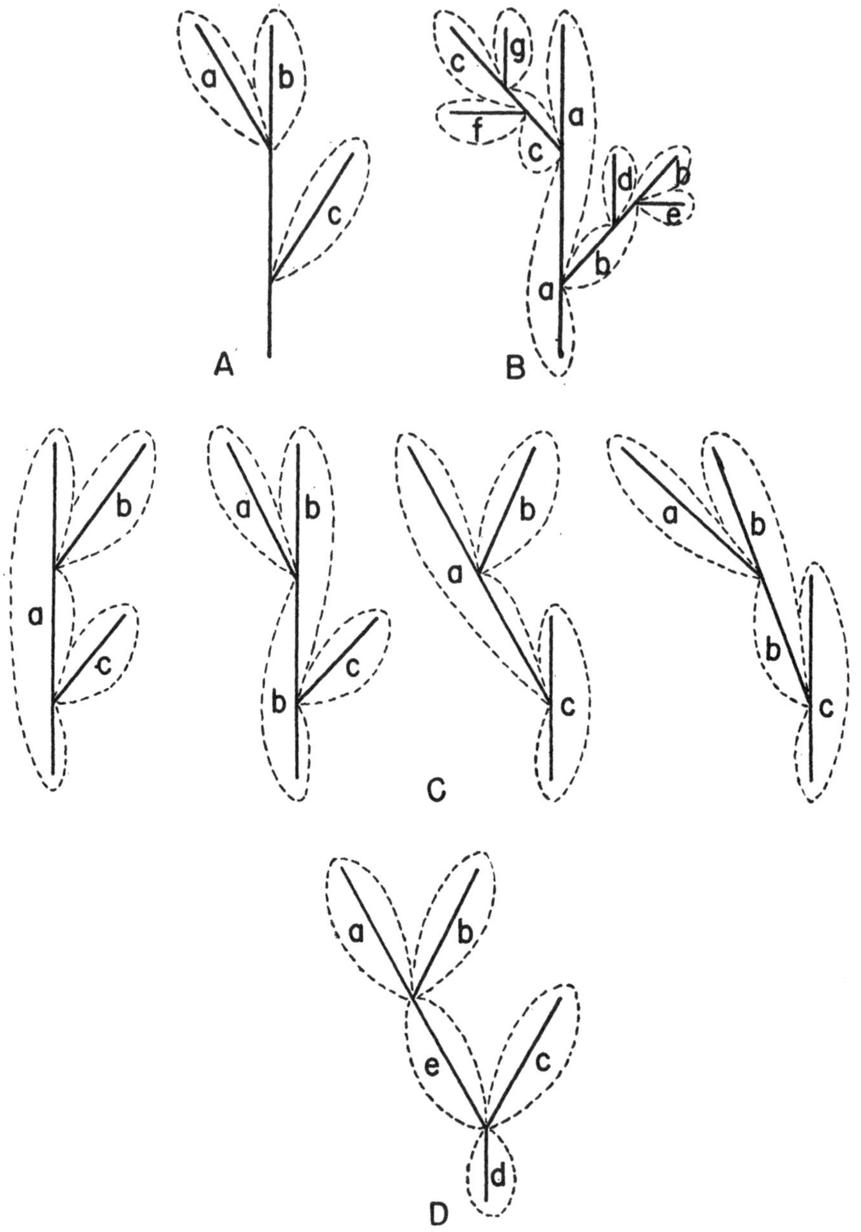

FIG. 2. Classification of successive populations with phyletic sequence and branching. In each diagram, the time sequence is from bottom to top and the solid lines represent phyletic descent. The broken lines enclose phyletic segments classified by various methods as distinct species. A, branching phyletic sequence with three clearly distinct evolutionary species, *a–c*. B, similar but more branched sequence with main lines and branches classified as species. C, four different possibilities of designating main lines and branches in the same phyletic sequence as in A. D, sequence as in A, with species boundaries set at points of branching. See discussion in text.

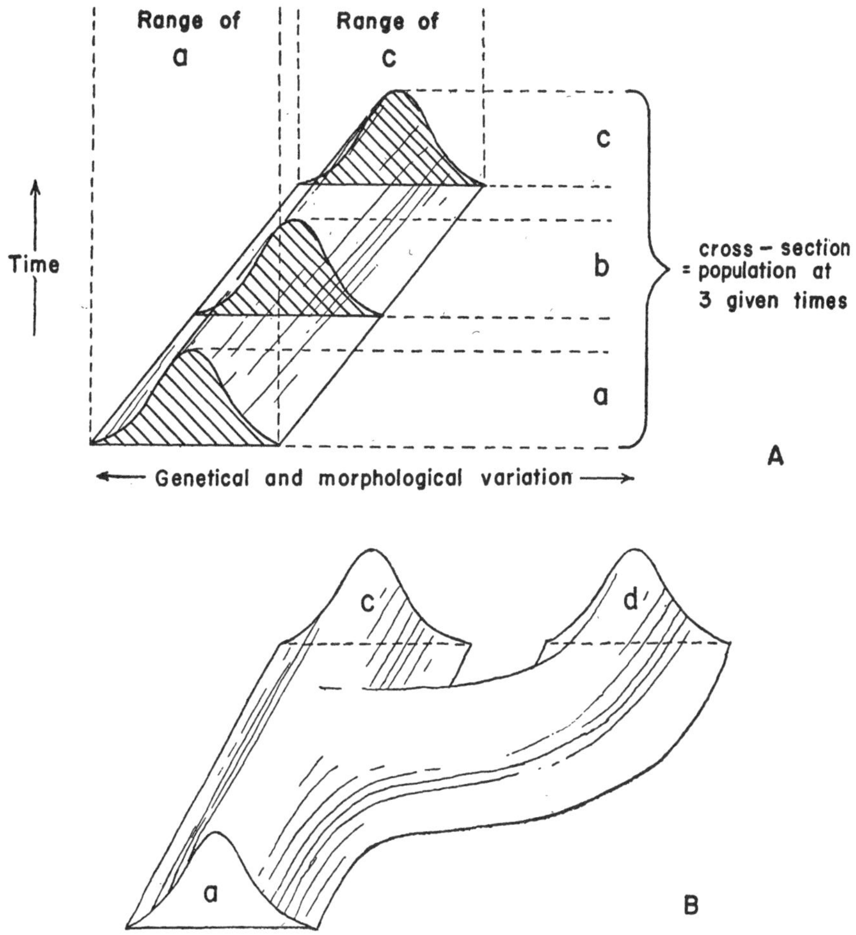

FIG. 3. Phyletic evolution and classification of successive populations. Phyletic sequence in a varying population is represented by a normal curve moving through time and changing in mean character, and by the solid generated thereby. A, sequence without branching. B, branched sequence. See discussion in text.

The whole sequence of populations, *a* to *c* in figure 3A, is genetically continuous and it fulfills the conditions of both genetical and evolutionary definitions of a species, as previously discussed. By these concepts, it is a single taxonomic group, defined as a species. Yet with the passage of time and continuation of progressive evolution, *c* has become quite different from *a*. For purposes of evolutionary study and of practical application to stratigraphy, it is essential that a distinction be made between these populations, which are different.

In practice, the paleontologist calls *a* and *c* different species if, as in figure 3A, the inferred ranges of variation do not overlap. They are *not* different species by the widely accepted genetical criteria or by the proposed evolutionary criteria discussed above. The comparison is clarified in figure 3B in which speciation (in the neontological sense) is represented as also having occurred. In this diagram, *c* and *d* are different species by any current usage; explicitly they are different genetical-evolutionary species and also different species in current paleontological practice.

But *a* and *c* are parts of a single genetical-evolutionary species, although called different species in paleontological practice.

The paleontologist thus uses the designation "species" for two sorts of entities which are radically and fundamentally incongruent. The only way in which the species category might be defined so as apparently to include both sorts of entities would be to abandon any evolutionary significance for taxonomy and to use purely morphological criteria. But this is not a useful solution. The general undesirability and impracticality of *purely* morphological taxonomic concepts have been sufficiently emphasized above. Moreover, the *whole* sequence of organisms represented in figure 3B cannot be classified at all, in morphological or any other terms, if the pattern in time, i.e. the evolutionary situation, is ignored. As static, separate pictures, the morphological difference between *a* and *c* and that between *c* and *d* are of the same sort, but within the pattern of the whole group in time, even the morphological relationships are not the same in the two cases, for *a* and *c* are morphologically (as well as genetically) continuous through intervening populations and *c* and *d* are not. (One might say here, in line with Dunbar, 1950, that *c* and *d* are continuous through the sequence *c–a–d*, but the fact that this involves a reversal in direction of time still makes an essential difference from the sequence *a–c* or *c–a*, which is consistent in the direction of time.)

In the situation represented in figure 3A, the desirable and indeed necessary taxonomic separation of *a* and *c*, whether they are called species or by some other category term, is arbitrary, because through intervening stages they are continuous by all meaningful criteria. The placing of an intermediate population, such as *b*, in one category or the other is, of course, also arbitrary. When the data really reflect the continuity of the sequence, intermediate populations must often be placed by rule of thumb rather than by any more positive and meaningful criterion. It is in such situations that the frequent occurrence of discontinuities of record, the absence of part of the sequence *a–c*, is practically useful in providing a means of separating *a* and *c*. This is still a separation in what *was* a continuum, but it is non-arbitrary (by the special definition of that word in this paper) as regards the actually available materials being classified.

Since paleontologists are applying the designation "species" to two fundamentally dissimilar sorts of taxonomic categories, it would appear logical that they confine that name to one of them and use a different name for the other. This has also been suggested, but it runs up against another serious practical difficulty: the paleontologist often does not know and has no way to determine which of the two basically different sorts of groups called "species" he has before him.

It is a common situation to have two discontinuous paleontological samples such as *a* and *c* of figure 4A. (It has been noted that if *a* and *c* are discontinuous in space, the possibility that they are also different in time can seldom be ruled out.) By applying the practical methods previously summarized, the paleontologist can readily draw population inferences from these samples, find that variation probably did not overlap in the populations, and define them as different "species." However, he does not know in what sense they are different species, because he does not know whether the relationship is as in figure 4B or as in figure 4C, and unless other crucial populations can be sampled he may have no conclusive way of finding out. In dealing with different samples from closely similar populations, this is one of the commonest situations in the practice of paleontology.

In such cases, a distinction cannot be made in practice between "species" in the basic genetical or evolutionary sense and in the sense of subdivisions in a continuous ancestral-descendent line. I do not

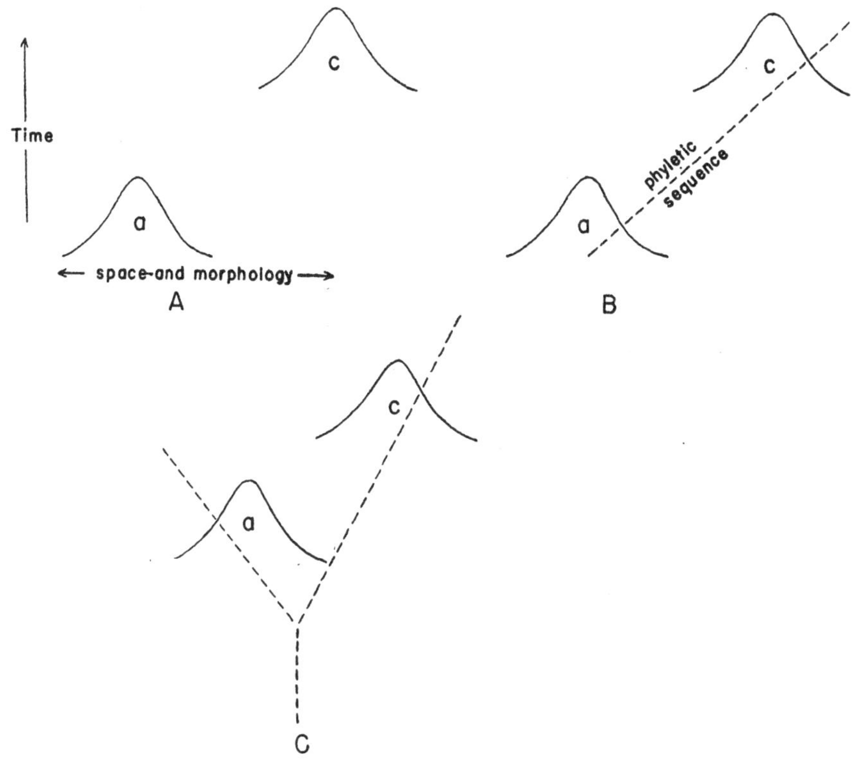

FIG. 4. Diagram illustrating problems of interpretation of two samples of related fossil organisms separated in space and (or, and possibly) in time. The variable populations represented by the samples are represented by normal curves. A, the given situation. B, interpretation as a single phyletic sequence, in which *a* and *c* represent the same species by genetic definition. C, interpretation as a branching sequence, in which *a* and *c* are different species by genetic or evolutionary definition. See discussion in text.

here favor or propose a special term for the latter sort of taxonomic group. I do maintain that it is desirable and useful to realize that these are two quite different things, and that the "species" of paleontological taxonomy may be of either sort.

There are many other pertinent and interesting points that might be considered, such as the problem of dual, partly coincident evolutionary species *a–c* and *a–d* in figure 3B, the uses of evolutionary acceleration and appearance of key characters for separating ancestral and descendent species, or the relationships of subdivisions of choroclines and chronoclines. This paper is, however, sufficiently long already, and the possible extension of its general point of view and method of approach to other points will probably be evident to most taxonomists.

LITERATURE CITED

ARKELL, W. J. 1950. A classification of the Jurassic ammonites. Jour. Paleont., **24**: 354–364.
BELL, W. C. 1950. Stratigraphy: a factor in paleontologic taxonomy. Jour. Paleont., **24**: 492–496.
BURMA, B. H. 1949a. The species concept: a semantic review. Evolution, **3**: 369–370.
——. 1949b. Postscriptum. Evolution, **3**: 372–373.
CAMP, W. H. 1951. Biosystematy. Brittonia, **7**: 113–127.
DUNBAR, C. O. 1950. The species concept: further discussion. Evolution, **4**: 175–176.
ELIAS, M. K. 1950. Paleontologic versus neontologic species and genera. Evolution **4**: 176–177.

GREGG, J. R. 1950. Taxonomy, language and reality. Amer. Nat., **84**: 419–435.
JELETZKY, J. A. 1950. Some nomenclatorial and taxonomic problems in paleozoology. Jour. Paleont., **24**: 19–38.
MAYR, E. 1942. Systematics and the origin of species. Columbia Univ. Press, New York.
——. 1946. The naturalist in Leidy's time and today. Proc. Acad. Nat. Sci. Philadelphia, **98**: 271–276.
——. 1949. The species concept: semantics versus semantics. Evolution, **3**: 371–372.
MILLER, H. 1949. The community of man. Macmillan, New York.
NEWELL, N. D. 1948. Infraspecific categories in invertebrate paleontology. Jour. Paleont., **22**: 225–232.
SCHINDEWOLF, O. H. 1950. Grundfragen der Paläontologie. Schweizerbart, Stuttgart.
SIMPSON, G. G. 1943. Criteria for genera, species, and subspecies in zoology and paleontology. Ann. New York Acad. Sci., **44**: 145–178.
STEBBINS, G. L., JR. 1950. Variation and evolution in plants. Columbia Univ. Press, New York.
WELLER, J. M. 1949. Paleontologic classification. Jour. Paleont., **23**: 680–690.
WRIGHT, C. W. 1950. Paleontologic classification. Jour. Paleont., **24**: 746–748.

Watson J. D., and F. H. C. Crick. 1953. Molecular structure of nucleic acids: a structure for deoxyribose nucleic acid. *Nature* 171:737–738.

Evolution can be defined most simply as alterations through time in the genetic composition of a population of organisms. Thus, it is ironic that the fields of evolutionary biology and population genetics were able to achieve so much in the first half of the twentieth century without understanding that DNA was the genetic material of life. That situation began to change in midcentury with the demonstration that nucleic acids (rather than proteins or some other class of molecules) were the genuine stuff of heredity (e.g., Avery et al. 1944; Hershey and Chase 1952). These discoveries set the stage for the next logical step, which was to determine the precise molecular structure of DNA. The breakthrough took place in 1953, when James Watson and Francis Crick (following a collaboration with Maurice Wilkins) announced their discovery of the double-helical structure of DNA, a feat for which the trio was awarded a Nobel Prize in 1962. (Rosalind Franklin in Wilkins' laboratory had provided the X-ray crystallography photographs from which Watson and Crick developed their physical models of DNA's molecular structure. She died in 1958, when she was 38 years old, well before the Nobel Prize award of 1962.)

We include a reprint of the original article in this volume, because it laid the foundation for all that has been accomplished in molecular and evolutionary genetics ever since that time. We also include this paper in the current collection because it contains a sentence toward the end that stands among the most famous understatements in the history of science: "It has not escaped our notice that the specific pairing we have postulated immediately suggests a possible copying mechanism for the genetic material."

RELATED READING

Avery, O. T., C. M. MacLeod, and M. McCarty. 1944. Studies on the chemical nature of the substance inducing transformation of Pneumococcal types: I. Induction of transformation by a DNA fraction isolated from *Pneumococcus* type III. Journal of Experimental Medicine 79:137–158.

Hershey, A. D., and M. C. Chase. 1952. Independent functions of viral protein and nucleic acid growth of bacteriophage. *Journal of General Physiology* 36:39–56.

Watson, J. D. 1968. The double helix. Atheneum Press, New York, New York, USA.

MOLECULAR STRUCTURE OF NUCLEIC ACIDS

A Structure for Deoxyribose Nucleic Acid

WE wish to suggest a structure for the salt of deoxyribose nucleic acid (D.N.A.). This structure has novel features which are of considerable biological interest.

A structure for nucleic acid has already been proposed by Pauling and Corey[1]. They kindly made their manuscript available to us in advance of publication. Their model consists of three intertwined chains, with the phosphates near the fibre axis, and the bases on the outside. In our opinion, this structure is unsatisfactory for two reasons: (1) We believe that the material which gives the X-ray diagrams is the salt, not the free acid. Without the acidic hydrogen atoms it is not clear what forces would hold the structure together, especially as the negatively charged phosphates near the axis will repel each other. (2) Some of the van der Waals distances appear to be too small.

Another three-chain structure has also been suggested by Fraser (in the press). In his model the phosphates are on the outside and the bases on the inside, linked together by hydrogen bonds. This structure as described is rather ill-defined, and for this reason we shall not comment on it.

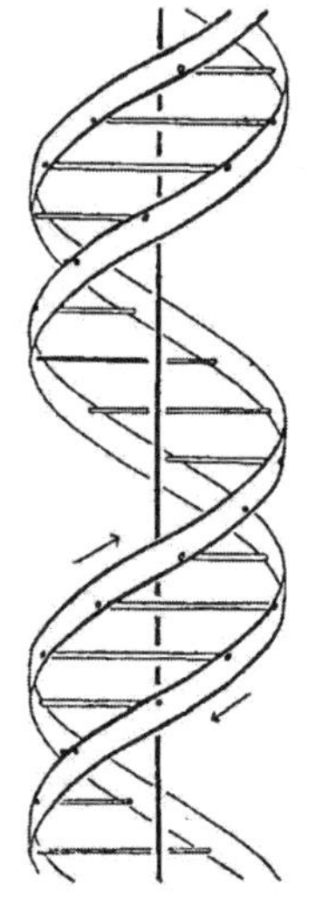

This figure is purely diagrammatic. The two ribbons symbolize the two phosphate—sugar chains, and the horizontal rods the pairs of bases holding the chains together. The vertical line marks the fibre axis

We wish to put forward a radically different structure for the salt of deoxyribose nucleic acid. This structure has two helical chains each coiled round the same axis (see diagram). We have made the usual chemical assumptions, namely, that each chain consists of phosphate diester groups joining β-D-deoxyribofuranose residues with 3′,5′ linkages. The two chains (but not their bases) are related by a dyad perpendicular to the fibre axis. Both chains follow right-handed helices, but owing to the dyad the sequences of the atoms in the two chains run in opposite directions. Each chain loosely resembles Furberg's[2] model No. 1; that is, the bases are on the inside of the helix and the phosphates on the outside. The configuration of the sugar and the atoms near it is close to Furberg's 'standard configuration', the sugar being roughly perpendicular to the attached base. There is a residue on each chain every 3·4 A. in the z-direction. We have assumed an angle of 36° between adjacent residues in the same chain, so that the structure repeats after 10 residues on each chain, that is, after 34 A. The distance of a phosphorus atom from the fibre axis is 10 A. As the phosphates are on the outside, cations have easy access to them.

The structure is an open one, and its water content is rather high. At lower water contents we would expect the bases to tilt so that the structure could become more compact.

The novel feature of the structure is the manner in which the two chains are held together by the purine and pyrimidine bases. The planes of the bases are perpendicular to the fibre axis. They are joined together in pairs, a single base from one chain being hydrogen-bonded to a single base from the other chain, so that the two lie side by side with identical z-co-ordinates. One of the pair must be a purine and the other a pyrimidine for bonding to occur. The hydrogen bonds are made as follows: purine position 1 to pyrimidine position 1; purine position 6 to pyrimidine position 6.

If it is assumed that the bases only occur in the structure in the most plausible tautomeric forms (that is, with the keto rather than the enol configurations) it is found that only specific pairs of bases can bond together. These pairs are: adenine (purine) with thymine (pyrimidine), and guanine (purine) with cytosine (pyrimidine).

In other words, if an adenine forms one member of a pair, on either chain, then on these assumptions the other member must be thymine; similarly for guanine and cytosine. The sequence of bases on a single chain does not appear to be restricted in any way. However, if only specific pairs of bases can be formed, it follows that if the sequence of bases on one chain is given, then the sequence on the other chain is automatically determined.

It has been found experimentally[3,4] that the ratio of the amounts of adenine to thymine, and the ratio of guanine to cytosine, are always very close to unity for deoxyribose nucleic acid.

It is probably impossible to build this structure with a ribose sugar in place of the deoxyribose, as the extra oxygen atom would make too close a van der Waals contact.

The previously published X-ray data[5,6] on deoxyribose nucleic acid are insufficient for a rigorous test of our structure. So far as we can tell, it is roughly compatible with the experimental data, but it must be regarded as unproved until it has been checked against more exact results. Some of these are given in the following communications. We were not aware of the details of the results presented there when we devised our structure, which rests mainly though not entirely on published experimental data and stereochemical arguments.

It has not escaped our notice that the specific pairing we have postulated immediately suggests a possible copying mechanism for the genetic material.

Full details of the structure, including the conditions assumed in building it, together with a set of co-ordinates for the atoms, will be published elsewhere.

We are much indebted to Dr. Jerry Donohue for constant advice and criticism, especially on interatomic distances. We have also been stimulated by a knowledge of the general nature of the unpublished experimental results and ideas of Dr. M. H. F. Wilkins, Dr. R. E. Franklin and their co-workers at

King's College, London. One of us (J. D. W.) has been aided by a fellowship from the National Foundation for Infantile Paralysis.

J. D. WATSON
F. H. C. CRICK

Medical Research Council Unit for the Study of the Molecular Structure of Biological Systems,
Cavendish Laboratory, Cambridge.
April 2.

[1] Pauling, L., and Corey, R. B., *Nature*, **171**, 346 (1953); *Proc. U.S. Nat. Acad. Sci.*, **39**, 84 (1953).
[2] Furberg, S., *Acta Chem. Scand.*, **6**, 634 (1952).
[3] Chargaff, E., for references see Zamenhof, S., Brawerman, G., and Chargaff, E., *Biochim. et Biophys. Acta*, **9**, 402 (1952).
[4] Wyatt, G. R., *J. Gen. Physiol.*, **36**, 201 (1952).
[5] Astbury, W. T., Symp. Soc. Exp. Biol. 1, Nucleic Acid, 66 (Camb. Univ. Press, 1947).
[6] Wilkins, M. H. F., and Randall, J. T., *Biochim. et Biophys. Acta*, **10**, 192 (1953).

Anderson, E., and G. L. Stebbins Jr. 1954. Hybridization as an evolutionary stimulus. *Evolution* 8:378–388.

Recent years have witnessed a resurgence of interest in the role of hybridization in evolution. Traditionally, the genetic variation that underlies evolutionary changes in populations was attributed to mutation and—in sexually reproducing species—to genetic recombination. Mutations are the ultimate source of novel genetic variants that recombination then shuffles through meiosis and syngamy (fertilization) into multitudinous new amalgams during each generation of sexual reproduction. But in the background lingered a nagging question: might introgressive hybridization, the crossing of a hybrid with one of its parent species, also be a significant wellspring of genetic variation upon which natural selection acts in at least some species? Evolution might occur disproportionately in bursts of high activity following introgressive hybridization events between disparate taxa. Introgression—even if it were relatively common—would not negate the preeminent roles for mutation and genetic recombination during the evolutionary process, but it would add an important ingredient or flavor to many discussions about exactly how evolution transpires.

Edgar Anderson and G. Ledyard Stebbins construct an eloquent and clear argument for the thesis that hybridization may indeed give an added boost to creative diversification in particular clades, especially when ecological circumstances are such that related organisms with distinct evolutionary histories are suddenly thrown into direct contact (such as might happen in highly disturbed habitats or perhaps through long-distance anthropogenic dispersal). Work continues today on empirically evaluating the extent to which this intriguing hypothesis applies in a wide variety of taxonomic groups ranging from plants to fishes.

RELATED READING

Anderson, E. 1949. Introgressive hybridization. Wiley, New York, New York, USA.
Arnold, M. L. 1997. Natural hybridization and evolution. Oxford University Press, New York, New York, USA.
Arnold, M. L. 2006. Evolution through genetic exchange. Oxford University Press, New York, New York, USA.

HYBRIDIZATION AS AN EVOLUTIONARY STIMULUS

E. Anderson and G. L. Stebbins, Jr.
Missouri Botanical Garden and University of California, Davis

Received July 1, 1954

One of the most spectacular facets of the newer studies of evolution has been the demonstration that evolution has not proceeded by slow, even steps but that seen in the large there have been bursts of creative activity. Some of the evidence for these bursts is from paleontology; Simpson (1953) has recently assembled a wealth of data concerning them and has discussed in detail their possible causes. Paleobotanists are equally aware of such events as the great upsurge of angiosperms in the Cretaceous, and of primitive vascular plants in the Devonian period. Other evidence for evolutionary bursts comes from the existence of large clusters of related endemic species and genera in the modern fauna and flora of certain regions, particularly oceanic islands and fresh water lakes. The snails (Achatinellidae) and honey sucker birds (Drepanidae) of Hawaii are classical examples, as are also the Gammarid crustaceans of Lake Baikal, and the fishes of Lakes Tanganyika and Nyasa in Africa, and particularly of Lake Lanao in the Philippines (see Brooks, 1950 for a summary and discussion of the data). It is true that some of these examples may represent normal rates of evolution occurring in a restricted area which has been isolated for a very long time, but there can be little doubt that in the case of others evolution has been phenomenally rapid.

As Simpson (1944, 1953) has clearly stated, the cause of this rapid evolution is to be sought in the organism-environment relationship. Along with most authors, however, he has tended to emphasize the peculiar environment present during these evolutionary bursts, and has suggested that one need not postulate any unusual type of population structure as a contributing factor. Zimmermann (1948) has given a plausible account of the environmental factors operating in the case of oceanic islands; reduction of competition, frequent migration to new habitats, and populations repeatedly reduced to a very few individuals, giving a maximum opportunity for the operation of chance as well as for the rapid action of selection.

To the student of hybridization, however, another factor which may have contributed largely to these evolutionary bursts presents itself. Hybridization between populations having very different genetic systems of adaptation may lead to several different results. If the reproductive isolation between the populations is slight enough so that functional, viable and fertile individuals can result from segregation in the F_2 and later generations, then new adaptive systems, adapted to new ecological niches, may arise relatively quickly in this fashion. If, on the other hand, the populations are well isolated from each other so that the hybrids between them are largely sterile, then one of two things may happen. The hybrids may become fertile and genetically stabilized through allopolyploidy, and so become adapted to more or less exactly intermediate habitats, or they may back cross to one or both parents, and so modify the adjoining populations of the parental species through introgression. This latter phenomenon has now been abundantly documented in the higher plants, and several good examples are known in animals (see bibliographies in Anderson, 1949; Heiser, 1949; Anderson, 1953). By introgressive hybridization elements of an entirely foreign genetic adaptive system can be carried over into a previously stabilized one, permitting the rapid reshuffling of varying adaptations and complex modifier systems. Natural selection is presented not with one or two new alleles but with segregating blocks of genic material belonging to entirely different adaptive systems. A simple analogy will show the comparative effectiveness of introgression.

Let us imagine an automobile industry in which new cars are produced only by copying old cars one part at a time and then putting them together on an assembly line. New models can be produced only by changing one part at a time. They cannot be produced *de novo* but must be built up from existing assembly lines. Imagine one factory producing only model 'T' Fords and another producing model 'T' Fords and also modern station wagons. It will be clear that if changes could only be brought about by using existing assembly lines these would have to proceed slowly in the factory which had only one assembly line to choose from. In the other factory, however, an ingenious mechanic, given two whole assembly lines to work with, could use different systems out of either and quickly produce a whole set of new models to suit various new needs when they arose.

Just as in the example of the two assembly lines, hybrids between the same two species could produce various different recombinations, each of which could accommodate itself to a different niche. When a big fresh water lake was formed *de novo,* hybrids between the same two species could rapidly differentiate into various new types suitable for the various new niches created in the big new lake. A technical point of much significance is that each of the various heterozygous introgressive segments brought in by hybridization would (by crossing over) be capable of producing increased variation generation after generation for periods running into whole geological eras (Anderson, 1939). The enhanced plasticity due to crossing over in introgressed segments has been shown on theoretical grounds to be present for many generations. Such studies as those of Woodson on Asclepias (1947, 1952), of Hall on Juniperus (1952) and of Dansereau on Cistus (1941) indicate that this does actually happen and that introgressive segments may persist for geological periods and produce effects of continental magnitude.

To students of introgressive hybridization it would seem like an excellent working hypothesis to suppose that when Lake Baikal was formed, and when each new island of the Hawaiian archipelago arose from the ocean, species belonging to different faunas and floras were brought together and that physical and biological barrier systems were broken down. There were increased chances for hybridization in an environment full of new ecological niches in which some new recombinations would be at selective advantages. There is a growing body of experimental data to support such an hypothesis. These data fall largely in two groups (1) Evolution under domestication, (2) Evolution in disturbed habitats.

(1) For evolution under domestication the evidence is overwhelming that by conscious and unconscious selection, man has created forms of plants and animals which are specifically distinct from their wild progenitors. This large body of evidence demonstrates that given a habitat in which novelties (or at least some of them) are at a great selective advantage, evolution may proceed very rapidly. There is presumptive evidence that many of these domesticates originated through introgression but the process began so early that getting exact experimental evidence for the history of any one of them will entail long-continued cooperative research (see, however, Mangelsdorf and Smith (1949), Alava (1952), and Nickerson (1953) for evidence that modern Zea is greatly different from the maize of five thousand

years ago and that much of this differentiation may well be the result of introgression from Tripsacum). For some ornamentals, domestication is such a recent event that critical evidence is easier to assemble. Anderson (1952) has presented in elementary detail the case of *Tradescantia virginiana.* He shows that in four hundred years by introgression from *T. ohiensis* and *T. subaspera* (unconsciously encouraged by man) it has evolved under cultivation into a variable complex quite distinct from *T. virginiana* as a genuinely wild species.

(2) Evolution in disturbed habitats. It has been repeatedly shown (Anderson, 1949; Heiser, 1949; Epling, 1947) that species which do not ordinarily produce hybrids and backcrosses may readily do so when man or any other agent disturbs the habitat. This phenomenon was referred to as "Hybridization of the Habitat" by Anderson (1948). After citing the work of several authors who have emphasized the role of man in promoting and creating habitats favorable for the perpetuation of hybrids and hybrid derivatives, he reached the following conclusion (1948, p. 6). "Does this mean that introgression as a phenomenon is limited to the areas disturbed by man and that its results are mere artifacts and not genuine natural phenomena? I think not. Though freely admitting that nearly all the introgression which has been studied experimentally (for one exception see Dansereau, 1941) is of the nature of an artifact, I believe that at particular times, and in particular places, introgression may have been a general evolutionary factor of real importance."

The great frequency of hybrid derivatives in disturbed habitats is only in part due to the breaking down of barrier systems, allowing previously isolated species to cross. It can and does occur when the barrier systems are not broken down (see for instance Heiser, 1951). Much more important is the production of new and varying ecological niches; more or less open habitats in which some of the almost infinitely various backcrosses and occasional types resulting directly from segregation in F_2 and later generations will be at a selective advantage. A particularly significant example was investigated by Anderson (unpublished) who studied *Salvia apiana* and *Salvia mellifera* in the San Gabriel mountains, confirming and extending Epling's (1947) previous studies. He found hybrid swarms not in the chaparral itself where both of these species are native but adjacent to it in cutover live oaks amidst an abandoned olive orchard. In this greatly disturbed area, new niches were created for the hybrid progeny, which are apparently always being produced in the chaparral but at a very low frequency. In this strange new set of various habitats some of the mongrels were at a greater selective advantage and the population of the deserted olive orchard was composed of hybrids and back-crosses to the virtual exclusion of *S. apiana* and *S. mellifera.*

It has been customary to dismiss the evidence of introgression under the influence of man as relatively unimportant to general theories of evolution because nothing quite like it had previously occurred. A little reflection will show that this is not so. Man at the moment is having a catastrophic effect upon the world's faunas and floras. He is, in Carl Sauer's phrase, an ecological dominant but he is not the first organism in the world's history to achieve that position. When the first land vertebrates invaded terrestrial vegetation they must have been quite as catastrophic to the flora which had been evolved in the absence of such creatures. When the large herbivorous reptiles first appeared, and also when the first large land mammals arrived in each new portion of the world there must have been violent readjustments and the creation of new ecological niches.

The last of these (the arrival of the large land mammals) is close enough to us in geological time so that we have witnessed the very end of the process. The vegetation of New Zealand had had no experience with mammals until the arrival of the Maori in the fourteenth century followed by Europeans in the 18th and 19th centuries. Man, pigs, horses, cattle, rats, sheep, goats, and rabbits were loosed upon a vegetation which had had no previous experience with simians or herbivores. The effect was catastrophic. Hybrid swarms were developed upon the most colossal scale known in modern times. A succession of New Zealand naturalists have occupied themselves with the problem and it has been treated monographically by Cockayne (1923) and by Allen (1937).

The extent to which disturbance of the habitat combined with reorganization of adaptive systems through hybridization could have been responsible for evolutionary bursts, "proliferation," "tachytely" or "quantum evolution" (Simpson 1953) can best be estimated by summarizing the geological and paleontological evidence concerning the time of occurrence of habitat disturbances, and comparing this with probable evolutionary changes in certain groups of organisms which were most likely initiated by hybridization. In such a survey, all three of the possible results of hybridizations—introgression, segregation of new types without backcrossing and allopolyploidy—must be considered. Reference to allopolyploids is particularly important, since hybrid derivatives of this type can easily be distinguished from their parental species by their chromosome numbers, and the time and place of hybridization can often be indicated with a high degree of probability (Stebbins, 1950, Chap. 9).

Preceding the advent of man, the most revolutionary event in the history of the northern continents was the Pleistocene glaciation and the contemporary pluvial periods of regions south of the ice sheet. This involved not only radical oscillations in climate, but also great disturbances of the soil, both in the glaciated regions and in areas to the south of them. In the latter, the extensive deposits of loess immediately south of the ice margin and the masses of alluvium carried for miles down the river valleys must have disturbed these areas almost as much as the ice sheets churned up the areas which they covered.

The activity of hybridization in developing plant populations adapted to these new habitats is amply evident from the frequency of allopolyploids in them. Specific examples are *Iris versicolor* and *Oxycoccus quadripetalus* (Stebbins, 1950); the polyploid complexes of *Salix, Betula, Vaccinium, Antennaria, Poa, Calamagrostis,* and many others can also be cited. The best example of introgression among species which have invaded the ice-free areas in post-Pleistocene time is in the complex *Acer saccharophorum* (Dansereau and Desmarais, 1947). The numerous examples cited by Anderson (1953) of hybrid and introgressant types which occupy the central Mississippi Valley between the Appalachian, Ozark, and central Texas highlands probably represent late Pleistocene or post-Pleistocene invasion of these areas which were strongly affected by outwash from the ice sheet and from the postglacial lakes. The origin of *Potentilla glandulosa* subsp. *Hanseni* in the post-Pleistocene meadows of the Sierra Nevada is discussed by Stebbins (1950, p. 279).

During the Tertiary and earlier geological periods three types of changes in the inanimate environment can be singled out which probably gave rise to disturbed habitats favorable to the establishment of hybrid derivatives. These were mountain building movements, advance and retreat of epicontinental seas, and radical changes in the earth's climate.

Some of the direct effects of mountain buildings are the rapid creation of raw, unoccupied habitats (such as lava flows, for instance), in which plants belonging to very different ecological associations may temporarily mingle and gain a chance to hybridize. In central California the canyon of the Big Sur River is a typical example of the mixing together of species belonging to very different floras in a region of recent uplift which has a rugged, youthful topography. Here yuccas and redwoods grow within a stone's throw of each other. An example of hybridization in this area is between two species of *Hieracium; H. albiflorum,* which is typical of northern California, the Pacific Northwest, and the Rocky Mountains, and *H. argutum,* a Southern California endemic which here reaches its northern limit except for one known station in the Sierra foothills. Examples such as this could undoubtedly be multiplied by a careful study of any youthful mountain region.

The retreat of epicontinental seas in the latter part of the Pliocene period, plus faulting in the Pleistocene and recent times, has been largely responsible for the present topography of coastal California with its flat valleys and abrupt mountain ridges. One hybrid polyploid which appears to have spread as a result of these changes is the octoploid *Eriogonum fasciculatum* var. *foliolosum* (Stebbins, 1942); another is probably the tetraploid *Zauschneria californica* (Clausen, Keck, and Hiesey, 1940). A series of hybrid swarms which may have arisen in response to the same topographical changes is that of *Quercus Alvordi* (Tucker, 1952). *Delphinium gypsophilum* is a relatively well stabilized species, probably of hybrid origin, endemic to this same recently emerged area of California (Epling, 1947), and the species of *Gilia* considered by Grant (1953) to be of hybrid origin have the same general distribution. In the Old World, Dansereau (1941) has suggested that *Cistus ladaniferus* var. *petiolatus,* which occupies the recently emerged coast of North Africa, is a product of hybridization between typical *C. ladaniferus* and *C. laurifolius* both of which occur in the more ancient land mass of the Iberian Peninsula.

Among the radical changes in the earth's climate which occurred recently enough so that their effect on the vegetation can be recorded, is the advent of the Mediterranean type of climate with its wet winters and dry summers in most of California. The time of this climatic change is now fully documented by the fossil record; it took place during the middle part of the Pliocene period. It was preceded by a general decrease in precipitation, with biseasonal maxima (Axelrod, 1944, 1948).

The effects of this climatic change on the woody vegetation of the area are also well documented by the fossil record. One very probable example of a hybrid swarm exists in a fossil flora. In the Remington Hill Flora, which was laid down in the Sierra foothills at the beginning of the Pliocene, there is a great abundance of oak leaves corresponding to the modern *Q. morehus,* a hybrid between the mesophytic, deciduous *Q. Kelloggii,* and the xerophytic, evergreen *Q. Wislizenii* (Condit, 1944). That these fossil leaves were borne by hybrid trees is evidenced not only by their very unusual and characteristic shape, but also by their great variability and the fact that no similar leaves are found in any of the numerous Miocene and Pliocene floras of California. Furthermore, the Remington Hill is the only one of these fossil floras which contains the counterparts of both parental species. At present, the *Q. Kelloggii* × *Wislizenii* hybrid is frequent in the Sierra foothills, but it usually grows as single trees in company with dense stands of *Q. Wislizenii* and *Q. Kelloggii.* The populations of the parental species growing in the vicinity of the hybrids appear little or not at all different from those occurring by themselves, far from any other species of this complex. On the other hand, *Q. Wislizenii* shows considerable geographic variation, with the more northernly and more coastal variants, i.e., those adapted to increasingly mesic climates, possessing an increasingly greater resemblance to *Q. Kelloggii* in habit, leaves, buds, and fruits. This suggests that the present variation pattern in *Q. Wislizenii* is the result chiefly of extensive introgression from *Q. Kelloggii,* which began with the hybrid swarms of Mio-Pliocene time, and has since been ordered into a regular, clinical series of variants by the selective action of the changing Pliocene and Pleistocene climates. Tucker (oral comm.) has suggested that *Q. Douglasii,* a completely unrelated oak with a similar geographical distribution, may have also originated from one or more hybrid swarms of a similar geological age. The modern variation pattern of the common chaparral species *Adenostoma fasciculatum* (Anderson, 1952) could be interpreted on the same basis, while less thorough observations by the junior author suggest that several other examples can be found in the California flora.

Conditions favorable for the origin and spread of hybrid derivatives are made not only by changes in the inanimate environment, but also by the advent and disappearance of various types of animals. Previous to man and his associated domesticates, some of these disturbances were as follows. In the Eocene and Oligocene periods, large grazing mammals made their first appearance on the earth. Their effect on the woody vegetation cannot be detected in the fossil record, and probably was not great. The herbaceous plants, however, must have been greatly affected by their inroads, and if these smaller plants had been abundantly preserved as fossils, we might be able to record a burst of evolution in them during these early Tertiary epochs. Babcock (1947, p. 132), after careful consideration of all lines of evidence, has suggested the latter part of the Oligocene as the time of origin of the genus *Crepis,* one of the larger, more specialized, and probably more recent genera of Compositae. On this basis, one might suggest that the greatest period of evolution of genera in this largest of plant families was during late Eocene and Oligocene time. The junior author, from his studies of various grass genera of temperate North America, believes that many facts about their present distribution patterns could best be explained on the assumption that they began their diversification during the Oligocene epoch. They appear to have attained much of their present diversity by the middle of the Miocene, by which time many of the now extensive polyploid complexes, such as those in *Bromus, Agropyron,* and *Elymus,* had begun to be formed. The extensive Miocene record of species belonging to the relatively advanced tribe Stipeae (Elias, 1942) would support such an assumption.

At an earlier period, namely the beginning of the Cretaceous, the world saw for a relatively short time the dominance of the largest land animals which have ever existed, the great herbivorous dinosaurs. These monsters must have consumed huge quantities of the fern and gymnosperm vegetation which prevailed at the time, and it is difficult to see how these plants, with their relatively slow growth and reproduction, could have kept up with such inroads. It is very tempting, in fact, to speculate that over grazing on the part of giant dinosaurs contributed toward the extinction of the Mesozoic gynmospermous vegetation, as well as of the larger dinosaurs themselves, during the middle of the Cretaceous period. At the same time, shallow epicontinental seas were advancing and retreating, leaving coastal plain areas open for plant colonization; other significant events during this period were the rise of modern birds and of Hymenoptera, particularly bees.

The writers venture to suggest that these four nearly or quite concurrent events—retreat of seas, overgrazing by dinosaurs, advent of a diversified avifauna which transported seeds long distances, and rise of flower pollinating bees and other insects—all contributed to the greatest revolution in vegetation which the world has even seen; the replacement of gymnosperms by the predominant angiosperm flora of the upper part of the Cretaceous period. One should note that all of these conditions would favor hybridization and the spread of hybrid derivatives, by giving unusual opportunities for previously separated types to be brought together by wide seed dispersal, by permitting cross pollination between types previously isolated from each other, and by opening up new areas for colonization by the hybrid derivatives. The suggestion has been made elsewhere (Stebbins, 1950, p. 363) that differentation of genera and sub-families among primitive angiosperms took place partly via allopolyploidy; the time of origin of this polyploidy may well have been during the Cretaceous period. Evidence of introgression at so remote a time is probably impossible to obtain; by an analogy we should assume that in the past as now, conditions favorable for allopolyploidy also promoted introgression.

Going still further into the past, let us speculate on the events which must have taken place at the time when vascular plants and vertebrates first spread over the land. The principal geological period involved is the Devonian. At the beginning of this period comes the first extensive fossil record of vascular plants, all belonging to the primitive order Psilophytales. By the end of the Devonian, forms recognizable as club mosses (Lycopsida), ferns (primitive Filicales), and seed plants (Pteridospermae) were already widespread. We shall, of course, never know what chromosome numbers existed in these extinct groups of primitive vascular plants. But their nearest living descendants are nearly all very high polyploids, as has now been most elegantly demonstrated by Manton (1950, 1953). She has suggested that the living Psilotales, which have gametic numbers of about 52, 104, and over 200, "are the end-products of very ancient polyploid series which date back to simple beginnings. . . ." The relationship between the modern Psilotales and the Devonian Psilophytales is not clear, but to the present authors they appear to resemble each other nearly enough so that they could belong to the same complex network of allopolyploids, which developed its greatest diversity in the Devonian period. In the genus *Ophioglossum,* generally regarded as one of the three most primitive genera of true ferns, we have the highest chromosome numbers known to the plant kingdom, namely n = ca. 256 in the northern *O. vulgatum* and n = ca. 370 in the tropical *O. pendulum.* These ferns are not preserved in the fossil record because of their soft texture, but their origin during the Devonian period is a fair inference. There is very good reason to believe, therefore, that the great proliferation of genera and families of vascular plants during this earliest period of their dominance was accompanied by allopolyploidy just as it has been in the more recent periods of very active evolution. Where allopolyploidy was widespread, we can also suspect abundant introgression.

The reader may well ask at this point whether any of this evidence contributes to the central theme of the present discussion, namely the hypothesis that these extensive hybridizations, both ancient and relatively modern, gave rise to really new types, which formed the beginnings of families, orders, and classes having different adaptive complexes from any plants previously existing. It is undoubtedly true that the results of introgression and allopolyploidy are chiefly the blurring of previously sharp distinctions between separate evolutionary lines, and the multiplication of variants on adaptive types which were already established during previous cycles of evolution. Nevertheless, the fact must not be overlooked that conditions favorable for introgression and allopolyploidy, namely the existence of widely different and freely recombining genotypes in a variety of new habitats, also favor the establishment and spread of new variants. Establishment of new adaptive systems is under any circumstances a relatively rare event; in any group of organisms we have hundreds of species and subspecies which are variants of old adaptive types to one which represents a really new departure. Hence we cannot expect to recognize introgressive or polyploid complexes which have given rise to such new types until we have carefully analyzed hundreds of those which have not. Furthermore, our methods of recognizing these complexes almost preclude the chance of identifying the new types which have arisen from them. We make the assumption that hybrid derivatives, whether introgressants or allopolyploids, have characteristics which can all be explained on the basis of intermediacy between or recombination of the characteristics of the putative parents, and then devise methods of verifying this assumption. The new types, falling outside of this assumption, would be rejected by our methods.

The junior author can suggest two examples known to him of new and distinctive morphological characteristics which may have arisen in recent hybrid derivatives. One of these is the presence in *Ceanothus Jepsonii,* a species narrowly endemic to the serpentine areas of northern California, of flowers with six and seven sepals, petals, and anthers (Nobs, 1951). This characteristic is not known anywhere else in the family Rhamnaceae or even in the entire order Rhamnales, an order which almost unquestionably dates back to the Cretaceous period. Mason (1942) has given strong evidence for the recent origin of *Ceanothus Jepsonii.* It inhabits an environment which is certainly recent, since the mountains on which it occurs were covered by a thick layer of volcanic rocks even as late as the end of the Pliocene epoch, and the serpentine formations to which it is endemic were not exposed until after the faulting which occurred at the beginning of the Pleistocene (Mason, 1942). It belongs to a complex of closely related species and subspecies, among which hybridization is still very actively taking place (Nobs, 1951). In characteristics other than sepal and petal number, it is intermediate between various ones of its relatives rather than an extreme type. Hence there is a good reason to suspect that *Ceanothus Jepsonii* represents a species of relatively recent (i.e. Pleistocene) hybrid origin which has evolved a morphological characteristic previously unknown in its family, and in fact one which is relatively uncommon in the entire subclass of dicotyledons.

The other example is in the grass species *Sitanion jubatum.* This species is distinguished by possessing glumes which are divided into a varying number of linear segments, a characteristic not found elsewhere in the tribe Hordeae, and one which is the basis of a distinctive mechanism for seed dispersal (Stebbins, 1950, p. 141). *Sitanion jubatum* is endemic to Pacific North America, being most abundant in the coast ranges and foothills of central California. Its nearest relative is *S. hystrix,* a species found in the montane areas of the same region, and extending far eastward and southeastward. The two species are distinguished only by the degree of division of the glumes, and in fact appear to grade into each other. Field observations suggest that they actually consist of a large swarm of genetically isolated microspecies, such as has been demonstrated experimentally to exist in the related *Elymus glaucus* (Snyder, 1950).

Cytogenetically, both species of *Sitanion* are allotetraploids (Stebbins, Valencia, and Valencia, 1946). Extensive chromosome counts from various parts of the ranges of both species plus still more numerous measurements of sizes of pollen and stomata have failed to reveal any form of *Sitanion* which could be diploid. Furthermore, the chromosomes of both species are strongly homologous with those of *Elymus glaucus,* as evidenced by complete pairing in the F_1 hybrid. All of this evidence suggests that *S. jubatum* did not have any diploid ancestors which possessed its distinctive glumes, but has evolved out of a complex of allopolyploids which has existed in western North America for a long time, probably since the middle of the Tertiary period.

The forms of *S. jubatum* which have the most extreme division of the glumes occupy habitats which are recent, and which in some ways are intermediate between the most extreme habitats occupied by *Sitanion* and those characteristic of *Elymus glaucus*. They are known from the shore of San Francisco Bay, in northeastern Marin County, from the eastern edge of the Sacramento Valley north and east of Sacramento and from the Sierra foothills in Mariposa county. In growth habit, these races of *S. jubatum* could be regarded as intermediate between the most extreme xerophytes found in *Sitanion* on the one hand, and *E. glaucus* on the other.

Experimental evidence (Stebbins, unpublished) has now indicated that the complex of microspecies within the taxonomic species *Elymus glaucus* originated partly if not entirely through introgression. The probability is strong that *Sitanion* consists of a similar swarm of microspecies which originated also by introgression. The extensive subdivision of the glumes in some of these microspecies, therefore, may well have originated through the establishment of new mutations, or of new types of gene interaction, in genotypes produced by hybridization and introgression between morphologically very different and genetically well isolated species.

When all of this evidence has been considered, the writers can hardly escape the conclusion that hybridization in disturbed habitats has produced the conditions under which the more familiar processes of evolution, mutation, selection, and the origin of reproductive isolation barriers, have been able to proceed at maximum rates. Far from being insignificant because much of it is in habitats greatly disturbed by man, the recent rapid evolution of weeds and semi-weeds is an indication of what must have happened again and again in geological history whenever any species or group of species became so ecologically dominant as greatly to upset the habitats of their own times.

Summary

(1) It has been established by recent work in Palaeontology and Systematics that evolution has not proceeded at a slow even rate. There have instead been bursts of evolutionary activity as for example when large fresh water lakes (Baikal, Tanganyika, and Lanao) were created *de novo*.

(2) Recent studies of introgression (hybridization and subsequent back-crossing) have demonstrated that under the influence of man evolution has been greatly accelerated. There has been a rapid evolution of plants and animals under domestication and an almost equally rapid evolution of weed species and strains in greatly disturbed habitats.

(3) The rapidity of evolution in these bursts of creative evolution may well have been due to hybridization. At such times diverse faunas and floras were brought together in the presence of new or greatly disturbed habitats where some hybrid derivates would have been at a selective advantage. Far from being without bearing on general theories of evolution, the repeated demonstrations of accelerated introgression in disturbed habitats are of tremendous significance, showing how much more rapidly evolution can proceed under the impact of a new ecological dominant (in this case, Man). Such an agent may bring diverse faunas and floras into contact. Even more important is the creation of various new, more or less open habitats in which novel deviates of partially hybrid ancestry are at a selective advantage. The enhanced evolution which we see in our own gardens, dooryards, dumps and roadsides may well be typical of what happened during the rise of previous ecological dominants. The first vertebrates to enter isolated continents or islands, the first great herbivorous reptiles, the first herbiv-

orous mammals must have created similar havoc upon the biotae of their own times. Introgression must have played the same predominant role in these disturbed habitats as it does today under the impact of man. These arguments are supported by a homely analogy (page 379) and by various kinds of experimental and taxonomic data.

LITERATURE CITED

ALAVA, REINO O. 1952. Spikelet variation in *Zea Mays L.* Ann. Mo. Bot. Gar., **39**: 65–96.

ALLAN, H. H. 1937. Wild species-hybrids in the phanerogams. Bot. Rev., **3**: 593.

ANDERSON, E. 1939. Recombination in species crosses. Genetics, **24**: 688.

——. 1948. Hybridization of the habitat. Evolution, **2**: 1–9.

——. 1949. Introgressive Hybridization. Wiley & Sons, New York, 109 pp.

——. 1952. The ecology of introgression in Adenostoma. Nat. Acad. Sci.: Abstracts of papers presented at the autumn meeting, Nov. 10–12, 1952 (Sci., **116**: 515–516).

——. 1952. Plants, Man and Life. Little, Brown & Co., Boston, 245 pp.

——. 1953. Introgressive hybridization. Biol. Rev., **28**: 280–307.

AXELROD, D. I. 1944. The Pliocene sequence in central California. Carnegie Inst. Wash. Publ., 553: 207–224.

——. 1948. Climate and evolution in western North America during Middle Pliocene time. Evolution, **2**: 127–144.

BABCOCK, E. B. 1947. The genus *Crepis*. Part I. The taxonomy, phylogeny, distribution and evolution of *Crepis*. Univ. Calif. Publ. Bot., **21**: 1–198.

BROOKS, J. L. 1950. Speciation in ancient lakes. Quart. Rev. Biol., **25**: 131–176.

CLAUSEN, J., D. D. KECK AND W. HIESEY. 1940. Experimental studies on the nature of species. I. Effect of varied environment on western North American plants. Carnegie Inst. Wash. Publ., 520, vii, 452 pp., figs. 1–155.

COCKAYNE, L. 1923. Hybridism in the New Zealand flora. New Phytol., **22**: 105–127.

CONDIT, C. 1944. The Remington Hill flora. Carnegie Inst. Wash. Publ., 553: 21–55.

DANSEREAU, P. 1941. Etudes sur les hybrides de Cistes. VI. Introgression dans la section Ladanium. Can. Jour. Research, **19**: 59–67.

—— AND Y. DESMARAIS. 1947. Introgression in sugar maples. II. Amer. Midl. Nat., **37**: 146–161.

ELIAS, M. K. 1942. Tertiary prairie grasses and other herbs from the high plains. Spec. Papers Geol. Soc. Amer., 41: 176 pp.

EPLING, C. 1947. Actual and potential gene flow in natural populations. Am. Nat., **81**: 81–113.

——. 1947. Natural hybridization of *Salvia apiana* and *S. mellifera*. Evolution, **1**: 69–78.

GRANT, V. 1953. The role of hybridization in the evolution of the leafy-stemmed *Gilias*. Evolution, **7**: 51–64.

HALL, M. T. 1952. Variation and hybridization in *Juniperus*. Ann. Mo. Bot. Gard., **39**: 1–64.

HEISER, C. B., JR. 1949. Natural hybridization with particular reference to introgression. Bot. Rev., **15**: 645–687.

——. 1951. A comparison of the flora as a whole and the weed flora of Indiana as to polyploidy and growth habits. Indiana Acad. Sci., Proc., **59**: 64–70.

MANGELSDORF, P. C., AND C. E. SMITH, JR. 1949. New archeological evidence on evolution in maize. Bot. Mus. Leaflets, Harvard Univer., **13**: 213–247.

MANTON, I. 1950. Problems of Cytology and Evolution in the Pteridophyta. 316 pp., Cambridge University Press.

——. 1953. The cytological evolution of the fern flora of Ceylon. Soc. Exp. Biol., Symp., **7**: Evolution, 174–185.

MASON, H. L. 1942. Distributional history and fossil record of *Ceanothus*. pp. 281–303 in Van Rensselaer, M., and H. E. McMinn. *Ceanothus*, publ. by Santa Barbara Bot. Gard., Santa Barbara.

NICKERSON, N. H. 1953. Variation in cob morphology among certain archaeological and ethnological races of maize. Ann. Mo. Bot. Gard., **40**: 79–111.

NOBS, M. 1951. Ph.D. Thesis, Univ. California, Library.

SAUER, C. O. 1952. Agricultural origins and dispersals. Am. Geogr. Soc., Bowman memorial lectures Ser. II, v. 110 pp., 4 pls. New York.

SIMPSON, G. G. 1944. Tempo and Mode in Evolution. xiii, 237 pp., New York.

——. 1953. The Major Features of Evolution. Columbia University Press, New York, 434 pp.

SNYDER, L. A. 1950. Morphological variability and hybrid development in *Elymus glaucus*. Amer. Jour. Bot., **37**: 628–636.

STEBBINS, G. L., JR. 1942. Polyploid complexes in relation to ecology and the history of floras. Am. Nat., **76**: 36–45, figs. 1–2.

——. 1950. Variation and Evolution in Plants. Columbia University Press, New York, 643 pp.

——, J. I. Valencia and R. M. Valencia. 1946. Artificial and natural hybrids in the Gramineae, tribe Hordeae I. *Elymus, Sitanion* and *Agropyron*. Am. Jour. Bot., **33**: 338–351.

Tucker, M. 1952. Evolution of the California oak *Quercus alvordiana*. Evolution, **6**: 162–180.

Woodson, R. E., Jr. 1947. Some dynamics of leaf variation in *Asclepias tuberosa*. Ann. Mo. Bot. Gard., **34**: 353–432.

——. 1952. A biometric analysis of natural selection in *Asclepias tuberosa*. Nat. Acad. Sci.: Abstracts of papers presented at the autumn meeting, Nov. 10–12, 1952 (Sci., **116**: 531).

Zimmerman, E. C. 1948. Insects of Hawaii. Vol. I. Introduction. University of Hawaii Press, Honolulu, 206 pp.

Dobzhansky, T., and O. Pavlovsky. 1957. An experimental study of interaction between genetic drift and natural selection. *Evolution* 11:311–319.

Random genetic drift can be defined as stochastic processes that occur because of accidents of sampling, especially in small populations. Fluctuations in the frequency of mutations, in numbers of migration between populations, and in selection rates are also stochastic processes. In the absence of these processes, the Hardy-Weinberg law (see Hardy 1908 in this volume) asserts that, with random mating, gene frequencies in populations remain constant from generation to generation. This constancy is expected only in ideal, infinitely large populations, which is never the case in reality. In small populations, the frequencies of different genes in the next generation may be higher, or lower, than in the preceding one. The smaller the number of breeding individuals in a population, the higher the probability that gene frequencies will change from one to another generation.

Random genetic drift in very small populations may lead to the fixation of one or another of the alleles or chromosomal arrangements present in the original population, but different populations may become fixed for one or other of the initial genetic variants (see also Kimura and Ohta 1971 in this volume). In populations that are not extremely small, gene frequencies will oscillate differently in separate populations. Natural selection, however, will interact with drift, favoring one or another genetic variant so that the effects of drift may not lead to genetic fixation even in small populations. Assessing the interactions between genetic drift and natural selection is the objective of the experiments in the present paper. In addition to his fundamental contributions to the theory of evolution, Dobzhansky was a leading experimentalist.

RELATED READING

Dobzhansky, T. 1970. Genetics of the evolutionary process. Columbia University Press, New York, New York, USA.

Kerr, W. E., and S. Wright. 1954. Experimental studies of the distribution of gene frequencies in very small populations of *Drosophila melanogaster.* Evolution 8:293–302.

Kimura, M. 1968. Evolutionary rate at the molecular level. Nature 217:624–626.

Neel, J. V., and E. A. Thompson. 1978. Founder effect and the number of private polymorphisms observed in Amerindian tribes. Proceedings of the National Academy of Sciences, USA 75:1904–1908.

O'Brien, S. J., D. E. Wildt, M. Bush, T. M. Caro, C. FitzGibbon, I. Aggundey, and R. E. Leakey. 1987. East African cheetahs: evidence for two population bottlenecks. Proceedings of the National Academy of Sciences, USA 84:508–511.

Wright, S. 1931. Evolution in Mendelian populations. Genetics 16:97–159.

AN EXPERIMENTAL STUDY OF INTERACTION BETWEEN GENETIC DRIFT AND NATURAL SELECTION

Theodosius Dobzhansky and Olga Pavlovsky [1]
Department of Zoology, Columbia University, New York City

Received January 31, 1957

Introduction

The role of random genetic drift in the evolutionary process has, for about two decades, been one of the controversial issues in population genetics. Some authors have appealed to "drift" as a convenient explanation of the origin of differences among organisms for which no other explanations seemed to be available. But one's inability to discover the adaptive significance of a trait does not mean that it has none (cf. Dobzhansky, 1956). The hypothesis of random genetic drift should not be used as a loophole; to be accepted it requires a firmer basis than suspicion. Other authors seem to think that drift and natural selection are alternatives. As soon as a gene is shown to have any effect whatever on fitness, the conclusion is drawn that its distribution in populations must be determined solely by selection and cannot be influenced by random drift. But this is a logical non-sequitur. The important work of Aird *et al.* (1954) and of Clarke *et al.* (1956) disclosed that the incidence of certain types of gastro-intestinal ulceration is significantly different in persons with different blood groups. This is, however, far from a convincing demonstration that the observed diversity in the frequencies of the blood group genes in human populations is governed wholly, or even partially, by selection for resistance to ulcers. To make such a conclusion tenable it would have to be demonstrated that the environments in which human racial differences have evolved actually favored greater resistance in certain parts of the world and lesser resistance in certain other parts. Thus far no evidence has been adduced to substantiate any such claim.

As defined by Wright (1949) random genetic drift includes all variations in gene frequencies which are indeterminate in direction. Such variations are caused by accidents in gene sampling in populations of finite genetically effective size, as well as by fluctuations in the intensity or in the direction of selection, mutation and gene exchange between populations. Wright (1932, 1948, 1948, 1951) as well as the present writer (Dobzhansky, 1937–1941–1951) have stressed that random drift by itself is not likely to bring about important evolutionary progress. Indeed, variations in gene frequencies induced by random drift in small isolated populations are apt to be inadaptive, and hence likely to result in extinction of such populations. However, random drift may be important in conjunction with systematic pressures on the gene frequencies, particularly with natural selection. What is most necessary, then, is the type of experimental evidence that would permit analysis of the interactions between random drift and selection. Such evidence, although difficult to obtain, should be within the range of what is possible. Kerr and Wright (1954) and Wright and Kerr (1954) studied models of Drosophila populations in which the number of the progenitors in every generation was fixed arbitrarily, and in which classical laboratory mutants were used as traits subject to drift and to selection. In the experimental Drosophila populations described in the following pages naturally occurring genetic variants, inversions in the third

[1] The work reported in this article has been carried out under Contract No. AT-(30-1)-1151, U. S. Atomic Energy Commission.

chromosomes of *Drosophila pseudoobscura* were used. Severe limitation of the population sizes was introduced in some of the populations in only a single generation, at the beginning of the experiments. Experiments so conducted may to some extent reproduce genetic events which occur in natural populations.

Preliminary Experiments

It has been shown (see Dobzhansky, 1949 and 1954, for reviews) that heterozygotes of *Drosophila pseudoobscura* which carry two third chromosomes with different gene arrangements derived from the same locality are, as a rule, superior in Darwinian fitness to the corresponding homozygotes. The situation is more complex when flies of different geographic origins are hybridized. Chromosomal heterozygotes which carry two third chromosomes derived from different geographic regions may or may not exhibit heterosis. Experimental populations, bred in the laboratory in so-called population cages, behave differently depending upon whether the foundation stock of the population consists of flies of geographically uniform or of geographically mixed origin. In the former case, the chromosomes with different gene arrangements usually reach certain equilibrium frequencies. Replicate experiments, conducted with reasonable precautions to make the environments uniform, give results repeatable within the limits of sampling errors. With geographically mixed populations the results do not obey simple rules. The course of natural selection in such populations is often erratic; equilibrium may or may not be reached, or may be reached and then lost; replicate experiments do not give uniform results; heterosis may or may not be present at the start of the experiments, and may or may not develop in the course of selection in the experimental populations.

The indeterminacy observed in the populations of geographically mixed origin is however understandable (Dobzhansky and Pavlovsky, 1953; Dobzhansky, 1954). Race hybridization releases a flood of genetic variability; the number of potentially possible gene combinations far exceeds the number of the flies in the experimental populations; natural selection perpetuates the genotypes which possess high adaptive values under experimental conditions, but it is a matter of chance which of the possible adaptive genotypes will be formed first in a given population. In some populations these genotypes will happen to be structural heterozygotes, and in others homozygotes.

We have tested about thirty experimental populations of mixed geographic origins, using different combinations of flies from diverse localities (Dobzhansky and Pavlovsky, 1953, and much unpublished data). Among them were two replicate populations, Nos. 119 and 120, which are relevant here. They were started on February 8, 1954, in wood-and-glass population cages used in our laboratory and described previously. The foundation stocks consisted of F_1 hybrids between 12 strains derived from flies collected near Austin, Texas in 1953 and 10 strains derived from Mather, California, in 1947. The Texas strains were homozygous for the Pikes Peak (PP) gene arrangement, and the California strains for the Arrowhead (AR) gene arrangement in their third chromosomes. In each of the two cages 2,395 flies of both sexes, taken from the same F_1 culture bottles of Texas by California crosses, were introduced. The populations were kept in an incubator at 25° C., samples of eggs deposited in the population cages were taken at desired intervals, larvae hatching from these eggs were grown under optimal conditions in regular culture bottles, and their salivary glands were dissected and stained in acetic orcein.

The course of the events in the populations Nos. 119 and 120 is shown in table 1 and figure 1. The percentage frequencies of PP chromosomes are given in this table, the frequencies of AR chromosomes are the balance to 100 per cent. Each sample is based on determination of the

TABLE 1. *Changes in the frequencies (in per cent) of PP chromosomes in two replicate experimental populations of* Drosophila pseudoobscura *of mixed geographic origin (Texas PP by California AR)*

Days from start	Population No. 119	Population No. 120	Chi-Square	P
0	50.0	50.0	—	—
35	49.3	48.7	0.02	0.90
70	39.0	40.7	0.08	0.75
105	42.3	36.7	1.01	0.35
250	30.0	43.7	6.01	0.01
300	29.0	40.7	4.50	0.03
365	26.3	42.0	15.60	0.001
425	25.0	41.7	9.37	0.002

gene arrangement in 300 third chromosomes (150 larvae, taken in 6 subsamples on 6 successive days). The first samples, 35 days from the start, showed little change from the original frequencies, 50 per cent, of the chromosomes. At 70 and 105 days the frequencies of PP diminished, about equally in both populations, as shown by the low chi-square (each chi-square has one degree of freedom). But at 250 days the frequency of PP diminished in the population No. 119, while it failed to change, or even increased, in No. 120. This situation persisted until April 9, 1955, about 425 days from the start, when the last samples were taken and the populations were discarded. The chi-squares shown in table 1 attest that the outcomes of natural selection in these two experimental populations were clearly unlike. It should be noted that the magnitude of the divergence between the replicate populations Nos. 119 and 120 is not exceptionally great for the type of experiments in which flies from geographically remote localities are involved.

MAIN EXPERIMENTS

Certain consequences should follow from the above interpretation of the indeterminacy observed in populations of geographically mixed parentage. The indeterminacy should be a function of the genetic variability in the foundation stock

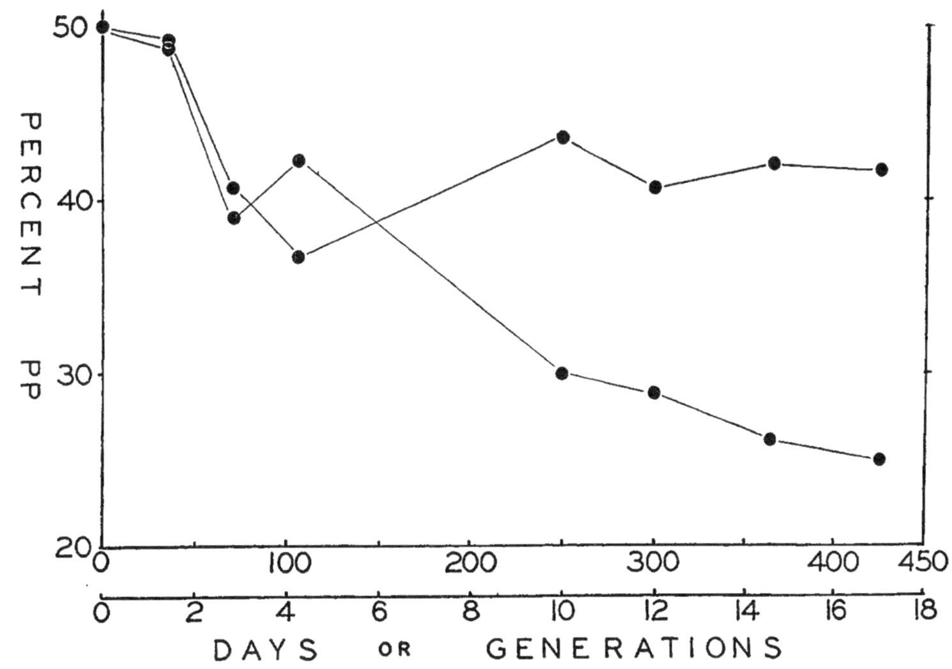

FIG. 1. Changes in the frequencies of PP chromosomes in two replicate experimental populations of mixed geographic origin (Texas by California).

of the populations. Chromosomes with PP and AR gene arrangements are recognizable under the microscope; their frequencies are made uniform in the foundation stock of all populations, and we observe changes in their frequencies as the experiment progresses. However, we infer that, apart from this overt variability in the frequency of the gene arrangements, there must exist also a large amount of genic variability released owing to gene recombination in the F_2 and later generations of interracial hybrids. Although there is no way of telling by how many genes the races differ, the number of the possible gene combinations must be several to many orders of magnitude greater than the number actually realized. The outcome of selection in the experimental populations should, then, be more variable in small than in large populations.

This working hypothesis is open to experimental test, but the experimental technique must be carefully thought through. One could make some experimental populations smaller than others by keeping them in cages of different sizes and with different amounts of food. The drawback of this would be that the environments of the populations of different sizes would be dissimilar. Therefore, we have chosen to vary the sizes of the foundation stocks of our populations, but to permit them to expand to equal size, which, because of the high fecundity of the flies, they do within a little more than a single generation.

The same 12 Texas PP and 10 California AR strains were used in the main as in the preliminary experiments (see above). F_1 hybrids between them, which were necessarily heterozygous PP/AR, were raised in regular culture bottles, and so were the F_2 hybrids. In June 1955, 4,000 F_2 flies, about equal numbers being females and males and derived equally from the different crosses, were placed in a population cage. Between June 15 and 27, 15 cups with yeasted culture medium were inserted in the cage daily. The flies covered the medium with eggs overnight. The cups with the eggs were then withdrawn and placed in another population cage containing no adult flies. In this manner ten population cages, Nos. 145–154, were obtained on ten successive days. They were descended, then, from the same foundation stock of 4,000 F_2 interlocality hybrids. The frequencies of PP and AR chromosomes in the foundation stock are evidently 50–50. These are the "large" populations.

Ten groups of 20 F_2 flies each, 10 ♀♀ and 10 ♂♂, were taken from the same F_2 cultures which served as the source of the foundation stock for the "large" populations, care being taken to include in each group flies from all the F_2 cultures.

TABLE 2. *Frequencies (in per cent) of PP chromosomes in the experimental populations*

Large populations			Small populations		
No.	Oct. '55	Nov. '56	No.	Oct. '55	Nov. '56
145	39.3	31.7	155	37.7	18.0
146	42.3	29.0	156	30.7	32.0
147	29.3	34.7	157	31.0	46.0
148	38.0	34.0	158	32.3	46.7
149	33.3	22.7	159	34.3	32.7
150	36.0	20.3	160	41.7	47.3
151	40.3	32.0	161	37.3	16.3
152	41.0	22.3	162	25.3	34.3
153	37.0	25.7	163	37.7	32.0
154	42.0	22.0	164	25.3	22.0
Mean	37.85	27.44	Mean	33.33	32.73
Variance	15.30	26.96	Variance	26.73	118.91

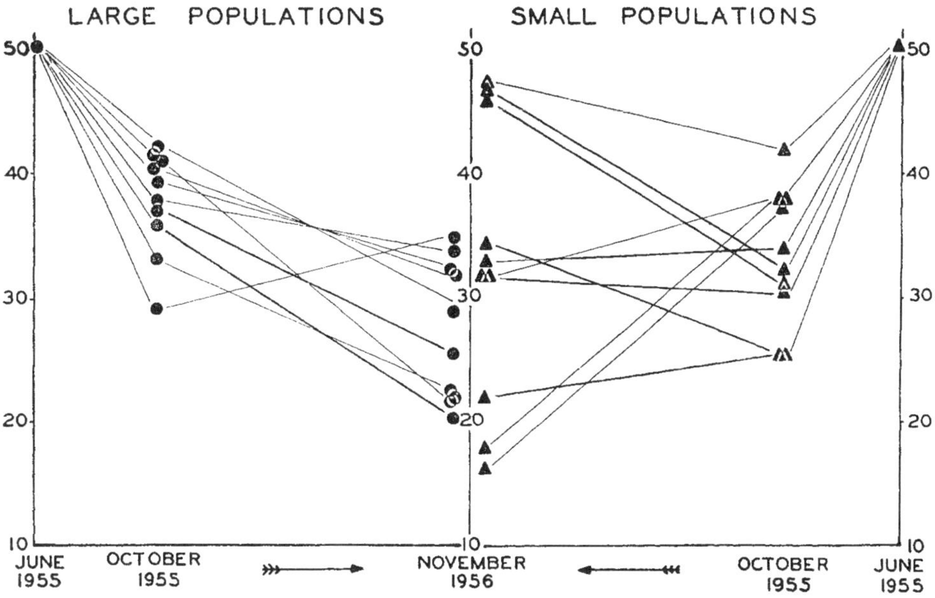

FIG. 2. The frequencies (in percent) of PP chromosomes in twenty replicate experimental populations of mixed geographic origin (Texas by California).

These groups of F_2 flies were placed in regular culture bottles and allowed to produce progenies. Each progeny was then transferred to population cages of the same type as those used for the "large" populations. Ten population cages, Nos. 155–164, were thus obtained. They are the "small" populations. It should be reiterated that the "large" and the "small" differed only in the foundation stocks, these being 4,000 and 20 flies respectively. All populations were kept at 25° C. and treated similarly in every way.

In October and November 1955, about 4 generations after the start, the populations were examined; egg samples were taken and the chromosomes in the salivary gland cells of the larvae that hatched from these eggs were studied. The gene arrangements in the chromosomes were determined and scored by Dr. Louis Levine. The results are summarized in table 2 and figure 2. As usual, each sample consisted of 300 chromosomes. The frequencies of PP varied from 29.3% to 42.3% in the "large" populations, and from 25.3% to 41.7% in the "small" ones. The heterogeneity is significant in both; the chi-square for the "large" is 19.5 and for the "small" ones 35.9, which correspond to probabilities of about 0.02 and of much less than 0.001 respectively. The heterogeneity among the replicate experiments is, of course, not surprising in view of the outcome of the preliminary experiments (table 1), although in these latter a significant heterogeneity first appeared after somewhat more than 4 generations from the start. It may be noted (table 2) that the variance for the "small" populations (26.7) is ostensibly greater than for the "large" ones (15.3), but the F value is not significant. It may also be noted that the frequencies of PP chromosomes have declined from the 50% value in the foundation stock, the decline being somewhat greater in the "small" (33.3%) than in the "large" (37.8%) populations.

The next, and the final, test of the populations was made in November 1956, i.e., more than a year after the first test and about 19 generations after the populations were placed in the population cages. The preliminary experiments (table 1) show that the populations reach equilibria in the frequencies of PP and AR chromosomes within less than a year from the start. The samples, 300 chromosomes per cage, were taken in the usual manner and scored by one of us (Th. D). The results are reported in table 2 and figure 2.

It may be noted that the mean frequency of PP in the "large" populations is now 27.4% and in the "small" ones 32.7%. These means are not significantly different from each other, but the 1956 mean for the "large" populations is significantly lower than the 1955 mean. The outcomes of natural selection in the "large" and the "small" populations are, then, similar on the average. It is otherwise when the outcomes in the individual populations are considered. As shown in table 2 and figure 2, the frequencies of PP in the "large" populations range from 20.3% to 34.7%, and in the "small" ones from 16.3% to 47.3%. In both instances the heterogeneity is highly significant (the chi-square for the "large" population is 40.4 which, for 9 degrees of freedom, has a negligible probability of being due to chance. Both in the "large" and especially in the "small" populations the variance has increased during the year intervening between the two tests (1955–1956).

Most important of all is, however, that the "small" populations show a heterogeneity significantly greater than the "large" ones. The variances, 118.9 and 27.0, now give an F ratio of 4.4 which is significant at between the 0.025 and 0.010 levels. The greater heterogeneity is evidently due to the different magnitudes of the foundation stocks in these populations. This heterogeneity was indicated already by the tests in October 1955, but it has become significant as the selection continued during the year between the two tests. Finally, it may be pointed out that there appears to be no significant correlation between the status of a given population in 1955 and 1956. For example, No. 160 had the highest frequency of PP in both tests (Table 2), but No. 161 which had the lowest frequency in 1956 had an above average frequency in 1955.

Discussion

The results of the present investigation can be stated very simply: Although the trait studied (the gene arrangement in the third chromosome) is subject to powerful selection pressure, the outcome of the selection in the experimental populations is conditioned by random genetic drift. The either-selection-or-drift point of view is a fallacy.

In our experiments, the heterozygotes which carry two third chromosomes with different gene arrangements are heterotic; natural selection in the experimental populations establishes equilibrium states at which both gene arrangements occur with certain frequencies; these frequencies are determined by the relative fitness of the homozygotes and heterozygotes. Now, the environments being reasonably uniform in all experimental populations, the outcome of the selection processes in the replicate experiments should also be uniform. And so it is, in experimental populations of geographically uniform origin. But it is not so in geographically mixed populations. In the latter, the selective fates of the chromosomal gene arrangements become dependent upon the polygenic genetic background, which is highly complex and variable because of the gene recombination that is bound to occur in populations descended from race hybrids. Here random drift becomes operative and important. It becomes important despite the populations being small only at the beginning of the experiments, because the foundation stocks in some populations consisted of small numbers of individuals. Thereafter, all the populations expand to equal sizes, fluctuating roughly between 1,000 and 4,000 adult individuals. Such populations can be regarded as small only in relation to the number of gene recombinations which are possible in populations of hybrid origin.

For reasons that are not far to seek, geneticists visualize the evolutionary process usually in terms of the destinies of single genes. With the notable exception of the contributions of Wright (1932 and subsequent work), this is the frame of reference of most of the mathematical theory of population genetics. This makes manageable an otherwise impossibly complex topic, and yet the oversimplified models usually suffice for understanding of microevolutionary processes. But as we move into the realm of mesoevolution (Dobzhansky, 1954), not to speak of macroevolution, it becomes indispensable to consider not only the destinies of single genes but also of integrated genotypes, and finally of the gene pool of Mendelian populations. In our experiments, the foundation stock of the populations consisted of F_2 hybrids between rather remote geographic races; a highly variable gene pool arose owing to the hybridization; random drift caused different segments of this gene pool to be included in the foundation stocks of each population, especially in the small ones; natural selection then produced divergent results in different populations, especially again amongst the small ones.

It is now logical to inquire whether the events observed in our experimental populations resemble situations which occur in nature. The excellent work of Dowdeswell and Ford (1952, 1953) and Ford (1954) has disclosed a most suggestive case. Populations of the butterfly *Maniola jurtina* are rather uniform throughout southern England, despite some obvious environmental diversity in different parts of this territory. In contrast to this, the populations of the same species show quite appreciable divergence on the islands of the Scilly archipelago, although these islands are within only a few miles of each other and their environments appear rather uniform. Especially remarkable is the divergence observed between the populations of certain small islands, while larger islands have more nearly similar populations. The small islands happen, however, to be situated between the larger ones. The investigators have estimated that the populations of the small islands consist of numbers of individuals of the order of 15,000 and that the populations of the large islands must be considerably greater. The authors conclude that the genetic divergence between these populations must be produced entirely by selection, random drift being inconsequential. The evidence is, however, weighed in favor of the view that the genetic divergence was initiated by the island populations being derived from small numbers of immigrants from the mainland or from other islands. These immigrants introduced somewhat different sets of genes on each island, whereupon natural selection built different genetic systems in the different populations.

The divergence between island and mainland populations has been studied also by Kramer and Mertens (1938) and by Eisentraut (1950) in lizards, and by Lowe (1955) in mammals and reptiles. Most data agree in showing that the divergence is greater on smaller than on larger islands, and greater on islands more remote from the mainland than on those which are apt to receive immigrants most frequently. Many authors, including this writer (Dobzhansky, 1937, 1941), interpreted these situations as arising through random drift in populations of continuously small size, or frequently passing through narrow "bottlenecks." This interpretation need no longer be sustained. It is more probable, especially in the light of the experiments described in the present article, that in the island populations we are observing the emergence of novel genetic systems moulded by interaction of random drift with natural selection.

Mayr (1954) has pointed out that conspicuous divergence of peripherally isolated populations of a species is a fairly general phenomenon, well known to systematists. He rightly concludes that this divergence cannot be due entirely to random drift "in the ordinary sense," i.e., to fluctuations of the gene frequencies in populations of persistently small size. Indeed, some of the peripheral populations consist of thousands or even millions of individuals. Mayr's interpretation can be stated best in his own words: "Isolating a few individuals (the 'founders') from a variable population which is situated in the midst of a stream of genes which flows ceaselessly through every widespread species will produce a sudden change of the genetic environment of most loci. This change, in fact, is the most drastic genetic change (except for polyploidy and hybridization) which may occur in a population, since it may effect all loci at once. Indeed, it may have the character of a veritable 'genetic revolution.' Furthermore, this 'genetic revolution,' released by the isolation of the founder population, may well have the character of a chain reaction. Changes in any locus will in turn affect the selective values at many other loci, until finally the system has reached a new state of equilibrium." The outcome of our experiments described above may, in a sense, be regarded as experimental verification of Mayr's hypothesis.

Summary

Twenty replicate experimental populations of *Drosophila pseudoobscura* were kept in a uniform environment for approximately 18 months. The foundation stocks of all the populations consisted of F_2 hybrids between flies of Texas origin which had the PP gene arrangement in their third chromosomes, and flies of California origin with the AR gene arrangement in the same chromosome. In ten of the populations the founders numbered 4,000 individuals; the other ten populations descended from only 20 founders each.

The frequencies of PP and AR chromosomes in all the populations were originally 50 per cent. Eighteen months later, the frequencies of PP varied from about 20 to 35 per cent in the populations descended from the large numbers of founders, and from 16 to 47 per cent in those descended from small numbers of the founders. The heterogeneity of these frequencies of PP chromosomes observed in the replicate populations is statistically highly significant. More important still, the heterogeneity is significantly greater in the populations descended from small numbers of founders than in those descended from large numbers of founders.

Heterozygotes which carry a PP and an AR third chromosome are superior in adaptive value to the PP and AR homozygotes. Therefore, the frequencies of PP and AR chromosomes in the experimental populations are controlled by natural selection. However, the heterogeneity of the results in the replicate populations is conditioned by random genetic drift.

Only some of the possible combinations of the genes of the Texas and California genomes are actually realized in the populations. The segments of the gene pool which arise from race hybridization are smaller, and therefore less uniform, in the populations descended from small than in those descended from large numbers of founders. It may reasonably be inferred that evolutionary changes involving interactions of natural selection and random drift of the kind observed in our experiments are not infrequent in nature.

Acknowledgments

We take pleasure in acknowledging our obligations to our colleague Professor Howard Levene, and to Dr. Bruce Wallace of the Long Island Biological Laboratories, for their counsels regarding the statistical and experimental procedures; to Professor Louis Levine for the examination of the chromosomal constitution of the experimental populations in October of 1955; to Mr. and Mrs. B. Spassky for the maintenance of the experimental populations during the two months when both authors were absent from New York; and to many colleagues and friends for many discussions of the problems and issues dwelt upon in the present article.

LITERATURE CITED

AIRD, I., H. H. BENTALL, J. H. MEHIGAN, AND J. A. FRASER ROBERTS. 1954. The blood groups in relation to peptic ulceration and carcinoma of the colon, rectum, breast, and bronchus. Brit. Med. Journ., **2**: 315–321.

CLARKE, C. H., J. W. EDWARDS, D. R. W. HADDOCK, A. W. HOWELL EVANS, R. B. McCONNELL, AND P. M. SHEPPARD. 1956. ABO blood groups and secretor character in duodenal ulcer. Brit. Med. Journ., **2**: 725–730.

DOBZHANSKY, TH. 1937–1941–1951. Genetics and the Origin of Species. Columbia University Press, New York (1st, 2nd and 3rd editions).

——. 1949. Observations and experiments on natural selection in *Drosophila*. Proc. 8th Internat. Cong. Genetics: 210–224.

——. 1954. Evolution as a creative process. Proc. 9th Internat. Cong. Genetics, **1**: 435–449.

——. 1956. What is an adaptive trait? Amer. Naturalist, **20**: 337–347.

——. 1957. Genetics of natural populations. XXVI. EVOLUTION (in press).

——, and O. PAVLOVSKY. 1953. Indeterminate outcome of certain experiments on *Drosophila* populations. EVOLUTION, **7**: 198–210.

DOWDESWELL, W. H., AND E. B. FORD. 1952. Geographical variation in the butterfly *Maniola jurtina*. Heredity, **6**: 99–109.

——, AND E. B. FORD. 1953. The influence of isolation on variability in the butterfly *Maniola jurtina*. Symp. Soc. Exp. Biol., **7**: 254–273.

EISENTRAUT, M. 1950. Die Eidechsen der spanischen Mittelmeerinseln und ihre Rassenaufspaltung in Lichte der Evolution. Berlin.

FORD, E. B. 1954. Problems in the evolution of geographical races. In: Evolution as a Process: 99–108.

KERR, W., AND S. WRIGHT, 1954. Experimental studies of the distribution of gene frequencies in very small populations of *Drosophila melanogaster*. EVOLUTION **8**: 225–240.

KRAMER, G., AND R. MERTENS. 1938. Rassenbildung bei west-istrianischen Inseleidechsen in Abhangigkeit von Isolierungsalter und Arealgrosse. Arch. Naturgesch., **7**: 189–234.

LOWE, CH. H. 1955. An evolutionary study of island faunas in the Gulf of California, Mexico, with a method of comparative analysis. EVOLUTION, **9**: 339–344.

MAYR, E. 1954. Change of genetic environment and evolution. In: Evolution as a Process: 157–180.

WRIGHT, S. 1932. The roles of mutation, inbreeding, crossbreeding and selection in evolution. Proc. 6th Internat. Congress Genetics, **1**: 356–366.

——. 1948. On the roles of directed and random changes in gene frequency in the genetics of populations. EVOLUTION, **2**: 279–294.

——. 1949. Population structure in evolution. Proc. Amer. Philos. Soc., **93**: 471–478.

——. 1951. Fisher and Ford on "The Sewall Wright Effect." Amer. Scientist, **39**: 452–479.

——, AND W. KERR. 1954. Experimental studies of the distribution of gene frequencies in very small populations of *Drosophila melanogaster*. EVOLUTION, **8**: 225–240.

MacArthur, R. H., and E. O. Wilson. 1963. An equilibrium theory of insular zoogeography. *Evolution* 17:373–387.

This paper was a synoptic prelude to Robert H. MacArthur and Edward O. Wilson's (1967) landmark book on island biogeography. Their treatment introduced graphical models in which rates of immigration and extinction were balanced to derive numerical statements about equilibrium levels of species' diversity in insular habitats. Ever since the time of Alfred Russel Wallace (the co-discoverer of natural selection, acknowledged as the "father of biogeography") in the late 1800s, biogeography (like much of ecology) had been largely a descriptive science with little experimentation or quantification. This paper and its book-length version were seminal, because they helped to transform biogeography into a more rigorous, quantitative discipline, a scientific tradition that has continued into the modern era (e.g., Hubbell 2001).

The conceptual framework that MacArthur and Wilson introduced was an overtly equilibrium theory, thus placing it in contradistinction to some later developments in biogeography (notably the rise of phylogeography, which explicitly acknowledges that various taxon-specific idiosyncrasies and historical contingencies underlie the present-day abundances and distributions of organisms).

RELATED READING

Andrewatha, H. G., and L. C. Birch. 1954. The distribution and abundance of animals. University of Chicago Press, Chicago, Illinois, USA.

Hubbell, S. P. 2001. The unified neutral theory of biodiversity and biogeography. Princeton University Press, Princeton, New Jersey, USA.

MacArthur, R. H., and E. O. Wilson. 1967. The theory of island biogeography. Princeton University Press, Princeton, New Jersey, USA.

Wallace, A. R. 1876. The geographical distribution of animals. Harper & Brothers, New York, New York, USA.

EVOLUTION

INTERNATIONAL JOURNAL OF ORGANIC EVOLUTION

PUBLISHED BY

THE SOCIETY FOR THE STUDY OF EVOLUTION

Vol. 17 | DECEMBER, 1963 | No. 4

AN EQUILIBRIUM THEORY OF INSULAR ZOOGEOGRAPHY

Robert H. MacArthur[1] and Edward O. Wilson[2]

Received March 1, 1963

The Fauna–Area Curve

As the area of sampling A increases in an ecologically uniform area, the number of plant and animal species s increases in an approximately logarithmic manner, or

$$s = bA^k, \tag{1}$$

where $k < 1$, as shown most recently in in the detailed analysis of Preston (1962). The same relationship holds for islands, where, as one of us has noted (Wilson, 1961), the parameters b and k vary among taxa. Thus, in the ponerine ants of Melanesia and the Moluccas, k (which might be called the *faunal coefficient*) is approximately 0.5 where area is measured in square miles; in the Carabidae and herpetofauna of the Greater Antilles and associated islands, 0.3; in the land and freshwater birds of Indonesia, 0.4; and in the islands of the Sahul Shelf (New Guinea and environs), 0.5.

The Distance Effect in Pacific Birds

The relation of number of land and freshwater bird species to area is very orderly in the closely grouped Sunda Islands (fig. 1), but somewhat less so in the islands of Melanesia, Micronesia, and Polynesia taken together (fig. 2). The greater variance of the latter group is attributable primarily to one variable, distance between the islands. In particular, the distance effect can be illustrated by taking the distance from the primary faunal "source area" of Melanesia and relating it to faunal number in the following manner. From fig. 2, take the line connecting New Guinea and the nearby Kei Islands as a "saturation curve" (other lines would be adequate but less suitable to the purpose), calculate the predicted range of "saturation" values among "saturated" islands of varying area from the curve, then take calculated "percentage saturation" as $s_i \times 100/B_i$, where s_i is the real number of species on any island and B_i the saturation number for islands of that area. As shown in fig. 3, the percentage saturation is nicely correlated in an inverse manner with distance from New Guinea. This allows quantification of the rule expressed qualitatively by past authors (see Mayr, 1940) that island faunas become progressively "impoverished" with distance from the nearest land mass.

[1] Division of Biology, University of Pennsylvania, Philadelphia, Pennsylvania.

[2] Biological Laboratories, Harvard University, Cambridge, Massachusetts.

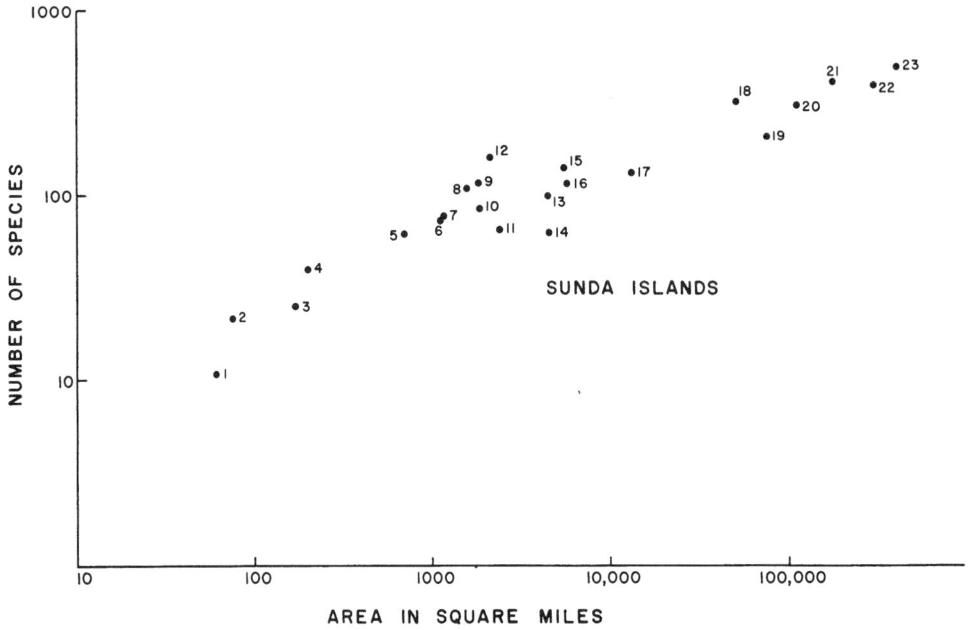

FIG. 1. The numbers of land and freshwater bird species on various islands of the Sunda group, together with the Philippines and New Guinea. The islands are grouped close to one another and to the Asian continent and Greater Sunda group, where most of the species live; and the distance effect is not apparent. (1) Christmas, (2) Bawean, (3) Engano, (4) Savu, (5) Simalur, (6) Alors, (7) Wetar, (8) Nias, (9) Lombok, (10) Billiton, (11) Mentawei, (12) Bali, (13) Sumba, (14) Bangka, (15) Flores, (16) Sumbawa, (17) Timor, (18) Java, (19) Celebes, (20) Philippines, (21) Sumatra, (22) Borneo, (23) New Guinea. Based on data from Delacour and Mayr (1946), Mayr (1940, 1944), Rensch (1936), and Stresemann (1934, 1939).

An Equilibrium Model

The impoverishment of the species on remote islands is usually explained, if at all, in terms of the length of time species have been able to colonize and their chances of reaching the remote island in that time. According to this explanation, the number of species on islands grows with time and, given enough time, remote islands will have the same number of species as comparable islands nearer to the source of colonization. The following alternative explanation may often be nearer the truth. Fig. 4 shows how the number of new species entering an island may be balanced by the number of species becoming extinct on that island. The descending curve is the rate at which *new* species enter the island by colonization. This rate does indeed fall as the number of species on the islands increases, because the chance that an immigrant be a new species, not already on the island, falls. Furthermore, the curve falls more steeply at first. This is a consequence of the fact that some species are commoner immigrants than others and that these rapid immigrants are likely, on typical islands, to be the first species present. When there are no species on the island ($N = 0$), the height of the curve represents the number of species arriving per unit of time. Thus the intercept, I, is the rate of immigration of species, new or already present, onto the island. The curve falls to zero at the point $N = P$ where all of the immigrating species are already present so that no new ones are arriving. P is thus the number of species in the "species pool" of immigrants. The shape of the rising curve in the same figure, which represents the

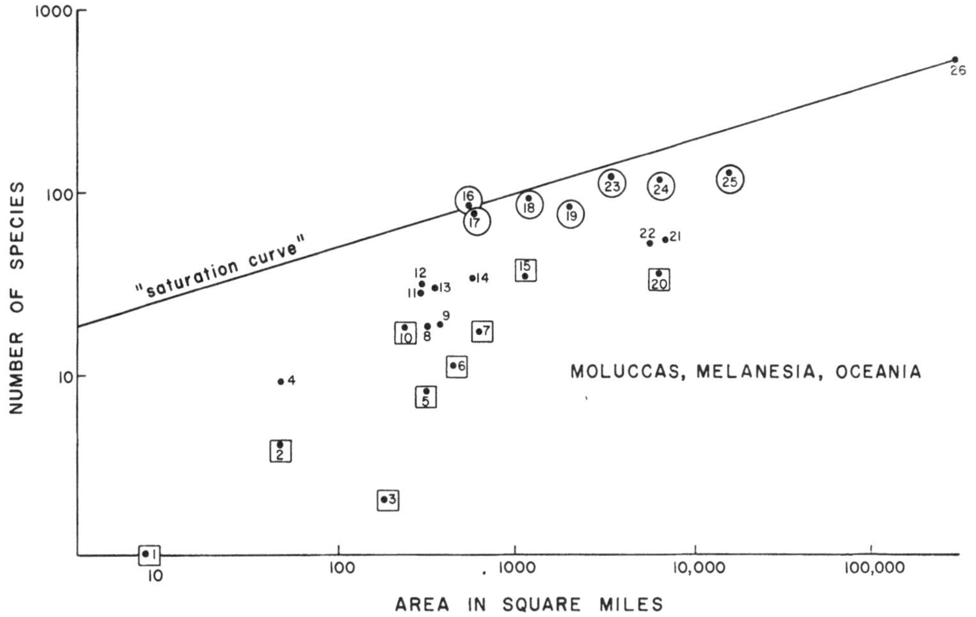

FIG. 2. The numbers of land and freshwater bird species on various islands of the Moluccas, Melanesia, Micronesia, and Polynesia. Here the archiplagoes are widely scattered, and the distance effect is apparent in the greater variance. Hawaii is included even though its fauna is derived mostly from the New World (Mayr, 1943). "Near" islands (less than 500 miles from New Guinea) are enclosed in circles, "far" islands (greater than 2,000 miles) in squares, and islands at intermediate distances are left unenclosed. The saturation curve is drawn through large and small islands at source of colonization. (1) Wake, (2) Henderson, (3) Line, (4) Kusaie, (5) Tuamotu, (6) Marquesas, (7) Society, (8) Ponape, (9) Marianas, (10) Tonga, (11) Carolines, (12) Palau, (13) Santa Cruz, (14) Rennell, (15) Samoa, (16) Kei, (17) Louisiade, (18) D'Entrecasteaux, (19) Tanimbar, (20) Hawaii, (21) Fiji, (22) New Hebrides, (23) Buru, (24) Ceram, (25) Solomons, (26) New Guinea. Based on data from Mayr (1933, 1940, 1943) and Greenway (1958).

rate at which species are becoming extinct on the island, can also be determined roughly. In case all of the species are equally likely to die out and this probability is independent of the number of other species present, the number of species becoming extinct in a unit of time is proportional to the number of species present, so that the curve would rise linearly with N. More realistically, some species die out more readily than others and the more species there are, the rarer each is, and hence an increased number of species increases the likelihood of any given species dying out. Under normal conditions both of these corrections would tend to increase the slope of the extinction curve for large values of N. (In the rare situation in which the species which enter most often as immigrants are the ones which die out most readily—presumably because the island is atypical so that species which are common elsewhere cannot survive well—the curve of extinction may have a steeper slope for small N.) If N is the number of species present at the start, then $E(N)/N$ is the fraction dying out, which can also be interpreted crudely as the probability that any given species will die out. Since this fraction cannot exceed 1, the extinction curve cannot rise higher than the straight line of a 45° angle rising from the origin of the coordinates.

It is clear that the rising and falling curves must intersect and we will denote by $\hat{s}$ the value of N for which the rate of immigration of new species is balanced by

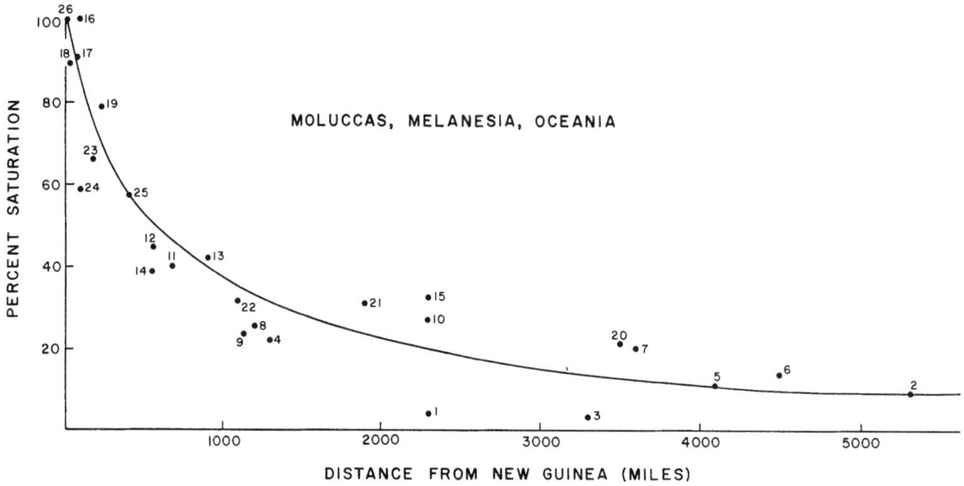

FIG. 3. Per cent saturation, based on the "saturation curve" of fig. 2, as a function of distance from New Guinea. The numbers refer to the same islands identified in the caption of fig. 2. Note that from equation (4) it is an oversimplification to take distances solely from New Guinea. The abscissa should give a more complex function of distances from all the surrounding islands, with the result that far islands would appear less "distant." But this representation expresses the distance effect adequately for the conclusions drawn.

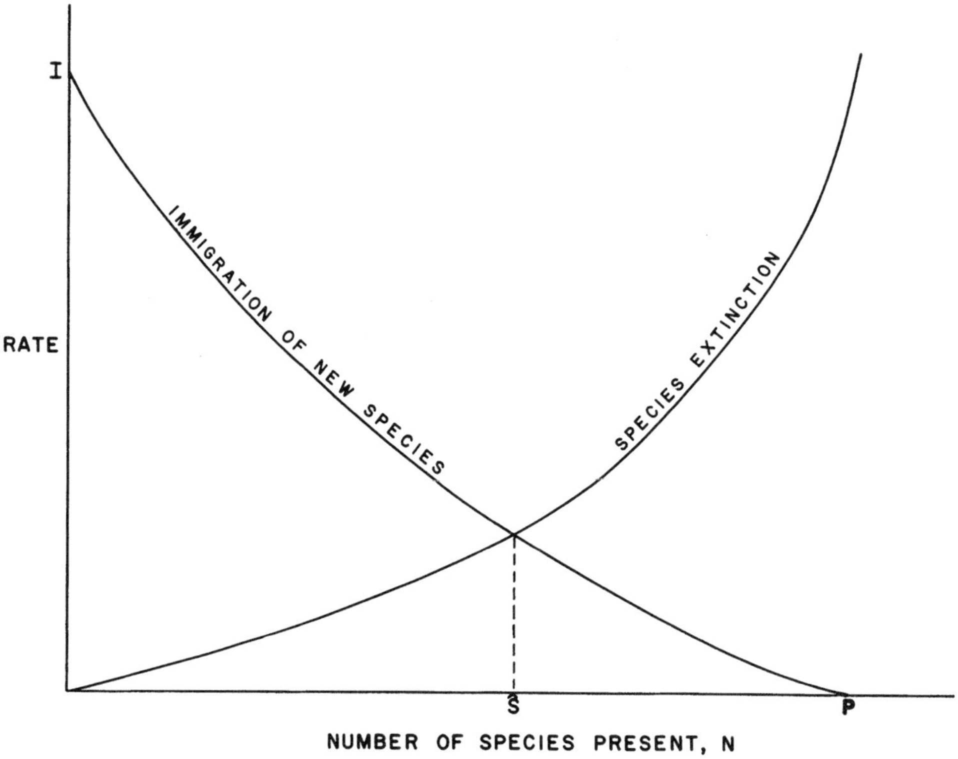

FIG. 4. Equilibrium model of a fauna of a single island. See explanation in the text.

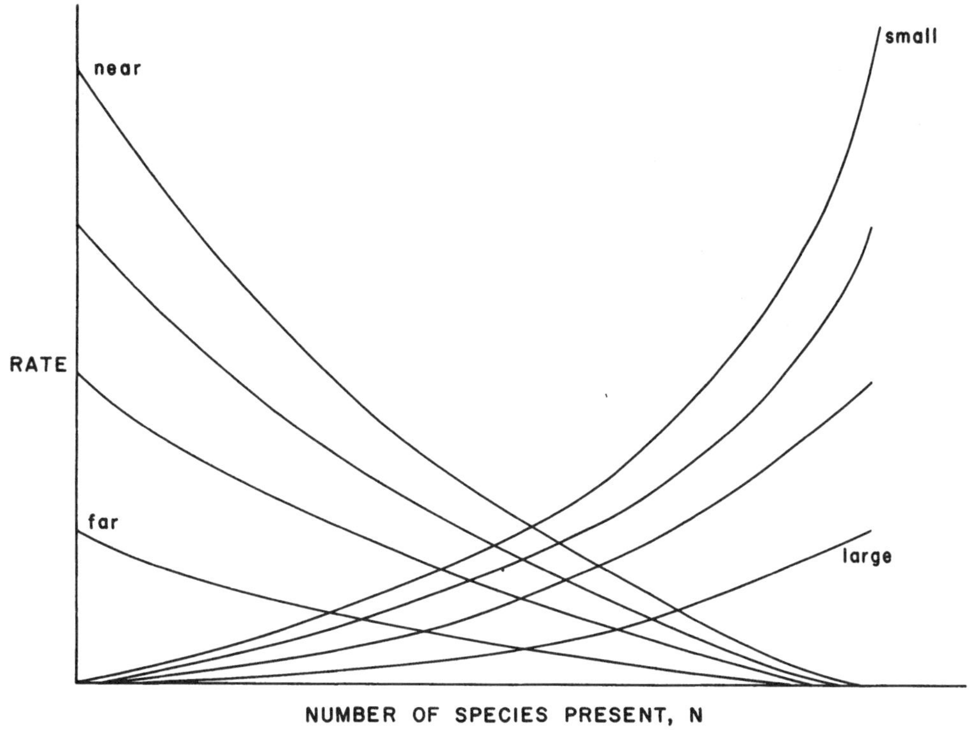

FIG. 5. Equilibrium model of faunas of several islands of varying distances from the source area and varying size. Note that the effect shown by the data of fig. 2, of faunas of far islands increasing with size more rapidly than those of near islands, is predicted by this model. Further explanation in text.

the rate of extinction. The number of species on the island will be stabilized at $\hat{s}$, for a glance at the figure shows that when N is greater than $\hat{s}$, extinction exceeds immigration of new species so that N decreases, and when N is less than $\hat{s}$, immigration of new species exceeds extinction so that N will increase. Therefore, in order to predict the number of species on an island we need only construct these two curves and see where they intersect. We shall make a somewhat oversimplified attempt to do this in later paragraphs. First, however, there are several interesting qualitative predictions which we can make without committing ourselves to any specific shape of the immigration and extinction curves.

A. An island which is farther from the source of colonization (or for any other reason has a smaller value of I) will, other things being equal, have fewer species, because the immigration curve will be lower and hence intersect the mortality curve farther to the left (see fig. 5).

B. Reduction of the "species pool" of immigrants, P, will reduce the number of species on the island (for the same reason as in A).

C. If an island has smaller area, more severe climate (or for any other reason has a greater extinction rate), the mortality curve will rise and the number of species will decrease (see fig. 5).

D. If we have two islands with the same immigration curve but different extinction curves, any given species on the one with the higher extinction curve is more likely to die out, because $E(N)/N$ can be seen to be higher [$E(N)/N$ is the slope of the line joining the intersection point to the origin].

E. The number of species found on islands far from the source of colonization will grow more rapidly with island area than will the number on near islands. More precisely, if the area of the island is denoted by A, and $\hat{s}$ is the equilibrium number of species, then d^2s/dA^2 is greater for far islands than for near ones. This can be verified empirically by plotting points or by noticing that the change in the angle of intersection is greater for far islands.

F. The number of species on large islands decreases with distance from source of colonization faster than does the number of species on small islands. (This is merely another way of writing E and is verified similarly.)

Further, as will be shown later, the variance in $\hat{s}$ (due to randomness in immigrations and extinctions) will be lower than that expected if the "classical" explanation holds. In the classical explanation most of those species will be found which have at any time succeeded in immigrating. At least for distant islands this number would have an approximately Poisson distribution so that the variance would be approximately equal to the mean. Our model predicts a reduced variance, so that if the observed variance is significantly smaller than the mean for distant islands, it is evidence for the equilibrium explanation.

The evidence in fig. 2, relating to the insular bird faunas east of Weber's Line, is consistent with all of these predictions. To see this for the non-obvious prediction *E,* notice that a greater slope on this log-log plot corresponds to a greater second derivative, since A becomes sufficiently large.

The Form of the Immigration and Extinction Curves

If the equilibrium model we have presented is correct, it should be possible eventually to derive some quantitative generalizations concerning rates of immigration and extinction. In the section to follow we have deduced an equilibrium equation which is adequate as a first approximation, in that it yields the general form of the empirically derived fauna–area curves without contradicting (for the moment) our intuitive ideas of the underlying biological processes. This attempt to produce a formal equation is subject to indefinite future improvements and does not affect the validity of the graphically derived equilibrium theory. We start with the statement that

$$\Delta s = M + G - D, \qquad (2)$$

where s is the number of species on an island, M is the number of species successfully immigrating to the island per year, G is the number of new species being added per year by local speciation (not including immigrant species that merely diverge to species level without multiplying), and D is the number of species dying out per year. At equilibrium,

$$M + G = D.$$

The immigration rate M must be determined by at least two independent values: (1) the rate at which propagules reach the island, which is dependent on the size of the island and its distance from the source of the propagules, as well as the nature of the source area, but not on the condition of the recipient island's fauna; and (2) as noted already, the number of species already resident on the island. Propagules are defined here as the minimum number of individuals of a given species needed to achieve colonization; a more exact explication is given in the Appendix. Consider first the source region. If it is climatically and faunistically similar to other potential source regions, the number of propagules passing beyond its shores per year is likely to be closely related to the size of the population of the taxon living on it, which in turn is approximately a linear function of its area. This notion is supported by the evidence from Indo-Australian ant zoogeography, which indicates that the ratio of faunal exchange is about equal to the ratio of the areas of the source regions (Wilson, 1961). On the other hand, the number of propagules reaching the recipient island probably varies linearly with the angle it subtends with reference to the center of the source region. Only near islands will vary much because of this factor. Finally, the number of propagules reaching the recipient island is most likely to be an exponential function of its distance from the source region. In the simplest case, if the probability that a given propagule ceases its overseas voyage (e.g., it falls into the sea and dies) at any given instant in time remains constant, then the fraction of propagules reaching a given distance fits an exponential holding-time distribution. If these assumptions are correct, the number of propagules reaching an island from a given source region per year can be approximated as

$$\alpha A_i \frac{\text{diam}_i e^{-\lambda d_i}}{2\pi d_i}, \tag{3}$$

where A_i is the area of the source region, d_i is the mean distance between the source region and recipient island, diam_i is the diameter of the recipient island taken at a right angle to the direction of d_i, and α is a coefficient relating area to the number of propagules produced. More generally, where more than one source region is in position, the rate of propagule arrival would be

$$\frac{\alpha}{2\pi}\sum_i \frac{\text{diam}_i}{d_i} A_i e^{-\lambda d_i}, \tag{4}$$

where the summation is of contributions from each of the ith source regions. Again, note that a propagule is defined as the minimum number of individuals required to achieve colonization.

Only a certain fraction of arriving propagules will add a new species to the fauna, however, because except for "empty" islands at least some ecological positions will be filled. As indicated in fig. 4, the rate of immigration (i.e., rate of propagule arrival times the fraction colonizing) declines to zero as the number of resident species (s) approaches the limit P. The curve relating the immigration rate to degree of unsaturation is probably a concave one, as indicated in fig. 4, for two reasons: (1) the more abundant immigrants reach the island earlier, and (2) we would expect otherwise randomly arriving elements to settle into available positions according to a simple occupancy model where one and only one object is allowed to occupy each randomly placed position (Feller, 1958). These circumstances would result in the rate of successful occupation decelerating as positions are filled. While these are interesting subjects in themselves, a reasonable approximation is obtained if it is assumed that the rate of occupation is an inverse linear function of the number of occupied positions, or

$$\left(1 - \frac{s}{P}\right). \tag{5}$$

Then

$$M = \frac{\alpha(1 - s/P)}{2\pi}\sum_i \frac{\text{diam}_i}{d_i} A_i e^{-\lambda d_i}. \tag{6}$$

We know the immigration line in fig. 4 is not straight; to take this into account we must modify formula 5 by adding a term in s^2. However, this will not be necessary for our immediate purposes.

Now let us consider G, the rate of new productions on the island by local speciation. Note that this rate does not include the mere divergence of an island endemic to a specific level with reference to the stock species in the source area; that species is still counted as contributing to M, the immigration rate, no matter how far it evolves. Only new species generated from it and in addition to it are counted in G. First, consider an archipelago as a unit and the increase of s by divergence of species on the various islands to the level of allopatric species, i.e., the production of a local archipelagic superspecies. If this is the case, and no exchange of endemics is yet achieved among the islands of the archipelago, the number of species in the archipelago is limited to

$$\sum_{i=1}^{\infty} n_i \hat{s}_i, \tag{7}$$

where n_i is the number of islands in the archipelago of ith area and $\hat{s}_i$ is the number of species occurring at equilibrium on islands of ith area. But the generation of allopatric species in superspecies does not multiply species on single islands or greatly change the fauna of the archipelago as a whole from the value predicted by the fauna–area curve, as can be readily seen in figs. 2 and 3. G, the increase of s by local speciation on single islands and exchange of autochthonous species between islands, probably becomes significant only in the oldest, largest, and most isolated archipelagoes, such as Hawaii and the Galápagos. Where it occurs, the exchange among the islands can be predicted from (6), with individual islands in the archipelago serving as both source regions and recipient islands. However, for most cases it is probably safe to omit G from the model, i.e., consider only source regions outside the archipelago, and hence

$$\Delta s = M - D. \qquad (8)$$

The extinction rate D would seem intuitively to depend in some simple manner on (1) the mean size of the species populations, which in turn is determined by the size of the island and the number of species belonging to the taxon that occur on it; and (2) the yearly mortality rate of the organisms. Let us suppose that the probability of extinction of a species is merely the probability that all the individuals of a given species will die in one year. If the deaths of individuals are unrelated to each other and the population sizes of the species are equal and nonfluctuating,

$$D = sP^{N_r/s}, \qquad (9)$$

where N_r is the total number of individuals in the taxon on the recipient island and P is their annual mortality rate. More realistically, the species of a taxon, such as the birds, vary in abundance in a manner approximating a Barton–Davis distribution (MacArthur, 1957) although the approximation is probably not good for a whole island. In s nonfluctuating species ordered according to their rank (K) in relative rareness,

$$D = \sum_{i=1}^{s} p^{(N_r/s)\sum_{1=i}^{K} 1/(s-i+1)}. \qquad (10)$$

This is still an oversimplification, if for no other reason than the fact that populations do fluctuate, and with increased fluctuation D will increase. However, both models, as well as elaborations of them to account for fluctuation, predict an exponential increase of D with restriction of island area. The increase of D which accompanies an increase in number of resident species is more complicated but is shown in fig. 4.

Model of Immigration and Extinction Process on a Single Island

Let $P_s(t)$ be the probability that, at time t, our island has s species, λ_s be the rate of immigration of new species onto the island, when s are present, μ_s be the rate of extinction of species on the island when s are present; and λ_s and μ_s then represent the intersecting curves in fig. 4. This is a "birth and death process" only slightly different from the kind most familiar to mathematicians (cf. Feller, 1958, last chapter). By the rules of probability

$$P_s(t+h) = P_s(t)(1 - \lambda_s h - \mu_s h) + P_{s-1}(t)\lambda_{s-1}h + P_{s+1}(t)\mu_{s+1}h,$$

since to have s at time $t+h$ requires that at a short time preceding one of the following conditions held: (1) there were s and that no immigration or extinction took place, or (2) that there were $s-1$ and one species immigrated, or (3) that there $s+1$ and one species became extinct. We take h to be small enough that probabilities of two or more extinctions and/or immigrations can be ignored. Bringing $P_s(t)$ to the left-hand side, dividing by h, and passing to the limit as $h \to 0$

$$\frac{dP_s(t)}{dt} = -(\lambda_s + \mu_s)P_s(t) + \lambda_{s-1}P_{s-1}(t) + \mu_{s+1}P_{s+1}(t). \qquad (11)$$

For this formula to be true in the case where $s = 0$, we must require that $\lambda_{-1} = 0$ and $\mu_0 = 0$. In principle we could solve

(11) for $P_s(t)$; for our purposes it is more useful to find the mean, $M(t)$, and the variance, $\mathrm{var}(t)$, of the number of species at time t. These can be estimated in nature by measuring the mean and variance in numbers of species on a series of islands of about the same distance and area and hence of the same λ_s and μ_s. To find the mean, $M(t)$, from (11) we multiply both sides of (11) by s and then sum from $s =$ 0 to $s = \infty$. Since $\sum_{s=0}^{\infty} sP_s(t) = M(t)$, this gives us

$$\frac{dM(t)}{dt} = -\sum_{s=0}^{\infty} (\lambda_s + \mu_s) sP_s(t) + \sum_{s-1=0}^{\infty} \lambda_{s-1}[(s-1)+1]P_{s-1}(t) + \sum_{s+1=0}^{\infty} \mu_{s+1}[(s+1)-1]P_{s+1}(t).$$

(Here terms $\lambda_{-1}\cdot 0 \cdot P_{-1}(t) = 0$ and $\mu_0 \cdot (-1)P_0(t) = 0$ have been subtracted or added without altering values.) This reduces to

$$\frac{dM(t)}{dt} = \sum_{s=0}^{\infty} \lambda_s P_s(t) - \sum_{s=0}^{\infty} \mu_s P_s(t) = \overline{\lambda_s(t)} - \overline{\mu_s(t)}. \tag{12}$$

But, since λ_s and μ_s are, at least locally, approximately straight, the mean value of λ_s at time t is about equal to $\lambda_{M(t)}$ and similarly $\overline{\mu_s(t)} \sim \mu_{M(t)}$. Hence, approximately

$$\frac{dM(t)}{dt} = \lambda_{M(t)} - \mu_{M(t)}, \tag{13}$$

or the expected number of species in Fig. 4 moves toward $\hat{s}$ at a rate equal to the difference in height of the immigration and extinction curves. In fact, if $d\mu/ds - d\lambda/ds$, evaluated near $s = \hat{s}$ is abbreviated by F, then, approximately $dM(t)/dt = F(\hat{s} - M(t))$ whose solution is $M(t) = \hat{s}(1 - e^{-Ft})$. Finally, we can compute the time required to reach 90% (say) of the saturation value $\hat{s}$ so that $M(t)/\hat{s} = 0.9$ or $e^{-Ft} = 0.1$.
Therefore,

$$t = \frac{2.303}{F}. \tag{13a}$$

A similar formula for the variance is obtained by multiplying both sides of (11) by $(s - M(t))^2$ and summing from $s = 0$ to $s = \infty$. As before, since $\mathrm{var}(t) = \sum_{s=0}^{\infty} (s - M(t))^2 P_s(t)$, this results in

$$\begin{aligned}\frac{d\,\mathrm{var}(t)}{dt} &= -\sum_{s=0}^{\infty} (\lambda_s + \mu_s)(s - M(t))^2 P_s(t) \\ &+ \sum_{s-1=0}^{\infty} \lambda_{s-1}[(s-1-M(t))+1]^2 P_{s+1}(t) \\ &+ \sum_{s+1=0}^{\infty} \mu_{s+1}[(s+1-M(t))-1]^2 P_{s+1}(t) \\ &= 2\sum_{s=0}^{\infty} \lambda_s (s - M(t)) P_s(t) \\ &- 2\sum_{s=0}^{\infty} \mu_s (s - M(t)) P_s(t) \\ &+ \sum_{s=0}^{\infty} \lambda_s P_s(t) + \sum_{s=0}^{\infty} \mu_s P_s(t).\end{aligned} \tag{14}$$

Again we can simplify this by noting that the λ_s and μ_s curves are only slowly curving and hence in any local region are approximately straight. Hence, where derivatives are now evaluated near the point $s = M(t)$,

$$\lambda_s = \lambda_{M(t)} + [s - M(t)]\frac{d\lambda}{ds}$$
$$\mu_s = \mu_{M(t)} + [s - M(t)]\frac{d\mu}{ds}. \tag{15}$$

Substituting (15) into (14) we get

$$\begin{aligned}\frac{d\,\mathrm{var}(t)}{dt} &= 2(\lambda_{M(t)} - \mu_{M(t)}) \sum_{s=0}^{\infty} (s - M(t)) P_s(t) \\ &+ 2\left(\frac{d\lambda}{ds} - \frac{d\mu}{ds}\right) \sum_{s=0}^{\infty} (s - M(t))^2 P_s(t) \\ &+ [\lambda_{m(t)} + \mu_{M(t)}] \sum_{s=0}^{\infty} P_s(t) \\ &+ \left(\frac{d\lambda}{ds} + \frac{d\mu}{ds}\right) \sum_{s=0}^{\infty} (s - M(t)) P_s(t),\end{aligned}$$

which, since $\sum_{s=0}^{\infty} P_s(t) = 1$ and $\sum (s - M(t)) P_s(t) = M(t) - M(t) = 0$, becomes,

$$\frac{d\,\mathrm{var}(t)}{dt} = -2\left(\frac{d\mu}{ds} - \frac{d\lambda}{ds}\right)\mathrm{var}(t) + \lambda_{M(t)} + \mu_{M(t)}\,. \qquad (16)$$

This is readily solved for var(t):

$$\mathrm{var}(t) = e^{-2[(d\mu/ds)-(d\lambda/ds)]t} \times \int_0^t (\lambda_{M(t)} + \mu_{M(t)})\, e^{2[(d\mu/ds)-(d\lambda/ds)]t}\, dt\,. \qquad (16a)$$

However, it is more instructive to compare mean and variance for the extreme situations of saturation and complete unsaturation, or equivalently of $t =$ near ∞ and $t =$ near zero.

At equilibrium, $\frac{d\,\mathrm{var}(t)}{dt} = 0$, so by (16) ·

$$\mathrm{var}(t) = \frac{\lambda_{\hat{s}} + \mu_{\hat{s}}}{2\left(\frac{d\mu}{ds} - \frac{d\lambda}{ds}\right)}. \qquad (17)$$

At equilibrium $\lambda_{\hat{s}} = \mu_{\hat{s}} = x$ say and we have already symbolized the difference of the derivatives at $s = \hat{s}$ by F (cf. eq. [13a]). Hence, at equilibrium

$$\mathrm{var} = \frac{X}{F}. \qquad (17a)$$

Now since μ_s has non-decreasing slope $X/s \leqslant d\mu/ds\,|_{s=\hat{s}}$ or $X \leqslant \hat{s}\, d\mu/ds\,|_{\hat{s}}$.

Therefore, variance $\leqslant \frac{\hat{s}\, d\mu/ds}{d\mu/ds - d\lambda/ds}$ or, at equilibrium

$$\frac{\mathrm{variance}}{\mathrm{mean}} \leqslant \frac{d\mu/ds}{d\mu/ds - d\lambda/ds}. \qquad (18)$$

In particular, if the extinction and immigration curves have slopes about equal in absolute value, (variance/mean) $\leqslant$ ½. On the other hand, when t is near zero, equation (16) shows that var$(t) \sim \lambda_0 t$. Similarly, when t is near zero, equations (13) or (14) show that $M(t) \sim \lambda_0 t$. Hence, in a very unsaturated situation, approximately,

$$\frac{\mathrm{variance}}{\mathrm{mean}} = 1\,. \qquad (19)$$

Therefore, we would expect the variance/mean to rise from somewhere around ½ to 1, as we proceed from saturated islands to extremely unsaturated islands farthest from the source of colonization.

Finally, if the number of species dying out per year, X (at equilibrium), is known, we can estimate the time required to 90% saturation from equations (13a) and (17a):

$$\frac{2.303}{t} = \frac{X}{\mathrm{variance}}$$

$$t = \frac{2.303\ \mathrm{variance}}{X} \doteq \frac{2.303}{2}\,\frac{\mathrm{mean}}{X}. \qquad (19a)$$

The above model was developed independently from an equilibrium hypothesis just published by Preston (1962). After providing massive documentation of the subject that will be of valuable assistance to future biogeographers, Preston draws the following particular conclusion about continental versus insular biotas: "[The depauperate insular biotas] are not depauperate in any absolute sense. They have the correct number of species for their area, provided that each area is an isolate, but they have far fewer than do equal areas on a mainland, because a mainland area is merely a 'sample' and hence is greatly enriched in the Species/Individuals ratio." To illustrate, "in a sample, such as the breeding birds of a hundred acres, we get many species represented by a single pair. Such species would be marked for extinction with one or two seasons' failure of their nests were it not for the fact that such local extirpation can be made good from outside the 'quadrat,' which is not the case with the isolate." This point of view agrees with our own. However, the author apparently missed the precise distance effect and his model is consequently not predictive in the direction we are attempting. His model is, however, more accurate in its account of

TABLE 1. *Number of species of land and freshwater birds on Krakatau and Verlaten during three collection periods together with losses in the two intervals (from Dammerman, 1948)*

	1908			1919–1921			1932–1934			Number "lost"	
	Non-migrant	Migrant	Total	Non-migrant	Migrant	Total	Non-migrant	Migrant	Total	1908 to 1919–1921	1919–1921 to 1932–1934
Krakatau	13	0	13	27	4	31	27	3	30	2	5
Verlaten	1	0	1	27	2	29	29	5	34	0	2

relative abundance, corresponding to our equation (10).

The Case of the Krakatau Faunas

The data on the growth of the bird faunas of the Krakatau Islands, summarized by Dammerman (1948), provide a rare opportunity to test the foregoing model of the immigration and extinction process on a single island. As is well known, the island of Krakatau proper exploded in August, 1883, after a three-month period of repeated eruptions. Half of Krakatau disappeared entirely and the remainder, together with the neighboring islands of Verlaten and Lang, was buried beneath a layer of glowing hot pumice and ash from 30 to 60 meters thick. Almost certainly the entire flora and fauna were destroyed. The repopulation proceeded rapidly thereafter. Collections and sight records of birds, made mostly in 1908, 1919–1921, and 1932–1934, show that the number of species of land and freshwater birds on both Krakatau and Verlaten climbed rapidly between 1908 and 1919–1921 and did not alter significantly by 1932–1934 (see table 1). Further, the number of non-migrant land and freshwater species on both islands in 1919–1921 and 1932–1934, i.e., 27–29, fall very close to the extrapolated fauna–area curve of our fig. 1. Both lines of evidence suggest that the Krakatau faunas had approached equilibrium within only 25 to 36 years after the explosion.

Depending on the exact form of the immigration and extinction curves (see fig. 4), the ratio of variance to mean of numbers of species on similar islands at or near saturation can be expected to vary between about ¼ and ¾. If the slopes of the two curves are equal at the point of intersection, the ratio would be near ½. Then the variance of faunas of Krakatau-like islands (same area and isolation) can be expected to fall between 7 and 21 species. Applying this estimate to equation (19a) and taking t (the time required to reach 90% of the equilibrium number) as 30 years, X, the annual extinction rate, is estimated to lie between 0.5 and 1.6 species per year.

This estimate of annual extinction rate (and hence of the acquisition rate) in an equilibrium fauna is surprisingly high; it is of the magnitude of 2 to 6% of the standing fauna. Yet it seems to be supported by the collection data. On Krakatau proper, 5 non-migrant land and freshwater species recorded in 1919–1921 were not recorded in 1932–1934, but 5 other species were recorded for the first time in 1932–1934. On Verlaten 2 species were "lost" and 4 were "gained." This balance sheet cannot easily be dismissed as an artifact of collecting technique. Dammerman notes that during this period, "The most remarkable thing is that now for the first time true fly catchers, *Muscicapidae,* appeared on the islands, and that there were no less than four species: *Cyornis rufigastra, Gerygone modigliani, Alseonax latirostris* and *Zanthopygia narcissina.* The two last species are migratory and were therefore only accidental visitors, but the sudden appearance of the *Cyornis* species in great numbers is noteworthy. These birds, first observed in May 1929, had already colonized three islands and may now be called common there. Moreover the *Gerygone,* unmistakable from his gentle note and common along the coast and in the mangrove forest, is certainly a new acquisition." Extinctions are less susceptible of proof but the following evidence is suggestive. "On the other hand two species mentioned by Jacobson (1908) were not found in 1921 and have not been observed since, namely the small kingfisher *Alcedo coerulescens* and the familiar bulbul *Pycnonotus aurigaster*." Between 1919–1921 and 1932–1934 the conspicuous *Demiegretta s. sacra* and *Accipter* sp. were "lost," although these species may not have been truly established as breeding populations. But "the well-known grey-backed shrike (*Lanius schach bentet*), a bird conspicuous in the open field, recorded in 1908 and found breeding in 1919, was not seen in 1933. Whether the species had really completely disappeared or only diminished so much in numbers that it was not noticed, the future must show." Future research on the Krakatau fauna would indeed be of great interest, in view of the very dynamic equilibrium suggested by the model we have presented. If the "losses" in the data represent true extinctions, the rate of extinction would be 0.2 to 0.4 species per year, closely approaching the predicted rate of 0.5 to 1.6. This must be regarded as a minimum figure, since it is likely that species could easily be lost and regained all in one 12-year period.

Such might be the situation in the early history of the equilibrium fauna. It is not possible to predict whether the rate of turnover would change through time. As other taxa reached saturation and more species of birds had a chance at colonization, it is conceivable that more "harmonic" species systems would accumulate within which the turnover rate would decline.

Prediction of a "Radiation Zone"

On islands holding equilibrium faunas, the ratio of the number of species arriving from other islands in the same archipelago (G in equation no. 2) to the number arriving from outside the archipelago (M in no. 2) can be expected to increase with distance from the major extra-archipelagic source area. Where the archipelagoes are of approximately similar area and configuration, G/M should increase in an orderly fashion with distance. Note that G provides the best available measure of what is loosely referred to in the literature as adaptive radiation. Specifically, adaptive radiation takes place as species are generated within archipelagoes, disperse between islands, and, most importantly, accumulate on individual islands to form diversified associations of sympatric species. In equilibrium faunas, then, the following prediction is possible: adaptive radiation, measured by G/M, will increase with distance from the major source region and after corrections for area and climate, reach a maximum on archipelagoes and large islands located in a circular zone close to the outermost range of the taxon. This might be referred to as the "radiation zone" of taxa with equilibrium faunas. Many examples possibly conforming to such a rule can be cited: the birds of Hawaii and the Galápagos, the murid rodents of Luzon, the cyprinid fish of Mindanao, the frogs of the Seychelles, the gekkonid lizards of New Caledonia, the Drosophilidae of Hawaii, the ants of Fiji and New Caledonia, and many others (see especially in Darlington, 1957; and Zimmerman, 1948). But there are conspicuous exceptions: the frogs just reach New Zealand but have not radiated there; the same is true of the insectivores of the Greater Antilles, the terrestrial mammals of the Solomons, the snakes of Fiji, and the lizards of Fiji and Samoa. To say that the latter taxa have only recently reached the islands in question, or that they are not in equilibrium, would be a premature if not facile explanation. But it is worth considering as a working hypothesis.

Estimating the Mean Dispersal Distance

A possible application of the equilibrium model in the indirect estimation of the mean dispersal distance, or λ in equation

(3). Note that if similar parameters of dispersal occur within archipelagoes as well as between them,

$$\frac{G}{M} = \frac{A_1 \,\text{diam}_1\, d_2}{A_2 \,\text{diam}_2\, d_1} e^{\lambda(d_2 - d_1)}, \qquad (20)$$

and

$$\lambda = \ln \frac{A_2 \,\text{diam}_2\, d_1 G}{A_1 \,\text{diam}_1\, d_2 M} \bigg/ (d_2 - d_1)\,, \qquad (21)$$

where, in a simple case, A_1, diam_1, and d_1 refer to the relation between the recipient island and some single major source island within the same archipelago; and A_2, diam_2, and d_2 refer to the relation between the recipient island and the major source region outside the archipelago.

Consider the case of the Geospizinae of the Galápagos. On the assumption that a single stock colonized the Galápagos (Lack, 1947), G/M for each island can be taken as equal to G, or the number of geospizine species. In particular, the peripherally located Chatham Island, with seven species, is worth evaluating. South America is the source of M and Indefatigable Island can probably be regarded as the principal source of G for Chatham. Given G/M as seven and assuming that the Geospizinae are in equilibrium, λ for the Geospizinae can be calculated from (21) as 0.018 mile. For birds as a whole, where G/M is approximately unity, λ is about 0.014 mile.

But there are at least three major sources of error in making an estimate in this way:

1. Whereas M is based from the start on propagules from an equilibrium fauna in South America, G increased gradually in the early history of the Galápagos through speciation of the Geospizinae on islands other than Chatham. Hence, G/M on Chatham is actually higher than the ratio of species drawn from the Galápagos to those drawn from outside the archipelago, which is our only way of computing G/M directly. Since λ increases with G/M, the estimates of λ given would be too low, if all other parameters were correct.

2. Most species of birds probably do not disperse according to a simple exponential holding-time distribution. Rather, they probably fly a single direction for considerable periods of time and cease flying at distances that can be approximated by the normal distribution. For this reason also, λ as estimated above would probably be too low.

3. We are using $\hat{S}_G/\hat{S}_M$ for G/M, which is only approximate.

These considerations lead us to believe that 0.01 mile can safely be set as the lower limit of λ for birds leaving the eastern South American coast. Using equation no. 12 in another case, we have attempted to calculate λ for birds moving through the Lesser Sunda chain of Indonesia. The Alor group was chosen as being conveniently located for the analysis, with Flores regarded as the principal source of western species and Timor as the principal source of eastern species. From the data of Mayr (1944) on the relationships of the Alor fauna, and assuming arbitrarily an exponential holding-time dispersal, λ can be calculated as approximately 0.3 mile. In this case the first source of error mentioned above with reference to the Galápagos fauna is removed but the second remains. Hence, the estimate is still probably a lower limit.

Of course these estimates are in themselves neither very surprising nor otherwise illuminating. We cite them primarily to show the possibilities of using zoogeographic data to set boundary conditions on population ecological phenomena that would otherwise be very difficult to assess.

Finally, while we believe the evidence favors the hypothesis that Indo-Australian insular bird faunas are at or near equilibrium, we do not intend to extend this conclusion carelessly to other taxa or even other bird faunas. Our purpose has been to deal with general equilibrium criteria, which might be applied to other faunas, together with some of the biological implications of the equilibrium condition.

Summary

A graphical equilibrium model, balancing immigration and extinction rates of species, has been developed which appears fully consistent with the fauna–area curves and the distance effect seen in land and freshwater bird faunas of the Indo-Australian islands. The establishment of the equilibrium condition allows the development of a more precise zoogeographic theory than hitherto possible.

One new and non-obvious prediction can be made from the model which is immediately verifiable from existing data, that the number of species increases with area more rapidly on far islands than on near ones. Similarly, the number of species on large islands decreases with distance faster than does the number of species on small islands.

As groups of islands pass from the unsaturated to saturated conditions, the variance-to-mean ratio should change from unity to about one-half. When the faunal buildup reaches 90% of the equilibrium number, the extinction rate in species/year should equal 2.303 times the variance divided by the time (in years) required to reach the 90% level. The implications of this relation are discussed with reference to the Krakatau faunas, where the buildup rate is known.

A "radiation zone," in which the rate of intra-archipelagic exchange of autochthonous species approaches or exceeds extra-archipelagic immigration toward the outer limits of the taxon's range, is predicted as still another consequence of the equilibrium condition. This condition seems to be fulfilled by conventional information but cannot be rigorously tested with the existing data.

Where faunas are at or near equilibrium, it should be possible to devise indirect estimates of the actual immigration and extinction rates, as well as of the times required to reach equilibrium. It should also be possible to estimate the mean dispersal distance of propagules overseas from the zoogeographic data. Mathematical models have been constructed to these ends and certain applications suggested.

The main purpose of the paper is to express the criteria and implications of the equilibrium condition, without extending them for the present beyond the Indo-Australian bird faunas.

Acknowledgments

We are grateful to Dr. W. H. Bossert, Prof. P. J. Darlington, Prof. E. Mayr, and Prof. G. G. Simpson for material aid and advice during the course of the study. Special acknowledgment must be made to the published works of K. W. Dammerman, E. Mayr, B. Rensch, and E. Stresemann, whose remarkably thorough faunistic data provided both the initial stimulus and the principal working material of our analysis. The work was supported by NSF Grant G-11575.

Literature Cited

Dammerman, K. W. 1948. The fauna of Krakatau 1883–1933. Verh. Kon. Ned. Akad. Wet. (Nat.), (2) **44**: 1–594.

Darlington, P. J. 1957. Zoogeography. The geographical distribution of animals. Wiley.

Delacour, J., and E. Mayr. 1946. Birds of the Philippines. Macmillan.

Feller, W. 1958. An introduction to probability theory and its applications. Vol. 1, 2nd ed. Wiley.

Greenway, J. G. 1958. Extinct and vanishing birds of the world. Amer. Comm. International Wild Life Protection, Special Publ. No. 13.

Lack, D. 1947. Darwin's finches, an essay on the general biological theory of evolution. Cambridge University Press.

MacArthur, R. H. 1957. On the relative abundance of bird species. Proc. Nat. Acad. Sci. [U. S.], **43**: 293–294.

Mayr, E. 1933. Die Vogelwelt Polynesiens. Mitt. Zool. Mus. Berlin, **19**: 306–323.

——. 1940. The origin and history of the bird fauna of Polynesia. Proc. Sixth Pacific Sci. Congr., **4**: 197–216.

——. 1943. The zoogeographic position of the Hawaiian Islands. Condor, **45**: 45–48.

——. 1944. Wallace's Line in the light of recent zoogeographic studies. Quart. Rev. Biol., **19**: 1–14.

Preston, F. W. 1962. The canonical distribution of commonness and rarity: Parts I, II. Ecology, **43**: 185–215, 410–432.

Rensch, B. 1936. Die Geschischte des Sundabogens. Borntraeger, Berlin.

STRESEMANN, E. 1934. "Aves." *In* Handb. Zool., W. Kukenthal, ed. Gruyter, Berlin.

———. 1939. Die Vögel von Celebes. J. für Ornithologie, **87**: 299–425.

WILSON, E. O. 1961. The nature of the taxon cycle in the Melanesian ant fauna. Amer. Nat., **95**: 169–193.

ZIMMERMAN, E. C. 1948. Insects of Hawaii. Vol. 1. Introduction. University of Hawaii Press.

APPENDIX: MEASUREMENT OF A PROPAGULE

A rudimentary account of how many immigrants are required to constitute a propagule may be constructed as follows. Let η be the average number of individuals next generation per individual this generation. Thus, for instance, if $\eta = 1.03$, the population is increasing at 3% interest rate.

Let us now suppose that the number of descendants per individual has a Poisson distribution. If it has not, due to small birth rate, the figures do not change appreciably. Then, due to chance alone, the population descended from immigrants may vanish. This subject is well known in probability theory as "Extinction probabilities in branching processes" (cf. Feller 1958, p. 274). The usual equation for the probability ζ of eventual extinction (Feller's equation 5.2 with $P(\zeta) = e^{-\eta(1-\zeta)}$, for a Poisson distribution), gives

$$\zeta = e^{-\eta(1-\zeta)}.$$

Solving this by trial and error for the probability of eventual extinction ζ, given a variety of values of η, we get the array shown in table 2. From this we can calculate how large a number of simultaneous immigrants would stand probability just one-half of becoming extinct during the initial stages of population growth following the introduction. In fact, if r pairs immigrate simultaneously, the probability that all will eventually be without descendants is ζ^r. Solving $\zeta^R = 0.5$ we find the number, R, of pairs of immigrants necessary to stand half a chance of not becoming extinct as given in table 3. From this it is clear that when η is 1, the propagule has infinite size, but that as η increases, the propagule size decreases rapidly, until, for a species which increases at 38.5% interest rate, one pair is sufficient to stand probability 1.2 of effecting a colonization. With sexual species which hunt for mates, η may be very nearly 1 initially.

TABLE 2. *Relation of replacement rate (η) of immigrants to probability of extinction (ζ)*

η	1	1.01	1.1	1.385
ζ	1	0.98	0.825	0.5

TABLE 3. *Relation of replacement rate (η) to the number of pairs (R) of immigrants required to give the population a 50% chance of survival*

η	1	1.01	1.1	1.385
R	∞	34	3.6	1

Hamilton, W. D. 1964. The genetical evolution of social behaviour: I. *Journal of Theoretical Biology* 7:1–16.

This was the first of two conceptually challenging papers in which the British evolutionary theorist W. D. Hamilton laid out a revolutionary framework for viewing selection and altruism, topics that are central to the field of sociobiology (the biology of social interactions). Hamilton's theory proposed that the fitness of an organism's genes is a function not merely of the individual's progeny but rather its *inclusive fitness*, which also incorporates the probability that an individual's genes are transmitted by genetic relatives. These latter transmission probabilities are influenced by the coefficients of relatedness (proportions of genes shared) by the relevant kin. Hamilton's papers introduced the concept of kin selection, now widely accepted as essential in sociobiology. The notion that kin selection underlies altruism led to Hamilton's rule, which states that an allele encoding a social behavior will tend to increase in frequency if r (the coefficient of genetic relatedness between two agents) is greater than the ratio of the cost (C) of the behavior (loss in personal fitness through self-sacrifice) compared with the benefit (B) received through increased reproduction by relatives ($r > C/B$). Hamilton's rule has been invoked to explain a wide range of seemingly altruistic phenomena in nature, ranging from aspects of apparent self-sacrifice in humans to the highly eusocial behaviors of ants and bees.

A few years before his death in 2000 at the age of 63, Hamilton edited two volumes (entitled *Narrow Roads of Gene Land*) of his collected publications.

RELATED READING

Dawkins, R. 1989. The selfish gene, 2nd ed. Oxford University Press, Oxford, England.
Hamilton, W. D. 1996a. Narrow roads of gene land: I. evolution of social behaviour. Oxford University Press, Oxford, England.
Hamilton, W. D. 1996b. Narrow roads of gene land: II. evolution of sex. Oxford University Press, Oxford, England.
Wilson, E. O. 1975. Sociobiology. Belknap Press, Cambridge, Massachusetts, USA.

J. Theoret. Biol. (1964) 7, 1–16

The Genetical Evolution of Social Behaviour. I

W. D. Hamilton

The Galton Laboratory, University College, London, W.C.2

(*Received* 13 *May* 1963, *and in revised form* 24 *February* 1964)

A genetical mathematical model is described which allows for interactions between relatives on one another's fitness. Making use of Wright's Coefficient of Relationship as the measure of the proportion of replica genes in a relative, a quantity is found which incorporates the maximizing property of Darwinian fitness. This quantity is named "inclusive fitness". Species following the model should tend to evolve behaviour such that each organism appears to be attempting to maximize its inclusive fitness. This implies a limited restraint on selfish competitive behaviour and possibility of limited self-sacrifices.

Special cases of the model are used to show (a) that selection in the social situations newly covered tends to be slower than classical selection, (b) how in populations of rather non-dispersive organisms the model may apply to genes affecting dispersion, and (c) how it may apply approximately to competition between relatives, for example, within sibships. Some artificialities of the model are discussed.

1. Introduction

With very few exceptions, the only parts of the theory of natural selection which have been supported by mathematical models admit no possibility of the evolution of any characters which are on average to the disadvantage of the individuals possessing them. If natural selection followed the classical models exclusively, species would not show any behaviour more positively social than the coming together of the sexes and parental care.

Sacrifices involved in parental care are a possibility implicit in any model in which the definition of fitness is based, as it should be, on the number of adult offspring. In certain circumstances an individual may leave more adult offspring by expending care and materials on its offspring already born than by reserving them for its own survival and further fecundity. A gene causing its possessor to give parental care will then leave more replica genes in the next generation than an allele having the opposite tendency. The selective advantage may be seen to lie through benefits conferred indifferently on a set of relatives each of which has a half chance of carrying the gene in question.

From this point of view it is also seen, however, that there is nothing special about the parent-offspring relationship except its close degree and a certain fundamental asymmetry. The full-sib relationship is just as close. If an individual carries a certain gene the expectation that a random sib will carry a replica of it is again one-half. Similarly, the half-sib relationship is equivalent to that of grandparent and grandchild with the expectation of replica genes, or genes "identical by descent" as they are usually called, standing at one quarter; and so on.

Although it does not seem to have received very detailed attention the possibility of the evolution of characters benefitting descendants more remote than immediate offspring has often been noticed. Opportunities for benefitting relatives, remote or not, in the same or an adjacent generation (i.e. relatives like cousins and nephews) must be much more common than opportunities for benefitting grandchildren and further descendants. As a first step towards a general theory that would take into account all kinds of relatives this paper will describe a model which is particularly adapted to deal with interactions between relatives of the same generation. The model includes the classical model for "non-overlapping generations" as a special case. An excellent summary of the general properties of this classical model has been given by Kingman (1961b). It is quite beyond the author's power to give an equally extensive survey of the properties of the present model but certain approximate deterministic implications of biological interest will be pointed out.

As is already evident the essential idea which the model is going to use is quite simple. Thus although the following account is necessarily somewhat mathematical it is not surprising that eventually, allowing certain lapses from mathematical rigour, we are able to arrive at approximate principles which can also be expressed quite simply and in non-mathematical form. The most important principle, as it arises directly from the model, is outlined in the last section of this paper, but a fuller discussion together with some attempt to evaluate the theory as a whole in the light of biological evidence will be given in the sequel.

2. The Model

The model is restricted to the case of an organism which reproduces once and for all at the end of a fixed period. Survivorship and reproduction can both vary but it is only the consequent variations in their product, net reproduction, that are of concern here. All genotypic effects are conceived as increments and decrements to a basic unit of reproduction which, if possessed by all the individuals alike, would render the population both stationary and non-evolutionary. Thus the fitness $a^{\bullet}$ of an individual is treated as the sum of his basic unit, the effect δa of his personal genotype and the total e° of effects on him due to his neighbours which will depend on their genotypes:

$$a^{\bullet} = 1+\delta a+e^{\circ}. \quad (1)$$

The index symbol $^{\bullet}$ in contrast to $^{\circ}$ will be used consistently to denote the inclusion of the personal effect δa in the aggregate in question. Thus equation (1) could be rewritten

$$a^{\bullet} = 1+e^{\bullet}.$$

In equation (1), however, the symbol $^{\bullet}$ also serves to distinguish this neighbour modulated kind of fitness from the part of it

$$a = 1+\delta a$$

which is equivalent to fitness in the classical sense of individual fitness.

The symbol δ preceding a letter will be used to indicate an effect or total of effects due to an individual treated as an addition to the basic unit, as typified in

$$a = 1+\delta a.$$

The neighbours of an individual are considered to be affected differently according to their relationship with him.

Genetically two related persons differ from two unrelated members of the population in their tendency to carry replica genes which they have both inherited from the one or more ancestors they have in common. If we consider an autosomal locus, not subject to selection, in relative B with respect to the same locus in the other relative A, it is apparent that there are just three possible conditions of this locus in B, namely that both, one only, or neither of his genes are identical by descent with genes in A. We denote the respective probabilities of these conditions by c_2, c_1 and c_0. They are independent of the locus considered; and since

$$c_2+c_1+c_0 = 1,$$

the relationship is completely specified by giving any two of them. Li & Sacks (1954) have described methods of calculating these probabilities adequate for any relationship that does not involve inbreeding. The mean number of genes per locus i.b.d. (as from now on we abbreviate the phrase "identical by descent") with genes at the same locus in A for a hypothetical population of relatives like B is clearly $2c_2+c_1$. One half of this number, $c_2+\frac{1}{2}c_1$, may therefore be called the expected fraction of genes i.b.d. in a relative. It can be shown that it is equal to Sewall Wright's Coefficient of Relationship r (in a non-inbred population). The standard methods of calculating r without obtaining the complete distribution can be found in Kempthorne (1957). Tables of

$$f = \tfrac{1}{2}r = \tfrac{1}{2}(c_2+\tfrac{1}{2}c_1) \quad \text{and} \quad F = c_2$$

1—2

for a large class of relationships can be found in Haldane & Jayakar (1962).

Strictly, a more complicated metric of relationship taking into account the parameters of selection is necessary for a locus undergoing selection, but the following account based on use of the above coefficients must give a good approximation to the truth when selection is slow and may be hoped to give some guidance even when it is not.

Consider now how the effects which an arbitrary individual distributes to the population can be summarized. For convenience and generality we will include at this stage certain effects (such as effects on parents' fitness) which must be zero under the restrictions of this particular model, and also others (such as effects on offspring) which although not necessarily zero we will not attempt to treat accurately in the subsequent analysis.

The effect of A on specified B can be a variate. In the present deterministic treatment, however, we are concerned only with the means of such variates. Thus the effect which we may write $(\delta a_{\text{father}})_A$ is really the expectation of the effect of A upon his father but for brevity we will refer to it as the effect on the father.

The full array of effects like $(\delta a_{\text{father}})_A$, $(\delta a_{\text{specified sister}})_A$, etc., we will denote

$$\{\delta a_{\text{rel.}}\}_A.$$

From this array we can construct the simpler array

$$\{\delta a_{r,\,c_2}\}_A$$

by adding together all effects to relatives who have the same values for the pair of coefficients (r, c_2). For example, the combined effect $\delta a_{\frac{1}{4},0}$ might contain effects actually occurring to grandparents, grandchildren, uncles, nephews and half-brothers. From what has been said above it is clear that as regards changes in autosomal gene-frequency by natural selection all the consequences of the full array are implied by this reduced array—at least, provided we ignore (a) the effect of previous generations of selection on the expected constitution of relatives, and (b) the one or more generations that must really occur before effects to children, nephews, grandchildren, etc., are manifested.

From this array we can construct a yet simpler array, or vector,

$$\{\delta a_r\}_A,$$

by adding together all effects with common r. Thus $\delta a_{\frac{1}{4}}$ would bring together effects to the above-mentioned set of relatives and effects to double-first cousins, for whom the pair of coefficients is $(\frac{1}{4}, \frac{1}{16})$.

Corresponding to the effect which A causes to B there will be an effect of similar type on A. This will either come from B himself or from a person who stands to A in the same relationship as A stands to B. Thus corresponding to an effect by A on his nephew there will be an effect on A by his uncle. The similarity between the effect which A dispenses and that which he receives is clearly an aspect of the problem of the correlation between relatives. Thus the term e° in equation (1) is not a constant for any given genotype of A since it will depend on the genotypes of neighbours and therefore on the gene-frequencies and the mating system.

Consider a single locus. Let the series of allelomorphs be $G_1, G_2, G_3, \ldots, G_n$, and their gene-frequencies $p_1, p_2, p_3, \ldots, p_n$. With the genotype G_iG_j associate the array $\{\delta a_{\text{rel.}}\}_{ij}$; within the limits of the above-mentioned approximations natural selection in the model is then defined.

If we were to follow the usual approach to the formulation of the progress due to natural selection in a generation, we should attempt to give formulae for the neighbour modulated fitnesses $a^\bullet_{ij}$. In order to formulate the expectation of that element of e°_{ij} which was due to the return effect of a relative B we would need to know the distribution of possible genotypes of B, and to obtain this we must use the double measure of B's relationship and the gene-frequencies just as in the problem of the correlation between relatives. Thus the formula for e°_{ij} will involve all the arrays $\{\delta a_{r,c_2}\}_{ij}$ and will be rather unwieldy (see Section 4).

An alternative approach, however, shows that the arrays $\{\delta a_r\}_{ij}$ are sufficient to define the selective effects. Every effect on reproduction which is due to A can be thought of as made up of two parts: an effect on the reproduction of genes i.b.d. with genes in A, and an effect on the reproduction of unrelated genes. Since the coefficient r measures the expected fraction of genes i.b.d. in a relative, for any particular degree of relationship this breakdown may be written quantitatively:

$$(\delta a_{\text{rel.}})_{\text{A}} = r(\delta a_{\text{rel.}})_{\text{A}} + (1-r)(\delta a_{\text{rel.}})_{\text{A}}.$$

The total of effects on reproduction which are due to A may be treated similarly:

$$\sum_{\text{rel.}} (\delta a_{\text{rel.}})_{\text{A}} = \sum_{\text{rel.}} r(\delta a_{\text{rel.}})_{\text{A}} + \sum_{\text{rel.}} (1-r)(\delta a_{\text{rel.}})_{\text{A}},$$

or

$$\sum_{r} (\delta a_r)_{\text{A}} = \sum_{r} r(\delta a_r)_{\text{A}} + \sum_{r} (1-r)(\delta a_r)_{\text{A}},$$

which we rewrite briefly as

$$\delta T^\bullet_{\text{A}} = \delta R^\bullet_{\text{A}} + \delta S_{\text{A}},$$

where $\delta R^\bullet_{\text{A}}$ is accordingly the total effect on genes i.b.d. in relatives of A, and δS_{A} is the total effect on their other genes. The reason for the omission of an index symbol from the last term is that here there is, in effect, no question of whether or not the self-effect is to be in the summation, for if it is included it has to be multiplied by zero. If index symbols were used we should have $\delta S^{\bullet}_{\mathrm{A}} = \delta S^{\circ}_{\mathrm{A}}$, whatever the subscript; it therefore seems more explicit to omit them throughout.

If, therefore, all effects are accounted to the individuals that cause them, of the total effect $\delta T^{\bullet}_{ij}$ due to an individual of genotype G_iG_j a part $\delta R^{\bullet}_{ij}$ will involve a specific contribution to the gene-pool by this genotype, while the remaining part δS_{ij} will involve an unspecific contribution consisting of genes in the ratio in which the gene-pool already possesses them. It is clear that it is the matrix of effects $\delta R^{\bullet}_{ij}$ which determines the direction of selection progress in gene-frequencies; δS_{ij} only influences its magnitude. In view of this importance of the $\delta R^{\bullet}_{ij}$ it is convenient to give some name to the concept with which they are associated.

In accordance with our convention let

$$R^{\bullet}_{ij} = 1 + \delta R^{\bullet}_{ij};$$

then $R^{\bullet}_{ij}$ will be called the *inclusive fitness*, $\delta R^{\bullet}_{ij}$ the *inclusive fitness effect* and δS_{ij} the *diluting effect*, of the genotype G_iG_j.

Let

$$T^{\bullet}_{ij} = 1 + \delta T^{\bullet}_{ij}.$$

So far our discussion is valid for non-random mating but from now on for simplicity we assume that it is random. Using a prime to distinguish the new gene-frequencies after one generation of selection we have

$$p'_i = \frac{\sum_j p_i p_j R^{\bullet}_{ij} + p_i \sum_{j,k} p_j p_k \delta S_{jk}}{\sum_{j,k} p_j p_k T^{\bullet}_{jk}} = p_i \frac{\sum_j p_j R^{\bullet}_{ij} + \sum_{j,k} p_j p_k \delta S_{jk}}{\sum_{j,k} p_j p_k T^{\bullet}_{jk}}.$$

The terms of this expression are clearly of the nature of averages over a part (genotypes containing G_i, homozygotes G_iG_i counted twice) and the whole of the existing set of genotypes in the population. Thus using a well known subscript notation we may rewrite the equation term by term as

$$p'_i = p_i \frac{R^{\bullet}_{i.} + \delta S_{..}}{T^{\bullet}_{..}}$$

$$\therefore \quad p'_i - p_i = \Delta p_i = \frac{p_i}{T^{\bullet}_{..}}(R^{\bullet}_{i.} + \delta S_{..} - T^{\bullet}_{..})$$

or

$$\Delta p_i = \frac{p_i}{R^{\bullet}_{..} + \delta S^{\bullet}_{..}}(R^{\bullet}_{i.} - R^{\bullet}_{..}). \tag{2}$$

This form clearly differentiates the roles of the $R^{\bullet}_{ij}$ and $\delta S^{\bullet}_{ij}$ in selective progress and shows the appropriateness of calling the latter diluting effects.

For comparison with the account of the classical case given by Moran (1962), equation (2) may be put in the form

$$\Delta p_i = \frac{p_i}{T^{\bullet}_{..}}\left(\frac{1}{2}\frac{\partial R^{\bullet}_{..}}{\partial p_i} - R^{\bullet}_{..}\right)$$

where $\partial/\partial p_i$ denotes the usual partial derivative, written d/dp_i by Moran.

Whether the selective effect is reckoned by means of the $a^{\bullet}_{ij}$ or according to the method above, the denominator expression must take in all effects occurring during the generation. Hence $a^{\bullet}_{..} = T^{\bullet}_{..}$.

As might be expected from the greater generality of the present model the extension of the theorem of the increase of mean fitness (Scheuer & Mandel, 1959; Mulholland & Smith, 1959; a much shorter proof by Kingman, 1961a) presents certain difficulties. However, from the above equations it is clear that the quantity that will tend to maximize, if any, is $R^{\bullet}_{..}$, the mean inclusive fitness. The following brief discussion uses Kingman's approach.

The mean inclusive fitness in the succeeding generation is given by

$$R^{\bullet\prime}_{..} = \sum_{i,j} p'_i p'_j R^{\bullet}_{ij} = \frac{1}{T^{\bullet 2}_{..}} \sum_{i,j} p_i p_j R^{\bullet}_{ij}(R^{\bullet}_{i.} + \delta S_{..})(R^{\bullet}_{.j} + \delta S_{..}).$$

$$\therefore \quad R^{\bullet\prime}_{..} - R^{\bullet}_{..} = \Delta R^{\bullet}_{..} = \frac{1}{T^{\bullet 2}_{..}}\Big\{\sum_{i,j} p_i p_j R^{\bullet}_{ij} R^{\bullet}_{i.} R^{\bullet}_{.j} + 2\delta S_{..} \sum_{i,j} p_i p_j R^{\bullet}_{ij} R^{\bullet}_{i.} + + R^{\bullet}_{..}\,\delta S^2_{..} - R^{\bullet}_{..} T^{\bullet 2}_{..}\Big\}.$$

Substituting $R^{\bullet}_{..} + \delta S_{..}$ for $T^{\bullet}_{..}$ in the numerator expression, expanding and rearranging:

$$\Delta R^{\bullet} = \frac{1}{T^{\bullet 2}_{..}}\left\{\left(\sum_{i,j} p_i p_j R^{\bullet}_{ij} R^{\bullet}_{i.} R^{\bullet}_{.j} - R^{\bullet 3}_{..}\right) + + 2\delta S_{..}\left(\sum_{i,j} p_i p_j R^{\bullet}_{ij} R^{\bullet}_{i.} - R^{\bullet 2}_{..}\right)\right\}.$$

We have ( ) $\geqslant 0$ in both cases. The first is the proven inequality of the classical model. The second follows from

$$\sum_{i,j} p_i p_j R^{\bullet}_{ij} R^{\bullet}_{i.} = \sum_i p_i R^{\bullet 2}_{i.} \geqslant \left(\sum_i p_i R^{\bullet}_{i.}\right)^2 = R^{\bullet 2}_{..}.$$

Thus a sufficient condition for $\Delta R^{\bullet}_{..} \geqslant 0$ is $\delta S_{..} \geqslant 0$. That $\Delta R^{\bullet}_{..} \geqslant 0$ for positive dilution is almost obvious if we compare the actual selective changes with those which would occur if $\{R^{\bullet}_{ij}\}$ were the fitness matrix in the classical model.

It follows that $R^{\bullet}_{..}$ certainly maximizes (in the sense of reaching a local maximum of $R^{\bullet}_{..}$) if it never occurs in the course of selective changes that $\delta S_{..} < 0$. Thus $R^{\bullet}_{..}$ certainly maximizes if all $\delta S_{ij} \geqslant 0$ and therefore also if all $(\delta a_{\text{rel.}})_{ij} \geqslant 0$. It still does so even if some or all δa_{ij} are negative, for, as we have seen δS_{ij} is independant of δa_{ij}.

Here then we have discovered a quantity, inclusive fitness, which under the conditions of the model tends to maximize in much the same way that fitness tends to maximize in the simpler classical model. For an important class of genetic effects where the individual is supposed to dispense benefits to his neighbours, we have formally proved that the average inclusive fitness in the population will always increase. For cases where individuals may dispense harm to their neighbours we merely know, roughly speaking, that the change in gene frequency in each generation is aimed somewhere in the direction of a local maximum of average inclusive fitness,† but may, for all the present analysis has told us, overshoot it in such a way as to produce a lower value.

As to the nature of inclusive fitness it may perhaps help to clarify the notion if we now give a slightly different verbal presentation. Inclusive fitness may be imagined as the personal fitness which an individual actually expresses in its production of adult offspring as it becomes after it has been first stripped and then augmented in a certain way. It is stripped of all components which can be considered as due to the individual's social environment, leaving the fitness which he would express if not exposed to any of the harms or benefits of that environment. This quantity is then augmented by certain fractions of the quantities of harm and benefit which the individual himself causes to the fitnesses of his neighbours. The fractions in question are simply the coefficients of relationship appropriate to the neighbours whom he affects: unity for clonal individuals, one-half for sibs, one-quarter for half-sibs, one-eighth for cousins, ... and finally zero for all neighbours whose relationship can be considered negligibly small.

Actually, in the preceding mathematical account we were not concerned with the inclusive fitness of individuals as described here but rather with certain averages of them which we call the inclusive fitnesses of types. But the idea of the inclusive fitness of an individual is nevertheless a useful one. Just as in the sense of classical selection we may consider whether a given character expressed in an individual is adaptive in the sense of being in the interest of his personal fitness or not, so in the present sense of selection we may consider whether the character or trait of behaviour is or is not adaptive in the sense of being in the interest of his inclusive fitness.

3. Three Special Cases

Equation (2) may be written

$$\Delta p_i = p_i \frac{\delta R^{\bullet}_{i.} - \delta R^{\bullet}_{..}}{1 + \delta T^{\bullet}_{..}}. \tag{3}$$

† That is, it is aimed "uphill": that it need not be at all directly towards the local maximum is well shown in the classical example illustrated by Mulholland & Smith (1959).

Now $\delta T^{\bullet}_{ij} = \sum_r (\delta a_r)_{ij}$ is the sum and $\delta R^{\bullet} = \sum_r r(\delta a_r)_{ij}$ is the first moment about $r = 0$ of the array of effects $\{\delta a_{\text{rel.}}\}_{ij}$ cause by the genotype G_iG_j; it appears that these two parameters are sufficient to fix the progress of the system under natural selection within our general approximation.

Let

$$r^{\bullet}_{ij} = \frac{\delta R^{\bullet}_{ij}}{\delta T^{\bullet}_{ij}}, \qquad (\delta T^{\bullet}_{ij} \neq 0); \tag{4}$$

and let

$$r^{\circ}_{ij} = \frac{\delta R^{\circ}_{ij}}{\delta T^{\circ}_{ij}}, \qquad (\delta T^{\circ}_{ij} \neq 0). \tag{5}$$

These quantities can be regarded as average relationships or as the first moments of reduced arrays, similar to the first moments of probability distributions.

We now consider three special cases which serve to bring out certain important features of selection in the model.

(a) The sums $\delta T^{\bullet}_{ij}$ differ between genotypes, the reduced first moment $r^{\bullet}$ being common to all. If all higher moments are equal between genotypes, that is, if all arrays are of the same "shape", this corresponds to the case where a stereotyped social action is performed with differing intensity or frequency according to genotype.

Whether or not this is so, we may, from equation (4), substitute $r^{\bullet}\delta T^{\bullet}_{ij}$ for $\delta R^{\bullet}_{ij}$ in equation (3) and have

$$\Delta p_i = p_i r^{\bullet} \frac{\delta T^{\bullet}_{i.} - \delta T^{\bullet}_{..}}{1 + \delta T^{\bullet}_{..}}.$$

Comparing this with the corresponding equation of the classical model,

$$\Delta p_i = p_i \frac{\delta a_{i.} - \delta a_{..}}{1 + \delta a_{..}}. \tag{6}$$

we see that placing genotypic effects on a relative of degree $r^{\bullet}$ instead of reserving them for personal fitness results in a slowing of selection progress according to the fractional factor $r^{\bullet}$.

If, for example, the advantages conferred by a "classical" gene to its carriers are such that the gene spreads at a certain rate the present result tells us that in exactly similar circumstances another gene which conferred similar advantages to the sibs of the carriers would progress at exactly half this rate.

In trying to imagine a realistic situation to fit this sort of case some concern may be felt about the occasions where through the probabilistic nature of things the gene-carrier happens not to have a sib, or not to have one suitably placed to receive the benefit. Such possibilities and their frequencies of realization must, however, all be taken into account as the effects $(\delta a_{\text{sibs}})_A$, etc., are being evaluated for the model, very much as if in a classical case allowance were being made for some degree of failure of penetrance of a gene.

(b) The reduced first moments $r_{ij}^{\bullet}$ differ between genotypes, the sum $\delta T^{\bullet}$ being common to all. From equation (4), substituting $r_{ij}^{\bullet}\delta T^{\bullet}$ for $\delta R_{ij}^{\bullet}$ in equation (3) we have

$$\Delta p_i = p_i \frac{\delta T^{\bullet}}{T^{\bullet}}(r_{i.}^{\bullet} - r_{..}^{\bullet}).$$

But it is more interesting to assume δa is also common to all genotypes. If so it follows that we can replace $^{\bullet}$ by $^{\circ}$ in the numerator expression of equation (3). Then, from equation (5), substituting $r_{ij}^{\circ}\delta T^{\circ}$ for δR_{ij}°, we have

$$\Delta p_i = p_i \frac{\delta T^{\circ}}{T^{\bullet}}(r_{i.}^{\circ} - r_{..}^{\circ}).$$

Hence, if a giving-trait is in question (δT° positive), genes which restrict giving to the nearest relative ($r_{i.}^{\circ}$ greatest) tend to be favoured; if a taking-trait (δT° negative), genes which cause taking from the most distant relatives tend to be favoured.

If all higher reduced moments about $r = r_{ij}^{\circ}$ are equal between genotypes it is implied that the genotype merely determines whereabouts in the field of relationship that centres on an individual a stereotyped array of effects is placed.

With many natural populations it must happen that an individual forms the centre of an actual local concentration of his relatives which is due to a general inability or disinclination of the organisms to move far from their places of birth. In such a population, which we may provisionally term "viscous", the present form of selection may apply fairly accurately to genes which affect vagrancy. It follows from the statements of the last paragraph but one that over a range of different species we would expect to find giving-traits commonest and most highly developed in the species with the most viscous populations whereas uninhibited competition should characterize species with the most freely mixing populations.

In the viscous population, however, the assumption of random mating is very unlikely to hold perfectly, so that these indications are of a rough qualitative nature only.

(c) $\delta T_{ij}^{\bullet} = 0$ for all genotypes.

$$\therefore \quad \delta T_{ij}^{\circ} = -\delta a_{ij}$$

for all genotypes, and from equation (5)

$$\delta R^{\circ}_{ij} = -\delta a_{ij} r^{\circ}_{ij}.$$

Then, from equation (3), we have

$$\Delta p_i = p_i(\delta R^{\bullet}_{i.} - \delta R^{\bullet}_{..}) = p_i\{(\delta a_{i.} + \delta R^{\circ}_{i.}) - (\delta a_{..} + \delta R^{\circ}_{..})\}$$
$$= p_i\{\delta a_{i.}(1 - r^{\circ}_{i.}) - \delta a_{..}(1 - r^{\circ}_{..})\}.$$

Such cases may be described as involving transfers of reproductive potential. They are especially relevant to competition, in which the individual can be considered as endeavouring to transfer prerequisites of survival and reproduction from his competitors to himself. In particular, if $r^{\circ}_{ij} = r^{\circ}$ for all genotypes we have

$$\Delta p_i = p_i(1 - r^{\circ})(\delta a_{i.} - \delta a_{..}).$$

Comparing this to the corresponding equation of the classical model (equation (6)) we see that there is a reduction in the rate of progress when transfers are from a relative.

It is relevant to note that Haldane (1923) in his first paper on the mathematical theory of selection pointed out the special circumstances of competition in the cases of mammalian embryos in a single uterus and of seeds both while still being nourished by a single parent plant and after their germination if they were not very thoroughly dispersed. He gave a numerical example of competition between sibs showing that the progress of gene-frequency would be slower than normal.

In such situations as this, however, where the population may be considered as subdivided into more or less standard-sized batches each of which is allotted a local standard-sized pool of reproductive potential (which in Haldane's case would consist almost entirely of prerequisites for pre-adult survival), there is, in addition to a small correcting term which we mention in the short general discussion of competition in the next section, an extra overall slowing in selection progress. This may be thought of as due to the wasting of the powers of the more fit and the protection of the less fit when these types chance to occur positively assorted (beyond any mere effect of relationship) in a locality; its importance may be judged from the fact that it ranges from zero when the batches are indefinitely large to a halving of the rate of progress for competition in pairs.

4. Artificialities of the Model

When any of the effects is negative the restrictions laid upon the model hitherto do not preclude certain situations which are clearly impossible from the biological point of view. It is clearly absurd if for any possible set of gene-frequencies any $a^{\bullet}_{ij}$ turns out negative; and even if the magnitude of δa_{ij} is sufficient to make $a^{\bullet}_{ij}$ positive while $1+e^{\circ}_{ij}$ is negative the situation is still highly artificial, since it implies the possibility of a sort of overdraft on the basic unit of an individual which has to be made good from his own takings. If we call this situation "improbable" we may specify two restrictions: a weaker, $e^{\circ}_{ij} > -1$, which precludes "improbable" situations; and a stronger, $e^{\bullet}_{ij} > -1$, which precludes even the impossible situations, both being required over the whole range of possible gene-frequencies as well as the whole range of genotypes.

As has been pointed out, a formula for $e^{\bullet}_{ij}$ can only be given if we have the arrays of effects according to a double coefficient of relationship. Choosing the double coefficient (c_2, c_1) such a formula is

$$e^{\bullet}_{ij} = \sum_{c_2, c_1}{}^{\bullet}[c_2 \operatorname{Dev}(\delta a_{c_2, c_1})_{ij} + \tfrac{1}{2}c_1\{\operatorname{Dev}(\delta a_{c_2, c_1})_{i.} + \operatorname{Dev}(\delta a_{c_2, c_1})_{.j}\}] + \delta T^{\circ}_{..}$$

where

$$\operatorname{Dev}(\delta a_{c_2, c_1})_{ij} = (\delta a_{c_2, c_1})_{ij} - (\delta a_{c_2, c_1})_{..} \quad \text{etc.}$$

Similarly

$$e^{\circ}_{ij} = \sum{}^{\circ}['']+\delta T^{\circ}_{..},$$

the self-effect $(\delta a_{1,\,0})_{ij}$ being in this case omitted from the summations.

The following discussion is in terms of the stronger restriction but the argument holds also for the weaker; we need only replace $^{\bullet}$ by $^{\circ}$ throughout.

If there are no dominance deviations, i.e. if

$$(\delta a_{\text{rel.}})_{ij} = \tfrac{1}{2}\{(\delta a_{\text{rel.}})_{ii} + (\delta a_{\text{rel.}})_{jj}\} \quad \text{for all } ij \text{ and rel.,}$$

it follows that each ij deviation is the sum of the $i.$ and the $j.$ deviations. In this case we have

$$e^{\bullet}_{ij} = \sum{}^{\bullet} r \operatorname{Dev}(\delta a_r)_{ij} + \delta T^{\bullet}_{..}.$$

Since we must have $e^{\bullet}_{..} = \delta T^{\bullet}_{..}$, it is obvious that some of the deviations must be negative.

Therefore $\delta T^{\bullet}_{..} > -1$ is a necessary condition for $e^{\bullet}_{ij} > -1$. This is, in fact, obvious when we consider that $\delta T^{\bullet}_{..} = -1$ would mean that the aggregate of individual takings was just sufficient to eat up all basic units exactly. Considering that the present use of the coefficients of relationships is only valid when selection is slow, there seems little point in attempting to derive mathematically sufficient conditions for the restriction to hold;

intuitively however it would seem that if we exclude over- and under-dominance it should be sufficient to have no homozygote with a net taking greater than unity.

Even if we could ignore the breakdown of our use of the coefficient of relationship it is clear enough that if $\delta T_{..}^{\bullet}$ approaches anywhere near -1 the model is highly artificial and implies a population in a state of catastrophic decline. This does not mean, of course, that mutations causing large selfish effects cannot receive positive selection; it means that their expression must moderate with increasing gene-frequency in a way that is inconsistent with our model. The "killer" trait of *Paramoecium* might be regarded as an example of a selfish trait with potentially large effects, but with its only partially genetic mode of inheritance and inevitable density dependance it obviously requires a selection model tailored to the case, and the same is doubtless true of most "social" traits which are as extreme as this.

Really the class of model situations with negative neighbour effects which are artificial according to a strict interpretation of the assumptions must be much wider than the class which we have chosen to call "improbable". The model assumes that the magnitude of an effect does not depend either on the genotype of the effectee or on his current state with respect to the prerequisites of fitness at the time when the effect is caused. Where taking-traits are concerned it is just possible to imagine that this is true of some kinds of surreptitious theft but in general it is more reasonable to suppose that following some sort of an encounter the limited prerequisite is divided in the ratio of the competitive abilities. Provided competitive differentials are small however, the model will not be far from the truth; the correcting term that should be added to the expression for Δp_i can be shown to be small to the third order. With giving-traits it is more reasonable to suppose that if it is the nature of the prerequisite to be transferable the individual can give away whatever fraction of his own property that his instincts incline him to. The model was designed to illuminate altruistic behaviour; the classes of selfish and competitive behaviour which it can also usefully illuminate are more restricted, especially where selective differentials are potentially large.

For loci under selection the only relatives to which our metric of relationship is strictly applicable are ancestors. Thus the chance that an arbitrary parent carries a gene picked in an offspring is $\frac{1}{2}$, the chance that an arbitrary grandparent carries it is $\frac{1}{4}$, and so on. As regards descendants, it seems intuitively plausible that for a gene which is making steady progress in gene-frequency the true expectation of genes i.b.d. in a n-th generation descendant will exceed $\frac{1}{2}^n$, and similarly that for a gene that is steadily declining in frequency the reverse will hold. Since the path of genetic connection with a simple same-generation relative like a half-sib includes an "ascending part" and a "descending part" it is tempting to imagine that the ascending part can be treated with multipliers of exactly $\frac{1}{2}$ and the descending part by multipliers consistently more or less than $\frac{1}{2}$ according to which type of selection is in progress. However, a more rigorous attack on the problem shows that it is more difficult than the corresponding one for simple descendants, where the formulation of the factor which actually replaces $\frac{1}{2}$ is quite easy at least in the case of classical selection, and the author has so far failed to reach any definite general conclusions as to the nature and extent of the error in the foregoing account which his use of the ordinary coefficients of relationship has actually involved.

Finally, it must be pointed out that the model is not applicable to the selection of new mutations. Sibs might or might not carry the mutation depending on the point in the germ-line of the parent at which it had occurred, but for relatives in general a definite number of generations must pass before the coefficients give the true—or, under selection, the approximate—expectations of replicas. This point is favourable to the establishment of taking-traits and slightly against giving-traits. A mutation can, however, be expected to overcome any such slight initial barrier before it has recurred many times.

5. The Model Limits to the Evolution of Altruistic and Selfish Behaviour

With classical selection a genotype may be regarded as positively selected if its fitness is above the average and as counter-selected if it is below. The environment usually forces the average fitness $a_{..}$ towards unity; thus for an arbitrary genotype the sign of δa_{ij} is an indication of the kind of selection. In the present case although it is $T^{\bullet}_{..}$ and not $R^{\bullet}_{..}$ that is forced towards unity, the analogous indication is given by the inclusive fitness effect $\delta R^{\bullet}_{ij}$, for the remaining part, the diluting effect δS_{ij}, of the total genotypic effect $\delta T^{\bullet}_{ij}$ has no influence on the kind of selection. In other words the kind of selection may be considered determined by whether the inclusive fitness of a genotype is above or below average.

We proceed, therefore, to consider certain elementary criteria which determine the sign of the inclusive fitness effect. The argument applies to any genotype and subscripts can be left out.

Let

$$\delta T^{\circ} = k\,\delta a. \tag{7}$$

According to the signs of δa and δT° we have four types of behaviour as set out in the following diagram:

		Neighbours gain; δT° +ve	Neighbours lose; δT° −ve
Individual	gains; δa +ve	k +ve *Selected*	k −ve Selfish behaviour ?
	loses; δa −ve	k −ve Altruistic behaviour ?	k +ve *Counter-selected*

The classes for which k is negative are of the greatest interest, since for these it is less obvious what will happen under selection. Also, if we regard fitness as like a substance and tending to be conserved, which must be the case in so far as it depends on the possession of material prerequisites of survival and reproduction, k −ve is the more likely situation. Perfect conservation occurs if $k = -1$. Then $\delta T^{\bullet} = 0$ and $T^{\circ} = 1$: the gene-pool maintains constant "volume" from generation to generation. This case has been discussed in Case (c) of section 3. In general the value of k indicates the nature of the departure from conservation. For instance, in the case of an altruistic action $|k|$ might be called the ratio of gain involved in the action: if its value is two, two units of fitness are received by neighbours for every one lost by an altruist. In the case of a selfish action, $|k|$ might be called the ratio of diminution: if its value is again two, two units of fitness are lost by neighbours for one unit gained by the taker.

The alarm call of a bird probably involves a small extra risk to the individual making it by rendering it more noticeable to the approaching predator but the consequent reduction of risk to a nearby bird previously unaware of danger must be much greater.† We need not discuss here just how risks are to be reckoned in terms of fitness: for the present illustration it is reasonable to guess that for the generality of alarm calls k is negative but $|k| > 1$. How large must $|k|$ be for the benefit to others to outweigh the risk to self in terms of inclusive fitness?

† The alarm call often warns more than one nearby bird of course—hundreds in the case of a flock—but since the predator would hardly succeed in surprising more than one in any case the total number warned must be comparatively unimportant.

$$\begin{aligned}\delta R^{\bullet} &= \delta R^{\circ} + \delta a \\ &= r^{\circ}\,\delta T^{\circ} + \delta a && \text{from (5)} \\ &= \delta a(kr^{\circ}+1) && \text{from (7).}\end{aligned}$$

Thus of actions which are detrimental to individual fitness (δa −ve) only those for which $-k > \frac{1}{r^{\circ}}$ will be beneficial to inclusive fitness ($\delta R^{\bullet}$ +ve).

This means that for a hereditary tendency to perform an action of this kind to evolve the benefit to a sib must average at least twice the loss to the individual, the benefit to a half-sib must be at least four times the loss, to a cousin eight times and so on. To express the matter more vividly, in the world of our model organisms, whose behaviour is determined strictly by genotype, we expect to find that no one is prepared to sacrifice his life for any single person but that everyone will sacrifice it when he can thereby save more than two brothers, or four half-brothers, or eight first cousins . . . Although according to the model a tendency to simple altruistic transfers ($k = -1$) will never be evolved by natural selection, such a tendency would, in fact, receive zero counter-selection when it concerned transfers between clonal individuals. Conversely selfish transfers are always selected except when from clonal individuals.

As regards selfish traits in general (δa +ve, k −ve) the condition for a benefit to inclusive fitness is $-k < \frac{1}{r^{\circ}}$. Behaviour that involves taking too much from close relatives will not evolve. In the model world of genetically controlled behaviour we expect to find that sibs deprive one another of reproductive prerequisites provided they can themselves make use of at least one half of what they take; individuals deprive half-sibs of four units of reproductive potential if they can get personal use of at least one of them; and so on. Clearly from a gene's point of view it is worthwhile to deprive a large number of distant relatives in order to extract a small reproductive advantage.

REFERENCES

HALDANE, J. B. S. (1923). *Trans. Camb. phil. Soc.* **23,** 19.

HALDANE, J. B. S. & JAYAKAR, S. D. (1962). *J. Genet.* **58,** 81.

KEMPTHORNE, O. (1957). "An Introduction to Genetical Statistics". New York: John Wiley & Sons, Inc.

KINGMAN, J. F. C. (1961a). *Quart. J. Math.* **12,** 78.

KINGMAN, J. F. C. (1961b). *Proc. Camb. phil. Soc.* **57,** 574.

LI, C. C. & SACKS, L. (1954). *Biometrics,* **10,** 347.

MORAN, P. A. P. (1962). *In* "The Statistical Processes of Evolutionary Theory", p. 54. Oxford: Clarendon Press.

MULHOLLAND, H. P. & SMITH, C. A. B. (1959). *Amer. math. Mon.* **66,** 673.

SCHEUER, P. A. G. & MANDEL, S. P. H. (1959). *Heredity,* **31,** 519.

Ehrlich, P. R., and P. H. Raven. 1964. Butterflies and plants: a study in coevolution. *Evolution* 18:586–608.

Coevolution can be defined as the joint evolution of two or more ecologically interacting species, each of which evolves in response to selective pressures imposed by the other. Predator-prey, host-parasite, pollinator-host, and numerous symbioses are a few examples in which species that interact in ecological time impact one another across evolutionary time. We have included in this volume the study by Paul Ehrlich and Peter Raven in which the term was introduced and defined. The authors focus on interactions between plants and phytophagous (plant-eating) insects, emphasizing both the intimacy and consequentiality of the reciprocal coevolutionary dance between these two species-rich sets of participants. In this paper, coevolution was defined only rather loosely, a fact that has allowed for a variety of interpretations by subsequent authors. For example, coevolution has been used to describe a broad set of interactions ranging from the intimate connections between pairs of intracellular symbionts to the manifold and often diffuse species' interactions that might shape the structure of natural communities.

Despite this definitional vagueness, the message from Ehrlich and Raven's paper rings loud and clear. Coevolution is a key point of contact between evolution and ecology, in the sense that organisms must be considered an important part of the environment that exerts selective pressures upon all species.

RELATED READING

Futuyma, D. J., and M. Slatkin, editors. 1983. Coevolution. Sinauer, Sunderland, Massachusetts, USA.

BUTTERFLIES AND PLANTS: A STUDY IN COEVOLUTION[1]

Paul R. Ehrlich and Peter H. Raven
Department of Biological Sciences, Stanford University, Stanford, California

Accepted June 15, 1964

One of the least understood aspects of population biology is community evolution—the evolutionary interactions found among different kinds or organisms where exchange of genetic information among the kinds is assumed to be minimal or absent. Studies of community evolution have, in general, tended to be narrow in scope and to ignore the reciprocal aspects of these interactions. Indeed, one group of organisms is all too often viewed as a kind of physical constant. In an extreme example a parasitologist might not consider the evolutionary history and responses of hosts, while a specialist in vertebrates might assume species of vertebrate parasites to be invariate entities. This viewpoint is one factor in the general lack of progress toward the understanding of organic diversification.

One approach to what we would like to call coevolution is the examination of patterns of interaction between two major groups of organisms with a close and evident ecological relationship, such as plants and herbivores. The considerable amount of information available about butterflies and their food plants make them particularly suitable for these investigations. Further, recent detailed investigations have provided a relatively firm basis for statements about the phenetic relationships of the various higher groups of Papilionoidea (Ehrlich, 1958, and unpubl.). It should, however, be remembered that we are considering the butterflies as a model. They are only one of the many groups of herbivorous organisms coevolving with plants. In this paper, we shall investigate the relationship between butterflies and their food plants with the hope of answering the following general questions:

1. Without recourse to long-term experimentation on single systems, what can be learned about the coevolutionary responses of ecologically intimate organisms?

2. Are predictive generalities about community evolution attainable?

3. In the absence of a fossil record can the patterns discovered aid in separating the rate and time components of evolutionary change in either or both groups?

4. Do studies of coevolution provide a reasonable starting point for the understanding of community evolution in general?

Factors Determining Food Choice

Before proceeding to a consideration of the relationships between butterfly groups and their food plants throughout the world, it is necessary briefly to consider some of the factors that determine the choice of food plants in this group and in phytophagous insects in general. Any group of phytophagous animals must draw its food supply from those plants that are available in its geographical and ecological range (Dethier, 1954). For instance, the butterflies are primarily a tropical group, and therefore there is a relatively greater utilization of primarily tropical than of temperate families of plants. The choice of oviposition site by the imago is also important. Many adult butterflies and moths lay their eggs on certain food plants with great precision as stressed by Merz (1959), but on the other hand, numerous "mistakes" have been recorded (e.g., Remington, 1952; Dethier, 1959). In such cases, larvae have either to find an appropriate plant or perish. There is an obvious selective advantage in oviposition on suitable

[1] This work has been supported in part by National Science Foundation Grants GB-123 (Ehrlich) and GB-141 (Raven).

plants, but inappropriate choices can be overcome by movement of the larvae. Furthermore, larvae feeding on herbs often consume the entire plant, and then must move even if the adult originally made an appropriate choice.

Larval choice therefore plays an important role in food plant relationships. An excellent review of a long series of experiments pertinent to this subject has recently been presented by Merz (1959); much of the following is based on his account. The condition of a given larva often has an effect on what foods it will or will not accept. In addition, many structural and mechanical characteristics of plants modify these relationships, mostly by limiting the acceptability of those plants in which they occur. For example, Merz (1959) found that larvae of *Lasiocampa quercus*, a moth that normally feeds along the edge of leaves, could not eat the sharply toothed leaves of holly (*Ilex*, Aquifoliaceae). When these same leaves were cut so that untoothed margins were presented, the larvae ate them voraciously. In other cases, larvae eat the young, soft leaves of plants but not the old, tough leaves of the same plants. Many Lycaenidae feed on flowers, and these butterflies may be unable to utilize the tough foliage of the same plants. Numerous similar examples could be given, but it must be borne in mind that chemical factors are operative in the same plants that present mechanical difficulties to larvae (Thorsteinson, 1960), and actually may be more important.

Chemical factors are of great general importance in determining larval food choice. In the first place, potential food sources are probably all nutritionally unbalanced to some extent (Gordon, 1961). The exploitation of a particular plant as a source of food thus involves metabolic adjustments on the part of an insect. These render the insect relatively inefficient in utilizing other sources of food and tend to restrict its choice of food plants. Secondly, many plants are characterized by the presence of secondary metabolic substances. These substances are repellent to most insects and may often be decisive in patterns of food plant selection (Thorsteinson, 1960). It has further been demonstrated that the chemical compounds that repel most animals can serve as trigger substances that induce the uptake of nutrients by members of certain oligophagous groups (Dethier, 1941, 1954; Thorsteinson, 1953, 1960). Presence of such repellent compounds may be correlated with the presence of the nutrients. Both odor and taste seem to be important.

The chemical composition of plants often changes with age, exposure to sunlight, or other environmental factors (Merz, 1959; Flück, 1963), and this may be critical for phytophagous insects (Dethier, 1954). For example, insects that feed on Umbelliferae prefer the old leaves, which appear to us less odorous than the young ones. Some insects that feed on alkaloid-rich species of *Papaver* (Papaveraceae) prefer the young leaves, which are relatively poor in alkaloids. Diurnal chemical cycles, influenced by exposure of the plant to sunlight, may be of prime importance in determining the habits of night-feeding groups, such as Argynnini.

Merz (1959, p. 159) has given a particularly interesting case of chemical repellents at the specific level. The larvae of the moth *Euchelia jacobaeae* feed on many species of *Senecio* (Compositae), but not on the densely glandular-hairy *S. viscosus*. When the glandular substance was dissolved in methyl alcohol, the larvae ate *S. viscosus*. When the same substance was painted on the leaves of other normally acceptable species of *Senecio*, these were refused. In an extensive study of the food plants of *Plebejus icarioides* (Lycaeninae), Downey (1961, 1962) showed that larvae would feed on any species of *Lupinus* (Leguminosae) in captivity, but populations in the field normally utilized only one or a few of the possible range of *Lupinus* species growing locally. This work suggests the subtle interaction of ecological, chemical, and mechanical factors that doubtless characterizes most natural situations. Relationships with predators (Brower, 1958), parasites (Downey, 1962), or, at least in the case of Lycaenidae, ants (Downey, 1962), may further modify patterns of food plant choice.

Despite all of these modifying factors, there is a general and long-recognized pattern running through the food plants of various groups of butterflies, and it is this pattern with which we shall be concerned. It certainly should not be inferred from anything that follows that all members of a family or genus of plants are equally acceptable to a given butterfly (for example, see Remington, 1952). We have placed our main emphasis on positive records, especially at the level of plant species and genera.

The Diversity of Butterflies

The butterflies comprise a single superfamily of Lepidoptera, the Papilionoidea. In comparison with many other superfamilies of insects they are uniform morphologically and behaviorally. Table 1 gives a rough idea of the taxonomic diversity of this superfamily.

Papilionoidea are divided into five families. Two of these, Nymphalidae and Lycaenidae, contain at least three-quarters of the genera and species; it is uncertain which family is the larger. Two smaller families, Pieridae and Papilionidae, include virtually all remaining butterflies. Pieridae, although containing many fewer genera and species than either Lycaenidae or Nymphalidae, form a prominent part of the butterfly fauna in many parts of the world, making up in number of individuals what they lack in number of kinds. Papilionidae are a group about half the size of Pieridae, but gain prominence through the large size of the included forms. The tiny family Libytheidae, closely related to Nymphalidae, is obscure to everyone except butterfly taxonomists.

The Papilionidae lead the butterflies in morphological diversity. Nymphalidae probably take second place, with Pieridae and Lycaenidae about tied for third. Difficult as this diversity is to estimate, it is clear that Papilionidae are a more heterogeneous group of organisms than any of the other families, whereas, considering the number of species and genera included, Lycaenidae are remarkably uniform. A rough idea of the phenetic relationships of the major groups of butterflies is given by Ehrlich (1958).

With food plant records from between 46 and 60% of all butterfly genera (table 1), it seems highly unlikely that future discoveries will necessitate extensive revisions of the conclusions drawn in this paper. The food plants of Riodininae are very poorly known, however, and it will be interesting to have more information about them and records for other outstanding "unknowns" such as *Styx* and *Pseudopontia*. It is, however, difficult to imagine any additional food plant record that would seriously distort the patterns outlined here.

Butterfly Food Plants

Sources of Information

The food plant information abstracted in this paper is derived principally from two sources. First, we have examined all the extensive and scattered literature that we could uncover with a library search and through the recommendations of various lepidopterists. Particularly helpful have been the volumes of Barrett and Burns (1951), Corbet and Pendlebury (1956), Costa Lima (1936), Ehrlich and Ehrlich (1961), Lee (1958), Seitz (1906–1927), van Son (1949, 1955), Wiltshire (1957), and Wynter-Blyth (1957), as well as the *Journal of the Entomological Society of South Africa*, the *Journal of the Lepidopterists' Society* (formerly *Lepidopterists' News*), and the *Journal of Research on the Lepidoptera*.

Our second major source of information has been provided by the following scientists, who have aided us in this ambitious undertaking not only by sending unpublished data and reprints of their works, but by helping to evaluate the validity of cer-

TABLE 1. *Summary of the taxonomic diversity of Papilionoidea*

Taxon	Approximate number of Genera*	Approximate number of Species	Distribution
Papilionidae	24(22)	575–700	
Baroniinae	1(1)	1	Central Mexico
Parnassiinae	8(8)	45–55	Holarctic and Oriental; greatest diversity, Asia
Papilioninae	15(13)	480–640	Worldwide; mainly tropical. Greatest diversity, Old World tropics
Pieridae	58(40)	950–1,150	
Coliadinae	11(8)	225–250	Cosmopolitan; greatest diversity tropics outside of Africa
Pierinae	43(29)	650–750	Cosmopolitan; greatest diversity tropics
Dismorphiinae	3(3)	80–120	Primarily Neotropical; one small Palearctic genus
Pseudopontiinae	1(0)	1	West equatorial Africa
Nymphalidae	325–400 (ca. 202)	4,800–6,200	
Ithomiinae	30–40(10)	300–400	Neotropical; *Tellervo* Australian
Danainae	10–12(10)	140–200	Cosmopolitan; greatest diversity Old World tropics
Satyrinae	120–150 (ca. 70)	1,200–1,500	Cosmopolitan; greatest diversity extratropical
Morphinae	23–26(12)	180–250	Indomalayan and Neotropical
Charaxinae	8–10(8)	300–400	Tropicopolitan, few temperate
Calinaginae	1(1)	1	Oriental
Nymphalinae	125–150 (ca. 85)	2,500–3,000	Cosmopolitan
Acraeinae	8(6)	225–275	Tropical; greatest diversity, Africa
Libytheidae	1(1)	10	Cosmopolitan
Lycaenidae	325–425 (ca. 167)	5,800–7,200	
Riodininae	75–125(17)	800–1,200	Tropical, few Nearctic and Palearctic. Metropolis, Neotropical
Styginae	1(0)	1	Peruvian Andes
Lycaeninae	250–300 (ca. 150)	5,000–6,000	Cosmopolitan; greatest diversity, Old World tropics
Total	730–930 (ca. 432)	12,000–15,000	

* Number in parentheses indicates number of genera for which food plant records are available.

tain published records and commenting on other aspects of the work. The cooperation of these people has been truly extraordinary, and we are particularly indebted to them: Remauldo F. d'Almeida (Brazil), Peter Bellinger (USA), C. M. de Biezanko (Brazil), L. P. Brower (USA), C. A. Clarke (England), H. K. Clench (USA), J. A. Comstock (USA), C. G. C. Dickson (Africa), J. C. Downey (USA), Maria Etcheverry (Chile), K. J. Hayward (Argentina), T. G. Howarth (England), Taro Iwase (Japan), T. W. Langer (Denmark), C. D. MacNeill (USA), D. P. Murray (England), G. Sevastopulo (Africa), Takashi Shirozu (Japan), E. M. Shull (India), Henri Stempffer (France), V. G. L. van Someren (Africa), G. van Son (Africa).

John C. Downey (Southern Illinois University), Gordon H. Orians (University of Washington), T. A. Geissman (University of California, Los Angeles), and Robert F. Thorne (Rancho Santa Ana Botanic Garden) have been so kind as to read the entire manuscript. Their advice has been invaluable.

To our knowledge, the data assembled here represent the most extensive body of information ever assembled on the interactions between a major group of herbivorous animals and their food plants.

Evaluation of the Literature

Extreme care has been taken in associating insects with particular food plants, as the literature is replete with errors and unverified records. In evaluating records, preference has been given to those which are concerned with the entire life cycle of a particular insect on a wild plant. Laboratory experiments and records from cultivated plants demonstrate only potentialities, not necessarily natural associations. In the laboratory, larvae may be starved or plants abnormal. In the wild, larvae are often misidentified, especially if not reared to maturity (cf. Brower, 1958b). Even more serious is the lack of precise plant identifications, or their identification in the vernacular only, which almost inevitably leads to confusion (Jörgensen, 1932). Any serious student of phytophagous animals should preserve adequate herbarium specimens of the plants with which he is concerned (cf. Remington, 1952, p. 62); only by doing this can the records be verified. Despite the extremely erratic oviposition behavior often shown by butterflies (Dethier, 1959), oviposition records have all too frequently been accepted as being equivalent to food plant records. Finally, nomenclatural difficulties, including changes in name and careless misspellings (e.g., "Oleaceae" for Olacaceae), have given rise to serious errors. In the literature on butterfly food plants, errors have often been compounded when copied from one source to another, and they are difficult to trace back to their origins. All of these problems make quantitative comparisons unreasonable. We therefore have been exceedingly conservative about accepting records, and focused our attention primarily on broad, repeatedly verified patterns of relationship.

The Food Plants of Butterflies

In this section, we will first outline the main patterns of food plant choice for each family, and then discuss what bearing these patterns have on our interpretation of relationships within the various butterfly families. It is necessary to give the data in considerable detail, as no comprehensive survey on a world basis is available elsewhere.

Papilionidae.—There are three subfamilies. *Baronia brevicornis*, the only species of Baroniinae, occurs in Mexico and feeds on *Acacia* (Leguminosae; Vazquez and Perez, 1961). In Parnassiinae, all five genera of Zerynthiini (Munroe, 1960; Munroe and Ehrlich, 1960) feed on Aristolochiaceae, as does *Archon* (Parnassiini). *Hypermnestra* (Parnassiini) is recorded from *Zygophyllum* (Zygophyllaceae). *Parnassius* feeds on Crassulaceae and herbaceous Saxifragaceae, two closely related families, with one small group on Fumariaceae. In view of the discussion below, it is of interest that Zygophyllaceae are close relatives of Rutaceae, and Fumariaceae are rich in alkaloids similar to those of woody Ranales (Hegnauer, 1963).

The third and last subfamily, Papilioninae, is cosmopolitan but best developed in the Old World, and consists of three tribes: Troidini, Graphiini, and Papilionini. The eight genera of Troidini feed mostly on Aristolochiaceae, with individual species of *Parides* also recorded from Rutaceae, Menispermaceae, Nepenthaceae, and Piperaceae. *Parides* (*Atrophaneura*) *daemonius* is reported to feed on *Osteomeles* (Rosaceae), and *P.* (*A.*) *antenor*, a Malagasy butterfly that is the only representative of its tribe in the Ethiopian region, feeds on *Combretum* (Combretaceae). Both of these last-mentioned records need confirmation. At least two species of *Battus* have been recorded from Rutaceae in addition to the usual Aristolochiaceae. Records available for five of the seven genera of Graphiini

(*Eurytides, Graphium, Lamproptera, Protographium, Teinopalpus*) are mostly from Annonaceae, Hernandiaceae, Lauraceae, Magnoliaceae, and Winteraceae. This is clearly a closely allied group of plant families referable to the woody Ranales. In addition, some species of *Graphium* feed on Rutaceae, and others both on Apocynaceae (*Landolphia*) and Annonaceae (one of the latter also on *Sphedamnocarpus*, Malpighiaceae). Several species of *Eurytides* feed on *Vitex* (Verbenaceae) and one on *Jacobinia* (Acanthaceae). *Eurytides lysithous* feeds both on Annonaceae and on *Jacobinia*, and *E. helios* both on *Vitex* and Magnoliaceae. The bitypic Palearctic *Iphiclides* departs from the usual pattern for the group in feeding on a number of Rosaceae–Pomoideae.

The third and last tribe, Papilionini, consists only of the enormous cosmopolitan genus *Papilio*. Two of the five sections recognized by Munroe (1960; II, IV) are primarily on Rutaceae, with occasional records from Canellaceae, Lauraceae, and Piperaceae. Members of the circumboreal *Papilio machaon* group are not only on Rutaceae but also on Umbelliferae and *Artemisia* (Compositae). The African *P. demodocus*, in another group, is known to feed on Rutaceae and Umbelliferae, as well as *Pseudospondias* (Anacardiaceae), *Ptaeroxylon* (Meliaceae), and *Hippobromus* (Sapindaceae). Another African species, *P. dardanus*, is recorded from Rutaceae and also from *Xymalos* (Flacourtiaceae; Dickson, pers. comm.). The Asian and Australian *P. demoleus* is mostly on Rutaceae but also locally on *Salvia* (Labiatae) and *Psoralea* (Leguminosae). The other three sections (Munroe's I, III, V) are primarily associated with Annonaceae, Canellaceae, Hernandiaceae, Lauraceae, and Magnoliaceae, with a few records from Berberidaceae, Malvaceae (*Thespea*), and Rutaceae. The North American temperate *Papilio glaucus* group (sect. II) feeds not only on Lauraceae and Magnoliaceae like its more southern relatives but also on Aceraceae, Betulaceae, Oleaceae, Platanaceae, Rhamnaceae, Rosaceae, Rutaceae (*Ptelea*), and Salicaceae (Brower, 1958b).

In Papilionidae, Munroe and Ehrlich (1960) have argued that the red-tuberculate, Aristolochiaceae-feeding larvae of Papilioninae–Troidini and Parnassiinae–Zerynthiini, plus *Archon* (Parnassiinae–Parnassini) are so similar, and the likelihood of their converging on Aristolochiaceae so remote, that these probably represent the remnants of the stock from which the rest of Papilioninae and Parnassiinae were derived. Viewed in this context other food plants of these groups are secondary. The two remaining tribes of Papilioninae (Papilionini, Graphiini) are above all associated with the group of dicotyledons known as the woody Ranales. This is a diverse assemblage of plant families showing many unspecialized characteristics. Thorne (1963) has used the food plant relationships of Papilionidae as a whole to support his suggestion of affinity between Aristolochiaceae and Annonaceae (one of the woody Ranales). He appears to have established the existence of this similarity on morphological evidence. Likewise, similar alkaloids are shared by Aristolochiaceae and woody Ranales (Hegnauer, 1963; Alston and Turner, 1963, p. 170). Recently, Vazquez and Perez (1961) described the life cycle of *Baronia brevicornis*, the only member of the third subfamily of the group. *Baronia* feeds on *Acacia* (Leguminosae) and has tuberculate larvae like those of the forms that feed on Aristolochiaceae. Considering its morphological distinctness, and in accordance with the scheme of relationships presented by Munroe and Ehrlich (1960, p. 175), it appears likely that *Baronia* represents a phylogenetic line which diverged early from that leading to the rest of Papilionidae. It may thus be the only member of the family neither feeding on Aristolochiaceae nor descended from forms that did.

Following this reasoning, we suggest that the original transition to Aristolochiaceae opened a new adaptive zone for Papilionidae. Their further spread and the multiplication of species was accompanied by the exploitation of other presumably chemically similar plant groups, such as woody Ranales, in areas where Aristolochiaceae were poorly represented, like Africa today. The site of greatest diversity for both Aristolochiaceae and Papilionidae is Asia. It is likely that the major diversification of Papilionidae (involving differentiation into Parnassiinae and Papilioninae) took place after the evolution of Aristolochiaceae. When this might have been is entirely uncertain, despite the unfounded speculations of Forbes (1958).

Another interesting problem is posed by the many representatives of Papilionini and Graphiini feeding on Rutaceae, in addition to woody Ranales. Rutaceae are morphologically very different from woody Ranales, and have not been closely associated with them taxonomically. Recently, however, Hegnauer (1963) has pointed out that some Rutaceae possess the alkaloids widespread in woody Ranales, in addition to an unusually rich repertoire of other alkaloids. Earlier, Dethier (1941) showed the similarity between the attractant essential oils in Rutaceae and Umbelliferae. Some Rutaceae-feeding groups of *Papilio* seem to have shifted to Umbelliferae, especially outside of the tropics. Dethier also implicated some of the similar-scented species of *Artemisia* (Compositae), another plant group fed on by at least one species of *Papilio*. Although species of *Papilio* link these groups of plants, none is known to feed on Burseraceae, Cneoraceae, Simarubaceae, or Zygophyllaceae, families thought to be related to Rutaceae but not known to contain alkaloids or coumarins (Price, 1963). On the other hand, the record of *Papilio demodocus* on *Ptaeroxylon* (Meliaceae), in addition to numerous Rutaceae, would seem to indicate a promising plant to investigate for the alkaloids suspected (Price, 1963) in Meliaceae.

Pieridae.—Our discussion of Pieridae is based taxonomically on the generic review of Klots (1933) as modified by Ehrlich (1958). There are four subfamilies, but nothing is known of the biology of the monobasic West African Pseudopontiinae. Of the remaining three, Dismorphiinae, including the Neotropical *Dismorphia* and *Pseudopieris* and the Palearctic *Leptidia,* are recorded only on Leguminosae. Larval food plants are known for 7 of the 11 genera of Coliadinae. *Catopsilia, Phoebis, Anteos, Eurema,* and *Colias* are mostly associated with Leguminosae, but there are a few records from Sapindaceae, Guttiferae, Euphorbiaceae, Simarubaceae, Oxalidaceae, Salicaceae, Ericaceae, and Gentianaceae (the last three with northern and montane species of *Colias*). On the other hand, three genera are associated with non-leguminous plants: *Gonepteryx* with *Rhamnus* (Rhamnaceae), *Nathalis* with Compositae, and *Kricogonia* with *Guaiacum* (Zygophyllaceae). Nonetheless, Leguminosae are decidedly the most important food plants of Coliadinae.

Pierinae, the third subfamily, are divided into two tribes, Pierini (36 genera) and Euchloini (7). In Euchloini, the temperate *Anthocharis, Euchloë, Zegris,* and *Hesperocharis* (also from *Phrygilanthus,* Loranthaceae) feed on Cruciferae, the tropical *Pinacopteryx* and *Hebemoia* on Capparidaceae. For Pierini, 14 of the 23 genera for which we know something of the food plants (namely, *Appias, Ascia, Belenois, Ceporis, Colotis, Dixeia, Elodina, Eronia, Ixias, Leptosia, Pareronia, Pieris, Prioneris,* and *Tatochila*) are primarily on Capparidaceae in the tropics and subtropics and on Cruciferae in temperate regions. Some have occasionally been reported to feed on Resedaceae, Salvadoraceae, and Tropaeolaceae. There are also a very few scattered records from other plants, including one or two from Leguminosae. The basis for selecting Capparidaceae, Cruciferae, Resedaceae, Salvadoraceae, and Tropaeolaceae is relatively easy to comprehend, since all of these plants are known to contain mustard oil glucosides (thioglucosides) and the associated enzyme myrosinase which acts in the hydrolysis of glucosides to release mustard oils (Alston and Turner,

1963, p. 284–288). In an early series of food choice experiments, Verschaeffelt (1910) found that larvae of *Pieris rapae* and *P. brassicae* would feed on Capparidaceae, Cruciferae, Resedaceae, and Tropaeolaceae, as well as another family which contains mustard oils but upon which Pierinae are not known to feed in nature: Moringaceae. Verschaeffelt also found that these larvae would eat flour, starch, or even filter paper if it was smeared with juice expressed from *Bunias* (Cruciferae), and Thorsteinson (1953, 1960) showed that the larvae would eat other kinds of leaves treated with sinigrin or sinalbin (two common mustard oil glucosides) if the leaves were not too tough and did not contain other kinds of repellents. Very few butterflies outside Pierdinae feed on these plants, but there is one example in Lycaeninae. In addition, there are at least two records of *Phoebis* (Coliadinae) from Capparidaceae and Cruciferae. Numerous groups of insects other than butterflies are characteristically associated with this same series of plant families (Fraenkel, 1959).

It is not so easy to interpret scattered records of these pierine genera feeding on other plant families: *Belenois raffrayi* on *Rhus* (Anacardiaceae); *Nepheronia argyia* on Capparidaceae, but also *Cassipourea* (Rhizophoraceae) and *Hippocratea* (Hippocrateaceae), with *N. thalassina* reported only from *Hippocratea*; *Ascia monuste* on Rhamnaceae and *Cassia* (Leguminosae), as well as Capparidaceae; and *Tatochila autodice* on *Cestrum* (Solanaceae) and also *Medicago* (Leguminosae). Several species of *Appias* have been reported from different genera of Euphorbiaceae, whereas others feed both on Capparidaceae and Euphorbiaceae, but this probably can be explained somewhat more simply, since mustard oils have been reported in some genera of Euphorbiaceae (Alston and Turner, 1963, p. 285).

The remaining genera of Pierini fall mostly into what has been called the *Delias* group. Of these, *Catasticta* and *Archonias* have been recorded from *Phrygilanthus* (Loranthaceae) in South America, and *Delias*, a large Indo-Malaysian genus, from "*Loranthus*" (Loranthaceae) and *Exocarpus* of the closely related Santalaceae, with *D. aglaja* on *Nauclea* (Rubiaceae). *Aporia*, a large genus of temperate regions of the Old World, has several species on *Berberis* (Berberidaceae), and one on woody Rosaceae. *Pereute*, South American, feeds on *Ocotea* (Lauraceae; Jörgensen, 1932), Tiliaceae, and Polygonaceae. Two very peculiar genera of the *Delias* group are the monotypic Mexican *Eucheira*, which feeds on woody hard-leaved Ericaceae, and the bitypic western North American *Neophasia*, which feeds on various genera of Pinaceae. It is very interesting that *Cepora*, which falls into the *Delias* group morphologically, feeds on *Capparis* like many other Pierini.

Finally, the large, taxonomically isolated, Ethiopian *Mylothris* feeds on Loranthaceae and Santalaceae (*Osyris*), with *M. bernice rubricosta* on *Polygonum* (Polygonaceae).

It is difficult to understand the reasons for large groups of Pierinae being associated both with plants that possess mustard oils and with Loranthaceae–Santalaceae; neither morphological nor biochemical evidence has been adduced to link these two groups of plants. Perhaps the Loranthaceae-feeders represent an old offshoot of Pierinae; in any case it would appear that the main diversification of this group occurred after it became associated with Capparidaceae–Cruciferae.

Nymphalidae.—This enormous family is divided into eight subfamilies which will be discussed one by one in the succeeding paragraphs.

Ithomiinae are primarily American, and there feed only on Solanaceae (many genera). The Indo-Malaysian *Tellervo*, only Old World representative of the group, which is segregated as a distinct tribe Tellervini, has been recorded from *Aristolochia* (Aristolochiaceae). The identity of the plant was inferred from the fact that papilionid larvae normally associated with *Aristolochia* were found on it with *Tellervo*. Solanaceae are rich in alkaloids (as are

Aristolochiaceae), and are very poorly represented among the food plants of butterflies as a whole. The diversification of the ithomiines that feed on them has probably followed a pattern similar to that of Papilionidae on Aristolochiaceae and Pierinae on Capparidaceae–Cruciferae. Many other groups of insects feed primarily on Solanaceae (Fraenkel, 1959).

Danainae are a rather uniform cosmopolitan group, obviously related to Ithomiinae. The danaines feed primarily and apparently interchangeably on Apocynaceae and Asclepiadaceae. In addition, there are records of *Euploea, Ituna,* and *Lycorella* on Moraceae and of the last occasionally on *Carica* (Caricaceae). All of these plants have milky juice. There is also a single record of *Ituna ilione,* which normally feeds on *Ficus* (Moraceae), from *Myoporum* (Myoporaceae). Apocynaceae and Asclepiadaceae form a virtual continuum in their pattern of variation and can scarcely be maintained as distinct (Safwat, 1962). Both are noted for their abundant bitter glycosides and alkaloids (Alston and Turner, 1963, p. 258), and share at least some alkaloids (Price, 1963, p. 431) and pyridines with Moraceae. Thus it appears very likely that here too the acquisition of the ability to feed on Apocynaceae and Asclepiadaceae has constituted for Danainae the penetration of a new adaptive zone, in which they have radiated. Numerous distinctive segments of other insect orders and groups likewise feed on these two plant families.

Eleven genera of Morphinae are recorded from a variety of monocotyledons: Bromeliaceae, Gramineae (mostly bamboos), Marantaceae, Musaceae, Palmae, Pandanaceae, and Zingiberaceae. In contrast, most species of *Morpho* feed on dicotyledons, including Canellaceae, Erythroxylaceae, Lauraceae, Leguminosae, Menispermaceae, Myrtaceae, Rhamnaceae, and Sapindaceae, but *M. aega* feeds on bamboos (Gramineae) and *M. hercules* on Musaceae. Whether the progenitors of *Morpho* fed on dicotyledons or monocotyledons cannot be determined.

Closely related to Morphinae are the more temperate Satyrinae, an enormous group that feeds mostly on Gramineae (including bamboos and canes) and Cyperaceae, occasionally on Juncaceae. *Pseudonympha vigilans* feeds on *Restio* (Restionaceae), a family close to Gramineae, *Physcaneura pione* on Zingiberaceae, and *Elymnias* on Palmae. There are no records of this group from dicotyledons. Thus the phenetically similar Morphinae–Satyrinae assemblage is the outstanding example in butterflies of a group associated primarily with monocotyledons.

The distinctive tropicopolitan Charaxinae are often associated with woody Ranales (Annonaceae, Lauraceae, Monimiaceae, Piperaceae), but also with such diverse families as Anacardiaceae, Araliaceae (*Schefflera*), Bombacaceae, Celastraceae, Connaraceae, Convolvulaceae, Euphorbiaceae, Flacourtiaceae, Hippocrateaceae, Leguminosae, Linaceae, Malvaceae, Meliaceae, Melianthaceae, Myrtaceae, Proteaceae, Rhamnaceae, Rutaceae, Salvadoraceae (*Charaxes hansali*), Sapindaceae, Sterculiaceae, Tiliaceae, Ulmaceae, and Verbenaceae. Records (largely Sevastopulo and van Someren, pers. comm.) are available for about 50 African species of *Charaxes*, most of which are associated with dicotyledons. At least three feed on grasses (Gramineae), two of these on dicotyledons also.

The Oriental *Calinaga buddha*, the only species of Calinaginae, feeds on *Morus* (Moraceae).

Nymphalinae are a huge cosmopolitan group with relatively few "gaps" in their pattern of variation which would permit the recognition of meaningful subgroups (cf. Reuter, 1896; Chermock, 1950). The tribes do, however, display some significant patterns in their choice of food plants, with Heliconiini (Michener, 1942) and Argynnini feeding mostly on the Passifloraceae–Flacourtiaceae–Violaceae–Turneraceae complex of families, a closely related group of plants also important for Acraeinae. Acraeinae (see below), Heliconiini, and Argynnini are closely related phenetically, and their diversification may have taken place from a common ancestor associated with this particular assemblage of plants. No biochemical basis is known for the association of this series of four plant families, but we confidently predict that one eventually will be found (cf. also Gibbs, 1963, p. 63). Melitaeini are often associated with Acanthaceae, Scrophulariaceae and their wind-pollinated derivatives Plantaginaceae, and with Compositae and Verbenaceae. Nymphalini feed on plants of the same families as Melitaeini, but also very prominently on the Ulmaceae–Urticaceae–Moraceae group and the Convolvulaceae, Labiatae, Portulacaceae, and Verbenaceae. A single species in this group, however, *Nymphalis canace,* feeds on Liliaceae and Dioscoreaceae. Apaturini are associated chiefly with Ulmaceae, especially *Celtis.* Cyrestini (*Chersonesia, Cyrestis,* and *Marpesia*) and Gynaeciini (*Gynaecia* and *Historis,* but not *Callizona* and *Smyrna*) are often associated with Moraceae, and Hamadryini (*Ectima, Hamadryas*), Didonini (*Didonis*), Ergolini (*Byblia, Byblis, Ergolis, Eurytela, Mestra*), Eunicini (*Asterope, Catonephele, Eunica, Myscelia*), and Dynamini (*Dynamine*) mostly on the related Euphorbiaceae, which, like Moraceae, have milky sap. In addition to the Eunicini just mentioned, *Diaethria* (*Callicore, Catagramma*), *Epiphile, Haematera, Pyrrhogyra,* and *Temenis* feed almost exclusively on Sapindaceae. There is no obvious dominant theme for the last tribe, Limenitini, but it is interesting to note that two species of *Euphaedra* (*Najas*), a group that is mostly on Sapindaceae, are on *Cocos* and other Palmae. Additional families represented among the food plants of Nymphalinae are: Aceraceae, Amaranthaceae, Anacardiaceae, Annonaceae, Berberidaceae, Betulaceae, Bignoniaceae, Bombacaceae, Boraginaceae, Caprifoliaceae, Combretaceae, Corylaceae, Crassulaceae, Curcurbitaceae, Dilleniaceae, Dipterocarpaceae, Ebenaceae, Eleagnaceae, Ericaceae, Fagaceae, Gentianaceae, Geraniaceae, Guttiferae, Icacinaceae, Leguminosae (very uncommonly), Loranthaceae, Malvaceae, Melastomaceae, Melianthaceae, Menispermaceae, Myrtaceae, Oleaceae, Ranunculaceae (*Vanessa* on *Delphinium*), Rhamnaceae, Rosaceae, Rubiaceae, Sabiaceae, Salicaceae, Sapotaceae, Saxifragaceae, Sterculiaceae, Thymeleaceae, Tiliaceae, and Vitaceae.

Some of the butterflies in this group feed on a very wide range of plants, and most of the families mentioned in the above list are represented by one or at most a very few records. For example, *Euptoieta claudia* is known to feed on Berberidaceae (*Podophyllum*), Crassulaceae, Leguminosae, Linaceae, Menispermaceae, Nyctaginaceae, Passifloraceae, Portulacaceae, Violaceae, and even Asclepiadaceae (*Cyanchum*), and *Precis lavinia* is recorded from, among others, Acanthaceae, Bignoniaceae, Compositae, Crassulaceae, Onagraceae (*Ludwigia*), Plantaginaceae, Scrophulariaceae, and Verbenaceae.

Acraeinae, a rather small tropical group, are often associated with Passifloraceae–Flacourtiaceae–Violaceae–Turneraceae, as noted above, but also with Amaranthaceae, Compositae, Convolvulaceae, Leguminosae, Lythraceae, Moraceae, Polygonaceae, Rosaceae, Sterculiaceae, Urticaceae, and Vitaceae. In addition, *Acraea encedon* is reported from *Commelina* (Commelinaceae).

For the very diverse Nymphalidae as a whole, the following groups of plants are especially important: (1) Passifloraceae–Flacourtiaceae–Violaceae–Turneraceae; (2) Ulmaceae–Urticaceae–Moraceae, as well as the closely related (Thorne, pers. comm.) Euphorbiaceae; (3) Acanthaceae–Scrophulariaceae–Plantaginaceae. The second and third of these groups are represented among the food plants of other butterflies, such as Lycaeninae, but not abundantly. Conversely, as will be seen, the groups of food plants commonly represented in Lycaenidae —for example, Fagaceae, Leguminosae, Oleaceae, Rosaceae—are rare in Nymphalidae. Although each of these two families of butterflies is very wide in its choice of food plants, there is a distinctiveness to the two patterns which suggests a history of selection along different lines.

Libytheidae.—This small family consists of a single widespread genus, *Libythea,* which feeds almost exclusively on *Celtis* (Ulmaceae), but in southern Japan on *Prunus* (Rosaceae). *Libythea* is obviously closely related to Nymphalidae (Ehrlich, 1958), as has recently been confirmed by a quantitative study of adult internal anatomy (Ehrlich, unpubl.).

Lycaenidae.—An enormous group, the family Lycaenidae may be larger even than the Nymphalidae. Lycaenidae are in general poorly known from the standpoint of food plants (Downey, 1962). Our discussion is based largely on the classification of Clench (1955 and pers. comm.). Nothing is known of the life history of the Peruvian *Styx infernalis,* only member of Styginae. Of the two remaining subfamilies, Riodininae, divided into three tribes, will be discussed first. *Euselasia* (Euselasiini) has been recorded from *Mammea* (Guttiferae) and three genera of Myrtaceae. The Old World Hamearini consist of three genera, with *Dodona* and *Zemeros* on *Maesa* (Myrsinaceae) and *Hamearis* on *Primula* (Primulaceae). The two plant families are very closely related, with Myrsinaceae being primarily tropical and woody, Primulaceae primarily temperate and herbaceous. The third and largest tribe, Riodinini, is divided into four subtribes. *Abisara* (Abisariti) feeds, like the Hamearini, on Myrsinaceae; *Theope* (Theopiti) is on *Theobroma* (Sterculiaceae); and *Helicopus* (Helicopiti) is one of two members of the subfamily known to feed on a monocotyledon, in this case *Montrichardia* (Araceae). The remaining genera of Riodinini are in the exclusively New World Riodiniti, with very few records for a great many species. Plant families represented are Acanthaceae, Anacardiaceae, Aquifoliaceae, Chenopodiaceae, Compositae, Euphorbiaceae, Leguminosae, Moraceae, Myrtaceae, Polygonaceae, Ranunculaceae (*Clematis*), Rosaceae, Rutaceae, Sapindaceae, and Sapotaceae. Deserving special mention are the records of *Cariomathus* and *Rhetus* from Loranthaceae, *Napaea nepos* from *Oncidium* (Orchidaceae), and *Stalachtis* from *Oxpetalum* (Asclepiadaceae; cf. Jörgensen, 1932, p. 43, however, where it is suggested that associations of larvae of this group with ants may determine the food plant on which they are found). The scanty food plant records for this group are thus sufficiently diverse to suggest that further studies of food plants will be of considerable interest. The most salient feature is the occurrence of Hameaerini and Abisariti on Myrsinaceae and Primulaceae, two closely related families that are fed on by very few other butterflies.

Lycaèninae likewise consist of three tribes. Of these, Leptinini are African and feed on lichens, some of them (*Durbania, Durbaniopsis,* and *Durbaniella*) even on the low crustose lichens that grow on rocks. Liphyrini, almost entirely confined to the Old World tropics, are predaceous on aphids, coccids, ant larvae, membracids, and jassids. There are no reliable records of phytophagy in this group.

The largest of the three tribes, Lycaenini, presents a bewildering array of forms that can be separated only informally at present. Many of these larvae are closely associated with and tended by ants, and this association may modify their food plant relationships (Downey, 1962; Stempffer, pers. comm.). For the large *Plebejus* group (the "blues"), we have records of the food plants of 45 genera, and 33 of these are known to feed, at least in part, on Leguminosae. Records of special interest in this group include *Nacaduba* on several genera of Myrsinaceae and *Agriades* on Primulaceae; in this way they are like Hamaerini and Abisariti of Riodininae. *Chilades* and *Neopithecops* are recorded from Rutaceae. Four genera (*Philotes, Scolitantides, Talicada,* and *Tongeia*) are known to feed, at least in part, on Crassulaceae. *Catachrysops pandava* feeds not only on *Wagatea* and *Xylia* (Leguminosae) but also on *Cycas revoluta* (Cycadaceae), a cycad to which it does harm in gardens. *Hemiargus ceraunus* feeds on Marantaceae. Although most species of *Jamides* feed on Leguminosae, *J. alecto* feeds on Zingiberaceae (a monocotyledon).

In the Strymon group (Clench *in* Ehrlich and Ehrlich, 1961, plus *Strymonidia*), there is no obvious pattern, but there are several records of interest: *Dolymorpha* on *Solanum* (Solanaceae; Clench, unpubl.); *Eumaeus*, with *E. debora* on both *Dioon edule* (Cycadaceae) and *Amaryllis* (Liliaceae) and *E. atala* on both *Manihot* (Euphorbiaceae; Comstock, unpubl.) and on *Zamia integrifolia* (Cycadaceae); *Strymon melinus*, which feeds on a variety of dicotyledonous plants, but also on the flowers of *Nolina* (Liliaceae); and *Tmolus echion*, which feeds not only on *Lantana* (Verbenaceae), *Cordia* (Boraginaceae), *Datura* and *Solanum* (Solanaceae), *Hyptis* (Labiatae), and *Mangifera* (Anacardiaceae), but also on *Ananas* (Bromeliaceae). The impressive pattern of food plant radiation among the four subgenera of *Callophrys* deserves special mention, for subg. *Callophrys* and *Incisalia* feed mostly on angiosperms—Leguminosae, Polygonaceae, Rosaceae, and Ericaceae—but three species of *Incisalia* have switched to conifers, feeding on *Picea* and *Pinus* (Pinaceae). A third subgenus, *Mitoura*, feeds primarily on another group of conifers, Cupressaceae, with two species surprisingly on the pine mistletoes, *Arceuthobium* (Loranthaceae). Finally, *Callophrys* (*Sandia*) *macfarlandi*, the only species of its group, feeds on the flowers of *Nolina* (Liliaceae) in the southwestern United States.

Lycaena and *Heliophorus*, closely related, feed primarily on Polygonaceae throughout the nearly cosmopolitan but largely extratropical range of both groups.

The theclines, narrowly defined (Shirôzu and Yamamoto, 1956), have recently been treated by Shirôzu (1962), who has demonstrated that Fagaceae are the most important food plants, with a number of genera associated with Oleaceae. One genus (*Shirozua*) has become predaceous on aphids.

Among the remaining genera of Lycaeninae, a few points are especially noteworthy. The morphologically diverse South American group referred to "*Thecla*" is also extraordinarily diverse in its choice of food plants: Bromeliaceae, Celastraceae, Compositae, Euphorbiaceae, Leguminosae, Liliaceae, Malpighiaceae, Malvaceae, Sapotaceae, Solanaceae, and Ulmaceae. In addition to *Callophrys*, already mentioned, many distinctive and in some cases large genera feed primarily on Loranthaceae and the closely related Santalaceae: *Charana, Deudorix, Hypochrysops, Iolaus* s. str. (also often on *Ximenia*, Olacaceae), *Ogyris, Pretapa, Pseudodipsas, Rathinda*, and *Zesius*. It would appear that the epiphytic mistletoes and their relatives have constituted an important adaptive zone for a number of genera of Lycaeninae (as suggested by Clench, pers. comm.). Some species of *Iolaus* are on *Colocasia* (Araceae). Olacaceae, Loranthaceae, and Santalaceae are presumably closely related (Hutchinson, 1959), and interestingly share some acetylinic fatty acids (Sørensen, 1963) and lipids (Shorland, 1963). *Chliaria* feeds on the buds and flowers of a number of genera of Orchidaceae, and *Eooxylides, Loxura*, and *Yasoda* feed on *Smilax*, a hard-leaved member of Liliaceae, and the superficially similar *Dioscorea* (Dioscoreaceae). *Artipe* lives inside the fruits of *Punica* (Punicaceae), and *Bindahara* inside the fruits of *Salacia* (Celastraceae). Finally, *Aphnaeus* inhabits galleries hollowed out by ants in the twigs of *Acacia* (Leguminosae), where it feeds on fungi (van Son, pers. comm.)!

In summary, the plant families that are best represented among the food plants of Lycaenini are Ericaceae, Labiatae, Polygonaceae, Rhamnaceae, and Rosaceae. Other records from families not hitherto mentioned are: Aizoaceae, Amaranthaceae, Araliaceae, Betulaceae, Boraginaceae, Bruniaceae, Burseraceae, Caprifoliaceae, Caryophyllaceae, Chenopodiaceae, Cistaceae, Combretaceae, Convolvulaceae, Coriaraceae, Cornaceae, Diapensiaceae, Dipterocarpaceae, Ebenaceae, Eleagnaceae, Gentianaceae, Geraniaceae, Hamamelidaceae, Juglandaceae, Lauraceae, Lecithydaceae, Lythraceae, Meliaceae, Melianthaceae, Myricaceae, Oxalidaceae, Pittosporaceae, Plantaginaceae, Plumbaginaceae, Proteaceae, Rubiaceae, Saxifragaceae, Sterculiaceae, Styracaceae, Symplocaceae, Theaceae, Thymeleaceae, and Zygophyllaceae. As before it must be borne in mind that many of these listings represent single records only; for example, the widespread Holarctic *Celastrina argiolus* has been recorded from food plants belonging to at least 14 families of dicotyledons. Nonetheless, it should be evident that the pattern is very different from that of Nymphalinae, the only subfamily comparable to Lycaeninae in size.

Discussion

What generalities can be drawn from these observed patterns? We shall approach this question from the standpoint of the utilization of different plant groups by butterflies and see what light this throws on patterns of evolution in the two groups. Butterflies, of course, are only one of many phytophagous groups of organisms affecting plant evolution.

Within the appropriate ecological framework, our view of the immediate potentialities of studies of phytophagy has been stated clearly and succinctly by Bourgogne (1951, p. 330), who, speaking of the patterns of food plant choice in Lepidoptera, said: "Ces anomalies apparentes peuvent parfois démontrer l'existence, entre deux végétaux, d'une affinité d'ordre chimique, quelquefois même d'une parenté systématique. . . ." Thus, the choices exercised by phytophagous organisms may provide approximate but nevertheless useful indications of biochemical similarities among groups of plants. These do *not* necessarily indicate the plants' overall phenetic or phylogenetic relationships. The same can be said of the choice of arrow poisons by primitive human groups (Alston and Turner, 1963, p. 293) and of patterns of parasitism by fungi (Saville, 1954). In many of these cases, biochemists have not yet worked out the bases for the observed patterns, but as Merz (1959, p. 181) points out, we should nevertheless assume that they probably do have a chemical basis.

Now let us consider the groups of organisms utilized as food by butterfly larvae, starting with the most unusual diets. Two tribes of Lycaeninae have departed completely from the usual range of foods: Liptenini feed on lichens, Liphyrini are carnivorous. Many other Lycaeninae, however, are tended by ants and in some cases the larvae are brought into the ant nests. It would seem to be a relatively small step for such larvae to switch and feed on the ant grubs or fungi present in these nests. A number of species of the group exhibit well-developed cannibalism (Downey, 1962). Several Lycaenini, such as *Shirozua*, are carnivorous, and at least one species of *Aphnaeus* feeds on fungi in ant galleries. These transitional steps suggest the evolutionary pathways to the most divergent of butterfly larval feeding habits.

Among those groups of butterflies that feed on plants, none is known to feed on bryophytes or on Psilopsida, Lycopsida, or Sphenopsida, nor is any known from ferns. In fact, very few insects feed on ferns at all (cf. Docters van Leeuwen, 1958), a most surprising and as yet unexplained fact with no evident chemical or mechanical basis. At least one genus of moths, *Papaipema*, is known to feed on ferns, however (Forbes, 1958).

There are a few groups of butterflies that feed on gymnosperms. Two genera of Lycaenidae (*Catachrysops*; *Eumaeus*, two species) feed on Cycadaceae, but all three species involved also feed on angiosperms. *Neophasia* (Pierinae) and three species of *Callophrys* subg. *Incisalia* (Lycaeninae) feed on Pinaceae, while *Callophrys* subg. *Mitoura* feeds on Cupressaceae (and also on *Arceuthobium*, Loranthaceae, a mistletoe that grows on pines). It is well established that Cupressaceae and Pinaceae are chemically quite distinct (Erdtman, 1963, p. 120). Judging from the taxonomic distance between these butterfly groups, it can be assumed that butterflies feeding on gymnosperms had ancestors that fed on angiosperms.

An overwhelmingly greater number of butterfly larvae feed on dicotyledons than on monocotyledons. The only two groups primarily associated with monocotyledons are Satyrinae and Morphinae, closely related subfamilies of Nymphalidae. One genus of morphines (*Morpho*) is more often associated with dicotyledons, but we can think of no way to determine whether this represents a switch from previous monocotyledon feeding. No member of Papilionidae, Pieridae, or Libytheidae is known to feed on monocotyledons, but in Nymphalidae and Lycaenidae numerous genera do so in whole or in part. Among the Nymphalidae, several species of *Charaxes*, one of *Acraea*, one of *Nymphalis*, and two of *Euphaedra* (*Najas*) are known to feed on monocotyledons; all of these genera and some of the same species feed on dicotyledons also. In Lycaenidae, a number of very diverse groups, including at least two genera of Riodininae (*Helicopis* on Araceae, *Napaea* on Orchidaceae) and 11 genera of Lycaeninae (*Jamides* on Zingiberaceae; *Iolaus* on Araceae; *Tmolus* on Bromeliaceae; *Chliaria* on Orchidaceae; *Hemiargus* on Marantaceae; and *Callophrys, Eooxylides, Eumaeus, Strymon,* "*Thecla*," and *Yasoda* on Liliaceae) feed on monocotyledons. Representatives of many of these genera and in some cases the same species feed on dicotyledons also. This pattern strongly suggests that butterflies of two families have switched to monocotyledons from dicotyledons in a number of independent lines (probably at least 18).

A corollary to the observations presented above is that the diversity we see in modern butterflies has been elaborated against a dicotyledonous background. Indeed this is probably true for Lepidoptera as a whole (cf. Forbes, 1958). The dominant themes in this particular coevolutionary situation are therefore of considerable interest. We conclude from this relationship that the appearance of dicotyledons, as yet undated but surely pre-Cretaceous, must have antedated the evolutionary radiation that produced the modern lines of diversification in Lepidoptera and specifically in Papilionoidea. All utilization of foods other than dicotyledons by butterfly larvae (and probably by any Lepidoptera) is assumed to be the result of changes from an earlier pattern of feeding on dicotyledons.

In general, the patterns of utilization by butterflies of dicotyledonous food plants show a great many regularities. Certain relationships are very constant; the plants are usually fed upon by a single, phenetically coherent group of butterflies or several very closely related groups. As examples we have the Aristolochiaceae-feeding Papilionidae; Pierinae on Capparidaceae and Cruciferae; Ithomiinae on Solanaceae; Danainae on Apocynaceae and Asclepiadaceae; Acraeinae, Heliconiini, and Argynnini on Passifloraceae, Flacourtiaceae, Violaceae, and Turneraceae; and Riodininae–Hamearini and Abisariti on Myrsinaceae and Primulaceae. In many of these cases, the broad patterns observed probably support suggestions of overall phenetic similarity among the plants utilized and among the groups of butterflies concerned. Other clusterings on the basis of food plant choice like that of Ulmaceae, Urticaceae, Moraceae, and Euphorbiaceae by certain groups of Nymphalidae, probably also reflect phylogenetic relationship among the plants concerned. In several instances, the patterns of food plant choice of butterfly groups underscore the close relationship between certain sets of tropical woody and temperate herbaceous families elaborated by Bews (1927). Examples are Danainae, feeding interchangeably on Apocynaceae and Asclepiadaceae; Pierinae, on Capparidaceae and Cruciferae; and part of Riodininae, on Myrsinaceae and Primulaceae. In the first case, the families are generally thought to be closely related. On the other hand, Hutchinson (1959) widely separated the members of the second and third pairs of plant families in his system, but this disposition is considered inappropriate by almost all botanists since it is based upon his primary division of flowering plants into woody and herbaceous lines. In making such decisions based on larval food plants, it must be remembered that we are dealing only with an indirect measure of biochemical similarity. For example, Pierinae not only feed on Capparidaceae and Cruciferae, which most botanists would agree are closely related, but also on Salvadoraceae, which contain mustard oil glucosides but otherwise seem totally different from Capparidaceae and Cruciferae. More equivocal cases likewise occur. For example, Pierinae feed on Tropaeolaceae. Not only do Tropaeolaceae share mustard oil glucosides with Capparidaceae and Cruciferae, they likewise have in common the rare fatty acid, erucic acid. Can we, with Alston and Turner (1963, p. 287), dismiss this as coincidence, or do these groups of plants have more in common than is generally assumed? Finally, is there biochemical similarity between Loranthaceae–Santalaceae and Capparidaceae–Cruciferae, both common food plants of different groups of Pierinae (and of the genus *Hesperocharis*).

Whatever conclusions are drawn about the biochemical affinities of plants from the habits of phytophagous or parasitic organisms, little or no weight should be given to individual records. This is true not only because of the numerous sources of error enumerated earlier, but also because of the multiple explanations possible for such switches. For example, *Atella* (Nymphalinae) feeds on Flacourtiaceae and Salicaceae, among other plants. It is quite possible that these two families are fairly closely related, despite the greatly reduced anemophilous flowers of the latter (Thorne, pers. comm.). But to assume that the few records involved indicate biochemical similarity between the groups would be an unwarranted extension of the data; it would be far simpler and safer at that point to make comparative investigations of the biochemistry of the two plant families.

Patterns of food plant utilization provide evidence bearing on the relationship of Araliaceae and Umbelliferae. Some groups of Papilioninae, normally associated with Rutaceae, feed interchangeably on Umbelliferae, or in some cases have switched entirely to this family. As we have seen, these two plant families are chemically similar. But Araliaceae are close relatives of Umbelliferae (Rodríguez, 1957) despite their wide separation in the system of Hutchinson (1959), and are common in many regions where Papilioninae feed on Rutaceae. Despite this, there is not a single record of a papilionid butterfly (indeed very few butterflies of any kind) feeding on Araliaceae. An even more interesting relationship hinges on the suggestion that the three subfamilies of Umbelliferae—Apioideae, Hydrocotyloideae, and Saniculoideae—may represent three phylogenetic lines, derived independently from a group like the present-day Araliaceae. All records of Papilioninae from Umbelliferae are concerned with Apioideae, and indeed Dethier (1941) found that Umbelliferae-feeding papilionine larvae refused *Hydrocotyle* (Hydrocotyloideae). This very strongly suggests that biochemical analysis may go far in elucidating relationships within the Araliaceae–Umbelliferae complex, and that the chemical properties generally ascribed to Umbelliferae as a whole may be characteristic only of one subfamily, Apioideae. Araliaceae and Umbelliferae are known to share certain distinctive fatty acids (Alston and Turner, 1963, p. 121) and acetylinic compounds (Sørensen, 1963), and it might be very instructive to see how these were distributed in Umbelliferae outside of Apioideae.

In further evaluating the patterns of food plant choice in butterflies, it is important to consider those plant families, especially dicotyledons, which are absent or very poorly represented. One outstanding group is that partly characterized by Merz (1959, p. 169) as "Sphingidpflanzen"—plants fed on by moths of the family Sphingidae. These include, among others,

Onagraceae, Lythraceae, Balsaminaceae, Vitaceae, Rubiaceae, and Caprifoliaceae. The first two, and probably the third and fourth, are generally regarded as fairly closely related. Each one of the first five families (Rubiaceae only in part) is characterized by the abundant presence of raphides, bundles of needlelike crystals of calcium oxalate (see discussion in Gibbs, 1963). In a very interesting experiment, Merz (1959) offered mature leaves of *Vitis* (Vitaceae) to larvae of *Pterogon proserpina* (Sphingidae). Young larvae ate these leaves, their pointlike bites falling between the clusters of raphides. Older larvae, which make large slashing bites, could not avoid the raphides and did not eat the leaves. After the raphides were dissolved in very dilute hydrochloric acid, the leaves were accepted by larvae of all sizes. Although it cannot be proven that some other chemical repellent was not removed by this treatment, it is obvious that raphides offer considerable mechanical difficulty for phytophagous insects. A number of families of moths other than Sphingidae feed on this same series of plant families (Forbes, 1958).

Rubiaceae, one of the families mentioned above, is perhaps the most prominent family that is nearly absent from the records of butterfly food plants. Probably the third largest family of dicotyledons, with nearly 10,000 species, it is, like the butterflies themselves, mostly tropical. One can only speculate that some chemical factor, perhaps the rich representation of alkaloids, sharply restricts the ability of butterfly larvae to feed on plants of this family. In this respect, the similarities between the alkaloids of Apocynaceae (which however have milky juice) and Rubiaceae are of interest. Other dicotyledonous families that are very poorly represented or not represented at all among butterfly food plants include Begoniaceae, Bignoniaceae, Boraginaceae, Celastraceae, Cornaceae, Curcurbitaceae (with curcurbitacins, bitter-tasting terpenes), Gesneriaceae, Hydrophyllaceae, Loasaceae, Menispermaceae (rich in alkaloids), Myrtaceae, Polemoniaceae, Ranunculaceae (rich in alkaloids), and Theaceae. In addition, very few butterflies feed on Centrospermae, a group characterized both by its morphological and biochemical traits (summary in Alston and Turner, 1963, p. 141–143, 276–279). This group includes such large families as Amaranthaceae, Cactaceae, Caryophyllaceae, Chenopodiaceae, Nyctaginaceae, and Portulacaceae. Although no biochemical basis for this lack of utilization is known at present, one probably exists. For a family such as the enormous Compositae, poorly represented among the food plants of butterflies, the explanation may lie either in their chemical composition or largely extratropical distribution, or most likely a combination of these. One prominent family of monocotyledons that is practically unrepresented among butterfly food plants is Araceae (*Helicopis*, Riodininae, and species of *Iolaus*, Lycaeninae, are exceptions).

One can conclude only that at least some of the plant groups enumerated above have chemical or mechanical properties that render them unpalatable to butterfly larvae. Thus far the combination of circumstances permitting a shift into the adaptive zones represented by these groups has not occurred. The assumption that such a shift is theoretically possible is strengthened by the observation that nearly every one of these plant groups is fed upon by one or more families of moths.

Conclusions

A systematic evaluation of the kinds of plants fed upon by the larvae of certain subgroups of butterflies leads unambiguously to the conclusion that secondary plant substances play the leading role in determining patterns of utilization. This seems true not only for butterflies but for all phytophagous groups and also for those parasitic on plants. In this context, the irregular distribution in plants of such chemical compounds of unknown physiological function as alkaloids, quinones, essential oils (including terpenoids), glycosides (including cyanogenic substances and saponins), flavonoids, and even raphides (needlelike calcium oxalate crystals) is immediately explicable (Dethier, 1954; Fraenkel, 1956, 1959; Lipke and Fraenkel, 1956; Thorsteinson, 1960; Gordon, 1961).

Angiosperms have, through occasional mutations and recombination, produced a series of chemical compounds not directly related to their basic metabolic pathways but not inimical to normal growth and development. Some of these compounds, by chance, serve to reduce or destroy the palatability of the plant in which they are produced (Fraenkel, 1959). Such a plant, protected from the attacks of phytophagous animals, would in a sense have entered a new adaptive zone. Evolutionary radiation of the plants might follow, and eventually what began as a chance mutation or recombination might characterize an entire family or group of related families. Phytophagous insects, however, can evolve in response to physiological obstacles, as shown by man's recent experience with commercial insecticides. Indeed, response to secondary plant substances and extreme nutritional imbalances and the evolution of resistance to insecticides seem to be intimately connected (Gordon, 1961). If a recombinant or mutation appeared in a population of insects that enabled individuals to feed on some previously protected plant group, selection could carry the line into a new adaptive zone. Here it would be free to diversify largely in the absence of competition from other phytophagous animals. Thus the diversity of plants not only may tend to augment the diversity of phytophagous animals (Hutchinson, 1959), the converse may also be true.

Changes in food plant choice would be especially favored in situations where the supply of the "preferred" plant is sufficiently limited to be an important factor in the survival of the larvae. Such situations have been described by Dethier (1959), who showed that the density of *Aster umbellatus* (Compositae) plants was critical to the success of the larvae of *Chlosyne harrisii*. Similarly, in western Colorado, the density of the small plants of *Lomatium eastwoodiae* (Umbelliferae) is an important factor limiting population size in *Papilio indra* (T. and J. Emmel, pers. comm.). In these, and many similar situations, it is logical to assume that genetic variants able to utilize another food plant successfully would be relatively favored. This advantage would be much enhanced if genotypes arose that permitted switching to a new food plant sufficiently novel biochemically that it was not utilized, or little utilized, by herbivores in general.

The degree of physiological specialization acquired in genetic adjustment to feeding on a biochemically unusual group of plants would very likely also act to limit the choice of food available to the insect group in the general flora (Merz, 1959, p. 187; Gordon, 1961). As stressed by Brower (1958a), moreover, close relationships between insects and a narrow range of food plants may be promoted by the evolution of concealment from predators in relation to a single background. The food plant provides the substrate for the larvae, not just their food (Dethier, 1954, p. 38).

After the restriction of certain groups of insects to a narrow range of food plants, the formerly repellent substances of these plants might, for the insects in question, become chemical attractants. Particularly interesting is the work of Thorsteinson (1953), who found that certain mustard oil glucosides from Cruciferae would elicit feeding responses from larvae that fed on these plants if these glucosides were smeared on other, normally unacceptable, leaves. But if these glucosides were smeared on the alkaloid-rich leaves of *Lycopersicum* (Solanaceae), the larvae still refused them. Similarly, Sevastopulo (pers. comm.) was unable to induce the larvae of *Danais chrysippus* to eat anything but Asclepiadaceae even by smearing the leaves of other plants with the juice of *Calotropis* (Asclepiadaceae).

This illustrates clearly that the choice of

a particular food plant or of a spectrum of food plants may be governed by repellents present in other plants (Thorsteinson, 1960) as well as by attractants in the normal food plants; this fully accords with the model outlined above. It should not, however, be assumed without experimental verification that a particular secondary plant substance is an attractant or feeding stimulant for the insects feeding on plants that contain it. Indeed, for the beetle *Leptinotarsa,* the alkaloids of the Solanaceae on which it feeds serve as repellents (summary in Fraenkel, 1959, p. 1467–1468).

In view of these considerations, we propose a comparable pattern of adaptive radiation for each of the more or less strictly limited groups of butterflies enumerated above. It is likewise probable that the elaboration of biochemical defenses has played a critical role in the radiation of those groups of plants characterized by unusual accessory metabolic products.

Further, it can be pointed out that all groups of butterflies that are important in furnishing models in situations involving mimicry are narrowly restricted in food plant choice: Papilioninae–Troidini; Ithomiinae; Nymphalinae–Heliconiini; Acraeinae; and Danainae. This is in accordance with a long-standing supposition of naturalists and students of mimicry that the physiological shifts that enabled the butterfly groups to feed on these plants conferred a double advantage by making the butterflies in question unpalatable. These groups of butterflies have been selected for warning coloration, and once established, this conspicuousness would tend to put anything that would maintain their distastefulness at a selective premium.

Conversely, those groups of butterflies that furnish most of the Batesian mimics—Papilioninae–Papilionini and Graphiini; Satyrinae; Nymphalinae except Heliconiini; Pierinae; and Dismorphiinae—feed mostly on plant groups that are shared with other dissimilar groups of butterflies. It is somewhat surprising that Nymphalinae–Argynnini, which feed on the same plants as Heliconiini and Acraeinae, do not play any prominent role as models for mimetic forms. This may be in part because Argynnini are best developed in temperate regions, where butterflies and therefore mimetic complexes are less common. It may further be suggested that the mimicry supposed to exist between dark female forms of various species of *Speyeria* and the model *Battus philenor* may well be a case of Müllerian rather than Batesian mimicry. Indeed, the results of Brower (1958) with the model *Danais plexippus* and its mimic *Limenitis archippus* suggest that there is in fact no sharp line between Batesian and Müllerian mimicry. These should be thought of as the extremes of a continuum. Thus *Speyeria* females may be somewhat distasteful but can only acquire warning coloration when the selective balance is tipped by the presence of other distasteful forms. This follows logically from numerous experiments and observations on the behavior of predators with mimetic complexes, which rarely reveal an "either/or" type of response (cf. Swynnerton, 1919). Assuming a balance of this sort would resolve some of the difficulties of interpretation concerning the two kinds of mimicry (e.g., Sheppard, 1963, p. 145).

Numerous unusual feeding patterns scattered among butterfly families attest to the frequency of radiation into new groups of food plants. We can, however, only guess at the probability of future radiation in the new adaptive zone. For example, does *Stalachtis susanae* (Riodininae), which is known to feed on *Oxypetalum campestre* (Asclepiadaceae) in Argentina, represent the start of a new phylogenetic series of butterflies restricted to this group of plants? Probably not, since the examples of this sort of unusual feeding habit today far exceed the number of radiations observed in the past. Nevertheless, the close patterns of coadaptation we have discussed above must have started in a similar fashion. Comparable patterns can be found among plants with biochemical innovations, for example, *Senecio viscosus*, mentioned earlier, or the relationship of *Sedum acre* to other species of its genus (Merz, 1959, p. 160).

In viewing present-day patterns of food plant utilization, however, the historical aspect of the situation must not be neglected. A biochemical innovation might have had a considerable selective advantage for a group of plants in the Cretaceous. Such an advantage would, of course, have been in terms of the phytophagous animals and parasites present in the Cretaceous, and not necessarily those of the present day. The crossing of an adaptive threshold by a member of a living group of phytophagous animals would have an entirely different significance now than that which it would have had in the Cretaceous.

For example, even though a species of *Stalachtis* is able to feed on Asclepiadaceae, it shares the available supply of these plants with representatives of numerous other groups of phytophagous animals. These have, somewhere in the course of geological time, acquired the ability to feed on asclepiads. Further, if the phytophagous organisms switching early to milkweeds became protected from predators by their ingestion of distasteful plant juices, this initial advantage might have been overcome in the intervening years by corresponding changes in prospective predators. A species of bird long selected to like milkweed bugs might find milkweed-feeding *Stalachtis* a gourmet's delight.

As in the occupation of any adaptive zone, the first organisms to enter it have a tremendous advantage and are apt to have the opportunity to become exceedingly diverse before evolution in other organisms sharply restricts their initial advantage. In short, the nature of any adaptive zone is altered by the organisms that enter it. From our vantage point in time we view only the remnants, doubtless often disarranged if not completely shattered by subsequent events, of the great adaptive radiations of the past.

In view of these considerations, we cannot accept the theoretical picture of a generalized group of polyphagous insects from which specialized oligophagous forms were gradually derived. Just as there is no truly "panphagous" insect (cf. Fraenkel, 1959), so there is no universally acceptable food plant; and this doubtless has always been true. This statement is based on the chemical variation observed in plants and the physiological variation observed in insects. Leguminosae are important food plants for several groups of Lycaenidae and Pieridae, and woody Ranales are well represented among the food plants of Papilionidae and Nymphalidae; but this should not be taken to prove that these groups of plants are "inert" chemically or readily available to other phytophagous groups of insects. The initial radiation of butterfly taxa onto these groups may for a time have produced a pattern just as spectacular as, for example, the close association between Troidini and Aristolochiaceae seen today. We hold that plants and phytophagous insects have evolved in part in response to one another, and that the stages we have postulated have developed in a stepwise manner.

As suggested by Fraenkel (1956, 1959), secondary plant substances must have been formed early in the history of angiosperms. At the present day, many classes of organic compounds are nearly or quite restricted to this group of plants (for example, see Alston and Turner, 1963, p. 164; Harborne, 1963, p. 360; Paris, 1963, p. 357). We suggest that some of these compounds may have been present in early angiosperms and afforded them an unusual degree of protection from the phytophagous organisms of the time, relative to other contemporary plant groups. Behind such a biochemical shield the angiosperms may have developed and become structurally diverse. Such an assumption of the origin of angiosperms provides a cogent reason why one of many structurally modified groups of gymnosperms would have been able to give rise to the bewildering diversity of modern angiosperms, while most other lines became extinct. It seems at least as convincing to us as do theories based on the structural peculiarities of angiosperms. Although the chemical basis for the success of early angiosperms may no longer be discernible, it can be mentioned that woody Ranales, generally accepted as the most "primitive" assemblage of living angiosperms on other grounds, are as a group characterized by many alkaloids as well as by essential oils. Of course this might also be interpreted only to mean that the development of alkaloids has permitted this group to persist despite its many generalized features.

In turn, the fantastic diversification of modern insects has developed in large measure as the result of a stepwise pattern of coevolutionary stages superimposed on the changing pattern of angiosperm variation. With specific reference to the butterflies, one is tempted in terms of present-day patterns to place more emphasis on the full exploitation of diurnal feeding habits by the adults than on the penetration of any particular biochemical barrier by the larvae. On the other hand, phenetic relationships suggest that Papilionoidea (with Hesperioidea, the skippers) are representatives of a line that is amply distinct from all other living Lepidoptera (Ehrlich, 1958). Thus it is entirely possible that radiation onto a new food plant was decisive at the time Papilionoidea first diverged, even though the feeding habits of the order as a whole are now much wider than those of the butterflies alone. The impossibility of deciding objectively which groups of Papilionoidea are more primitive than others (cf. Ehrlich, 1958, p. 334–335) relegates the task of identifying the original group of food plants for butterflies to the realm of profitless speculation.

We would like to return now to the four general questions posed at the beginning of this paper. First, what have we learned of the reciprocal responses of butterflies and their food plants? The observed patterns clearly point to the critical importance of plant biochemistry in governing the relationships between the two groups. The degree of plasticity of chemoreceptive response and the potential for physiological adjustment to various plant secondary substances in butterfly populations must in large measure determine their potential for evolutionary radiation. Of secondary, but still possibly major importance, are mechanical plant defenses, and the butterflies' responses to them.

With respect to the second question on the generation of predictions the answer also seems clear. We cannot predict the results of any given interaction with precision—*Stalachtis* on Asclepiadaceae or *Neophasia* on pines may or may not form the basis for further patterns of radiation. On the other hand, the basis for a probabilistic statement of "Further radiation unlikely" seems to have been developed. A great many minor predictions can be made, such as the probable presence of alkaloids in *Ptaeroxylon* (Meliaceae), the solanaceous character of the food plants of unknown larvae of Ithomiinae, and so forth.

Although the data we have gathered permit us to make some reasonable sequence predictions about phylogenetic patterns (e.g., diversification of Apocynaceae and Solanaceae before Danainae and Ithomiinae, respectively), these predictions cannot be tested and the relationships cannot be specified further in the absence of a fossil record. The reconstruction of phylogenies on the basis of this sort of information would seem an unwarranted imposition on the data, since evolutionary rate and time are still inseparable.

In response to the fourth question, it seems to us that studies of coevolution provide an excellent starting point for understanding community evolution. Indeed the seeming ease with which our conclusions have been extended to include the complex interactions among plants, phytophagous organisms, mimics, models, and predators leads us to believe that population biologists should pursue similar studies of other systems. Many examples come to mind such as parasitoid–caterpillar, *Plasmodium*–

hemoglobin, tree–mycorrhizal fungus, in which stepwise reciprocal selective response is to be expected. Studying most of these systems experimentally tends to be difficult, and may be complicated by lack of repeatability in the results.

An approach to biology that is concerned with broad patterns quite possibly will lead to a better understanding of some other problems of community ecology. For example, biologists have long been interested in the reasons for the differences in species diversity between tropical and temperate areas. An important factor in maintaining these differences may be the sort of synergistic interactions between plants and herbivores we have been discussing. The selective advantage of living in a tropical climate is evident for insects, which are poikilothermal. Insects are much more abundant in the tropics than elsewhere and doubtless constitute the major class of herbivorous animals. The penetration of relatively cold environments or other environments requiring diapause is probably a rather recent occurrence in most insect groups. That these environments are not always readily entered is attested to by the repeated failure of insects such as the butterfly *Mestra amymone* to survive the winter in localities at the northern fringes of their ranges where summer colonies have been established (Ehrlich and Ehrlich, 1961).

The abundance of phytophagous insects in tropical regions would be expected to accentuate the pace of evolutionary interactions with plants. These interactions may have been the major factor in promoting the species diversity of both plants and animals observed in the tropics today. As this diversity was being produced, it became arrayed in richly varied mixtures of species with relatively great distances between individuals of any one plant species. As Grant (1963, p. 420–422) has suggested, this arrangement would have the additional advantage of providing a maximum degree of protection from epidemic outbreaks of plant diseases and plant pests. It must, however, also be mentioned that the relatively permissive tropical climate presumably allows a greater diversity of plant life forms and therefore secondarily of animals (Hutchinson, 1959, p. 150).

Probably our most important overall conclusion is that the importance of reciprocal selective responses between ecologically closely linked organisms has been vastly underrated in considerations of the origins of organic diversity. Indeed, the plant–herbivore "interface" may be the major zone of interaction responsible for generating terrestrial organic diversity.

Summary

The reciprocal evolutionary relationships of butterflies and their food plants have been examined on the basis of an extensive survey of patterns of plant utilization and information on factors affecting food plant choice. The evolution of secondary plant substances and the stepwise evolutionary responses to these by phytophagous organisms have clearly been the dominant factors in the evolution of butterflies and other phytophagous groups. Furthermore, these secondary plant substances have probably been critical in the evolution of angiosperm subgroups and perhaps of the angiosperms themselves. The examination of broad patterns of coevolution permits several levels of predictions and shows promise as a route to the understanding of community evolution. Little information useful for the reconstruction of phylogenies is supplied. It is apparent that reciprocal selective responses have been greatly underrated as a factor in the origination of organic diversity. The paramount importance of plant–herbivore interactions in generating terrestrial diversity is suggested. For instance, viewed in this framework the rich diversity of tropical communities may be traced in large part to the hospitality of warm climates toward poikilothermal phytophagous insects.

Literature Cited

Alston, R. E., and B. L. Turner. 1963. Biochemical systematics. Prentice-Hall, Englewood Cliffs, N. J.

BARRETT, C., AND A. N. BURNS. 1951. Butterflies of Australia and New Guinea. N. H. Seward, Melbourne.

BEWS, J. W. 1927. Studies in the ecological evolution of the angiosperms. New Phytol., **26**: 1–21, 65–84, 129–148, 209–248, 273–294.

BOURGOGNE, J. 1951. Ordre des Lépidoptères. Traité Zool., **10**: 174–448.

BROWER, JANE VAN ZANDT. 1958. Experimental studies of mimicry in some North American butterflies. Part I. The monarch, *Danaus plexippus*, and viceroy, *Limenitis archippus archippus*. EVOLUTION, **12**: 32–47.

BROWER, L. P. 1958a. Bird predation and food plant specificity in closely related procryptic insects. Amer. Nat., **92**: 183–187.

——. 1958b. Larval food plant specificity in butterflies of the *Papilio glaucus* group. Lepidop. News, **12**: 103–114.

CHERMOCK, R. L. 1950. A generic revision of the Limenitini of the world. Am. Midl. Nat., **43**: 513–569.

CLENCH, H. K. 1955. Revised classification of the butterfly family Lycaenidae and its allies. Ann. Carnegie Mus., **33**: 261–274.

CORBET, A. S., AND H. M. PENDLEBURY. 1956. The butterflies of the Malay Peninsula. Ed. 2. Oliver and Boyd, London.

COSTA LIMA, A. M. DA. 1936. Terceiro Catalogo dos Insectos que vivem nas plantas do Brasil. Directoria de Estatistica da Producção Secção de Publicade, Rio de Janeiro, p. 201–231.

DETHIER, V. G. 1941. Chemical factors determining the choice of food plants by *Papilio* larvae. Amer. Nat., **75**: 61–73.

——. 1954. Evolution of feeding preferences in phytophagous insects. EVOLUTION, **8**: 33–54.

——. 1959. Food-plant distribution and density and larval dispersal as factors affecting insect populations. Canad. Entom., **91**: 581–596.

DOCTERS VAN LEEUWEN, W. M. 1958. Zoocecidia. *In* Verdoorn, F., ed., Manual of pteridology. Martinus Nijhoff, The Hague, p. 192–195.

DOWNEY, J. C. 1962. Host-plant relations as data for butterfly classification. Syst. Zool., **11**: 150–159.

—— AND W. C. FULLER. 1961. Variation in *Plebejus icarioides* (Lycaenidae). I. Food plant specificity. J. Lepidop. Soc., **15**: 34–42.

EHRLICH, P. R. 1958. The comparative morphology, phylogeny and higher classification of the butterflies (Lepidoptera: Papilionoidea). Univ. Kansas Sci. Bull., **39**: 305–370.

—— AND A. H. EHRLICH. 1961. How to know the butterflies. Wm. C. Brown, Dubuque.

ERDTMAN, H. 1963. Some aspects of chemotaxonomy. *In* Swain, T., ed., Chemical plant taxonomy. Academic Press, London, p. 89–125.

FLÜCK, H. 1963. Intrinsic and extrinsic factors affecting the production of secondary plant products. *In* Swain, T., ed., Chemical plant taxonomy. Academic Press, London, p. 167–186.

FORBES, W. T. M. 1958. Caterpillars as botanists. Proc. Tenth Int. Congr. Ent., **1**: 313–317.

FRAENKEL, G. 1956. Insects and plant biochemistry. The specificity of food plants for insects. Proc. 14th Int. Congr. Zool., p. 383–387.

——. 1959. The raison d'etre of secondary plant substances. Science, **129**: 1466–1470.

GIBBS, R. D. 1963. History of chemical taxonomy. *In* Swain, T., ed., Chemical plant taxonomy. Academic Press, London, p. 41–88.

GORDON, H. T. 1961. Nutritional factors in insect resistance to chemicals. Ann. Rev. Entom., **6**: 27–54.

GRANT, V. 1963. The origin of adaptations. Columbia University Press, New York and London.

HARBORNE, J. B. 1963. Distribution of anthocyanins in higher plants. *In* Swain, T., ed., Chemical plant taxonomy. Academic Press, London, p. 359–388.

HEGNAUER, R. 1963. The taxonomic significance of alkaloids. *In* Swain, T., ed., Chemical plant taxonomy. Academic Press, London, p. 389–427.

HUTCHINSON, G. E. 1959. Homage to Santa Rosalia *or* why are there so many kinds of animals. Amer. Nat., **93**: 145–159.

HUTCHINSON, J. 1959. The families of flowering plants. Ed. 2. 2 vols. Clarendon Press, Oxford.

JÖRGENSEN, P. 1932. Lepidopterologisches aus Südamerika. Deutsch. Ent. Zeitschr. Iris, Dresden, **46**: 37–66.

KLOTS, A. B. 1933. A generic revision of the Pieridae (Lepidoptera). Entom. Amer. n.s., **12**: 139–242.

LEE, C. L. 1958. Butterflies. Academia Sinica (in Chinese).

LIPKE, H., AND G. FRAENKEL. 1956. Insect nutrition. Ann. Rev. Ent., **1**: 17–44.

MERZ, E. 1959. Pflanzen und Raupen. Über einige Prinzipien der Futterwahl bei Grossschmetterlingsraupen. Biol. Zentr., **78**: 152–188.

MICHENER, C. D. 1942. A generic revision of the Heliconiinae (Lepidoptera, Nymphalidae). Am. Mus. Novitates, **1197**: 1–8.

MUNROE, E. 1960. The generic classification of the Papilionidae. Canad. Ent., Suppl., **17**: 1–51.

—— AND P. R. EHRLICH. 1960. Harmonization of concepts of higher classification of the Papilionidae. J. Lepid. Soc., **14**: 169–175.

PARIS, R. 1963. The distribution of plant glycosides. *In* Swain, T., ed., Chemical plant taxonomy. Academic Press, London, p. 337–358.

PINHEY, E. C. G. 1949. Butterflies of Rhodesia. Rhodesia Sci. Association, Salisbury.

PRICE, J. R. 1963. The distribution of alkaloids in the Rutaceae. *In* Swain, T., ed., Chemical plant taxonomy. Academic Press, London, p. 429–452.

REMINGTON, C. L. 1952. The biology of nearctic Lepidoptera I. Food plants and life-histories of Colorado Papilionoidea. Psyche, **59**: 61–70.

REUTER, E. 1896. Über die Palpen der Rhopaloceren. Acta Soc. Sci. Fennica, **22**: i–xvi, 1–577.

RODRÍGUEZ, R. L. 1957. Systematic anatomical studies on *Myrrhidendron* and other woody Umbellales. Univ. Calif. Publ. Bot., **29**: 145–318, pls. 36–47.

SAFWAT, FUAD M. 1962. The floral morphology of *Secamone* and the evolution of the pollinating apparatus in Asclepiadaceae. Ann. Missouri Bot. Gard., **49**: 95–129.

SAVILLE, D. B. O. 1954. The fungi as aids in the taxonomy of flowering plants. Science, **120**: 583–585.

SEITZ, A. 1906–1927. The macrolepidoptera of the world. Vols. 1, 5, 9, 13. Fritz Lehman Verlag and Alfred Kerner Verlag, Stuttgart.

SHEPPARD, P. M. 1963. Some genetic studies of Müllerian mimics in butterflies of the genus *Heliconius*. Zoologica, **48**: 145–154, pls. 1–2.

SHIRÔZU, T. 1962. Evolution of the food-habits of larvae of the thecline butterflies. Tyô to Ga (Trans. Lepidop. Soc. Japan), **12**: 144–162.

—— AND H. YAMAMOTO. 1956. A generic revision and the phylogeny of the tribe Theclini (Lepidoptera: Lycaenidae). Sieboldia, **1**: 329–421.

SHORLAND, F. B. 1963. The distribution of fatty acids in plant lipids. *In* Swain, T., ed., Chemical plant taxonomy. Academic Press, London, p. 253–303.

SØRENSEN, N. A. 1963. Chemical taxonomy of acetylinic compounds. *In* Swain, T., ed., Chemical plant taxonomy. Academic Press, London, p. 219–252.

SWYNNERTON, C. M. F. 1919. Experiments and observations bearing on the explanation of form and colouring, 1908–1913, Africa. J. Linn. Soc. Zool., **33**: 203–385.

THORNE, R. F. 1963. Some problems and guiding principles of angiosperm phylogeny. Amer. Nat., **97**: 287–306.

THORSTEINSON, A. J. 1953. The chemotactic responses that determine host specificity in an oligophagous insect (*Plutella maculipennis* (Curt.) Lepidoptera). Canad. J. Zool., **31**: 52–72.

——. 1960. Host selection in phytophagous insects. Ann. Rev. Ent., **5**: 193–218.

VAN SON, G. 1949. The butterflies of southern Africa. Part I, Papilionidae and Pieridae. Part II (1955), Danainae and Satyrinae.

VAZQUEZ G., LEONILA, AND PEREZ R., H. 1961. Observaciones sobre la biología de *Baronia brevicornis* Salv. (Lepidoptera: Papilionidae-Baroniinae). An. Inst. Biol. Mex., **22**: 295–311.

VERSCHAEFFELT, E. 1910. The cause determining the selection of food in some herbivorous insects. Proc. Acad. Sci., Amsterdam, **13**: 536–542.

WILTSHIRE, E. P. 1957. The Lepidoptera of Iraq. Rev. ed. Nicholas Kaye, London.

WYNTER-BLYTH, M. A. 1957. Butterflies of the Indian region. Bombay Nat. Hist. Soc., Bombay.

Lewontin, R. C., and J. L. Hubby. 1966. A molecular approach to the study of genic heterozygosity in natural populations: II. amount of variation and degree of heterozygosity in natural populations of *Drosophila pseudoobscura*. *Genetics* 54:595–609.

In 1966, protein-electrophoretic techniques were introduced into population biology by several research teams. The article reprinted here is one of these original studies, the scientific significance of which can hardly be overstated. Prior to the mid-1960s, evolutionary biologists could do little more than guess—based on phenotypic observations—what might be the magnitudes of genetic variation in natural populations. Thus, ever since Darwin and Mendel, a fundamental but mostly unanswered empirical challenge had always been to quantify population genetic variability (the underlying basis of evolutionary change). The extent of uncertainty about genetic variation was evidenced by two opposing schools of thought that had emerged by the mid-twentieth century. Proponents of the "balance school" argued that many, perhaps most, loci were polymorphic (contained multiple alleles) and that most individuals were heterozygous at a substantial fraction of gene loci. By contrast, advocates of the "classical school" maintained that genetic variability was low, such that conspecific individuals must be homozygous for a selectively favored "wild-type" allele at nearly every genetic locus. Central to the classical school was the concept of genetic load—the idea that genetic variation produces a heavy burden of diminished fitness. In short, the balance and classical schools viewed genetic variability as a blessing or as a curse, respectively.

The paper by Richard Lewontin and John Hubby had a profound impact in several regards: it introduced a user-friendly molecular method to a field (population biology) that was unaccustomed to having access to abundant population genetic data, it overthrew extreme versions of classical thought by showing that natural populations house extensive genetic variation at the molecular level, and it set the research agenda in population genetics for at least the next quarter century (Lewontin 1974). The compelling question shifted from "How much genetic variation?" to "What is the adaptive significance of genetic variation?" Although the toolkits of molecular biology have expanded dramatically since the age of protein electrophoresis, the question of the adaptive significance of genetic variation continues today to be a prime motivator of research effort in the fields of molecular ecology and evolution.

RELATED READING

Lewontin, R. C. 1974. The genetic basis of evolutionary change. Columbia University Press, New York, New York, USA.

A MOLECULAR APPROACH TO THE STUDY OF GENIC HETEROZYGOSITY IN NATURAL POPULATIONS. II. AMOUNT OF VARIATION AND DEGREE OF HETEROZYGOSITY IN NATURAL POPULATIONS OF *DROSOPHILA PSEUDOOBSCURA*[1]

R. C. LEWONTIN AND J. L. HUBBY

Department of Zoology, University of Chicago, Chicago, Illinois

Received March 30, 1966

As pointed out in the first paper of this series (HUBBY and LEWONTIN 1966), no one knows at the present time the kinds and frequencies of variant alleles present in natural populations of any organism, with the exception of certain special classes of genes. For human populations we know a good deal about certain polymorphisms for blood cell antigens, serum proteins, and metabolic disorders of various kinds but we can hardly regard these, *a priori*, as typical of the genome as a whole. Clearly we need a method that will randomly sample the genome and detect a major proportion of the individual allelic substitutions that are segregating in a population. In our previous paper, we discussed a method for accomplishing this end by means of a study of electrophoretic variants at a large number of loci and we showed that the variation picked up by this method behaves in a simple Mendelian fashion so that phenotypes can be equated to homozygous and heterozygous genotypes at single loci.

It is the purpose of this second paper to show the results of an application of the method to a series of samples chosen from natural populations of *Drosophila pseudoobscura*. In particular, we will show that there is a considerable amount of genic variation segregating in all of the populations studied and that the real variation in these populations must be greater than we are able to demonstrate. This study does not make clear what balance of forces is responsible for the genetic variation observed, but it does make clear the kind and amount of variation at the genic level that we need to explain.

An exactly similar method has recently been applied by HARRIS (1966) for the enzymes of human blood. In a preliminary report on ten randomly chosen enzymes, HARRIS describes two as definitely polymorphic genetically and a third as phenotypically polymorphic but with insufficient genetic data so far. Clearly these methods are applicable to any organism of macroscopic dimensions.

The Populations Studied

We have chosen populations of *D. pseudoobscura* for a number of reasons. This species is not commensal with man, as is *D. melanogaster*, and so can be

[1] The work reported here was supported in part by grants from the National Science Foundation (GB 3112 and GB 3213) and the Public Health Service (GM 11216).

Genetics 54: 595–609 August 1966.

said to be truly "wild." It has a wide distribution in Western North and Central America from British Columbia to Guatemala with a recently discovered outlier as far south as Bogotá, Colombia. *D. pseudoobscura* is genetically well known, at least to the extent of having marker genes and inversions on all of its four major chromosomes, and there exists a vast literature on the population genetics of the inversion systems on chromosome 3 of this species by DOBZHANSKY and his school. No species of Drosophila is really well understood in its ecological aspects, but for *D. pseudoobscura* 30 years of study of natural populations has led to a fair knowledge of population size fluctuation, kind of vegetation with which the species is associated, diurnal activity and temperature tolerance. Numerous samples from wild populations exist in the laboratory, and new samples are constantly becoming available. All of these reasons suggested to us that *D. pseudoobscura* would be a good species for our first survey of natural genic variation. It seemed to us that the variation found within and between populations of this species ought to be typical of a common, relatively widespread, sexually reproducing organism.

The populations in this study are represented by a number of separate lines each stemming from a single fertilized female caught in nature. For example, nine separate single-female lines maintained separately in the laboratory since 1957 represent the population from Flagstaff, Arizona. Because we were unable to get fresh samples (except for one case) we preferred these separate lines to any mixed population. Such separate lines may each suffer homozygosis because of inbreeding, but the differences *between* lines will preserve some portion of the original population variance. If the lines had been pooled and kept since 1957 as a mixture, more of the variability originally introduced would have been lost. As our results will show, most, but not all, lines are in fact homozygous but differences between lines have been preserved. Nevertheless, the loss of variation because of inbreeding needs to be kept in mind when we analyze the results.

The population samples in the study were as follows: (1) Flagstaff, Arizona. Nine lines collected in a ponderosa pine forest above 5,000 feet elevation in 1957. The natural population is virtually pure for the Arrowhead gene arrangement on the third chromosome and all lines are Arrowhead homozygotes (see DOBZHANSKY and EPLING 1944). (2) Mather, California. Seven lines collected between 1957 and 1960 in a Transition Zone forest at 4,600 feet elevation. This population is highly polymorphic for third chromosome inversions in nature. All strains used were homozygous Arrowhead (see DOBZHANSKY, 1948). (3) Wildrose, California. Ten strains collected in 1957 in the Panamint Range at 8,000 feet elevation in a piñon Juniper forest. The population is highly polymorphic for inversions, but the strains tested were all homozygous Arrowhead (see DOBZHANSKY and EPLING 1944). (4) Cimarron, Colorado. Six lines collected in a *Quercus gambelii* grove at about 7,000 feet elevation in 1960. All lines are homozygous Arrowhead. (5) Strawberry Canyon (Berkeley), California. Ten strains from a much larger collection made in 1965 at an elevation of 800 feet. This population is highly polymorphic for third chromosome inversions, and the strains used were also polymorphic, being the F_2 and F_3 from the wild females. (6) A single strain from Bogotá, Colombia. A much larger sample is planned for this extreme outlier of the species range, but the single strain collected in 1960 was included since it was available. The population occurs between 8,000 and 10,000 feet elevation and has two inversions, Santa Cruz and Treeline in proportions 65:35 (see DOBZHANSKY *et al.* 1963).

The natural and laboratory history of these various strains is thus rather different. Two, Cimarron and Flagstaff, are from the eastern part of the species distribution where chromosomal (inversion) variability is low. All but Strawberry Canyon have been in the laboratory for 5 to 8 years as separate strains, while Strawberry Canyon is a fresh sample from nature, and is polymorphic for inversions. One strain, Bogotá, represents a geographically remote population that surely represents the extreme southern part of the species distribution. All in all, the sample was chosen to give a diversity of histories so that the results could be given some generality.

The laboratory maintenance of all strains was the same. They were kept at 18°C in half-pint culture bottles with an average of about 50 parents each generation, but with considerable variation in size. At times in their culture, most, if not all, suffered one or more extreme breeding size bottlenecks. Thus, there has been inbreeding to an unknown extent. At the culture temperature of 18°C, there is little or no difference in selective values among third chromosome inversion types, although nothing can be said in this respect about other segregating gene systems.

RESULTS

The methods of electrophoretic separation and detection of enzyme systems are fully explained by HUBBY and LEWONTIN (1966) and we will take it as demonstrated in that paper that the phenotypes we see are reflective of simple allelic substitutions at single genetic loci. Therefore, in what follows in this paper, we will refer to "alleles" and "loci" without again referring to the phenotypic appearance of the electrophoretic gels.

In every case, five or more individuals were tested from each strain. A strain is classified as homozygous for an allele if all individuals tested were homozygous, while the strain is classified as segregating for two alleles if any of the individuals was heterozygous or if homozygotes of two different kinds were found. The notation *.95/1.07*, for example, means that the allele *.95* and the allele *1.07* for a gene were found segregating among the tested individuals of the strain. Throughout we use the relative electrophoretic mobilities as names of alternate alleles (see HUBBY and LEWONTIN 1966).

The observations are summarized in Table 1. The body of the table shows the number of strains (not individuals) either homozygous or segregating for various alleles at various loci. Of the ten enzyme systems discussed in HUBBY and LEWONTIN (1966), two (*ap-1* and *ap-2*) are not included here because they appeared on the gels infrequently and are not sufficiently reliable to be used in a population study. For the same reason, only ten of the 13 larval proteins are included in the present study. The decision whether to include a band in the study was made solely on the basis of reliability, and independently of whether it showed electrophoretic variants.

The entry in Table 1 for Leucine aminopeptidase (*lap*) is different in meaning from the others. The relative mobilities of the variant forms are so close for this locus that it is not possible to make the proper cross assignments between populations. There are at least four alleles at the locus, but we do not at present know unambiguously which are present in which populations. Therefore, in Table 1 we have simply indicated how many alleles are present among the strains of that population.

Table 1 shows some remarkable results. First, of the 18 loci represented, there is some genetic variation in some population for nine of them. Second, genetic variation is found in more than one population for seven of the loci: malic dehydrogenase (*mdh*), esterase-5 (*e-5*), leucine aminopeptidase (*lap*), alkaline phosphatase-7 (*ap-7*), *pt-7*, *pt-8* and *pt-10*. This variation in more than one popu-

TABLE 1

Number of strains from each population either homozygous or segregating for various alleles at different loci

Locus	Allele	Strawberry Canyon	Wildrose	Cimarron	Mather	Flagstaff	Bogotá
esterase-5	*.85*	0	0	0	1	0	0
	.95	0	1	0	1	1	0
	1.00	0	3	3	0	4	1
	1.03	0	1	0	2	0	0
	1.07	0	0	2	1	4	0
	1.12	0	1	0	2	0	0
	.95/1.00	1	0	0	0	0	0
	.95/1.07	1	0	0	0	0	0
	.95/1.12	0	0	1	0	0	0
	1.00/1.07	4	1	0	0	0	0
	1.00/1.12	3	1	0	0	0	0
	1.03/1.07	1	1	0	0	0	0
	1.03/1.12	0	1	0	0	0	0
	1.07/1.12	1	0	0	0	0	0
malic dehydrogenase	*.90*	0	0	0	1	0	0
	1.00	6	10	6	4	8	1
	1.11	2	0	0	0	0	0
	1.22	0	0	0	0	1	0
	.90/1.00	0	0	0	2	0	0
	1.00/1.11	2	0	0	0	0	0
glucose-6-phospate dehydrogenase	*1.00*	9	10	4	6	9	1
alkaline phosphatase-4	*.93*	0	0	0	0	1	.
	1.00	9	11	6	7	8	.
alkaline phosphatase-6	+	9	10	5	7	9	.
	−/+	0	0	1*	0	0	.
alkaline phosphatase-7	+	9	9	5	7	9	.
	−/+	0	1	1*	0	0	.
α-glycerophosphate dehydrogenase	*1.00*	10	10	6	6	8	1
leucine aminopepidase	*.95*, *.97*, *1.00*, *1.02*	2† alleles	3‡ alleles	2 alleles	2§ alleles	3 alleles	1 allele
pt-4	*.45*	10	10	6	6	8	1
pt-5	*.55*	1	4	4	6	2	1
pt-6	*.62*	10	10	6	6	8	1
pt-7	*.73*	0	0	0	0	1	0
	.75	9	10	5	5	6	1
	.77	0	0	0	0	0	0
	.73/.75	0	0	0	0	1	0
	.75/.77	1	0	1	1	0	0
pt-8	*.80*	0	0	0	0	0	1
	.81	2	2	3	2	1	0
	.83	1	4	1	1	5	0
	.81/83	7	4	2	3	2	0
pt-9	*.90*	3	8	4	1	0	0

TABLE 1—Continued

Number of strains from each population either homozygous or segregating for various alleles at different loci

Locus	Allele	Strawberry Canyon	Wildrose	Cimarron	Mather	Flagstaff	Bogotá
pt-10	*1.02*	0	0	0	0	0	0
	1.04	4	9	6	4	8	0
	1.06	0	0	0	0	0	1
	1.02/1.04	0	1	0	0	0	0
	1.04/1.06	6	0	0	2	0	0
pt-11	*1.12*	4	10	6	6	8	.
pt-12	*1.18*	5	10	6	6	8	1
pt-13	*1.30*	7	10	6	6	8	1

* Both loci segregating in the same strain.
† Three strains segregating.
‡ One strain segregating.
§ Two strains segregating.

lation must be characterized as polymorphism in the usual sense because variant alleles occur with some appreciable frequency in more than an isolated case.

Third, and most remarkable of all, is the widespread occurrence of segregation in strains that have been in the laboratory for as many as seven years. As might be expected, the Strawberry Canyon strains are segregating at those loci that are polymorphic. In fact, not a single strain of Strawberry Canyon is homozygous for an allele of *e-5*. But four strains of Wildrose are also segregating for alleles at this locus, as is one strain of Cimarron. Most striking of all is the case of the *.81/.83* polymorphism at the *pt-8* locus where there are segregating strains in every population (not including the single strain from Bogotá). Despite the segregation at many of these loci, Table 1 definitely gives the impression of an effect of inbreeding over the many generations during which the strains have been maintained in the laboratory. The Strawberry Canyon strains segregate far more frequently than any of the others, and, in general, more of the genetic variation in the other populations is between homozygous strains.

Fourth, the genotype of the single strain from Bogotá is sometimes unusual. In most cases, the Bogotá strain is homozygous for the allele most commonly found in other localities. This is not the case for *pt-8*, however, where Bogotá is homozygous for an allele not found elsewhere, and *pt-10* where Bogotá is homozygous for one of the less common alleles.

In order to make the pattern of genic variation simpler to perceive, Table 2 has been constructed from the data in Table 1. In Table 2 *very approximate* gene frequencies are calculated for the alleles shown in Table 1 by using the following convention. Each of the original strains carried four independent doses of each gene when it was brought into culture. A large proportion of the strains still have more than one of these original doses since so many strains are still polymorphic and therefore carry at least two of the original four alleles. How many of the original alleles are still represented in any strain can only be guessed at, however. We make an arbitrary convention that each line shall be counted

TABLE 2

Approximate gene frequencies calculated from the data of Table 1

Locus	Allele	Strawberry Canyon	Wildrose	Cimarron	Mather	Flagstaff	Bogotá
esterase-5	*.85*	0	0	0	.14	0	0
	.95	.09	.10	.08	.14	.11	0
	1.00	.36	.40	.50	0	.44	x
	1.03	.05	.20	0	.29	0	0
	1.07	.32	.10	.33	.14	.44	0
	1.12	.18	.20	.08	.29	0	0
malic dehydrogenase	*.90*	0	0	0	.29	0	0
	1.00	.70	1.00	1.00	.71	.89	x
	1.11	.30	0	0	0	0	0
	1.22	0	0	0	0	.11	0
glucose-6-phosphate dehydrogenase	*1.00*	1.00	1.00	1.00	1.00	1.00	x
alkaline phosphatase-4	*.93*	0	0	0	0	.11	.
	1.00	1.00	1.00	1.00	1.00	.88	.
alkaline phosphatase-6	+	1.00	1.00	.92	1.00	1.00	.
	—	0	0	.08	0	0	.
alkaline phosphatase-7	+	1.00	.95	.92	1.00	1.00	.
	—	0	.05	.08	0	0	.
α-glycerophosphate dehydrogenase	*1.00*	1.00	1.00	1.00	1.00	1.00	x
leucine aminopeptidase	*.95* *.97* *1.00* *1.02*	2 alleles	3 alleles	2 alleles	2 alleles	3 alleles	1 allele
pt-4	*.45*	1.00	1.00	1.00	1.00	1.00	x
pt-5	*.55*	1.00	1.00	1.00	1.00	1.00	x
pt-6	*.62*	1.00	1.00	1.00	1.00	1.00	x
pt-7	*.73*	0	0	0	0	.19	0
	.75	.95	1.00	.92	.92	.81	x
	.77	.05	0	.08	.08	0	0
pt-8	*.80*	0	0	0	0	0	x
	.81	.55	.40	.67	.58	.25	0
	.83	.45	.60	.33	.42	.75	0
pt-9	*.90*	1.00	1.00	1.00	1.00	1.00	x
	1.02	0	.05	0	0	0	0
pt-10	*1.04*	.70	.95	1.00	.83	1.00	0
	1.06	.30	0	0	.17	0	x
pt-11	*1.12*	1.00	1.00	1.00	1.00	1.00	x
pt-12	*1.18*	1.00	1.00	1.00	1.00	1.00	x
pt-13	*1.30*	1.00	1.00	1.00	1.00	1.00	x

One strain=2 alleles. No gene frequency estimate can be made for Bogotá, so the allele present is marked with an x.

equally and, since many of the strains are segregating, each allele in such lines is given a weight of one half. So, for example, in Strawberry Canyon, for locus *pt-8*, there are two strains homozygous for allele *.81*, seven strains segregating *.81/.83*, and one strain homozygous *.83*. Then the gene frequency of allele *.81* is $q_{.81} = (2 + 7/2)/(2 + 7 + 1) = .55$. Such a method can give only a very crude estimate of the frequency of alleles in the original sample brought into the laboratory, except for Strawberry Canyon where the sample was examined in the F_2 and F_3 generations from the wild. Since these original samples were themselves small, we cannot take our gene frequency estimation in Table 2 too seriously. They are meant only to give a qualitative picture of the variation, yet they show certain patterns and on the basis of these crude estimates we can characterize the variation at each locus as falling into certain broad categories.

1. *Monomorphism.* In a sufficiently large population, no locus can be completely without variant alleles. However, we class as monomorphic those loci that are without variation in our sample and those with only a single variant allele in a single strain. It might be argued that the presence of even a single variant allele in such a small sample as ours is evidence that in the population this variant is at reasonably high frequency. Nevertheless, we prefer to err on the side of conservatism and class such isolated variants as newly arisen mutations that have not yet been eliminated from the population by natural selection or genetic drift. Using the criterion that a variant must be present in more than one strain in more than one population in order for a population to be considered polymorphic, we find 11 out of 18 loci monomorphic. Of these, nine are completely without variation in our sample: glucose-6-phosphate dehydrogenase, α-glycerol phosphate dehydrogenase, *pt-4*, *pt-5*, *pt-6*, *pt-9*, *pt-11*, *pt-12*, and *pt-13*. The other two, alkaline phosphatase-4 and alkaline phosphatase-6 each have a single variant allele in a single strain. In the case of alkaline phosphatase-4, the strain is homozygous for the variant allele so it is likely that it has been in the strain for some time, probably from the original sample from the wild. Nevertheless, we do not count this locus as polymorphic.

2. *Widespread polymorphism with one allele in high frequency.* In this class there are three loci in our sample: *ap-7* which has the same variant allele in two different geographical regions but in low frequency, *pt-7* which is similar, but has the polymorphism more widespread and which also has a second variant allele restricted to one population, and *pt-10* which is like *pt-7* except that the rarer allele is found fixed in the Bogotá strain. These three loci are clearly polymorphic, but one allele in each case is found in high frequency in every population and so may be considered the "typical" allele. For *pt-10* the "type" concept is shaky since in Strawberry Canyon the atypical allele is in a frequency of 30% and the allele is fixed in the single Bogotá strain.

3. *Ubiquitous polymorphism with no wild type.* This class includes three loci. The most extreme case is the esterase-5 gene which has six alleles so far recovered. Populations are segregating for between three and five of these and no one allele is most common. Allele *1.00* comes close to being most common, but it is completely lacking in the Mather sample. Only one allele, *.85*, is restricted to a single population, all others being found in a minimum of three populations. *pt-8* has about a 50:50 polymorphism of alleles *.81* and *.83* in all the populations and this is related to the fact that all populations had some strains still segregating for these two alleles. In addition, *pt-8* has a unique allele in Bogotá. *Leucine amino-*

peptidase appears to fall in this group, although there is some suspicion, not yet confirmed, that allele *1.00* is most common in all populations.

4. *Local indigenous polymorphism*. Only one locus is completely of this sort, malic dehydrogenase. Three of the five populations have a local variant in high frequency, but it is a different variant in each case. Allele *1.00* would appear to be a "type" allele or at least a most common form. In addition to *mdh*, we have already noted an occasional local variant, such as the allele *.80* of *pt-8* in the Bogotá strain, the allele *.73* of *pt-7* found only in Flakstaff, and the allele *.85* of esterase-5 known only from Mather. In these last two cases, it is impossible to distinguish them from the single homozygous variant of alkaline phosphatase-4 which we have classed as nonpolymorphic.

5. *Local pure races*. A class of variation that is completely lacking in our sample of loci is the local pure race. In no case do we find some populations homozygous for one allele and other populations homozygous for a different one. We expect such a pattern if the alleles were functionally equivalent isoalleles not under any natural selection pressure. The failure to find such cases is important to our hypotheses about the forces responsible for the observed variation.

To sum up these classes, out of 18 loci included in the population study, seven are clearly polymorphic in more than one population and two are represented by rare local variants in a single population which, to be conservative, are not considered polymorphic. Thus, conservatively 39% of loci are polymorphic. This takes account of all populations and does not give an estimate of the polymorphism in any given population, which will be less. Table 3 is a summary of the information for each population separately. The populations are very similar to each other in their degree of polymorphism with an average of 30% of the loci varying in each. It is interesting that Strawberry Canyon, a fresh sample from the wild, is not different from the others. We can assume that most of the variation from nature has been preserved in the laboratory stocks but has been converted to variation between strains by the inbreeding attendant on laboratory culture. Another point of interest is that the great similarity in *proportion* of loci polymorphic in each population is not entirely a result of identity of poly-

TABLE 3

Proportion of loci, out of 18, polymorphic and proportion of the genome estimated to be heterozygous in an average individual for each population studied

Population	No. of loci polymorphic	Proportion of loci polymorphic	Proportion of genome heterozygous per individual	Maximum proportion of genome heterozygous
Strawberry Canyon	6	.33	.148	.173
Wildrose	5	.28	.106	.156
Cimarron	5	.28	.099	.153
Mather	6	.33	.143	.173
Flagstaff	5	.28	.081	.120
Average		.30	.115	.155

morphisms. Thus, although Wildrose and Flagstaff are both polymorphic at five out of 18 loci, only three of these are common to both populations. Flagstaff is polymorphic at two loci, *mdh* and *pt-7*, for which Wildrose is monomorphic, but Wildrose is polymorphic for *ap-7* and *pt-10*, while Cimarron is monomorphic at these loci.

Yet another question that can be asked from the data is, "At what proportion of his loci will the average individual in a population be heterozygous?" In fact, this can be described without exaggeration as the central problem of experimental population genetics at the present time. A complete discussion of the conflicting results on this question is not possible here, but the issue is very clearly drawn by WALLACE (1958). The results reported by WALLACE in that paper, in previous papers (WALLACE 1956) and in subsequent works by WALLACE (1963), WALLACE and DOBZHANSKY (1962), DOBZHANSKY, KRIMBAS and KRIMBAS (1960), and many others, all point, although indirectly, toward a high level of heterozygosity in natural populations. On the other hand, theoretical considerations by KIMURA and CROW (1964) and experiments of HIRAIZUMI and CROW (1960), GREENBERG and CROW (1960), MULLER and FALK (1961) and FALK (1961) among others, point in the opposite direction. These latter authors interpret their results as showing that the proportion of loci heterozygous in a typical individual from a population will be quite small and that polymorphic loci will represent a small minority of all genes.

Our data enable us to estimate the proportion of heterozygosity per individual directly. This is estimated in the next to the last column of Table 3 for each population separately. This estimate is made by taking the gene frequencies of all the alleles at a locus in a population, calculating the expected frequencies of heterozygotes from the Hardy-Weinberg proportions, and then averaging over all loci for each population separately. For example, at the *e-5* locus in Flagstaff there are three alleles at frequency .44, .44, and .11, respectively. The expected frequency of heterozygotes at this locus in Flagstaff is then given by:

$$\textit{Proportion heterozygotes} = 2(.11)(.44) + 2(.11)(.44) + 2(.44)(.44) = .581.$$

This value is then averaged with similarly derived values from each of the other loci for Flagstaff, including the monomorphic ones which contribute no heterozygosity. Obviously, for a given number of alleles the proportion of heterozygosity is maximized when all are in equal frequency. In such a case

$$\textit{maximum proportion heterozygosity} = (n-1)/n$$

where n is the number of alleles present. This value is given for comparison in the last column of Table 3.

As Table 3 shows, between 8% and 15% of the loci in an average individual from one of these populations will be in a heterozygous state and this is not very different from the maximum heterozygosity expected from the number of alleles actually segregating in the population. It is interesting that the two populations with the lowest amount of chromosomal polymorphism, Flagstaff and Cimarron (DOBZHANSKY and EPLING 1944) also have a slightly lower genic heterozygosity than the chromosomally highly polymorphic populations of Mather, Strawberry Canyon, and Wildrose. More extensive data on chromosomally polymorphic and monomorphic populations are being taken now.

DISCUSSION

Biases: Before we attempt to explain the amount of polymorphism shown in Table 3, we need to ask what the biases in our experiment are. There are four sources of bias in our estimates and they are all in the same direction.

1. The method of electrophoretic separation detects only some of the differences between proteins. Many amino acid substitutions may occur in a protein without making a detectable difference in the net charge. We do not know what proportion of substitutions we are detecting but it is probably on the order of one half. Depending upon the protein, different results have been observed. For tryptophan synthetase about 7/9 of all mutations tested are electrophoretically detectable (HENNING and YANOFSKY 1963), but none of the forms of cytochrome-c are electrophoretically separable despite extensive amino-acid substitution over the plant and animal kingdoms (MARGOLIASH, personal communication). Presumably in the latter case, net charge is critical to proper function. At any rate, our estimate of the number of variant alleles is clearly on the low side.

2. Our lines have preserved only a portion of the variation originally present in them when they were taken from nature. Because of the inbreeding effect of maintaining small populations with occasional bottle necks in breeding size, some of the alleles originally present must have been lost. This causes our estimate of variation to be on the low side.

3. The original lines were only a small sample of the natural populations. We have tested very few lines, as few as six in the case of Cimarron, so that we are only sampling a portion of the natural variation. Alleles at frequencies of say 5% or 10% may easily be lacking in such samples. Again our experiment underestimates the variation within each population.

4. We have deliberately excluded as polymorphic two loci in which only a single variant allele was found. This coupled with the fact that only five individuals were surveyed in each strain will leave out of account real polymorphisms at low frequencies. Had we included the two rare variants in Table 3, both Cimarron and Flagstaff would have had 33% of loci polymorphic which would change the overall average to 32%. The proportion of loci heterozygous per individual in these populations would be increased from .09 and .081 to .107 and .092, respectively, bringing the average over all populations to 12%, a very small change.

All these sources of bias cause us to underestimate the proportion of loci polymorphic and the proportion of heterozygous loci per individual, but by how much we cannot say. At present we are studying a large sample of over 100 F_1 lines from females caught in Strawberry Canyon over the course of a year. This study will eliminate biases 2 and 3 above and give us an appropriate correction for our present estimates.

One other possible source of bias is in the choice of enzyme assays. If there were some subtle reason that the enzymes we have chosen to use tended to be more or less genetically variable than loci in general, our results would not be referable to the genome as a whole. Our chief protection against this sort of bias is in the use of the larval proteins in addition to the specific enzyme assays. Both of these classes of genes give about the same degree of polymorphism: three out of ten polymorphic loci for larval proteins and four out of eight for the enzymes. While it might be argued that the very existence of a published method for the detection of an enzyme on a gel is a bias in favor of variable enzymes, no such argument can be made for the larval proteins, all of which are developed on the same gel by a general protein stain. Moreover, two of the enzymes, malic dehydrogenase and α-glycerophosphate dehydrogenase, were developed in this laboratory simply because suitable coupling methods are known for dehydrogenases.

In order to avoid the bias that might arise from considering only a particular enzyme function, we have deliberately not assayed a large number of proteins associated with similar functions. For example, there are ten different sites of esterase activity, presumably representing ten different genes, but we have only assayed the one with the greatest activity. To load our sample with more esterases might introduce a bias if there were some reason why esterase loci were more or less polymorphic than other genes.

The source of the variation: It is not possible in this paper to examine in detail all of the alternative explanations possible for the large amount of genic variation we have observed in natural populations. Our observations do require explanation and we already have some evidence from the observations themselves.

Genetic variation is destroyed by two forces: genetic drift in populations under going periodic size reductions and selection against recessive or partly dominant deleterious genes. Genetic variation is increased or maintained by three factors: mutation, migration between populations with different gene frequencies, and balancing selection usually of the form of selection in favor of heterozygotes. On the basis of combination of these factors, we can distinguish three main possibilities to explain the variation we have seen.

(1) The alleles we have detected have no relevance to natural selection but are adaptively equivalent isoalleles. In such a case, genetic drift will drive populations to homozygosity, but will be resisted by recurrent mutation and migration. We have some idea of the effective breeding size, N, in populations of *D. pseudoobscura* from the experiments of DOBZHANSKY and WRIGHT (1941, 1943) and WRIGHT, DOBZHANSKY, and HOVANITZ (1942). Various estimates agree that "panmictic unit" has an effective size, N, of between 500 and 1,000 in the Mount San Jacinto populations where the species is most dense and successful. At Wildrose the population size is between one-fifth and one-tenth of that at Mount San Jacinto and, although there is no published evidence, the same is true at Cimarron where flies are rare even in summer. For the dense populations the conclusion of DOBZHANSKY and WRIGHT (1943) is that "the effective size of the panmictic unit in *D. pseudoobscura* turns out to be so large that but little permanent differentiation can be expected in a continuous population of this species owing to

genetic drift alone." For Cimarron and Wildrose, however, this is not true, yet we find these populations with the same average heterozygosity as other populations. The lack of any loci showing pure local races in nature is against the selective equivalence of isoalleles. It can be argued, however, that genetic drift in the marginal populations is producing local pure races but that migration from the other populations and mutation (of unknown magnitude for these alleles) is preventing differentiation. As a matter of fact, very little migration, of the order of one individual per generation, will effectively prevent homozygosis by drift. We must also take account of the observation that many lines in the laboratory are still segregating for several loci and that effective population size of these lines has been very small and migration (contamination) close to nil. The continued segregation of alleles in the laboratory might be caused by mutation rates much higher for isoalleles than for dysgenic alleles, and we are checking the mutation rate for a few alleles. All in all, however, complete selective neutrality is not a satisfactory explanation of all the observations.

(2) Selection tends to eliminate alternative alleles but mutation restores them. This hypothesis comes close to the neutral isoallele theory because our observed gene frequencies of alternate alleles would require that mutation rates and selection coefficients be of the same order of magnitude. That is, the equilibrium gene frequency for an allele selected against with intensity t in heterozygotes (we can ignore the rarer homozygotes) is approximately equal to u/t, where u is the mutation rate. Since our rarer alleles at each locus vary in frequency from 5% to 45%, u and t must be of about the same order of magnitude. This in turn suggests extraordinarily high mutation rates or very, very weak selection *on the average*. But an average selection coefficient of .001 implies that in some populations at some times the gene in question is selected for rather than against so that local pure race formation should be promoted. Again we must check to see that mutation rates are not higher than 10^{-3}.

(3) Selection is in favor of heterozygotes. This hypothesis satisfies all the objections to (1) and (2) above, since heterosis, if strong enough, can maintain genic variation in any size population, irrespective of mutation and migration. However, two different problems are raised by the assumption of nearly universal heterosis. First, unless we assume that the two homozygotes are very weakly selected against, in which case we are back effectively to alternatives (1) and (2), the total amount of differential selection in a population with many heterotic loci is tremendous. For example, suppose two alleles are maintained by selecting against both homozygotes to the extent of 10% each. Since half of all individuals are homozygotes at such a locus, there is a loss of 5% of the population's reproductive potential because of the locus alone. If our estimate is correct that one third of all loci are polymorphic, then something like 2,000 loci are being maintained polymorphic by heterosis. If the selection at each locus were reducing population fitness to 95% of maximum, the population's reproductive potential would be only $(.95)^{2000}$ of its maximum or about 10^{-46}. If each homozygote were 98% as fit as the heterozygote, the population's reproductive potential would be cut to 10^{-9}. In either case, the value is unbelievably low. While we cannot assign

an exact maximum reproductive value to the most fit multiple heterozygous genotype, it seems quite impossible that only one billionth of the reproductive capacity of a Drosophila population is being realized. No Drosophila female could conceivably lay two billion eggs in her lifetime.

There is a strong possibility that the intensity of heterosis decreases as the number of loci heterozygous increases (VANN 1966). This does not really solve the problem, however, since drift will fix loci until the heterosis per locus still segregating is high enough to resist random fixation.

We then have a dilemma. If we postulate weak selective forces, we cannot explain the observed variation in natural populations unless we invoke much larger mutation and migration rates than are now considered reasonable. If we postulate strong selection, we must assume an intolerable load of differential selection in the population.

Some most interesting numerical calculations have been made by KIMURA and CROW (1964) relating the mutation rate, population size, heterozygosity, and genetic load of isoallelic systems. Their conclusions on the theoretical implications of widespread heterosis are similar to ours. One possible resolution of this dilemma is to suppose that in any given environment, only a portion, say 10% or less, of the polymorphisms are actually under selection so that most polymorphisms are relics of previous selection. If this is coupled with a small amount of migration between populations sufficient to retard genetic drift between periods of selection, we might explain very large amounts of variation without intolerable genetic loads. Such a process needs to be explored theoretically, while tests for heterosis need to be made under controlled conditions in the laboratory for a variety of loci and environments. Such tests are now under way. One such test by MACINTYRE and WRIGHT (1966) on esterase alleles in *D. melanogaster* was ambiguous in its result, but pointed in the direction of selective neutrality for the alleles tested.

Second, if we are to postulate heterosis on such a wide scale, we must be able to explain the adaptive superiority of heterozygotes for so many different functions. Heterozygotes differ from homozygotes in an important respect: they have present in the same organism both forms of the protein, and, in some cases they also have a third form, the hybrid protein. Only some of our enzyme proteins and none of our larval proteins show hybrid enzyme formation, so that hybrid enzyme *per se* cannot lie at the basis of general heterosis. But variation in physicochemical characteristics of the same functional protein might very well enhance the flexibility of an organism living in a variable environment. One of the best evidences that such heteromorphy of protein structure is adaptive in evolution is the occurrence of polymeric proteins made up of very similar but not identical subunits. Obviously the genes responsible for the α and β subunits of hemoglobin or the subunits of lactic dehydrogenase tetramers must have arisen by a process of gene duplication since the polypeptides they produce are so similar in amino acid sequence. The advantage of duplicate genes with slight differentiation over a single gene with different alleles is that in the former case every individual in the population can have the advantage of polymorphism. Gene duplication provides the opportunity for fixed "heterozygosity" at the functional level while allelic variation always suffers from segregation of less fit homozygotes. Heterozygosis, then, is a suboptimal solution to the problem that duplicate genes solve optimally. An excellent presentation of this argument may be found in the last chapter of FINCHAM (1966).

We are greatly indebted to DR. SUMIKO NARISE and MR. ALAN NOVETSKY for their contribution to the survey of the strains. DR. CHRISTOPHER WILLS and MR. ALAN WICK have provided us most generously with flies from Strawberry Canyon. A number of illuminating comments and criticisms of the ideas were provided by BRUCE WALLACE, ROSS MACINTYRE, JAMES CROW, and HERMAN LEWIS, to all of whom we are most grateful.

SUMMARY

Using genetic differences in electrophoretic mobility, demonstrated by HUBBY and LEWONTIN (1966) to be single Mendelian alternatives, we have surveyed the allelic variation in samples from five natural populations of *D. pseudoobscura*. Out of 18 loci randomly chosen, seven are shown to be clearly polymorphic in more than one population and two loci were found to have a rare local variant segregating. Thus, 39% of loci in the genome are polymorphic over the whole species. The average population is polymorphic for 30% of all loci. The estimates of gene frequency at these loci enable us to estimate the proportion of all loci in an individual's genome that will be in heterozygous state. This value is between 8% and 15% for different populations, with an average of 12%. A suggestion of a relationship has been observed between the extent of this heterogeneity and the amount of inversion polymorphism in a population.—An examination of the various biases in the experiment shows that they all conspire to make our estimate of polymorphism and heterozygosity lower than the true value. There is no simple explanation for the maintenance of such large amounts of genic heterozygosity.

LITERATURE CITED

DOBZHANSKY, TH., 1948 Genetics of natural populations. XVI. Altitudinal and seasonal changes produced by natural selection in certain populations of *D. pseudoobscura* and *D. persimilis*. Genetics **33**: 158–176.

DOBZHANSKY, TH., and C. EPLING, 1944 Contributions to the genetics, taxonomy and ecology of *Drosophila pseudoobscura* and its relatives. Carnegie Inst. Wash. Publ. **554**:

DOBZHANSKY, TH., A. S. HUNTER, O. PAVLOSKY,, B. SPASSKY, and BRUCE WALLACE, 1963 Genetics of natural populations. XXXI. Genetics of an isolated marginal population of *Drosophila pseudoobscura*. Genetics **48**: 91–103.

DOBZHANSKY, TH., C. KRIMBAS, and M. G. KRIMBAS, 1960 Genetics of natural populations. XXX. Is the genetic load in *Drosophila pseudoobscura* a mutational or balanced load? Genetics **45**: 741–753.

DOBZHANSKY, TH., and S. WRIGHT, 1941 Genetics of natural populations. V. Relations between mutation rate and accumulation of lethals in populations of *Drosophila pseudoobscura*. Genetics **26**: 23–51. —— 1943 Genetics of natural populations. X. Dispersion rates in *Drosophila pseudoobscura*. Genetics **28**: 304–340.

FALK, R., 1961 Are induced mutations in Drosophila overdominant? II. Experimental results. Genetics **46**: 737–757.

FINCHAM, J. R. S., 1966 *Genetic Complementation.* Benjamin, New York.

GREENBERG, R., and J. F. CROW, 1960 A comparison of the effect of lethal and detrimental chromosomes from Drosophila populations. Genetics **45**: 1153–1168.

HARRIS, H., 1966 Enzyme polymorphisms in man. Proc. Roy. Soc. Lond. B **164**: 298–310.

HENNING, V., and C. YANOFSKY, 1963 An electrophoretic study of mutationally altered A proteins of the tryptophan synthetase of *Escherichia coli.* J. Mol. Biol. **6**: 16–21.

HIRAIZUMI, Y., and J. F. CROW, 1960 Heterozygous effects on viability, fertility, rate of development, and longevity of Drosophila chromosomes that are lethal when homozygous. Genetics **45**: 1071–1083.

HUBBY, J. L., and R. C. LEWONTIN, 1966 A molecular approach to the study of genic heterozygosity in natural populations. I. The number of alleles at different loci in *Drosophila pseudoobscura.* Genetics **54**: 577–594.

KIMURA, M., and J. F. CROW, 1964 The number of alleles that can be maintained in a finite population. Genetics **49**: 725–738.

MACINTYRE, ROSS, and T. R. F. WRIGHT, 1966 Responses of esterase 6 alleles of *Drosophila melanogaster* and *D. simulans* to selection in experimental populations. Genetics **53**: 371–387.

MULLER, H. J., and R. FALK, 1961 Are induced mutations in Drosophila overdominant? I. Experimental design. Genetics **46**: 727–735.

VANN, E., 1966 The fate of X-ray induced chromosomal rearrangements introduced into laboratory populations of *D. melanogaster.* Am. Naturalist (in press)

WALLACE, B., 1956 Studies on irradiated populations of *D. melanogaster.* J. Genet. **54**: 280–293. —— 1958 The average effect of radiation induced mutations on viability in *D. melanogaster.* Evolution **12**: 532–552. —— 1963 Further data on the overdominance of induced mutations. Genetics **48**: 633–651.

WALLACE, B., and TH. DOBZHANSKY, 1962 Experimental proof of balanced genetic loads in Drosophila. Genetics **47**: 1027–1042.

WRIGHT, S., TH. DOBZHANSKY, and W. HOVANITZ, 1942 Genetics of natural populations. VII. The allelism of lethals in the third chromosome of *Drosophila pseudoobscura.* Genetics **27**: 363–394.

Fitch, W. M., and E. Margoliash. 1967. Construction of phylogenetic trees. *Science* 155:279–284.

Biologists have used several evocative metaphors to describe the genealogical history of life on Earth: rungs on life's ladder, hereditary threads in the fabric of existence, streams or watersheds of heredity, and—most commonly—phylogenetic trees. In the first 100 years after Darwin, scientists devised phylogenetic trees for various taxa by comparing visible organismal phenotypes—e.g., morphological, physiological, or behavioral characteristics—that they could only presume reflected underlying genetic relationships. Then something remarkable happened—molecular technologies were introduced that gave relatively quick and direct access to the voluminous genetic information registered in DNA and protein sequences. Ever since then, biologists have tapped the sap of heredity flowing through evolutionary trees to reconstruct phylogeny.

The initial transition to molecular information in phylogenetic biology took place in the early 1960s, and this article by Walter Fitch and Emanuel Margoliash was a model of clarity that helped pave the way for this revolutionary adjustment. Although phenotypes continue to be an extremely important source of information in the field of systematics, molecular data are of special significance because they are: (1) unambiguously genetic; (2) ubiquitously distributed across all forms of life; (3) copious in information content; and (4) arguably less prone than many adaptive phenotypes to homoplasy (evolutionary convergences or reversals) that otherwise complicate phylogeny estimation. Molecular phylogenies can serve as historical backdrop for interpreting the evolution of alternative phenotypes, thus enriching scientific understanding of the latter as well.

Although the general tree-building philosophy introduced by Fitch and Margoliash was illustrated using molecular information from only one small gene, the broader evolutionary implications were enormous and immediately apparent. Today, phylogeneticists have access to vast molecular data and refined tree-building algorithms beyond what was true in the 1960s, but their efforts trace their roots to the foundation laid in the paper by Fitch and Margoliash.

RELATED READING

Avise, J. C. 2006. Evolutionary pathways in nature. Cambridge University Press, Cambridge, England.

Ayala, F. J., editor. 1976. Molecular evolution. Sinauer, Sunderland, Massachusetts, USA.

Felsenstein, J. 2004. Inferring phylogenies. Sinauer, Sunderland, Massachusetts, USA.

Hedges, S. B., and S. Kumar, editors. 2009. The timetree of life. Oxford University Press, Oxford, England.

Pietsch, T. W. 2012. Trees of life: a visual history of evolution. Johns Hopkins University Press, Baltimore, Maryland, USA.

Zuckerkandl, E., and L. Pauling. 1965. Evolutionary divergence and convergence in proteins." Pages 97–166, *in* Evolving genes and proteins, V. Bryson and H. J. Vogel, editors. Academic Press, New York, New York, USA.

Construction of Phylogenetic Trees

A method based on mutation distances as estimated from cytochrome *c* sequences is of general applicability.

Walter M. Fitch and Emanuel Margoliash

Biochemists have attempted to use quantitative estimates of variance between substances obtained from different species to construct phylogenetic trees. Examples of this approach include studies of the degree of interspecific hybridization of DNA (*1*), the degree of cross reactivity of antisera to purified proteins (*2*), the number of differences in the peptides from enzymic digests of purified homologous proteins, both as estimated by paper electrophoresis-chromatography or column chromatography and as estimated from the amino acid compositions of the proteins (*3*), and the number of amino acid replacements between homologous proteins whose complete primary structures had been determined (*4*). These methods have not been completely satisfactory because (i) the portion of the genome examined was often very restricted, (ii) the variable measured did not reflect with sufficient accuracy the mutation distance between the genes examined, and (iii) no adequate mathematical treatment for data from large numbers of species was available. In this paper we suggest several improvements under categories (ii) and (iii) and, using cytochrome *c*, for which much precise information on amino acid sequences is available, construct a tree which, despite our examining but a single gene, is remarkably like the classical phylogenetic tree that has been obtained from purely biological data (*5*). We also show that the analytical method employed has general applicability, as exemplified by the derivation of appropriate relationships among ethnic groups from data on their physical characteristics (*6, 7*).

Dr. Fitch is an assistant professor of physiological chemistry at the University of Wisconsin Medical School in Madison, Dr. Margoliash is head of the Protein Section in the Department of Molecular Biology, Abbott Laboratories, North Chicago, Illinois.

Determining the Mutation Distance

The *mutation distance* between two cytochromes is defined here as the minimal number of nucleotides that would need to be altered in order for the gene for one cytochrome to code for the other. This distance is determined by a computer making a pairwise comparison of homologous amino acids (*8*). For each pair a *mutation value* is taken from Table 1 which gives the minimum number of nucleotide changes required to convert the coding from one amino acid to the other. The table is derived from Fig. 2 of Fitch (*9*) except that, as a result of the work of Weigert and Garen (*10*) and Brenner, Stretton, and Kaplan (*11*), the uridyl-adenosylpurine trinucleotide is now treated as a chain-terminating codon. This change of codon meaning, although it does not affect the method of calculation, does cause the mutation values for amino acid pairs involving glutamine with cysteine, phenylalanine, tyrosine, serine, and tryptophan to become 1 greater than in the table previously published (*12*). Also, misprints involving the leucine-glycine and valine-cysteine pairs have been corrected. To maintain homology, deletions, all of which occur near the ends of the chains, are represented by X's. The amino- and carboxyl-terminal sequences in which deletions occur are shown in Table 2. Thus all cytochromes are regarded as being 110 amino acids long. If the homologous pairing includes an X, no mutation value is assigned.

For each possible pairing of cytochromes, the 110 mutation values found are summed to obtain the minimal mutation distance. For purposes of calculation, these mutation distances are proportionally adjusted to compensate for variable numbers of pairs of residue positions in which at least one member contains an X. For example, the number of X-containing amino acid pairs occurring between the *Saccharomyces* and *Candida* cytochromes *c* is 1, whereas that between two mammalian cytochromes *c* is 6. Thus the known mutation distance of the former pairing is multiplied by 110/109 whereas that of the latter is multiplied by 110/104. The results for 20 known cytochromes *c*, rounded off to the nearest whole number, are shown in the lower left half of Table 3.

The basic approach to the construction of the tree is illustrated in Fig. 1, which shows three hypothetical proteins, A, B, and C, and their mutation distances. There are two fundamental problems: (i) Which pair does one join together first? (ii) What are the lengths of legs *a*, *b*, and *c*?

As a first approximation, one solves problem (i) simply by choosing the pair with the smallest mutation distance, which in this case is A and B, with a distance of 24. Hence A and B are shown connected at the lower apex in Fig. 1. To solve the second problem, one notes that the distance from A to C, 28, is 4 less than the distance from B to C. Hence there must have been at least 4 more countable mutations in the descent of B from the lower apex than in the descent of A. Thus if $a + b = 24$ and $b - a = 4$, then $a = 10$, $b = 14$, and therefore $c = 18$. Note that an exact solution is obtained from which a reconstruction of the mutation distances precisely matches the input data.

Table 1. Mutation values for amino acid pairs. Each value is the minimum number of nucleotides that would need to be changed in order to convert a codon for one amino acid into a codon for another. The table is symmetrical about the diagonal of zeros. Letters across the top represent the amino acids in the same order as in the first column and conform to the single-letter code of Keil, Prusik, and Sŏrm (*21*).

	A	C	E	F	G	H	I	L	M	N	O	P	Q	R	S	T	U	V	W	Y
Aspartic acid	0	2	2	2	1	1	2	1	3	1	1	2	2	2	2	3	2	1	2	1
Cysteine	2	0	2	1	3	2	3	2	3	2	1	2	3	1	1	1	2	2	2	1
Threonine	2	2	0	2	2	2	1	1	1	1	2	1	2	1	1	2	2	2	1	2
Phenylalanine	2	1	2	0	3	2	3	2	2	2	1	2	3	2	1	2	1	1	1	2
Glutamic acid	1	3	2	3	0	2	1	1	2	2	2	2	1	2	2	2	2	1	3	1
Histidine	1	2	2	2	2	0	2	2	3	1	1	1	1	1	2	3	1	2	2	2
Lysine	2	3	1	3	1	2	0	2	1	1	2	2	1	1	2	2	2	2	2	2
Alanine	1	2	1	2	1	2	2	0	2	2	2	1	2	2	1	2	2	1	2	1
Methionine	3	3	1	2	2	3	1	2	0	2	3	2	2	1	2	2	1	1	1	2
Asparagine	1	2	1	2	2	1	1	2	2	0	1	2	2	2	1	3	2	2	1	2
Tyrosine	1	1	2	1	2	1	2	2	3	1	0	2	2	2	1	2	2	2	2	2
Proline	2	2	1	2	2	1	2	1	2	2	2	0	1	1	1	2	1	2	2	2
Glutamine	2	3	2	3	1	1	1	2	2	2	2	1	0	1	2	2	1	2	3	2
Arginine	2	1	1	2	2	1	1	2	1	2	2	1	1	0	1	1	1	2	2	1
Serine	2	1	1	1	2	2	2	1	2	1	1	1	2	1	0	1	1	2	1	1
Tryptophan	3	1	2	2	2	3	2	2	2	3	2	2	2	1	1	0	1	2	3	1
Leucine	2	2	2	1	2	1	2	2	1	2	2	1	1	1	1	1	0	1	1	2
Valine	1	2	2	1	1	2	2	1	1	2	2	2	2	2	2	2	1	0	1	1
Isoleucine	2	2	1	1	3	2	2	2	1	1	2	2	3	2	1	3	1	1	0	2
Glycine	1	1	2	2	1	2	2	1	2	2	2	2	2	1	1	1	2	1	2	0

Table 2. Areas of cytochromes *c* involving deletions. The first seven and the last four amino acids of the cytochromes *c* for the 20 species studied are shown. Deletions are represented by X's. Sequences are reported in the single-letter code of Keil *et al.* (*21*), a key to which is provided in Table 1.

Amino terminal positions 1–7	Organism	Carboxyl terminal positions 107–110
PLPFGQY	*Candida*	LXSI
XEGFILY	*Saccharomyces*	LXCG
XXYFSLY	*Neurospora*	LELX
XXYVPLY	Moth	SEXI
XXYVPLY	Screwworm fly	LSEI
XXXXXXY	Tuna	LESX
XXXXXXY	All other vertebrates	No deletions

When information from more than three proteins is utilized, the basic procedure is the same, except that initially each protein is assigned to its own subset. One then simply joins two subsets to create a single, more comprehensive, subset. This process is repeated according to the rules set forth below until all proteins are members of a single subset. A phylogenetic tree is but a graphical representation of the order in which the subsets were joined.

In the present case, we start with 20 subsets, each subset consisting of a single cytochrome *c* amino acid sequence. To determine which two subsets should be joined, all possible pairwise combinations of subsets are in turn assigned to sets *A* and *B*, with all remaining subsets in each case assigned to set *C*. In each alternative test all proteins are thus a part of one of the three sets. The three sets are treated exactly as in the preceding example, except that now the mutation distances used are averages determined from every possible pairing of proteins, one from each of the two sets whose average mutation distance is being calculated.

One arbitrarily accepts, from among all the possible pairings examined, that assignment of protein subsets to sets *A*, *B*, and *C* which provides the lowest average mutation distance from *A* to *B*. The leg lengths are then calculated and recorded. Henceforth the proteins of *A* and *B* so joined are treated as a single subset, and the entire procedure described in the preceding paragraph is repeated. Thus the number of subsets, originally equal to the number of proteins (*N*), is reduced by 1 with each cycle. In this fashion, after *N*–1 joinings of subsets, the initial phylogenetic tree will have been produced. Because average mutation distances are now being used, the solutions obtained are very unlikely to permit an exact reconstruction of the input data.

Testing Alternative Trees

Because of the arbitrary nature of the rule by which proteins are assigned to sets *A* and *B*, the initial tree will not necessarily represent the best use of the information. To examine reasonable alternatives, one simply constructs another tree by assigning an alternative pair of protein subsets to sets *A* and *B* whenever the mutation distance between the two subsets is not greater by some arbitrary amount than that between the members of the initial pair used in constructing the initial phylogenetic tree (*13*). The tree that is less satisfactory on the basis of criteria set forth below is discarded, and other alternatives are tested.

The best of 40 phylogenetic trees so far examined is presented in Fig. 2. Each juncture is located on the ordinate at a point representing the average of all distances between the juncture and the species descendant from it. The mutation distance to any one descendant may be more or less than the ordinate value.

By summing distances over the tree, it is possible to reconstruct values (upper right half of Table 3) comparable to the original input mutation distances (lower left half of Table 3).

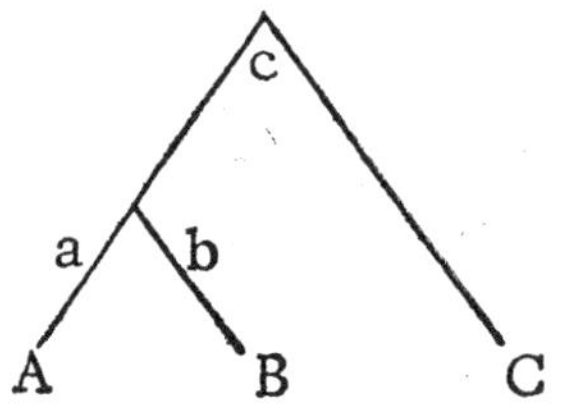

Mutation Distances

	B	C
A	24	28
B		32

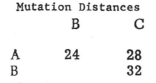

Fig. 1. Calculation of observed mutation distances. The upper apex represents a hypothetical ancestral organism that divided into two descending lines, one of which subsequently also divided. Thus we have three present-day species, A, B, and C. The number of observable mutations that have occurred in a particular gene since the A and B lines of descent diverged are represented respectively by *a* and *b*. The number of mutations that separate the lower apex and C is represented by *c*. The sums of $a + b$, $a + c$, and $b + c$, then, are the mutation distances of the three species as currently observed.

The 20 species are indicated in the last column; the identifying numbers in the first column and the top row of the table may be used as coordinates. Thus the tabulated values interrelating the human and horse cytochromes at coordinates (1,4) and (4,1) are mutation distances of 17 and 15 respectively, the former being the input datum, the latter having been obtained from the tree by reconstruction. If the absolute difference between two such mutation distances $|(i,j) - (j,i)|$ is multiplied by 100 and divided by (i,j), the result is the percentage of change from the input data. If such values are squared and the squares are summed over all values of $i < j$, the resultant sum (Σ) may be used to obtain the percent "standard deviation" (*4*) of the reconstructed values from the input mutation distances. The number of mutation distances summed is $N(N-1)/2$, or 190 for our case. If this number is reduced by 1, divided into the sum Σ, and the square root taken, the result is the percent "standard deviation." Since the standard deviation is a larger number than the standard error, the probable error, or the average deviation, the percent "standard deviation" is used here, it being less likely to create overconfidence in the significance of a result (*4*).

The Statistically Optimal Tree

In testing phylogenetic alternatives, one is seeking to minimize the percent "standard deviation." The scheme shown in Fig. 2 has a percent "standard deviation" of 8.7, the lowest of the 40 alternatives so far tested. The percent "standard deviation" for the initial tree was 12.3.

In addition to using a gene product to discover evolutionary relationships among several species, one can similarly delineate evolutionary relationships among different genes. Our procedure constructs, from the amino acid sequences of human alpha, beta, gamma, and delta hemoglobin chains and whale myoglobin (*15*), the gene phylogeny

Table 3. Minimum numbers of mutations required to interrelate pairs of cytochromes *c*. Values in the lower left half of the table are mutation distances as determined from the amino acid sequences and, prior to rounding off, were used to derive Fig. 2. Values in the upper right half of the table are reconstructed distances found by summing the leg lengths in Fig. 2. The references cited in the last column are to studies of the amino acid sequences of the cytochromes *c* of the indicated species.

Protein	1	2	3	4	5	6	7	8	9	10	11	12	13	14	15	16	17	18	19	20	
1		1	13	15	15	13	11	14	15	15	16	16	17	29	29	30	33	64	62	68	Man (*22*)
2	1		12	15	14	12	11	13	15	14	15	15	16	28	29	29	32	63	61	67	Monkey (*Macacus mulatta*) (*23*)
3	13	12		9	8	6	7	8	13	13	13	14	15	26	27	27	30	61	59	65	Dog (*24*)
4	17	16	10		1	5	10	11	15	15	16	16	17	29	29	30	33	64	62	68	Horse (*25*)
5	16	15	8	1		4	9	10	14	14	15	15	16	28	28	29	32	63	61	67	Donkey (*26*)
6	13	12	4	5	4		7	8	13	12	13	13	14	26	27	27	30	61	59	65	Pig (*27*)
7	12	11	6	11	10	6		7	11	11	12	12	13	24	25	25	29	60	57	63	Rabbit (*30*)
8	12	13	7	11	12	7	7		13	13	14	14	15	27	27	28	31	62	60	66	Kangaroo (*Canopus canguru*) (*28*)
9	17	16	12	16	15	13	10	14		3	3	3	8	26	27	27	30	61	59	65	Pekin duck (*29*)
10	16	15	12	16	15	13	8	14	3		4	4	8	26	27	27	30	61	59	65	Pigeon (*29*)
11	18	17	14	16	15	13	11	15	3	4		2	9	27	27	28	31	62	60	66	Chicken (*17*)
12	18	17	14	17	16	14	11	13	3	4	2		9	27	27	28	31	62	60	66	King penguin (*Aptenodytes patagonica*) (*29*)
13	19	18	13	16	15	13	11	14	7	8	8	8		28	29	29	32	63	61	67	Snapping turtle (*Chelydra serpentina*) (*31*)
14	20	21	30	32	31	30	25	30	24	24	28	28	30		33	34	37	68	66	72	Rattlesnake (*Crotalus adamanteus*) (*32*)
15	31	32	29	27	26	25	26	27	26	27	26	27	27	38		35	38	69	67	73	Tuna (*33*)
16	33	32	24	24	25	26	23	26	25	26	26	28	30	40	34		16	59	56	63	Screwworm fly (*Haematobia irritans*) (*29*)
17	36	35	28	33	32	31	29	31	29	30	31	30	33	41	41	16		62	60	66	Moth (*Samia cynthia*) (*34*)
18	63	62	64	64	64	64	62	66	61	59	61	62	65	61	72	58	59		56	62	*Neurospora* (*crassa*) (*35*)
19	56	57	61	60	59	59	59	58	62	62	62	61	64	61	66	63	60	57		41	*Saccharomyces* (*oviformis*) iso-1 (*36*)
20	66	65	66	68	67	67	67	68	66	66	66	65	67	69	69	65	61	61	41		*Candida* (*krusei*) (*37*)

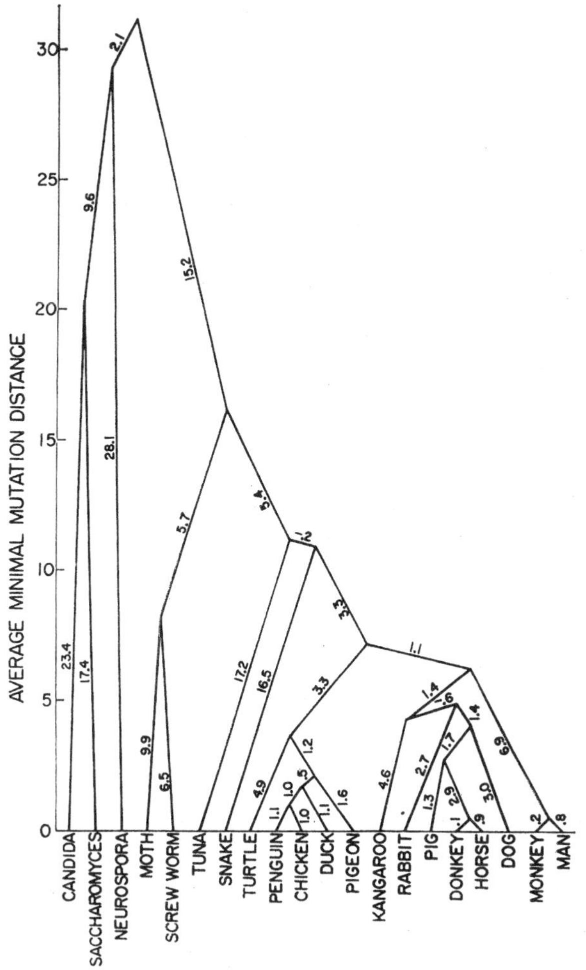

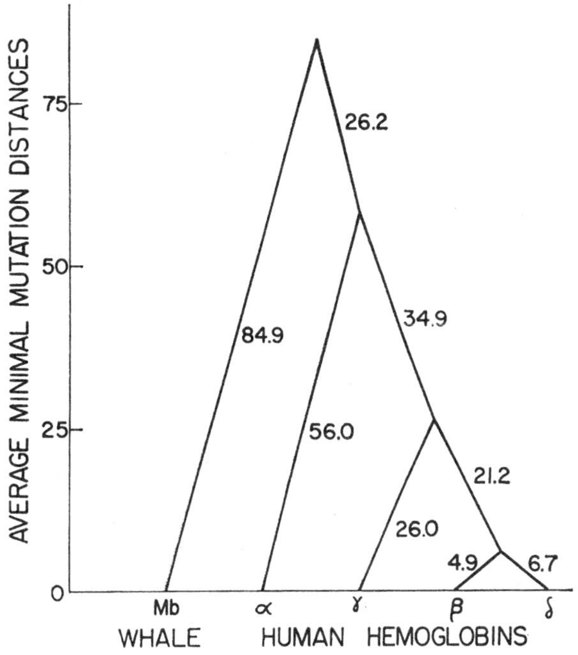

Fig. 2 (left). Phylogeny as reconstructed from observable mutations in the cytochrome *c* gene. Each number on the figure is the corrected mutation distance (see text) along the line of descent as determined from the best computer fit so far found. Each apex is placed at an ordinate value representing the average of the sums of all mutations in the lines of descent from that apex.

Fig. 3 (right above). A gene phylogeny as reconstructed from observable mutations in several heme-containing globins. See Fig. 2 for details. The percent "standard deviation" (*7*) for this tree is 1.33.

Table 4. Descent of the mammalian cytochromes. Changes in amino acids are shown in large capitals, with subscripts to indicate the number of mutations that had to occur to produce the indicated change. In general, unchanging amino acids are not repeated, but occasionally it has been necessary to relist an unchanged amino acid because a mutation appearing in one line of descent did not apply to other lines listed further down the page. Such unchanged amino acids are shown in small capitals. The lines of descent are shown on either side of the table. The last two columns give the sum of the mutations indicated in that row and the corresponding value from Fig. 2. The following rules were used in formulating each amino acid position of the ancestral sequences: Choose the amino acid so that the changes in the codon during descent require (i) the smallest overall number of mutations; (ii) the fewest segments containing multiple mutations (that is, two lines with one mutation each are preferred to one line with two mutations); (iii) the fewest sequential mutations (that is, one mutation in each of two lines following a branch point is preferred to one mutation before and one after the branch point); (iv) the fewest back mutations; (v) the fewest kinds of amino acids. Rule (i), where applicable, took priority over all others and rule (ii) took priority over the remainder. It was not found necessary to choose among the last three rules. The ancestral mammalian cytochrome *c* sequence shown was derived from the amino acid sequences of all 20 cytochromes *c*.

Amino acid No.	17	18	21	39	41	50	52	53	56	64	66	68	89	94	95	98	109	Mutuations (No.) Listed in this table	Mutuations (No.) From leg lengths in Fig. 2
Ancestral mammal	V	Q	L	H	U	P	O	S	A	E	Y	A	L	I	G	L	N		
Ancestral primate	W_1	M_2	S_1	.	.	.	.	.	L_1	.	.	.	V_1	.	.	.	.	6	6.9
Monkey	.	.	.	.	.	.	.	.	.	.	.	.	.	.	.	.	.	0	0.2
Man	.	.	.	.	.	.	.	.	.	W_1	.	.	.	.	.	.	.	1	.8
	V	Q	L	.	.	.	F_1	.	A	E	.	.	L	.	Y_1	.	.	2	1.4
Kangaroo	.	.	.	N_1	W_1	.	.	E_1	.	W_1	.	.	.	.	.	.	.	4	4.6
	.	.	.	H	U	.	.	S	.	E	.	.	.	.	.	.	.	0	−.6
Rabbit	.	.	.	.	.	V_2	.	.	.	.	.	.	.	.	A_1	.	.	3	2.7
	.	.	.	.	.	P	.	.	.	.	.	G_1	.	.	Y	.	.	1	1.4
Dog	.	.	.	.	.	.	.	.	.	.	.	.	.	E_1	.	.	I_4	2	3.0
Ancestral ungulate	.	.	.	.	.	.	.	.	.	.	.	.	.	I	.	Q_2	N	2	1.7
Pig	.	.	.	.	.	.	.	.	.	.	.	.	.	.	.	.	.	0	1.3
Ancestral perissodactyl	.	.	.	.	.	.	.	.	.	.	I_2	.	.	.	E_2	.	.	4	2.9
Donkey	.	.	.	.	.	.	.	.	.	.	.	.	.	.	.	.	.	0	0.1
Horse	.	.	.	.	.	.	.	E_1	.	.	.	.	.	.	.	.	.	1	.9

shown in Fig. 3. The overall result is as Ingram had previously indicated (*15*). A cautionary note may be derived from this. A wildly incorrect result could easily be obtained if the presence of multiple, homologous genes were not recognized and a phylogeny were constructed from sequences which were coded for, say, half by genes for alpha hemoglobin chains and half by genes for beta hemoglobin chains. This results from the speciation having occurred more recently than the gene duplication which permitted the separate evolution of the alpha and beta genes.

The method described can also be used to develop treelike relationships by employing data which are very different in character from mutation distances. For example, the physical characteristics of human beings have been used to construct a tree relating several ethnic groups (Fig. 4; *6*).

Although we are examining the product of but a single gene, and a rather small one at that, the phylogenetic scheme in Fig. 2 is remarkably like that constructed in accord with classical zoological comparisons (*5*). There are only three noticeable deviations, discussed below, and these may well be changed as more species are added to the list. Of even greater value would be sequences from other genes, since special environmental effects may easily cause the convergence of one or several genes in phylogenetically disparate organisms. Hemoglobin amino acid sequences may soon be available in great enough numbers to prove useful in this respect.

Almost all the alternative phylogenetic schemes tested involved rearrangements within the groups birds (*16, 17*) and nonprimate mammals (*14, 18, 19*). With respect to the birds, it will be noticed that the penguin is closely associated with the chicken, whereas one might have expected that all the "birds of flight" (Neognathae) would be more closely related to each other than to the penguin (Impennae). This discrepancy is probably related to the very small numbers of mutations involved. In this regard, it is interesting to note that on the basis of a micro-complement-fixation technique using antisera to several purified enzymes, Wilson *et al.* (*2*) found that the duck is more closely related to the chicken than is the pigeon. This agrees with our findings.

In the second group, the kangaroo is shown closely associated with the nonprimate mammals, whereas most zoologists would maintain that the placental mammals, including the primates, are more closely related to each other than to the marsupials.

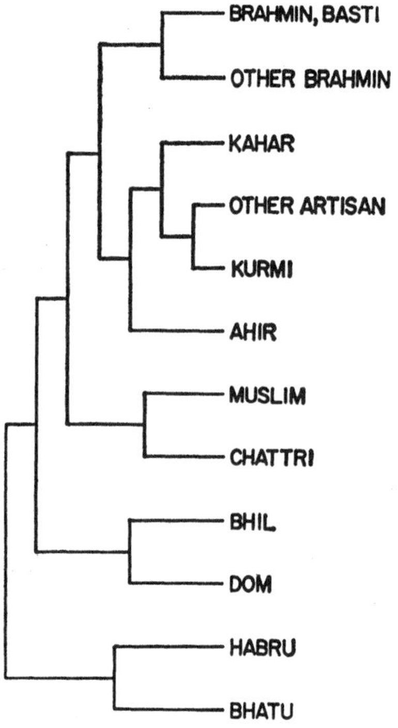

Fig. 4. Relations among various tribes and castes of India. The data used to construct this scheme are the D^2 values given by Rao (*6*). This figure is in principle like Fig. 2 except that, to prevent misinterpretation of the physical significance of the numbers one obtains, branching is shown as a rather uniform step function which preserves the relationships but obliterates the quantitative distances of the ordinate.

A third anomaly is that the turtle appears more closely associated with the birds than to its fellow reptile the rattlesnake. Although it is true that the snake is involved in seven of the nine instances where the reconstructed values differ from the input mutation distances by more than 4 mutations, this cannot account for the anomaly, which in fact results from the close similarity of the turtle's cytochrome *c* amino acid sequence to those of the birds.

Thus the phylogenetic tree in Fig. 2 is imperfect. Nevertheless, considering that only one gene product was analyzed and that no choices were made other than those dictated by the statistical analysis, the results are very promising, and a phylogeny based upon a quantitative determination of those very events which permit speciation, namely mutations, must ultimately be capable of providing the most accurate phylogenetic trees.

Elapsed Time and Evolutionary Change

It should be pointed out that the ordinate of Fig. 2 represents the minimum number of mutations observable. Since multiple mutations in a single codon are not likely to produce mutation values as large as the actual number of mutations sustained, Fig. 2 is greatly foreshortened with respect to the actual number of mutations (*20*). The possibility of obtaining an ordinate scale denoted as actual mutations by applying a correction factor, using the relative frequencies of codons observed to have sustained one, two, and three nucleotide changes, must await reliable statistical information on the relative probabilities that given amino acid substitutions will permit the progeny to compete successfully in their environment. Any meaningful correction of this sort is precluded at present by the lack of such statistical information, but its importance may be emphasized by noting that such a correction would yield an ordinate in Fig. 2 in which equal numbers of mutations would correspond to equal intervals of time, as long as the rate at which mutations are fixed, averaged for many lines of descent over very long periods of evolutionary history, does not vary appreciably (*20*).

It should be noted that the method does not assume any particular value for the rate at which mutations have accumulated during evolution. Indeed, from any phylogenetic ancestor, today's descendants are equidistant with respect to time but not, as computations show, equidistant genetically. Thus the method indicates those lines in which the gene has undergone the more rapid changes. For example, from the point at which the primates separate from the other mammals, there are, on the average, 7.5 mutations in the descent of the former and 5.8 in that of the latter, indicating that the change in the cytochrome *c* gene has been much more rapid in the descent of the primates than in that of the other mammals.

The method allows negative mutation distances, and a few were observed in some of the discarded phylogenetic schemes. Their absence from the best-fitting scheme would indicate that there were no significant evolutionary reversals in this gene.

One highly desirable goal is the reconstruction of the ancestral cytochrome *c* amino acid sequences. The procedure, though not difficult, is dependent upon the phylogenetic tree on which these sequence data are arranged. Given the present scheme (see Fig. 2) one can reconstruct the ancestral proteins. A reconstruction of the ancestral amino acid sequences for the mammalian portion of tree is shown in Table 4. One can then ask such a question as "What are the mutations required to account for the difference between the cytochromes *c* of the ancestral primate and of the ancestral mammal?" The data in Table 4 clearly identify the mutations as occurring in positions 17, 18, 21, 56, and 89. In a similar manner, the monkey and human lines are distinguished by a single mutation in the human line which resulted in the substitution of isoleucine for threonine at position 64.

There is presently no detectable relationship between the primary structures of cytochrome *c* and those of hemoglobins (*12*). Nevertheless, the reconstruction and comparison of the ancestral amino acid sequences may reveal a homology that cannot be detected in present-day proteins. The employment of such ancestral sequences may be generally useful for detecting common ancestry not otherwise observable.

Note added in proof. Since this article was accepted our attention has been called to several earlier papers which present some of the important concepts discussed here. Sokal and his collaborators (*38*) have for several years been studying various ways of producing treelike relationships from quantitative taxonomic information. In an interesting application of this type of technique, using the amino acid sequences of fibrinopeptides from several ungulates, R. F. Doolittle and B. Blombäck (*39*) constructed such a tree and specifically indicated how knowledge of the genetic code would be useful for more precise constructions.

Jukes (*40*, fig. 3) has presented the Ingram scheme of the hemoglobin gene duplications and placed upon the various legs estimates of the numbers of nucleotide substitutions. His figure is not essentially different from Fig. 3 of this article.

References and Notes

1. B. J. McCarthy and E. T. Bolton, *Proc. Natl. Acad. Sci. U.S.* **50**, 156 (1963).
2. A. C. Wilson, N. O. Kaplan, L. Levine, A. Pesce, M. Reichlin, W. S. Allison, *Fed. Proc.* **23**, 1258 (1964); C. A. Williams, Jr., in *Peptides of Biological Fluids*, H. Peeters, Ed. (Elsevier, New York, 1965), p. 62; M. Goodman, *ibid.*, p. 70; A. S. Hafleigh and C. A. Williams, Jr., *Science* **151**, 1530 (1966).
3. R. L. Hill, J. Buettner-Janusch, V. Buettner-Janusch, *Proc. Natl. Acad. Sci. U.S.* **50**, 885 (1963); R. L. Hill and J. Buettner-Janusch, *Fed. Proc.* **23**, 1236 (1964).
4. E. Margoliash, *Proc. Natl. Acad. Sci. U.S.* **50**, 672 (1963); E. L. Smith and E. Margoliash, *Fed. Proc.* **23**, 1243 (1964); E. Margoliash and E. L. Smith, in *Evolving Genes and Proteins*, V. Bryson and H. Vogel, Eds. (Academic Press, New York, 1965), p. 221.
5. A. S. Romer, *Vertebrate Paleontology* (Univ. of Chicago Press, Chicago, ed. 2, 1945).
6. Our procedure may be compared with the "cluster analysis" approach as formulated by A. W. F. Edwards and L. L. Cavalli-Sforza [*Biometrics* **21**, 362 (1965)]; their approach is, in one sense, the reverse of that we have used, since cluster analysis starts with all the elements as members of the same subset and proceeds to subdivide that subset into successively smaller but more numerous subsets until each element is the sole member of its own subset. In terms of Fig. 2, Edwards and Cavalli-Sforza constructed their tree from the top down, whereas we built ours from the bottom up. Edwards and Cavalli-Sforza report testing their method on C. R. Rao's data [*Advanced Statistical Methods in Biometric Research* (Wiley, New York, 1952)] on physical characteristics of 12 Indian castes and tribes. Rao had used these data to postulate relationships among the castes and tribes. Although the nature of these data is quite different from that of ours, the formal mathematical problems are very much alike, and we have used the D^2 values of Rao, as did Edwards and Cavalli-Sforza, to find the best tree. Edwards and Cavalli-Sforza's tree has a percent "standard deviation" (*7*) of 32.6. Our result, shown in Fig. 4, has a percent "standard deviation" of 29.2 and, except that it possesses greater detail, conforms to the conclusions drawn by Rao.
7. The quotation marks are placed around "standard deviation" because the data used in its formulation here are not statistically independent as is generally required. This is evident in that only 20 amino acid sequences determine the 190 mutation distances utilized.
8. The homology may be found by aligning the cysteine residues which bind the heme. Excellent examples of this may be seen in Fig. 10 of E. Margoliash and A. Schejter, *Advan. Protein Chem.* **20**, 114 (1965).
9. W. M. Fitch, *J. Mol. Biol.* **16**, 1 (1966).
10. M. G. Weigert and A. Garen, *Nature* **206**, 992 (1965).
11. S. Brenner, A. O. W. Stretton, S. Kaplan, *ibid.*, p. 994.
12. W. M. Fitch, *J. Mol. Biol.* **16**, 9 (1966).
13. It will be recognized that once the first tree is calculated, the number of computations required for alternatives becomes greatly reduced. For example, if instead of the tree shown in Fig. 2 one wishes to test a tree which differs only in the order in which the chicken, duck, and penguin are joined, the only legs in need of recalculation are those five descending to these birds from the avian apex.
14. The cow (*18*) and sheep (*19*) cytochromes *c* are identical with that of the pig (*27*).
15. V. M. Ingram, *The Hemoglobins in Genetics and Evolution* (Columbia Univ. Press, New York, 1963); A. B. Edmundson, *Nature* **205**, 883 (1965).
16. The cytochrome *c* of the turkey (*29*) is identical with that of the chicken (*16*).
17. S. K. Chan and E. Margoliash, *J. Biol. Chem.* **241**, 507 (1966).
18. K. T. Yasunobu, T. Nakashima, H. Higo, H. Matsubara, A. Benson, *Biochim. Biophys. Acta* **78**, 791 (1963).
19. S. K. Chan, S. B. Needleman, J. W. Stewart, E. Margoliash, unpublished results.
20. This is analogous to the relationship between numbers of amino acid replacements and the evolutionary time scale discussed by E. Margoliash and E. L. Smith in *Evolving Genes and Proteins*, V. Bryson and H. Vogel, Eds. (Academic Press, New York, 1965), p. 221.
21. B. Keil, Z. Prusik, F. Sorm, *Biochim. Biophys. Acta* **78**, (1963).
22. H. Matsubara and E. L. Smith, *J. Biol. Chem.* **238**, 2732 (1963).
23. J. A. Rothfus and E. L. Smith, *ibid.* **240**, 4277 (1965).
24. M. A. McDowall and E. L. Smith, *ibid.* p. 4635.
25. E. Margoliash, E. L. Smith, G. Kreil, H. Tuppy, *Nature* **192**, 1125 (1961).
26. O. F. Walasek and E. Margoliash, unpublished results.
27. J. W. Stewart and E. Margoliash, *Can. J. Biochem.* **43**, 1187 (1965).
28. C. Nolan and E. Margoliash, *J. Biol. Chem.* **241**, 1049 (1966).
29. S. K. Chan, I. Tulloss, E. Margoliash, unpublished results.
30. S. B. Needleman and E. Margoliash, *J. Biol. Chem.* **241**, 853 (1966).
31. S. K. Chan, I. Tulloss, E. Margoliash, *Biochemistry* **5**, 2586 (1966).
32. O. P. Bahl and E. L. Smith, *J. Biol. Chem.* **240**, 3585 (1965).
33. G. Kreil, *Z. Physiol. Chem.* **334**, 154 (1963).
34. S. K. Chan and E. Margoliash, *J. Biol. Chem.* **241**, 335 (1966).
35. J. Heller and E. L. Smith, *Proc. Natl. Acad. Sci. U.S.* **54**, 1621 (1965).
36. Y. Yaoi, K. Titani, K. Narita, *J. Biochem. Tokyo* **59**, 247 (1966).
37. K. Narita and K. Titani, *Proc. Japan Acad.* **41**, 831 (1965).
38. R. R. Sokal, *Syst. Zool.* **10**, 70 (1961); F. J. Rohlf and R. R. Sokal, *Univ. Kansas Sci. Bull.* **45**, 3 (1965); J. H. Camin and R. R. Sokal, *Evolution* **19**, 311 (1965).
39. R. F. Doolittle and B. Blombäck, *Nature* **202**, 147 (1964).
40. T. H. Jukes, *Advan. Biol. Med. Phys.* **9**, 1 (1963).

41. This project received support from grants from NIH (NB-04565) and NSF (GB-4017) to W.M.F. We thank Peter Guetter and Daniel Brick for valuable technical assistance.

Britten, R. J., and E. H. Davidson. 1969. Gene regulation for higher cells: a theory. *Science* 165:349–357.

This paper is a fine example of path-breaking conceptual science. By the early 1960s, molecular discoveries had revealed the startling fact that genomes in eukaryotic cells harbor hordes of repetitive DNA sequences. Also emerging was the provocative notion that changes in gene expression (i.e., gene regulation) might play far more important evolutionary roles than do simple alterations in protein-coding genes. In other words, changes in gene regulation must be important both with respect to the ontogenetic differentiation of body parts within an individual and the phylogenetic differentiation of body plans across diverse forms of life.

Roy Britten and Eric Davidson formulated an innovative hypothesis that wove these ideas into a grand schema for how genetic regulation generally might transpire mechanistically. The authors envisioned batteries of producer loci (protein-coding or other housekeeping genes) being coordinately regulated by activator RNA molecules synthesized by integrator genes whose effect was to induce joint transcription from multiple producer genes in response to environmental stimuli (such as hormones) acting on particular sensor genes. The model is worthy of study today, not because it has proved to be entirely correct in its details (it has not) but rather because it offers a prime example of integrative and imaginative thinking in evolutionary genomics. Today, the study of gene regulation remains one of the hottest topics in evolutionary biology (ENCODE Project Consortium 2012), with diverse regulatory mechanisms routinely being illuminated at transcriptional, translational, and post-translational levels. Many of these modern discoveries can trace their intellectual roots to the powerful conceptual framework laid down with clarity and verve by Britten and Davidson.

RELATED READING

Carroll, S. B. 2006. Endless forms most beautiful: the new science of evo devo and the making of the animal kingdom. Norton, New York, New York, USA.

Carroll, S. B., J. K. Grenier, and S. D. Weatherbee. 2001. From DNA to diversity: molecular genetics and the evolution of animal design. Blackwell, London, England.

ENCODE Project Consortium. 2012. An integrated encyclopedia of DNA elements in the human genome. *Nature* 489:57–74.

Raff, R. A., and T. C. Kaufman. 1983. Embryos, genes, and evolution: the developmental genetic basis of evolutionary change. Macmillan, New York, New York, USA.

Gene Regulation for Higher Cells: A Theory

New facts regarding the organization of the genome provide clues to the nature of gene regulation.

Roy J. Britten and Eric H. Davidson

Cell differentiation is based almost certainly on the regulation of gene activity, so that for each state of differentiation a certain set of genes is active in transcription and other genes are inactive. The establishment of this concept (*1*) has depended on evidence indicating that the cells of an organism generally contain identical genomes (*2*). Direct support for the idea that regulation of gene activity underlies cell differentiation comes from evidence that much of the genome in higher cell types is inactive (*3*) and that different ribonucleic acids (RNA) are synthesized in different cell types (*4*).

Little is known, however, of the molecular mechanisms by which gene expression is controlled in differentiated cells. As far as we are aware no theoretical concepts have been advanced which provide an interpretation of certain of the salient features of genomic structure and function in higher organisms. We consider here experimental evidence relating to these features. (i) Change in state of differentiation in higher cell types is often mediated by simple external signals, as, for example, in the action of hormones or embryonic inductive agents. (ii) A given state of differentiation tends to require the integrated activation of a very large number of noncontiguous genes. (iii) There exists a significant class of genomic sequences which are transcribed in the nuclei of higher cell types but appear to be absent from cytoplasmic RNA's. (iv) The genome present in higher cell types is extremely large, compared to that in bacteria. (v) This genome differs strikingly from the bacterial genome due to the presence of large fractions of repetitive nucleotide sequences which are scattered throughout the genome. (vi) Furthermore, these repetitive sequences are transcribed in differentiated cells according to cell type-specific patterns.

In this article we propose a new set of regulatory mechanisms for the cells of higher organisms such that multiple changes in gene activity can result from a single initiatory event. These proposals are presented in the form of a specific, relatively detailed model at the level of complexity which appears to us to be required for the genomic regulatory machinery of higher cells. We make no attempt to arrive at definitive statements regarding these proposed mechanisms; obviously evidence is not now available to support any model in detail. Our purpose in presenting an explicit theory is to describe the regulatory system proposed in terms of elements and processes which are capable of facing direct experimental test. It is hoped that our relatively detailed commitment will induce discussion and experiment, and it is expected that major modifications in concept will result.

Undoubtedly important regulatory processes occur at all levels of biological organization. We emphasize that this theory is restricted to processes of cell regulation at the level of genomic transcription.

We begin by describing our usage of certain terms and their role in the model, and then present the model itself. We then consider relevant experimental observations and certain testable implications of the model. Finally, some general implications of the model for evolutionary theory are mentioned.

Elements of the Model

The following definitions are intended only to clarify the usage of certain terms in our discussion of this model.

Gene: A region of the genome with a narrowly definable or elementary

Dr. Britten is on the staff of the Department of Terrestrial Magnetism, Carnegie Institution of Washington, Washington, D.C. 20015, and Dr. Davidson is on the staff of Rockefeller University, New York 10021.

function. It need not contain information for specifying the primary structure of a protein.

Producer gene: A region of the genome transcribed to yield a template RNA molecule or other species of RNA molecules, except those engaged directly in genomic regulation. We are using this term in a manner analogous to that in which the term "structural gene" has been used in the context of certain bacterial regulation systems (*5*). Products of the producer gene include all RNA's other than those exclusively performing genomic regulation by recognition of a specific sequence. Among producer genes, for example, are the genes on which the messenger RNA template for a hemoglobin subunit is synthesized, and also the genes on which transfer RNA molecules are synthesized.

Receptor gene: A DNA sequence linked to a producer gene which causes transcription of the producer gene to occur when a sequence-specific complex is formed between the receptor sequence and an RNA molecule called an activator RNA. We do not, in this model, wish to specify a mode of action for the receptor gene—that is, the nature of the molecular events occurring between the DNA, histones, polymerases, and so forth, present in the receptor complex. This model is concerned primarily with interrelations among the DNA sequences present in the genome.

Activator RNA: The RNA molecules which form a sequence-specific complex with receptor genes linked to producer genes. The complex suggested here is between native (double-stranded) DNA and a single-stranded RNA molecule (*6*). The role proposed for activator RNA could well be carried out by protein molecules coded by these RNA's without changing the formal structure of the model (*7*). Decisive evidence is lacking in higher cells, and we have chosen the simpler alternative (*8*). As the discussion of the evolutionary implications of this model will indicate, however, the probability of formation of new batteries of genes in evolution appears to differ greatly between these two alternatives.

Integrator gene: A gene whose function is the synthesis of an activator-RNA. The term integrator is intended to emphasize the role of these genes in leading, by way of their activator RNA's, to the coordinated activity of a number of producer genes. A set of linked integrator genes is activated together in response to a specific initiating event, resulting in the concerted activity of a number of producer genes not sharing a given receptor gene sequence.

Sensor gene: A sequence serving as a binding site for agents which induce the occurrence of specific patterns of activity in the genome. Binding of these inducing agents is a sequence-specific phenomenon dependent on the sensor gene sequence, and it results in the activation of the integrator gene or genes linked to the sensor gene. Such agents include, for example, hormones and other molecules active in intercellular relations as well as in intracellular control. Most will not bind to sensor gene DNA, and an intermediary structure such as a specific protein molecule will be required. This structure must complex with the inducing agent and must bind to the sensor gene DNA in a sequence-specific way.

Battery of genes: The set of producer genes which is activated when a particular sensor gene activates its set of integrator genes. A particular cell state will usually require the operation of many batteries.

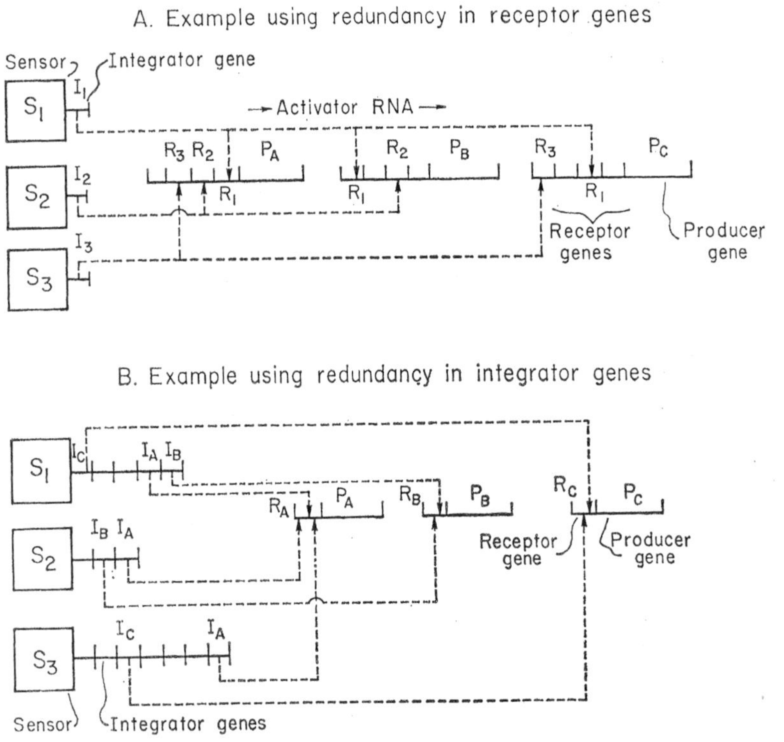

Fig. 1. Types of integrative system within the model. (A) Integrative system depending on redundancy among the regulator genes. (B) Integrative system depending on redundancy among the integrator genes. These diagrams schematize the events that occur after the three sensor genes have initiated transcription of their integrator genes. Activator RNA's diffuse (symbolized by dotted line) from their sites of synthesis—the integrator genes—to receptor genes. The formation of a complex between them leads to active transcription of the producer genes P_A, P_B, and P_C.

Integrative Function of the Model

The concerted activation of one or more batteries of producer genes is considered to underlie the existence of diverse states of differentiation. Examples of two basic aspects of the proposed integrative function appear in Fig. 1. In each case, the producer genes shown are integrated into three different, very small batteries. Sensor gene S_1 and its integrator specify the activation of producer genes P_A, P_B, and P_C; S_2 that of P_A and P_B; and S_3 that of P_A and P_C.

In Fig. 1A, the control pattern depends on the existence of redundant receptor sequences in the receptor gene sets of the three producer genes. Inclusion of a particular producer gene in each of the batteries calling on it depends on the presence of the appropriate receptor gene adjacent to the producer gene. Thus, in the case where there is only one integrator gene per sensor as in Fig. 1A, there will be as many copies of a given receptor gene sequence as there are producer genes in a battery.

In the case shown in Fig. 1B, however, redundancy is present between the integrator genes of different integrator sets. A particular producer gene, in this example, is included in each of several batteries calling on it by virtue of the inclusion of the same integrator gene adjacent to each of the appropriate sensor genes. Here there will be as many copies of a given integrator-gene as there are batteries that call on its producer gene. For certain commonly required genes, for example those used in the fundamental biochemistry of each cell, this could be a very large number indeed.

Systems of the type portrayed in Fig. 1A might be most useful in the case where the producer genes to be integrated direct the synthesis of enzymes whose function is tightly coordinated physiologically, for example, the ten enzymes of the urea synthesis system. Where the system is needed, all the genes would be needed. The system portrayed in Fig. 1B is a more powerful integrative system since it can govern a larger diversity of producer genes. The number of receptor sequences governing each producer sequence is probably small since transcription of a producer gene sequence is not likely to be activated from a great distance along the DNA strand. There is no reason a priori, on the other hand, to restrict the number of integrator genes per integrator set, except for the requirement that the integrator genes not be so distant that there is a high probability of their being separated by translocation.

In this model, regulation is accomplished by sequence-specific binding of an activator RNA and not by sequence recognition on the part of histones. The latter seem clearly to be the general inhibitors of transcription in the genome, but evidently these general inhibitors do not possess sufficient diversity to be considered as sequence-specific regulatory elements themselves (*9, 10*). We have assumed that, unless otherwise specified, the state of the higher cell genome is histone-mediated repression and that regulation is accomplished by specific activation of otherwise repressed sites, rather than by repression of otherwise active sites.

Figure 2 combines the elements and systems we have thus far described. In the remainder of our discussion we consider various properties and consequences of the minimal model, as portrayed in this figure. The magnitude of the producer gene batteries is only suggested by the diagram in Fig. 2, and of course no attempt has been made to portray the actual complexity of the system, that is, to illustrate the number of elements whose function is likely to be integrated in a living cell. Obviously, the coordination of many batteries of genes is required in order to account for massive changes in differentiated state, such as the neogenesis of a tissue during development. We visualize such phenomena as being mediated by sensor genes sensitive to the products of integrator genes in other integrative sets. In other words, a single inducing agent could lead to the activation of a number of sensor-integrator sets, activating a vast number of producer genes.

Sequential patterns of gene activation, as in development, could result if certain sensors respond to the products of producer genes. In addition, the protein of a newly effective sensor assembly is, in the model, a product of a previously activated producer gene. Stabilization of a cell type in a given state of differentiation might also be explained in this way. Living systems continuously adjust their activities in accordance with their internal state, and it is evident that a requirement for sensors sensitive to feedback control by certain producer-gene products exists as well.

Fraction of the Genome Utilized for Regulation

Broadly speaking, genome size increases with the grade of organization of eukaryotes, as first pointed out by Mirsky and Ris in 1951 (*11, 12*). The wide range of genome sizes often observed among closely related creatures obscures the correlation. Organisms with large genomes presumably have a requirement for genomic information similar to that of their relatives with smaller genomes. This implies the evolutionary multiplication of the genome of ancestors possessing the minimum amount of DNA required to effect each grade of organization. It is thus useful to consider the minimum amount of DNA observable at each grade of organization. Figure 3 shows the minimum genome size (*13*) for some major steps in evolution between viruses and the higher chordates.

A reasonable explanation has not been advanced for the large genome sizes occurring at the higher organizational levels. Most of the known biosynthetic pathways are already represented in unicellular organisms. It is not possible to estimate the increase in number of producer genes required to specify structure and chemistry at the higher levels of organization. Nonetheless, it seems unlikely that the 30-fold

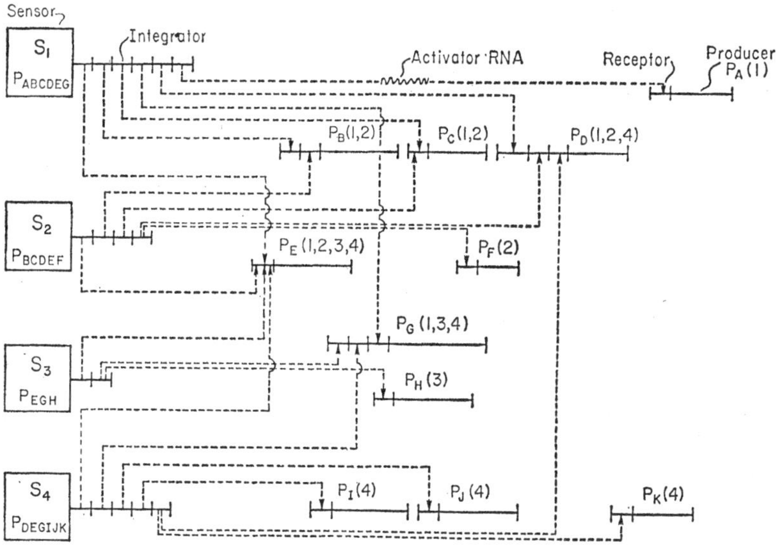

Fig. 2. This diagram is intended to suggest the existence of overlapping batteries of genes and to show how, according to the model, control of their transcription might occur. The dotted lines symbolize the diffusion of activator RNA from its sites of synthesis, the integrator genes, to the receptor genes. The numbers in parentheses show which sensor genes control the transcription of the producer genes. At each sensor the battery of producer genes activated by that sensor is listed. In reality many batteries will be much larger than those shown and some genes will be part of hundreds of batteries.

Table 1. Several of the functionally linked enzyme systems present in liver (*17*, chapter 12; *36*). Uridine monophosphate, UMP; adenosine monophosphate, AMP.

System	Number of enzymes
Glycogen synthesis	5
Galactose synthesis	6
Phosphogluconate oxidation	11
Glycolysis	12
Citric acid cycle	17
Lecithin synthesis	8
Fatty acid breakdown	5
Lanosterol synthesis	10
Phenylalanine oxidation	8
Methionine to cysteine	10
Methionine to aspartic acid	10
Urea formation	10
Coenzyme A synthesis	6
Heme synthesis	9
Pyrimidine synthesis (to UMP)	6
Purine synthesis (to AMP)	14

increase from poriferan to mammal can be attributed to a 30-fold increase in the number of producer genes. This problem cannot be escaped by attributing the large genome size to redundancy. Fifty-five percent of the DNA of the calf, for example, occurs in nonrepetitive sequences (*14*). This is enough DNA to provide almost 10^7 diverse producer-gene sequences the size of the gene coding for the beta chain of hemoglobin. A few other measurements have been made which indicate that such diversity in DNA sequence is general (*15*).

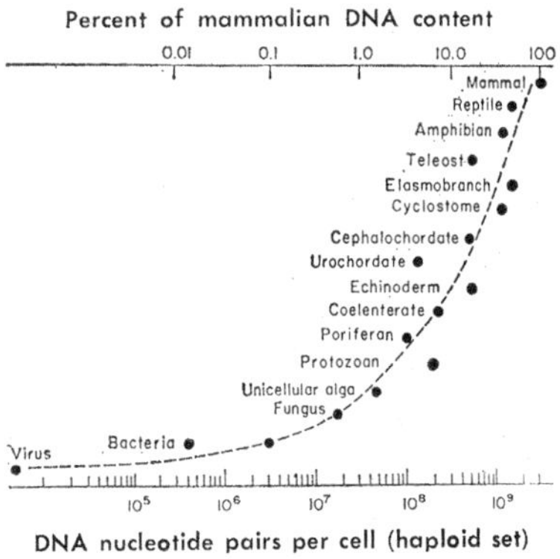

Fig. 3. The minimum amount of DNA that has been observed for species (*13*) at various grades of organization. Each point represents the measured DNA content per cell for a haploid set of chromosomes. In the cases of mammals, amphibians, teleosts, bacteria, and viruses enough measurements exist to give the minimum value meaning. However for the intermediate grades few measurements are available, and the values shown may not be truly minimal. No measurements were unearthed for acoela, pseudocoela and mesozoa. The ordinate is not a numerical scale, and the exact shape of the curve has little significance. The figure shows that a great increase in DNA content is a necessary concomitant to increased complexity of organization.

Quite possibly, the principal difference between a poriferan and a mammal could lie in the degree of integrated cellular activity, and thus in a vastly increased complexity of regulation rather than a vastly increased number of producer genes (*16*). Much of the DNA accumulating in the genomes toward the upper end of the curve in Fig. 3 might then have a regulative function. The model also suggests that a large amount of DNA could be devoted to regulatory function: consider integrator and receptor sequences which are not redundant. In this case a battery of producer genes would require a distinct integrator gene for each producer gene. Producer genes occurring in several batteries would require receptor genes corresponding to each battery. The resulting multiplicity of integrator and receptor genes might result in a much larger quantity of DNA in regulatory sequences than in producer sequences. It is likely that an ever-growing library of different combinations of groups of producer genes is needed as more complex organisms evolve. An effective way of storing the information specifying these combinations in the genome is to make use of sensors responsive to the activator RNA's of other integrative sets. Thus we propose that a higher level of integrator gene sets is accumulated. Each of these, when activated, could specify a very large program of producer gene activations by specifying the activity of a network of other sensor-integrator sets. Thereby many batteries of genes of the sort shown in Fig. 2 could be activated.

Experimental Justification of the Elements of the Model

There are five important classes of elements in this model: sensor genes, integrator genes, activator RNA, receptor genes, and producer genes. Is this degree of complexity really necessary? The particular set of elements we have postulated may of course not be the required ones. Five, however, is the minimum number of classes of elements which can carry out the following formally described process: (i) response to an external signal; (ii) production of a second signal; (iii) transmission of the second signal to a number of receptors unresponsive to the original signal; (iv) reception of the second signal; and (v) response to this event by activation of a producer gene and its transcription to provide the cell with the producer gene product. In the following sections we examine evidence that such a description is applicable to gene regulation in higher organisms, and explore evidence that suggests the existence of the elements of the model.

Integration of Physically Unlinked Producer-Gene Activity

We have assumed that a given state of differentiation depends on the coordinated activity of a number of biochemical systems. Each of these systems will probably contain a number of components. As an example, Table 1 lists some of the enzyme systems operating in one cell type, mammalian liver.

An underlying principle of this model is that producer genes active in any given tissue need not be physically linked in the genome. For physically adjacent producer genes, integration of activity could be based on the operation of gigantic polycistronic tissue-specific operons. There are good reasons for believing that this is not the case in eukaryotes. Some producer genes are called into activity in a number of different tissues, as illustrated in Table 2.

Table 2 shows the overlapping pattern of activity for 17 enzymes in 8 tissues. Direct contiguity of active producer genes could not produce this set of patterns if a single copy of each gene were present in the genome. Genetic evidence does not at present indicate the presence of multiple producer genes yielding identical products, except for ribosomal RNA and transfer RNA. An equally strong point can be made that control of the producer gene sets for the systems listed in Table 1 cannot be based on physical linkage of one set to the next in the liver genome. In other tissues, some but not all of these systems are functional (*17*). In other words, even where the producer genes within a physiologically coordinated enzyme system (Table 1) are linked, the same formal problem remains: a mechanism is required for coordinating the activity of the noncontiguous systems of producer genes characteristic of each state of differentiation. In at least some instances the

Table 2. Distribution of various enzymes in tissues of one organism, rat (*17*, Table XIII). + Means enzyme is present in amount 40 to 100 percent of that in the tissue where it is most plentiful; 0 means enzyme is essentially absent, that is, less than 8 percent the level of the tissue with the highest activity. If the level falls between 8 and 40 percent, or if data are lacking, space is left blank.

E.C. No.	Enzyme	Liver	Kidney	Spleen	Heart	Skeletal muscle	Small intestine	Pancreas	Brain
1.1.1.30	3-Hydroxybutyrate dehydrogenase	+		0	0	0		0	0
1.1.1.37	Malate dehydrogenase		0		+				
1.5.1.1	Pyrroline-2-carboxylate reductase		+	0	0				+
1.11.1.6	Catalase	+	+			0		0	0
1.11.1.7	Peroxidase	0	0	+	0	0	+		0
1.13.1.5	Homogentisate oxidase	+	+	0	0	0	0		0
2.1.1.6	Catechol methyltransferase	+		0	0	0	0		0
2.1.1.3	Dimethythetin-homocysteine methyltransferase	+		0		0			0
2.7.7.16	Ribonuclease	0	0		0	0		+	
3.1.1.1	Carboxylesterase		0	0		0		+	
3.1.1.5	Phospholipase	+		+	0	0	+		0
3.1.1.7	Acetylcholinesterase	0	0			0			+
3.1.1.8 } 3.1.1.9 }	Cholinesterases		0		+	0	+		
3.1.3.1	Alkaline phosphatase	0	+	0		0	+	0	0
3.1.3.2	Acid phosphatase			+	0	0			
3.1.3.9	Glucose-6-phosphatase	+		0	0	0			0
3.2.1.25	β-Mannosidase	+	+		0	0		+	0
3.2.1.30	β-Acetylamino deoxyglucosidase		+		0	0		0	0
3.2.1.31	β-Glucuronidase	+		+		0		0	
3.5.3.1	Arginase	+		0	0	0	0	0	0
3.5.4.3	Guanine deaminase	+	+	+		0		+	0
4.1.2.7	Aldolase	0	0	0	0	+			0
4.1.3.7	Citrate synthase	0	0		+	0			
4.2.1.3	Aconitate hydratase		+		+				0
6.3.1.2	Glutamine synthetase	+	0	0		0		0	+

integrated producer genes within each physiologically coordinated set are known to be noncontiguous in higher organisms. As an example, in the human the producer genes coding for the alpha and beta subunits of hemoglobin are unlinked (*18*). Another case concerns two of the enzymes of the phosphogluconic acid oxidation pathway (system No. 3 of Table 1) in *Drosophila melanogaster*. These are glucose-6-phosphate dehydrogenase (E.C.1.1.1.49) and 6-phosphogluconate dehydrogenase (E.C.1.1.1.43) whose genes are located on separate linkage groups (*19*). Evidently producer genes whose activity must be functionally integrated in the most intimate way can be located far apart in the genome. We conclude that within at least some functionally integrated producer-gene systems as well as among these systems, specification of particular patterns of activity requires a method of control other than one depending on contiguity of the producer loci.

The data considered so far provide instances of the type of pattern which our model is designed to interpret, but they do not indicate the extensiveness of the producer-gene batteries called forth in given conditions of differentiation. Table 3 lists some of the effects of estrogen on the uterus, an estrogen target tissue. Although we are ignorant of the diverse proteins involved in effecting these changes it is obvious that there must be many. Though in Table 3, of course, we present only a partial list, and the number of diverse producer genes required for each item on the list can only be guessed at present, this table provides a more realistic description of the magnitude of the problem of producer-gene integration than Tables 1 or 2 do. Analogous problems exist in explaining the integration of a multitude of producer genes in every cell lineage during development. As diverse cell lineages differentiate, a huge variety of qualitatively novel properties appear together. As in the case of the hormones, such processes of differentiation appear to require a mechanism for the simultaneous activation of many systems, such as proposed in this model.

Evidence for the Existence of Sensor Elements

There are many chemically defined agents that have the evident property of inducing large-scale changes in the producer-gene activity of specific target tissues. These agents now include steroid hormones, polypeptide hormones, several plant hormones, several vitamins, and several embryonic inductive agents (*20*). Frequently, the responsible agents also produce an alteration in the spectrum of RNA's being transcribed in the target tissues, as indicated by data obtained with RNA-DNA hybridization and studies in vitro of chromatin template activity (*20*); and these agents have been identified in the nuclear apparatus of the target cells (*20*). The most intensively studied system is perhaps estrogen response (Table 3). All of the above-mentioned forms of evidence exist for this hormone (*20*).

In addition, Maurer and Chalkey (*21*) have isolated from calf endometrial chromatin a protein that binds 17β-estradiol. The binding is stereospecific, noncovalent, and strong (the Michaelis constant, K_m, for binding is $2 \times 10^{-8} M$); and the responsible protein appears not to be a histone. It does not bind steroids as closely related as 17 α-estradiol or diethylstilbestrol. Such a protein, in combination with the specific external agent for which it is the receptor, must interact with the genome in a sequence-specific way, since this interaction results in the activation of only a certain group of genes. Consider a system in which the genomic binding sites are simply adjacent to all the producer genes activated by the external chemical agent. Such a system would appear to possess a limited integrative function which might be utilized for certain small gene batteries. However, the binding of an

Table 3. Some effects of estrogen on uterine cells.

Effect	Ref.
Increase in total cell protein	(*37*)
Increase in transport of amino acids into cell	(*37, 38*)
Increase in protein synthesis activity per unit amount of polyribosomes	(*37*)
Increased synthesis of new ribosomes	(*37*)
Alteration of amounts of nuclear protein to nucleus	(*37*)
Increased amount of polyribosomes per cell	(*37*)
Increase in nucleolar mass and number	(*39*)
Increase in activity of two RNA polymerases	(*37*)
Increase in synthesis of contractile proteins	(*40*)
Imbibition of water	(*41*)
Increased synthesis of many phospholipids	(*42*)
Increased de novo synthesis of purines (dependent on new enzyme synthesis)	(*43*)
Alteration in membrane excitability	(*40*)
Alteration in glucose metabolism	(*44*)
Increase in synthesis of various mucopolysaccharides	(*45*)

external agent to a sequence-specific site on the genome could lead to the activation of a large number of distant producer genes. This is exactly the role the sensor elements of this model carry out. Implicit in the available data on hormone action are genomic elements performing some of the functions of the producer and integrator genes of the model.

Evidence Suggesting the Existence of Activator RNA's

Many of the properties attributed to the RNA in our model are actually those of a certain class of RNA molecules already described extensively; yet no known function has so far been attributed to this class of RNA's. The activator RNA molecules of the model have the following properties that can be tested. (i) They will, in the main, be confined to the nucleus, that is, they are not precursors of cytoplasmic polysomes. (ii) When observed in their functional role, they would be found in chromatin, bound to DNA in a sequence-specific manner. (iii) They are often the product of the redundant fraction of the genome. (iv) They include sequences not present in the polysomes carrying producer-gene templates, that is, most or all cytoplasmic polysomes. Table 4 summarizes some recent studies of RNA's which seem to fulfill condition (i). These RNA's have the suggestive properties of nuclear location, heterogeneity, and probable lack of "precursor" relation to cytoplasmic polysomal templates.

The hybridization experiments of McCarthy and Shearer (*22*) (Table 4) were performed at relatively low concentrations of nucleic acid and at short incubation time. Therefore, the RNA's they describe are the products of the redundant fraction of the genome. The presence of sequences specific to the nucleus and their absence from the cytoplasm is indicated by competition experiments. Furthermore, the nuclear RNA's contain sequences binding as much as five times more DNA than the cytoplasmic RNA at empirical saturation of the DNA with RNA. These and the other data of Table 4 show that RNA's are already known which might fulfill the functions we have assigned to activator RNA's.

At the heart of the model regulation system lies nucleus-confined RNA which determines the pattern of cellular gene activity, and this remains a key area of uncertainty. Bekhor, Kung, and Bonner (*10*), and also Huang and Huang (*10*), have presented evidence suggesting that sequence-specific binding between chromosomal RNA and genomic DNA determines the sites at which the transcription-inhibiting chromatin proteins bind to the DNA (*23*). Thus, according to these experiments, sequence recognition between the special chromosomal RNA's and the DNA specifies the pattern of gene activity (*10, 23*). Furthermore, the chromosomal RNA's have the property of binding, in what is apparently a sequence-specific way, to double-stranded native DNA (*7*). The significance of this line of investigation for experimental test of the idea of activator RNA's is obvious.

Large Changes in Transcription of Redundant Sequences

It is a striking fact that very large changes in the spectrum of RNA's deriving from repetitive sequences are observed when the state of differentiation alters. This knowledge is derived from RNA–DNA hybridization experiments carried out at relatively low concentrations of nucleic acid and short annealing times, so that reaction of RNA with any but the repetitive sequences in the genome is precluded. The spectrum of RNA's present or in the process of being synthesized in different tissues (*4, 24*), both in hormone response (*25*) and in embryonic development and differentiation (*26, 27*), has been investigated with competition procedures. In these experiments RNA from a cell type in one state of differentiation is used to compete with RNA from a cell type in another state of differentiation for binding sites in the repetitive fraction of the DNA. This type of analysis has shown that different families of repetitive genomic sequences are represented in the RNA of cells in diverse states of differentiation. Changes as large as 100 percent (apparent complete lack of homology) in the measured RNA's have been observed—for example, in successive stages of the embryogenesis of *Xenopus* (*27*). It is not particularly obvious why such changes should be detected, since the populations of producer genes active in each state of differentiation might be expected in general to be strongly overlapping. One possible explanation would be that much of the pulse-labeled RNA monitored in these studies is the rapidly turning over product of different regulatory genes such as the integrator genes of this model.

Regulatory Genes Known in Higher Organisms

The model suggests that a sizable portion of the functional genes in differentiated cell types may be regulatory genes (integrator and receptor genes). If this is so, it might be expected that, despite the difficulty of detecting such genes with classical genetic procedures, a certain number of apparent regulatory mutations would be known in higher organisms. The distinguishing characteristic of such regulatory loci would be pleiotropic effects on the activity of a number of producer genes, particularly with reference to a pattern of integration on the part of the latter. A number of good cases of this genre actually exist, particularly for drosophila and maize. A notable example is the Notch series of x-chromosome deficiencies (*28*), some of which are sharply localized. Notch mutants display a very large variety of developmental abnormalities—all affecting early embryonic organization—for example, failure to form a complete gut, failure of meso-

Table 4. Nucleus-confined, apparently heterogeneous RNA's of unknown function.

Source	Size of RNA	Compositional peculiarity if any	Turnover	Evidence against precursor relation with cytoplasmic RNA	Reference
HeLa cell nuclei	Heterodisperse 10*S* to 65*S*	29 to 32% uridine	Rapid	Composition; size; absence of association with nascent proteins	(*46*)
L-cell nuclei			Rapid	Sequences present in nuclear RNA absent in cytoplasmic RNA	(*22*)
Kidney and liver cell nuclei				Sequences present in nuclear RNA absent in cytoplasmic RNA	(*47*)
Reticulocyte nuclei	Heterodisperse 30*S* to 80*S*	29 to 31% uridine	Rapid	Kinetics (cytoplasmic mRNA does not turn over at all); size; base composition	(*48*)
HeLa cell nuclei	100 to 180 nucleotides	Extensive methylation	Extremely low	Composition; small size	(*49*)
Pea seedling nuclei	40 to 60 nucleotides	Presence of dihydro-pyrimidines		Sequences present in nuclear RNA absent from cytoplasmic RNA; composition	(*50*)

dermal differentiations to occur, overly large neural structure, and subnormal ectodermal skin production. Their effects are clearly pléiotropic. The multiplicity of the actual primary failures of these mutants is unknown. That is, no comparison can be made of the number of diverse producer genes affected simultaneously, as opposed to the array of sequential effects that follow the initial primary effects. Nonetheless, the effect of the Notch genes on the organization of the embryo is consistent with what would be expected of mutations in integrator gene sets. Many similar cases are known in which specific organizational lesions result from simple mutations affecting a small region of the genome (*29*). Studies with drosophila imaginal disk cell determination and transdetermination carried out by Hadorn and his associates (*30*) also demonstrate the existence of an apparatus in the genome for specifying integrated patterns of activity in the various cell types deriving from the disk cells. In experimental imaginal disk systems, highly exact specification of the patterns of producer gene activity is heritable through many cell divisions and is separated in time from producer gene function per se (that is, manifest differentiation).

Genes are known in maize which display control over producer genes and are located in the genome at sites distant from the producer genes that they control (*31*). In addition, McClintock and others have demonstrated the presence of other control sites adjacent in the genome to the same producer genes as those controlled by the distant regulatory elements (*31*). Control of the expression of the producer gene is accomplished through the interaction of the distant regulatory gene with the contiguous regulatory gene. This point has been demonstrated by insertion of the contiguous regulatory genes at different sites in the genome, near known genes, which then respond to the same control system governed by the distant regulatory unit. An example is the system termed Ac-Ds. Here the distant regulatory element Ac (which behaves as an integrator gene of this model) can be made to govern producer genes in other chromosomes such as the gene series for synthesis of anthocyanin pigment. Establishment of Ac control over the pigment synthesis system is accomplished by transposing the contiguous regulatory element responsive to Ac (Ds) to the loci of the anthocyanin producer genes (Ds thus behaves like a receptor gene of this model). In several ways, these and other data presented by McClintock (*32*) would seem easily to fit a model such as that presented here.

DNA Sequence Repetition

The existence of repeated sequences in higher organisms led us independently to consider models of gene regulation of the type we describe here. This model depends in part on the general presence of repeated DNA sequences. The model suggests a present-day function for these repeated DNA sequences in addition to their possible evolutionary role as the raw material for creation of novel producer gene sequences. The apparently universal occurrence of large quantities of sequence repetition in the genomes of higher organisms (*14*) suggests strongly that they have an important current function.

The quantity of DNA in repeated sequences, the frequency of repetition (that is, number of times a given sequence is present per genome), and the precision of the repetition show great variation among species. Frequencies from 100 to 1,000,000 have been observed, and the quantities of DNA involved range from 15 to 80 percent of the total DNA. The usual relation between repeated sequences is not that of a perfect copy (*33*), but the sharing of most of the nucleotides in a sequence extending for at least a few hundred nucleotides. Repeated sequence families in the DNA are observed, with degrees of similarity varying from perfect matching to matching of perhaps only two-thirds of the nucleotides. Expression of families of repeated sequences by transcription into RNA shows tissue specificity (as mentioned above) in spite of the fact that the individual families contain these widely divergent sequences.

In the cases studied there is good evidence that the repeated sequences are scattered throughout the DNA. For example in bovine DNA, 75 percent of all fragments about 5000 nucleotides long contain a segment of repeated DNA (*34*). When the fragment size is reduced to about 500 nucleotides, only 45 percent contain repeated sequences. Therefore, the typical bovine DNA fragment of 5000 nucleotides is a composite of lengths of repeated sequence and nonrepeated sequence. For longer fragments (20,000 or so nucleotides), there is suggestive evidence (*14*) that more than 95 percent contain repetitive sequences. Therefore, for bovine DNA (and probably that of other organisms) repeated sequences are intimately interspersed with nonrepeated sequences, throughout the length of the genome. This is precisely the pattern required in our model if repeated sequences are usually or often regulatory in function.

Evolutionary Implications of the Model

Any evolutionary changes in the phenotype of an organism require, in addition to changes in the producer genes, consistent changes in the regulatory system. Not only must the changes be compatible with the interplay of regulatory processes in the adult, but also during the events of development and differentiation. At higher grades of organization, evolution might indeed be considered principally in terms of changes in the regulatory systems. It is therefore a requirement of a theory of genetic regulation that it supply a means of visualizing the process of evolution.

Inactivity of New Genetic Material

A characteristic of this model is that DNA sequences are inactive in transcription, unless specifically activated. Thus the genome of an organism can accommodate new and even useless or dangerous segments of DNA sequence such as might result from a saltatory replication (*35*). Initially these sequences would not be transcribed, and thus would not be subject to adverse selection. Only by inclusion in integrated producer gene batteries (through translocation of receptor genes) would their usefulness as producer genes be tested.

Formation of New Integrative Relations

A peculiar combination of conservatism and flexibility is supplied by the model system. Preexisting useful batteries of genes will tend to remain integrated in function. At the same time, there is the potentiality of formation of new integrative combinations of preexisting producer genes. These combinations would be the result of translocations, principally among the integrator gene sets. Less often, new producer gene batteries would result from events in which receptor genes are translocated into positions contiguous to other producer genes.

We visualize many of the integrator genes and receptor genes as being members of families of repeated DNA sequences. It is known that new repeated sequence families have originated periodically in the course of evolution (*35*). The new families of repeated sequences might well be utilized to form integrator and receptor gene sets specifying novel batteries of producer genes. Thus saltatory replications can be considered the source of new regulatory DNA. All that is required for regulatory function in this model is sequence complementarity (translocation of members of the same repetitive sequence family to integrator and receptor positions). Almost any set of nucleotide sequences would suffice. *The likelihood of utilization of new DNA for regulation is thus far greater than the likelihood of invention of a new and useful amino acid sequence*, since for the latter case great restrictions on the nucleotide sequence exist.

Changes in the integrator systems make possible the origin of new functions and possibly even of new tissues and organs. In other words, the model supplies an avenue for the appearance of novelty in evolution by combining into new systems the already functioning parts of preexisting systems.

Divergence within Repeated Sequence Families

Individual sequences may differ from others in a family as a result of many base changes. We presume that binding of activator RNA to the receptor genes will occur for a degree of sequence homology far short of perfect complementarity. However, at some degree of divergence, binding would be lost, and a producer gene would fail to be activated as a part of its previous battery. Eventually, the process of divergence might yield regulatory DNA in which the original patterns of repetition are no longer observable. In this way, nonrepeated (unique) regulatory DNA could arise, leading to the situation discussed earlier with respect to the fraction of the genome utilized for regulation.

The possibility of increasing sequence divergence among integrator and receptor genes suggests a novel evolutionary mechanism. The divergence of regulatory sequences can be expected to be reversible. If the degree of complementarity required for binding between activator RNA and receptor sequence is fairly low then a reasonably good probability would exist for a subsequent base change to restore the complementarity lost by an earlier change. Intermediate degrees of transcription of certain producer genes will probably result since sequences with a degree of complementarity near some critical value will bind only part of the time. Natural selection could then reversibly affect the integration of individual producer genes into batteries. The potentiality for smoothly changing patterns of integration among many sets of producer genes supplies a mechanism for direct adjustment by natural selection of the organization of systems of cellular activity. In other words, the model implies that selective factors can influence the integrative configurations in which an organism uses its genes.

The families of repeated sequences that appear and remain in the genome of a species affect the rate at which newly integrated systems of producer genes will arise. Thereby, the rate of evolution is affected. It follows that the rate of evolution will be acted on by natural selection.

The issues raised in considering the evolution of the regulatory systems themselves are of a magnitude which is really out of reach in this brief discussion. However, the model offers interesting and surprising predictions. The properties of the model regulatory system suggest that *both the rate and the direction of evolution (for example, toward greater or lesser complexity) may be subject to control by natural selection.*

Summary

A theory for the genomic regulation systems of higher organisms is described. Batteries of producer genes are regulated by activator RNA molecules synthesized on integrator genes. The effect of the integrator genes is to induce transcription of many producer genes in response to a single molecular event. Current evidence suggesting the existence of elements of this model is summarized. Some evolutionary implications are indicated.

References and Notes

1. The variable gene activity theory of cell differentiation was explicitly proposed in the early 1950's, by A. E. Mirsky [in *Genetics in the Twentieth Century*, L. C. Dunn, Ed. (Macmillan, New York, 1951), p. 127] and by E. Stedman and E. Stedman [*Nature* **166**, 780 (1950)]. T. H. Morgan, among several others, had earlier considered this idea [*Embryology and Genetics* (Columbia Univ. Press, New York, 1934)].
2. Evidence for the equivalence of differentiated cell genomes comes from a variety of sources, including regeneration experiments, nuclear transplantation experiments, early embryological studies in which nuclei were positioned in cells other than those normally receiving them, measurements of DNA content per cell, and so forth. [See E. H. Davidson, *Gene Activity in Early Development* (Academic Press, New York, 1968), pp. 3–9.] Critical recent evidence has been provided by Gurdon's demonstration that differentiated intestinal cell nuclei from *Xenopus* tadpoles can direct the development of whole frogs when reimplanted into enucleate eggs [J. B. Gurdon, *Develop. Biol.* **4**, 256 (1962)], and by DNA reassociation studies in which the DNA of different tissues was shown to be indistinguishable in sequence content [B. J.

McCarthy and B. H. Hoyer, *Proc. Nat. Acad. Sci. U.S.* **52**, 915 (1964)].
3. V. G. Allfrey and A. E. Mirsky, *Proc. Nat. Acad. Sci. U.S.* **44**, 981 (1958); *ibid.* **48**, 1590 (1960); V. C. Littau, V. G. Allfrey, A. E. Mirsky, *ibid.* **52**, 93 (1964); J. Bonner, M. E. Dahmus, D. Fambrough, R. C. Huang, K. Marushige, Y. H. Yuan, *Science* **159**, 47 (1968).
4. J. Paul and R. S. Gilmour, *Nature* **210**, 992 (1966); B. J. McCarthy and B. H. Hoyer, *Proc. Nat. Acad. Sci. U.S.* **52**, 915 (1964).
5. F. Jacob and J. Monod, *J. Mol. Biol.* **3**, 318 (1961).
6. I. Bekhor, J. Bonner, G. K. Dahmus, *Proc. Nat. Acad. Sci. U.S.* **62**, 271 (1969).
7. In this case it would be desirable to use a distinctive term other than producer gene for the DNA sequences coding for such regulative protein molecules. The producer genes would then be defined as those sequences coding for RNA's other than those translated to produce proteins which recognize the receptor sequences.
8. J. H. Frenster [*Nature* **206**, 1269 (1965)] has proposed that RNA could act as a "derepressor." There is only a formal similarity to one part of this model—the activator RNA. An RNA molecule, fully complementary to the "unread" strand was supposed to maintain the DNA of a gene in a strand-separated condition and thus permit its transcription.
9. E. W. Johns and J. A. V. Butler, *Nature* **204**, 853 (1964); K. Murray, *Ann. Rev. Biochem.* **34**, 209 (1965); L. S. Hnilica, *Biochem. Biophys. Acta* **117**, 163 (1966); and H. A. Kappler, *Fed. Proc.* **24**, Pt. I (2), 601 (1965); J. L. Beeson and E. L. Triplett, *Exp. Cell. Res.* **48**, 61 (1967); D. M. Fambrough and J. Bonner, *J. Biol. Chem.* **243**, 4434 (1968); R. J. deLange, D. M. Fambrough, E. L. Smith, J. Bonner, *J. Biol. Chem.* **243**, 5906 (1968). In addition to lack of diversity of tissue specificity of histones shown by the above references, information also exists which indicates directly that histone-DNA recognition does not control the sequence-specific pattern of transcription in chromatin in vitro. [See J. Paul and R. S. Gilmour, *J. Mol. Biol.* **34**, 305 (1968)].
10. I. Bekhor, G. M. Kung, J. Bonner, *J. Mol. Biol.* **39**, 351 (1969); R. C. Huang and P. C. Huang, *ibid.*, p. 365.
11. A. E. Mirsky and H. Ris, *J. Gen. Physiol.* **34**, 451 (1951).
12. A. E. Mirsky, *Harvey Lec.* **Ser. 46**, 98 (1950–51).
13. Species whose genome sizes appear in Fig. 3 are as follows (from top to bottom): *Bos bos* [C. Leuchtenberger, R. Leuchtenberger, C. Vendrely, R. Vendrely, *Exp. Cell Res.* **3**, 240 (1952)]; *Chelonia mydas* [H. Ris and A. E. Mirsky, *J. Gen. Physiol.* **33**, 125 (1949)]; *Scaphiopus couchi* [E. Sexsmith, thesis, Univ. of Toronto (1968)]; *Tetraodon fluviatilis* [R. Hinegardner, *Amer. Natur.* **102**, 517 (1968)]; *Carcharias obscurus* (*11*); *Lampetra planeri* and *Amphioxus lanceolatus* [N. B. Atkin and S. Ohno, *Chromosoma* **23**, 10 (1967)]; *Acidea atra* (*11*); *Paracentrotus lividus* [R. Vendrely, *Compt. Rend. Soc. Biol.* **143**, 1386 (1949)]; *Cassiopea* and *Dysidea crawshagi* (*11*); *Amoeba hystolytica* (A. Gelderman, unpublished data); *Sacharomyces* [M. Ogur, S. Minckler, G. Lindegren, C. G. Lindergren, *Arch. Biochem. Biophys.* **40**, 175 (1952)]; *Escherichia coli* [J. Cairns, *J. Mol. Biol.* **4**, 407 (1962)]; *Mycoplasma* [H. Bode and H. J. Morowitz, *ibid.* **23**, 191 (1967)]; *Simian Virus* 40 [T. Ben-Porat and A. S. Kaplan, *Virology* **16**, 261 (1962)].
14. R. J. Britten and D. E. Kohne, *Science* **161**, 529 (1968).
15. In addition to calf and mouse (*14*), the following species have been shown to contain relatively large quantities of nonrepetitive DNA: *Xenopus laevis* (E. H. Davidson and B. R. Hough, *Proc. Nat. Acad. Sci. U.S.*, in press); Sea urchin and chicken (D. E. Kohne, unpublished data); *Drosophila*, 3 species (C. D. Laird and B. J. McCarthy, in press); Human (R. J. Britten, unpublished data). Even *Amphiuma* with a genome 30 times the size of the mammalian appears to have about 20 percent nonrepeated DNA [R. J. Britten, *Carnegie Inst. Wash. Year B.* **66**, 75 (1968)].
16. The choice of a mammal in this comparison could be objected to on the grounds that an enormous number of producer genes could be required for the construction of the super-complex central nervous system and immunoresponse systems of higher chordates. However, it seems unlikely to us that the enormous increase in genome size which has occurred in evolution could be explained in this way.
17. M. Dixon and E. C. Webb, *Enzymes* (Academic Press, New York, 1964).
18. C. J. Epstein and A. G. Motulsky, in *Progr. Med. Genet.* **4**, 97 (1965).
19. W. J. Young, *J. Hered.* **57**, 58 (1966).
20. Action of the steroid hormones aldosterone, cortisone, testosterone, and estrogen are known to depend on gene activity for their effects in target tissues: for testosterone, S. Liao, R. W. Barton, A. H. Lin [*Proc. Nat. Acad. Sci. U.S.* **55**, 1593 (1966)] and S. Liao [*J. Biol. Chem.* **240**, 1236 (1965)] demonstrate increase in prostate template-active RNA, and actinomycin-sensitive increase in RNA synthesis, with change in base composition. Template activity of muscle chromatin is also increased [C. B. Brewer and J. R. Florini, *Biochemistry* **5**, 3857 (1966)] and testosterone binds to a protein or proteins of prostate chromatin in vivo [N. Bruchovsky and J. D. Wilson, *J. Biol. Chem.* **243**, 5953 (1968)]. A similar range of evidence exists for the effect of estrogen on uterus [see T. H. Hamilton, *Science* **161**, 649 (1968)]. Striking increase in RNA synthesis in uterus occurs almost instantly with estrogen [A. R. Means and T. H. Hamilton, *Proc. Nat. Acad. Sci. U.S.* **56**, 1594 (1966)] and the spectrum of gene products made is different [B. O'Malley, W. L. McGuire, P. A. Middleton, *Nature* **218**, 1249 (1968)]. Corticosteroids cause many liver enzymes to be synthesized [G. Weber, S. K. Srivastava, L. Radhey, *J. Biol. Chem.* **240**, 750 (1965); D. Greengard and G. Acs, *Biochim. Biophys. Acta* **61**, 652 (1962)] and cause sharp increases in liver nuclear RNA synthesis [G. T. Kenny and F. J. Kull, *Proc. Nat. Acad. Sci. U.S.* **50**, 493 (1963)], resulting from both increased polymerase [O. Barnabei, B. Romano, G. Bitono, V. Tomasi, F. Sereni, *Biochim. Biophys. Acta* **113**, 478 (1966); W. Schmid, D. Gallwitz, C. E. Sekeris, *ibid.* **134**, 85 (1967)] and from increased template activity in isolated chromatin [M. E. Dahmus and J. Bonner, *Proc. Nat. Acad. Sci. U.S.* **54**, 1370 (1965)]. The specific effect of aldosterone on isolated bladder is abolished by actinomycin, and this hormone localizes in the bladder cell nuclei [G. A. Porter, R. Bogorach, I. S. Edelman, *Proc. Nat. Acad. Sci. U.S.* **52**, 1326 (1964); I. S. Edelman, R. Bogorach, G. A. Porter, *ibid.* **50**, 1169 (1963)]. The polypeptide gonadotropins follicular stimulating hormone and lutenizing hormone stimulate protein synthesis in ovary in an actinomycin-sensitive way [M. Civen, C. B. Brown, J. Hilliard, *Biochim. Biophys. Acta* **114**, 127 (1966)], and adrenocorticotropin (ACTH) promotes actinomycin-sensitive increase in adrenal RNA synthesis and steroidogenesis [R. B. Farese, *Fed. Proc.* **24** Pt. I (2), 306 (1965)]. Mobilization of Ca^{++} from bone, promoted by the polypeptide parathyroid hormone, is prevented by actinomycin [H. Rasmussen, D. Arnaud, C. Hawker, *Science* **144**, 1019 (1969)]. Several specific plant hormones, including giberellins, auxin, and cytokinins [reviewed by A. Trewas, in *Progr. Phytochem.* **1**, 114 (1968)] as well as antheridogens (B. R. Voeller, personal communication) also appear to act at the genomic level in tissues affected by them. Transport of Ca^{++} mediated by vitamin D in intestinal mucosa is blocked by actinomycin [J. E. Zull, E. Czarnowska-Misztall, H. F. deLuca, *Science* **149**, 183 (1965)]. Various embryonic processes known to occur as a result of specific inducing agents [see review by T. Yamada, in *Compr. Biochem.* **28**, 113 (1967)] are known to be sensitive to actinomycin treatment of the competent tissues, for example, in mouse pancreas induction [N. K. Wessels and F. H. Wilt, *J. Mol. Biol.* **13**, 767 (1965)] and neurulation in the frog [J. Brachet, H. Denis, F. deVitry, *Develop. Biol.* **9**, 398 (1964)].
21. H. R. Maurer and G. R. Chalkey, *J. Mol. Biol.* **27**, 431 (1967).
22. R. W. Shearer and B. J. McCarthy, *Biochemistry* **6**, 283 (1967).
23. These experiments are carried out by dissociating the chromatin in high salt and allowing the DNA and the proteins to reassociate by gradually lowering salt concentration in the presence of 5*M* urea. If the chromosomal RNA (RNA found initially bound to chromatin) is destroyed by ribonuclease or zinc hydrolysis, or if conditions are such as to prevent sequence-specific binding of the RNA to the DNA during the reconstitution procedure, the proteins reassociate in random positions and the reconstituted chromatin makes all the species of gene products made by totally deproteinized DNA. If the chromosomal RNA is allowed to associate with the DNA during the reconstitution process, on the other hand, the proteins of the chromatin evidently return to positions on the DNA close to those in the starting material, so that the reconstituted chromatin now synthesizes a spectrum of RNA's homologous to that of the original chromatin.
24. N. Miyagi, D. Kohl, R. A. Flickinger, *J. Exp. Zool.* **165**, 147 (1967); H. Ursprung, K. D. Smith, W. H. Sofer, D. T. Sullivan, *Science* **160**, 1075 (1968).
25. B. O'Malley, W. L. McGuire, P. A. Middleton, *Nature* **218**, 1249 (1968).
26. R. B. Church and B. J. McCarthy, *J. Mol. Biol.* **23**, 459 (1967); V. R. Glisin, M. V. Glisin, P. Doty, *Proc. Nat. Acad. Sci. U.S.* **56**, 285 (1966); A. H. Whitely, B. J. McCarthy, H. R. Whitely, *ibid.* **55**, 519 (1966).
27. H. Denis, *J. Mol. Biol.* **22**, 285 (1966); E. H. Davidson, M. Crippa, A. E. Mirsky, *Proc. Nat. Acad. Sci. U.S.* **60**, 152 (1968).
28. D. F. Poulson, *Amer. Natur.* **79**, 340 (1945).
29. Some of the most striking examples are the *t* mutant series in the mouse [L. C. Dunn and S. Gluecksohn-Waelsh, *Genetics* **38**, 261 (1953); P. Chesley and L. C. Dunn, *ibid.* **21**, 525 (1936)]; the creeper gene in the chicken [W. Landauer, *J. Genet.* **25**, 367 (1932)]; the control of a variety of morphogenetic patterns in insects such as bristle formation and wing color pattern [see review by D. H. Waddington, *New Patterns in Genetics and Development* (Columbia Univ. Press, New York, 1962), chap. 6]. A spectacular case of organizational control loci is provided by the "bithorax" series in drosophila [E. B. Lewis, in *The Role of Chromosomes in Development*, M. Locke, Ed. (Academic Press, New York, 1964), p. 231].
30. S. Hadorn, *Brookhaven Symp. Biol.* **18**, 148 (1965).
31. B. McClintock, *ibid.*, p. 162.
32. ———, *Develop. Biol.* (*Suppl.*) **1**, 84 (1967).
33. Since the relationship is commonly imperfect and nothing is yet known of the functional importance of sequence multiplicity, the phrase "redundant sequences" cannot be applied in its strict sense. Nevertheless redundant and repeated are used interchangeably since no properly applicable word exists.
34. R. J. Britten, in preparation.
35. ——— and D. E. Kohne, *Carnegie Inst. Wash. Year B.* **66**, 83 (1968).
36. C. Rouiller, Ed., *The Liver: Morphology, Biochemistry, Physiology* (Academic Press, New York, 1963), vol. 1, chaps. 9–11.
37. T. H. Hamilton, *Science* **161**, 649 (1968).
38. M. W. Noall and W. M. Allen, *J. Biol. Chem.* **236**, 2987 (1961).
39. R. J. Laquens, *Ultrastructures Res.* **10**, 578 (1964).
40. A. Csapo, *Ann. N.Y. Acad. Sci.* **75**, 740 (1959).
41. G. Pincus, Ed., *The Hormones: Physiology, Chemistry, and Applications* (Academic Press, New York, 1964), vol. 5, pp. 787–793.
42. Y. Aizawa and G. C. Mueller, *J. Biol. Chem.* **236**, 381 (1961).
43. G. C. Mueller, J. Gorski, Y. Aizawa, in *Mechanism of Action of Steroid Hormones*, C. A. Ville and L. L. Engel, Eds. (Pergamon, Oxford, 1961), pp. 181–184.
44. J. A. Nicolette and J. Gorski, *Arch. Biochem. Biophys.* **107**, 279 (1964).
45. H. Sinohara and H. H. Sky-peck, *ibid.* **106**, 138 (1964).
46. R. Soeiro, H. C. Birnboim, J. E. Darnell, *J. Mol. Biol.* **19**, 362 (1966).
47. R. B. Church and B. J. McCarthy, *Proc. Nat. Acad. Sci. U.S.* **58**, 1548 (1967).
48. G. Attardi, H. Parnas, M-I. H. Hwang, B. Attardi, *J. Mol. Biol.* **20**, 145 (1966).
49. R. A. Weinberg and S. Penman, *ibid.* **38**, 289 (1968).
50. J. Bonner and J. Widholm, *Proc. Nat. Acad. Sci. U.S.* **57**, 1379 (1967); I. Bekhor, J. Bonner, G. K. Dahmus, *ibid.*, in press.

Kimura, M., and T. Ohta. 1971. Protein polymorphism as a phase of molecular evolution. *Nature* 229:467–469.

In the 1960s and 1970s, numerous studies were published on all sorts of organisms concerning protein polymorphism, as well as rates of evolution and molecular phylogenies derived from the amino acid sequences of proteins. In the ensuing decades, these studies were supplemented by investigations of nucleotide sequences in DNA and eventually largely replaced by them. Obtaining DNA sequences had become more efficient, faster, and less expensive than obtaining protein sequences. Two startling discoveries came about from the investigation of protein and DNA polymorphisms. First, that organisms harbor an enormous amount of protein polymorphisms and, second, that the rates of evolution of any particular protein were seemingly constant through millions and even hundreds of millions of years.

The profusion of polymorphisms was apparently in contradiction with the so-called Haldane's dilemma (Haldane 1957), emerging from consideration of the huge number of genetic deaths required in a population as a mutant favored by natural selection replaces an alternative allele. Seemingly, the required number of genetic deaths per generation would be much greater than the number of individuals in the population. One way to resolve the dilemma was to postulate that evolution at the molecular level mostly involved adaptively neutral substitutions. That is, it involved the replacement of one allele for another that was molecularly different but had equal fitness and thus did not involve natural selection. This revolutionary proposal, the neutral theory of molecular evolution" was first advanced by Motoo Kimura (1968). This theory could also account for the apparently constant rate of evolution at the molecular level, because evolution would mostly involve the replacement of one amino acid or one nucleotide for another adaptively equivalent. These replacements would not involve natural selection but would depend on the rate of mutation and the time elapsed, which should be the same for a given gene or protein through time and in different species.

RELATED READING

Ayala, F. J. 1969. An evolutionary dilemma: fitness of genotypes versus fitness of populations. Canadian Journal of Genetics and Cytology 11:439–456.

Ayala, F. J., editor. 1976. Molecular evolution. Sinauer, Sunderland, Massachusetts, USA.

Haldane, J. B. S. 1957. The cost of natural selection. Journal of Genetics 55:511–524.

Kimura, M. 1968. Evolutionary rate at the molecular level. Nature 217:624–626.

Kimura, M. 1983. The neutral theory of molecular evolution. Cambridge University, Cambridge, England.

King, J. L., and T. H. Jukes. 1969. Non-Darwinian evolution. Science 164:788–798.

Nei, M. 1972. Genetic distance between populations. American Naturalist 106:283–291.

Rodríguez-Trelles, F., R. Tarrío, and F. J. Ayala. 2003. Molecular clocks: whence and whither? Pages 5–26 *in* Donoghue, P. C. J., and M. P. Smith, editors, Telling evolutionary time: molecular clocks and the fossil record. CRC Press, London, England.

Protein Polymorphism as a Phase of Molecular Evolution

MOTOO KIMURA & TOMOKO OHTA

National Institute of Genetics, Mishima, Shizuoka-ken

It is proposed that random genetic drift of neutral mutations in finite populations can account for observed protein polymorphisms.

SINCE one of us[1] put forward the theory that the chief cause of molecular evolution is the random fixation of selectively neutral mutants, some have supported the theory[2–5] and others have criticized it[6–8].

Rate of Evolution

Probably the strongest evidence for the theory is the remarkable uniformity for each protein molecule in the rate of mutant substitutions in the course of evolution. This is particularly evident in the evolutionary changes of haemoglobins[5], where, for example, the number of amino-acid substitutions is about the same in the line leading to man as in that leading to the carp from their common ancestor. Similar constancy is found on the whole for cytochrome C, although the rate is different from that of the haemoglobins. The observed rate of amino-acid substitution for the haemoglobins is very near to one Pauling (10^{-9}/amino-acid site/yr) over all vertebrate lines[5]. The rate for cytochrome C is roughly 0.3, while the average rate for several proteins is about 1.6 times this figure[2].

If we define the rate, k, of mutant substitution in evolution as the long term average of the number of mutants that are substituted in the population at a cistron per unit time (year, generation and so on), then under the neutral mutation–random drift theory, we have a simple formula

$$k = u \quad (1)$$

where u is the mutation rate per gamete for neutral mutants per unit time at this locus. Note that this rate k is different from the rate at which an individual mutant increases its frequency within a population. The latter depends on effective population size.

The uniformity of the rate of mutant substitution per year for a given protein may be explained by assuming constancy of neutral mutation rate per year over diverse lines. Moreover, the difference of the evolutionary rates among different molecules can be explained by assuming that the different fraction of mutants is neutral depending on the functional requirement of the molecules.

On the other hand, it can be shown that if the mutant substitution is carried out principally by natural selection

$$k = 4N_e s_1 u \quad (2)$$

where N_e is the effective population number of the species, s_1 is the selective advantage of the mutant and u is the rate at which the advantageous mutants are produced per gamete per unit time[9]. In this case we must assume that in the course of evolution three parameters N_e, s_1 and u are adjusted in such a way that their product remains constant per year over diverse lines. The mere assumption of constancy in the "internal environment" is, however, far from being satisfactory to explain such uniformity of evolutionary rate. In our example of carp–human divergence, we must assume that $N_e s_1 u$ is kept constant in two lines which have been separate for some 400 million years in spite of the fact that the evolutionary rates at the phenotypic level (likely to be governed by natural selection) are so different.

Polymorphism in Sub-populations

Kimura[1] also suggested that the widespread enzyme polymorphisms in *Drosophila*[10] and man[11] as detected by electrophoresis are selectively neutral and that the high level of heterozygosity at such loci can be explained by assuming that most mutations at these cistrons are neutral. This suggestion, however, has been much criticized[12–14]. One of the chief objections is that the same alleles are found in similar frequencies among different sub-populations of a species and that some kind of balancing selection must therefore be involved.

Robertson[15] suggested that if a large fraction of mutations at a locus is selectively neutral, we find either very many alleles segregating in large populations, or a small number of different set of alleles in different isolated small populations. He considered that because neither of these alternatives is found, most polymorphisms have at some time been actively maintained by selection.

Actually, both the situations suggested by Robertson are typical of the heterochromatic pattern of chromosomes in wild populations of the perennial plant *Trillium kamtschaticum*. Extensive cytological studies of this plant by Haga and his associates[16,17] have shown that several chromosome types are segregating within a large population, while different types are fixed in small isolated populations. Indeed, Robertson's suggestion is pertinent if isolation between sub-populations is nearly complete and if the mutation rate for neutral isoallelic variations is sufficiently high that more than one new mutant appears within a large population each generation. The chromosome polymorphism in *Trillium* can be explained by assuming a relatively high mutation rate per chromosome and very low migration rate per generation for this plant.

Mutation and Mobility

On the other hand, it is possible that in animal species such as *Drosophila*, mouse and man well able to migrate, no local population is sufficiently isolated to prevent the entire species or subspecies from forming effectively one panmictic population. In his study on "isolation by distance", Wright[18], using his model of continuum over an area, has concluded that the total species differ little from a single panmictic population if the size of the "neighbourhood" from which parents come is more than 200.

Recently, Maruyama[19,20] made an extensive mathematical analysis of the stepping stone model of finite size. He worked out the exact relationship between local differentiation of gene frequencies and the amount of migration. His results show that in the two dimensional stepping stone model, if N is the effective size of each colony and m is the rate at which each colony exchanges individuals with four surrounding colonies per generation, then marked local differentiation is possible only when Nm is smaller than unity (assuming a large number of colonies arranged on a torus). This is a very severe restriction for migration between colonies because the number of individuals which each colony exchanges with surrounding colonies must be less than an average of one per generation, irrespective of the size of each colony. For the model of continuous distribution of individuals over an area, this condition is equivalent to $N_\sigma < \pi$, where N_σ is the average number of individuals within a circle of radius σ, the standard deviation of the distance of individual migration in one direction per generation. If, on the other hand, there is more migration, the whole population tends to become effectively panmictic. The transition from marked local differentiation to practical panmixis is very rapid for the distribution over an area, and it can be shown that if $N_\sigma > 12$, the whole populations behave as if it were a single panmictic population.

This means that when two or more alleles happen to be segregating within a species, their frequencies among different localities far apart from each other are nearly the same. For animals with separate sexes, it is expected that at least several individual males and females usually exist within a circle of radius σ and the condition $N_\sigma > 12$ is therefore almost always met by widely distributed and actively moving animals. Maruyama also showed that when isolation is more complete and different alleles tend to fix in different local populations, they are connected by zones of intermediate frequencies. The overall pattern then mimics a gene frequency cline resulting from selection, even if alleles are in fact neutral.

Heterozygosity and Probability of Polymorphism

Let us now consider the number of neutral isoalleles maintained in a finite population. Kimura and Crow[21] have shown that if u is the mutation rate per locus (cistron) for neutral mutants, the effective number of alleles maintained in a population of effective size N_e at equilibrium is

$$n_e = 4N_eu + 1 \quad (3)$$

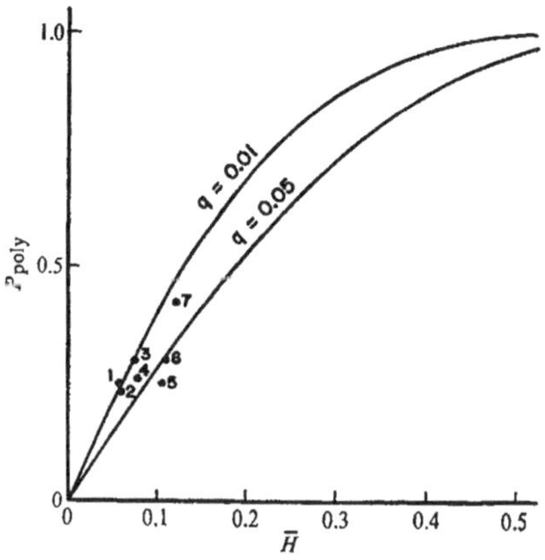

Fig. 1 Relationship between the probability of polymorphism (P_{poly}) and the average heterozygosity $\bar{H}$. The two curves represent the theoretical relationship based on the neutral polymorphism theory; the dots represent observed values. 1, *Limulus polyphemus*; 2, *Peromyscus polionotus*; 3, *Homo sapiens*; 4, *Mus musculus* (Denmark); 5, *Drosophila persimitis*; 6, *Mus musculus* (California); 7, *Drosophila pseudoobscura*.

In deriving this formula, it was assumed that the possible number of allelic states per locus is so large that whenever a mutant appears it represents a new, not pre-existing allele. The effective number of alleles given by formula (3) is equal to the reciprocal of the average homozygosity and is the number estimated by the ordinary procedure of allelism test. Then the average heterozygosity is

$$\bar{H} = 1 - 1/n_e = 4N_eu/(4N_eu + 1) \quad (4)$$

This is the mean frequency of the heterozygotes averaged over all cases including monomorphic and polymorphic cases.

We shall call a population "monomorphic" if the sum of the frequencies of "variant" alleles is q or less. Then it can be shown[22] that the probability of a population being monomorphic is

$$P_{mono} = q^{n_e - 1} \quad (5)$$

when n_e is the effective number of alleles. This formula has been derived under the same condition as formulae (3) and (4). Note that the value of q is arbitrary but a reasonable value is 0.01. If, however, a sample from each locality consists of only a dozen or so individuals, $q=0.05$ may be more appropriate. The probability that a population is polymorphic $(1 - P_{mono})$ is then

$$P_{poly} = 1 - q^{\bar{H}/(1-\bar{H})} \quad (6)$$

Fig. 1 illustrates the relationship between the probability of polymorphism and the average heterozygosity at two levels of q (0.01, 0.05) together with some observed values compiled by Selander *et al.*[23] in their Table 3. The agreement between theoretical and observed values is satisfactory.

From this discussion it can be seen that for actively moving animals such as *Drosophila*, mouse and man, the frequency and the observed pattern of polymorphism can be explained by assuming effective migration such that

$$Nm > 4 \quad (7)$$

between adjacent local populations and also assuming a low mutation rate per cistron for neutral isoalleles such that

$$4N_eu \approx 0.1 \quad (8)$$

In formulae (7) and (8), N refers to the effective size of the local population (colony) while N_e refers to the effective size of the total Mendelian population such as species or subspecies.

When these two conditions are met, the effective number of alleles is on the average 1.1 and at each of the polymorphic loci (constituting roughly 0.3 of all loci) a particular allele takes a high frequency such as 0.8 while the remaining alleles exist in frequencies less than 0.2 in all. Furthermore, frequencies of those alleles among different local populations are about the same throughout the species.

Relative Neutral Mutation Rate

As mentioned already, we know from studies of molecular evolution the average rate of amino-acid substitution. Then, using equation (1), we can infer that the mutation rate for neutral isoalleles is $u_{aa}=1.6\times 10^{-9}$ per amino-acid site per year. If the average cistron responsible for isozyme polymorphisms consists of 300 amino-acids and if 0.3 of amino-acid changes lead to change of the electric charge

$$u=1.6\times 10^{-9}\times 300\times(0.3)=1.6\times 10^{-7}$$

per year. If the fraction of neutral mutants is less among mutants that can be detected by electrophoresis than among those that cannot be so detected, this figure is an overestimate. The same applies if the changes that can be detected by electrophoresis are restricted to amino-acids that are exposed to the surface of the protein molecule. It is therefore possible that the true mutation rate is much lower than this and in the following treatment we take $u=10^{-7}$.

For species such as the mouse, with possibly two generations per year, the mutation rate per generation for neutral isoalleles detectable by electrophoresis is half as large, while for man it should be some twenty times as large. Then the effective population number that satisfies formula (8) is $N_e\approx 0.5\times 10^6$ for the mouse and $N_e\approx 1.3\times 10^4$ for man. The effective number here refers to the species or subspecies in the course of evolution.

We note that if the mutation rate u is constant per year, then the product $N_e u$ should be less variable among different organisms than its components N_e and u, because the species with short generation time tends to have small body size and attain a large population number, while the species which takes many years for one generation tends to have a small population number. At any rate, $u=10^{-7}$ per year is much lower than the standard figure of 10^{-5} per generation even for man and this suggests that, in general, neutral mutants constitute a small fraction of all the mutants at a cistron. Thus, we consider this as one important revision to earlier work[1] in which it was assumed that the neutral mutation rate per locus was high.

We must emphasize, however, that most mutants that spread into the species are neutral, even if the neutral mutants constitute a small fraction of the total mutants at the time of occurrence. Those mutants that are destined to spread to the species take a long time until fixation and on their way take the form of "protein polymorphism".

If most protein polymorphisms constitute a phase of molecular evolution, then the behaviour of molecular mutants in a population is crucial for an understanding of the polymorphism. It was shown by Kimura and Ohta[24] that for a selectively neutral mutant, it takes about $4N_e$ generations to reach fixation in the population (excluding the cases of eventual loss). We can also compute, using the solution of Kimura[25], the average number of generations that have elapsed since the appearance of a neutral allele which happens to have reached frequency 0.5. It turns out that this is about $(1.25)N_e$ generations.

One further factor we must consider is "associative overdominance". When truly overdominant loci are distributed over the genome, they will cause neutral loci to behave as if they were overdominant[26]. As we have shown (unpublished), this will somewhat prolong the time spent by a neutral mutant at intermediate frequencies, but this has no effect on the rate of mutant substitution in evolution. The associative overdominance, however, will play an important role when a small number of chromosomes are extracted from natural populations and rapidly multiplied for an experiment. In this case, spurious "balancing selection" will be observed.

Polymorphism in Living Fossils

Returning to the problems of evolutionary time, we note that the average number of generations between two consecutive fixations of mutants at a given locus (cistron) in the course of evolution is $1/u$. This is roughly ten times as long as the time taken for an individual mutant to reach fixation if $4N_e u=0.1$. It may be interesting to ask, then, how long it takes until new mutants accumulate in the species causing detectable change in the fraction P_d of proteins. If we denote by T_d the average length of time for such change, then

$$T_d=-(1/u)\log_e(1-P_d) \qquad (9)$$

According to Selander *et al.*[27], two Danish subspecies of the house mouse differ at 32% of their loci. Putting $P_d=0.32$ and assuming $u=10^{-7}$ per year, we obtain $T_d\approx 3.9\times 10^6$ yr. The time since divergence of these two subspecies from their common ancestor is given by $T_d/2$ or roughly 2 million years.

In our view, protein polymorphism and molecular evolution are not two separate phenomena, but merely two aspects of a single phenomenon caused by random frequency drift of neutral mutants in finite populations. If this view is correct, we should expect that not only genes in "living fossils" have undergone as many DNA base (and therefore amino-acid) substitutions as corresponding genes in more rapidly evolving species as predicted by Kimura[5], but also they are equally polymorphic and heterozygous at the protein level. A study by Selander *et al.*[23] on the variation of the horseshoe crab at the protein level seems to support this view.

At the moment, our observations are limited to a few organisms, and we do not know how typical their heterozygosities are. It is possible that for organisms with short generation time and small effective population number, the level of heterozygosity is much lower (because of a very small $N_e u$).

The neutral mutation-random drift theory allows us to make a number of definite quantitative as well as qualitative predictions by which the theory can be tested. We hope that through this process we will be able to gain deeper understanding of the mechanism of evolution at the molecular level and will be emancipated from a naive pan-selectionism.

Received October 29, 1970.

1 Kimura, M., *Nature*, **217**, 624 (1968).
2 King, J. L., and Jukes, T. H., *Science*, **164**, 788 (1969).
3 Crow, J. F., *Proc. Twelfth Intern. Cong. Genet.*, **3**, 105 (1969).
4 Arnheim, N., and Taylor, C. E., *Nature*, **223**, 900 (1969).
5 Kimura, M., *Proc. US Nat. Acad. Sci.*, **63**, 1181 (1969).
6 Maynard Smith, J., *Nature*, **219**, 1114 (1968).
7 Richmond, R. C., *Nature*, **225**, 1025 (1970).
8 Clarke, B., *Science*, **168**, 1009 (1970).
9 Kimura, M., and Ohta, T., *J. Mol. Evol.* (in the press).
10 Lewontin, R. C., and Hubby, J. L., *Genetics*, **54**, 595 (1966).
11 Harris, H., *Proc. Roy. Soc.*, B, **164**, 298 (1966).
12 Prakash, S., Lewontin, R. C., and Hubby, J. L., *Genetics*, **61**, 841 (1969).
13 Petras, M. L., Reimer, J. D., Biddle, F. G., Martin, J. E., and Linton, R. S., *Canad. J. Genet. Cytol.*, **11**, 497 (1969).
14 Maynard Smith, J., *Amer. Nat.*, **104**, 231 (1970).
15 Robertson, S., in *Population Biology and Evolution* (edit. by Lewontin, R.), 5 (Syracuse University Press, New York, 1968).
16 Haga, T., and Kurabayashi, M., *Cytologia*, **18**, 13 (1953).
17 Haga, T., in *Chromosomes Today*, **2** (edit. by Darlington, C. D., and Lewis, K. R.), 207 (1969).
18 Wright, S., *Annals of Eugenics*, **15**, 323 (1951).
19 Maruyama, T., *Theoretical Population Biology*, **1**, 101 (1970).
20 Maruyama, T., *Japan. J. Genet.* (in the press).
21 Kimura, M., and Crow, J. F., *Genetics*, **49**, 725 (1964).
22 Kimura, M., *Theoretical Population Biology* (in the press).
23 Selander, R. K., Yang, S. Y., Lewontin, R. C., and Johnson, W. E., *Evolution*, **24**, 402 (1970).
24 Kimura, M., and Ohta, T., *Genetics*, **61**, 763 (1969).
25 Kimura, M., *Proc. US Nat. Acad. Sci.*, **41**, 144 (1955).
26 Ohta, T., and Kimura, M., *Genet. Res.* (in the press).
27 Selander, R. K., Hunt, W. G., and Yang, S. Y., *Evolution*, **23**, 379 (1969).

Eldredge, N., and S. J. Gould. 1972. Punctuated equilibria: an alternative to phyletic gradualism. Pages 82–115 *in* Schopf, T. J. M., editor, *Models in paleobiology*. Freeman, Cooper & Co, San Francisco, California.

This paper had a huge impact on some branches (notably paleontology) of evolutionary biology. Niles Eldredge and Stephen Jay Gould argue that cladogenesis (i.e., speciation) is more important than anagenesis (single-lineage change) as a driver of phenotypic evolution. They suggest that apparent "gaps" in the fossil record are real rather than artifactual, being due to extremely rapid morphological changes (typically in allopatry) that presumably transpire during each speciation event. The authors view evolution as being characterized mostly by phenotypic stasis within established species, punctuated by bursts of rapid evolutionary change during speciation episodes. This paper was also an unabashed attempt by the authors to raise the status of paleobiology by suggesting that the fossil record has much to add to macroevolutionary thought, which had traditionally been dominated by extrapolations (phyletic gradualism) from standard microevolutionary processes such as those addressed in the field of population genetics.

Steven Stanley (1975) introduced the term *rectangular evolution* to characterize punctuated equilibrium, because, under the latter scenario, the branching points (speciation events) in phylogenetic trees might better be plotted as rectangular corners (right-angle lines) to denote that substantial phenotypic changes often accompany cladogenesis. Some critics argued that rectangular evolution is compromised by circular reasoning: if species in the fossil record are recognized by their morphological differences (as they are), then how could it be otherwise than that morphological change must appear to be associated with speciation (cladogenesis)? (For more on this topic, see Stebbins and Ayala 1981 in this volume.)

RELATED READING

Gould, S. J. 1989. Wonderful life: the Burgess Shale and the nature of history. Norton, New York, New York, USA.

Stanley, S. M. 1975. A theory of evolution above the species level. Proceedings of the National Academy of Sciences, USA 72:646–650.

5

PUNCTUATED EQUILIBRIA: AN ALTERNATIVE TO PHYLETIC GRADUALISM

Niles Eldredge • Stephen Jay Gould

Editorial introduction. Moving from populations to species, we recall that the process of speciation as seen through the hyperopic eyes of the paleontologist is an old and venerable theme. But the significance of "gaps" in the fossil record has been a recurrent "difficulty," used on the one hand to show that spontaneous generation is a "fact," and on the other hand to illustrate the "incompleteness" of the fossil record. Some have expressed a third interpretation, which views such gaps as the logical and expected result of the allopatric model of speciation.

Bernard's Eléments de Paléontologie (1895) discusses the existence of gaps in the fossil record as follows, p. 25, English edition:

> Still it remains an indisputable fact that in the most thoroughly explored regions, those where the fauna is best known, as, for instance, the Tertiary of the Paris basin, the species of one bed often differ widely from those of the preceding, even where no stratigraphic gap appears between them. This is easily explained. The production of new forms usually takes place within narrowly limited regions. It may happen in reality that one form evolves in the same manner in localities widely separated from each other, and farther on we shall see examples of this: but this is not generally the case, the area of the appearance of species is

usually very circumscribed. This fact has been established in the case of certain butterflies and plants. The diversity having once occurred, the new types spread often to great distances, and may be found near the present form without crossing with it or presenting any trace of transition.

The same phenomenon must have taken place in former epochs. It is then only by the merest chance that geologists are able to locate the origin of the species they have under consideration; if, furthermore, the phenomena of erosion or metamorphism have destroyed or changed the locality in question, direct observation will not furnish any means of supplying the missing links of the chain.

Although this has been pointed out nicely by Bernard—and moreover, any number of paleontologists will tell you that this is what they teach—comprehension and application are two different things. And indeed, the fossil record has been interpreted by many to show just the opposite. J. B. S. Haldane's classical *The Cause of Evolution*, published in 1932, contains the following passage (p. 213):

> But [Sewall] Wright's theory [that evolution is most rapid in populations large enough to be reasonably variable, but small enough to permit large changes in gene frequencies due to random drift] certainly supports the view taken in this book that the evolution in large random-mating populations, which is recorded by paleontology, is not representative of evolution in general, and perhaps gives a false impression of the events occurring in less numerous species.

Thus an extremely eminent student of the evolutionary process considered that the known fossil record supported the view of evolution proceeding as a stately unfolding of changes in large populations.

The interpretation supported by Eldredge and Gould is that allopatric speciation in small, peripheral populations *automatically* results in "gaps" in the fossil record. Throughout their essay, however, runs a larger and more important lesson: *a priori* theorems often determine the results of "empirical" studies, before the first shred of evidence is collected. This idea, that theory dictates what one sees, cannot be stated too strongly.

Statement

In this paper we shall argue:

(1) The expectations of theory color perception to such a degree that new notions seldom arise from facts collected under the influence of old pictures of the world. New pictures must cast their influence before facts can be seen in different perspective.

(2) Paleontology's view of speciation has been dominated by the picture of "phyletic gradualism." It holds that new species arise from the slow and steady transformation of entire populations. Under its influence, we seek unbroken fossil series linking two forms by insensible gradation as the only complete mirror of Darwinian processes; we ascribe all breaks to imperfections in the record.

(3) The theory of allopatric (or geographic) speciation suggests a different interpretation of paleontological data. If new species arise very rapidly in small, peripherally isolated local populations, then the great expectation of insensibly graded fossil sequences is a chimera. A new species does not evolve in the area of its ancestors; it does not arise from the slow transformation of all its forbears. Many breaks in the fossil record are real.

(4) The history of life is more adequately represented by a picture of "punctuated equilibria" than by the notion of phyletic gradualism. The history of evolution is not one of stately unfolding, but a story of homeostatic equilibria, disturbed only "rarely" (i.e., rather often in the fullness of time) by rapid and episodic events of speciation.

The Cloven Hoofprint of Theory

> Innocent, unbiased observation is a myth.
> P. B. Medawar (1969, p. 28)

Isaac Newton possessed no special flair for the turning of phrases. Yet two of his epigrams have been widely cited as guides for the humble and proper scientist—his remark in a letter of 1675 written to Hooke: "If I have seen farther, it is by standing on the shoulders of giants," and his confusing comment of the *Principia* (1726 edition, p. 530): "hypotheses non fingo"—[I frame no hypotheses]. The first is not his own; it has a pedigree extending back at least to Bernard of Chartres in 1126 (Merton, 1965). The second is his indeed, but modern philosophers have offered as many interpretations for it as the higher critics heaped upon Genesis 1 in their heyday (see Mandelbaum, 1964, p. 72 for a bibliography).

Although most scholars would now hold, with Hanson (1969, 1970, see also Koyré, 1968), that Newton meant only to eschew idle speculation and untestable opinion, his phrase has traditionally been interpreted in another light—as the credo of an inductivist philosophy that views "objective" fact as the primary input to science and theory as the generalization of this unsullied information. For example, Ernst Mach, the great physicist-philosopher, wrote (1893, p. 193): "Newton's reiterated and emphatic protestations that he is not concerned with hypotheses as to the causes of phenomena, but has simply to do with the investigation and transformed statement of *actual facts* . . . stamps him as a philosopher of the *highest* rank."

Today, most philosophers and psychologists would brand the inductivist credo as naive and untenable on two counts:

(1) We do not encounter facts as *data* (literally "given") discovered objectively. All observation is colored by theory and expectation. (See Vernon, 1966, on the relation between expectation and perception. For a radical view, read Feyerabend's (1970) claim that theories act as "party lines" to force observation in preset channels, unrecognized by adherents who think they perceive an objective truth.)

(2) Theory does not develop as a simple and logical extension of observation; it does not arise merely from the patient accumulation of facts. Rather, we observe in order to test hypotheses and examine their consequences. Thus, Hanson (1970, pp. 22–23) writes: "Much recent philosophy of science has been dedicated to disclosing that a 'given' or a 'pure' observation language is a myth-eaten fabric of philosophical fiction. . . . In any observation statement the cloven hoofprint of theory can readily be detected."

Yet, inductivist notions continue to control the methodology and ethic of practicing scientists raised in the tradition of British empiricism. In unguarded moments, great naturalists have correctly attributed their success to skill in hypothesizing and power in imagination; yet, in the delusion of conscious reflection, they have usually ascribed their accomplishments to patient induction. Thus, Darwin, in a statement that should be a motto for all of us (letter to Fawcett, September 18, 1861, quoted in Medawar, 1969), wrote:

> About thirty years ago there was much talk that geologists ought only to observe and not theorize; and I well remember someone saying that at this rate a man might as well go into a gravel-pit and count the pebbles and describe the colours. How odd it is that anyone should not see that all observation must be for or against some view if it is to be of any service.

Yet, in traditional obeisance to inductivist tenets, he wrote in his autobiography that he had "worked on true Baconian principles, and without any theory collected facts on a wholesale scale" (see discussion of this point in Ghiselin, 1969a; Medawar, 1969; and de Beer, 1970).

Almost all of us adhere, consciously or unconsciously, to the inductivist methodology. We do not recognize that all our perceptions and descriptions are made in the light of theory. Leopold (1969, p. 12), for example, claimed that he could describe and analyze the aesthetics of rivers "without introduction of any personal preference or bias." He began by generating "uniqueness" values, but abandoned that approach when the sluggish, polluted, murky Little Salmon River scored highest among his samples. He then selected a very small subset of his measures for a simplified type of multivariate scaling. As he must have known before he started, Hells Canyon of the Snake River now ranked best. It cannot be accidental that the article was written by an opponent to applications then before the Federal Power Commission for the damming of Hells Canyon. (It is no less fortuitous that so many philosophers, Hegel and Spencer in particular, generated ideal states by pure reason that mirrored their own so well.)

In paleontology, even the most "objective" undertaking, the "pure" description of fossils, is all the more affected by theory because that theory is unacknowledged. We describe part by part and are led, subtly but surely, to the view that complexity is irreducible. Such description stands against a developing science of form (Gould, 1970a, 1971a) because it both gathers different facts (static states rather than dynamic correlations) and presents contrary comparisons (compendia of differences rather than reductions of complexity to fewer generating factors). D'Arcy Thompson, with his usual insight, wrote of the "pure" taxonomist (1942, p. 1036), "when comparing one organism with another, he describes the differences between them point by point and 'character' by 'character.' If he is from time to time constrained to admit the existence of 'correlation' between characters . . . yet all the while he recognizes this fact of correlation somewhat vaguely, as a phenomenon due to causes which, except in rare instances, he can hardly hope to trace; and he falls readily into the habit of thinking and talking of evolution as though it had proceeded on the lines of his own description, point by point and character by character."

The inductivist view forces us into a vicious circle. A theory often compels us to see the world in its light and support. Yet, we think we see objectively and therefore interpret each new datum as an independent confirmation of our theory. Although our theory may be wrong, we cannot confute it. To extract ourselves from this dilemma, we must bring in a more adequate theory; it will not arise from facts collected in the old way. Paleontology supported creationism in continuing comfort, yet the imposition of Darwinism forced a new, and surely more adequate, interpretation upon old facts. Science progresses more by the introduction of new world-views or "pictures"* than by the steady accumulation of information.

This issue is central to the study of speciation in paleontology. We believe that an inadequate picture has been guiding our thoughts on speciation for 100 years. We hold that its influence has been all the more tenacious because paleontologists, in claiming that they see objectively, have not recognized its guiding sway. We contend that a notion developed elsewhere, the theory of allopatric speciation, supplies a more satisfactory picture for the ordering of paleontological data.

* We have no desire to enter the tedious debate over what is, or is not, a "model," "theory," or "paradigm" (Kuhnian, not Rudwickian). In using the neutral word "picture," we trust that readers will understand our concern with alternate ways of seeing the world that render the same facts in *different* ways.

Phyletic Gradualism: Our Old and Present Picture

Je mehr sich das palaeontologische Material vergrössert, desto zahlreicher und vollständiger werden dic Formenreihen.

Zittel, 1895, p. 11

Charles Darwin viewed the fossil record more as an embarrassment than as an aid to his theory. Why, he asked (1859, p. 310), do we not find the "infinitely numerous transitional links" that would illustrate the slow and steady operation of natural selection? "Why then is not every geological formation and every stratum full of such intermediate links? Geology assuredly does not reveal any such finely graduated organic chain; and this, perhaps, is the gravest objection which can be urged against my theory" (1859, p. 280). Darwin resolved this dilemma by invoking the great inadequacy of surviving evidence (1859, p. 342): "The geological record is extremely imperfect and this fact will to a large extent explain why we do not find interminable varieties, connecting together all the extinct and existing forms of life by the finest graduated steps. He who rejects these views on the nature of the geological record, will rightly reject my whole theory."

Thus, Darwin set a task for the new science of evolutionary paleontology: to demonstrate evolution, search the fossil record and extract the rare exemplars of Darwinian processes—insensibly graded fossil series, spared somehow from the ravages of decomposition, non-deposition, metamorphism, and tectonism. Neither the simple testimony of change nor the more hopeful discovery of "progress" would do, for anti-evolutionists of the catastrophist schools had claimed these phenomena as consequences of their own theories. The rebuttal of these doctrines and the test for (Darwinian) evolution could only be an *insensibly graded fossil sequence*—this discovery of all transitional forms linking an ancestor with its presumed descendant (*figure 5–1*). The task that Darwin set has guided our studies of evolution to this day.*

In titling his book *On the Origin of Species by Means of Natural Selection*, Darwin both identified this event as the keystone of evolution and stated his belief in its manner of occurrence. New species can arise in only two ways: by the transformation of an entire population from one state to another (phyletic evolution) or by the splitting of a lineage (speciation). The second process must occur: otherwise there could be no increase in numbers of taxa and life would cease as lineages became extinct. Yet, as Mayr (1959) noted, Darwin muddled this distinction and cast most of his discussion in terms of phyletic

* Beliefs in "saltative" evolution, buttressed by de Vries' "mutation theory," collapsed when population geneticists of the 1930's welded modern genetics and Darwinism into our "synthetic theory" of evolution. The synthetic theory is completely Darwinian in its identification of natural selection as the efficient cause of evolution.

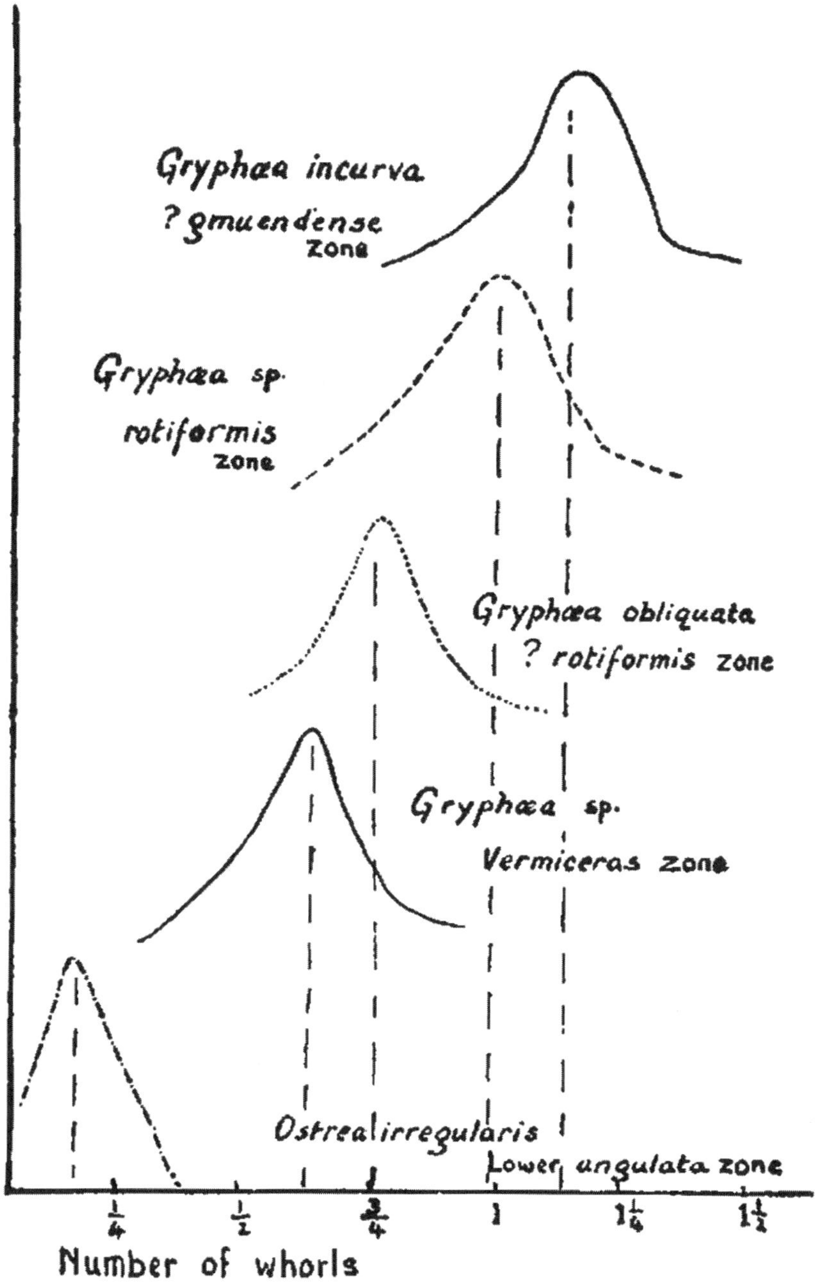

Figure 5–1: The classic case of postulated phyletic gradualism in paleontology. Slow, progressive, and gradual increase in whorl number in the basal Liassic oyster *Gryphaea*. From Trueman, 1922; figure 5.

evolution. His insistence on insensibly graded sequences among fossils reflects this emphasis, for if species arise by the gradual transformation of entire populations, an even sequence of intermediates should indeed be found. When Darwin did discuss speciation (the splitting of lineages), he

88

continued to look through the glasses of transformation: he saw splitting largely as a sympatric process, proceeding slowly and gradually, and producing progressive divergence between forms. To Darwin, therefore, speciation entailed the same expectation as phyletic evolution: a long and insensibly graded chain of intermediate forms. Our present texts have not abandoned this view (*figure 5–2*), although modern biology has.

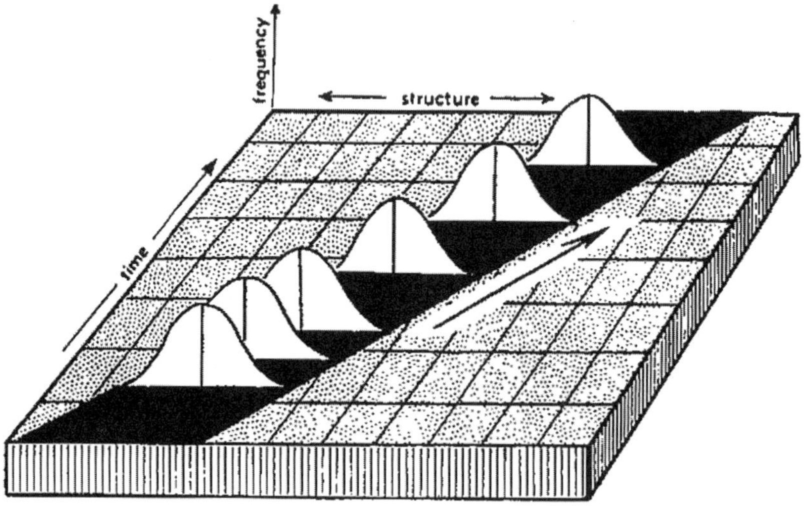

Figure 5–2: A standard textbook view of evolution *via* phyletic gradualism. From Moore, Lalicker, and Fischer, 1952; figure 1–14.

In this Darwinian perspective, paleontology formulated its picture for the origin of new taxa. This picture, though rarely articulated, is familiar to all of us. We refer to it here as "phyletic gradualism" and identify the following as its tenets:

(1) New species arise by the transformation of an ancestral population into its modified descendants.

(2) The transformation is even and slow.

(3) The transformation involves large numbers, usually the entire ancestral population.

(4) The transformation occurs over all or a large part of the ancestral species' geographic range.

These statements imply several consequences, two of which seem especially important to paleontologists:

(1) Ideally, the fossil record for the origin of a new species should consist of a long sequence of continuous, insensibly graded intermediate forms linking ancestor and descendant.

(2) Morphological breaks in a postulated phyletic sequence are due to imperfections in the geological record.

Under the influence of phyletic gradualism, the rarity of transitional series remains as our persistent bugbear. From the reputable claims of a Cuvier or an Agassiz to the jibes of modern cranks and fundamentalists, it has stood as the bulwark of anti-evolutionist arguments: "For evolution to be true, there had to be thousands, millions of transitional forms making an unbroken chain" (Anon., 1967—from a Jehovah's Witnesses pamphlet).

We have all heard the traditional response so often that it has become imprinted as a catechism that brooks no analysis: the fossil record is extremely imperfect. To cite but one example: "The connection of arbitrarily selected 'species' in a time sequence, in fact their complete continuity with one another, is to be expected in all evolutionary lineages. But, *fortunately*, because of the imperfect preservation of fossil faunas and floras, we shall meet relatively few examples of this, no matter how long paleontology continues" (Eaton, 1970, p. 23—our italics; we are amused by the absurdity of a claim that we should rejoice in a lack of data because of the taxonomic convenience thus provided).

This traditional approach to morphological breaks merely underscores what Feyerabend meant (see above) in comparing theories to party lines, for it renders the picture of phyletic gradualism virtually unfalsifiable. The picture prescribes an interpretation and the interpretation, viewed improperly as an "objective" rendering of data, buttresses the picture. We have encountered no dearth of examples, and cite the following nearly at random. Neef (1970) encountered "apparent saltation in the *Pelicaria* lineage" (p. 464), a group of Plio-Pleistocene snails. Although he cites no lithologic or geographic data favoring either interpretation, the picture of phyletic gradualism prescribes a preference: "It is likely that the discontinuity . . . is due to a period of non-deposition. . . . The possibility that the apparent saltations in the *Pelicaria* lineage are due to the migration of advanced forms from small nearby semi-isolated populations and that deposition of the Marima Sandstone was continuous cannot be entirely excluded" (1970, p. 454).

Moreover, the picture's influence has many subtle extensions. For instance:

(1) It colors our language. We are compelled to talk of "morphological breaks" in order to be understood. But the term is not a neutral descriptor; it presupposes the truth of phyletic gradualism, for a "break" is an interruption of something continuous. (Under a deVriesian picture, for example, "breaks" are "saltations"; they are real and expressive of evolutionary processes.)

(2) It prescribes the cases that are worthy of study. If breaks are artificial, the sequences in which they abound become, *ipso facto*, poor objects for evolutionary investigation. But surely there is something insidious here: if breaks are real and stand against the picture of phyletic gradualism, then the picture itself excludes an investigation of the very cases that could place it in jeopardy.

If we doubt phyletic gradualism, we should not seek to "disprove" it "in the rocks." We should bring a new picture from elsewhere and see if it provides a more adequate interpretation of fossil evidence. In the next section, we express our doubts, display a different picture, and attempt this interpretation.

But before leaving the picture of phyletic gradualism, we wish to illustrate its pervasive influence in yet another way. Kuhn (1962) has stressed the impact of textbooks in molding the thought of new professionals. The "normal science" that they inculcate is "a strenuous and devoted attempt to force nature into the conceptual boxes supplied by professional education" (1962, p. 5).

Before the "modern synthesis" of the 1930's and 40's, English-speaking invertebrate paleontologists were raised upon two texts—Eastman's translation of Zittel (1900) and that venerable *Gray's Anatomy* of British works, Woods' *Palaeontology* (editions from 1893 to 1946, last edition reprinted five times before 1958 and still very much in use). Both present an orthodox version of phyletic gradualism. In a classic statement, ending with the sentence that serves as masthead to this section, Zittel wrote (Eastman translation, 1900, p. 10):

> Weighty evidence for the progressive evolution of organisms is afforded by fossil transitional series, of which a considerable number are known to us, notwithstanding the imperfection of the palaeontological record. By transitional series are meant a greater or lesser number of similar forms occurring through several successive horizons, and constituting a practically unbroken morphic chain . . . With increasing abundance of palaeontological material, the more numerous and more complete are the series of intermediate forms which are brought to light.

The last edition of Woods (1946) devotes three pages to evolution; all but two paragraphs (one on ontogeny, the other on orthogenesis) to an exposition of phyletic gradualism (one page on the imperfection of the record, another on some rare examples of graded sequences).

Our current textbooks have changed the argument not at all. Moore, Lalicker and Fischer (1952, p. 30), in listing the fossil record among "evidences of evolution," have only this to say about it: "Although lack of knowledge is immeasurably greater than knowledge, many lineages among fossils of various groups have been firmly established. These demonstrate the transformation of one species or genus into another and thus constitute documentary evidence of gradual evolution." And Easton (1960, p. 34), citing the apotheosis of our achievements, writes: "An evolutionary series represents the peak of scientific accomplishment in organizing fossil invertebrates. It purports to show an orderly progression in morphologic changes among related creatures during successive intervals of time."

That these older texts hold so strongly to phyletic gradualism should surprise no one; harder to understand is the fact that virtually all modern texts repeat the same arguments even though their warrant had disappeared, as we shall now show, with the advent of the allopatric theory of speciation.

The Biospecies and Punctuated Equilibria: A Different Picture of Speciation

> Habits of thought in the tradition of a science are not readily changed, it is not easy to deviate from the customary channels of accumulated experience in conventionalized subjects.
>
> G. L. Jepsen, 1949, p. v

An irony. The formulation of the biological species concept was a major triumph of the synthetic theory (Mayr, 1963, abridged and revised 1970, remains the indispensable source on its meaning and implications). Since paleontology has always taken its conceptual lead from biology (with practical guidance from geology), it was inevitable that paleontologists should try to discover the meaning of the biospecies for their own science.

Here we meet an ironic situation: the taxonomic perspective—one of our persistent albatrosses—dictated an approach to the biospecies. Instead of extracting its insights about evolutionary processes, we sought only its prescriptions for classification. We learned that species are populations, that they are recognized in fossils by ranges of variability not by correspondence to idealized types. The "new systematics" ushered in the revolution in species-level classification that Darwin's theory had implied but not effected. In paleontology, its main accomplishment has been a vast condensation and elimination of spurious taxa established on typological criteria.

But the new systematics also rekindled a theoretical debate unsurpassed in the annals of paleontology for its ponderous emptiness: What is the nature of a paleontological species? In this reincarnation: can taxa designated as biospecies be recognized from fossils? Biologists insisted that the biospecies is a "real" unit of nature, a population of interacting individuals, reproductively isolated from all other groups. Yet its reality seemed to hinge upon what Mayr calls its "non-dimensional" aspect: species are distinct at any moment in time, but the boundaries between forms must blur in temporal extension—a continuous lineage cannot be broken into objective segments. Attempts to reconcile or divorce the non-dimensional biospecies and the temporal "paleospecies" creep on apace (Imbrie, 1957; Weller, 1961; McAlester, 1962; Shaw, 1969; and an entire symposium edited by Sylvester-Bradley, 1956); if obfuscation is any sign of futility, we offer the following as a plea for the termination of this discussion: "Such a plexiform lineage . . .

constitutes a chronospecies (or paleospecies), and it is composed of many successional polytypic morphospecies ('holomorphospecies'), each of which is in theory the paleontological equivalent of a neontological biospecies" (Thomas, 1956, p. 24).

The discussion is futile for a very simple reason: the issue is insoluble; it is not a question of fact (phylogeny proceeds as it does no matter how we name its steps), but a debate about ways of ordering information. When Whitehead said that all philosophy was a footnote to Plato, he meant not only that Plato had identified all the major problems, but also that the problems were still debated because they could not be solved. The point is this: the hierarchical system of Linnaeus was established for his world: a world of discrete entities. It works for the living biota because most species are discrete at any moment in time. It has no objective application to evolving continua, only an arbitrary one based on subjective criteria for division. Linnaeus would not have set up the same system for our world. As Vladimir Nabokov writes in *Ada* (1969, p. 406): "Man . . . will never die, because there may never be a taxonomical point in his evolutionary progress that could be determined as the last stage of man in the cline turning him into Neohomo, or some horrible throbbing slime."

Then does the biospecies offer us nothing but semantic trouble? On one level, the answer is no because it can be applied with great effectiveness to past time-planes. But on another level, and this involves our irony, we must avoid the narrow approach that embraces a biological concept only when it can be transplanted bodily into our temporal taxonomy. The biospecies abounds with implications for the operation of evolutionary processes. Instead of attempting vainly to name successional taxa objectively in its light (McAlester, 1962), we should be applying its concepts. In the following section, we argue that one of these concepts—the theory of allopatric speciation—might reorient our picture for the origin of taxa.

Implications of allopatric speciation for the fossil record. We wish to consider an alternate picture to phyletic gradualism; it is based on a theory of speciation that arises from the behavior, ecology, and distribution of modern biospecies. First, we must emphasize that mechanisms of speciation can be studied directly only with experimental and field techniques applied to living organisms. No theory of evolutionary mechanisms can be generated directly from paleontological data. Instead, theories developed by students of the modern biota generate predictions about the course of evolution in time. With these predictions, the paleontologist can approach the fossil record and ask the following question: Are observed patterns of geographic and stratigraphic distribution, and apparent rates and directions of morphological change, consistent with the consequences of a particular theory of speciation? We can apply and test, but we cannot generate new mechanisms. If discrepancies are found between paleontological data and the expected patterns,

93

we may be able to identify those aspects of a general theory that need improvement. But we cannot formulate these improvements ourselves.*

During the past thirty years, the allopatric theory has grown in popularity to become, for the vast majority of biologists, *the* theory of speciation. Its only serious challenger is the sympatric theory. Here we discuss only the implications of the allopatric theory for interpreting the fossil record of sexually-reproducing metazoans. We do this simply because it is the allopatric, rather than the sympatric, theory that is preferred by biologists. We shall therefore contrast the allopatric theory with the picture of phyletic gradualism developed in the last section.

Most paleontologists, of course, are aware of this theory, but the influence of phyletic gradualism remains so strong that discussions of geographic speciation are almost always cast in its light: geographic speciation is seen as the slow and steady transformation of two separated lineages—i.e., as *two* cases of phyletic gradualism (*figure 5–3*). Raup and Stanley (1971, p. 98), for example, write:

> Let us consider populations of a species living at a given time but not in geographic contact with each other. . . . Two or more segments of the species thus evolve and undergo *phyletic* speciation independently. . . . The distinction between phyletic and geographic speciation is to some extent artificial in that both processes depend on natural selection. The critical difference is that phyletic speciation is accomplished in the absence of geographic isolation and geographic speciation requires geographic isolation (italics ours).

The central concept of allopatric speciation is that new species can arise only when a small local population becomes isolated at the margin of the geographic range of its parent species. Such local populations are termed *peripheral isolates.* A peripheral isolate develops into a new species if *isolating mechanisms* evolve that will prevent the re-initiation of gene flow if the new form re-encounters its ancestors at some future time. As a consequence of the allopatric theory, new fossil species do not originate in the place where their ancestors lived. It is extremely improbable that we shall be able to trace the gradual splitting of a lineage merely by following a certain species up through a local rock column.

Another consequence of the theory of allopatric processes follows: since selection always maintains an equilibrium between populations and their local environment, the morphological features that distinguish the descendant

* The rate and direction of morphological change over long periods of time is the most obvious kind of evolutionary pattern that we can test against predictions based on processes observed over short periods of time by neontologists. We try to do this in the next section.

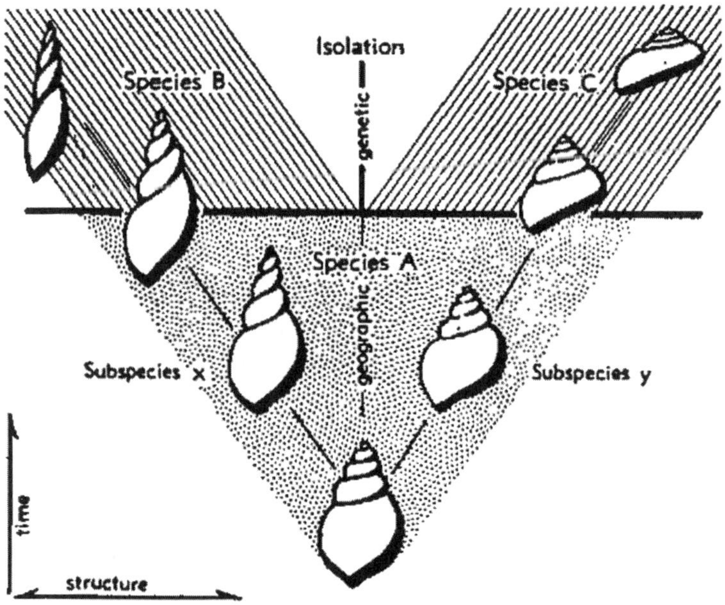

Figure 5–3: A hypothetical case of geographic speciation viewed from the perspective of phyletic gradualism—slow and gradual transformation in two lineages. From Moore, Lalicker, and Fischer, 1952; figure 1–15.

species from its ancestor are present close after, if not actually prior to, the onset of genetic isolation. These differences are often accentuated if the two species become sympatric at a later date (character displacement—Brown and Wilson, 1956). In any event, most morphological divergence of a descendant species occurs very early in its differentiation, when the population is small and still adjusting more precisely to local conditions. After it is fully established, a descendant species is as unlikely to show gradual, progressive change as is the parental species. Thus, in the fossil record, we should not expect to find gradual divergence between two species in an ancestral-descendant relationship. Most evolutionary changes in morphology occur in a short period of time relative to the total duration of species. After the descendant is established as a full species, there will be little evolutionary change except when the two species become sympatric for the first time.

These simple consequences of the allopatric theory can be combined into an expected pattern for the fossil record. Using stratigraphic, radiometric, or biostratigraphic criteria (for organisms other than those under study), we establish a regional framework of correlation. Starting with these correlations, patterns of geographic (not stratigraphic) variation among samples of fossils should appear. Tracing a fossil species through any local rock column, so long as no drastic changes occur in the physical environment, should produce *no* pattern of constant change, but one of oscillation in mean values. Closely related (perhaps descendant) species that enter the rock column should appear suddenly and show no intergradation with the "ancestral" species in morphological features that act as inter-specific differentia. There should be no gradual divergence between the two species when both persist for some time to higher stratigraphic levels. Quite the contrary—it is likely that the two species will display their greatest difference when the descendant first appears. Finally, in exceptional circumstances, we may be able to identify the general area of the ancestor's geographic range in which the new species arose.

Another conclusion is that time and geography, as factors in evolution, are not so comparable as some authors have maintained (Sylvester-Bradley, 1951). The allopatric theory predicts that most variation will be found among samples drawn from different geographic areas rather than from different stratigraphic levels in the local rock column. The key factor is adjustment to a heterogeneous series of micro-environments vs. a general pattern of stasis through time.

In summary, we contrast the tenets and predictions of allopatric speciation with the corresponding statements of phyletic gradualism previously given:

(1) New species arise by the splitting of lineages.

(2) New species develop rapidly.

(3) A small sub-population of the ancestral form gives rise to the new species.

(4) The new species originates in a very small part of the ancestral species' geographic extent—in an isolated area at the periphery of the range.

These four statements again entail two important consequences:

(1) In any *local* section containing the ancestral species, the fossil record for the descendant's origin should consist of a sharp morphological break between the two forms. This break marks the migration of the descendant, from the peripherally isolated area in which it developed, into its ancestral range. Morphological change in the ancestor, even if directional in time, should bear no relationship to the descendant's morphology (which arose in response to local conditions in its isolated area). Since speciation occurs rapidly in small populations occupying small areas far from the center of ancestral abundance, we will rarely discover the actual event in the fossil record.

(2) Many breaks in the fossil record are real; they express the way in which evolution occurs, not the fragments of an imperfect record. The sharp break in a local column accurately records what happened in that area through time. Acceptance of this point would release us from a self-imposed status of inferiority among the evolutionary sciences. The paleontologist's gut-reaction is to view almost any anomaly as an artifact imposed by our institutional millstone—an imperfect fossil record. But just as we now tend to view the rarity of Precambrian metazoans as a true reflection of life's history rather than a testimony to the ravages of metamorphism or the lacunae of Lipalian

intervals, so also might we reassess the smaller breaks that permeate our Phanerozoic record. We suspect that this record is much better (or at least much richer in optimal cases) than tradition dictates.

Problems of phyletic gradualism. In our alternate picture of phyletic gradualism, we are not confronted with a self-contained theory from modern biology. The postulated mechanism for gradual uni-directional change is "orthoselection," usually viewed as a constant adjustment to a uni-directional change in one or more features of the physical environment. The concept of orthoselection arose as an attempt to remove the explanation of gradual morphological change from the realm of metaphysics ("orthogenesis"). It does *not* emanate from *Drosophila* laboratories, but represents a hypothetical extrapolation of selective mechanisms observed by geneticists.

Extrapolation of gradual change under selection to a complete model for the origin of species fails to recognize that speciation is primarily an ecological and geographic process. Natural selection, in the allopatric theory, involves adaptation to local conditions and the elaboration of isolating mechanisms. Phyletic gradualism is, in itself, an insufficient picture to explain the origin of diversity in the present, or any past, biota.

Although phyletic gradualism prevails as a picture for the origin of new species in paleontology, very few "classic" examples purport to document it. A few authors (MacGillavry, 1968, Eldredge, 1971) have offered a simple and literal interpretation of this situation: *in situ*, gradual, progressive evolutionary change is a rare phenomenon. But we usually explain the paucity of cases by a nearly-ritualized invocation of the inadequacy of the fossil record. It *is* valid to point out the rarity of thick, undisturbed, highly fossiliferous rock sections in which one or more species occur continuously throughout the sequence. Nevertheless, if most species evolved according to the tenets of phyletic gradualism, then, no matter how discontinuous a species' occurrence in thick sections, there should be a shift in one or more variables from sample to sample up the section. This is, in fact, the situation in most cases of postulated gradualism: the "gradualism" is represented by dashed lines connecting known samples. This procedure provides an excellent example of the role of preconceived pictures in "objectively documented" cases. One of the early "classics" of phyletic gradualism, Carruthers' (1910) study of the Carboniferous rugose coral *Zaphrentites delanouei* (Milne-Edwards and Haime) and its reinterpretation by Sylvester-Bradley (1951), is of this kind. We do not say that the analysis is incorrect; the *Z. delanouei* stock may have evolved as claimed. We merely wish to show how the *a priori* picture of phyletic gradualism has imposed itself upon limited data.

How pervasive, then, is gradualism in these quasi-continuous sequences? A number of authors (including, *inter alia*, Kurtén, 1965, MacGillavry, 1968, and Eldredge, 1971) have claimed that most species show little or no change throughout their stratigraphic range. But though it is tempting to conclude that gradual, progressive morphological change is an illusion, we recognize that there is little hard evidence to support either view.

As a final, and admittedly extreme, example of *a priori* beliefs in phyletic gradualism, we cite the work of Brace (1967) on human evolution. This is all the more instructive since most paleoanthropologists, in reversing an older view that Brace still maintains, now claim that hominid evolution involves speciation by splitting as well as phyletic evolution by transformation (seen especially in the presumed coexistence of two australopithecine species in the African lower Pleistocene—Howell, 1967; Tobias, 1965; Pilbeam, 1968; Pilbeam and Simons, 1965). Brace (1967) has claimed that the fossil record of man includes four successive "stages" in direct ancestral-descendant relation. These are the Australopithecine (with two successive "phases"—the australopithecus and paranthropus), the Pithecanthropus, the Neanderthaloid, and, finally, the Modern Stage. In discussing the history of paleoanthropology, Brace shows that most denials of ancestral-descendant relationships among hominid fossils stem from a desire to avoid the conclusion that *Homo sapiens* evolved from some "lower," more "brutish" form. But Brace has lumped all such analyses under the catch phrase "hominid catastrophism." Hominid catastrophism, according to Brace, is the denial of ancestral-descendant relationships among fossils, with the invocation of extinction and subsequent migrations of new populations that arose by successive creation. Such views are, of course, absurd, but Brace would include *all* cladistic interpretations of the hominid record within "hominid catastrophism." To view hominid phylogeny as a gradual, progressive, unilineal process involving a series of stages, Brace claims, is the interpretation most consonant with evolutionary theory. His interpretation of phylogeny may be correct (though most experts deny it), but he is seriously wrong to claim that phyletic gradualism is the picture most consistent with modern biological thought. Quite apart from the issue of probable overlap in the ranges of his stages, it would be of great interest to determine the degree of stasis attained by them during any reasonably long period of time.

Application of allopatric concepts to paleontological examples. At this point, there is some justification for concluding that the picture of phyletic gradualism is poorly documented indeed, and that most analyses purporting to illustrate it directly from the fossil record are interpretations based on a preconceived idea. On the other hand, the alternative picture of stasis punctuated by episodic events of allopatric speciation rests on a few general statements in the literature and a wealth of informal data. The idea of *punctuated equilibria* is just as much a preconceived picture as that of phyletic gradualism. We readily admit our bias towards it and urge readers, in the ensuing discussion, to remember that our interpretations are as colored by our preconceptions as are the claims of the champions of phyletic gradualism by theirs. We merely reiterate: (1) that one must have some picture of speciation in mind, (2) that the data of paleontology cannot decide which picture is more adequate, and (3) that the picture of punctuated equilibria is more in accord with the process of speciation as understood by modern evolutionists.

We could cite any number of reported sequences that fare better under notions of allopatric processes than under the interpretation of phyletic gradualism that was originally applied. This is surely true for all or part of the three warhorses of the English literature: horses themselves, the Cretaceous echinoid *Micraster*, and the Jurassic oyster *Gryphaea*. Simpson (1951) has shown that the phylogeny of horses is a luxuriant, branching bush, not the ladder to one toe and big teeth that earlier authors envisioned (Matthew and Chubb, 1921). Nichols (1959) believes that *Micraster senonensis* was a migrant from elsewhere and that it did not arise and diverge gradually from *M. cortestudinarium* as Rowe (1899) had maintained. Hallam (1959, 1962) has argued that the transition from *Liostrea* to *Gryphaea* was abrupt and that *neither* genus shows *any* progressive change through the basal Liassic zones, contrary to Trueman's claim (1922, p. 258) that: "It is doubtful whether any better example of lineage of fossil forms could be found." Gould (1971b and in press) has confirmed Hallam's conclusions. Hallam interprets the sudden appearance of *Gryphaea* as the first entry into a local rock column of a species that had evolved rapidly elsewhere. He writes (1962, p. 574): "This interpretation is more in accord with the experience of most invertebrate paleontologists who, despite continued collecting all over the world and an ever-increasing amount of research, find 'cryptogenic' genera and species far more commonly than they detect gradual trends or lineages. The sort of evolution I tentatively propose for *Gryphaea* could in fact be quite normal among the invertebrates." We agree.

We choose, rather, to present two examples from our own work which we believe are interpreted best from the viewpoint of allopatric speciation. We prefer to emphasize our own work simply because we are most familiar with it and are naturally more inclined to defend our interpretations.

Gould (1969) has analyzed the evolution of *Poecilozonites bermudensis zonatus* Verrill, a pulmonate snail, during the last 300,000 years of the Bermudian Pleistocene. The specimens were collected from an alternating sequence of wind-blown sands and red soils. Formational names, dominant lithologies, and glacial-interglacial correlations are given in *table 5–1.*

The small area and striking differentiation of stratigraphic units in the Bermudian Pleistocene permit a high degree of geographic and temporal control. *P. bermudensis* (Pfeiffer) is plentiful in all post-Belmont formations; in addition, one subspecies, *P.b. bermudensis*, is extant and available for study in the laboratory.

Distinct patterns of color banding differentiate an eastern from a western population of *P. bermudensis zonatus.* The boundary between these two groups is sharp, and there are no unambiguous cases of introgression. *P.*

bermudensis zonatus was divided into two stocks, evolving in parallel with little gene flow between them, throughout the entire interval of Shore Hills to Southampton time. Both eastern and western *P.b. zonatus* became extinct sometime after the deposition of Southampton dunes; they were replaced by *P.b. bermudensis*, a derivative of eastern *P.b. zonatus* which had been evolving separately in the area of St. George's Island since St. George's time. Gould (1969, 1970b) has discussed the parallel oscillation of several morphological features in both stocks of *P.b. zonatus*; these are adaptive shifts in response to glacially-controlled variations in climate. Both stocks exhibit stability in other features that serve to distinguish them from their nearest relatives. There is no evidence for any gradual divergence between eastern and western *P.b. zonatus*.

Several samples of *P. bermudensis* share many features that distinguish them from *P. bermudensis zonatus*. These characters can be arranged in four categories: color, general form of the spire, thickness of the shell, and shape of the apertural lip. The ontogeny of *P.b. zonatus* illustrates the interrelation of these categories. Immature shells of *P.b. zonatus* are weakly colored, relatively wide, lack a callus, and have the lowest portion of the outer apertural lip at the umbilical border. This combination of character states is exactly repeated in the large *mature* shells of non-*zonatus* samples of *P. bermudensis*. Since every ontogenetic feature developed at or after the fifth whorl in non-*zonatus* samples is attained by whorls 3–4 in *P.b. zonatus*, Gould (1969) concludes that the non-*zonatus* samples of *P. bermudensis* are derived by paedomorphosis from *P.b. zonatus*.

These paedomorphic samples range through the entire interval of Shore Hills to Recent. The most obvious hypothesis would hold that they constitute a continuous lineage evolving separately from *P.b. zonatus*. Gould rejects this and concludes that paedomorphic offshoots arose from the *P.b. zonatus* stock at four different times; the arguments are based on details of stratigraphic and geographic distrubution, as well as on morphology.

Figure 5–4 summarizes the history of splitting in the *P.b. zonatus* lineage. The earliest paedomorph, *P.b. fasolti* Gould, occurs in the Shore Hills Formation within the geographic range of eastern *P.b. zonatus*. *P.b. fasolti* and the contemporary population of eastern *P.b. zonatus* share a unique set of morphological features including, *inter alia*, small size at any given whorl, low spire, relatively wide shell, and a wide umbilicus. These features unite the Shore Hills paedomorph and non-paedomorph, and set them apart from all post-Shore Hills *P. bermudensis*.

In the succeeding Harrington Formation, paedomorphic samples of *P. bermudensis* lived in both the eastern and western geographic regions of *P.b. zonatus*. The eastern paedomorph, *P.b. sieglindae* Gould, may have evolved from the Shore Hills paedomorph, *P.b. fasolti*. However, both *P.b. sieglindae*

Table 5–1. Stratigraphic column of Bermuda.

Formation	*Description*	*Interpretation*
Recent	Poorly developed brownish soil or crust	Interglacial
Southampton	Complex of eolianites and discontinuous unindurated zones	„
St. George's	Red paleosol of island wide extent	Glacial
Spencer's Point	Intertidal marine, beach and dune facies	Interglacial
Pembroke	Extensive eolianites and discontinuous unindurated zones	„
Harrington	Fairly continuous unindurated layer with shallow water marine and beach facies	„
Devonshire	Intertidal marine and poorly developed dune facies	„
Shore Hills	Well-developed red paleosol of island-wide extent	Glacial
Belmont	Complex shallow water marine, beach and dune facies	Interglacial
Soil (?)	A reddened surface rarely seen in the Walsingham district	Glacial?
Walsingham	Highly altered elolianites	Interglacial

and the contemporaneous population of eastern *P.b. zonatus* lack the distinctive features of all Shore Hills *P. bermudensis* and a more likely hypothesis holds that the features uniting all post-Shore Hills *P. bermudensis* were evolved only once. If this is the case, *P.b. sieglindae* is a second paedomorphic derivative of eastern *P.b. zonatus.*

P.b. sieglindae differs from its contemporary paedomorph *P.b. siegmundi* Gould in that each displays the color pattern of the local non-paedomorph. Very simply, *P.b. sieglindae* is found in eastern Bermuda and shares the banding pattern of eastern *P.b. zonatus*, while *P.b. siegmundi* is found in western Bermuda and has the same color pattern as western *P.b. zonatus.* In addition, both *P.b. sieglindae* and *P.b. siegmundi* evolved at the periphery of the known range of their putative ancestors. The independent derivation of the two Harrington paedomorphs from the two stocks of *P.b. zonatus* seems clear.

Finally, the living paedomorph, *P.b. bermudensis*, first appears in the St. George's Formation on St. George's Island. While St. George's Island is within the geographic range of eastern *P.b. zonatus*, it is far removed from the

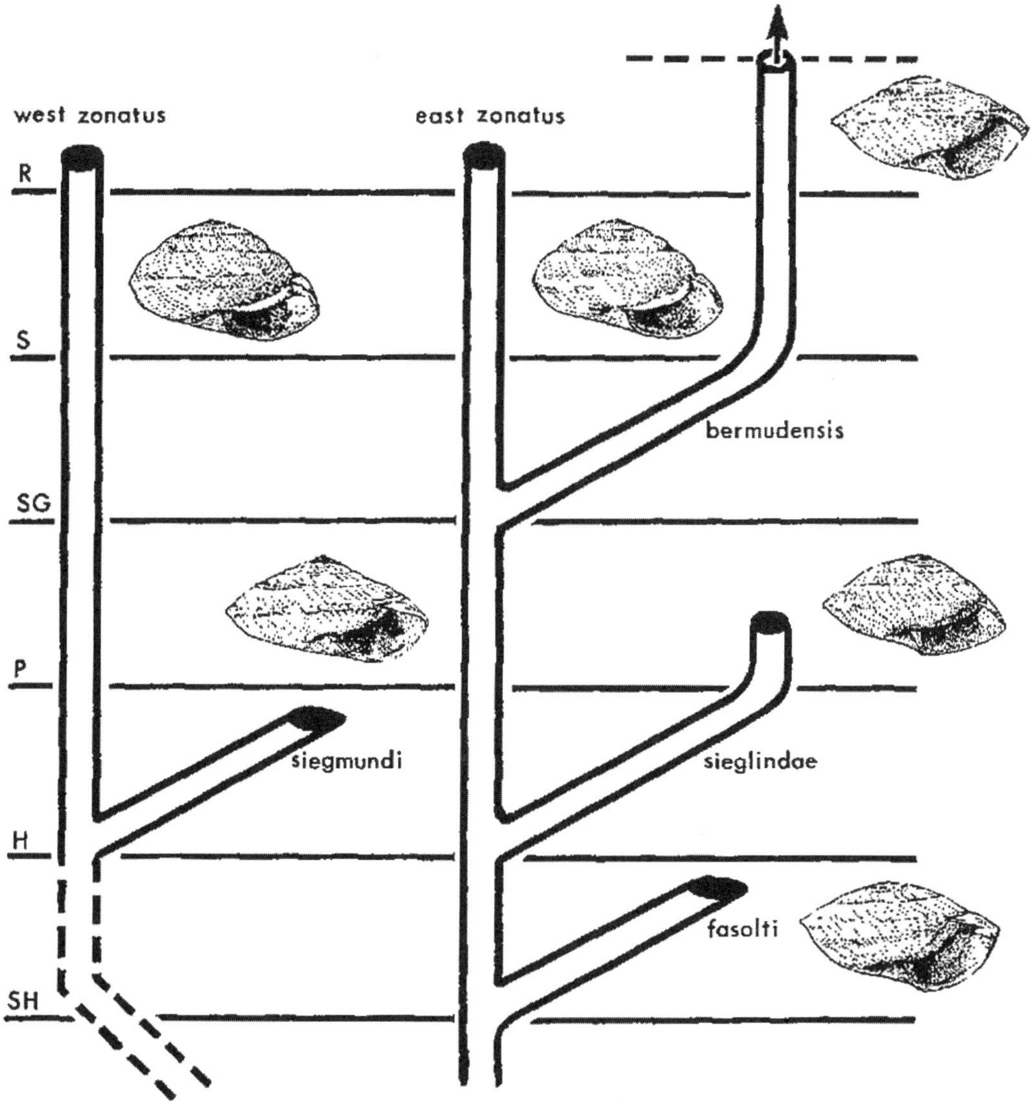

Figure 5–4: Reconstruction of the phylogenetic history of *P. bermudensis* showing iterative development of paedomorphic subspecies. SH—Shore Hills; H—Harrington; P—Pembroke; SG—St. George's; S—Southampton; R—Recent. From Gould, 1969; figure 20.

area in which *P.b. sieglindae* arose and lived. Gould concludes that *P.b. sieglindae* was a short-lived population that never enjoyed a wide geographic distribution; he estimates that the Pembroke population's range did not exceed 200 meters. Although there is little morphological evidence to support it, Gould recognizes a fourth paedomorphic subspecies, *P.b. bermudensis*, derived directly from (eastern) *P.b. zonatus*. The conclusion is based upon geographic and stratigraphic data.

102

Gould (1969) has advanced an adaptive explanation for the four separate origins of paedomorphic populations from *P.b. zonatus.* This explanation, based on the value of thin shells in lime-poor soils, need not be elaborated here. What is important, for our purposes, is to emphasize that the reconstruction of phylogenetic histories for the paedomorphs involves (1) attention to geographic data (the allopatric model), (2) discontinuous stratigraphic occurrence (a more literal interpretation of the fossil record), and (3) formal arguments based on morphology. It is entirely possible, from morphological data alone, to interpret the three paedomorphs of the eastern *zonatus* area as a gradational biostratigraphic series. *Figure 5–5* shows a tempting interpretation of phyletic gradualism for "lower eccentricity," an apertural

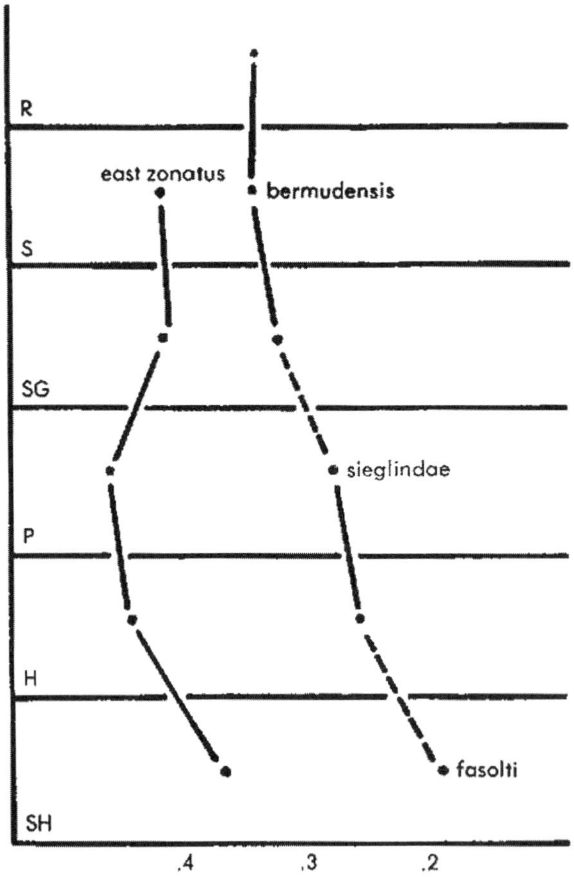

Figure 5–5: Plot of means of mean sample values of "lower eccentricity" in *P. bermudensis.* Dashed lines show the phylogeny of the three paedomorphs of eastern *zonatus* as a direct ancestral-descendant sequence, and offer a tempting instance of phyletic gradualism. Abbreviations as in *figure 5–4.*

variable. Values gradually increase through time. *Figure 5–6*, however, confounds this interpretation by showing that stratigraphic variability in "differential growth ratio" within both *P.b. sieglindae* and *P.b. bermudensis* varies in a direction *opposite* to the net stratigraphic "trend": *P.b. fasolti—P.b. sieglindae—P.b. bermudensis*: this could be read to indicate that each subspecies is unique. In fact, neither graph affords sufficient evidence to warrant either conclusion. Morphology, stratigraphy, and geography must all be evaluated.

The phylogenetic history of the trilobite *Phacops rana* (Green) from the Middle Devonian of North America (Eldredge, 1971; 1972) provides another example of the postulated operation of allopatric processes. As in *Poecilozonites bermudensis*, full genetic isolation was probably not established between "parent" and "daughter" taxa; this conclusion, based on

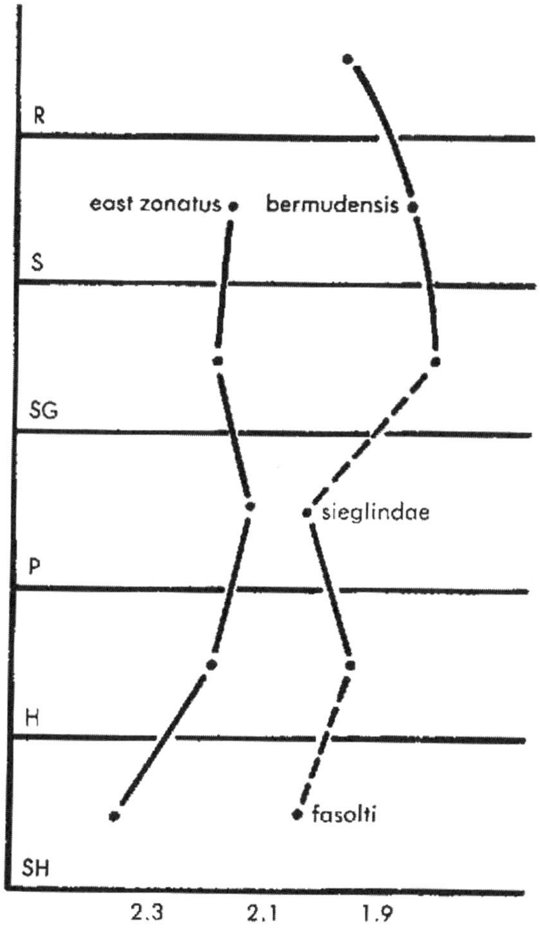

Figure 5–6: Plot of means of mean sample values for "differential growth ratio" in *P. bermudensis*. Dashed lines show the interpretation of the phylogeny of the three paedomorphs as a direct ancestral-descendant sequence. Abbreviations as in *figure 5–4*.

inferences from morphological variability, may be unwarranted. For our purposes, it does not matter whether we are dealing with four subspecies of *P. rana*, or four separate species of *Phacops*, including *P. rana* and its three closest relatives. The basic mode of evolution underlying the group's phylogenetic history as a whole is the same in either case.

Features of eye morphology exhibit the greatest amount of variation among samples of *P. rana*. Lenses are arranged on the visual surface of the eye in vertical dorso-ventral files (Clarkson, 1966). A stable number of dorso-ventral files, characteristic of the entire sample in any population, is reached early in ontogeny. The number of dorso-ventral (d.-v.) files is the most important feature of interpopulational variation in *P. rana*.

The closest known relative of *P. rana* is *P. schlotheimi* (Bronn) *s. l.*, from the Eifelian of Europe and Africa; this group has recently been revised by C. J. Burton (1969). In addition, several samples of *P. rana* have been found in the Spanish Sahara in northwestern Africa (Burton and Eldredge, in preparation). *P. schlotheimi* and the African specimens of *P. rana* are most similar to *P. rana milleri* Stewart and *P. rana crassituberculata* Stumm, the two oldest subspecies of *P. rana* in North America. All these taxa possess 18 dorso-ventral files. Eldredge (1972) concludes that 18 is the primitive number of d.-v. files for all North American *Phacops rana*.

Figure 5–7 summarizes relationships among the four subspecies of *P. rana* without regard to stratigraphic occurrence. The oldest North American *P. rana* occurs in the Lower Cazenovian Stage of Ohio and central New York State. All have 18 d.-v. files. Populations with 18 d.-v. files (*P. rana milleri* and *P. rana crassituberculata*) persist into the Upper Cazenovian Stage in the epicontinental seas west of the marginal basin in New York and the Appalachians.

Of the two samples the one that displays intra-populational variation in d.-v. file number occurs in the Lower Cazenovian of central New York. Some specimens have 18 d.-v. files, while others reduce the first d.-v. file to various degrees; a few lack it altogether. *All P. rana* from subsequent, younger horizons in New York and adjacent Appalachian states have 17 dorso-ventral files. Apparently, 17 d.-v. file *P. rana rana* arose from an 18 d.-v. file population on the northeastern periphery of the Cazenovian geographic range of *P. rana*. Seventeen d.-v. file *P. rana* persist, unchanged in most respects, through the Upper Cazenovian, Tioughniogan, and Taghanic Stages in the eastern marginal basin. Seventeen d.-v. file *P. rana rana* first appears in the shallow interior seas at the beginning of the Tioughniogan Stage, replacing the 18 d.-v. file populations that apparently became extinct during a general withdrawal of seas from the continental interior. All Tioughniogan *P. rana* possess 17 dorso-ventral files.

A second, similar event involving reduction in dorso-ventral files occurred during the Taghanic. Here again, a variable population inhabited the eastern

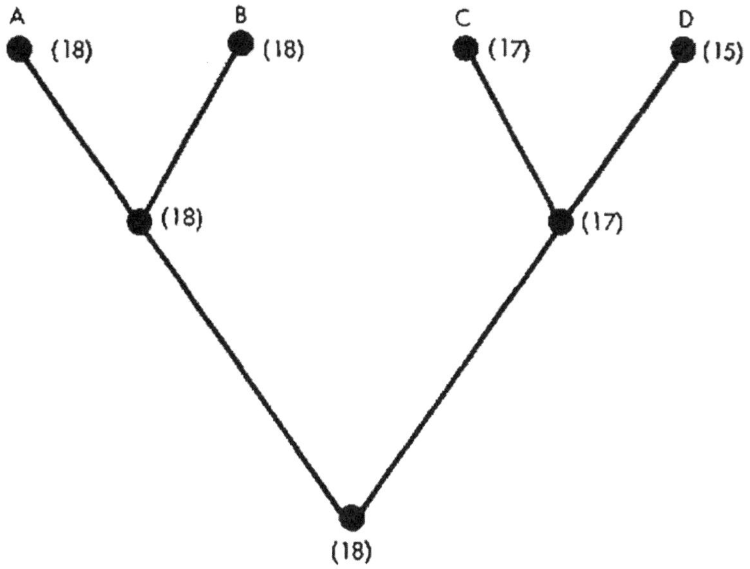

Figure 5–7: Outline of relationships of four subspecies of *Phacops rana*. A—*Phacops rana crassituberculata* Stumm; B—*Phacops rana milleri* Stewart; C—*Phacops rana rana* (Green); D—*Phacops rana norwoodensis* Stumm. Numbers in parentheses refer to number of dorso-ventral files typical of subspecies or hypothesized to characterize condition of common ancestor.

marginal basin in New York. This suggests that, once more, reduction in d.-v. files occurred allopatrically on the periphery of the known range of *P. rana rana*. The subsequent spread of stabilized, 15 d.-v. file *P. rana norwoodensis* through the Taghanic seas of the continental interior was instantaneous in terms of our biostratigraphic resolution. *Figure 5–8* summarizes this interpretation of the history of *P. rana*.

Under the tenets of phyletic gradualism, this story has a different (and incorrect) interpretation: the three successional taxa of the epeiric seas form an *in situ* sequence of gradual evolutionary modification. The sudden transitions from one form to the next are the artifact of a woefully incomplete fossil record. Most evolutionary change occurred during these missing intervals: fill in the lost pieces with an even dotted line.

If the interpreter pays attention to geographic detail, however, quite a different tale emerges, one that allows a more literal reading of the fossil record. Now the story is one of stasis: no variation in the most important feature of discrimination (number of d.-v. files—actually a complex of highly interrelated variables) through long spans of time. Two samples displaying intra-populational variation in numbers of d.-v. files identify relatively "sudden" events of reduction in files on the periphery of the species' geographic range. These two samples, moreover, have a very short stratigraphic, and very restricted geographic, distribution.

106

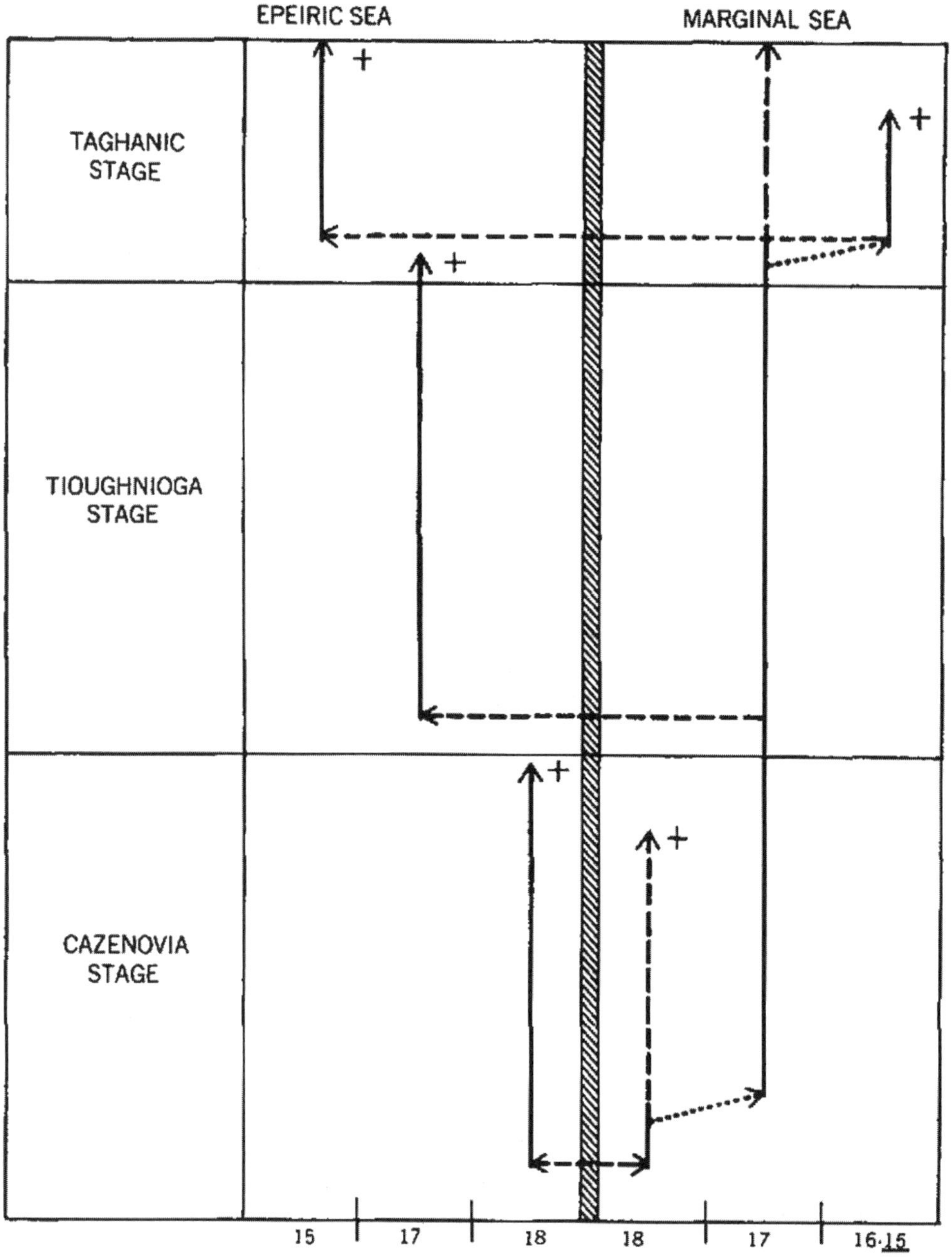

Figure 5–8: Hypothesized phylogeny of the *Phacops rana* stock in the Middle Devonian of North America. Numbers at the base of the diagram refer to the population number of dorso-ventral files. Dotted lines: origin of new (reduced) number of d.-v. files in a peripheral isolate; horizontal dashed lines: migration; vertical solid lines: presence of taxon in indicated area; dashed vertical lines: persistence of ancestral stock in a portion of the marginal sea other than that in which the derived taxon occurs. Crosses denote final disappearance; for fuller explanation, see text.

107

Our two examples, so widely separated in scale, age, and subject, have much in common as exemplars of allopatric processes. Both required an attention to details of *geographic* distribution for their elucidation. Both involved a *more literal* reading of the fossil record than is allowed under the unconscious guidance of phyletic gradualism. Both are characterized by *rapid* evolutionary events punctuating a history of stasis. These are among the expected consequences if most fossil species arose by allopatric speciation in small, peripherally isolated populations. This alternative picture merely represents the application to the fossil record of the dominant theory of speciation in modern evolutionary thought. We believe that the consequences of this theory are more nearly demonstrated than those of phyletic gradualism by the fossil record of the vast majority of Metazoa.

Some Extrapolations to Macroevolution

Before 1930, paleontology sought a separate theory for the causes of macroevolution. The processes of microevolution (including the origin of species) were deemed insufficient to generate the complexity and diversity of life, even under the generous constraint of geological time; a variety of special causes were proposed—vitalism, orthogenesis, racial "life" cycles, and universal acceleration in development to name just a few.

However, the advent of the "modern synthesis" inspired a reassessment that must stand as the major conceptual advance in 20th-century paleontology. Special explanations for macroevolution were abandoned for a simplifying theory of extrapolation from species-level processes. All evolutionary events, including those that seemed most strongly "directed" and greatly extended in time, were explained as consequences of mutation, recombination, selection, etc.—i.e., as consequences only of the phenomena that produce evolution in nature's real taxon, the species. (The modern synthesis received its name because it gathered under one theory—with population genetics at its core—the events in many subfields that had previously been explained by special theories unique to that discipline. Such an occurrence marks scientific "progress" in its truest sense—the replacement of special explanations carrying little power in prediction or extension with general theories, rich in implications and capable of unifying a diverse set of phenomena that had seemed unrelated. Thus Simpson (1944, 1953) did for paleontology what Dobzhansky (1937) had done for classical genetics, Mayr (1942) for systematics, de Beer (1940) for development, White (1954) for cytology, and Stebbins (1950) for systematic botany—he exemplified the phenomena of his field as the result of Darwinian processes acting upon species.)

We have discussed two pictures for the origin of species in paleontology. In the perspective of a species-extrapolation theory of macroevolution, we should now extend these pictures to see how macroevolution proceeds under their guidance. If actual events, as recorded by fossils, fit more comfortably with the predictions of either picture, this will be a further argument for that picture's greater adequacy.

Under phyletic gradualism, the history of life should be one of *stately unfolding*. Most changes occur slowly and evenly by phyletic transformation; splitting, when it occurs, produces a slow and very gradual divergence of forms (Weller's (1969) tree of life—reproduced as *figure 5–9*—records the extrapolation of this partisan view, not a neutral hatrack for the fossils themselves). We have already named our alternate picture for its predicted extrapolation—*punctuated equilibria*. The theory of allopatric speciation implies

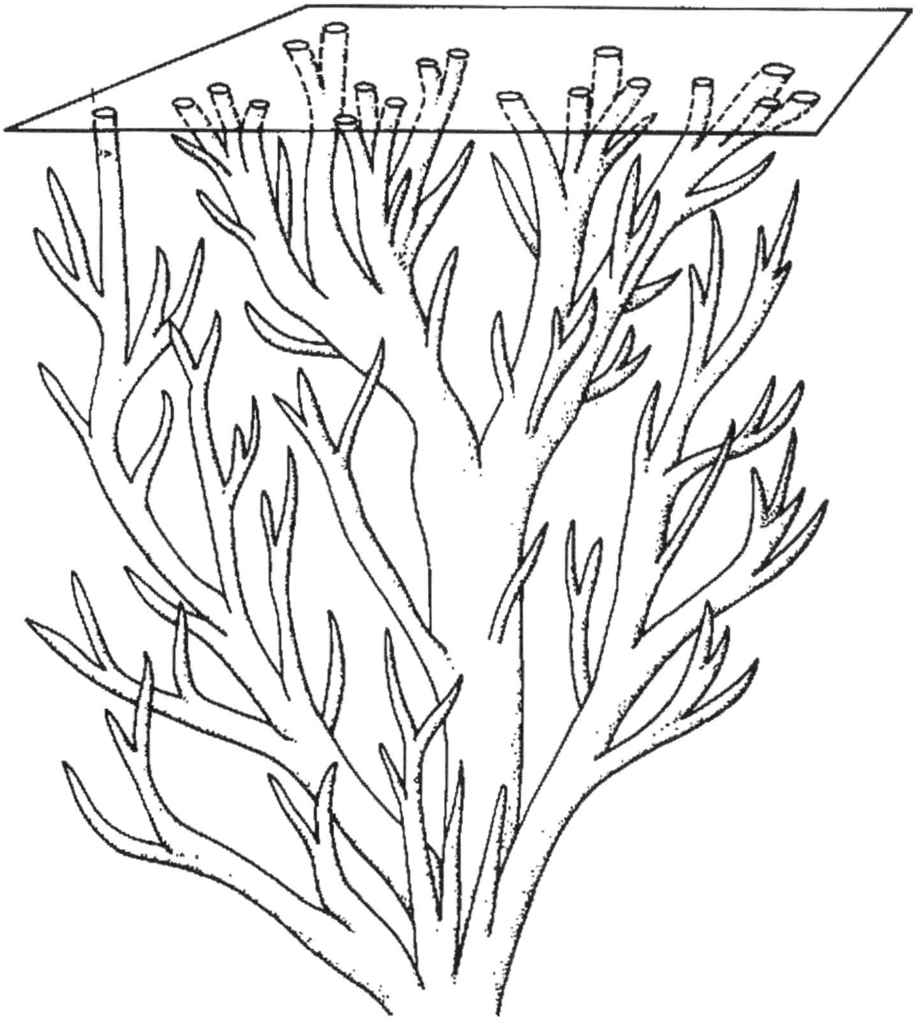

Figure 5–9: The "Tree of Life" viewed from the perspective of phyletic gradualism. Branches diverge gradually one from the other. A slow and relatively equal rate of evolution pervades the system. From Weller, 1969; figure 637.

that a lineage's history includes long periods of morphologic stability, punctuated here and there by rapid events of speciation in isolated subpopulations.

We now consider two phenomena of macroevolution as case studies of our extrapolated pictures. The first is widely recognized as anomalous under the unconscious guidance of stately unfolding; it emerges as an expectation under the notion of punctuated equilibria. The second phenomenon seems, superficially, to have an easier explanation under stately unfolding, but we shall argue that it has a more interesting interpretation when viewed with the picture of punctuated equilibria.

(1) *"Classes" of great number and low diversity*

To many paleontologists, nothing is more distressing than the current situation in echinoderm systematics. Ubaghs (1967), in his contribution to the *Treatise on Invertebrate Paleontology*, recognizes 20 classes and at least one has been added since then—Robison and Sprinkle's (1969) ctenocystoids. Yet, although all appeared by the Ordovician, only five survived the Devonian. Moreover, although each class has a distinct Bauplan, many display a diversity often considered embarrassingly small for so exalted a taxonomic rank—the *Treatise* describes eight classes with five or fewer genera; five of these include but a single genus (as does the new ctenocystoids).

There are two aspects to this tale that fit poorly with the traditional view of stately unfolding:

(1) The presence of 21 classes by the Ordovician, coupled with their presumed monophyletic descent, requires extrapolation to a common ancestor uncomfortably far back in the Precambrian if Ordovician diversity is the apex of a gradual unfolding. Yet current views of Precambrian evolution will not happily accommodate a complex metazoan so early (Cloud, 1968).

(2) We expect that successively higher ranks of the taxonomic hierarchy will contain more and more taxa: a class with one genus is anomalous and we are led either to desperate hopes for synonymy or, once again, to our old assumption—that we possess a fragmentary record of a truly diverse group. Yet this expectation is no consequence of the logic of taxonomy (which demands only that each taxon be *as* inclusive as the lower ones it incorporates); it arises, rather, from a picture of stately unfolding. In *figure 5–9*, a new higher taxon attains its rank *by virtue of* its diversity—an evenly progressing, evenly diverging set of branches cannot produce such a taxon with limited diversity, for a lineage "graduates" from family to order to class only as it persists to a tolerable age and branches an acceptable number of times.

With the picture of punctuated equilibria, however, classes of small membership are welcome and echinoderm evolution becomes more intriguing than bothersome. Since speciation is rapid and episodic, repeated splitting during short intervals is likely when opportunities for full speciation following isolation are good (limited dangers of predation or competition in peripheral

110

environments, for example—a likely Lower Cambrian situation). When these repeated splits affect a small, isolated lineage; when adaptation to peripheral environments involves new modes of feeding, protection, and locomotion; and when extinction of parental species commonly follows the migration of descendants to the ancestral area, then very distinct phenons with few species will develop. Since higher taxa are all "arbitrary" (they reflect no interacting group in nature, but rather a convenient arrangement of species that violates no rule of monophyly, hierarchical ordering, etc.), we believe that they should be defined by morphology. Criteria of diversity are too closely tied to partisan pictures; morphology, though not as "objective" as some numerical taxonomists claim, is at least more functional for information retrieval.

(2) *Trends*

Trends, or biostratigraphic character gradients, are frequently mentioned as basic features of the fossil record. Sequences of fossils, said to display trends, range from the infraspecific through the very highest levels of the taxonomic hierarchy. Trends at and below the species level were discussed in the previous section, but the relation between phyletic gradualism and trends among related clusters of species—families or orders—remains to be examined.

Many, if not most, trends involving higher taxa may simply reflect a selective rendering of elements in the fossil record, chosen because they seem to form a morphologically-graded series coincident with a progressive biostratigraphic distribution. In this sense, trends may represent simple extrapolations of phyletic gradualism.

But a claim that all documented trends are just unwarranted extrapolations based on a preconception would be altogether too facile an explanation for the large numbers of trends cited in the literature. For this discussion, we accept trends as a real and important phenomenon in evolution, and adopt the simple definition given by MacGillavry (1968, p. 72): "A trend is a direction which involves the *majority* of related lineages of a group" (our italics).

If trends are real and common, how can they be reconciled with our picture, in which speciation occurs in peripheral isolates by adaptation to local conditions and the perfection of isolating mechanisms? The problem may be stated in another way: Sewall Wright (1967, p. 120) has suggested that, just as mutations are stochastic with respect to selection within a population, so might speciation be stochastic with respect to the origin of higher taxa. As a slight extension of that statement, we might claim that adaptations to local conditions by peripheral isolates are stochastic with respect to long-term, net directional change (trends) within a higher taxon as a whole. We are left with a bit of a paradox: to picture speciation as an allopatric phenomenon, involving rapid differentiation within a general, long-term picture of stasis, is to deny the picture of directed gradualism in speciation. Yet, superficially at least, this directed gradualism is easier to reconcile with valid cases of long-term trends involving many species.

MacGillavry's definition of a trend removes part of the problem by using the expression "majority of related lineages." This frees us from the constraint of reconciling *all* events of adaptation to local conditions in peripheral isolates, with long-term, net directional change.

A reconciliation of allopatric speciation with long-term trends can be formulated along the following lines: we envision multiple "explorations" or "experimentations" (see Schaeffer, 1965)—i.e., invasions, on a stochastic basis, of new environments by peripheral isolates. There is nothing inherently directional about these invasions. However, a subset of these new environments might, in the context of inherited genetic constitution in the ancestral components of a lineage, lead to new and improved efficiency. Improvement would be consistently greater within this hypothetical subset of local conditions that a population might invade. The overall effect would then be one of net, apparently directional change: but, as in the case of selection upon mutations, the initial variations would be stochastic with respect to this change (*figure 5-10*). We postulate no "new" type of selection. We simply state a view of long-term, superficially "directed" phenomena that is in accord with the theory of allopatric speciation, and also avoids the largely untestable concept of orthoselection.

Conclusion: Evolution, Stately or Episodic?

Heretofore, we have spoken of the morphological stability of species in time without examining the reasons for it. The standard definition of a biospecies—as a group of actually or potentially reproducing organisms sharing a common gene pool—specifies the major reason usually cited: gene flow. Since the subpopulations of a species adapt to a range of differing local environments, we might expect these groups to differentiate, acquire isolating mechanisms and, eventually, to form new species. But gene flow exerts a homogenizing influence "to counteract local ecotypic adaptation by breaking up well-integrated gene complexes" (Mayr, 1963, p. 178). The role of gene flow is recognized in the central tenet of allopatric speciation: speciation occurs in *peripheral* isolates because only geographic separation from the parental species can reduce gene flow sufficiently to allow local differentiation to proceed to full speciation.

Recently, however, a serious challenge to the importance of gene flow in species' cohesion has come from several sources (Ehrlich and Raven, 1969, for example). Critics claim that, in most cases, gene flow is simply too restricted to exert a homogenizing influence and prevent differentiation. This

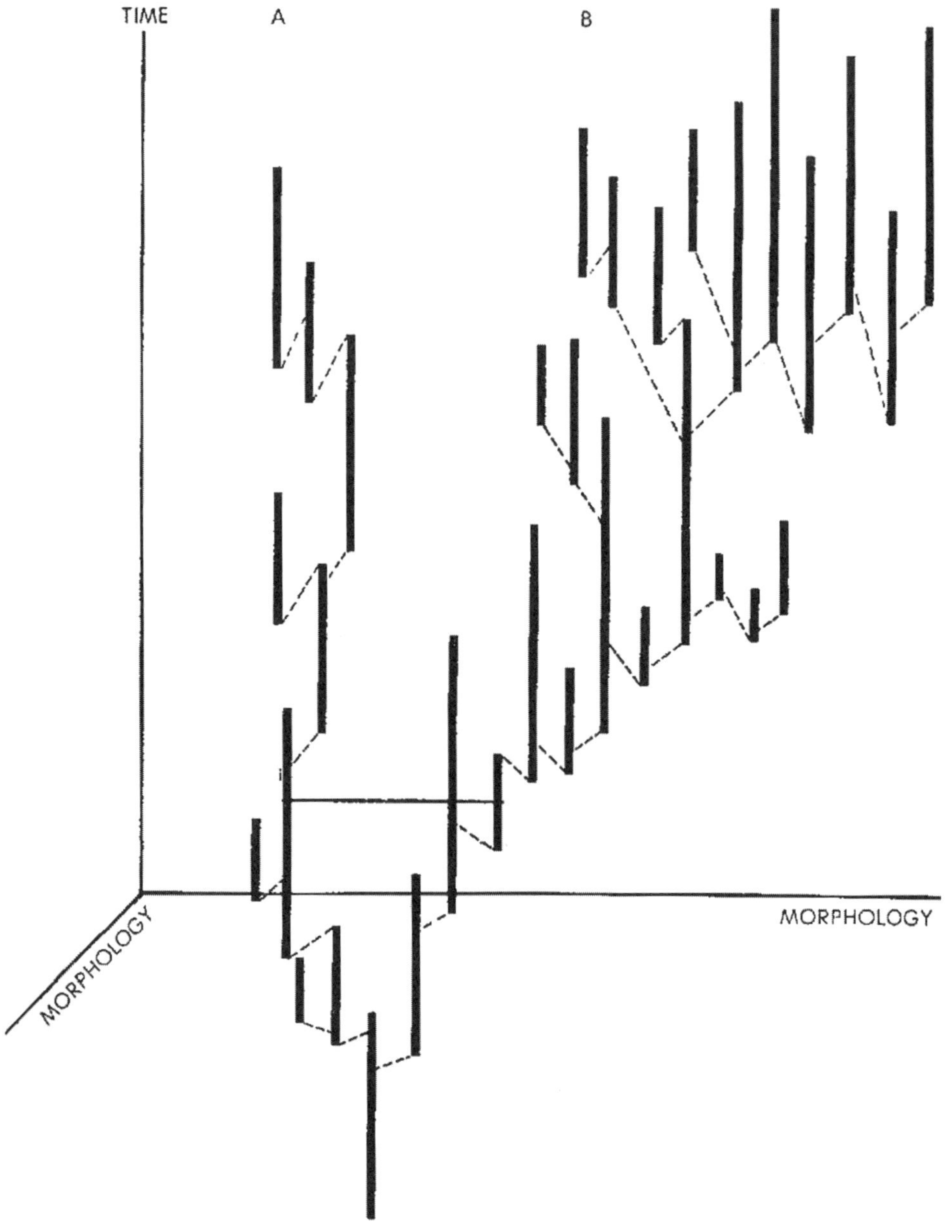

Figure 5–10: Three-dimensional sketch contrasting a pattern of relative stability (A) with a trend (B), where speciation (dashed lines) is occurring in both major lineages. Morphological change is depicted here along the horizontal axes, while the vertical axis is time. Though a retrospective pattern of directional selection might be fitted as a straight line in (B), the actual pattern is stasis within species, and differential success of species exhibiting morphological change in a particular direction. For further explanation, see text.

113

produces a paradox: why, then, are species coherent (or even recognizable)? Why do groups of (relatively independent) local populations continue to display a fairly consistent phenotype that permits their recognition as a species? Why does reproductive isolation not arise in every local population? Why is the local population itself not considered the "real" unit in evolution (as some would prefer—Sokal and Crovello, 1970, p. 151, for example)?

The answer probably lies in a view of species and individuals as homeostatic systems—as amazingly well-buffered to resist change and maintain stability in the face of disturbing influences. This concept has been urged particularly by Lerner (1954) and Mayr (1963), though the latter still gives more weight to gene flow than many will allow. Lerner (1954, p. 6) recognizes two types of homeostasis, mediated in both cases, he believes, by the generally higher fitness of heterozygous vs. homozygous genotypes: (1) ontogenetic self-regulation (developmental homeostasis) "based on the greater ability of the heterozygote to stay within the norms of canalized development" and (2) self-regulation of populations (genetic homeostasis) "based on natural selection favoring intermediate rather than extreme phenotypes." In this view, the importance of peripheral isolates lies in their small size and the alien environment beyond the species border that they inhabit—for only here are selective pressures strong enough and the inertia of large numbers sufficiently reduced to produce the "genetic revolution" (Mayr, 1963, p. 533) that overcomes homeostasis. The coherence of a species, therefore, is not maintained by interaction among its members (gene flow). It emerges, rather, as an historical consequence of the species' origin as a peripherally isolated population that acquired its own powerful homeostatic system. (We regard this idea as a serious challenge to the conventional view of species' reality that depends upon the organization of species as ecological units of *interacting* individuals in nature. If groups of nearly-independent local populations are recognized as species only because they share a set of homeostatic mechanisms developed long ago in a peripheral isolate that was "real" in our conventional sense of interaction, then some persistent anomalies are resolved. The arrangement of many asexual groups into good phenetic "species," quite inexplicable if interaction is the basis for coherence, receives a comfortable explanation under notions of homeostasis.)

Thus, the challenge to gene flow that seemed to question the stability of species in time ends by reinforcing that stability even more strongly. If we view a species as a set of subpopulations, all ready and able to differentiate but held in check only by the rein of gene flow, then the stability of species is a tenuous thing indeed. But if that stability is an inherent property both of individual development and the genetic structure of populations, then its power is immeasurably enhanced, for the basic property of homeostatic systems, of steady states, is that they resist change by self-regulation. That local popula tions do not differentiate into species, even though no external bar prevents it, stands as strong testimony to the inherent stability of species in time.

Paleontologists should recognize that much of their thought is conditioned by a peculiar perspective that they must bring to the study of life: they must look down from its present complexity and diversity into the past: their view must be retrospective. From this vantage point, it is very difficult to view evolution as anything but an easy and inevitable result of mere existence, as something that unfolds in a natural and orderly fashion. Yet we urge a different view. The norm for a species or, by extension, a community is stability. Speciation is a rare and difficult event that punctuates a system in homeostatic equilibrium. That so uncommon an event should have produced such a wondrous array of living and fossil forms can only give strength to an old idea: paleontology deals with a phenomenon that belongs to it alone among the evolutionary sciences and that enlightens all its conclusions—time.

Dobzhansky, T. 1973. Nothing in biology makes sense except in the light of evolution. *American Biology Teacher* 35:125–129.

Dobzhansky declares: "It is wrong to hold creation and evolution as mutually exclusive alternatives. I am a creationist *and* an evolutionist." We may see science and religion as two different windows for looking at the world. The two windows look at the same world, but they show different aspects of it. The scope of science is the world of nature, the reality that is observed, directly or indirectly, by our senses. Science advances explanations concerning the natural world: the composition of matter, the drift of the continents, the expansion of the galaxies, the origin and adaptations of organisms. Religion concerns the meaning and purpose of the world and of human life, the proper relation of people to the Creator and to each other, the moral values that inspire and govern people's lives. It is only when assertions are made beyond their legitimate boundaries that science and religious belief appear to be antithetical.

The evolutionary origin of animals and plants is a scientific conclusion beyond reasonable doubt. The evolution of organisms is considered by scientists to be as well confirmed as such established concepts like the roundness of the Earth, its revolution around the sun, and the atomic composition of matter. In recent decades, some people of faith have claimed that organisms come about by what they refer to as "intelligent design." Scientists have shown that the "design" of organisms comes about by natural selection combined with mutation and other evolutionary processes, without the intervention of some "designer" agent.

One difficulty with attributing the design of organisms to the Creator is that imperfections and defects pervade the living world. The human eye has a blind spot (which a squid's does not), the human jaw is not large enough for the teeth (wisdom teeth need to be removed), the human reproductive system is so poorly designed that about 20 percent of all recognized pregnancies end in spontaneous miscarriage during the first two months. This misfortune amounts to more than 20 million spontaneous abortions worldwide every year. People of faith might rather attribute this monumental calamity, as well as the imperfections and defects of living organisms, to the clumsy ways of the evolutionary process than to the incompetence of an "intelligent designer." And what to say about the waste of hundreds of millions of species that became extinct through the history of life on Earth?

RELATED READING

Avise, J. C. 2010. Inside the human genome: a case for non-intelligent design. Oxford University Press, Oxford, England.

Ayala, F. J. 2007. Darwin's gift to science and religion. Joseph Henry Press, Washington, D.C., USA.

Miller, K. R. 1999. Finding Darwin's god: a scientist's search for common ground. HarperCollins, New York, New York, USA.

National Academy of Sciences and the Institute of Medicine. 2008. Science, evolution, and creationism. National Academies Press, Washington, D.C. USA.

Polkinghorne, J. C. 1989. Science and providence: God's interaction with the world. Templeton Foundation Press, West Conshohocken, Pennsylvania, USA.

Young, M., and T. Edis, editors. 2004. Why intelligent design fails: a scientific critique of the new creationism. Rutgers University Press, New Brunswick, New Jersey, USA.

Nothing in Biology Makes Sense Except in the Light of Evolution

THEODOSIUS DOBZHANSKY

As RECENTLY AS 1966, sheik Abd el Aziz bin Baz asked the king of Saudi Arabia to suppress a heresy that was spreading in his land. Wrote the sheik:

"The Holy Koran, the Prophet's teachings, the majority of Islamic scientists, and the actual facts all prove that the sun is running in its orbit . . . and that the earth is fixed and stable, spread out by God for his mankind. . . . Anyone who professed otherwise would utter a charge of falsehood toward God, the Koran, and the Prophet."

The good sheik evidently holds the Copernican theory to be a "mere theory," not a "fact." In this he is technically correct. A theory can be verified by a mass of facts, but it becomes a proven theory, not a fact. The sheik was perhaps unaware that the Space Age had begun before he asked the king to suppress the Copernican heresy. The sphericity of the earth had been seen by astronauts, and even by many earth-bound people on their television screens. Perhaps the sheik could retort that those who venture beyond the confines of God's earth suffer hallucinations, and that the earth is really flat.

Parts of the Copernican world model, such as the contention that the earth rotates around the sun, and not vice versa, have not been verified by direct observations even to the extent the sphericity of the earth has been. Yet scientists accept the model as an accurate representation of reality. Why? Because it makes sense of a multitude of facts which are otherwise meaningless or extravagant. To nonspecialists most of these facts are unfamiliar. Why then do we accept the "mere theory" that the earth is a sphere revolving around a spherical sun? Are we simply submitting to authority? Not quite: we know that those who took time to study the evidence found it convincing.

The good sheik is probably ignorant of the evidence. Even more likely, he is so hopelessly biased that no amount of evidence would impress him. Anyway, it would be sheer waste of time to attempt to convince him. The Koran and the Bible do not contradict Copernicus, nor does Copernicus contradict them. It is ludicrous to mistake the Bible and the Koran for primers of natural science. They treat of matters even more important: the meaning of man and his relations to God. They are written in poetic symbols that were understandable to people of the age when they were written, as well as to peoples of all other ages. The king of Arabia did not comply with the sheik's demand. He knew that some people fear enlightenment, because enlightenment threatens their vested interests. Education is not to be used to promote obscurantism.

The earth is not the geometric center of the universe, although it may be its spiritual center. It is a mere speck of dust in cosmic spaces. Contrary to Bishop Ussher's calculations, the world did not appear in approximately its present state in 4004 B.C. The estimates of the age of the universe given by modern cosmologists are still only rough approximations, which are revised (usually upward) as the methods of estimation are refined. Some cosmologists take the universe to be about 10 billion years old; others suppose that it may have existed, and will continue to exist, eternally. The origin of life on earth is dated tentatively between 3 and 5 billion years ago; manlike beings appeared relatively quite recently, between 2 and 4 million years ago. The estimates of the age of the earth, of the duration of the geologic and paleontologic eras, and of the antiquity of man's ancestors are now based mainly on radiometric evidence—the proportions of isotopes of certain chemical elements in rocks suitable for such studies.

One of the world's leading geneticists, Theodosius Dobzhansky is professor emeritus, Rockefeller University, and adjunct professor of genetics, University of California, Davis 95616. Born in Russia, in 1900, he is a graduate of the University of Kiev and taught (with J. Philipchenko) at the University of Leningrad before coming to the U.S., in 1927; thereafter he taught at Columbia University and the California Institute of Technology before joining the Rockefeller faculty, in 1962. He has been president of the Genetics Society of America, the American Society of Naturalists, the Society for the Study of Evolution, the American Society of Zoologists, and the American Teilhard de Chardin Association. Among his many honors are the National Medal of Science (1964) and the Gold Medal Award for Distinguished Achievement in Science (1969). He holds 18 honorary doctorates from universities in this country and abroad. Among his well-known books are ***The Biological Basis of Human Freedom*** (1956) and ***Mankind Evolving*** (1963). The present paper was presented at the 1972 NABT convention.

Sheik bin Baz and his like refuse to accept the radiometric evidence, because it is a "mere theory." What is the alternative? One can suppose that the Creator saw fit to play deceitful tricks on geologists and biologists. He carefully arranged to have various rocks provided with isotope ratios just right to mislead us into thinking that certain rocks are 2 billion years old, others 2 million, while in fact they are only some 6,000 years old. This kind of pseudo-explanation is not very new. One of the early antievolutionists, P. H. Gosse, published a book entitled *Omphalos* ("the Navel"). The gist of this amazing book is that Adam, though he had no mother, was created with a navel, and that fossils were placed by the Creator where we find them now—a deliberate act on His part, to give the appearance of great antiquity and geologic upheavals. It is easy to see the fatal flaw in all such notions. They are blasphemies, accusing God of absurd deceitfulness. This is as revolting as it is uncalled for.

Diversity of Living Beings

The diversity and the unity of life are equally striking and meaningful aspects of the living world. Between 1.5 and 2 million species of animals and plants have been described and studied; the number yet to be described is probably about as great. The diversity of sizes, structures, and ways of life is staggering but fascinating. Here are just a few examples.

The foot-and-mouth disease virus is a sphere 8–12 mμ in diameter. The blue whale reaches 30 m in length and 135 t in weight. The simplest viruses are parasites in cells of other organisms, reduced to barest essentials—minute amounts of DNA or RNA, which subvert the biochemical machinery of the host cells to replicate their genetic information, rather than that of the host.

It is a matter of opinion, or of definition, whether viruses are considered living organisms or peculiar chemical substances. The fact that such differences of opinion can exist is in itself highly significant. It means that the borderline between living and inanimate matter is obliterated. At the opposite end of the simplicity–complexity spectrum you have vertebrate animals, including man. The human brain has some 12 billion neurons; the synapses between the neurons are perhaps a thousand times as numerous.

Some organisms live in a great variety of environments. Man is at the top of the scale in this respect. He is not only a truly cosmopolitan species but, owing to his technologic achievements, can survive for at least a limited time on the surface of the moon and in cosmic spaces. By contrast, some organisms are amazingly specialized. Perhaps the narrowest ecologic niche of all is that of a species of the fungus family Laboulbeniaceae, which grows exclusively on the rear portion of the elytra of the beetle *Aphenops cronei,* which is found only in some limestone caves in southern France. Larvae of the fly *Psilopa petrolei* develop in seepages of crude oil in California oilfields; as far as is known they occur nowhere else. This is the only insect able to live and feed in oil, and its adult can walk on the surface of the oil only as long as no body part other than the tarsi are in contact with the oil. Larvae of the fly *Drosophila carcinophila* develop only in the nephric grooves beneath the flaps of the third maxilliped of the land crab *Geocarcinus ruricola,* which is restricted to certain islands in the Caribbean.

Is there an explanation, to make intelligible to reason this colossal diversity of living beings? Whence came these extraordinary, seemingly whimsical and superfluous creatures, like the fungus *Laboulbenia,* the beetle *Aphenops cronei,* the flies *Psilopa petrolei* and *Drosophila carcinophila,* and many, many more apparent biologic curiosities? The only explanation that makes sense is that the organic diversity has evolved in response to the diversity of environment on the planet earth. No single species, however perfect and however versatile, could exploit all the opportunities for living. Every one of the millions of species has its own way of living and of getting sustenance from the environment. There are doubtless many other possible ways of living as yet unexploited by any existing species; but one thing is clear: with less organic diversity, some opportunities for living would remain unexploited. The evolutionary process tends to fill up the available ecologic niches. It does not do so consciously or deliberately; the relations between evolution and the environment are more subtle and more interesting than that. The environment does not impose evolutionary changes on its inhabitants, as postulated by the now abandoned neo-Lamarckian theories. The best way to envisage the situation is as follows: the environment presents challenges to living species, to which the latter may respond by adaptive genetic changes.

An unoccupied ecologic niche, an unexploited opportunity for living, is a challenge. So is an environmental change, such as the Ice Age climate giving place to a warmer climate. Natural selection may cause a living species to respond to the challenge by adaptive genetic changes. These changes may enable the species to occupy the formerly empty ecologic niche as a new opportunity for living, or to resist the environmental change if it is unfavorable. But the response may or may not be successful. This depends on many factors, the chief of which is the genetic composition of the responding species at the time the response is called for. Lack of successful response may cause the species to become extinct. The evidence of fossils shows clearly that the eventual end of most evolutionary lines is extinction. Organisms now living are successful descendants of only a minority of the species that lived in the past—and of smaller and smaller minorities the farther back you look. Nevertheless, the number of living species has not dwindled; indeed, it has probably grown with time. All this is understandable in the light of evolution theory; but what a senseless operation it would have been, on God's part, to fabricate a multitude of species ex nihilo and then let most of them die out!

There is, of course, nothing conscious or intentional in the action of natural selection. A biologic species does not say to itself, "Let me try tomorrow (or a million years from now) to grow in a different soil, or use a different food, or subsist on a different body part of a different crab." Only a human being could make such conscious decisions. This is why the species *Homo sapiens* is the apex of evolution. Natural selection is at one and the same time a blind and a creative process. Only a creative but blind process could produce, on the one hand, the tremendous biologic success that is the human species and, on the other, forms of adaptedness as narrow and as constraining as those of the overspecialized fungus, beetle, and flies mentioned above.

Antievolutionists fail to understand how natural selection operates. They fancy that all existing species were generated by supernatural fiat a few thousand years ago, pretty much as we find them today. But what is the sense of having as many as 2 or 3 million species living on earth? If natural selection is the main factor that brings evolution about, any number of species is understandable: natural selection does not work according to a foreordained plan, and species are produced not because they are needed for some purpose but simply because there is an environmental opportunity and genetic wherewithal to make them possible. Was the Creator in a jocular mood when he made *Psilopa petrolei* for California oil-fields and species of *Drosophila* to live exclusively on some body-parts of certain land crabs on only certain islands in the Caribbean? The organic diversity becomes, however, reasonable and understandable if the Creator has created the living world not by caprice but by evolution propelled by natural selection. It is wrong to hold creation and evolution as mutually exclusive alternatives. I am a creationist *and* an evolutionist. Evolution is God's, or Nature's, method of Creation. Creation is not an event that happened in 4004 B.C.; it is a process that began some 10 billion years ago and is still under way.

Unity of Life

The unity of life is no less remarkable than its diversity. Most forms of life are similar in many respects. The universal biologic similarities are particularly striking in the biochemical dimension. From viruses to man, heredity is coded in just two, chemically related substances: DNA and RNA. The genetic code is as simple as it is universal. There are only four genetic "letters" in DNA: adenine, guanine, thymine, and cytosine. Uracil replaces thymine in RNA. The entire evolutionary development of the living world has taken place not by invention of new "letters" in the genetic "alphabet" but by elaboration of ever-new combinations of these letters.

Not only is the DNA–RNA genetic code universal, but so is the method of translation of the sequences of the "letters" in DNA–RNA into sequences of amino acids in proteins. The same 20 amino acids compose countless different proteins in all, or at least in most, organisms. Different amino acids are coded by one to six nucleotide triplets in DNA and RNA. And the biochemical universals extend beyond the genetic code and its translation into proteins: striking uniformities prevail in the cellular metabolism of the most diverse living beings. Adenosine triphosphate, biotin, riboflavin, hemes, pyridoxin, vitamins K and B_{12}, and folic acid implement metabolic processes everywhere.

What do these biochemical or biologic universals mean? They suggest that life arose from inanimate matter only once and that all organisms, no matter how diverse in other respects, conserve the basic features of the primordial life. (It is also possible that there were several, or even many, origins of life; if so, the progeny of only one of them has survived and inherited the earth.) But what if there was no evolution, and every one of the millions of species was created by separate fiat? However offensive the notion may be to religious feeling and to reason, the antievolutionists must again accuse the Creator of cheating. They must insist that He deliberately arranged things exactly as if his method of creation was evolution, intentionally to mislead sincere seekers of truth.

The remarkable advances of molecular biology in recent years have made it possible to understand how it is that diverse organisms are constructed from such monotonously similar materials: proteins composed of only 20 kinds of amino acids and coded only by DNA and RNA, each with only four kinds of nucleotides. The method is astonishingly simple. All English words, sentences, chapters, and books are made up of sequences of 26 letters of the alphabet. (They can be represented also by only three signs of the Morse code: dot, dash, and gap.) The meaning of a word or a sentence is defined not so much by what letters it contains as by the sequence of these letters. It is the same with heredity: it is coded by the sequences of the genetic "letters"—the nucleotides—in the DNA. They are translated into the sequences of amino acids in the proteins.

Molecular studies have made possible an approach to exact measurements of degrees of biochemical similarities and differences among organisms. Some kinds of enzymes and other proteins are quasiuniversal, or at any rate widespread, in the living world. They are functionally similar in different living beings, in that they catalyze similar chemical reactions. But when such proteins are isolated and their structures determined chemically, they are often found to contain more or less different sequences of amino acids in different organisms. For example, the so-called alpha chains of hemoglobin have identical sequences of amino acids in man and the chimpanzee, but they differ in a single amino acid (out of 141) in the gorilla. Alpha chains of human hemoglobin differ from cattle hemoglobin in 17 amino acid substitutions, 18 from horse, 20 from donkey, 25 from rabbit, and 71 from fish (carp).

Cytochrome C is an enzyme that plays an important role in the metabolism of aerobic cells. It is found in the most diverse organisms, from man to molds. E. Margoliash, W. M. Fitch, and others have compared the amino acid sequences in cytochrome C in different branches of the living world. Most significant similarities as well as differences have been brought to light. The cytochrome C of different orders of mammals and birds differ in 2 to 17 amino acids, classes of vertebrates in 7 to 38, and vertebrates and insects in 23 to 41; and animals differ from yeasts and molds in 56 to 72 amino acids. Fitch and Margoliash prefer to express their findings in what are called "minimal mutational distances." It has been mentioned above that different amino acids are coded by different triplets of nucleotides in DNA of the genes; this code is now known. Most mutations involve substitutions of single nucleotides somewhere in the DNA chain coding for a given protein. Therefore, one can calculate the minimum numbers of single mutations needed to change the cytochrome C of one organism into that of another. Minimal mutational distances between human cytochrome C and the cytochrome C of other living beings are as follows:

Monkey	1	Chicken	18
Dog	13	Penguin	18
Horse	17	Turtle	19
Donkey	16	Rattlesnake	20
Pig	13	Fish (tuna)	31
Rabbit	12	Fly	33
Kangaroo	12	Moth	36
Duck	17	Mold	63
Pigeon	16	Yeast	56

It is important to note that amino acid sequences in a given kind of protein vary within a species as well as from species to species. It is evident that the differences among proteins at the levels of species, genus, family, order, class, and phylum are compounded of elements that vary also among individuals within a species. Individual and group differences are only quantitatively, not qualitatively, different. Evidence supporting the above propositions is ample and is growing rapidly. Much work has been done in recent years on individual variations in amino acid sequences of hemoglobins of human blood. More than 100 variants have been detected. Most of them involve substitutions of single amino acids—substitutions that have arisen by genetic mutations in the persons in whom they are discovered or in their ancestors. As expected, some of these mutations are deleterious to their carriers, but others apparently are neutral or even favorable in certain environments. Some mutant hemoglobins have been found only in one person or in one family; others are discovered repeatedly among inhabitants of different parts of the world. I submit that all these remarkable findings make sense in the light of evolution; they are nonsense otherwise.

Comparative Anatomy and Embryology

The biochemical universals are the most impressive and the most recently discovered, but certainly they are not the only vestiges of creation by means of evolution. Comparative anatomy and embryology proclaim the evolutionary origins of the present inhabitants of the world. In 1555 Pierre Belon established the presence of homologous bones in the superficially very different skeletons of man and bird. Later anatomists traced the homologies in the skeletons, as well as in other organs, of all vertebrates. Homologies are also traceable in the external skeletons of arthropods as seemingly unlike as a lobster, a fly, and a butterfly. Examples of homologies can be multiplied indefinitely.

Embryos of apparently quite diverse animals often exhibit striking similarities. A century ago these similarities led some biologists (notably the German zoologist Ernst Haeckel) to be carried by their enthusiasm so far as to interpret the embryonic similarities as meaning that the embryo repeats in its development the evolutionary history of its species: it was said to pass through stages in which it resembles its remote ancestors. In other words, early-day biologists supposed that by studying embryonic development one can, as it were, read off the stages through which the evolutionary development had passed. This so-called biogenetic law is no longer credited in its original form. And yet embryonic similarities are undeniably impressive and significant.

Probably everybody knows the sedentary barnacles which seem to have no similarity to free-swimming crustaceans, such as the copepods. How remarkable that barnacles pass through a free-swimming larval stage, the nauplius! At that stage of its development a barnacle and a *Cyclops* look unmistakably similar. They are evidently relatives. The presence of gill slits in human embryos and in embryos of other terrestrial vertebrates is another famous example. Of course, at no stage of its development is a human embryo a fish, nor does it ever have functioning gills. But why should it have unmistakable gill slits unless its remote ancestors did respire with the aid of gills? Is the Creator again playing practical jokes?

Adaptive Radiation: Hawaii's Flies

There are about 2,000 species of drosophilid flies in the world as a whole. About a quarter of them occur in Hawaii, although the total area of the archipelago is only about that of the state of New Jersey. All but 17 of the species in Hawaii are endemic (found nowhere else). Furthermore, a great majority of the Hawaiian endemics do not occur throughout the archipelago: they are restricted to single islands or even to a part of an island. What is the explanation of this extraordinary proliferation of drosophilid species in so small a territory? Recent work of H. L. Carson, H. T. Spieth, D. E. Hardy, and others makes the situation understandable.

The Hawaiian islands are of volcanic origin; they were never parts of any continent. Their ages are between 5.6 and 0.7 million years. Before man came their inhabitants were descendants of immigrants that had been transported across the ocean by air currents and other accidental means. A single drosophilid species, which arrived in Hawaii first, before there were numerous competitors, faced the challenge of an abundance of many unoccupied ecologic niches. Its descendants responded to this challenge by evolutionary adaptive radiation, the products of which are the remarkable Hawaiian drosophilids of today. To forestall a possible misunderstanding, let it be made clear that the Hawaiian endemics are by no means so similar to each other that they could be mistaken for variants of the same species; if anything, they are more diversified than are drosophilids elsewhere. The largest and the smallest drosophilid species are both Hawaiian. They exhibit an astonishing variety of behavior patterns. Some of them have become adapted to ways of life quite extraordinary for a drosophilid fly, such as being parasites in egg cocoons of spiders.

Oceanic islands other than Hawaii, scattered over the wide Pacific Ocean, are not conspicuously rich in endemic species of drosophilids. The most probable explanation of this fact is that these other islands were colonized by drosophilids after most ecologic niches had already been filled by earlier arrivals. This surely is a hypothesis, but it is a reasonable one. Antievolutionists might perhaps suggest an alternative hypothesis: in a fit of absent-mindedness, the Creator went on manufacturing more and more drosophilid species for Hawaii, until there was an extravagant surfeit of them in this archipelago. I leave it to you to decide which hypothesis makes sense.

Strength and Acceptance of the Theory

Seen in the light of evolution, biology is, perhaps, intellectually the most satisfying and inspiring science. Without that light it becomes a pile of sundry facts—some of them interesting or curious but making no meaningful picture as a whole.

This is not to imply that we know everything that can and should be known about biology and about evolution. Any competent biologist is aware of a multitude of problems yet unresolved and of questions yet unanswered. After all, biologic research shows no sign of approaching completion; quite the opposite is true. Disagreements and clashes of opinion are rife among biologists, as they should be in a living and growing science. Antievolutionists mistake, or pretend to mistake, these disagreements as indications of dubiousness of the entire doctrine of evolution. Their favorite sport is stringing together quotations, carefully and sometimes expertly taken out of context, to show that nothing is really established or agreed upon among evolutionists. Some of my colleagues and myself have been amused and amazed to read ourselves quoted in a way showing that we are really antievolutionists under the skin.

Let me try to make crystal clear what is established beyond reasonable doubt, and what needs further study, about evolution. Evolution as a process that has always gone on in the history of the earth can be doubted only by those who are ignorant of the evidence or are resistant to evidence, owing to emotional blocks or to plain bigotry. By contrast, the mechanisms that bring evolution about certainly need study and clarification. There are no alternatives to evolution as history that can withstand critical examination. Yet we are constantly learning new and important facts about evolutionary mechanisms.

It is remarkable that more than a century ago Darwin was able to discern so much about evolution without having available to him the key facts discovered since. The development of genetics after 1900—especially of molecular genetics, in the last two decades—has provided information essential to the understanding of evolutionary mechanisms. But much is in doubt and much remains to be learned. This is heartening and inspiring for any scientist worth his salt. Imagine that everything is completely known and that science has nothing more to discover: what a nightmare!

Does the evolutionary doctrine clash with religious faith? It does not. It is a blunder to mistake the Holy Scriptures for elementary textbooks of astronomy, geology, biology, and anthropology. Only if symbols are construed to mean what they are not intended to mean can there arise imaginary, insoluble conflicts. As pointed out above, the blunder leads to blasphemy: the Creator is accused of systematic deceitfulness.

One of the great thinkers of our age, Pierre Teilhard de Chardin, wrote the following: "Is evolution a theory, a system, or a hypothesis? It is much more —it is a general postulate to which all theories, all hypotheses, all systems must henceforward bow and which they must satisfy in order to be thinkable and true. Evolution is a light which illuminates all facts, a trajectory which all lines of thought must follow—this is what evolution is." Of course, some scientists, as well as some philosophers and theologians, disagree with some parts of Teilhard's teachings; the acceptance of his world view falls short of universal. But there is no doubt at all that Teilhard was a truly and deeply religious man and that Christianity was the cornerstone of his world view. Moreover, in his world view science and faith were not segregated in watertight compartments, as they are with so many people. They were harmoniously fitting parts of his world view. Teilhard was a creationists, but one who understood that the Creation is realized in this world by means of evolution. □

Maynard Smith, J., and G. R. Price. 1973. The logic of animal conflict. *Nature.* 246:15–18.

We include this paper in our collection for three reasons: (1) it illustrates the popular approaches of game theory and what Maynard Smith and Price refer to as "evolutionarily stable strategy"; (2) it contrasts "group selection" with arguments based on individual selection; and (3) it is *the* most frequently cited work by John Maynard Smith, one of the leading evolutionary biologists of the late twentieth century. The third point speaks for itself. We briefly elaborate on the first two points, in turn.

1. Game theory is an important branch of evolutionary analysis that deals with how animals might act when an individual's success depends on others' decisions as well as its own. In human social behavior, applications include bargaining encounters, ultimatum games, commitment choices, decisions about mutual aid versus defection, hawk-dove interactions, ownership issues, and truth versus deception. Game theory also finds many applications in the social behavior of animals in nature. The usual goal is to identify the behavioral tactics that benefit personal fitness and to discover whether particular tactics are evolutionarily stable (immune to invasion by alternative tactics) in a population. Each outcome immune to invasion is called an evolutionarily stable strategy, or ESS. To assess whether an ESS exists, game theorists begin by generating a payoff matrix showing changes in personal fitness to individuals who adopt one or another tactic when interacting with another. Computer simulations are then run or mathematical calculations made to determine whether a particular tactic or mixture of tactics is evolutionarily stable. In this paper, a "limited war" tactic proved to be an ESS for many animals in combat.

2. A recurring challenge in evolutionary biology is to identify (and then typically discredit) arguments that invoke benefits to groups of organisms (group selection), as opposed to more acceptable arguments that invoke natural selection operating at the level of individuals (or sometimes their constituent genes). In this case, Maynard Smith and Price argue that the benefits of a limited-war tactic apply to individual combatants as well as to the species involved.

RELATED READING

Maynard Smith, J. 1982. Evolution and the theory of games. Cambridge University Press, Cambridge, England.
Nowak, M. A., C. E. Tarnita, and E. O. Wilson. 2010. The evolution of eusociality. Nature 466:1057–1062.
Osborne, M. J. 2004. An introduction to game theory. Oxford University Press, New York, New York, USA.
Skyrms, B. 1996. Evolution of the social contract. Cambridge University Press, Cambridge, England.
Wilson, E. O. 2013. The social conquest of Earth. Liveright Publishing, New York, New York, USA.

The Logic of Animal Conflict

J. MAYNARD SMITH
School of Biological Sciences, University of Sussex, Falmer, Sussex BN1 9QG

G. R. PRICE
Galton Laboratory, University College London, 4 Stephenson Way, London NW1 2HE

Conflicts between animals of the same species usually are of "limited war" type, not causing serious injury. This is often explained as due to group or species selection for behaviour benefiting the species rather than individuals. Game theory and computer simulation analyses show, however, that a "limited war" strategy benefits individual animals as well as the species.

In a typical combat between two male animals of the same species, the winner gains mates, dominance rights, desirable territory, or other advantages that will tend toward transmitting its genes to future generations at higher frequencies than the loser's genes. Consequently, one might expect that natural selection would develop maximally effective weapons and fighting styles for a "total war" strategy of battles between males to the death. But instead, intraspecific conflicts are usually of a "limited war" type, involving inefficient weapons or ritualized tactics that seldom cause serious injury to either contestant. For example, in many snake species the males fight each other by wrestling without using their fangs[1,2]. In mule deer (*Odocoileus hemionus*) the bucks fight furiously but harmlessly by crashing or pushing antlers against antlers, while they refrain from attacking when an opponent turns away, exposing the unprotected side of its body[3]. And in the Arabian oryx (*Oryx leucoryx*) the extremely long, backward pointing horns are so inefficient for combat that in order for two males to fight they are forced to kneel down with their heads between their knees to direct their horns forward[4]. (For additional examples, see Collins[5], Darwin[4], Hingston[6], Huxley *et al.*[7], Lorenz[8] and Wynne-Edwards[9].)

How can one explain such oddities as snakes that wrestle with each other, deer that refuse to strike "foul blows", and antelope that kneel down to fight?

The accepted explanation for the conventional nature of contests is that if no conventional methods existed, many individuals would be injured, and this would militate against the survival of the species (see, for example, Huxley[7]). The difficulty with this type of explanation is that it appears to assume the operation of "group selection". Although one cannot rule out group selection as an agent producing adaptations, it is only likely to be effective in rather special circumstances[10–12]. Consequently it seems to us that group selection cannot by itself account for the complex anatomical and behavioural adaptations for limited conflict found in so many species, but there must also be individual selection for these, which means that a "limited war" strategy must be differentially advantageous for individuals.

We consider simple formal models of conflict situations, and ask what strategy will be favoured under individual selection. We first consider conflict in species possessing offensive weapons capable of inflicting serious injury on other members of the species. Then we consider conflict in species where serious injury is impossible, so that victory goes to the contestant who fights longest. For each model, we seek a strategy that will be stable under natural selection; that is, we seek an "evolutionarily stable strategy" or ESS. The concept of an ESS is fundamental to our argument; it has been derived in part from the theory of games, and in part from the work of MacArthur[13] and of Hamilton[14] on the evolution of the sex ratio. Roughly, an ESS is a strategy such that, if most of the members of a population adopt it, there is no "mutant" strategy that would give higher reproductive fitness.

A Computer Model

A main reason for using computer simulation was to test whether it is possible even in theory for individual selection to account for "limited war" behaviour.

We consider a species that possesses offensive weapons capable of inflicting serious injuries. We assume that there are two categories of conflict tactics: "conventional" tactics, *C*, which are unlikely to cause serious injury, and "dangerous" tactics, *D*, which are likely to injure the opponent seriously if they are employed for long. (Thus in the snake example, wrestling involves *C* tactics and use of fangs would be *D* tactics. In many species, *C* tactics are limited to threat displays at a distance, without any physical fighting. We consider a conflict between two individuals to consist of a series of alternate "moves". At each move, a contestant can employ *C* or *D* tactics, or retreat, *R*. If a contestant employs *D* tactics, there is a fixed probability that his opponent will be seriously injured: a contestant who is seriously injured always retreats. If a contestant retreats, the contest is at an end and his opponent is the winner. A possible conflict between contestants *A* and *B* can be represented in this way:

A's move *C C C C C C C C C C C D C C C C C C C D*
B's move *C C C C C C C C C C C D C C C C C C C R*

If a contestant plays *D* on the first move of a contest, or plays *D* in response to *C* by his opponent, this is called a "probe" or a "provocation". A probe made after the opening move is said to "escalate" a contest from *C* to *D* level. A contestant who plays *D* in reply to a probe is said to "retaliate". In the example shown above, *A* probes on his twelfth and twentieth moves; *B* retaliates after the first probe, but retreats after the second, leaving *A* the winner. At the end of a contest there are "pay-offs" to each contestant. The pay-offs are taken as measures of the contribution the contest has made to the reproductive success of the individual. They take account of three factors: the advantages of winning as compared with losing, the disadvantage of being seriously injured, and the disadvantage of wasting time and energy in the contest.

A "strategy" for a contestant is a set of rules which ascribe probabilities to the *C*, *D*, and *R* plays, as functions of what has previously happened in the course of the current contest. (No memory of what has happened in previous contests with the same or other opponents is assumed.) For computer simulation we programmed five possible strategies, each of which might be thought on *a priori* grounds to be optimal in certain circumstances. The strategies considered were as follows:

(1) "Mouse". Never plays *D*. If receives *D*, retreats at once before there is any possibility of receiving a serious injury. Otherwise plays *C* until the contest has lasted a preassigned number of moves.

(2) "Hawk". Always plays *D*. Continues the contest until he is seriously injured or his opponent retreats.

(3) "Bully". Plays *D* if making the first move. Plays *D* in response to *C*. Plays *C* in response to *D*. Retreats if opponent plays *D* a second time.

(4) "Retaliator". Plays *C* if making the first move. If opponent plays *C*, plays *C* (but plays *R* if contest has lasted a preassigned number of moves). If opponent plays *D*, with a high probability retaliates by playing *D*.

(5) "Prober-Retaliator". If making the first move, or after opponent has played *C*, with high probability plays *C* and with low probability plays *D* (but plays *R* if contest has lasted a preassigned number of moves). After giving a probe, reverts to *C* if opponent retaliates, but "takes advantage" by continuing to play *D* if opponent plays *C*. After receiving a probe, with high probability plays *D*.

The contestants were programmed as having identical fighting prowess, so that they differed only in the strategies they followed. The five strategies represent extremes, but from results with these it is possible to estimate the results likely to be found with intermediate types. The Hawk strategy is a "total war" strategy; Mouse, Retaliator, and Prober-Retaliator are "limited war" strategies. The question of main interest is whether individual selection will favour the former or one of the latter types.

The Simulation Test

The five strategies determine fifteen types of two-opponent contests. Two thousand contests of each type were simulated by computer, using pseudo-random numbers generated by an algorithm to vary the contests. The following probabilities were used: Probability of serious injury from a single *D* play=0.10. Probability that a Prober-Retaliator will probe on the opening move or after opponent has played *C*=0.05. Probability that Retaliator or Prober-Retaliator will retaliate against a probe (if not injured) by opponent=1.0. Pay-offs were calculated as follows: Pay-off for winning=+60. Pay-off for receiving serious injury=−100. Pay-off for each *D* received that does not cause serious injury (a "scratch")=−2. Pay-off for saving time and energy (awarded to each contestant not seriously injured) varied from 0 for a contest of maximum length, to +20 for a very short contest. The contest example shown earlier was one of the 2,000 Prober-Retaliator versus Prober-Retaliator contests.

Table 1 shows the average pay-off to each contestant in each type of contest. The number in a given row and column is the pay-off gained by the row strategy when the opponent uses the column strategy. For example, in contests between Mouse and Hawk, the average pay-offs are 19.5 to Mouse and 80.0 to Hawk.

To tell whether a strategy is evolutionarily stable against the other four strategies, we examine the corresponding column in Table 1. For example, for Hawk to be an ESS, it is necessary that it be the most profitable strategy in a population almost entirely of Hawks. In such a population, a given animal of any type will almost always have a Hawk as opponent. Therefore the pay-offs in the "Hawk" column apply. These show that Mouse and Bully are both more successful than Hawk. Therefore natural selection will cause alleles for Mouse and Bully behaviour to increase in frequency, and alleles giving Hawk behaviour to decrease. Thus Hawk is not an ESS.

Examining the other columns, we see that Mouse is not an ESS because Hawk, Bully, and Prober-Retaliator average higher pay-offs in a population almost entirely of Mouse. Nor is Bully an ESS. However, Retaliator is an ESS since no other strategy does better, though Mouse does equally well. And the last column shows that Prober-Retaliator is almost an ESS.

How would we expect such a population to evolve? It will come to consist mainly of Retaliators or Prober-Retaliators, with the other strategies maintained at a low frequency by mutation. The balance between the two main types will depend on the frequency of Mouse, since the habit of probing is only an advantage against Mouse. For the particular values in Table 1, it can be shown that if the frequency of Mouse is greater than 7%, Prober-Retaliator will replace Retaliator as the predominant type. It is worth noting that a real population would contain young, senile, diseased and injured individuals adopting the strategy Mouse for non-genetic reasons.

Thus the simulation shows emphatically the superiority, under individual selection, of "limited war" strategies in comparison with the Hawk strategy.

Briefly, the reason that conflict limitation increases individual fitness is that retaliation behaviour decreases the fitness of Hawks, while the existence of possible future mating opportunities reduces the loss from retreating uninjured.

This general result will not be altered by moderate changes in the program parameters, though very large changes will alter it. One way would be by changing the probability of serious injury from a single *D* from 0.10 to 0.90. This would give advantage to "Pre-emptive Strike" policies, making Hawk an ESS. (Such species are probably rare, because excessively dangerous weapons or tactics would be opposed by kin selection.) Another way to make selection favour "total war" behaviour would be by giving the same pay-off penalty for retreating uninjured as for serious injury. This would correspond to a species where an individual fights only a single battle in its lifetime, on which its reproductive success entirely depends. Our choice of +60 for winning, 0 for retreating uninjured, and −100 for serious injury represents a species where males have more than one opportunity to gain a mate. Changing these values to +60, −100, and −100 respectively, would make Hawk the optimal strategy. Conversely, +60, 0,

Table 1 Average Pay-offs in Simulated Intraspecific Contests for Five Different Strategies

		Opponent				
		"Mouse"	"Hawk"	"Bully"	"Retaliator"	"Prober-Retaliator"
Contestant receiving the pay-off	"Mouse"	29.0	19.5	19.5	29.0	17.2
	"Hawk"	80.0	−19.5	74.6	−18.1	−18.9
	"Bully"	80.0	4.9	41.5	11.9	11.2
	"Retaliator"	29.0	−22.3	57.1	29.0	23.1
	"Prober-Retaliator"	56.7	−20.1	59.4	26.9	21.9

−500 would represent a long-lived species with numerous opportunities to gain mates, where individual selection would still more strongly favour cautious strategies.

Real Animals

Real animal conflicts are vastly more complex than our simulated conflicts. (An interesting study by Dingle[15] shows that this holds true even at the lowly level of the mantis shrimp.) Probably our models are true to nature in emphasising a category distinction rather than an intensity distinction between "conventional" and "dangerous" tactics. In real animals, however, there exist not only the category distinction, but also individual differences in the intensity and skill with which each kind of tactic is employed. Also, in many species there are several categories of increasingly dangerous tactics, instead of only one.

The advantage from making a category distinction is that this simplifies behavioural requirements for limited conflict. It is probably easier for genetics to program a snake not to use fangs at all in certain situations than to program it to use fangs as intensively as possible up to intensity k, but not at intensities greater than k. Similarly, fair and foul blows are distinguished in boxing, and conventional and nuclear weapons in war.

Under the condition that any act of physical aggression is treated as a D act, the theoretical model will result in symbolic fighting by threat from a distance. This would be advantageous for a species that has an inherent difficulty in fighting physically at a safe level. For example, domestic and wild cattle, which have very dangerous horns and are somewhat clumsy in their charges, make much use of threat displays (stomping, pawing, bellowing). The model will not, however, give rise to conflict behaviour that is wholly symbolic and never backed up by physical aggression or other sanctions, since such behaviour would not be evolutionarily stable without some mechanism reducing the reproductive success of mutant individuals deficient in responding to the symbols. An interesting problem is how the felids, with their dangerous teeth and claws, limit their physical combats to non-fatal levels. Probably the explanation is that they have a hierarchy of many conflict categories and limit their probing to small escalations. Consequently, it takes repeated escalations to raise the conflict to the most dangerous level.

In most animal species there is probably a high correlation between prowess in C tactics and in D tactics. This means that C level conflict provides information to each animal about how its opponent is likely to perform if the conflict is escalated. This permits improvement in strategies over those used in the computer model. Instead of probing at random, an animal will be more likely to probe if its opponent is inferior in conventional fighting. On the other hand, if its opponent is very superior in conventional tactics, an animal will frequently retreat without waiting for its opponent to try a probe. Thus actual animals may combine Prober-Retaliator and Mouse capabilities.

If animals can adopt different strategies according to the opponent that confronts them, then an interesting possibility appears. The "Hawk" column of Table 1 shows that the best strategy against a Hawk is Mouse: that is, retreat immediately. If a species includes deviant individuals who follow the Hawk strategy and fight recklessly against every opponent, then it will be advantageous for ordinary members of the species to be able to estimate recklessness and avoid combat with Hawks. But if this happens, then it will be advantageous to simulate wild, incontrollable rage. And in fact the threat displays of some species do have an appearance of maniacal fury, hence there probably is some advantage in acting this way. However, if most species members simulate insane rage when actually their fighting is limited and controlled, then selection will favour individuals who partly discount the threat displays, and "call the bluff" of the pseudo-Hawks.

This leads to the suggestion that it might be advantageous for an individual animal to be maniacal in an easily recognisable way that could not be counterfeited. A possible instance of this is the phenomenon of going "on musth", which occurs periodically in adult male elephants[16,17]. The temporal glands secrete a dark brown fluid that runs down the face, giving a visual and olfactory sign that cannot be counterfeited. The madness of the animal "on musth" causes other elephants to avoid him, and this may give an increase in dominance status that persists for a time after the musth period is over.

Conflict in which Injury is Impossible

The previous section offers an explanation of why, in a species with offensive weapons capable of inflicting serious injury, escalated fighting may be rare or absent. In doing so, it raises a second problem. In a contest between opponents who are unable to inflict serious injury, victory goes to the one who is prepared to continue for a longer time. How are such contests decided?

Suppose that the pay-off to the victor is v. If a contest is ever to be settled, there must also be some disadvantage to the contestants in a long contest. If so, the only choice of strategy open to a contestant is of the period for which he is prepared to continue, and hence of the pay-off, say $-m$, he is prepared to accept. Thus if two contestants adopt strategies m_1 and m_2, where $m_1>m_2$, the pay-off to the first is $v-m_2$ and to the second is $-m_2$. Our problem then is how a contestant should choose a value of m, or, more precisely, whether there is a method of choosing m which is an ESS.

To answer this question, we need a more precise definition of an ESS. We define $E_J(I)$ as the expected pay-off to I played against J. Then I is an ESS if, for all J, $E_I(I)>E_I(J)$; if for any strategy J, $E_I(I)=E_I(J)$, then evolutionary stability requires that $E_J(I)>E_J(J)$. The relevance of the latter condition is as follows. If in a population adopting strategy I a mutant J arises whose expectation against I is the same as I's expectation against itself, then J will increase by genetic drift until meetings between two J's becomes a common event.

It is easy to show that no "pure" strategy (that is, no fixed value of m) is an ESS. Thus in a population adopting strategy m, a mutant adopting $m+\epsilon$ would always do better (and if $m>v$, a mutant adopting a zero strategy would also do better). It is, however, possible to find a mixed strategy which is an ESS. Let strategy I be a mixed strategy which selects a value of m between x and $x+\delta x$ with probability $p(x)\delta x$.

Then if

$$p(x)=(1/v)\exp(-x/v) \tag{1}$$

it can be shown that I is an ESS.

We conclude that an evolutionary stable population is either genetically polymorphic, the strategies of individuals being distributed as in equation (1), or that it consists of individuals whose behaviour differs from contest to contest as in (1). There is no stable pure strategy, and hence no behaviourally uniform population can be stable.

Conclusions

There are many complications left out of these simple models. The analysis is, however, sufficient to show that individual selection can explain why potentially dangerous offensive weapons are rarely used in intraspecific contests; a stable strategy does, however, require that contestants should respond to an "escalated" attack by escalating in return. Also, if contests are settled by a process of attrition, then evolutionary stability requires that the population be genetically polymorphic, or that individuals vary their behaviour from contest to contest.

A more detailed analysis will be published elsewhere.

Ideas similar to those described here have been applied to human neurotic behaviour by J. S. Price[18].

For suggestions, we thank Professors Hans Kalmus and R. C. Lewontin, and Drs W. D. Hamilton, Gerald Lincoln, T. B. Poole and M. J. A. Simpson. We thank the Science Research Council for support.

1 Shaw, C. E., *Herpetologica*, **4**, 137 (1948).
2 Shaw, C. E., *Herpetologica*, 7, 149 (1951).
3 Linsdale, J. M., and Tomich, P. Q., *A Herd of Mule Deer*, 511f (Univ. of California Press, Berkeley and Los Angeles, 1953).
4 Darwin, C., *The Descent of Man and Selection in Relation to Sex*, chap. 17 (Murray, London, 1882).
5 Collins, N. E., *Physiol. Zool.*, **17**, 83 (1944).
6 Hingston, R. W. G., *Character Person.*, **2**, 3 (1933).
7 Huxley, J. S., *Phil. Trans. R. Soc.*, **251B**, 249 (1956).
8 Lorenz, K., *On Aggression* (Methuen, London, 1966).
9 Wynne-Edwards, V. C., *Animal Dispersion in Relation to Social Behaviour*, chap. 8 (Oliver and Boyd, Edinburgh and London, 1962).
10 Maynard Smith, J., *Nature*, **201**, 1145 (1964).
11 Levins, R., in *Some Mathematical Questions in Biology* (American Mathematical Society, 1970).
12 Price, G. R., *Ann. hum. Genet.*, **35**, 485 (1972).
13 MacArthur, R. H., in *Theoretical and Mathematical Biology* (edit. by Waterman, T., and Horowitz, H.) (Blaisdell, New York, 1965).
14 Hamilton, W. D., *Science, N.Y.*, **156**, 477 (1967).
15 Dingle, H., *Anim. Behav.*, **17**, 561 (1969).
16 West, L. J., Pierce, C. M., and Thomas, W. D., *Science, N.Y.*, **138**, 1100 (1962).
17 Eisenberg, J. F., McKay, G. M., and Jainudeen, M. R., *Behaviour*, **38**, 193 (1971).
18 Price, J. S., *Proc. R. Soc. Med.*, **62**, 1107 (1969).

Trivers, R. L. 1974. Parent-offspring conflict. *American Zoologist* 14:249–264.

Robert Trivers, an American evolutionist and sociobiologist, has made fundamental theoretical contributions to the analysis of social evolution, notably in the areas of cooperation, conflict, and self-deception. His life story is fascinating in its own right, having included professional training in subjects ranging from mathematics and evolutionary biology to psychology and law. Trivers has been affiliated with organizations as different as Harvard University and the Black Panther Party. Trivers' eclectic career in evolutionary biology defies easy characterization, except perhaps for its consistent insight and creativity (as illustrated in several of his books, listed below).

We include Trivers' paper on parent-offspring conflict to provide one example of this researcher's unorthodoxy and evolutionary acumen. Traditionally, many biologists supposed that a parent and its offspring share mutualities of interest that should spawn the evolution of fully cooperative behaviors. Trivers shows that this expectation is an oversimplification; overt strategic conflicts between parent and child can be rampant even in such sacrosanct locations and times as within the mammalian womb during pregnancy.

RELATED READING

Avise, J. C. 2013. Evolutionary perspectives on pregnancy. New York: Columbia University Press.

Burt, A., and R. L. Trivers. 2006. Genes in conflict: the evolution of selfish genetic elements. Belknap Press, Cambridge, Massachusetts, USA.

Strassmann, J. E., D. C. Queller, J. C. Avise, and F. J. Ayala, editors. 2012. In the light of evolution, vol. 5: Cooperation and conflict. National Academies Press, Washington, D.C., USA.

Trivers, R. L. 1985. Social evolution. Benjamin/Cummings, Menlo Park, California, USA.

Trivers, R. L. 2011. The folly of fools: the logic of deceit and self-deception in human life. Basic Books, New York, New York, USA.

AMER. ZOOL., 14:249-264 (1974).

Parent-Offspring Conflict

ROBERT L. TRIVERS

Museum of Comparative Zoology, Harvard University, Cambridge, Massachusetts 02138

SYNOPSIS. When parent-offspring relations in sexually reproducing species are viewed from the standpoint of the offspring as well as the parent, conflict is seen to be an expected feature of such relations. In particular, parent and offspring are expected to disagree over how long the period of parental investment should last, over the amount of parental investment that should be given, and over the altruistic and egoistic tendencies of the offspring as these tendencies affect other relatives. In addition, under certain conditions parents and offspring are expected to disagree over the preferred sex of the potential offspring. In general, parent-offspring conflict is expected to increase during the period of parental care, and offspring are expected to employ psychological weapons in order to compete with their parents. Detailed data on mother-offspring relations in mammals are consistent with the arguments presented. Conflict in some species, including the human species, is expected to extend to the adult reproductive role of the offspring: under certain conditions parents are expected to attempt to mold an offspring, against its better interests, into a permanent nonreproductive.

In classical evolutionary theory parent-offspring relations are viewed from the standpoint of the parent. If parental investment (PI) in an offspring is defined as anything done by the parent for the offspring that increases the offspring's chance of surviving while decreasing the parent's ability to invest in other offspring (Trivers, 1972), then parents are classically assumed to allocate investment in their young in such a way as to maximize the number surviving, while offspring are implicitly assumed to be passive vessels into which parents pour the appropriate care. Once one imagines offspring as *actors* in this interaction, then conflict must be assumed to lie at the heart of sexual reproduction itself—an offspring attempting from the very beginning to maximize its reproductive success (RS) would presumably want more investment than the parent is selected to give. But unlike conflict between unrelated individuals, parent-offspring conflict is expected to be circumscribed by the close genetic relationship between parent and offspring. For example, if the offspring garners more investment than the parent has been selected to give, the offspring thereby decreases the number of its surviving siblings, so that any gene in an offspring that leads to an additional investment decreases (to some extent) the number of surviving copies of itself located in siblings. Clearly, if the gene in the offspring exacts too great a cost from the parent, that gene will be selected against even though it confers some benefit on the offspring. To specify precisely how much cost an offspring should be willing to inflict on its parent in order to gain a given benefit, one must specify how the offspring is expected to weigh the survival of siblings against its own survival.

The problem of specifying how an individual is expected to weigh siblings against itself (or any relative against any other) has been solved in outline by Hamilton (1964), in the context of explaining the evolution

I thank I. DeVore for numerous conversations and for detailed comments on the manuscript. For additional comments I thank W. D. Hamilton, J. Roughgarden, T. W. Schoener, J. Seger, and G. C. Williams. For help with the appendix I thank J. D. Weinrich. Finally, for help with the references I thank my research assistant, H. Hare, and the Harry Frank Guggenheim Foundation, which provides her salary. Part of this work was completed under an N.I.H. postdoctoral fellowship and partly supported by N.I.M.H. grant MH-13611 to I. DeVore.

of altruistic behavior. An altruistic act can be defined as one that harms the organism performing the act while benefiting some other individual, harm and benefit being defined in terms of reproductive success. Since any gene that helps itself spread in a population is, by definition, being selected for, altruistic behavior in the above sense can be selected only if there is a sufficiently large probability that the recipient of the act also has the gene. More precisely, the benefit/cost ratio of the act, times the chance that the recipient has the gene, must be greater than one. If the recipient of the act is a relative of the altruist, then the probability that the recipient has the gene by descent from a common ancestor can be specified. This conditional probability is called the *degree of relatedness*, r_0. For an altruistic act directed at a relative to have survival value its benefit/cost ratio must be larger than the inverse of the altruist's r_0 to the relative. Likewise an individual is expected to forego a selfish act if its cost to a relative, times the r_0 to that relative, is greater than the benefit to the actor.

The rules for calculating degrees of relatedness are straightforward for both diploid and haplodiploid organisms, even when inbreeding complicates the relevant genealogy (see the addendum in Hamilton, 1971). For example, in a diploid species (in the absence of inbreeding) an individual's r_0 to his or her full-siblings is $\frac{1}{2}$; to half-siblings, $\frac{1}{4}$; to children, $\frac{1}{2}$; to cousins, $\frac{1}{8}$. If in calculating the selective value of a gene one not only computes its effect on the reproductive success of the individual bearing it, but adds to this its effects on the reproductive success of related individuals, appropriately devalued by the relevant degrees of relatedness, then one has computed what Hamilton (1964) calls *inclusive fitness*. While Hamilton pointed out that the parent-offspring relationship is merely a special case of relations between any set of genetically related individuals, he did not apply his theory to such relations. I present here a theory of parent-offspring relations which follows directly from the key concept of inclusive fitness and from the assumption that the offspring is at all times capable of an active role in its relationship to its parents. The form of the argument applies equally well to haplodiploid species, but for simplicity the discussion is mostly limited to diploid species. Likewise, although many of the arguments apply to any sexually reproducing species showing parental investment (including many plant species), the arguments presented here are particularly relevant to understanding a species such as the human species in which parental investment is critical to the offspring throughout its entire prereproductive life (and often later as well) and in which an individual normally spends life embedded in a network of near and distant kin.

PARENT-OFFSPRING CONFLICT OVER THE CONTINUATION OF PARENTAL INVESTMENT

Consider a newborn (male) caribou calf nursing from his mother. The benefit to him of nursing (measured in terms of his chance of surviving) is large, the cost to his mother (measured in terms of her ability to produce additional offspring) presumably small. As time goes on and the calf becomes increasingly capable of feeding on his own, the benefit to him of nursing decreases while the cost to his mother may increase (as a function, for example, of the calf's size). If cost and benefit are measured in the same units, then at some point the cost to the mother will exceed the benefit to her young and the net reproductive success of the mother decreases if she continues to nurse. (Note that later-born offspring may contribute less to the mother's eventual RS than early-born, because their reproductive value may be lower [Fisher, 1930], but this is automatically taken into account in the cost function.)

The calf is not expected, so to speak, to view this situation as does his mother, for the calf is completely related to himself but only partially related to his future siblings, so that he is expected to devalue the cost of nursing (as measured in terms of future sibs) by his r_0 to his future sibs, when comparing the cost of nursing with its benefit to himself. For example, if future sibs are expected to be full-sibs, then the calf should nurse until the cost to the mother is more than twice the benefit to himself. Once the cost to the mother is more than twice the benefit to the calf, continued nursing is opposed by natural selection acting on both the mother and the calf. As long as one imagines that the benefit/cost ratio of a parental act changes continuously from some large number to some very small number near zero, then there must occur a period of time during which $\frac{1}{2}<B/C<1$. This period is one of expected conflict between mother and offspring, in the sense that natural selection working on the mother favors her halting parental investment while natural selection acting on the offspring favors his eliciting the parental investment. The argument presented here is graphed in Figure 1. (Note, as argued below, that there are specialized situations in which the offspring may be selected to consume *less* PI than the parent is selected to give.)

This argument applies to all sexually reproducing species that are not completely inbred, that is, in which siblings are not identical copies of each other. Conflict near the end of the period of PI over the continuation of PI is expected in all such species. The argument applies to PI in general or to any subcomponent of PI (such as feeding the young, guarding the young, carrying the young) that can be assigned a more or less independent cost-benefit function. Weaning conflict in mammals is an example of parent-offspring conflict explained by the argument given here. Such conflict is known to occur in a variety of mammals, in the field and in the laboratory: for example, baboons (DeVore, 1963), langurs (Jay, 1963), rhesus macaques (Hinde and Spencer-Booth, 1971), other macaques (Rosenblum, 1971), vervets (Struhsaker, 1971), cats (Schneirla et al., 1963), dogs (Rheingold, 1963), and rats (Rosenblatt and Lehrman, 1963). Likewise, I interpret conflict over parental feeding at the time of fledging in bird species as conflict explained by the present argument: for example, Herring Gulls (Drury and Smith, 1968), Red Warblers (Elliott, 1969), Verreaux's Eagles (Rowe, 1947), and White Pelicans (Schaller, 1964).

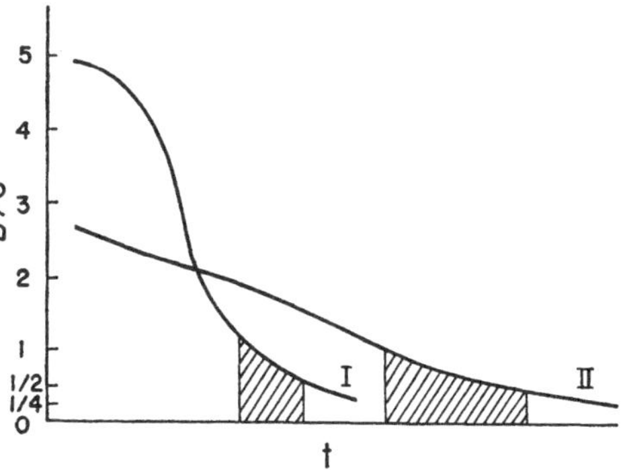

FIG. 1. The benefit/cost ratio (B/C) of a parental act (such as nursing) toward an offspring as a function of time. Benefit is measured in units of reproductive success of the offspring and cost in comparable units of reproductive success of the mother's future offspring. Two species are plotted. In species I the benefit/cost ratio decays quickly; in species II, slowly. Shaded areas indicate times during which parent and offspring are in conflict over whether the parental care should continue. Future sibs are assumed to be full-sibs. If future sibs were half-sibs, the shaded areas would have to be extended until $B/C = 1/4$.

Weaning conflict is usually assumed to occur either because transitions in nature are assumed always to be imperfect or because such conflict is assumed to serve the interests of both parent and offspring by informing each of the needs of the other. In either case, the marked inefficiency of weaning conflict seems the clearest argument in favor of the view that such conflict results from an underlying conflict in the way in which the inclusive fitness of mother and offspring are maximized. Weaning conflict in baboons, for example, may last for weeks or months, involving daily competitive interactions and loud cries from the infant in a species otherwise strongly selected for silence (DeVore, 1963). Interactions that inefficient *within* a multicellular organism would be cause for some surprise, since, unlike mother and offspring, the somatic cells within an organism are identically related.

One parameter affecting the expected length (and intensity) of weaning conflict is the offspring's expected r_o to its future siblings. The lower the offspring's r_0 to its future siblings, the longer and more intense the expected weaning conflict. This suggests a simple prediction. Other things being equal, species in which different, unrelated males commonly father a female's successive offspring are expected to show stronger weaning conflict than species in which a female's successive offspring are usually fathered by the same male. As shown below, however, weaning conflict is merely a special case of conflict expected throughout the period of parental investment, so that this prediction applies to the intensity of conflict prior to weaning as well.

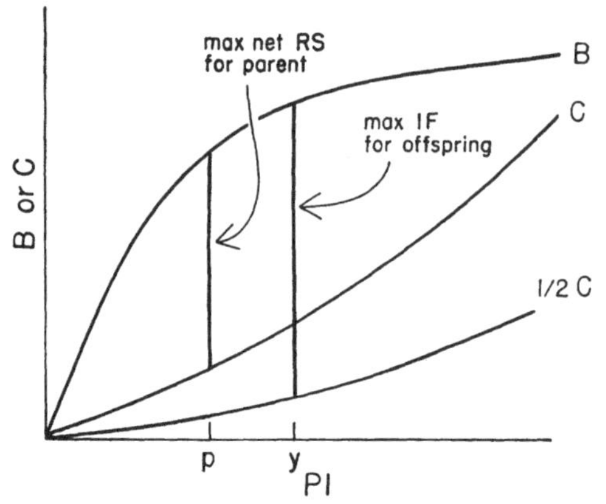

FIG. 2. The benefit, cost, and half the cost of a parental act toward an offspring at one moment in time as a function of the amount the parent invests in the act (PI). Amount of milk given during one day of nursing in a mammal would be an example of PI. At *p* the parent's inclusive fitness (B - C) is maximized; at *y* the offspring's inclusive fitness (B - C/2) is maximized. Parent and offspring disagree over whether *p* or *y* should be invested. The offspring's future siblings are assumed to be full-siblings. IF = inclusive fitness.

CONFLICT THROUGHOUT THE PERIOD OF PI OVER THE AMOUNT OF PI

In Figure 1 it was assumed that the amount of investment for each day (or moment in time) had already been established, and that mother and young were only selected to disagree over when such investment should be ended. But it can be shown that, in theory, conflict over the amount of investment that should at each moment be given, is expected throughout the period of PI.

At any moment in the period of PI the female is selected to invest that amount which maximizes the difference between the associated cost and benefit, where these terms are defined as above. The infant is selected to induce that investment which maximizes the difference between the benefit and a cost devalued by the relevant r_0. The different optima for a moment in time in a hypothetical species are graphed in Figure 2. With reasonable assumptions about the shape of the benefit and cost curves, it is clear that the infant will, at each instant in time, tend to favor greater parental investment than the parent is selected to give. The period of transition discussed in the previous section is a special case of this continuing competition, namely, the case in which parent and offspring compete over whether *any* investment should be given, as opposed to their earlier competition over *how much* should be given. Since parental investment begins before eggs are laid or young are born, and since there appears to be no essential distinction between parent-offspring conflict outside the mother (mediated primarily by behavioral acts) and parent-offspring conflict inside the mother (mediated primarily by chemical acts), I assume that parent-offspring conflict may in theory begin as early as meiosis.

It must be emphasized that the cost of parental investment referred to above (see Fig. 2) is measured *only* in terms of decreased ability to produce *future* offspring (or, when the brood size is larger than one, decreased ability to produce *other* offspring). To appreciate the significance of this definition, imagine that early in the period of PI the offspring garners more investment than the parent has been selected to give. This added investment may decrease the parent's later investment in the offspring at hand, either through an increased chance of parental mortality during the period of PI, or through a depletion in parental resources, or because parents have been selected to make the appropriate adjustment (that is, to reduce later investment below what otherwise would have been given). In short, the offspring may gain a temporary benefit but suffer a later cost. This self-inflicted cost is subsumed in the benefit function (B) of Figure 2, because it decreases the benefit the infant receives. It is not subsumed in the cost function (C) because this function refers only to the mother's future offspring.

THE TIME COURSE OF PARENT OFFSPRING CONFLICT

If one could specify a series of cost-benefit curves (such as Fig. 2) for each day of the period of PI, then the expected time course of parent-offspring conflict could be specified. Where the difference in the offspring's inclusive fitness at the parent's optimum PI (p in Fig. 2) and at the offspring's optimum PI (y) is large, conflict is expected to be intense. Where the difference is slight, conflict is expected to be slight or nonexistent. In general, where there is a strong difference in the offspring's inclusive fitness at the two different optima (p and y), there will also be a strong difference in the parent's inclusive fitness, so that both parent and offspring will simultaneously be strongly motivated to achieve their respective optimal values of PI. (This technique of comparing cost-benefit graphs can be used to make other predictions about parent-offspring conflict, for example that such conflict should decrease in intensity with increasing age, and hence decreasing reproductive value, of the parent; see Figure 3.) In the absence of such day-by-day graphs three factors can be identified, all of which will usually predispose parent and offspring to show greater conflict as the period of PI progresses.

1) *Decreased chance of self-inflicted cost.* As the period of PI progresses, the offspring faces a decreased chance of suffering a later self-inflicted cost for garnering additional investment at the moment. At the end of the period of PI any additional investment forced on the parent will only affect later offspring, so that at that time the interests of parent and offspring are maximally divergent. This time-dependent change in the offspring's chance of suffering a self-

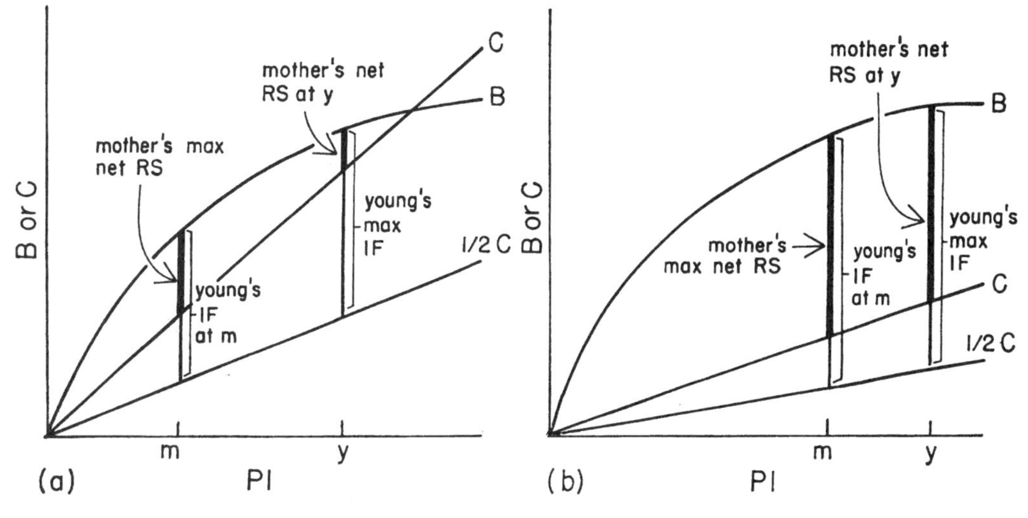

FIG. 3. The benefit and cost of a parental act (as in Fig. 2) toward (*a*) an offspring born to a young female and (*b*) an offspring born to an old female. One assumes that the benefit to the offspring of a given amount of PI does not change with birth order but that the cost declines as a function of the declining reproductive value (Fisher, 1930) of the mother: she will produce fewer future offspring anyway. The difference between the mother's inclusive fitness at m and y is greater for (*a*) than for (*b*). The same is true for the offspring. Conflict should be correspondingly more intense between early born young and their mothers than between late born young and their mothers.

inflicted cost will, other things being equal, predispose parent and offspring to increasing conflict during the period of PI.

2) *Imperfect replenishment of parental resources.* If the parent is unable on a daily basis to replenish resources invested in the offspring, the parent will suffer increasing depletion of its resources, and, as time goes on, the cost of such depletion should rise disproportionately, even if the amount of resources invested per day declines. For example, a female may give less milk per day in the first half of the nursing period than in the second half (as in pigs: Gill and Thomson, 1956), but if she is failing throughout to replenish her energy losses, then she is constantly increasing her deficit (although at a diminishing rate) and greater deficits may be associated with disproportionate costs. In some species a parent does not feed itself during much of the period of PI and at least during such periods the parent must be depleting its resources (for example, female elephant seals during the nursing period: LeBoeuf et al., 1972). But the extent to which parents who feed during the period of PI fail to replenish their resources is usually not known. For some species it is clear that females typically show increasing levels of depletion during the period of PI (e.g., sheep: Wallace, 1948).

3) *Increasing size of the offspring.* During that portion of the period of PI in which the offspring receives all its food from its parents, the tendency for the offspring to begin very small and steadily increase in size will, other things being equal, increase the cost to the parent of maintaining and enlarging it. (Whether this is always true will depend, of course, on the way in which the offspring's growth rate changes as a function of increasing size.) In addition, as the offspring increases in size the relative energetic expense to it of competing with its parents should decline.

The argument advanced here is only meant to suggest a general tendency for conflict to increase during the period of PI, since it is easy to imagine circumstances in which conflict might peak several times during the period of PI. It is possible, for example, that weight at birth in a mammal such as humans is strongly associated with the offspring's survival in subsequent weeks, but that the cost to the mother of bearing a large offspring is considerably greater than some of her ensuing investment. In such circumstances, conflict *prior* to birth over the offspring's weight at birth may be more intense than conflict over nursing in the weeks after birth.

Data from studies of dogs, cats, rhesus macaques, and sheep appear to support the arguments of this and the previous section. In these species, parent-offspring conflict begins well before the period of weaning and tends to increase during the period of PI. In dogs (Rheingold, 1963) and cats (Schneirla et al., 1963) postnatal maternal care can be divided into three periods according to increasing age of the offspring. During the first, the mother approaches the infant to initiate parental investment. No avoidance behavior or aggression toward the infant is shown by the mother. In the second, the offspring and the mother approach each other about equally, and the mother shows some avoidance behavior and some aggression in response to the infant's demands. The third period can be characterized as the period of weaning. Most contacts are initiated by the offspring. Open avoidance and aggression characterize the mother.

Detailed quantitative data on the rhesus macaque (Hinde and Spencer-Booth, 1967, 1971), and some parallel data on other macaques (Rosenblum, 1971), demonstrate that the behavior of both mother and offspring change during the period of postnatal parental care in a way consistent with theory. During the first weeks after she has given birth, the rhesus mother's initiative in setting up nipple contacts is high but it soon declines rapidly. Concurrently she begins to reject some of the infant's advances, and after her own initiatives toward nipple contact have ceased, she rejects her infant's advances with steadily increasing frequency until at the end of the period of investment all of the offspring's advances are rejected. Shortly after birth, the offspring leaves the mother more often than it approaches her, but as time goes on the initiative in maintaining mother-offspring proximity shifts to the offspring. This leads to the superficially paradoxical result that as the offspring becomes increasingly active and independent, spending more and more time away from its mother, its initiative in maintaining mother-offspring proximity *increases* (that is, it tends to approach the mother more often than it leaves her). According to the theory presented here, this result reflects the underlying tendency for parent-offspring conflict to increase during the period of PI. As the interests of mother and offspring diverge, the offspring must assume a greater role in inducing whatever parental investment is forthcoming.

Data on the production and consumption of milk in sheep (Wallace, 1948) indicate that during the first weeks of the lamb's life the mother typically produces more milk than the lamb can drink. The lamb's appetite determines how much milk is consumed. But after the fourth week, the mother begins to produce less than the lamb can drink, and from that time on it is the mother who is the limiting factor in determining how much milk is consumed. Parallel behavioral data indicate that the mother initially permits free access by her lamb(s) but after a couple of weeks begins to prevent some suckling attempts (Munro, 1956; Ewbank, 1967). Mothers who are in poor condition become the limiting factor in nursing earlier than do mothers in good condition, and this is presumably because the cost of a given amount of milk is considerably higher when the mother is in poor condition, while the benefit to the offspring remains more or less unchanged. Females who produce twins permit either twin to suckle on demand during the first three weeks after birth, but in the ensuing weeks they do not permit one twin to suckle unless the other is ready also (Ewbank, 1964; Alexander, 1960).

DISAGREEMENT OVER THE SEX OF THE OFFSPRING

Under certain conditions a potential offspring is expected to disagree with its parents over whether it should become a male or a female. Since one can not assume that potential offspring are powerless to affect their sex, sex ratios observed in nature should to some extent reflect the offspring's preferred value as well as the parents'.

Fisher (1930) showed that (in the absence of inbreeding) parents are selected to invest as much in the total of their daughters as in the total of their sons. When each son produced costs on average the same as each daughter, parents are selected to produce a sex ratio of 50/50. In such species, the expected reproductive success (RS) of a son is the same as that of a daughter, so that an offspring should be indifferent as to its sex. But if (for example) parents are selected to invest twice as much in a typical male as in a typical female, then they will be selected to produce twice as many females as males, and the expected RS of each son will be twice that of each daughter. In such a species a potential offspring would prefer to be a male, for it would then achieve twice the RS it would as a female, without suffering a comparable decrease in inclusive fitness through the cost forced on its parents, because the offspring is selected to devalue that cost by the offspring's expected r_0 to the displaced sibling. For the example chosen, the exact gain in the offspring's inclusive fitness can be specified as follows. If the expected RS of a female offspring is defined as one unit of RS, then, in being made male, the offspring gains one unit of RS, but it deprives its mother of an additional daughter (or half a son). This displaced sibling (whether a female or half of a male) would have achieved one unit of RS, but this unit is devalued from the offspring's standpoint by the relevant r_0. If the displaced sibling would have been a full sibling, then this unit of RS is devalued by $\frac{1}{2}$, and the offspring, in being made a male, achieves a $\frac{1}{2}$ unit net increase in inclusive fitness. If the displaced sibling would have been a half sibling, the offspring, in being made a male, achieves a $\frac{3}{4}$ unit net increase in inclusive fitness. The parent, on the other hand, experiences initially only a trivial decrease in RS, so that *initially* any gene in the offspring tending to make it a male against its parents' efforts would spread rapidly.

As a hypothetical gene for offspring control of sex begins to spread, the number of males produced increases, thereby lowering the expected RS of each male. This decreases the gain (in inclusive fitness) to the offspring of being made a male. If the offspring's equilibrial sex ratio is defined as that sex ratio at which an offspring is indifferent as to whether it becomes a male or a female, then this sex ratio can be calculated by determining the sex ratio at which the offspring's gain in RS in being made a male is exactly offset by its loss in inclusive fitness in depriving itself of a sister (or half a brother). The offspring's equilibrial sex ratio will depend on both the offspring's expected r_0 to the displaced siblings and on the extent to which parents invest more in males than in females (or vice versa). The general solution is given

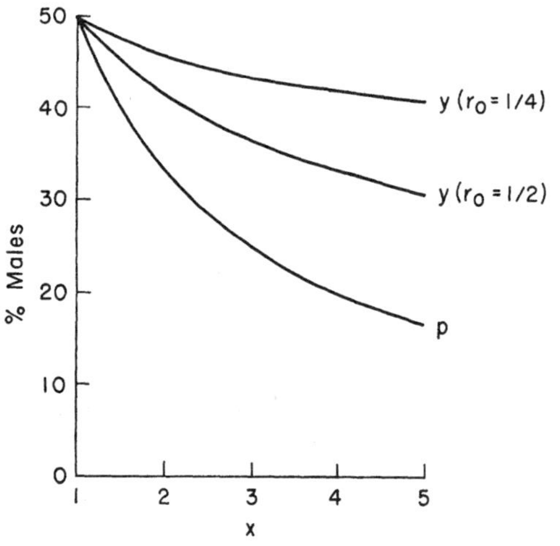

FIG. 4. The optimal sex ratio (per cent males) for the mother (m) and the young (y) where the mother invests more in a son than in a daughter by a factor of x (and assuming no paternal investment in either sex). Two functions are given for the offspring, depending on whether the siblings it displaces are full-siblings ($r_0 = 1/2$) or half-siblings ($r_0 = 1/4$). Note the initial rapid divergence between the mother's and the offspring's preferred sex ratio as the mother moves from equal investment in a typical individual of either sex ($x = 1$) to twice as much investment in a typical male ($x = 2$).

in the Appendix. Parent and offspring equilibrial sex ratios for different values of r_0 and different values of x (PI in a typical son/PI in a typical daughter) are plotted in Figure 4. For example, where the r_0 between siblings is ½ and where parents invest twice as much in a son as in a daughter ($x = 2$), the parents' equilibrial sex ratio is 1:2 (males:females) while that of the offspring is 1:1.414.

As long as all offspring are fathered by the same male, he will prefer the same sex ratio among the offspring that the mother does. But consider a species such as caribou in which the female produces only one offspring a year and assume that a female's successive offspring are fathered by different, unrelated males. If the female invests more in a son than in a daughter, then she will be selected to produce more daughters than sons. The greater cost of the son is not borne by the father, however, who invests nothing beyond his sperm, and who will not father the female's later offspring, so the father's equilibrial sex ratio is an equal number of sons and daughters. The offspring will prefer some probability of being a male that is intermediate between its parents' preferred probabilities, because (unlike the father) the offspring is related to the mother's future offspring but (unlike the mother) it is less related to them than to itself.

In a species such as just described (in which the male is heterogametic) the following sort of competitive interaction is possible. The prospective father produces more Y-bearing sperm than the female would prefer and she subjects the Y-bearing sperm to differential mortality. If the ratio of the sperm reaching the egg has been reduced to near the mother's optimal value, then the egg preferentially admits the Y-bearing sperm. If the mother ovulated more eggs than she intends to rear, she could then choose which to invest in, according to the sex of the fertilized egg, unless a male egg is able to deceive the mother about its sex until the mother has committed herself to investing in him. Whether such interactions actually occur in nature is at present unknown.

One consequence of the argument advanced here is that there is an automatic selective agent tending to keep maternal investment in a son similar to that in a daughter, for the greater the disparity between the investment in typical individuals of the two sexes, the greater the loss suffered by the mother in competitive interactions with her offspring over their preferred sex and in producing a sex ratio further skewed away from her preferred ratio (see Fig. 4). This automatic selection pressure may partly account for the apparent absence of strongly size-dimorphic young (at the end of PI) in species showing striking adult sexual dimorphism in size.

The argument presented here applies to any tendency of the parent to invest differentially in the young, whether according to sex or some other variable, except that in many species sex is irreversibly determined early in ontogeny and the offspring is expected at the very beginning to be able to discern its own sex and hence the predicted pattern of investment it will receive, so that, unlike other forms of differential investment, conflict is expected very early, namely, at the time of sex determination.

THE OFFSPRING AS PSYCHOLOGICAL MANIPULATOR

How is the offspring to compete effectively with its parent? An offspring can not fling its mother to the ground at will and nurse. Throughout the period of parental investment the offspring competes at a disadvantage. The offspring is smaller and less experienced than its parent, and its parent controls the resources at issue. Given this competitive disadvantage the offspring is expected to employ psychological rather than physical tactics. (Inside the mother the offspring is expected to employ chemical tactics, but some of the analysis presented below should also apply to such competition.) It should attempt to *induce* more investment than the parent wishes to give.

Since an offspring will often have better knowledge of its real needs than will its parent, selection should favor parental attentiveness to signals from its offspring that apprize the parent of the offspring's condition. In short, the offspring cries when hungry or in danger and the parent responds appropriately. Conversely, the offspring signals its parent (by smiling or wagging its tail) when its needs have been well met. Both parent and offspring benefit from this system of communication. But once such a system has evolved, the offspring can begin to employ it out of context. The offspring can cry not only when it is famished but also when it merely wants more food than the parent is selected to give. Likewise, it can begin to withhold its smile until it has gotten its way. Selection will then of course favor parental ability to discriminate the two uses of the signals, but still subtler mimicry and deception by the offspring are always possible. Parental experience with preceding offspring is expected to improve the parent's ability to make the appropriate discrimination. Unless succeeding offspring can employ more confusing tactics than earlier ones, parent-offspring interactions are expected to be increasingly biased in favor of the parent as a function of parental age.

In those species in which the offspring is more helpless and vulnerable the younger it is, its parents will have been more strongly selected to respond positively to signals of need emitted by the offspring, the younger that offspring is. This suggests that at any stage of ontogeny in which the offspring is in conflict with its parents, one appropriate tactic may be to revert to the gestures and actions of an earlier stage of development in order to induce the investment that would then have been forthcoming. Psychologists have long recognized such a tendency in humans and have given it the name of regression. A detailed functional analysis of regression could be based on the theory presented here.

The normal course of parent-offspring relations must be subject to considerable unpredictable variation in both the condition of the parent and (sometimes independently) the condition of the offspring. Both partners must be sensitive to such variation and must adjust their behavior appropriately. Low investment coming from a parent in poor condition has a different meaning than low investment coming from a parent in good condition. This suggests that from an early age the offspring is expected to be a psychologically sophisticated organism. The offspring should be able to evaluate the cost of a given parental act (which depends in part on the condition of the parent at that moment) and its benefit (which depends in part on the condition of the offspring). When the offspring's interests diverge from those of its parent, the offspring must be able to employ a series of psychological maneuvers, including the mimicry and regression mentioned above. Although it would be expected to learn appropriate information (such as whether its psychological maneuvers were having the desired effects), an important feature of the argument presented here is that the offspring cannot rely on its parents for disinterested guidance. One expects the offspring to be pre-programmed to resist some parental teaching while being open to other forms. This is particularly true, as argued below, for parental teaching that affects the altruistic and egoistic tendencies of the offspring.

If one event in a social relationship predicts to some degree future events in that relationship, the organism should be selected to alter its behavior in response to an initial event, in order to change the probability that the predicted events will occur. For example, if a mother's lack of love for her offspring early in its life predicts deficient future investment, then the offspring will be selected to be sensitive to such early lack of love, whether investment at that time is deficient or not, in order to increase her future investment. The best data relevant to these possibilities come from the work of Hinde and his associates on groups of caged rhesus macaques. In a series of experiments, a mother was removed from her 6-month-old infant, leaving the infant in the home cage with other group members. After 6 days, the mother was returned to the home cage. Behavioral data were gathered before, during, and after the separation (see points 1 and 2 below). In a parallel series of experiments, the infant was removed for 6 days from its mother, leaving her in the home cage, and the same behavioral data were gathered (see point 3 below). The main findings can be summarized as follows:

1) *Separation of mother from her offspring affects their relationship upon reunion.* After reunion with its mother, the infant spends more time on the mother than it did before separation—although, had the separation not occurred, the infant would have reduced its time on the mother. This increase is caused by the infant, and occurs despite an increase in the frequency of maternal rejection (Hinde and Spencer-Booth, 1971). These effects can be detected at least as long as 5 weeks after reunion. These data are consistent with the assumption that the infant has been selected to interpret its mother's disappearance as an event whose recurrence the infant can help prevent by devoting more of its energies to staying close to its mother.

2) *The mother-offspring relationship prior to separation affects the offspring's behavior on reunion.* Upon reunion with its mother, an infant typically shows distress, as measured by callings and immobility. The more frequently an infant was rejected *prior* to separation, the more distress it shows upon reunion. This correlation holds for at least 4 weeks after reunion. In addition, the more distressed the infant is, the greater is its role in maintaining proximity to its mother (Hinde and Spencer-Booth, 1971). These data support the assumption that the infant interprets its mother's disappearance in relation to her predeparture behavior in a logical way: the offspring should assume that a rejecting mother who temporarily disappears needs more offspring surveilance and intervention than does a nonrejecting mother who temporarily disappears.

3) *An offspring removed from its mother shows, upon reunion, different effects than an offspring whose mother has been removed.* Compared to an infant whose mother had been removed, an infant removed from its mother shows, upon reunion, and for up to 6 weeks after reunion, less distress and more time off the mother.

In addition, the offspring tends to play a smaller role in maintaining proximity to its mother, and it experiences less frequent maternal rejections (Hinde and Davies, 1972*a,b*). These data are consistent with the expectation that the offspring should be sensitive to the *meaning* of events affecting its relationship to its mother. The offspring can differentiate between a separation from its mother caused by its own behavior or some accident (infant removed from group) and a separation which may have been caused by maternal negligence (mother removed from group). In the former kind of separation, the infant shows less effects when reunited, because, from its point of view, such a separation does not reflect on its mother and no remedial action is indicated. A similar explanation can be given for differences in the mother's behavior.

PARENT OFFSPRING CONFLICT OVER THE BEHAVIORAL TENDENCIES OF THE OFFSPRING

Parents and offspring are expected to disagree over the behavioral tendencies of the offspring insofar as these tendencies affect related individuals. Consider first interactions among siblings. An individual is only expected to perform an altruistic act toward its full-sibling whenever the benefit to the sibling is greater than twice the cost to the altruist. Likewise, it is only expected to forego selfish acts when $C>2B$ (where a selfish act is defined as one that gives the actor a benefit, B, while inflicting a cost, C, on some other individual, in this case, on a full-sibling). But parents, who are equally related to all of their offspring, are expected to encourage all altruistic acts among their offspring in which $B>C$, and to discourage all selfish acts in which $C>B$. Since there ought to exist altruistic situations in which $C<B<2C$, parents and offspring are expected to disagree over the tendency of the offspring to act altruistically toward its siblings. Likewise, whenever for any selfish act harming a full-sibling $B<C<2B$, parents are expected to discourage such behavior and offspring are expected to be relatively refractory to such discouragement.

This parent-offspring disagreement is expected over behavior directed toward other relatives as well. For example, the offspring is only selected to perform altruistic acts toward a cousin (related through the mother) when $B>8C$. But the offspring's mother is related to her own nephews and nieces by $r_o = 1/4$ and to her offspring by $r_o = 1/2$, so that she would like to see any altruistic acts performed by her offspring toward their maternal cousins whenever $B>2C$. The same argument applies to selfish acts, and both arguments can be made for more distant relatives as well. (The father is unrelated to his mate's kin and, other things being equal, should not be distressed to see his offspring treat such individuals as if they were unrelated.)

The general argument extends to interactions with unrelated individuals, as long as these interactions have some effect, however remote and indirect, on kin. Assume, for example, that an individual gains some immediate benefit, B, by acting nastily toward some unrelated individual. Assume that the unrelated individual reciprocates in kind (Trivers, 1971), but assume that the reciprocity is directed toward both the original actor and some relative, e.g., his sibling. Assuming no other effects of the initial act, the original actor will be selected to perform the nasty act as long as $B>C_1 + 1/2(C_2)$, where C_1 is the cost to the original actor of the reciprocal nastiness he receives and C_2 is the cost to his sibling of the nastiness the sibling receives. The actor's parents viewing the interaction would be expected to condone the initial act only if $B>C_1 + C_2$. Since there ought to exist situations in which $C_1 + 1/2(C_2)<B<C_1 + C_2$, one expects conflict between offspring and parents over the offspring's tendency to perform the initial nasty act in the situation described. A similar argument can be made for altruistic behavior directed toward an unrelated individual if this behavior induces altruism in return, part of which benefits the original altruist's sibling. Parents are expected to encourage such altruism more often than the offspring is expected to undertake on his own. The

argument can obviously be extended to behavior which has indirect effects on kin other than one's sibling.

As it applies to human beings, the above argument can be summarized by saying that a fundamental conflict is expected during socialization over the altruistic and egoistic impulses of the offspring. Parents are expected to socialize their offspring to act more altruistically and less egoistically than the offspring would naturally act, and the offspring are expected to resist such socialization. If this argument is valid, then it is clearly a mistake to view socialization in humans (or in any sexually reproducing species) as only or even primarily a process of "enculturation," a process by which parents teach offspring their culture (e.g., Mussen et al., 1969, p. 259). For example, one is not permitted to assume that parents who attempt to impart such virtues as responsibility, decency, honesty, trustworthiness, generosity, and self-denial are merely providing the offspring with useful information on appropriate behavior in the local culture, for all such virtues are likely to affect the amount of altruistic and egoistic behavior impinging on the parent's kin, and parent and offspring are expected to view such behavior differently. That some teaching beneficial to the offspring transpires during human socialization can be taken for granted, and one would expect no conflict if socialization involved *only* teaching beneficial to the offspring. According to the theory presented here, socialization is a process by which parents attempt to mold each offspring in order to increase their own inclusive fitness, while each offspring is selected to resist some of the molding and to attempt to mold the behavior of its parents (and siblings) in order to increase its inclusive fitness. Conflict during socialization need not be viewed solely as conflict between the culture of the parent and the biology of the child; it can also be viewed as conflict between the biology of the parent and the biology of the child. Since teaching (as opposed to molding) is expected to be recognized by offspring as being in their own self-interest, parents would be expected to overemphasize their role as teachers in order to minimize resistance in their young. According to this view then, the prevailing concept of socialization is to some extent a view one would expect adults to entertain and disseminate.

Parent-offspring conflict may extend to behavior that is not on the surface either altruistic or selfish but which has consequences that can be so classified. The amount of energy a child consumes during the day, and the way in which the child consumes this energy, are not matters of indifference to the parent when the parent is supplying that energy, and when the way in which the child consumes the energy affects its ability to act altruistically in the future. For example, when parent and child disagree over when the child should go to sleep, one expects in general the parent to favor early bedtime, since the parent anticipates that this will decrease the offspring's demands on parental resources the following day. Likewise, one expects the parent to favor serious and useful expenditures of energy by the child (such as tending the family chickens, or studying) over frivolous and unnecessary expenditures (such as playing cards)—the former are either altruistic in themselves, or they prepare the offspring for future altruism. In short, we expect the offspring to perceive some behavior, that the parent favors, as being dull, unpleasant, moral, or any combination of these. One must at least entertain the assumption that the child would find such behavior more enjoyable if in fact the behavior maximized the offspring's inclusive fitness.

CONFLICT OVER THE ADULT REPRODUCTIVE ROLE OF THE OFFSPRING

As a special case of the preceding argument, it is clear that under certain conditions conflict is expected between parent and offspring over the adult reproductive role of the offspring. To take the extreme case, it follows at once from Hamilton's (1964) work that individuals who choose not to reproduce (such as celibate priests) are not necessarily acting counter to their genetic self-interest. One need merely assume that the nonreproducer thereby increases the reproductive success of relatives by an amount which, when devalued by the relevant degrees of relatedness, is greater than the nonreproducer would have achieved on his own. This kind of explanation has been developed in some detail to explain nonreproductives in the haplodiploid Hymenoptera (Hamilton, 1972). What is clear from the present argument, however, is that it is even more likely that the nonreproducer will thereby increase his *parents'* inclusive fitness than that he will increase his own. This follows because his parents are expected to value the increased reproductive success of kin relatively more than he is.

If the benefits of nonreproducing are assumed, for simplicity, to accrue only to full siblings and if the costs of nonreproducing are defined as the surviving offspring the nonreproducer would have produced had he or she chosen to reproduce, then parent-offspring conflict over whether the offspring should reproduce is expected whenever $C<B<2C$. Assuming it is sometimes possible for parents to predict while an offspring is still young what the cost and benefit of its not reproducing will be, the parents would be selected to mold the offspring toward not reproducing whenever $B>C$. Two kinds of nonreproductives are expected: those who are thereby increasing their own inclusive fitness ($B>2C$) and those who are thereby lowering their own inclusive fitness but increasing that of their parents ($C<B<2C$). The first kind is expected to be as happy and content as living creatures ever are, but the second is expected to show internal conflict over its adult role and to express ambivalence over the past, particularly over the behavior and influence of its parents. I emphasize that it is not necessary for parents to be conscious of molding an offspring toward nonreproduction in order for such molding to occur and to increase the parent's inclusive fitness. It remains to be explored to what extent the etiology of sexual preferences (such as homosexuality) which tend to interfere with reproduction can be explained in terms of the present argument.

Assuming that parent and offspring agree that the offspring should reproduce, disagreement is still possible over the form of that reproduction. Whether an individual attempts to produce few offspring or many is a decision that affects that individual's opportunities for kin-directed altruism, so that parent and offspring may disagree over the optimal reproductive effort of the offspring. Since in humans an individual's choice of mate may affect his or her ability to render altruistic behavior toward relatives, mate choice is not expected to be a matter of indifference to the parents. Parents are expected to encourage their offspring to choose a mate that will enlarge the offspring's altruism toward kin. For example, when a man marries his cousin, he increases (other things being equal) his contacts with relatives, since the immediate kin of his wife will also be related to him, and marriage will normally lead to greater contact with her immediate kin. One therefore might expect human parents to show a tendency to encourage their offspring to marry more closely related individuals (e.g., cousins) than the offspring would prefer. Parents may also use an offspring's marriage to cement an alliance with an unrelated family or group, and insofar as such an alliance is beneficial to kin of the parent in addition to the offspring itself, parents are expected to encourage such marriages more often than the offspring would prefer. Finally, parents will more strongly discourage marriage by their offspring to individuals the local society defines as pariahs, because such unions are likely to besmirch the reputation of close kin as well.

Because parents may be selected to employ parental investment itself as an incentive to induce greater offspring altruism, parent-offspring conflict may include situations in which the offspring attempts to terminate the period of PI *before* the parent wishes to. For example, where the parent is selected to retain one or more offspring as permanent "helpers at the nest" (Skutch, 1961), that is, permanent nonreproductives who help their parents raise additional offspring (or help those offspring to reproduce), the parent may be selected to give additional investment in order to tie the offspring to the parent. In this situation, selection on the offspring may favor any urge toward independence which overcomes the offspring's impulse toward additional investment (with its hidden cost of additional dependency). In short, in species in which kin-directed altruism is important, parent-offspring conflict may include situations in which the offspring wants *less* than the parent is selected to give as well as the more common situation in which the offspring attempts to garner *more* PI than the parent is selected to give.

Parent-offspring relations early in ontogeny can affect the later adult reproductive role of the offspring. A parent can influence the altruistic and egoistic tendencies of its offspring whenever it has influence over any variable that affects the costs and benefits associated with altruistic and egoistic behavior. For example, if becoming a permanent nonreproductive, helping one's siblings, is more likely to increase one's inclusive fitness when one is small in size relative to one's siblings (as appears to be true in some polistine wasps: Eberhard, 1969), then parents can influence the proportion of their offspring who become helpers by altering the size distribution of their offspring. Parent-offspring conflict over early PI may itself involve parent-offspring conflict over the eventual reproductive role of the offspring. This theoretical possibility may be relevant to human psychology if parental decision to mold an offspring into being a nonreproductive involves differential investment as well as psychological manipulation.

THE ROLE OF PARENTAL EXPERIENCE IN PARENT-OFFSPRING CONFLICT

It cannot be supposed that all parent-offspring conflict results from the conflict in the way in which the parent's and the offspring's inclusive fitnesses are maximized. Some conflict also results, ironically because of an overlap in the interests of parent and young. When circumstances change, altering the benefits and costs associated with some offspring behavior, both the parent and the offspring are selected to alter the offspring's behavior appropriately. That is, the parent is selected to mold the appropriate change in the offspring's behavior, and if parental molding is successful, it will strongly reduce the selection pressure on the offspring to change its behavior spontaneously. Since the parent is likely to discover the changing circumstances as a result of its own experience, one expects tendencies toward parental molding to appear, and spread, before the parellel tendencies appear in the offspring. Once parents commonly mold the appropriate offspring behavior, selection still favors genes leading toward voluntary offspring behavior, since such a developmental avenue is presumably more efficient and more certain than that involving parental manipulation. But the selection pressure for the appropriate offspring genes should be weak, and if circumstances change at a faster rate than this selection operates, there is the possibility of continued parent-offspring conflict resulting from the greater experience of the parent.

If the conflict described above actually occurs, then (as mentioned in an earlier section) selection will favor a tendency for parents to overemphasize their experience in all situations, and for the offspring to differentiate between those situations in which greater parental experience is real and those situations in which such experience is merely claimed in order to manipulate the offspring.

APPENDIX: THE OFFSPRING'S EQUILIBRIAL SEX RATIO

Let the cost of producing a female be one unit of investment, and let the cost of producing a male be x units, where x is larger than one. Let the expected reproductive success of a female be one unit of RS. Let the sex ratio produced be 1:y (males: females), where y is larger than one. At this sex ratio the expected RS of a male is y units of RS, so that, in being made a male instead of a female, an offspring gains y−1 units of RS. But the offspring also thereby deprives its mother of x−1 units of investment. The offspring's equilibrial sex ratio is that sex ratio at which the offspring's gain in RS in being made a male (y−1) is exactly offset by its loss in inclusive fitness which results because it thereby deprives its mother of x−1 units of investment. The mother would have allocated these units in such a way as to achieve a 1:y sex ratio, that is, she would have allocated x/(x+y) of the units to males and y/(x+y) of the units to females. In short, she would have produced (x−1)/(x+y) sons, which would have achieved RS of y(x−1)/(x+y), and she would have produced y(x−1)/(x+y) daughters, which would have achieved RS of y(x−1)/(x+y). The offspring is expected to devalue this loss by the offspring's r_o to its displaced siblings. Hence, the offspring's equilibrial sex ratio results when

$$y - 1 = \frac{r_o\, y\,(x - 1)}{x + y} + \frac{r_o\, y\,(x - 1)}{x + y}$$

$$= (2r_o y)\,\frac{x - 1}{x + y}$$

Rearranging gives

$$y^2 + y\,(x - 2r_o x + 2r_o - 1) - x = 0$$

$$y^2 + (x - 1)\,(1 - 2r_o)\,y - x = 0$$

The general solution for this quadratic equation is

$$y = \frac{-(x - 1)\,(1 - 2r_o)}{2} + \frac{\sqrt{(x - 1)^2\,(1 - 2r_o)^2 + 4x}}{2}$$

Where $r_o = \frac{1}{2}$, the equation reduces to $y = \sqrt{x}$. In other words, when the offspring displaces full siblings (as is probably often the case), the offspring's equilibrial sex ratio if $1 : \sqrt{x}$, while the parent's equilibrial sex ratio is 1:x. These values, as well as the offspring's equilibrial sex ratio where $r_o = \frac{1}{4}$, are plotted in Figure 4. The same general solution holds if parents invest more in females by a factor of x, except that the resulting sex ratios are then reversed (e.g., $\sqrt{x} : 1$ instead of $1 : \sqrt{x}$).

REFERENCES

Alexander, G. 1960. Maternal behaviour in the Merino ewe. Anim. Prod. 3:105-114.

DeVore, I. 1963. Mother-infant relations in free-ranging baboons, p. 305-335. *In* H. Rheingold [ed.], Maternal behavior in mammals. Wiley, N.Y.

Drury, W. H., and W. J. Smith. 1968. Defense of feeding areas by adult Herring Gulls and intrusion by young. Evolution 22:193-201.

Eberhard, M. J. W. 1969. The social biology of polistine wasps, Misc. Publ. Mus. Zool. Univ. Mich. 140:1-101.

Elliott, B. 1969. Life history of the Red Warbler. Wilson Bull. 81:184-195.

Ewbank, R. 1964. Observations on the suckling habits of twin lambs. Anim. Behav. 12:34-37.

Ewbank, R. 1967. Nursing and suckling behaviour amongst Clun Forest ewes and lambs. Anim. Behav. 15:251-258.

Fisher, R. A. 1930. The genetical theory of natural selection. Clarendon, Oxford.

Gill, J. C., and W. Thomson. 1956. Observations on the behavior of suckling pigs. Anim. Behav. 4: 46-51.

Hamilton, W. D. 1964. The genetical evolution of social behavior. J. Theoret. Biol. 7:1-52.

Hamilton, W. D. 1971. The genetical evolution of social behavior, p. 23-39. Reprinted, with addendum. *In* G. C. Williams [ed.], Group selection. Aldine-Atherton, Chicago.

Hamilton, W. D. 1972. Altruism and related phenomena, mainly in social insects. Annu. Rev. Ecol. Syst. 3:193-232.

Hinde, R. A., and Y. Spencer-Booth. 1967. The behaviour of socially living rhesus monkeys in their first two and a half years. Anim. Behav. 15:169-196.

Hinde, R. A., and Y. Spencer-Booth. 1971. Effects of brief separation from mother on rhesus monkeys. Science 173:111-118.

Hinde, R. A., and L. M. Davies. 1972*a*. Changes in mother-infant relationship after separation in rhesus monkeys. Nature 239:41-42.

Hinde, R. A., and L. M. Davies. 1972*b*. Removing infant rhesus from mother for 13 days compared with removing mother from infant. J. Child Psychol. Psychiat. 13:227-237.

Jay, P. 1963. Mother-infant relations in langurs, p. 282-304. *In* H. Rheingold [ed.], Maternal behaviour in mammals. Wiley, N.Y.

Le Boeuf, B. J., R. J. Whiting, and R. F. Gantt. 1972. Perinatal behavior of northern elephant seal females and their young. Behaviour 43:121-156.

Munro, J. 1956. Observations on the suckling behaviour of young lambs. Anim. Behav. 4:34-36.

Mussen, P. H., J. J. Conger, and J. Kagan. 1969. Child development and personality. 3rd ed. Harper and Row, N.Y.

Rheingold, H. 1963. Maternal behavior in the dog, p. 169-202. *In* H. Rheingold [ed.], Maternal behavior in mammals. Wiley, N.Y.

Rosenblatt, J. S., and D. S. Lehrman. 1963. Maternal behavior of the laboratory rat, p. 8-57. *In* H. Rheingold [ed.], Maternal behavior in mammals. Wiley, N.Y.

Rosenblum, L. A. 1971. The ontogeny of mother-infant relations in macaques, p. 315-367. *In* H. Moltz [ed.], The ontogeny of vertebrate behavior. Academic Press, N.Y.

Rowe, E. G. 1947. The breeding biology of *Aquila verreauxi* Lesson. Ibis 89:576-606.

Schaller, G. B. 1964. Breeding behavior of the White Pelican at Yellowstone Lake, Wyoming. Condor 66:3-23.

Schneirla, T. C., J. S. Rosenblatt, and E. Tobach. 1963. Maternal behavior in the cat, p. 122-168. *In* H. Rheingold [ed.], Maternal behavior in mammals. Wiley, N.Y.

Skutch, A. F. 1961. Helpers among birds. Condor 63:198-226.

Struhsaker, T. T. 1971. Social behaviour of mother and infant vervet monkeys (*Cercopithecus aethiops*). Anim. Behav. 19:233-250.

Trivers, R. L. 1971. The evolution of reciprocal altruism. Quart. Rev. Biol. 46:35-57.

Trivers, R. L. 1972. Parental investment and sexual selection, p. 136-179. *In* B. Campbell [ed.], Sexual selection and the descent of man, 1871-1971. Aldine-Atherton, Chicago.

Wallace, L. R. 1948. The growth of lambs before and after birth in relation to the level of nutrition. J. Agri. Sci. 38:93-153.

King, M.-C., and A. C. Wilson. 1975. Evolution at two levels in humans and chimpanzees. *Science* 188:107–116.

The main appeal of this article stems from its central thesis that morphological evolution and molecular evolution sometimes march to the beat of very different drummers. By the early 1970s, the molecular revolution in evolutionary biology was well under way, but a troubling empirical dilemma arose: measures of molecular distance and morphological distance between species often seemed to differ, even qualitatively. The relevant case in point involved humans and chimpanzees, which differ dramatically in appearance and biology and yet proved to be surprisingly similar with respect to their overall macromolecular makeup. Mary-Claire King and Allan Wilson address this paradox by first quantifying the magnitude of the disparity (see also Cherry et al. 1978) and then delving into its evolutionary causes. They speculate that changes in gene regulation, more so than standard mutational alterations in transcribed structural genes, may be at the heart of much of morphological evolution. This paper dovetailed nicely with emerging sentiments at the time (see for example Britten and Davidson 1969 in this volume) that gene regulation might be the real "stuff" of both genomic and phenotypic evolution.

RELATED READING

Cela-Conde, C., and F. J. Ayala. 2007. Human evolution: trails from the past. Oxford University Press, Oxford, England.

Cherry, L. M., S. M. Case, and A. C. Wilson. 1978. Frog perspective on the morphological divergence between humans and chimpanzees. Science 200:209–211.

11 April 1975, Volume 188, Number 4184 SCIENCE

Evolution at Two Levels in Humans and Chimpanzees

Their macromolecules are so alike that regulatory mutations may account for their biological differences.

Mary-Claire King and A. C. Wilson

Soon after the expansion of molecular biology in the 1950's, it became evident that by comparing the proteins and nucleic acids of one species with those of another, one could hope to obtain a quantitative and objective estimate of the "genetic distance" between species. Until then, there was no common yardstick for measuring the degree of genetic difference among species. The characters used to distinguish among bacterial species, for example, were entirely different from those used for distinguishing among mammals. The hope was to use molecular biology to measure the differences in the DNA base sequences of various species. This would be the common yardstick for studies of organismal diversity.

During the past decade, many workers have participated in the development and application of biochemical methods for estimating genetic distance. These methods include the comparison of proteins by electrophoretic, immunological, and sequencing techniques, as well as the comparison of nucleic acids by annealing techniques. The only two species which have been compared by all of these methods are chimpanzees (*Pan troglodytes*) and humans (*Homo sapiens*). This pair of species is also unique because of the thoroughness with which they have been compared at the organismal level—that is, at the level of anatomy, physiology, behavior, and ecology. A good opportunity is therefore presented for finding out whether the molecular and organismal estimates of distance agree.

The intriguing result, documented in this article, is that all the biochemical methods agree in showing that the genetic distance between humans and the chimpanzee is probably too small to account for their substantial organismal differences.

Indications of such a paradox already existed long ago. By 1963, it appeared that some of the blood proteins of humans were virtually identical in amino acid sequence with those of apes such as the chimpanzee or gorilla (*1*). In the intervening years, comparisons between humans and chimpanzees were made with many additional proteins and with DNA. These results, reported herein, are consistent with the early results. Moreover, they tell us that the genes of the human and the chimpanzee are as similar as those of sibling species of other organisms (*2*). So, the paradox remains. In order to explain how species which have such similar genes can differ so substantially in anatomy and way of life, we review evidence concerning the molecular basis of evolution at the organismal level. We suggest that evolutionary changes in anatomy and way of life are more often based on changes in the mechanisms controlling the expression of genes than on sequence changes in proteins. We therefore propose that regulatory mutations account for the major biological differences between humans and chimpanzees.

Similarity of Human and Chimpanzee Genes

To compare human and chimpanzee genes, one compares either homologous proteins or nucleic acids. At the protein level, one way of measuring the degree of genetic similarity of two taxa is to determine the average number of amino acid differences between homologous polypeptides from each population. The most direct method for determining this difference is to compare the amino acid sequences of the homologous proteins. A second method is microcomplement fixation, which provides immunological distances linearly correlated with amino acid sequence difference. A third method is electrophoresis, which is useful in analyzing taxa sufficiently closely related that they share many alleles. For the human-chimpanzee comparison all three methods are appropriate, and thus many human and chimpanzee proteins have now been compared by each method. We can therefore estimate the degree of genetic similarity between humans and chimpanzees by each of these techniques.

Sequence and immunological comparisons of proteins. During the last decade, amino acid sequence studies have been published on several human and chimpanzee proteins. As Table 1 indicates, the two species seem to have identical fibrinopeptides (*3*), cytochromes c (*4*), and hemoglobin chains [alpha (*4*), beta (*4*), and gamma (*5*, *6*)]. The structural genes for these proteins may therefore be identical in humans and chimpanzees. In other cases, for example, myoglobin (*7*) and the

Dr. King, formerly a graduate student in the Departments of Genetics and Biochemistry, University of California, Berkeley, is now a research geneticist at the Hooper Foundation and Department of International Health, University of California, San Francisco 94143. Dr. Wilson is a professor of biochemistry at the University of California, Berkeley 94720.

delta chain of hemoglobin (*5, 8*), the human polypeptide chain differs from that of the chimpanzee by a single amino acid replacement. The amino acid replacement in each case is consistent with a single base replacement in the corresponding structural gene.

Owing to the limitations of conventional sequencing methods, exactly comparable information is not available for larger proteins. Indeed, the sequence information available for the proteins already mentioned is not yet complete. By applying the microcomplement fixation method to large proteins, however, one can obtain an approximate measure of the degree of amino acid sequence difference between related proteins (*9*). This method indicates that the sequences of human and chimpanzee albumins (*10*), transferrins (*11*), and carbonic anhydrases (*4, 12*) differ slightly, but that lysozyme (*13*) is identical in the two species (Table 1) (*14*). Based on the proteins listed in Table 1, the average degree of difference between human and chimpanzee proteins is

$$\frac{19 \times 1000}{2633} = 7.2 \qquad (1)$$

amino acid sites per 1000 substitutions. That is, the sequences of human and chimpanzee polypeptides examined to date are, on the average, more than 99 percent identical.

Electrophoretic comparison of proteins. Electrophoresis can provide an independent estimate of the average amino acid sequence difference between closely related species. We have compared the human and chimpanzee polypeptide products of 44 different structural genes. Table 2 indicates the allelic frequencies and the estimated probability of identity at each locus. The symbol S_i represents the probability that human and chimpanzee alleles will be electrophoretically identical at a particular locus i, or

$$S_i = \sum_{j=1}^{A_i} x_{ij} y_{ij} \qquad (2)$$

where x_{ij} is the frequency of the jth allele at the ith locus in human populations, and y_{ij} the frequency of the jth allele at the ith locus in chimpanzee populations for all A_i alleles at that locus. For example, Table 2 indicates the frequencies of the three alleles (AP^a, AP^b, and AP^c) found at the acid phosphatase locus for human and chimpanzee populations. The probability of identity of human and chimpanzee alleles at this locus, that is, S_i is $(0.29 \times 0) + (0.68 \times 1.00) + (0.03 \times 0)$, or 0.68.

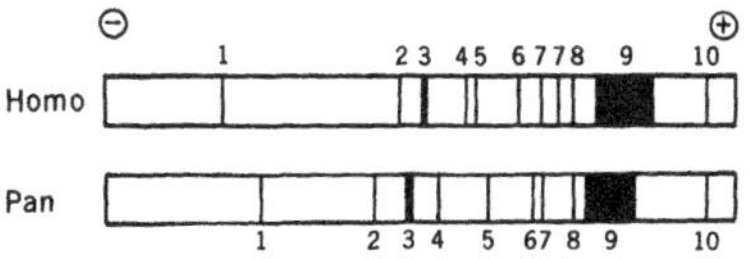

Fig. 1. Separation of human and chimpanzee plasma proteins by acrylamide electrophoresis at *p*H 8.9. The proteins are: 1, α_2-macroglobulin; 2, third component of complement; 3, transferrin; 4, haptoglobin; 5, ceruloplasmin; 6, α_{2HS}-glycoprotein; 7, Gc-globulins; 8, α_1-antitrypsin; 9, albumin; and 10, α_1-acid glycoprotein. The chimpanzee plasma has transferrin genotype *Pan* CC; the human plasma has transferrin genotype *Homo* CC and haptoglobin genotype 1-1. The direction of migration is from left to right.

Of the loci in Table 2, 31 code for intracellular proteins; 13 code for secreted or extracellular proteins. In general, the intracellular proteins were analyzed by starch gel electrophoresis of red blood cell lysates, with the buffer systems indicated in the table and stains specific for the enzymatic activity of each protein. For a few intracellular proteins (cytochrome c, the hemoglobin chains, and myoglobin), amino acid sequences have been published for both species, so that direct sequence comparison is also possible.

Most of the secreted proteins were compared by acrylamide gel electrophoresis of human and chimpanzee plasma (*15*). The electrode chamber contained tris(hydroxymethyl)aminomethane (tris) borate buffer, *p*H 8.9; acrylamide gel slabs were made with tris-sulfate buffer, *p*H 8.9. Gels were stained with amido black, a general protein dye. The identification of bands on a gel stained with this dye poses a problem, since it is not obvious, particularly for less concentrated proteins, which protein each band represents. We determined the electrophoretic mobilities of the plasma proteins by applying the same sample to several slots of the same gel, staining the outside columns, and cutting horizontal slices across the unstained portion of the gel at the position of each band. The protein was eluted separately from each band in 0.1 to 0.2 milliliter of an appropriate isotonic tris buffer (*9*) and tested for reactivity with a series of rabbit antiserums, each specific for a particular human plasma protein, by means of immunoelectrophoresis and immunodiffusion in agar (*15, 16*). The results of this analysis are shown in Fig. 1.

Some of the secreted proteins were compared by means of other electrophoretic methods as well. Albumin and transferrin were surveyed by cellulose acetate electrophoresis; and α_1-antitrypsin, Gc-globulin (group-specific component), the haptoglobin chains, lysozyme, and plasma cholinesterase were analyzed on starch gels, with the buffers indicated in Table 2.

The results of all electrophoretic comparisons are summarized in Fig. 2. About half of the proteins in this survey are electrophoretically identical for the two species, and about half of them are different. Only a few loci are highly polymorphic in both species (see *17*).

The proportion of alleles at an "average" locus that are electrophoretically identical in human and chimpanzee populations can be calculated from Table 2 and Eq. 3, where L is the number of loci observed:

$$\bar{S} = \frac{1}{L}(S_1 + S_2 + \ldots + S_L) = 0.52 \qquad (3)$$

Table 1. Differences in amino acid sequences of human and chimpanzee polypeptides. Lysozyme, carbonic anhydrase, albumin, and transferrin have been compared immunologically by the microcomplement fixation technique. Amino acid sequences have been determined for the other proteins. Numbers in parentheses indicate references for each protein.

Protein	Amino acid differences	Amino acid sites
Fibrinopeptides A and B (*3*)	0	30
Cytochrome c (*4*)	0	104
Lysozyme (*13*)	~0	130
Hemoglobin α (*4*)	0	141
Hemoglobin β (*4*)	0	146
Hemoglobin $^A\gamma$ (*5, 6*)	0	146
Hemoglobin $^G\gamma$ (*5, 6*)	0	146
Hemoglobin δ (*5, 8*)	1	146
Myoglobin (*7*)	1	153
Carbonic anhydrase (*4, 12*)	~3	264
Serum albumin (*10*)	~6	580
Transferrin (*11*)	~8	647
Total	~19	2633

In other words, the probability that human and chimpanzee alleles will be electrophoretically identical at a particular locus is about one-half.

Agreement between electrophoresis and protein sequencing. The results of electrophoretic analysis can be used to estimate the average number of amino acid differences per polypeptide chain for humans and chimpanzees, for comparison with the estimate based on amino acid sequences and immunological data. To calculate the average amino acid sequence difference between human and chimpanzee proteins, we need first an estimate of the proportion ($\bar{c}$) of amino acid substitutions detectable by electrophoresis. Electrophoretic techniques detect only amino acid substitutions that change the net charge of the protein observed. Four amino acid side chains are charged at *p*H 8.6: arginine, lysine, glutamic acid, and aspartic acid. The side chain of histidine is positively charged below approximately *p*H 6. The proportion of accepted point mutations that would be detectable by the buffer

Table 2. Electrophoretic comparison of chimpanzee and human proteins. In the first column, Enzyme Commission numbers are given in parentheses; *N* is the number of chimpanzees analyzed, both in this study and by other investigators. Abbreviations: MW, molecular weight; aa, amino acids; tris, tris(hydroxymethyl)aninomethane; EDTA, ethylenediaminetetraacetate. Secreted proteins differ more frequently for the two species than intracellular proteins *(93)*.

Locus (i) and allele (j)	Allele frequency: Human* (x_{ij})	Allele frequency: Chimpanzee (y_{ij})	Probability of identity† (S_i)	Comments and references‡
				Intracellular proteins
Acid phosphatase (3.1.3.2); $N = 86$				
AP^a	0.29	0	0.68	Red cells; 15,000 MW; 110 aa; citrate-phosphate, *p*H 5.9, starch electrophoresis *(54, 55)*
AP^b	0.68	1.00		
AP^c	0.03	0		
Adenosine deaminase (3.5.4.4); $N = 22$				
ADA^1	0.96	0	0	Red cells; 35,000 MW; 300 aa; chimpanzee protein faster on starch electrophoresis *(54)*; polymorphism in human populations *(16)*
ADA^2	0.04	0		
$ADA^{ape=5}$	0	1.00		
Adenylate kinase (2.7.4.3); $N = 86$				
AK^1	0.98	1.00	0.98	Red cells; 21,500 MW; 190 aa; well buffer is citrate-NaOH, *p*H 7.0; gel buffer is histidine-NaOH, pH 7.0, starch electrophoresis *(54, 56, 57)*
AK^2	0.02	0		
Carbonic anhydrase I or B (4.2.1.1); $N = 111$	1.00	1.00	1.00	Red cells; 28,000 MW; 264 aa; well buffer is borate-NaOH, *p*H 8.0; gel buffer is borate-NaOH, *p*H 8.6, starch electrophoresis *(56, 58)*
Cytochrome c	1.00	1.00	1.00	Mitochondria; 12,400 MW; 104 aa; sequence identity based on amino acid analysis *(5)*; possible heterogeneity in man *(59)*
Esterase A_1 (3.1.1.6); $N = 111$	1.00	0	0	Red cells; well buffer is lithium borate, *p*H 8.2; gel buffer is lithium-borate and tris-citrate, *p*H 7.3, starch electrophoresis *(58, 60)*
	0	1.00		
Esterase A_2§ (3.1.1.6); $N = 111$	1.00	Absent		See esterase A_1
Esterase A_3 (3.1.1.6); $N = 111$				
$EstA_3{}^a$	1.00	0	0	See esterase A_1
$EstA_3{}^b$	0	1.00		
Esterase B (3.1.1.1); $N = 111$	1.00	1.00	1.00	See esterase A_1
Glucose-6-phosphate dehydrogenase (1.1.1.49); $N = 86$				
Gd^A	0.01	0	0.99	Red cells; six subunits, each 43,000 MW; ~ 370 aa; phosphate, *p*H 7.0, starch electrophoresis *(56)*; A and B variants identical by microcomplement fixation *(61)*; sequences differ by one amino acid, aspartic acid in A variant, asparagine in B variant *(61)*
Gd^B	0.99	1.00		
Glutamate-oxalacetate transaminase (soluble form) (2.6.1.1); $N = 63$				
$sGOT^1$	1.00	0	0	Red cells; two subunits, each 50,000 MW; ~ 430 aa; tris-citrate, *p*H 7.0, starch electrophoresis *(62)*; chimpanzee protein faster *(63)*
$sGOT^2$	0	1.00		
Glutathione reductase (1.6.4.2); $N = 64$				
GSR^2 and GSR^3	0.97	1.00	0.97	Red cells; tris-EDTA, *p*H 9.6, starch electrophoresis; polymorphism in human populations *(64)*, possibly associated with gout; GSR^2 and GSR^3 not distinguishable at *p*H 9.6
GSR^5	0.01	0		
GSR^6	0.02	0		
Hemoglobin α chain; $N = 108$				
$Hb_\alpha{}^A$	1.00	0.99	0.99	Red cells; 15,100 MW; 141 aa; tris-glycine, *p*H 8.4, cellulose acetate electrophoresis *(15)*; tryptic peptides of human and chimpanzee α chains identical *(65)*; chimpanzee α chain variant is electrophoretically identical to human Hb^J *(66)*
$Hb_\alpha{}^J$	< 0.01	0.01		

(Table 2 is continued on pages 110 and 111)

Locus (i) and allele (j)	Allele frequency		Probability of identity† (S_i)	Comments and references‡
	Human* (x_{ij})	Chimpanzee (y_{ij})		
Hemoglobin β chain; $N = 108$				
Hb_β^A	0.99	0.99	0.98	Red cells; 16,000 MW; 146 aa; tris-glycine, *p*H 8.4, cellulose acetate electrophoresis (*15*); amino acid sequences of β^A chains identical (*65*); chimpanzee Hb^B electrophoretically identical to human Hb^S (*66*)
$Hb_\beta^{S=B}$	0.01	0.01		
Hemoglobin $^A\gamma$ chain	1.00	1.00	1.00	Fetal red cells; 16,000 MW; 146 aa; amino acid sequence of human and chimpanzee γ chains identical; $^A\gamma$ and $^G\gamma$ arep roducts of different structural genes, differ at residue 136; A, alanine; G, glycine (*67*)
Hemoglobin $^G\gamma$ chain	1.00	1.00	1.00	See hemoglobin $^A\gamma$
Hemoglobin δ chain	1.00	1.00	1.00	Red cells; 16,000 MW; 146 aa; human and chimpanzee electrophoretic mobilities identical, but one amino acid difference at position 125: humane δ, methionine; chimpanzee δ, valine (*8*)
Lactate dehydrogenase H (1.1.1.27); $N = 74$	1.00	1.00	1.00	Red cells; H and M subunits each 34,000 MW; 330 aa; citrate-phosphate, *p*H 6.0, starch electrophoresis (*69*); three intermediate bands of five-band, tetrameric electrophoretic pattern have different mobilities for humans and chimpanzees, because of difference in M polypeptide (*70*)
Lactate dehydrogenase M (1.1.1.27); $N = 74$				
$ldh\ M^a$	1.00	0	0	See lactate dehydrogenase H
$ldh\ M^b$	0	1.00		
Malate dehydrogenase (cytoplasmic) (1.1.1.37); $N = 88$	1.00	1.00	1.00	Red cells; two subunits, each 34,000 MW; 330 aa; see LDH for procedures; polymorphic in some human populations (*71*)
Methemglobin reductase (1.6.99); $N = 86$				
MR^1	1.00	0	0	Red cells; tris-citrate, *p*H 6.8, starch electrophoresis (*72*) distinguishes human and chimpanzee enzymes, no difference with tris-EDTA, *p*H 9.3, electrophoresis (*56, 73*)
MR^2	0	1.00		
Myoglobin	1.00	1.00	1.00	Muscle; 16,900 MW; 153 aa; tryptic and chymotryptic peptides of cyanmethemoglobin electrophoretically identical at *p*H 8.6 (*74*), but at position 116, human has glutamine, chimpanzee has histidine (*7*)
Peptidase A (3.4.3.2); $N = 63$				
$PepA^1$ and $PepA^8$	0.99	1.00	0.99	Red cells; two subunits, each 46,000 MW; ~ 400 aa; tris-maleate, *p*H 7.4 starch electrophoresis, leucyl-glycine substrate (*65*); $PepA^1$ and $PepA^8$ not distinguishable in red blood cell lysates (*75*)
$PepA^2$	0.01	0		
Peptidase C (3.4.3.2); $N = 63$				
$PepC^1$	0.99	1.00	0.99	Red cells; 65,000 MW; ~ 565 aa; see peptidase A for procedures; polymorphism in human populations (*76*)
$PepC^4$	0.01	0		
Phosphoglucomutase 1 ‖ (2.7.5.1); $N = 168$				
PGM_1^1	0.77	0.26	0.20	Red cells; subunits PGM_1 and PGM_2 each 62,000 MW; ~ 540 aa; tris-maleate-EDTA, *p*H 7.4, starch electrophoresis (*16, 55, 61, 77*)
PGM_1^2	0.23	0		
PGM_1^{Pan}	0	0.74		
Phosphoglucomutase 2 (2.7.5.1); $N = 168$				
PGM_2^1	1.00	1.00	1.00	See phosphoglucomutase 1
PGM_2^3	$<< 0.01$	< 0.01		
6-Phosphogluconate dehydrogenase (1.1.1.44); $N = 86$				
PGD^A	0.96	0	0.04	Red cells; two subunits, each 40,000 MW; 350 aa; see G6PD for procedures; chimpanzee allele electrophoretically identical to human "Canning" variant (*55*)
PGD^C	0.04	1.00		
Phosphohexose isomerase (5.3.1.9); $N = 86$				
PHI^1	1.00	0	0	Red cells; two subunits, each 66,000 MW; 580 aa; tris-citrate, *p*H 8.0, starch electrophoresis (*56*); chimpanzee protein has slower mobility, both cathodally migrating (*78*)
PHI^B	0	1.00		
Superoxide dismutase A (indophenol oxidase) (1.15.1.1); $N = 64$	1.00	1.00	1.00	Red blood cells; two subunits, each 16,300 MW; 158 aa (*68*); see phosphoglucomutase for procedure
Triosephosphate isomerase A (5.3.1.1)	1.00	1.00	1.00	Fibroblasts; dimers 48,000 MW; each polypeptide 248 aa (*79*); β polypeptide found only in hominoids.
Triosephosphate isomerase B (5.3.1.1)	1.00	1.00	1.00	See triosephosphate isomerase A

Locus (i) and allele (j)	Allele frequency		Probability of identity† (S_i)	Comments and references‡
	Human* (x_{ij})	Chimpanzee (y_{ij})		
			Secreted proteins	
α_1-Acid glycoprotein (orosomucoid); $N = 123$				
Or^S	0.32	0	0.68	Glycoprotein in plasma; carbohydrate > 50 percent; 44,100 MW; 181 aa; acrylamide electrophoresis, *p*H 8.9 (see text); polymorphism in human populations detectable at *p*H 2.9 (*80*); isoelectric point is 1.82 for human and chimpanzee proteins, but proteins differ by quantitative precipitin analysis (*81*)
Or^F	0.68	1.00		
Albumin; $N = 123$				
Alb^A	0	1.00	0	Plasma; 69,000 MW, ~ 580 aa; tris-citrate, *p*H 5.5, cellulose acetate electrophoresis; acrylamide electrophoresis, *p*H 8.9; chimpanzee protein slower mobility, immunological difference detected by microcomplement fixation (*10, 42*); rare polymorphic alleles in human populations (*82*)
Alb^{Pan}	1.00	0		
α_1-Antitrypsin; $N = 123$				
Pi^M	0.95	0	0	Plasma; 49,000 MW; ~ 380 aa; anodal well buffer is citrate-phosphate, *p*H 4.5; cathodal well buffer is borate-NaOH, *p*H 9.0; gel buffer is tris-citrate, *p*H 4.8; starch electrophoresis (*56*); acrylamide electrophoresis, *p*H 8.9; polymorphism in human populations (*83*)
Pi^F	0.03	0		
Pi^S	0.02	0		
Pi^{Pan}	0	1.00		
Ceruloplasmin; $N = 123$				
Cp^A and Cp^{Pan}	0.01	1.00	0.01	Plasma; eight subunits, each 17,000 MW; ~ 150 aa; acrylamide electrophoresis, *p*H 8.9; possible adaptive significance of polymorphism in human populations (*84*)
Cp^B	0.98	0		
Cp^C	0.01	0		
Third component of complement; $N = 123$				
$C'3^{1=F}$	0.12	0	0	Plasma; total MW 240,000; acrylamide electrophoresis, *p*H 8.9; polymorphism in human populations detectable by high voltage electrophoresis (*85*)
$C'3^{2=S}$	0.87	0		
$C'3^{S}$	0.01	0		
$C'3^{Pan}$	0	1.00		
Group-specific component; $N = 206$				
Gc^1	0.74	0	0	Plasma; two subunits, each 25,000 MW; ~ 220 aa; acrylamide electrophoresis, *p*H 8.9; human Gc 2-2 and chimpanzee protein similar on acrylamide, chimpanzee slightly faster on starch or immunoelectrophoresis (*86*)
Gc^2	0.26	0		
Gc^{Pan}	0	1.00		
α_{2HS}-Glycoprotein; $N = 123$				
Gly^A	1.00	0	0	Plasma; 49,000 MW; ~ 400 aa; acrylamide electrophoresis, *p*H 8.9 (*15*)
Gly^B	0	1.00		
Haptoglobin α chain; $N = 300$				
Hp_α^1	0.36	0	0	Plasma; α^1 chain is 8,900 MW, 83 aa; α^2 chain is 16,000 MW, 142 aa; β chain is 36,000 to 40,000 MW; ~ 330 aa; acrylamide electrophoresis, *p*H 8.9; borate-NaOH well buffer and tris-citrate gel buffer, *p*H 8.6, starch electrophoresis (*56*); chimpanzee Hp shares six human Hp 1-1 and eight Hp 2-2 antigenic determinants; Hp^2 evolved since human-chimpanzee divergence (*87*)
Hp_α^2	0.64	0		
Hp_α^{Pan}	0	1.00		
Haptoglobin β chain; $N = 300$	1.00	1.00	1.00	See haptoglobin α chain
Lysozyme				
lzm^A	1.00	0	0	Milk; 14,400 MW; 130 aa; starch gel electrophoresis, *p*H 5.3 (*88*)
lzm^B	0	1.00		
α_2-Macroglobulin; $N = 123$				
Xm^A	1.00	0	0	Plasma; four subunits, each 196,000 MW; acrylamide electrophoresis, *p*H 8.9; X-linked antigenic polymorphism observed in human populations (*89*) but not detectable by electrophoresis; human and chimpanzee proteins immunologically indistinguishable (*14*)
Xm^B	0	1.00		
Plasma cholinesterase (3.1.1.8); $N = 111$				
E_1^u	1.00	0	0	Plasma; four subunits, each ~ 87,000 MW; see esterase A_1 for procedures; chimpanzee protein has four components with faster mobilities than analogous human components (*15*)
E_1^{Pan}	0	1.00		
Transferrin; $N = 133$				
Homo: Tf^C	0.99	0	0	Plasma; 73,000 to 92,000 MW; ~ 650 aa; acrylamide electrophoresis, *p*H 8.9; tris-glycine, *p*H 8.4, cellulose acetate electrophoresis (*77, 90*)
Tf^{D_1}	0.01	0		
Pan: Tf^A	0	0.08		
Tf^B	0	0.06		
Tf^C	0	0.70		
Tf^D	0	0.15		
Tf^E	0	0.02		

* Allelic frequencies for human populations are calculated from data summarized by Nei and Roychoudhury (*28*). Sample sizes generally greater than 1000. Only alleles with frequency > 0.01 are listed. The relative sizes of racial groups were estimated to be Caucasian, 45 percent; Black African, 10 percent; and Mongoloid-Amerind (combined), 45 percent. † See Eq. 2 in text. ‡Given in this column are: the tissue used, polypeptide chain length, electrophoretic conditions, and references to previous studies on people and chimpanzees. Genetic, population, and physiological studies of most human red cell and plasma proteins are summarized by Giblett (*56*) or Harris (*91*); studies of plasma proteins are summarized by Schultze and Heremans (*92*). References are for additional studies of chimpanzee or human proteins. § Not included in identity calculations. ‖ Notation for the chimpanzee alleles at the PGM_1 locus differs in publishhed surveys. Ours is as follows: PGM_1^{Pan} (which is chimpanzee PGM_1^1, of Goodman and co-workers and PGM_1^{Pan} of Schmitt and co-workers) is the allele with slowest electrophoretic mobility; PGM_1^1 (which is human PGM_1^1, the chimpanzee PGM_1^1 of Schmitt, and the chimpanzee PGM_1^2 of Goodman) is intermediate; and PGM_1^2 (found only in human populations) has the fastest mobility.

systems used in this study is about 0.27 (*18*).

If we assume that, at a particular amino acid site on a given protein, amino acid substitutions have occurred (i) independently and (ii) at random with respect to species since the evolutionary divergence of humans and chimpanzees, then the number of proteins that have accumulated r amino acid substitutions since this divergence approximates a Poisson variate (*19*). That is, the probability that r substitutions have accumulated in a particular polypeptide is

$$P_r = \frac{(mc)^r \, e^{-mc}}{r!} \qquad (4)$$

where m is the expected number of amino acid substitutions per polypeptide (the mean of the Poisson distribution), and c is the proportion of those substitutions that are electrophoretically detectable. The probability that the polypeptides are electrophoretically identical (that is, that no electrophoretically detectable substitutions have occurred) is 0.52. Therefore,

$$P_0 = 0.52 = \frac{(mc)^0 \, e^{-mc}}{0!} = e^{-mc} \qquad (5)$$

Thus $mc = 0.65$ and the expected number of amino acid differences per polypeptide is

$$m = 0.65/0.27 = 2.41 \qquad (6)$$

For comparative purposes, this value can also be expressed in terms of the expected number of amino acid differences per 1000 amino acids. The average number of amino acids per polypeptide for all the proteins analyzed electrophoretically is 293 ± 27 (standard error). Therefore the expected degree of amino acid difference between human and chimpanzee is

$$\frac{2.41 \times 1000}{293} = 8.2 \qquad (7)$$

substitutions per 1000 sites, with a range (within one standard error) of 7.5 to 9.1 differences per 1000 amino acids. The estimate based on amino acid sequencing and immunological comparisons (Eq. 1) agrees well with this estimate. Both estimates indicate that the average human protein is more than 99 percent identical in amino acid sequence to its chimpanzee homolog (*20*).

Comparison of nucleic acids. Another method of comparing genomes is nucleic acid hybridization. Several workers have compared the thermostability of human-chimpanzee hybrid DNA formed in vitro with the thermostability of DNA from each species separately. By this criterion, human and chimpanzee mitochondrial DNA's appear identical (*21*). Working with "nonrepeated" DNA sequences, Kohne has estimated that human-chimpanzee hybrid DNA dissociates at a temperature (ΔT) 1.5°C lower than the dissociation temperature of reannealed human DNA (*22*). Hoyer *et al.*, on the other hand, have estimated that ΔT equals 0.7°C for human-chimpanzee hybrid DNA (*23*). If ΔT is the difference in dissociation temperature of reannealed human DNA and human-chimpanzee hybrid DNA prepared in vitro, then the percentage of nucleic acid sequence difference is $k \times \Delta T$ where the calibration factor k has been variously estimated as 1.5, 1.0, 0.9, or 0.45 (*22, 24*). Based on k being 1.0 and ΔT being 1.1°C, the nucleic acid sequence difference of human and chimpanzee DNA is about 1.1 percent. In a length of DNA 3000 bases long (representing 1000 amino acids), there will be about 0.011×3000, or 33 nucleotide sequence differences between the two species.

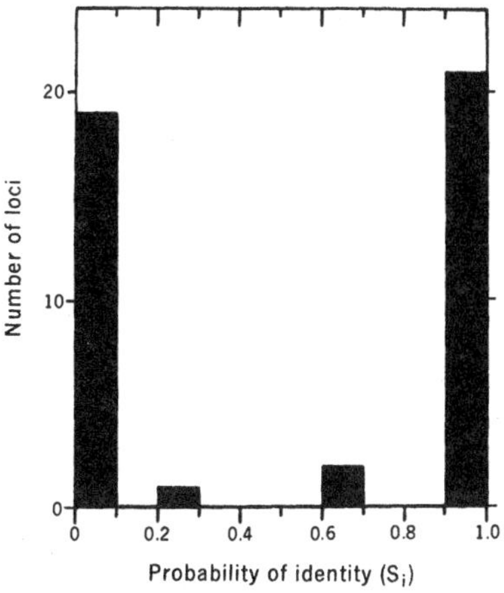

Fig. 2. Electrophoretic comparison of 43 proteins from humans and chimpanzees. The probability of identity (S_i) represents the likelihood that at locus i, human and chimpanzee alleles will appear electrophoretically identical.

The evidence from the DNA annealing experiments indicates that there may be more difference at the nucleic acid level than at the protein level in human and chimpanzee genomes. For every amino acid sequence difference observed, about four base differences are observed in the DNA. Li *et al.* (*25*) found the same distinction between amino acid and nucleic acid differences in the tryptophan synthetase of several bacterial species: the nucleic acid sequences were about three times as different as the amino acid sequences. A similar result has been observed in three related RNA bacteriophages, as well as in studies of the relative rates of DNA and protein evolution in cow, pig, and sheep (*26*).

There are a number of probable reasons for this discrepancy (*25, 26*). First, more changes may appear in DNA than in proteins because of the redundancy of the code and consequently the existence of third-position nucleotide changes which do not lead to amino acid substitutions. The nature of the code indicates that if first-, second-, and third-position substitutions were equally likely to persist, then about 30 to 40 percent of potential base replacements in a cistron would not be reflected in the coded protein; that is, 1.4 to 1.7 base substitutions would occur for each amino acid substitution (*27*). However, it is likely that a larger proportion of the actual base substitutions in a cistron are third-position changes, since base substitutions that do not affect amino acid sequence are more likely to spread through a population. In addition, many of the nucleic acid substitutions may have occurred in regions of the DNA that are not transcribed and are therefore not conserved during evolution. Proteins analyzed by electrophoresis, sequencing, or microcomplement fixation techniques, on the other hand, all have definite cellular functions and may therefore have been conserved to a greater extent during evolution.

Genetic Distance and the Evolution of Organisms

The resemblance between human and chimpanzee macromolecules has been measured by protein sequencing, immunology, electrophoresis, and nucleic acid hybridization. From each of these results we can obtain an estimate of the genetic distance between humans and chimpanzees. Some of the same approaches have been used to estimate the genetic distance between other taxa, so that these estimates may be compared to the human-chimpanzee genetic distance.

First, we consider genetic distance estimated from electrophoretic data, using the standard estimate of net codon differences per locus developed by Nei and Roychoudhury (*28*). Other indices have been suggested for handling electrophoretic data (*29*) and give the same qualitative results, though somewhat different underlying assumptions are required. Nei and Roychoudhury's standard estimate of genetic distance between humans and chimpanzees can be written:

$$D = D_{HC} - \frac{D_C + D_H}{2} \qquad (8)$$

where

$$D_{HC} = -\log_e \bar{S}$$

$$D_H = -\log_e \left(\frac{1}{L} \sum_{i=1}^{L} \sum_{j=1}^{A_i} x_{ij}^2 \right)$$

$$D_C = -\log_e \left(\frac{1}{L} \sum_{i=1}^{L} \sum_{j=1}^{A_i} y_{ij}^2 \right)$$

according to the notation of Table 2 and Eqs. 2 and 3. Therefore, D is an estimate of the variability between human and chimpanzee populations (D_{HC}), corrected for the variability within human populations (D_H) and within chimpanzee populations (D_C). D_C and D_H are also measurements of the degree of heterozygosity in human and chimpanzee populations (*30*). Based on the data of Table 2, D_{HC} is 0.65, D_C is 0.02, and D_H is 0.05, so that:

$$D = 0.62 \qquad (9)$$

In other words, there is an average of 0.62 electrophoretically detectable codon differences per locus between homologous human and chimpanzee proteins.

This distance is 25 to 60 times greater than the genetic distance between human races (*28, 31*). In fact, the genetic distance between Caucasian, Black African, and Japanese populations is less than or equal to that between morphologically and behaviorally identical populations of other species. In addition, these three human populations are equally distant from the chimpanzee lineage (Fig. 3).

However, with respect to genetic distances between species, the human-chimpanzee D value is extraordinarily small, corresponding to the genetic distance between sibling species of *Drosophila* or mammals (Fig. 4). Nonsibling species within a genus (referred to in the figure as congeneric species) generally differ more from each other, by electrophoretic criteria, than humans and chimpanzees. The genetic distances among species from different genera are considerably larger than the human-chimpanzee genetic distance.

The genetic distance between two species measured by DNA hybridization also indicates that human beings and chimpanzees are as similar as sibling species of other organisms. The difference in dissociation temperature, ΔT, between reannealed human DNA and human-chimpanzee hybrid DNA is about 1.1°C. However, for sibling species of *Drosophila*, ΔT is 3°C; for congeneric species of *Drosophila*, ΔT is 19°C; and for congeneric species of mice (*Mus*), ΔT is 5°C (*32*).

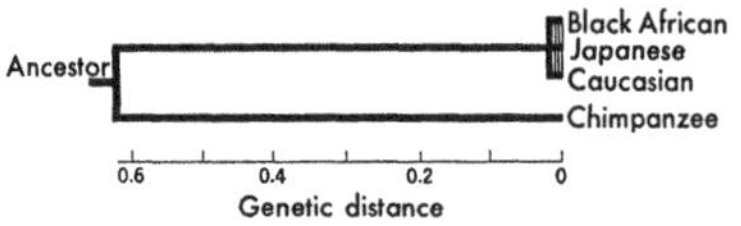

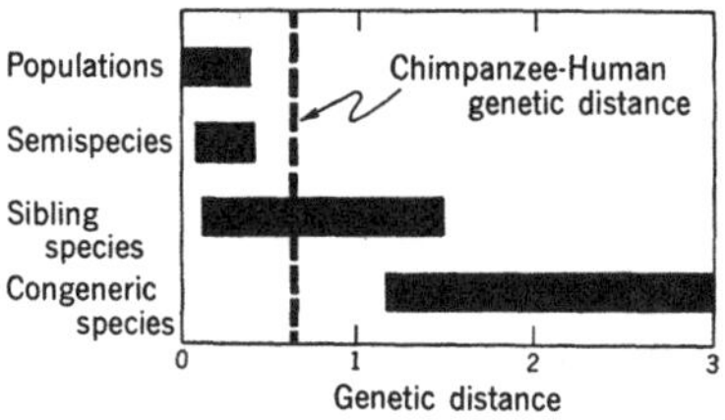

Fig. 3 (left). Phylogenetic relationship between human populations and chimpanzees. The genetic distances are based on electrophoretic comparison of proteins. The genetic distances among the three major human populations ($D = 0.01$ to 0.02) that have been tested are extremely small compared to those between humans and chimpanzees ($D = 0.62$). No human population is significantly closer than another to the chimpanzee lineage. The vertically hatched area between the three human lineages indicates that the populations are not really separate, owing to gene flow. Fig. 4 (right). The genetic distance, D, between humans and chimpanzees (dashed line) compared to the genetic distances between other taxa. Taxa compared include several species of *Drosophila* [*D. willistoni* (*94*), *D. paulistorum* (*95*), and *D. pseudoobscura* (*96*)], the horseshoe crab *Limulus polyphemus* (*97*), salamanders from the genus *Taricha* (*98*), lizards from the genus *Anolis* (*99*), the teleost fish *Astyanax mexicanus* (*100*), bats from the genus *Lasiurus* (*101*), and several genera of rodents [*Mus, Sigmodon, Dipodomys, Peromyscus,* and *Thomomys* (*99*), *Geomys* (*101*), and *Apodemus* (*102*)]. Selander and Johnson (*99*) summarize most of the data used in this figure. The great majority of proteins in these studies are intracellular.

Immunological and amino acid sequence comparisons of proteins lead to the same conclusion. Antigenic differences among the serum proteins of congeneric squirrel species are several times greater than those between humans and chimpanzees (*33*). Moreover, antigenic differences among the albumins of congeneric frog species (*Rana* and *Hyla*) are 20 to 30 times greater than those between the two hominoids (*34, 35*). In addition, the genetic distances among *Hyla* species, estimated electrophoretically, are far larger than the chimpanzee-human genetic distance (*36*). Finally, the human and chimpanzee β chains of hemoglobin appear to have identical sequences (Table 1), while the β chains of two *Rana* species differ by at least 29 amino acid substitutions (*37*). In summary, the genetic distance between humans and chimpanzees is well within the range found for sibling species of other organisms.

The molecular similarity between chimpanzees and humans is extraordinary because they differ far more than sibling species in anatomy and way of life. Although humans and chimpanzees are rather similar in the structure of the thorax and arms, they differ substantially not only in brain size but also in the anatomy of the pelvis, foot, and jaws, as well as in relative lengths of limbs and digits (*38*). Humans and chimpanzees also differ significantly in many other anatomical respects, to the extent that nearly every bone in the body of a chimpanzee is readily distinguishable in shape or size from its human counterpart (*38*). Associated with these anatomical differences there are, of course, major differences in posture (see cover picture), mode of locomotion, methods of procuring food, and means of communication. Because of these major differences in anatomy and way of life, biologists place the two species not just in separate genera but in separate families (*39*). So it appears that molecular and organismal methods of evaluating the chimpanzee-human difference yield quite different conclusions (*40*).

An evolutionary perspective further illustrates the contrast between the results of the molecular and organismal approaches. Since the time that the ancestor of these two species lived, the chimpanzee lineage has evolved slowly relative to the human lineage, in terms of anatomy and adaptive strategy. According to Simpson (*41*):

> *Pan* is the terminus of a conservative lineage, retaining in a general way an anatomical and adaptive facies common to all recent hominoids *except Homo. Homo* is both anatomically and adaptively the most radically distinctive of all hominoids, divergent to a degree considered familial by all primatologists.

This concept is illustrated in the left-hand portion of Fig. 5. However, at the macromolecular level, chimpanzees and humans seem to have evolved

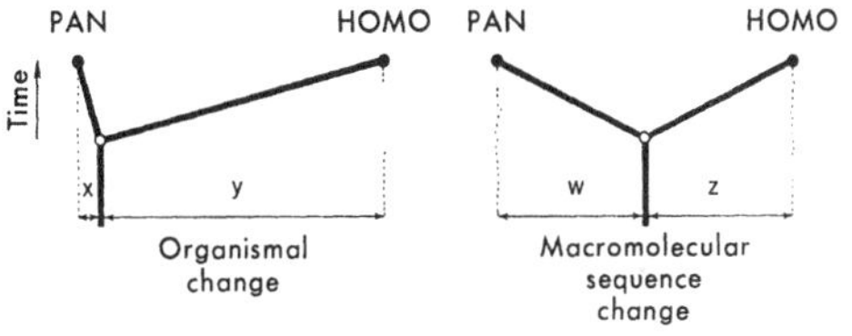

Fig. 5. The contrast between biological evolution and molecular evolution since the divergence of the human and chimpanzee lineages from a common ancestor. As shown on the left, zoological evidence indicates that far more biological change has taken place in the human lineage (y) than in the chimpanzee lineage ($y \gg x$); this illustration is adapted from that of Simpson (*41*). As shown on the right, both protein and nucleic acid evidence indicate that as much change has occurred in chimpanzee genes (w) as in human genes (z).

at similar rates (Fig. 5, right). For example, human and chimpanzee albumins are equally distinct immunologically from the albumins of other hominoids (gorilla, orangutan, and gibbon) (*10, 42, 43*), and human and chimpanzee DNA's differ to the same degree from DNA's of other hominoids (*21, 22*). Construction of a phylogenetic tree for primate myoglobins shows that the single amino acid difference between the sequences of human and chimpanzee myoglobin occurred in the chimpanzee lineage (*7*). Analogous reasoning indicates that the single amino acid difference between the sequences of human and chimpanzee hemoglobin δ chains arose in the human lineage (*8*). It appears that molecular change has accumulated in the two lineages at approximately equal rates, despite a striking difference in rates of organismal evolution. Thus, the major adaptive shift which took place in the human lineage was probably not accompanied by accelerated protein or DNA evolution.

Such an observation is by no means peculiar to the case of hominid evolution. It appears to be a general rule that anatomically conservative lineages, such as frogs, have experienced as much sequence evolution as have lineages that have undergone rapid evolutionary changes in anatomy and way of life (*34, 35, 44*).

Molecular Basis for the Evolution of Organisms

The contrasts between organismal and molecular evolution indicate that the two processes are to a large extent independent of one another. Is it possible, therefore, that species diversity results from molecular changes other than sequence differences in proteins? It has been suggested by Ohno (*45*) and others (*46*) that major anatomical changes usually result from mutations affecting the expression of genes. According to this hypothesis, small differences in the time of activation or in the level of activity of a single gene could in principle influence considerably the systems controlling embryonic development. The organismal differences between chimpanzees and humans would then result chiefly from genetic changes in a few regulatory systems, while amino acid substitutions in general would rarely be a key factor in major adaptive shifts.

Regulatory mutations may be of at least two types. First, point mutations could affect regulatory genes. Nucleotide substitutions in a promoter or operator gene would affect the production, but not the amino acid sequence, of proteins in that operon. Nucleotide substitutions in a structural gene coding for a regulatory protein such as a repressor, hormone, or receptor protein, could bring about amino acid substitutions, altering the regulatory properties of the protein. However, we suspect that only a minor fraction of the substitutions which accumulate in regulatory proteins would be likely to alter their regulatory properties.

Second, the order of genes on a chromosome may change owing to inversion, translocation, addition or deletion of genes, as well as fusion or fission of chromosomes. These gene rearrangements may have important effects on gene expression (*47*), though the biochemical mechanisms involved are obscure. Evolutionary changes in gene order occur frequently. Microscopic studies of *Drosophila* salivary chromosomes show, as a general rule, that no two species have the same gene order and that inversions are the commonest type of gene rearrangement (*48*). Furthermore, there is a parallel between rate of gene rearrangement and rate of anatomical evolution in the three major groups of vertebrates that have been studied in this respect, namely birds, mammals, and frogs (*46*). Hence gene rearrangements may be more important than point mutations as sources for evolutionary changes in gene regulation.

Although humans and chimpanzees have rather similar chromosome numbers, 46 and 48, respectively, the arrangement of genes on chimpanzee chromosomes differs from that on human chromosomes. Only a small proportion of the chromosomes have identical banding patterns in the two species. The banding studies indicate that at least 10 large inversions and translocations and one chromosomal fusion have occurred since the two lineages diverged (*49*). Further evidence for the possibility that chimpanzees and humans differ considerably in gene arrangement is provided by annealing studies with a purified DNA fraction. An RNA which is complementary in sequence to this DNA apparently anneals predominantly at a cluster of sites on a single human chromosome, but at widely dispersed sites on several chimpanzee chromosomes (*50*). The arrangement of chromosomal sites at which ribosomal RNA anneals may also differ between the two species (*50*).

Biologists are still a long way from understanding gene regulation in mammals (*51*), and only a few cases of regulatory mutations are now known (*52*). New techniques for detecting regulatory differences at the molecular level are required in order to test the hypothesis that organismal differences between individuals, populations, or species result mainly from regulatory differences. When the regulation of gene expression during embryonic development is more fully understood, molecular biology will contribute more significantly to our understanding of the evolution of whole organisms. Most important for the future study of human evolution would be the demonstration of differences between apes and humans in the timing of gene expression during development, particularly during the development of adaptively crucial organ systems such as the brain.

Summary and Conclusions

The comparison of human and chimpanzee macromolecules leads to several inferences:

1) Amino acid sequencing, immunological, and electrophoretic methods of protein comparison yield concordant estimates of genetic resemblance. These approaches all indicate that the average human polypeptide is more than 99 per-

cent identical to its chimpanzee counterpart.

2) Nonrepeated DNA sequences differ more than amino acid sequences. A large proportion of the nucleotide differences between the two species may be ascribed to redundancies in the genetic code or to differences in nontranscribed regions.

3) The genetic distance between humans and chimpanzees, based on electrophoretic comparison of proteins encoded by 44 loci is very small, corresponding to the genetic distance between sibling species of fruit flies or mammals. Results obtained with other biochemical methods are consistent with this conclusion. However, the substantial anatomical and behavioral differences between humans and chimpanzees have led to their classification in separate families. This indicates that macromolecules and anatomical or behavioral features of organisms can evolve at independent rates.

4) A relatively small number of genetic changes in systems controlling the expression of genes may account for the major organismal differences between humans and chimpanzees. Some of these changes may result from the rearrangement of genes on chromosomes rather than from point mutations (*53*).

References and Notes

1. S. L. Washburn, Ed., *Classification and Human Evolution* (Aldine, Chicago, 1963). That there were striking similarities in blood proteins between apes and humans was known in 1904 [G. H. F. Nuttall, *Blood Immunity and Blood Relationships* (Cambridge Univ. Press, London, 1904)].
2. Sibling species are virtually identical morphologically.
3. R. F. Doolittle, G. L. Wooding, Y. Lin, M. Riley, *J. Mol. Evol.* **1**, 74 (1971).
4. M. O. Dayhoff, Ed., *Atlas of Protein Sequence and Structure* (National Biomedical Research Foundation, Georgetown Univ. Medical Center, Washington, D.C., 1972), vol. 5.
5. S. H. Boyer, E. F. Crosby, A. N. Noyes, G. F. Fuller, S. E. Leslie, L. J. Donaldson, G. R. Vrablik, E. W. Schaefer, T. F. Thurmon, *Biochem. Genet.* **5**, 405 (1971).
6. W. W. W. DeJong, *Biochim. Biophys. Acta* **251**, 217 (1971).
7. A. E. Romero Herrera and H. Lehmann, *ibid.* **278**, 62 (1972).
8. W. W. W. DeJong, *Nat. New Biol.* **234**, 176 (1971).
9. E. M. Prager and A. C. Wilson, *J. Biol. Chem.* **246**, 5978 and 7010 (1971).
10. V. M. Sarich and A. C. Wilson, *Science* **158**, 1200 (1967).
11. A. C. Wilson and V. M. Sarich, *Proc. Natl. Acad. Sci. U.S.A.* **63**, 1088 (1969); J. E. Cronin and V. M. Sarich, personal communication; R. Palmour, personal communication.
12. L. Nonno, H. Herschman, L. Levine, *Arch. Biochem. Biophys.* **136**, 361 (1970).
13. N. Hanke, E. M. Prager, A. C. Wilson, *J. Biol. Chem.* **248**, 2824 (1973).
14. A variety of immunological techniques have been used to compare chimpanzee proteins with their human counterparts [N. Mohagheghpour and C. A. Leone, *Comp. Biochem. Physiol.* **31**, 437 (1969); M. Goodman and G. W. Moore, *Syst. Zool.* **20**, 19 (1971); K. Bauer, *Humangenetik* **17**, 253 (1973)]. The immunodiffusion techniques employed in these studies are less sensitive to small differences in amino acid sequence than is microcomplement fixation [E. M. Prager and A. C. Wilson, *J. Biol. Chem.* **246**, 5978 (1971)]. Nevertheless, their results are generally consistent with those in Table 1. The few cases of large antigenic differences between human and chimpanzee proteins are probably not indicative of large sequence differences. For example, the haptoglobin difference reported by Mohagheghpour and Leone is due mainly to the fact that the haptoglobin 2 polypeptide is nearly twice the length of the haptoglobin 1 polypeptide [J. A. Black and G. H. Dixon, *Nature (Lond.)* **218**, 736 (1968)]. Human haptoglobin 1 is immunologically very similar to chimpanzee haptoglobin [J. Javid and H. H. Fuhrmann, *Am. J. Hum. Genet.* **207**, 496 (1971)]. The immunoglobulin differences reported by Bauer may be due to comparison of peptide chains that are not strictly homologous. In addition, Bauer's kappa chain results conflict with quantitative studies which detected no immunological difference [A. C. Wang, J. Shuster, A. Epstein, H. H. Fudenberg, *Biochem. Genet.* **1**, 347 (1968)]. Finally, the large Xh factor difference that Bauer reported might result from the fact that the chimpanzees in his studies were not pregnant and thus lacked Xh factor.
15. M.-C. King, thesis, University of California, Berkeley (1973).
16. D. Stollar and L. Levine, in *Methods in Enzymology*, S. P. Colowick and N. O. Kaplan, Eds. (Academic Press, New York, 1967), vol. 6, p. 928.
17. F. J. Ayala, M. L. Tracey, L. G. Barr, J. F. McDonald, S. Perez-Salas, *Genetics* **77**, 343 (1974).
18. See figure 9.3 in Dayhoff (*4*). We determined the proportion of amino acid substitutions causing a charge change during vertebrate evolution for several additional proteins: cytochrome c, lysozyme, myoglobin, α and β hemoglobin chains, triosephosphate dehydrogenase, α-lactalbumin, growth hormone, trypsin, and insulin. The average for these proteins is also 0.27. Our estimate of 0.16 for hemoglobin alone is very similar to that of Boyer *et al.* [S. H. Boyer, A. N. Noyes, C. F. Timmons, R. A. Young, *J. Hum. Evol.* **1**, 515 (1972)], who calculated that the ratio between electrophoretically silent and electrophoretically detectable hemoglobin alleles in primates is about 5.5; that is, about 15 percent of amino acid substitutions in primate hemoglobin evolution would be electrophoretically detectable.

    A change in charge at a single amino acid site may have little or no effect on the net charge of a protein unless the substituted amino acid is on the exposed surface of the protein. Lee and Richards [B. Lee and F. M. Richards, *J. Mol. Biol.* **55**, 379 (1971)] determined the degree of exposure of each of the amino acid residues of lysozyme, ribonuclease, and myoglobin, based on the three-dimensional structure of these molecules. Their data indicate that 100 percent of the lysine residues, 100 percent of the arginine residues, 95 percent of the aspartic acid residues, 100 percent of the glutamic acid residues, and 70 percent of the histidine residues are on exposed surfaces of the proteins. Thus more than 90 percent of the substitutions involving charged amino acids would have affected the net charge of the protein and would, therefore, be detectable by electrophoresis.
19. A negative binomial variable may better describe the distribution of amino acid substitutions along lineages, since substitutions occur in proteins which are subject to varying selective pressures. That is, since different proteins evolve at different rates, the probability of a particular protein accepting a mutation varies from protein to protein [T. Uzzell and K. W. Corbin, *Science* **172**, 1089 (1971)]. However, for small values of m the negative binomial distribution agrees substantially with the Poisson distribution [C. I. Bliss and R. A. Fisher, *Biometrics* **9**, 176 (1953)]. Thus, for this calculation, the Poisson distribution should provide a very good estimate of the true m.
20. Further evidence regarding the correlation between electrophoretic and immunological measures of genetic distance has been obtained in studies involving many taxa of mammals, reptiles, and amphibians, by S. M. Case, H. C. Dessauer, G. C. Gorman, P. Haneline, K. H. Keeler, L. R. Maxson, V. M. Sarich, D. Shochat, A. C. Wilson, and J. S. Wyles.
21. W. M. Brown and R. L. Hallberg, *Fed. Proc.* **31** (2), Abstr. 1173 (1972).
22. D. E. Kohne, *Q. Rev. Biophys.* **3**, 327 (1970); ———, J. A. Chiscon, B. H. Hoyer, *J. Hum. Evol.* **1**, 627 (1972).
23. B. H. Hoyer, N. W. van de Velde, M. Goodman, R. B. Roberts, *J. Hum. Evol.* **1**, 645 (1972).
24. N. R. Kallenbach and S. D. Drost, *Biopolymers* **11**, 1613 (1972); J. R. Hutton and J. G. Wetmur, *Biochemistry* **12**, 558 (1973); J. S. Ullman and B. J. McCarthy, *Biochim. Biophys. Acta* **294**, 416 (1973).
25. S. L. Li, R. M. Denney, C. Yanofsky, *Proc. Natl. Acad. Sci. U.S.A.* **70**, 1112 (1973).
26. H. D. Robertson and P. G. N. Jeppesen, *J. Mol. Biol.* **68**, 417 (1972); C. Laird, B. L. McConaughy, B. J. McCarthy, *Nature (Lond.)* **224**, 149 (1969).
27. *Cold Spring Harbor Symp. Quant. Biol.* **31**, 1 (1966).
28. M. Nei and A. K. Roychoudhury, *Am. J. Hum. Genet.* **26**, 431 (1974).
29. L. L. Cavalli-Sforza and A. W. F. Edwards, *ibid.* **19**, 233 (1967); N. E. Morton, *Annu. Rev. Genet.* **3**, 53 (1969); V. Balakrishnan and L. D. Sangvhi, *Biometrics* **24**, 859 (1968); T. W. Kurczynski, *ibid.* **26**, 525 (1970); P. W. Hedrick, *Evolution* **25**, 276 (1971); R. R. Sokal and P. H. A. Sneath, *Principles of Numerical Taxonomy* (Freeman, San Francisco, 1973); J. S. Rogers, "Studies in Genetics," *Univ. Texas Publ. No. 7213* (1972), vol. 7, p. 145.
30. The average heterozygosity estimates for the loci in this study are quite low, especially for chimpanzee populations. To obtain comparable heterozygosity estimates for humans and chimpanzees, we included only loci that have been surveyed for both species and only polymorphisms detectable by techniques used for surveying both species. Thus some confirmed polymorphisms in human populations were excluded. There are at least three reasons for the difference between the heterozygosity estimates for human and chimpanzee populations. First, many more humans than chimpanzees have been surveyed at each locus, so that the variability estimate for humans is biased insofar as it is based on alleles present at low frequency in human populations. Second, there are many more humans than chimpanzees alive today, living in a greater variety of environments and with a larger number of gene pools. As a result, more mutants reach appreciable frequencies in human populations. Third, and probably most important, the chimpanzees in colonies available for study are based on even fewer gene pools and are highly inbred in many cases. The discrepancies in real population size and sampling technique between human and chimpanzee populations probably account for the greater number of polymorphic loci, the larger number of alleles at polymorphic loci, and the higher average heterozygosity estimates in human populations.
31. M. Nei and A. K. Roychoudhury, *Science* **177**, 434 (1972).
32. N. R. Rice, *Brookhaven Symp. Biol.* **23**, 44 (1972); C. D. Laird, *Annu. Rev. Genet.* **3**, 177 (1973).
33. M. E. Hight, M. Goodman, W. Prychodko, *Syst. Zool.* **23**, 12 (1974).
34. D. G. Wallace, L. R. Maxson, A. C. Wilson, *Proc. Natl. Acad. Sci. U.S.A.* **68**, 3127 (1971); D. G. Wallace, M.-C. King, A. C. Wilson, *Syst. Zool.* **22**, 1 (1973).
35. L. R. Maxson, and A. C. Wilson, *Syst. Zool.*, in press.
36. R. K. Selander, personal communication; L. R. Maxson and A. C. Wilson, *Science* **185**, 66 (1974).
37. T. O. Baldwin and A. Riggs, *J. Biol. Chem.* **249**, 6110 (1974).
38. G. H. Bourne, Ed., *The Chimpanzee* (Karger, New York, 1970).
39. Th. Dobzhansky, in *Classification and Human Evolution*, S. L. Washburn, Ed. (Aldine, Chicago, 1963), p. 347; E. Mayr, in *ibid.*, p. 332; E. L. Simons, *Primate Evolution* (Mac-

millan, New York, 1972); G. G. Simpson, *Principles of Animal Taxonomy* (Columbia Univ. Press, New York, 1961). L. Van Valen [*Am. J. Phys. Anthropol.* **30**, 295 (1969)] has suggested that, based on differences in their adaptive zones, humans and chimpanzees be placed in separate suborders.
40. On the basis of some protein evidence available in 1970, Goodman and Moore proposed that humans and African apes be placed in the same subfamily [M. Goodman and G. W. Moore, *Syst. Zool.* **20**, 19 (1971)]. By analogy, the protein evidence now available would lead to placement of chimpanzees and humans in the same genus. However, as protein evolution and organismal evolution apparently can proceed independently, it is questionable whether organismal classifications should be revised on the basis of protein evidence alone.
41. G. G. Simpson, in *Classification and Human Evolution*, S. L. Washburn, Ed. (Aldine, Chicago, 1963).
42. V. M. Sarich, in *Old World Monkeys*, J. R. Napier and P. H. Napier, Eds. (Academic Press, New York, 1970), p. 175.
43. ——— and A. C. Wilson, *Proc. Natl. Acad. Sci. U.S.A.* **58**, 142 (1967).
44. V. M. Sarich, *Syst. Zool.* **18**, 286 and 416 (1969); *Nature (Lond.)* **245**, 218 (1973).
45. S. Ohno, *J. Hum. Evol.* **1**, 651 (1972).
46. A. C. Wilson, L. R. Maxson, V. M. Sarich, *Proc. Natl. Acad. Sci. U.S.A.* **71**, 2843 (1974); A. C. Wilson, V. M. Sarich, L. R. Maxson, *ibid.*, p. 2028; E. M. Prager and A. C. Wilson, *ibid.* **72**, 200 (1975).
47. E. Bahn, *Hereditas* **67**, 79 (1971); B. Wallace and T. L. Kass, *Genetics* **77**, 541 (1974).
48. M. J. D. White, *Annu. Rev. Genet.* **3**, 75 (1969).
49. J. de Grouchy, C. Turleau, M. Roubin, F. C. Colin, *Nobel Symp.* **23**, 124 (1973); B. Dutrillaux, M.-O. Rethoré, M. Prieur, J. Lejeune, *Humangenetik* **20**, 343 (1973); D. Warburton, I. L. Firschein, D. A. Miller, F. E. Warburton, *Cytogenet. Cell Genet.* **12**, 453 (1973); C. C. Lin, B. Chiarelli, L. E. M. de Boer, M. M. Cohen, *J. Hum. Evol.* **2**, 311 (1973); J. Ecozcue, M. R. Caballin, C. Goday, *Humangenetik* **18**, 77 (1973); M. Bobrow and K. Madan, *Cytogenet. Cell Genet.* **12**, 107 (1973).
50. In situ annealing studies have been performed with RNA complementary to purified human satellite DNA [K. W. Jones, J. Prosser, G. Carneo, E. Ginelli, M. Bobrow, *Symp. Med. Hoechst* **6**, 45 (1973)] and with human ribosomal RNA [A. Henderson, D. Warburton, K. C. Atwood, *Chromosoma* **46**, 435 (1974)].
51. J. E. Darnell, W. R. Jelinek, G. R. Molloy, *Science* **181**, 1214 (1973); E. H. Davidson and R. J. Britten, *Q. Rev. Biol.* **48**, 565 (1973); C. A. Thomas, Jr., in *Regulation of Transcription and Translation in Eukaryotes*, E. K. F. Bautz, Ed. (Springer-Verlag, Berlin, 1973).
52. D. J. Weatherall and J. B. Clegg, *The Thalassaemia Syndromes* (Blackwell, Oxford, ed. 2, 1972).
53. Additional inferences can be drawn from the comparison of human and chimpanzee macromolecules; some of these will be discussed elsewhere.
54. I. N. H. White and P. J. Butterworth, *Biochim. Biophys. Acta* **229**, 193 (1971).
55. J. Schmitt, K. H. Lichte, W. Fuhrmann, *Humangenetik* **10**, 138 (1970); G. Tariverdian, H. Ritter, J. Schmitt, *ibid.* **11**, 323 (1971).
56. E. R. Giblett, *Genetic Markers in Human Blood* (Blackwell, Oxford, 1969).
57. J. Schmitt, G. Tariverdian, H. Ritter, *Humangenetik* **11**, 100 (1971).
58. R. E. Tashian, *Am. J. Hum. Genet.* **17**, 257 (1965).
59. H. Matsubara and E. L. Smith, *J. Biol. Chem.* **237**, 3575 (1962).
60. R. Schiff and C. Stormont, *Biochem. Genet.* **4**, 11 (1970).
61. A. Yoshida, *Proc. Natl. Acad. Sci. U.S.A.* **57**, 835 (1967); *Biochem. Genet.* **1**, 81 (1967); J. Kömpf, H. Ritter, J. Schmitt, *Humangenetik* **11**, 342 (1971); M. Goodman and M. D. Poulik, "Genetic variations and phylogenetic properties of protein macromolecules of chimpanzees" (6571st Aeromedical Research Laboratory, Rep. ARL-TR-68-3, Holloman Air Force Base, New Mexico, 1968).
62. J. L. Brewbaker, M. D. Upadhya, Y. Mäkinen, T. Macdonald, *Physiol. Plant* **21**, 930 (1968); C. R. Shaw and R. Prasad, *Biochem. Genet.* **4**, 297 (1970).
63. J. Kömpf, H. Ritter, J. Schmitt, *Humangenetik* **13**, 72 (1971).
64. J. C. Kaplan, *Nature (Lond.)* **217**, 256 (1968); W. K. Long, *Science* **155**, 712 (1967).
65. D. Rifkin and W. Konigsberg, *Biochim. Biophys. Acta* **104**, 457 (1965).
66. H. Harris, *J. Med. Genet.* **8**, 444 (1971).
67. R. G. Davidson, J. A. Cortner, M. C. Rattazzi, F. H. Ruddle, H. A. Lubs, *Science* **196**, 391 (1970).
68. G. Beckman, E. Lundgren, A. Tärnvik, *Hum. Hered.* **23**, 338 (1973); B. B. Keele, Jr., J. M. McCord, I. Fridovich, *J. Biol. Chem.* **246**, 2875 (1971); H. M. Steinman and R. L. Hill, *Proc. Natl. Acad. Sci. U.S.A.* **70**, 3725 (1973).
69. G. S. Bailey and A. C. Wilson, *J. Biol. Chem.* **243**, 5843 (1968).
70. A. L. Koen and M. Goodman, *Biochem. Genet.* **3**, 457 (1969).
71. R. L. Kirk, E. M. McDermid, N. M. Blake, R. L. Wight, E. H. Yap, M. J. Simons, *Humangenetik* **17**, 345 (1973); G. Tariverdian, H. Ritter, J. Schmitt, *ibid.* **11**, 339 (1971).
72. G. Tariverdian, H. Ritter, G. G. Wendt, *ibid.*, p. 75.
73. J. Schmitt, G. Tariverdian, H. Ritter, *ibid.*, p. 95.
74. P. C. Hudgins, C. M. Whorton, T. Tomoyoshi, A. J. Riopelle, *Nature (Lond.)* **212**, 693 (1966).
75. W. H. P. Lewis, *Ann. Hum. Genet.* **36**, 267 (1973).
76. S. Povey, G. Corney, W. H. P. Lewis, E. B. Robson, J. M. Parrington, H. Harris, *ibid.* **35**, 455 (1972).
77. M. Goodman and R. E. Tashian, *Hum. Biol.* **41**, 237 (1969).
78. G. Tariverdian, H. Ritter, J. Schmitt, *Humangenetik* **12**, 105 (1971).
79. H. Rubinson, M. C. Meienhofer, J. C. Dreyfus, *J. Mol. Evol.* **2**, 243 (1973); P. H. Corran and S. G. Waley, *Biochem. J.* **139**, 1 (1974).
80. W. E. Marshall, *J. Biol. Chem.* **241**, 4731 (1966).
81. Y. T. Li and S. C. Li, *ibid.* **245**, 825 (1970).
82. A. L. Tarnoky, B. Dowding, A. L. Lakin, *Nature (Lond.)* **225**, 742 (1970).
83. G. Kellermann and H. Walter, *Humangenetik* **10**, 145 (1970); G. Kellermann and H. Walter, *ibid.*, p. 191.
84. M. H. K. Shokeir and D. C. Shreffler, *Biochem. Genet.* **4**, 517 (1970).
85. C. A. Alper and F. S. Rosen, *Immunology* **14**, 251 (1971); E. A. Azen, O. Smithies, O. Hiller, *Biochem. Genet.* **3**, 214 (1969).
86. H. Cleve, *Hum. Hered.* **20**, 438 (1970); F. D. Kitchin and A. G. Bearn, *Am. J. Hum. Genet.* **17**, 42 (1965).
87. B. S. Blumberg, *Proc. Soc. Exp. Biol. Med.* **104**, 25 (1960); S. H. Boyer and W. J. Young, *Nature (Lond.)* **187**, 1035 (1960); M. Cresta, *Riv. Antrop. Roma* **47**, 225 (1961); V. Lange and J. Schmitt, *Folia Primatol.* **1**, 208 (1963); O. Mäkelä, O. V. Rekonen, E. Salonen, *Nature (Lond.)* **185**, 852 (1960); W. C. Parker and A. G. Bearn, *Ann. Hum. Genet.* **25**, 227 (1961); J. Javid and M. H. Fuhrman, *Am. J. Hum. Genet.* **23**, 496 (1971); B. S. Shim and A. G. Bearn, *ibid.* **16**, 477 (1964); J. Planas, *Folia Primatol.* **13**, 177 (1970).
88. Although human and chimpanzee lysozymes have been reported to be electrophoretically identical [N. Hanke, E. M. Prager, A. C. Wilson, *J. Biol. Chem.* **248**, 2824 (1973)], more refined techniques indicate that their mobilities in fact differ (E. Prager, personal communication).
89. K. Berg and A. G. Bearn, *Annu. Rev. Genet.* **2**, 341 (1968).
90. M. Goodman, R. McBride, E. Poulik, E. Reklys, *Nature (Lond.)* **197**, 259 (1963); M. Goodman and A. J. Riopelle, *ibid.*, p. 261; M. Goodman, W. G. Wisecup, H. H. Reynolds, C. H. Kratochvil, *Science* **150**, 98 (1967).
91. H. Harris, *The Principles of Human Biochemical Genetics* (Elsevier, New York, 1970).
92. H. E. Schultze and J. F. Heremans, *Molecular Biology of Human Proteins* (Elsevier, New York, 1966), vol. 1.
93. M.-C. King and A. C. Wilson, in preparation.
94. F. J. Ayala, J. R. Powell, M. L. Tracey, C. A. Mourão, S. Pérez-Salas, *Genetics* **70**, 113 (1972).
95. R. C. Richmond, *ibid.*, p. 87.
96. S. Prakash, R. C. Lewontin, J. L. Hubby, *ibid.* **61**, 841 (1969).
97. R. K. Selander, S. Y. Yang, R. C. Lewontin, W. E. Johnson, *Evolution* **24**, 402 (1970).
98. D. Hedgecock and F. J. Ayala, *Copeia* (1974), p. 738.
99. R. K. Selander and W. E. Johnson, *Annu. Rev. Ecol. Syst.* **4**, 75 (1973).
100. C. R. Shaw, *Biochem. Genet.* **4**, 275 (1970).
101. R. K. Selander, D. W. Kaufman, R. J. Baker, S. L. Williams, *Evolution* **28**, 557 (1974).
102. W. Engel, W. Vogel, I. Voiculescu, H. Ropers, M. T. Zenges, K. Bender, *Comp. Biochem. Physiol.* **44B**, 1165 (1973).
103. Samples of chimpanzee blood for this study were obtained from the Laboratory for Experimental Medicine and Surgery in Primates, New York University Medical Center, P.O. Box 575, Tuxedo, N.Y. 10987; M. Goodman, Wayne State University School of Medicine, Detroit, Mich. 48201; and H. Hoffman, National Institutes of Health, Bethesda, Md. 20014. This work was supported by grant GM-18578 from NIH. Many colleagues helped us with this project. We thank S. S. Carlson, R. K. Colwell, L. R. Maxson, J. Maynard Smith, E. M. Prager, V. M. Sarich, and G. S. Sensabaugh for advice and ideas; M. Nei for unpublished data; and E. Bradley, D. Healy, P. Lozar, K. Pippen, and R. Wayner for expert technical assistance.

Jacob, F. 1977. Evolution and tinkering. *Science* 196:1161–1166.

Organisms consist exhaustively of material components—atoms and molecules—but it does not follow that they are nothing but heaps of these tiny parts. The world, and not only the world of life, is hierarchically structured. A steam engine may consist only of iron and other materials, but it is something more than iron and its other components. Similarly, an electronic computer is not only a pile of semiconductors, wires, plastic, and other materials. Organisms are made up of atoms and molecules, but they are highly complex systems—and systems of systems—of these atoms and molecules. Cellular, physiological, developmental, and other living processes are highly special and highly improbable patterns of physical and chemical processes, which come about gradually over the eons. Time adds one additional dimension of the evolutionary hierarchy, with the significant consequence that transitions from one level to another occur. As time proceeds, the descendants of a single cell may include multicellular organisms, and the descendants of a single species may include separate genera, families, and so forth.

A goal of physics is said to be to formulate a "theory of everything." Such a theory, if it is ever discovered, may account simultaneously for the phenomena of quantum mechanics, relativity, and other physical theories. But it will never be truly a theory of everything. Will the behavior of a cheetah chasing a deer be ever predicted from knowledge about the atoms and molecules making up these animals?

A different sort of difficulties arises for those who seek to explain the complexity of organisms as the intentional outcome of an "intelligent designer." Jacob, a Nobel Prize–winning biologist, points out (p. 1163) that "natural selection does not work as an engineer works. It works like a tinkerer—a tinkerer who does not know exactly what he is going to produce but uses whatever he finds around him."

RELATED READING

Carroll, S. B. 2005. From DNA to diversity: molecular genetics and the evolution of animal design, 2nd ed. Blackwell, Malden, Massachusetts, USA.

Gould, S. J. 1996. Full house: the spread of excellence from Plato to Darwin. Harmony Books, New York, New York, USA.

Mayr, E. 2001. What evolution is. Basic Books, New York, New York, USA.

Simpson, G. G. 1967. The meaning of evolution. Yale University Press, New Haven, Connecticut, USA.

10 June 1977, Volume 196, Number 4295 SCIENCE

Evolution and Tinkering

François Jacob

Some of the 16th-century books devoted to zoology and botany are illustrated by superb drawings of the various animals that populate the earth. Certain contain detailed descriptions of such creatures as dogs with fish heads, men with chicken legs, or even women without heads. The notion of monsters that blend the characteristics of different species is not itself surprising: everyone has imagined or sketched such hybrids. What is disconcerting today is that in the 16th century these creatures belonged, not to the world of fantasies, but to the real world. Many people had seen them and described them in detail. The monsters walked alongside the familiar animals of everyday life. They were within the limits of the possible.

When looking at present-day science fiction books, one is struck by the same phenomenon: the abominable animals that hunt the poor astronaut lost on a distant planet are products of recombinations between the organisms living on the earth. The creatures coming from outer space to explore the earth are depicted in the likeness of man. You can watch them emerging from their unidentified flying objects (UFO's); they are vertebrates, mammals without any doubt, walking erect. The only variations concern body size and the number of eyes. Generally these creatures have larger skulls than humans, to suggest bigger brains, and sometimes one or two radioantennae on the head, to suggest very sophisticated sense organs. The surprising point here again is what is considered possible. It is the idea, more than a hundred years after Darwin, that, if life occurs anywhere, it is bound to produce animals not too different from the terrestrial ones; and above all to evolve something like man.

The interest in these monsters is that they show how a culture handles the possible and marks its limits. It is a requirement of the human brain to put order in the universe. It seems fair to say that all cultures have more or less succeeded in providing their members with a unified and coherent view of the world and of the forces that run it. One may disagree with the explanatory systems offered by myths or magic, but one cannot deny them unity and coherence. In fact, they are often charged with too much unity and coherence because of their capacity to explain anything by the same simple argument. Actually, despite their differences, whether mythic, magic, or scientific, all explanatory systems operate on a common principle. In the words of the physicist Jean Perrin, the heart of the problem is always "to explain the complicated visible by some simple invisible" (*1*). A thunderstorm can be viewed as a consequence of Zeus' anger or of a difference of potential between the clouds and the earth. A disease can be seen as the result of a spell cast on the patient or of an infection by a virus. In all cases, however, one watches the visible effect of some hidden cause related to the whole set of invisible forces that are supposed to run the world.

The World View of Science

Whether mythic or scientific, the view of the world that man constructs is always largely a product of imagination. For the scientific process does not consist simply in observing, in collecting data, and in deducing from them a theory. One can watch an object for years and never produce any observation of scientific interest. To produce a valuable observation, one has first to have an idea of what to observe, a preconception of what is possible. Scientific advances often come from uncovering a hitherto unseen aspect of things as a result, not so much of using some new instrument, but rather of looking at objects from a different angle. This look is necessarily guided by a certain idea of what the so-called reality might be. It always involves a certain conception about the unknown, that is, about what lies beyond that which one has logical or experimental reasons to believe. In the words of Peter Medawar, scientific investigation begins by the "invention of a possible world or of a tiny fraction of that world" (*2*). So also begins mythical thought. But it stops there. Having constructed what it considers as the only possible world, it easily fits reality into its scheme. For scientific thought, instead, imagination is only a part of the game. At every step, it has to meet with experimentation and criticism. The best world is the one that exists and has proven to work already for a long time. Science attempts to confront the possible with the actual.

The price to be paid for this outlook, however, turned out to be high. It was, and is perhaps more than ever, renouncing a unified world view. This results from the very way science proceeds. Most other systems of explanation—mythic, magic, or religious—generally encompass everything. They apply to every domain. They answer any possible question. They account for the origin, the present, and the end of the universe. Science proceeds differently. It operates by detailed experimentation with nature and thus appears less ambitious, at least at first glance. It does not aim at reaching at once a complete and definitive explanation of the whole universe, its beginning, and its present form. Instead, it looks for partial and provisional answers about those phenomena that can be isolated and well defined. Actually, the beginning of modern science can be dated from the time when such general questions as, "How was the universe created?

The author is a professor of cell genetics at the Institut Pasteur, 28 Rue du Dockteur Roux, 75015, Paris, France. This article is the text of a lecture delivered at the University of California, Berkeley, in March 1977.

What is matter made of? What is the essence of life?" were replaced by such limited questions as "How does a stone fall? How does water flow in a tube? How does blood circulate in vessels?" This substitution had an amazing result. While asking general questions led to limited answers, asking limited questions turned out to provide more and more general answers.

At the same time, however, this scientific method could hardly avoid a parceling out of the world view. Each branch of science investigates a particular domain that is not necessarily connected with the neighboring ones. Scientific knowledge thus appears to consist of isolated islands. In the history of sciences, important advances often come from bridging the gaps. They result from the recognition that two hitherto separate observations can be viewed from a new angle and seen to represent nothing but different facets of one phenomenon. Thus, terrestrial and celestial mechanisms became a single science with Newton's laws. Thermodynamics and mechanics were unified through statistical mechanics, as were optics and electromagnetism through Maxwell's theory of magnetic field, or chemistry and atomic physics through quantum mechanics. Similarly different combinations of the same atoms, obeying the same laws, were shown by biochemists to compose both the inanimate and the living worlds.

The Hierarchy of Objects

Despite such generalizations, however, large gaps remain, some of which probably will not be bridged for a long time, if ever. Today, there exists a series of sciences that differ, not only by the nature of the objects that are studied, but also by the concepts and the language that are used. These sciences can be arranged in a certain order—physics, chemistry, biology, psychosociology—an order that corresponds to the hierarchy of complexity found in the objects of these sciences. Following the line from physics to sociology, one goes from the simpler to the more complex objects and also, for obvious reasons, from the older to the younger science, from the poorer to the richer empirical content, as well as from the harder to the softer system of hypotheses and experimentation. In order to obtain a unified world view through science, the question has repeatedly been raised as to the possibility of making bridges between adjacent disciplines. Because of the hierarchy of objects, the problem is always to explain the more complex in terms and concepts applying to the simpler. This is the old problem of reduction, emergence, whole and parts, and so forth. Is it possible to reduce chemistry to physics, biology to physics plus chemistry, and so forth? Clearly an understanding of the simple is necessary to understand the more complex, but whether it is sufficient is questionable.

This type of question has resulted in endless arguments. Obviously, the two critical events of evolution—first the appearance of life and later that of thought and language—led to phenomena that previously did not exist on the earth. To describe and to interpret these phenomena, new concepts, meaningless at the previous level, are required. What can the notions of sexuality, of predator, or of pain represent in physics or chemistry? Or the ideas of justice, of increase in value or of democratic power in biology? At the limit, total reductionism results in absurdity. For the pretention that every level can be completely reduced to a simpler one would result, for example, in explaining democracy in terms of the structure and properties of elementary particles; and this is clearly nonsense.

This problem can be considered in a different way. One can look at the series of objects, moving from the simpler to the more complex. Molecules are made of atoms. They therefore obey the laws that determine the behavior of atoms. But, in addition, two statements can be made about molecules. First, they can exhibit new properties, such as isomerization, racemization, and so forth. Second, the subject matter of chemistry, the molecules found in nature or produced in the laboratory, represents only a small fraction of all the possible interactions between atoms. Chemistry constitutes, therefore, a special case of physics. This is even more so with biology that deals with a complex hierarchy of objects ranging from cells to populations and ecosystems. The objects which exist at each level constitute a limitation of the total possibilities offered by the simpler level. For instance, the set of molecules found in living organisms represents a very restricted range of chemical objects. At the next level, the number of animal species amounts to several millions; however, this is small relative to the number that could exist. All vertebrates are composed of a very limited number of cellular types, at most 200, such as muscle cells, skin cells, and nerve cells. The great diversity of vertebrates results from differences in the arrangement, in the number, and in the proportion of these 200 types. Similarly, the human societies with which ethnology and sociology deal represent only a restricted group of all possible interactions between human beings.

Constraints and History

Nature functions by integration. Whatever the level, the objects analyzed by natural sciences are always organizations, or systems. Each system at a given level uses as ingredients some systems of the simpler level, but some only. The hierarchy in the complexity of objects is thus accompanied by a series of restrictions and limitations. At each level, new properties may appear which impose new constraints on the system. But these are merely additional constraints. Those that operate at any given level are still valid at all more complex levels. Every proposition that is true for physics is also true for chemistry, biology, or sociology. Similarly every proposition that is valid for biology holds true in sociology. But as a general rule, the statements of greatest importance at one level are of no interest at the more complex ones. The law of perfect gases is no less true for the objects of biology or sociology than for those of physics. It is simply irrelevant in the context of the problems with which biologists, and even more so sociologists, are concerned.

This hierarchy of successive integrations, characterized by restrictions and by the appearance of new properties at each level, has several consequences. The first is the necessity of analyzing complex objects at all levels. If molecular biology, which presents a strong reductionist attitude, yielded such a successful analysis of heredity, it was mainly because, at every step, the analysis was carried out simultaneously at the level of the molecules and at the level of the black box, the bacterial cell. This applies also to recent developments in immunology. And it seems likely that such a convergence of analysis will play an important role in the study of human beings and their societies.

The second point concerns predictability. Is it possible to make predictions at one level on the basis of what is known at a simpler one? Only to a very limited extent. The properties of a system can be explained by the properties of its components. They cannot be deduced from them. Starting from fundamental laws of physics, there is no way of reconstructing the universe. This means that a particular system, say a cell, has only a certain probability of appearing. All predictions about its existence can only be statistical. Molecular biology has shown that ultimately the characteristics of a cell rest on the structure of its molecular components. But the appearance of life on the earth was not the necessary consequence of the presence of certain molecular structures in prebiotic times. In fact, there is absolutely no way of estimating what was the probability for life appearing on earth. It may very well have appeared only once.

The third point concerns the nature of the restrictions and limitations found at every step of increasing complexity. Can one explain why, among all the possible interactions at one level, only certain are actually observed at the more complex one? How is it that only some types of molecular structures are present, for instance, in living organisms? Or only some interactions in human societies? There is no general answer to such questions, and it seems doubtful that there will ever be a specific answer for any one particular level of complexity. Complex objects are produced by evolutionary processes in which two factors are paramount: the constraints that at every level control the systems involved, and the historical circumstances that control the actual interactions between the systems. The combination of constraints and history exists at every level, although in different proportions. Simpler objects are more dependent on constraints than on history. As complexity increases, history plays a greater part. But history has always to be introduced into the picture, even in physics. According to present theories, heavier nuclei are composed of lighter ones and ultimately of hydrogen nuclei and neutrons. The transformation of heavy hydrogen into helium occurs during the fusion process, which is the main source of energy in the sun as well as in hydrogen bombs. Helium and all the heavier elements are thus the result of a cosmological evolution. According to present views, the heavier elements are considered as products of supernovae explosions. They seem to be very rare and not to exceed 1 or 2 percent by mass of all matter, while helium represents one-fifth and hydrogen four-fifths of all matter. The earth and the other planets of the solar system have thus been made of very rare material under conditions that seem to be rarely encountered in the cosmos. The source of hydrogen itself is left to theories and speculations concerning the origin of the universe.

Natural Selection

The constraints to which systems are subjected vary with the level of complexity. There are always some constraints imposed by stability and thermodynamics. But as complexity increases, additional constraints appear—such as reproduction for living systems, or economic requirements for social systems. Consequently, there cannot be any general law of evolution, any recipe that accounts for increasing complexity at all levels. Since Darwin, biologists have progressively elaborated a reasonable, although still incomplete, picture of the mechanism that operates in the evolution of the living world, namely, natural selection. For many, it has been tempting to invoke a similar mechanism of selection to describe any possible evolution, whether cosmological, chemical, cultural, ideological, or social. But this seems condemned to fail from the outset. The rules of the game differ at each level. New principles have, therefore, to be worked out at each level.

Natural selection is the result of two constraints imposed on every living organism: (i) the requirement for reproduction, which is fulfilled through genetic mechanisms carefully adjusted by special devices such as mutation, recombination, and sex to produce organisms similar, but not identical, to their parents; and (ii) the requirement for a permanent interaction with the environment because living beings are what thermodynamicists call open systems and persist only by a constant flux of matter, energy, and information. The first of these factors generates random variations and produces populations in which all individuals are different. The interplay of the two factors results in differential reproduction and consequently in populations that evolve progressively as a function of environmental circumstances, of behavior, and of new ecological niches. But natural selection does not act merely as a sieve eliminating detrimental mutations and favoring reproductions of beneficial ones as is often suggested. In the long run, it integrates mutations, and it orders them into adaptatively coherent patterns adjusted over millions of years, and over millions of generations as a response to environmental challenges. It is natural selection that gives direction to changes, orients chance, and slowly, progressively produces more complex structures, new organs, and new species. Novelties come from previously unseen association of old material. To create is to recombine.

Engineer and Tinkerer

The action of natural selection has often been compared to that of an engineer. This, however, does not seem to be a suitable comparison. First, because in contrast to what occurs in evolution, the engineer works according to a preconceived plan in that he foresees the product of his efforts. Second, because of the way the engineer works: to make a new product, he has at his disposal both material specially prepared to that end and machines designed solely for that task. Finally, because the objects produced by the engineer, at least by the good engineer, approach the level of perfection made possible by the technology of the time. In contrast, evolution is far from perfection. This is a point which was repeatedly stressed by Darwin who had to fight against the argument of perfect creation. In the *Origin of Species*, Darwin emphasizes over and over again the structural or functional imperfections of the living world. For instance, when he discusses natural selection (*3*, p. 472):

> Nor ought we to marvel if all the contrivances in nature be not, as far as we can judge, absolutely perfect. We need not marvel at the sting of the bee causing the bee's own death; at drones being produced in such vast numbers for one single act, and being then slaughtered by their sterile sisters; at the astonishing waste of pollen by our firtrees; at the instinctive hatred of the queen bee for her own fertile daughters; at ichneumonidae feeding within the live bodies of caterpillars; and at other such cases. The wonder indeed is, on the theory of natural selection, that more cases of the want of absolute perfection have not been observed.

There are innumerable statements of this type in the *Origin of Species*. In fact, one of the best arguments against perfection comes from extinct species. While the number of living species in the animal kingdom can be estimated to be around a few million, the number of extinct ones since life existed on earth has been estimated by Simpson (*4*) at around five hundred million.

Natural selection has no analogy with any aspect of human behavior. However, if one wanted to play with a comparison, one would have to say that natural selection does not work as an engineer works. It works like a tinkerer—a tinkerer who does not know exactly what he is going to produce but uses whatever he finds around him whether it be pieces of string, fragments of wood, or old cardboards; in short it works like a tinkerer who uses everything at his disposal to produce some kind of workable object. For the engineer, the realization of his task depends on his having the raw materials and the tools that exactly fit his project. The tinkerer, in contrast, always manages with odds and ends. What he ultimately produces is generally related to no special project, and it results from a series of contingent events, of all the opportunities he had to enrich his stock with leftovers. As was discussed by Levi-Strauss (*5*), none of the materials at the tinkerer's disposal has a precise and definite function. Each can be used in a number of different ways. In contrast with the engineer's tools, those of the tinkerer cannot be defined by a project. What these objects have in common is "it might well be of some use." For what? That depends on the opportunities.

Evolution as Tinkering

This mode of operation has several aspects in common with the process of evolution. Often, without any well-defined long-term project, the tinkerer gives his materials unexpected functions to produce a new object. From an old bicycle wheel, he makes a roulette; from a broken chair the cabinet of a radio. Similarly evolution makes a wing from a leg or a part of an ear from a piece of jaw. Naturally, this takes a long time. Evolution behaves like a tinkerer who, during eons upon eons, would slowly modify his work, unceasingly retouching it, cutting here, lengthening there, seizing the opportunities to adapt it progressively to its new use. For instance, the lung of terrestrial vertebrates was, according to Mayr (*6*), formed in the following way. Its development started in certain freshwater fishes living in stagnant pools with insufficient oxygen. They adopted the habit of swallowing air and absorbing oxygen through the walls of the esophagus. Under these conditions, enlargement of the surface area of the esophagus provided a selective advantage. Diverticula of the esophagus appeared and, under continuous selective pressure, enlarged into lungs. Further evolution of the lung was merely an elaboration of this theme—enlarging the surface for oxygen uptake and vascularization. To make a lung with a piece of esophagus sounds very much like tinkering.

Unlike engineers, tinkerers who tackle the same problem are likely to end up with different solutions. This also applies to evolution, as exemplified by the variety of eyes found in the living world [see (*7*)]. It is obviously a great advantage under many conditions to possess light receptors, and the variety of photoreceptors in the living world is amazing. The most sophisticated are the image-forming eyes that provide information, not only on the intensity of incoming light, but also on the objects light comes from, on their shape, color, position, motion, speed, distance, and the like. Such sophisticated structures are necessarily complex. They can develop only in organisms already complex themselves. One might suppose, therefore, that there is just one way of producing such a structure. This is not the case. Eyes appeared a great many times in the course of evolution, based on at least three principles—pinhole, lens, and multiple tubes. Lens eyes, like ours, appeared both in mollusks and vertebrates. Nothing looks so much like our eye as the octopus eye. Both work in almost exactly the same way. Yet they did not evolve in the same way. Whereas in vertebrates photoreceptor cells of the retina point away from light, in mollusks they point toward light. Among all solutions found to the problem of photoreceptors, these two are similar but not identical. In each case, natural selection did what it could with the materials at its disposal.

Evolution does not produce novelties from scratch. It works on what already exists, either transforming a system to give it new functions or combining several systems to produce a more elaborate one. This happened, for instance, during one of the main events of cellular evolution: namely, the passage from unicellular to multicellular forms. This was a particularly important transition because it carried an enormous potential for a specialization of the parts. Such a transition, which probably occurred several times, did not require the creation of new chemical species, for there are no major differences between molecular types of uni- and multicellular organisms. It was mainly a reorganization of what already existed.

Molecular Tinkering

It is at the molecular level that the tinkering aspect of natural selection is perhaps most apparent. What characterizes the living world is both its diversity and its underlying unity. The living world contains bacteria and whales, viruses and elephants, organisms living at −20°C in polar areas and others at 70°C in hot springs. All these objects, however, exhibit a remarkable unity of chemical structures and functions. Similar polymers, nucleic acids or proteins, always made of the same basic elements, the four bases and the 20 amino acids, play similar roles. The genetic code is the same and the translating machineries are very nearly so. The same coenzymes mediate similar reactions. Many metabolic steps remain essentially the same, from bacteria to man. Obviously, for life to emerge, a number of new molecular types had first to be formed. During chemical evolution in prebiotic times and at the beginning of biological evolution, all those molecules of which every living being is built had to appear. But once life had started in the form of some primitive self-reproducing organism, further evolution had to proceed mainly through alterations of already existing compounds. New functions developed as new proteins appeared. But these were merely variations on previous themes. A sequence of a thousand nucleotides codes for a medium-sized protein. The probability that a functional protein would appear de novo by random association of amino acids is practically zero. In organisms as complex and integrated as those that were already living a long time ago, creation of entirely new nucleotide sequences could not be of any importance in the production of new information.

The appearance of new molecular structures during much of biological evolution must, therefore, have rested on alteration of preexisting ones. This is exemplified by the finding that large segments of genetic information, that is, of DNA, turn out to be homologous, not only in the same organism, but also among different organisms, even among those that are phylogenetically distant. Similarly, as more is known about amino acid sequences in proteins, it appears not only that proteins fulfilling similar functions in different organisms have frequently similar sequences, but also that proteins with different functions often exhibit rather large segments in common. The hypothesis most generally envisaged to account for these similarities was proposed by Horowitz (*8*), by Ingram (*9*), and by Ohno (*10*). A segment of DNA, corresponding to one or several genes, is assumed to be duplicated by some genetic mechanism. When a gene exists in more than one copy in a cell or a gamete, it is released from the contraints imposed on functions by natural selection. Mutations can then accumulate more or less freely and result in modified protein structures, some of which can eventually fulfill new functions. Since natural selection exerts a continual pressure on organisms, an alteration in a protein can be further improved by other, later changes. It can also lead to a perturbation in the interactions with other proteins and eventually favor modifications of these proteins. A large fraction of the genome of complex organisms might actually derive from a few ancestral genes.

Biochemical changes do not seem, therefore, to be a main driving force in the diversification of living organisms. The really creative part in biochemistry must have occurred very early. For the biochemical unity that underlies the living world makes sense only if most of the important molecular types found in organisms, that is, most of the metabolic pathways involved in the production of energy and in biosynthesis or degradation of the essential building blocks already existed in very primitive organisms such as bacteria. Once this stage passed, biochemical evolution continued as more complex organisms appeared. But it is not biochemical novelties that generated diversification of organisms. In all likelihood, it worked the other way around. It is the selective pressure resulting from changes in behavior or in ecological niches that led to biochemical adjustments and changes in molecular types. What distinguishes a butterfly from a lion, a hen from a fly, or a worm from a whale is much less a difference in chemical constituents than in the organization and the distribution of these constituents. The few big steps of evolution required acquisition of new information. But specialization and diversification occurred by using differently the same structural information. Among neighboring groups, vertebrates for instance, chemistry is the same. What makes one vertebrate different from another is a change in the time of expression and in the relative amounts of gene products rather than the small differences observed in the structure of these products. It is a matter of regulation rather than of structure [see (*11*)].

After egg fertilization, embryonic development occurs in a fixed order and according to a precise schedule set by the genetic program contained in the chromosomes. This program determines when and where lines of differentiated cells will emerge, when and where different proteins will be made and in what amounts. Both the quality and quantity of the different proteins vary in time and space during development. Thus in the adult, the various types of cells or tissues contain different repertoires of molecular types in agreement with their functions. The genetic program is executed through complex regulatory circuits that switch the different biochemical activities of the organism on or off. Very little is known as yet about the regulatory circuits that operate in the development of complex organisms. It is known, however, that, among related organisms such as mammals, the first steps of embryonic development are remarkably similar, with divergences showing up only progressively as development proceeds. These divergences concern much less the actual structure of cellular or molecular types than their number and position. It seems likely that divergence and specialization of mammals, for instance, resulted from mutations altering regulatory circuits rather than chemical structures. Small changes modifying the distribution in time and space of the same structures are sufficient to affect deeply the form, the functioning, and the behavior of the final product—the adult animal. It is always a matter of using the same elements, of adjusting them, of altering here or there, of arranging various combinations to produce new objects of increasing complexity. It is always a matter of tinkering.

Consequences of Tinkering

Marks of this tinkering are thus found at every level thoughout the living world. Of course, they can be found in human beings as shown by the following few examples. In humans, as in many mammals, there exist very complex processes responsible for such functions as blood coagulation, inflammatory reactions against foreign bodies, and the immunological defenses mediated by the so-called complement system. These three processes have been independently analyzed in some detail during recent years. Each one exhibits an unexpected complexity. Each involves about ten proteins, none of which initially has enzymatic activity. Conversion of the first protein into a catalytically active form triggers a cascade of reactions. The first protein cleaves the second one at a specific point; a product of this reaction cleaves the third protein, and so on. In this series of reactions, the individual proteins are thus split in sequence and the released fragments serve as activators, or inhibitors, in other reactions of the chain. Furthermore, these three chains of reactions are not wholly independent. A product of cleavage in one chain can suddenly become an active element in another chain or even play a role in a completely different process. These products may serve as signals to connect chemically unrelated, but physiologically dependent, systems. It is as though some protein molecules, which happened to be formed, were used here or there as a source of smaller but active peptides as new functions were taking shape. Recently, a number of peptides of different sizes have been found to participate in a variety of physiological processes. Some of them, such as hormone peptides or brain peptides, are known not to be chemically transformed in the reaction they activate or inhibit. They appear just to bind to some protein to favor an allosteric transition, thus acting as simple chemical signals. For the biologist, it is thus generally impossible to make a prediction, or even an inspired guess, about the nature of such molecules and their structural relations with other constituents. All he can do is to detect them, purify them, and analyze them. Later, as the structures of more proteins become known, there will perhaps be a chance to define the functional interrelations and evolutionary relationship among such molecules.

Another example of tinkering can be found in early human embryonic development. Embryonic development is a tremendously complicated process of which little is known at present. Studies of the past 10 or 20 years have revealed an amazing phenomenon. In various human populations, 50 percent of all conceptions are estimated to result in spontaneous abortion [see (*12*)]. A large fraction of these abortions occur during the first 3 weeks of pregnancy and generally pass unnoticed. Thus, in half of the total conceptions, something is wrong to begin with. Many of these spontaneous abortions appear to be due to an odd number of chromosomes; instead of having one set of chromosomes derived from its mother and one from its father, the embryo lacks a chromosome, or has an extra one, or even has three sets instead of two. As a result, some functions necessary to embryonic development are not performed correctly. The fetus dies and is expelled. Thus many potentially malformed fetuses disappear; not all, unfortunately, since some of them still come to term. This reveals the imperfections of a mechanism that is at the very core of any living system and that has been refined over millions of years.

A third example of tinkering which is very intriguing when one thinks about it is the association between reproduction and what is generally called pleasure. Sex is one of the most efficient inventions of evolution. In lower organisms which apparently reproduce asexually by fission, the genetic program is scrupulously recopied at every generation. Within a population, it always remains the same, except for rare mutations. Division of the organism is an automatic process resulting from growth. When something resembling sexuality exists, as in bacteria, it is a luxury. In such populations, adaptation necessarily involves the selection of rare mutants under environmental conditions. In contrast, sexual reproduction, which probably occurred early in evolution, compels reassortment of genetic programs in interbreeding populations. As a result, every genetic program (that is, every individual) is different from the others. This permanent reshuffling of genetic elements provides tremendous potentialities of adaptation. But once sexuality had become a necessary condition for reproduction, it required special mechanisms: one, allowing individuals of opposite sexes to recognize and meet each other and a second, driving them to unite. The first of these requirements has been fulfilled by a variety of specific signaling systems—visual, auditory, or olfactory—of amazing precision and efficiency. The second has been met through the development of genetically determined and very rigid programs of behavior. For instance, in birds, at the proper season, the view of an individual of the opposite sex initiates a whole process of rituals, courtship, and parade leading almost automatically to copulation, nidation, and progeny care. The course of evolution, however, is characterized by a trend to greater flexibility in the execution of the genetic program. As this program became more open, so to speak, the behavior became less rigidly determined by the genes. Reactions to sexual signals were no longer completely automatic. In order to drive the individuals toward reproduction, sexuality had therefore to be associated with some other devices. Among these was pleasure. In the Oxford dictionary, pleasure is defined as "the opposite of pain," obviously, but also as "the condition of consciousness induced by the enjoyment of what is felt or viewed as good or desirable." It seems likely that feelings of discomfort and pleasure must already have existed for a long time in complex animals. An animal is more likely to have progeny if a feeling of discomfort dissuades it from entering harmful situations. It is clear that the existence of nervous centers, connected with sense organs and able to correlate what is felt as pleasant or unpleasant with what is actually good or bad for survival, is of great selective value. In fact, such centers are now known to exist. Some 20 years ago, neurobiologists detected in the brain, first in the rat and later in many vertebrates, the presence of two remarkable centers—one called the center of aversion and the other called the center of autostimulation. Fitted with correctly implanted electrodes and given the means of activating at will the latter center, a rat gives himself pleasure until it collapses from sheer exhaustion. Experiments performed during brain surgery and descriptions of feelings by the patients leave very little doubt as to the existence of such centers in man and to its association with sexual activity. Thus pleasure appears as a mere expedient to push individuals to indulge in sex and therefore to reproduce. A rather successful expedient indeed, as judged by the state of the world population.

A Final Example of Tinkering: The Human Brain

Although our brain represents the main adaptive feature of our species, what it is adapted to is not clear at all. What is clear, however, is that, like the rest of our body, our brain is a product of natural selection, that is, of differential reproductions accumulated over millions of years under the pressure of various environmental conditions. Our brain has therefore evolved at our gonad's service, as already emphasized by Freud many years ago. But curiously enough, brain development in mammals was not as integrated a process as, for instance, the transformation of a leg into a wing. The human brain was formed by superposition of new structures on old ones. To the old rhinencephalon of lower mammals a neocortex was added that rapidly, perhaps too rapidly, took a most important role in the evolutionary sequence leading to man. For some neurobiologists, especially McLean (*13*), these two types of structures correspond to two types of functions but have not been completely coordinated or hierarchized. The recent one, the neocortex, controls intellectual, cognitive activity. The old one, derived from the rhinencephalon, controls emotional and visceral activities. In contrast to the former, the latter does not seem to possess any power of specific discrimination, or any capacity for symbolization, language, or self-consciousness. The old structure which, in lower mammals, was in total command has been relegated to the department of emotions. In man, it constitutes what McLean calls "the visceral brain." Perhaps because development is so prolonged and maturity so delayed in man, these centers maintain strong connections with lower autonomic centers and continue to coordinate such fundamental drives as obtaining food, hunting for a sexual partner, or reacting to an enemy. This evolutionary procedure—the formation of a dominating neocortex coupled with the persistence of a nervous and hormonal system partially, but not totally under the rule of the neocortex—strongly resembles the tinkerer's procedure. It is somewhat like adding a jet engine to an old horse cart. It is not surprising, in either case, that accidents, difficulties, and conflicts can occur.

It is hard to realize that the living world as we know it is just one among many possibilities; that its actual structure results from the history of the earth. Yet living organisms are historical structures: literally creations of history. They represent, not a perfect product of engineering, but a patchwork of odd sets pieced together when and where opportunities arose. For the opportunism of natural selection is not simply a matter of indifference to the structure and operation of its products. It reflects the very nature of a historical process full of contingency.

As Simpson (*4*) pointed out, the interplay of local opportunities—physical, ecological, and constitutional—produces a net historical opportunity which in turn determines how genetic opportunities will be exploited. It is this net historical opportunity that mainly controls the direction and pace of adaptive evolution. This is why the probability is practically zero that living systems, which might well exist elsewhere in the cosmos, would have evolved into something looking like human beings. Even if life in outer space uses the same material as on the earth, even if the environment is not too different from ours, even if the nature of life and of its chemistry strongly limits the way to fulfill certain functions, the sequence of historical opportunities there could not be the same as here. A different play had to be performed by different actors. Despite science fiction, Martians cannot look like us. And we might as well have looked like one of those 16th-century monsters.

References and Notes

1. J. Perrin, *Les Atomes* (Alcan, Paris, 1914).
2. P. B. Medawar, *The Hope of Progress* (Doubleday, New York, 1973).
3. C. Darwin, *On the Origin of Species* (London, 1859).
4. G. G. Simpson, *Evolution* **6**, 342 (1952).
5. C. Levi-Strauss, *La Pensée Sauvage* (Plon, Paris, 1962).
6. E. Mayr. *Fed. Proc. Fed. Am. Soc. Exp. Biol.* **23**, 1231 (1964).
7. G. G. Simpson, *The Meaning of Evolution* (Yale Univ. Press, New Haven, Conn., 1967).
8. N. Horowitz, *Adv. Genet.* **3**, 33 (1950).
9. V. M. Ingram, *Hemoglobins in Genetics and Evolution* (Columbia Univ. Press, New York, 1963).
10. S. Ohno, *Evolution by Gene Duplication* (Springer-Verlag, New York, 1970).
11. M. C. King and A. C. Wilson, *Science* **188**, 107 (1975).
12. A. Boue and J. G. Boue, in *Physiology and Genetics of Reproduction*, E. Coutinho and F. Fuchs, Eds. (Plenum, New York, 1975), vol. 4b, p. 317.
13. P. McLean, *Psychosom. Med.* **11**, 338 (1949).

Woese, C. M., and G. E. Fox. 1977. Phylogenetic structure of the prokaryotic domain: the primary kingdoms. *Proceedings of the National Academy of Sciences, USA* 74:5088–5090.

From Greek antiquity until recent times, a common notion was that life on Earth could be divided into two primary kingdoms: animals and plants. Following invention of the microscope and the discovery of microbes, a different view gained traction, namely, that life could be subdivided into prokaryotes (microorganisms lacking a membrane-bound nucleus) and eukaryotes (organisms composed of cells with membrane-bound nuclei).

Another scientific breakthrough followed the introduction of molecular techniques during the 1960s and ensuing decades, as evolutionary biologists used the vast genealogical information in DNA sequences and proteins to reconstruct phylogenetic trees for many kinds of organisms. Of special interest were conservative genes, such as those encoding subunits of ribosomal RNA, because these genes are ubiquitous across nearly all forms of life and because they evolve so slowly at the nucleotide sequence level that scientists can use them to decipher even the most ancient branches in the Tree of Life.

This paper by Carl Woese and George Fox was arguably the single most impactful of deep phylogenetic analyses. Using sequences from the 16S rDNA subunit isolated from organisms ranging from unicellular microbes to the cytoplasmic compartments of multicellular eukaryotes, the authors demonstrated that life on Earth appears to be organized into three major phylogenetic units or domains, which are now named eubacteria (or Bacteria), archaebacteria (or Archaea), and eukaryotes (or Eucarya). There could no longer be any doubt that major phylogenetic lineages, previously unappreciated, exist among prokaryotic life-forms. Indeed, the apparent lengths of many branches in the microbial Tree of Life eventually proved to be much longer than those that separate (for example) humans and other vertebrates from worms or even from fungi and plants.

Conclusions from the pioneering work of Woese and his colleagues largely have withstood the test of time. Today, however, most textbooks and reviews summarize the results in the form of evolutionary trees, as done, for example, in Woese (2000), rather than in the tabular format employed in this early classic.

SOURCES AND FURTHER READING

Hall, B. G. 2001. Phylogenetic trees made easy. Sinauer, Sunderland, Massachusetts.

Hillis, D. M., C. Moritz, and B. K. Mable. 1996. Molecular systematics, 2nd ed. Sinauer, Sunderland, Massachusetts.

Page, R. D. M., and E. C. Holmes. 1998. Molecular evolution: a phylogenetic approach. Blackwell Science, Oxford, England.

Tudge, C. 2000. The variety of life: a survey and a celebration of all the creatures that have ever lived. Oxford University Press, Oxford, England.

Woese, C. R. 2000. Interpreting the universal phylogenetic tree. Proceedings of the National Academy of Sciences USA 97:8392–8396.

Proc. Natl. Acad. Sci. USA
Vol. 74, No. 11, pp. 5088–5090, November 1977
Evolution

Phylogenetic structure of the prokaryotic domain: The primary kingdoms

(archaebacteria/eubacteria/urkaryote/16S ribosomal RNA/molecular phylogeny)

CARL R. WOESE AND GEORGE E. FOX*

Department of Genetics and Development, University of Illinois, Urbana, Illinois 61801

Communicated by T. M. Sonneborn, August 18, 1977

ABSTRACT A phylogenetic analysis based upon ribosomal RNA sequence characterization reveals that living systems represent one of three aboriginal lines of descent: (*i*) the eubacteria, comprising all typical bacteria; (*ii*) the archaebacteria, containing methanogenic bacteria; and (*iii*) the urkaryotes, now represented in the cytoplasmic component of eukaryotic cells.

The biologist has customarily structured his world in terms of certain basic dichotomies. Classically, what was not plant was animal. The discovery that bacteria, which initially had been considered plants, resembled both plants and animals less than plants and animals resembled one another led to a reformulation of the issue in terms of a yet more basic dichotomy, that of eukaryote versus prokaryote. The striking differences between eukaryotic and prokaryotic cells have now been documented in endless molecular detail. As a result, it is generally taken for granted that all extant life must be of these two basic types.

Thus, it appears that the biologist has solved the problem of the primary phylogenetic groupings. However, this is not the case. Dividing the living world into *Prokaryotae* and *Eukaryotae* has served, if anything, to obscure the problem of what extant groupings represent the various primeval branches from the common line of descent. The reason is that eukaryote/prokaryote is not primarily a phylogenetic distinction, although it is generally treated so. The eukaryotic cell is organized in a different and more complex way than is the prokaryote; this probably reflects the former's composite origin as a symbiotic collection of various simpler organisms (1–5). However striking, these organizational dissimilarities do not guarantee that eukaryote and prokaryote represent phylogenetic extremes.

The eukaryotic cell *per se* cannot be directly compared to the prokaryote. The composite nature of the eukaryotic cell makes it necessary that it first be conceptually reduced to its phylogenetically separate components, which arose from ancestors that were noncomposite and so individually are comparable to prokaryotes. In other words, the question of the primary phylogenetic groupings must be formulated solely in terms of relationships among "prokaryotes"—i.e., noncomposite entities. (Note that in this context there is no suggestion *a priori* that the living world is structured in a dichotomous way.)

The organizational differences between prokaryote and eukaryote and the composite nature of the latter indicate an important property of the evolutionary process: Evolution seems to progress in a "quantized" fashion. One level or domain of organization gives rise ultimately to a higher (more complex) one. What "prokaryote" and "eukaryote" actually represent are two such domains. Thus, although it is useful to define phylogenetic patterns within each domain, it is not meaningful to construct phylogenetic classifications between domains: Prokaryotic kingdoms are not comparable to eukaryotic ones. This should be recognized by an appropriate terminology. The highest phylogenetic unit in the prokaryotic domain we think should be called an "urkingdom"—or perhaps "primary kingdom." This would recognize the qualitative distinction between prokaryotic and eukaryotic kingdoms and emphasize that the former have primary evolutionary status.

The passage from one domain to a higher one then becomes a central problem. Initially one would like to know whether this is a frequent or a rare (unique) evolutionary event. It is traditionally assumed—without evidence—that the eukaryotic domain has arisen but once; all extant eukaryotes stem from a common ancestor, itself eukaryotic (2). A similar prejudice holds for the prokaryotic domain (2). [We elsewhere argue (6) that a hypothetical domain of lower complexity, that of "progenotes," may have preceded and given rise to the prokaryotes.] The present communication is a discussion of recent findings that relate to the urkingdom structure of the prokaryotic domain and the question of its unique as opposed to multiple origin.

Phylogenetic relationships cannot be reliably established in terms of noncomparable properties (7). A comparative approach that can measure degree of difference in comparable structures is required. An organism's genome seems to be the ultimate record of its evolutionary history (8). Thus, comparative analysis of molecular sequences has become a powerful approach to determining evolutionary relationships (9, 10).

To determine relationships covering the entire spectrum of extant living systems, one optimally needs a molecule of appropriately broad distribution. None of the readily characterized proteins fits this requirement. However, ribosomal RNA does. It is a component of all self-replicating systems; it is readily isolated; and its sequence changes but slowly with time—permitting the detection of relatedness among very distant species (11–13). To date, the primary structure of the 16S (18S) ribosomal RNA has been characterized in a moderately large and varied collection of organisms and organelles, and the general phylogenetic structure of the prokaryotic domain is beginning to emerge.

A comparative analysis of these data, summarized in Table 1, shows that the organisms clearly cluster into several primary kingdoms. The first of these contains all of the typical bacteria so far characterized, including the genera *Acetobacterium*, *Acinetobacter*, *Acholeplasma*, *Aeromonas*, *Alcaligenes*, *Anacystis*, *Aphanocapsa*, *Bacillus*, *Bdellovibrio*, *Chlorobium*, *Chromatium*, *Clostridium*, *Corynebacterium*, *Escherichia*, *Eubacterium*, *Lactobacillus*, *Leptospira*, *Micrococcus*, *Mycoplasma*, *Paracoccus*, *Photobacterium*, *Propionibacterium*,

The costs of publication of this article were defrayed in part by the payment of page charges. This article must therefore be hereby marked "*advertisement*" in accordance with 18 U. S. C. §1734 solely to indicate this fact.

* Present address: Department of Biophysical Sciences, University of Houston, Houston, TX 77004.

Table 1. Association coefficients (S_{AB}) between representative members of the three primary kingdoms

	1	2	3	4	5	6	7	8	9	10	11	12	13
1. *Saccharomyces cerevisiae*, 18S	—	0.29	0.33	0.05	0.06	0.08	0.09	0.11	0.08	0.11	0.11	0.08	0.08
2. *Lemna minor*, 18S	0.29	—	0.36	0.10	0.05	0.06	0.10	0.09	0.11	0.10	0.10	0.13	0.07
3. L cell, 18S	0.33	0.36	—	0.06	0.06	0.07	0.07	0.09	0.06	0.10	0.10	0.09	0.07
4. *Escherichia coli*	0.05	0.10	0.06	—	0.24	0.25	0.28	0.26	0.21	0.11	0.12	0.07	0.12
5. *Chlorobium vibrioforme*	0.06	0.05	0.06	0.24	—	0.22	0.22	0.20	0.19	0.06	0.07	0.06	0.09
6. *Bacillus firmus*	0.08	0.06	0.07	0.25	0.22	—	0.34	0.26	0.20	0.11	0.13	0.06	0.12
7. *Corynebacterium diphtheriae*	0.09	0.10	0.07	0.28	0.22	0.34	—	0.23	0.21	0.12	0.12	0.09	0.10
8. *Aphanocapsa* 6714	0.11	0.09	0.09	0.26	0.20	0.26	0.23	—	0.31	0.11	0.11	0.10	0.10
9. Chloroplast (*Lemna*)	0.08	0.11	0.06	0.21	0.19	0.20	0.21	0.31	—	0.14	0.12	0.10	0.12
10. *Methanobacterium thermoautotrophicum*	0.11	0.10	0.10	0.11	0.06	0.11	0.12	0.11	0.14	—	0.51	0.25	0.30
11. *M. ruminantium* strain M-1	0.11	0.10	0.10	0.12	0.07	0.13	0.12	0.11	0.12	0.51	—	0.25	0.24
12. *Methanobacterium* sp., Cariaco isolate JR-1	0.08	0.13	0.09	0.07	0.06	0.06	0.09	0.10	0.10	0.25	0.25	—	0.32
13. *Methanosarcina barkeri*	0.08	0.07	0.07	0.12	0.09	0.12	0.10	0.10	0.12	0.30	0.24	0.32	—

The 16S (18S) ribosomal RNA from the organisms (organelles) listed were digested with T1 RNase and the resulting digests were subjected to two-dimensional electrophoretic separation to produce an oligonucleotide fingerprint. The individual oligonucleotides on each fingerprint were then sequenced by established procedures (13, 14) to produce an oligonucleotide catalog characteristic of the given organism (3, 4, 13–17, 22, 23; unpublished data). Comparisons of all possible pairs of such catalogs defines a set of association coefficients (S_{AB}) given by: $S_{AB} = 2N_{AB}/(N_A + N_B)$, in which N_A, N_B, and N_{AB} are the total numbers of nucleotides in sequences of hexamers or larger in the catalog for organism A, in that for organism B, and in the interreaction of the two catalogs, respectively (13, 23).

Pseudomonas, Rhodopseudomonas, Rhodospirillum, Spirochaeta, Spiroplasma, Streptococcus, and *Vibrio* (refs. 13–17; unpublished data). The group has three major subdivisions, the blue-green bacteria and chloroplasts, the "Gram-positive" bacteria, and a broad "Gram-negative" subdivision (refs. 3, 4, 13–17; unpublished data). It is appropriate to call this urkingdom the *eubacteria*.

A second group is defined by the 18S rRNAs of the eukaryotic cytoplasm—animal, plant, fungal, and slime mold (unpublished data). It is uncertain what ancestral organism in the symbiosis that produced the eukaryotic cell this RNA represents. If there had been an "engulfing species" (1) in relation to which all the other organisms were endosymbionts, then it seems likely that 18S rRNA represents that species. This hypothetical group of organisms, in one sense the major ancestors of eukaryotic cells, might appropriately be called *urkaryotes*. Detailed study of anaerobic amoebae and the like (18), which seem not to contain mitochondria and in general are cytologically simpler than customary examples of eukaryotes, might help to resolve this question.

Eubacteria and urkaryotes correspond approximately to the conventional categories "prokaryote" and "eukaryote" when they are used in a phylogenetic sense. However, they do not constitute a dichotomy; they do not collectively exhaust the class of living systems. There exists a third kingdom which, to date, is represented solely by the methanogenic bacteria, a relatively unknown class of anaerobes that possess a unique metabolism based on the reduction of carbon dioxide to methane (19–21). *These "bacteria" appear to be no more related to typical bacteria than they are to eukaryotic cytoplasms.* Although the two divisions of this kingdom appear as remote from one another as blue-green algae are from other eubacteria, they nevertheless correspond to the *same* biochemical phenotype. The apparent antiquity of the methanogenic phenotype plus the fact that it seems well suited to the type of environment presumed to exist on earth 3–4 billion years ago lead us tentatively to name this urkingdom the *archaebacteria*. Whether or not other biochemically distinct phenotypes exist in this kingdom is clearly an important question upon which may turn our concept of the nature and ancestry of the first prokaryotes.

Table 1 shows the three urkingdoms to be equidistant from one another. Because the distances measured are actually proportional to numbers of mutations and not necessarily to time, it cannot be proven that the three lines of descent branched from the common ancestral line at about the same time. One of the three may represent a far earlier bifurcation than the other two, making there in effect only two urkingdoms. Of the three possible unequal branching patterns the case for which the initial bifurcation defines urkaryotes vs. all bacteria requires further comment because, as we have seen, there is a predilection to accept such a dichotomy.

The phenotype of the methanogens, although ostensibly "bacterial," on close scrutiny gives no indication of a specific phylogenetic resemblance to the eubacteria. For example, methanogens do have cell walls, but these do not contain peptidoglycan (24). The biochemistry of methane formation appears to involve totally unique coenzymes (23, 25, 26). The methanogen rRNAs are comparable in size to their eubacterial counterparts, but resemble the latter specifically in neither sequence (Table 1) nor in their pattern of base modification (23). The tRNAs from eubacteria and eukaryotes are characterized by a common modified sequence, TΨCG; methanogens modify this tRNA sequence in a quite different and unique way (23). It must be recognized that very little is known of the general biochemistry of the methanogens—and almost nothing is known regarding their molecular biology. Hence, although the above points are few in number, they represent most of what is now known. There is no reason at present to consider methanogens as any closer to eubacteria than to the "cytoplasmic component" of the eukaryote. Both in terms of rRNA sequence measurement and in terms of general phenotypic differences, then, the three groupings appear to be distinct urkingdoms.

If a third urkingdom exists, does this suggest that many more such will be found among yet to be characterized organisms? We think not, although the matter clearly requires an exhaustive search. As seen above, the number of species that can be classified as eubacteria is moderately large. To this list can be added *Spirillum* and *Desulfovibrio*, whose rRNAs appear typically eubacterial by nucleic acid hybridization measurements (27). Because the list is also phenotypically diverse, it seems unlikely that many, if any, of the yet uncharacterized prokaryotic groups will be shown to have coequal status with the present three. Conceivably the halophiles whose cell walls contain no peptidoglycan, are candidates for this distinction (28, 29).

Eukaryotic organelles, however, could be a different matter. There can be no doubt that the chloroplast is of specific eubacterial origin (3, 4). A question arises with the remaining organelles and structures. Mitochondria, for example, do not conform well to a "typically prokaryotic" phenotype, which has led some to conclude that they could not have arisen as endosymbionts (30). By using "prokaryote" in a phylogenetic sense, this formulation of the issue does not recognize a third alternative—that the organelle in question arose endosymbiotically from a separate line of descent whose phenotype is not "typically prokaryotic" (i.e., eubacterial). It is thus conceivable that some endosymbiotically formed structures represent still other major phylogenetic groups; some could even be the only extant representation thereof.

The question that remains to be answered is whether the common ancestor of all three major lines of descent was itself a prokaryote. If not, each urkingdom represents an independent evolution of the prokaryotic level of organization. Obviously, much more needs to be known about the general properties of all the urkingdoms before this matter can be definitely settled. At present we can point to two arguments suggesting that each urkingdom does represent a separate evolution of the prokaryotic level of organization.

The first argument concerns the stability of the general phenotypes. The general eubacterial phenotype has been stable for at least 3 billion years—i.e., the apparent age of blue-green algae (31). The methanogenic phenotype seems to be at least this old in that branchings within the two urkingdoms are comparably deep (see Table 1). The time available to form each phenotype (from their common ancestor) is then short by comparison, which seems paradoxical in that the two phenotypes are so fundamentally different. We think that this ostensible paradox implies that the common ancestor in this case was not a prokaryote. It was a far simpler entity; it probably did not evolve at the "slow" rate characteristic of prokaryotes; it did not possess many of the features possessed by prokaryotes, and so these evolved independently and differently in separate lines of descent.

The second argument concerns the quality of the differences in the three general phenotypes. It seems highly unlikely, for example, that differences in general patterns of base modification in rRNAs and tRNAs are related to the niches that organisms occupy. Rather, differences of this nature imply independent evolution of the properties in question. It has been argued elsewhere that features such as RNA base modification generally represent the final stage in the evolution of translation (32). If these features have evolved separately in two lines of descent, their common ancestor, lacking them, had a more rudimentary version of the translation mechanism and consequently, could not have been as complex as a prokaryote (6).

With the identification and characterization of the urkingdoms we are for the first time beginning to see the overall phylogenetic structure of the living world. It is not structured in a bipartite way along the lines of the organizationally dissimilar prokaryote and eukaryote. Rather, it is (at least) tripartite, comprising (*i*) the typical bacteria, (*ii*) the line of descent manifested in eukaryotic cytoplasms, and (*iii*) a little explored grouping, represented so far only by methanogenic bacteria.

The ideas expressed herein stem from research supported by the National Aeronautics and Space Administration and the National Science Foundation. We are grateful to a number of colleagues who have helped to generate the yet unpublished data that make these speculations possible: William Balch, Richard Blakemore, Linda Bonen, Tristan Dyer, Jane Gibson, Ramesh Gupta, Robert Hespell, Bobby Joe Lewis, Kenneth Luehrsen, Linda Magrum, Jack Maniloff, Norman Pace, Mitchel Sogin, Stephan Sogin, David Stahl, Ralph Tanner, Thomas Walker, Ralph Wolfe, and Lawrence Zablen. We thank Linda Magrum and David Nanney for suggesting the name "archaebacteria."

1. Stanier, R. Y. (1970) *Symp. Soc. Gen. Microbiol.* **20,** 1–38.
2. Margulis, L. (1970) *Origin of Eucaryotic Cells* (Yale University Press, New Haven).
3. Zablen, L. B., Kissel, M. S., Woese, C. R. & Buetow, D. E. (1975) *Proc. Natl. Acad. Sci. USA* **72,** 2418–2422.
4. Bonen, L. & Doolittle, W. F. (1975) *Proc. Natl. Acad. Sci. USA* **72,** 2310–2314.
5. Bonen, L., Cunningham, R. S., Gray, M. W. & Doolittle, W. F. (1977) *Nucleic Acid Res.* **4,** 663–671.
6. Woese, C. R. & Fox, G. E. (1977) *J. Mol. Evol.,* in press.
7. Sneath, P. H. A. & Sokal, R. R. (1973) *Numerical Taxonomy* (W. H. Freeman, San Francisco).
8. Zuckerkandl, E. & Pauling, L. (1965) *J. Theor. Biol.* **8,** 357–366.
9. Fitch, W. M. & Margoliash, E. (1967) *Science* **155,** 279–284.
10. Fitch, W. M. (1976) *J. Mol. Evol.* **8,** 13–40.
11. Sogin, S. J., Sogin, M. L. & Woese, C. R. (1972) *J. Mol. Evol.* **1,** 173–184.
12. Woese, C. R., Fox, G. E., Zablen, L., Uchida, T., Bonen, L., Pechman, K., Lewis, B. J. & Stahl, D. (1975) *Nature* **254,** 83–86.
13. Fox, G. E., Pechman, K. R. & Woese, C. R. (1977) *Int. J. Syst. Bacteriol.* **27,** 44–57.
14. Uchida, T., Bonen, L., Schaup, H. W., Lewis, B. J., Zablen, L. B. & Woese, C. R. (1974) *J. Mol. Evol.* **3,** 63–77.
15. Zablen, L. B. & Woese, C. R. (1975) *J. Mol. Evol.* **5,** 25–34.
16. Doolittle, W. F., Woese, C. R., Sogin, M. L., Bonen, L. & Stahl, D. (1975) *J. Mol. Evol.* **4,** 307–315.
17. Pechman, K. J., Lewis, B. J. & Woese, C. R. (1976) *Int. J. Syst. Bacteriol.* **26,** 305–310.
18. Bovee, E. C. & Jahn, T. L. (1973) in *The Biology of Amoeba,* ed. Jeon, K. W. (Academic Press, New York), p. 38.
19. Wolfe, R. S. (1972) *Adv. Microbiol. Phys.* **6,** 107–146.
20. Zeikus, J. G. (1977) *Bacteriol. Rev.* **41,** 514–541.
21. Zeikus, J. G. & Bowen, V. G. (1975) *Can. J. Microbiol.* **21,** 121–129.
22. Balch, W. E., Magrum, L. J., Fox, G. E., Wolfe, R. S. & Woese, C. R. (1977) *J. Mol. Evol.,* in press.
23. Fox, G. E., Magrum, L. J., Balch, W. E., Wolfe, R. S. & Woese, C. R. (1977) *Proc. Natl. Acad. Sci. USA,* **74,** 4537–4541.
24. Kandler, O. & Hippe, H. (1977) *Arch. Microbiol.* **113,** 57–60.
25. Taylor, C. D. & Wolfe, R. S. (1974) *J. Biol. Chem.* **249,** 4879–4885.
26. Cheeseman, P., Toms-Wood, A. & Wolfe, R. S. (1972) *J. Bacteriol.* **112,** 527–531.
27. Pace, B. & Campbell, L. L. (1971) *J. Bacteriol.* **107,** 543–547.
28. Brown, A. D. & Cho, K. Y. (1970) *J. Gen. Microbiol.* **62,** 267–270.
29. Reistad, R. (1972) *Arch. Mikrobiol.* **82,** 24–30.
30. Raff, R. A. & Mahler, H. R. (1973) *Science* **180,** 517–521.
31. Shopf, J. W. (1972) *Exobiology—Frontiers of Biology* (North Holland, Amsterdam), Vol. 23, pp. 16–61.
32. Woese, C. R. (1970) *Symp. Soc. Gen. Microbiol.* **20,** 39–54.

Avise, J. C., C. Giblin-Davidson, J. Laerm, J. C. Patton, and R. A. Lansman. 1979. Mitochondrial DNA clones and matriarchal phylogeny within and among geographic populations of the pocket gopher, *Geomys pinetis*. *Proceedings of the National Academy of Sciences, USA* 76:6694–6698.

Traditional population genetics was anchored in the study of Mendelian nuclear genes: genes that segregate and recombine during each generation of sexual reproduction. That is why the earliest studies of mitochondrial (mt) DNA had such a revisionary influence on evolutionary studies. Unlike most coding genes that reside in the genome of a cell's nucleus, cytoplasmic mtDNA is transmitted without recombination through the matrilineal descendants of an organismal pedigree. Thus, mtDNA is clonally inherited, even in species that otherwise reproduce sexually. Animal mtDNA is comprised of about 35 different functional genes, but, from a genealogical or evolutionary vantage, it behaves as if it were a single genetic locus. Mitochondrial DNA's non-recombining mode of transmission and its surprisingly rapid pace of nucleotide substitution make it suitable for phylogenetic reconstruction across recent (microevolutionary) time scales. These and other favorable properties have combined to make this tiny molecule the single most popular genetic marker for microevolutionary studies of multicellular animal populations.

The paper is one of a handful of articles from the late 1970s that introduced population biologists to animal mtDNA by demonstrating this molecule's exceptional utility for genealogical appraisals at the intraspecific scale. This paper can be deemed the first substantive foray into the evolutionary realm of what would later be termed *intraspecific phylogeography*. For a personal account of the author's serendipitous involvement with mtDNA (and pocket gophers), curious readers are directed to Avise (2001).

RELATED READING

Avise, J. C. 2001. Captivating life: a naturalist in the age of genetics. Smithsonian Institution Press, Washington, D.C., USA.

Ayala, F. J. 1995. The myth of Eve: molecular biology and human origins. Science 270:1930–1936.

Ruvolo, M., S. Zehr, M. von Dornum, D. Pan, B. Chang, and J. Lin. 1993. Mitochondrial COII sequences and modern human origins. Molecular Biology and Evolution 10:1115–1135.

Proc. Natl. Acad. Sci. USA
Vol. 76, No. 12, pp. 6694–6698, December 1979
Population Biology

Mitochondrial DNA clones and matriarchal phylogeny within and among geographic populations of the pocket gopher, *Geomys pinetis**

(population structure/genetic variation/restriction enzymes)

JOHN C. AVISE[†‡], CECILIA GIBLIN-DAVIDSON[†], JOSHUA LAERM[†], JOHN C. PATTON[†‡], AND ROBERT A. LANSMAN[§,‡]

Departments of [†]Zoology and [§]Biochemistry, and ‡Program in Genetics, University of Georgia, Athens, Georgia 30602

Communicated by Norman H. Giles, August 27, 1979

ABSTRACT Restriction endonuclease assay of mitochondria DNA (mtDNA) and standard starch-gel electrophoresis of proteins encoded by nuclear genes have been used to analyze phylogenetic relatedness among a large number of pocket gophers (*Geomys pinetis*) collected throughout the range of the species. The restriction analysis clearly distinguishes two populations within the species, an eastern and a western form, which differ by at least 3% in mtDNA sequence. Qualitative comparisons of the restriction phenotypes can also be used to identify mtDNA "clones" within each form. The mtDNA clones interconnect in a phylogenetic network which represents an estimate of matriarchal phylogeny for *G. pinetis*. Although the protein electrophoretic data also differentiate the eastern and western forms, the data are of limited usefulness in establishing relationships among more local subpopulations. The comparison between these two data sets suggests that restriction analysis of mtDNA is probably unequalled by other techniques currently available for determining phylogenetic relationships among conspecific organisms.

Type II restriction endonucleases cleave duplex DNA at specific recognition sites usually containing four, five, or six nucleotides (1, 2). The fragment patterns produced by the digestion of two homologous DNA molecules can differ because they are differentially modified (i.e., by methylation of bases within the recognition sites), nucleotide substitutions have abolished cleavage sites or created new ones, or major sequence rearrangements have altered the relative positions of cleavage sites within the molecules. Restriction endonucleases have been used to compare mtDNA sequences from a variety of mammalian sources. These studies have shown that the population of mtDNA molecules within an individual animal appears homogeneous in sequence (3, 4), mtDNA does not appear to be modified in ways that affect digestion (3, 5), mtDNA is maternally inherited (3, 6–8), and mtDNA evolves rapidly enough to produce easily detectable sequence heterogeneity within species (3, 4, 6, 9, 10). These observations provide the rationale for the use of restriction analysis of mtDNA to identify mtDNA "clones" in nature and to estimate their evolutionary geneology (phylogeny).

Critical assessment of the utility of any new information used for phylogenetic reconstruction is hampered by the fact that the true evolutionary history of the populations or species being studied is seldom if ever known. In the absence of an absolute calibration, the potential of a new approach can be evaluated only by comparing results with those obtained by independent methods. Gel electrophoresis of proteins is the simplest and strongest technique in current use for estimating the genetic relatedness of closely similar organisms. We have, therefore, performed both protein electrophoretic and mtDNA restriction analyses with a large number of individuals of the pocket gopher, *Geomys pinetis*, live-trapped throughout the range of the species (11).

Unless local populations are completely isolated, we do not expect the two methods of analysis to yield exactly the same kind of information. First, the level of allozyme differentiation among conspecific populations is usually low and consists largely of differing frequencies of the same electromorphs (12). Recently introduced techniques for protein survey analysis, such as gel sieving and heat lability, may, however, prove to refine discrimination of local populations (13, 14). Second, a given electromorph may often be polyphyletic, in fact representing an assemblage of different allelic products sharing electrophoretic mobility (15–17). We have suggested elsewhere that convergence to a common mtDNA restriction phenotype from unrelated phenotypes is unlikely (3). Third, in contrast to mtDNA, nuclear alleles encoding allozymes are segregated and recombined during each generation of sexual reproduction. For these reasons, allozymes rarely provide unambiguous information about the phylogeny of conspecific populations, much less of individuals within these populations. We will show that analysis of mtDNA by restriction enzymes can partially or totally circumvent many of these difficulties.

MATERIALS AND METHODS

Gophers were live-trapped from the localities listed in Table 1. mtDNA was purified as described (3) from livers of individual animals and digested with six restriction endonucleases. *Hinc*II, *Hind*III, and *Bam*HI were purchased from Bethesda Research (Rockville, MD). *Bgl* II and *Bst*EII were prepared and provided by Richard Meagher. *Eco*RI was purified by published procedures (18). Samples of mtDNA were electrophoresed through 1.1% agarose gels (19) with *Hind*III-digested phage λ DNA as molecular weight standards. Fragments smaller than 500 base pairs (200 base pairs in the *Hinc*II) were not included in the data analysis because they would not be detectable in the more dilute DNA samples. Proteins were electrophoresed on horizontal starch gels according to standard procedures (20, 21).

RESULTS

mtDNA divergence

The mtDNA fragment phenotypes produced by digestion with three of the six restriction endonucleases are pictured in Fig. 1. Estimates of the number of mtDNA base substitutions per

The publication costs of this article were defrayed in part by page charge payment. This article must therefore be hereby marked "*advertisement*" in accordance with 18 U. S. C. §1734 solely to indicate this fact.

Abbreviations: *PGD*, 6-phosphogluconate dehydrogenase gene; *ALB*, albumin gene.

* This is paper no. 2 in the series "The use of restriction endonucleases to measure mitochondrial DNA sequence relatedness in natural populations." Paper no. 1 in the series is ref. 3.

Table 1. mtDNA digestion phenotypes observed in geographic samples of *G. pinetis*

Type	Composite mtDNA digestion phenotype*	Counties	No. of individuals: mtDNA	No. of individuals: Proteins	Mean heterozygosity†
1	1N2N3N4N5N6N	Pierce, Camden, Charlton, GA; Nassau, FL	20		
2	1N2N3N4N5M6N	Camden, GA	1	44	0.001
3	1Q2N3N4N5N6N	Pierce, GA	1		
4	1N2N3N4N5P6N	Camden, GA	3		
5	1N2N3N4N5O6N	Screven, Jenkins, Burke, Richmond, GA	16	22	0.003
6	1N2N3N4N5N6P	Alachua, FL	2		
7	1N2N3M4N5-6P	Alachua, FL	1	12	0.030
8	1N2N3N4N5S6L	Alachua, FL	1		
9	1N2N3N4N5L6M	Levy, FL	1		
10	1N2N3N4N5N6M	Citrus, FL	1	12	0.058
11	1N2N3N4M5N6M	Marion, FL	2		
12	1N2N3N4N5Q6P	Grady, GA	2	2	
13	1N2O3N4N5Q6P	Gadsden, FL	3	3	
14	1N2N3N4I5-6K	Leon, FL	1	2	
15	1N2N3N4K5-6P	Thomas, GA; Wakulla, FL	2	8	0.051
16	1M2M3O4O5R6Q	Walton, FL	1		
17	1L2M3O4O5R6Q	Walton, FL	1	13	0.022
18	1M2M3N4O5R6Q	Walton, FL			
		Macon, Autauga, AL;		9	0.009
		Crenshaw, AL;	19		
		Taylor, Talbot, GA		5	0.024
19	1M2M3N4O5R6O	Taylor, GA	1	15	0.020
20	1L2M3N4O5R6R	Taylor, GA	1		
21	1J2M3N4O5R6Q	Russell, AL	2	12	0.033
22	1M2M3N4P5R6Q	Escambia, AL	3	8	0.020
23	1K2M3N4O5R6Q	Baldwin, AL	2	3	
		Totals	87	171	$\overline{H}$ = 0.025

* Numbers refer to restriction enzymes: 1, *Eco*RI; 2, *Bam*HI; 3, *Bst*EII; 4, *Hin*dIII; 5, *Hin*cII; 6, *Bgl* II. Letters refer to digestion phenotypes, as exemplified by Fig. 1.

† Estimates of mean genic heterozygosity (for samples $n \geq 5$) at 25 protein loci.

nucleotide (p) differentiating individual organisms were calculated from the total counted fraction of shared mtDNA digestion fragments produced by all restriction enzymes using Nei and Li's approach (22). Estimates of p calculated by an approach suggested by Upholt (23) were virtually identical. To divide the samples into natural subdivisions according to estimated mtDNA sequence divergence, we subjected this matrix of p values to cluster analysis (24). Results yielded two distinct subdivisions, largely homogeneous internally but significantly different from one another in percentages of shared mtDNA fragments. All gophers collected in northwest Georgia, Alabama, and the western panhandle of Florida belong to one subdivision, whereas samples collected from the remainder of Georgia and Florida belong to the other cluster. From now on we will refer to these two sets of populations as the "western" and "eastern" forms of *G. pinetis*, respectively. Estimates of

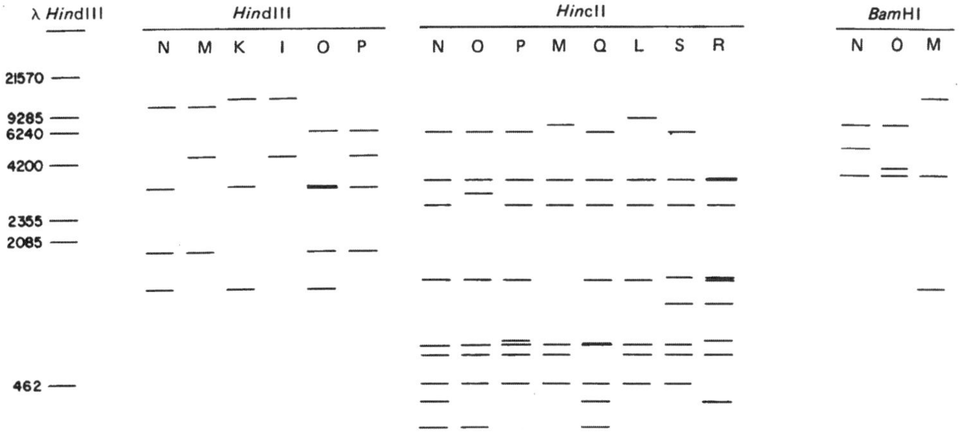

FIG. 1. Diagram of digestion phenotypes of mtDNA observed in 86 specimens of *G. pinetis*. The molecular weight markers, for which sizes are given in base pairs, are the fragments generated by a *Hin*dIII digestion of phage λ DNA.

mtDNA base substitutions per nucleotide within and among these two forms of *G. pinetis* are summarized in Table 2. There is no overlap in *p* values in comparisons within compared to between the two forms.

A qualitative method of data analysis that does not submerge the digestion phenotypes in a quantitative summary statistic provides much additional information. An example of the geographic distribution of mtDNA phenotypes, produced by digestion with enzyme *Bam*HI, is shown in Fig. 2. All composite mtDNA digestion phenotypes produced by the six restriction enzymes are listed in Table 1. These composite phenotypes were connected into a most-parsimonious phylogenetic network. The basic procedure is similar to that used for estimation of phylogenies from another kind of qualitative data base—chromosomal inversions in *Drosophila* (25). For example, composite type 1 (abbreviated 1N2N3N4N5N6N) was found in 20 gophers collected from Pierce, Camden, and Charlton Cos., GA, and from Nassau Co., FL. This composite type differs from type 2 (observed in Camden Co., GA; Table 1) only in digestion phenotype produced by enzyme no. 5, *Hin*cII. Similarly, type 3 (observed in Pierce Co., GA) differs from type 1 only in digestion phenotype produced by *Eco*RI. Thus, types 2 and 3 connect separately to type 1 by a single fragment pattern change. In this fashion, and without any regard to geographic collection site, all 23 composite phenotypes listed in Table 1 were added to the phylogenetic network.

This unoriented network was subsequently superimposed on the geographic source of the collections, with results shown in Fig. 3. Each circle encompasses the geographic area within which a given composite phenotype was observed. Composite types are interconnected by branches of the network. In cases where two fragment patterns (e.g., *Bam*HI-O and *Bam*HI-M) cannot be interconverted by a single base substitution, we have added a solid line and a dashed line crossing the branches to indicate that the patterns differ in at least two restriction sites. Thus, the number of solid plus dashed lines indicates the minimum number of base substitutions required to account for the differences in the composite digestion phenotypes.

We have suggested earlier (3) that one advantage of a qualitative analysis of mtDNA digestion phenotypes lies in the low probability that complex phenotypes arise independently in the evolutionary process. In this study, we have observed only a single instance of probable parallelism: clones 17 and 20 both possess the digestion phenotype "L" produced by *Eco*RI, although they link in the phylogenetic network to clones 16 and 18, respectively, both of which exhibit phenotype 1M. 1M and 1L are relatable by a single base substitution; thus, parallelism in this case is not too surprising.

Most of the composite phenotypes in Fig. 3, on both a local and regional basis, are clearly related by one or a few restriction site changes recognized by the restriction enzymes used. For example, type 18, which is found in a wide region within the range of the western form of *G. pinetis*, appears to have given rise by independent mutations to types 19, 20, and 16, which were observed in different localities within the range of 18. Type 17 may subsequently have evolved from type 16 (or conceivably vice versa; the tree is nondirectional) by an additional base change. If type 16 had not been included in our limited samples, the relationship of types 17 and 18 (two assayed base changes) would have appeared similar to the current connectedness of types 18 and 20.

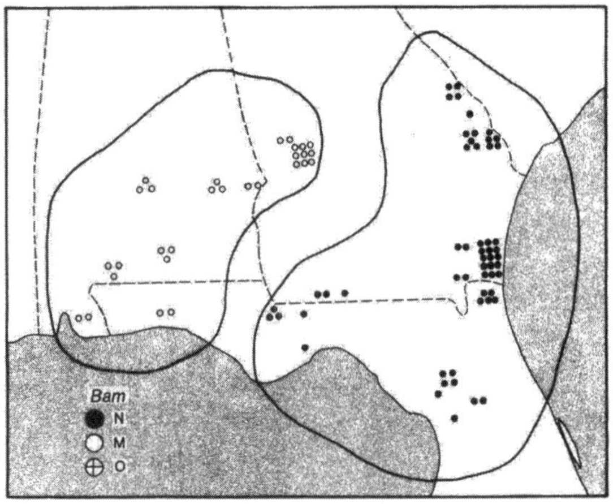

FIG. 2. Geographic distribution of *Bam*HI digestion phenotypes. Each dot represents a single individual.

The eastern and western forms of *G. pinetis* differ in digestion phenotypes produced by five of the six restriction enzymes used, and these phenotype differences represent a *minimum* of nine base changes. With the present data it is impossible to determine which particular populations of the eastern and western forms are best directly connected into the network, although a few populations (exhibiting composite types such as 7, 16, and 17) can be effectively eliminated from consideration by possession of different rare digestion phenotypes produced by *Bst*EII.

Protein divergence

There was relatively little allozymic variation, within or among populations, at proteins encoded by most of the 25 loci examined in this study. Counted heterozygosities (*H*, mean pro-

Table 2. Genetic distances between eastern and western forms of *G. pinetis*

	mtDNA		25 protein loci	
Comparison	No. pairwise comparisons of individuals	p,* mean (range)	No. pairwise comparisons of populations†	$\bar{D}$,‡ mean (range)
Eastern	1378	0.005 (0.000–0.018)	10	0.009 (0.000–0.020)
Western	435	0.002 (0.000–0.010)	15	0.024 (0.001–0.049)
Eastern vs. western	1590	0.034 (0.025–0.047)	30	0.065 (0.040–0.098)
Totals	3403	0.018 (0.000–0.047)	55	0.044 (0.000–0.098)

* Base substitutions per nucleotide. To greatly simplify calculations, *Hin*cII was counted as a "six-base" enzyme; this will have only trivial effects on absolute values of *p* and should not alter relative values.

† $n \geq 5$, see Table 1.

‡ Codon substitutions per locus.

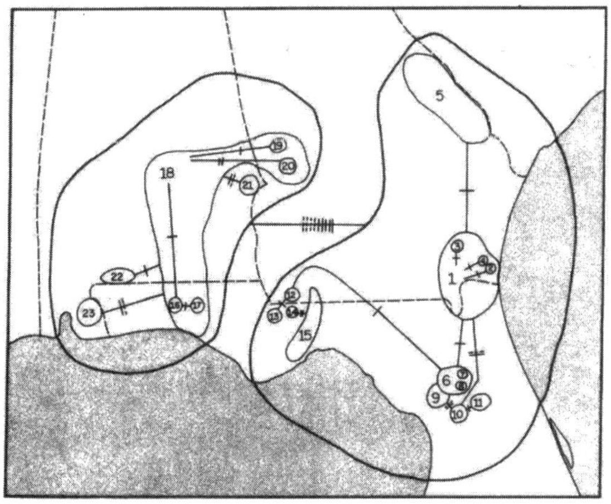

FIG. 3. Phylogenetic network of *G. pinetis* inferred from mtDNA fragment phenotypes produced by digestion with six different restriction endonucleases. Numbers refer to the composite mtDNA phenotypes listed in Table 1. Phenotypes 7, 14, and 15 are not connected to the network because information from *Hin*cII was not obtained. Phenotypes 8 and 23 were equally related to more than one other phenotype, and their placement in the network was decided by geographic contiguity.

portion of individuals heterozygous per locus in a local population) range from 0.000 to 0.058, with overall unweighted $\overline{H}$ equal to 0.025 (Table 1). Similar levels of electrophoretically assayed genic variation have previously been noted in various species and genera of pocket gophers (26–29).

Protein electromorphs encoded by two loci, 6-phosphogluconate dehydrogenase (*PGD*) and albumin (*ALB*), do show marked regional differentiation. Allele frequencies at these two loci are shown in Figs. 4 and 5. The "100" electromorph for albumin (ALB^{100}) is fixed in all populations of eastern *G. pinetis* defined by mtDNA and is also a common allele in a related congener *G. bursarius* (26). Hence, it is probably plesiomorphic (ancestral) to the *G. pinetis* group of gophers, distinct from ALB^{95}, which may be a derived allele unique to and nearly fixed in western *G. pinetis*. With the exception of peninsular

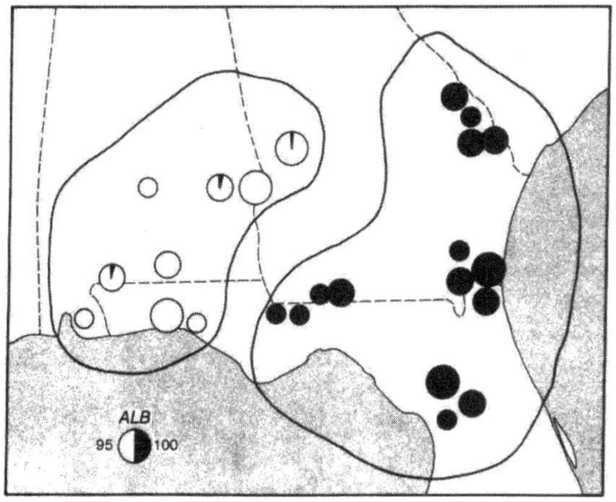

FIG. 4. Geographic distribution of electromorph frequencies of albumin in *G. pinetis*. Large circles represent samples of $n \geq 12$; medium circles, $5 \leq n \leq 11$; small circles, $n \leq 4$. Heavy lines encircle eastern and western forms of *G. pinetis* as defined by the mtDNA analysis.

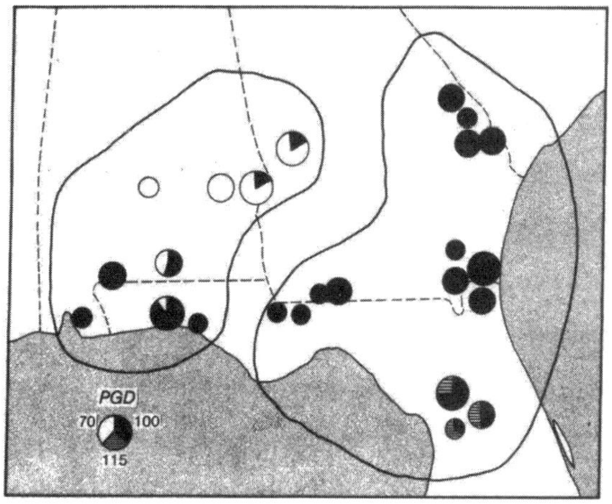

FIG. 5. Geographic distribution of electromorph frequencies of 6-phosphogluconate dehydrogenase, as in Fig. 4.

Florida populations, which exhibit both PGD^{115} and PGD^{100} in high frequency, all eastern *G. pinetis* populations (as well as several western populations) appear monomorphic for PGD^{100}. PGD^{70} was observed only in western populations, where it sometimes reaches high frequency. Because PGD^{100}, PGD^{70}, and PGD^{115} are also observed in the congener *G. bursarius* (26), they are likely ancestral to the entire *G. pinetis* assemblage. Hence, none may have arisen after the evolutionary separation of western from eastern *G. pinetis* stock.

Mean genetic distances (30) at 25 loci were calculated between all samples of $n > 5$ with results shown in Table 2. All distances are well within the range of values typical of conspecific populations in other vertebrates and invertebrates (12). Distances were subjected to a cluster analysis (24), and two major subdivisions corresponding perfectly to the eastern and western forms of *G. pinetis* previously defined by mtDNA were apparent. Apart from the distinctness of these two macrogeographic assemblages, these protein data were largely uninformative about possible phylogenetic relationships among local populations of *G. pinetis*.

DISCUSSION

The heart of the mtDNA data is summarized in Fig. 3. Each circled number represents one or a group of individuals sharing an identical mtDNA genotype as assayed by our restriction enzymes, distinct from all other such groups. If current belief is correct, that mitochondria are strictly maternally inherited, each group then represents a mtDNA clone. Organisms belonging to this clone must have evolved from a common female parent at some time in the past. Interconnectedness among clones, depicted by branches of the network in Fig. 3, provide strong estimates of matriarchal phylogeny.

The power and precision of this approach to natural population analysis are clear. Among 87 gophers examined, 23 clones were identified. Some of these clones (such as 5 and 18) are widespread, while others (such as 2 and 19) appear more local. The local clones are usually relatable by one or two assayed base substitutions to the widespread clone of that region. Clones in different geographic regions can also be readily related to one another. Because mtDNA is inherited through females, segregation and recombination during sexual reproduction do not confound attempts at reconstruction of mtDNA phylogeny. For these reasons also, an individual animal carries within its mitochondrial genome relatively unambiguous information about the female lineage to which it belongs. In contrast to nuclear gene analyses, which are typically based on *population* allele frequencies, the natural and basic unit of analysis for mtDNA is the individual organism.

A striking contrast of the different kinds of information yielded by protein and mtDNA analyses is apparent in the data presented in Figs. 3 and 5 for samples collected from the Florida peninsula. This group of samples contains several identifiable mtDNA clones in two groups. Clones 9, 10, and 11 appear to be more closely related to the southeast Georgia assemblage (clone 1) than they are to clones 6 and 7, which appear in adjacent locales. One possible interpretation is that this area of Florida has been successfully colonized separately by gophers of rather distinct maternal lineages which now coexist and interbreed. Animals from both maternal lineages share a *PGD* electromorph (115) that has not been found in any other area. This allele probably arose *in situ* by mutation or represents a plesiomorph that has been spread by sexual reassortment so that it now appears in animals of distinct maternal phylogeny.

The choice of the restriction endonucleases used in this study was arbitrary. We expect that the use of a different group of enzymes would resolve a different set of mtDNA clones and would probably produce a slightly different representation of the relatedness of the samples we have used. Fig. 3 is, therefore, an estimation rather than an absolute evaluation of matriarchal phylogeny within *G. pinetis*. Clearly, more refined determinations of clonal diversity and interrelatedness could be obtained by including data from larger numbers of restriction endonucleases. The only limitation on the resolution obtained is the amount of mtDNA available from each sample. Even the rather limited data set presented here illustrates the wealth of phylogenetic information present in the distribution of mtDNA clones within a single species. It is clear to us that restriction analysis of mtDNA, either alone or in concert with conventional data on nuclear genes, will find a wide range of applications in population and evolutionary biology.

Work was supported by National Science Foundation Grant DEB7814195, by the Theodore Roosevelt Memorial Fund of the American Museum of Natural History, and by a grant from the Georgia Department of Natural Resources. J.C.P. was a predoctoral trainee of the U.S. Public Health Service.

1. Boyer, H. W. (1971) *Annu. Rev. Microbiol.* **25,** 153–176.
2. Boyer, H. W. (1974) *Fed. Proc. Fed. Am. Soc. Exp. Biol.* **33,** 1125–1127.
3. Avise, J. C., Lansman, R. A. & Shade, R. O. (1979) *Genetics* **92,** 279–295.
4. Upholt, W. B. & Dawid, I. B. (1977) *Cell* **11,** 571–583.
5. Chang, A. C. Y., Lansman, R. A., Clayton, D. A. & Cohen, S. N. (1975) *Cell* **6,** 231–244.
6. Kroon, A. M., deVos, W. M. & Bakker, H. (1978) *Biochim. Biophys. Acta* **519,** 269–273.
7. Dawid, I. B. & Blacker, A. W. (1972) *Dev. Biol.* **29,** 152–161.
8. Conde, M. J., Pring, D. R. & Levings, C. S., III (1979) *J. Hered.* **70,** 2–4.
9. Brown, W. M., George, M., Jr. & Wilson, A. C. (1979) *Proc. Natl. Acad. Sci. USA* **76,** 1967–1971.
10. Brown, W. M. & Wright, J. W. (1979) *Science* **203,** 1247–1249.
11. Hall, E. R. & Kelson, K. R. (1959) *The Mammals of North America* (Ronald, New York), Vol. 2.
12. Avise, J. C. (1974) *Syst. Zool.* **23,** 465–481.
13. Johnson, G. B. (1976) *Genetics* **83,** 149–167.
14. Singh, R. S., Lewontin, R. C. & Felton, A. A. (1976) *Genetics* **84,** 609–629.
15. King, J. L. & Ohta, T. (1975) *Genetics* **79,** 681–691.
16. Coyne, J. A. (1976) *Genetics* **84,** 593–607.
17. Singh, A. S., Lewontin, R. C. & Felton, A. A. (1976) *Genetics* **84,** 609–629.
18. Greene, P. J., Heyneker, H. L., Bolivar, F., Rodriguez, R. L., Betlach, M. C., Covarrbias, A. A., Backman, K., Russel, D. J., Tait, R. & Boyer, H. W. (1978) *Nucleic Acids Res.* **5,** 2373–2380.
19. Helling, R. B., Goodman, H. M. & Boyer, H. M. (1974) *J. Virol.* **14,** 1235–1244.
20. Ayala, F. J., Powell, J. R., Tracey, M. L., Mourao, C. A. & Perez-Salas, S. (1972) *Genetics* **70,** 113–139.
21. Selender, R. K., Smith, M. H., Yang, S. Y., Johnson, W. E. & Gentry, J. B. (1971) *Stud. Genet. VI Univ. Texas Publ.* **7103,** 49–90.
22. Nei, M. & Li, W.-H. (1979) *Proc. Natl. Acad. Sci. USA* **76,** 5269–5273.
23. Upholt, W. B. (1977) *Nucleic Acids Res.* **4,** 1257–1265.
24. Sneath, P. H. A. & Sokal, R. R. (1973) *Numerical Taxonomy* (Freeman, San Francisco).
25. Carson, H. L. & Kaneshiro, K. Y. (1976) *Annu. Rev. Ecol. Syst.* **7,** 311–345.
26. Penney, D. F. & Zimmerman, E. G. (1976) *Evolution* **30,** 473–483.
27. Nevo, E., Kim, Y. J., Shaw, C. R. & Thaeler, C. S. (1974) *Evolution* **28,** 1–23.
28. Selander, R. K. Kaufman, D. W., Baker, R. J. & Williams, S. L. (1974) *Evolution* **28,** 557–564.
29. Patton, J. L., Selander, R. K. & Smith, M. H. (1972) *Syst. Zool.* **21,** 263–270.
30. Nei, M. (1972) *Am. Nat.* **106,** 283–292.

Gould, S. J., and R. C. Lewontin. 1979. The spandrels of San Marco and the Panglossian paradigm: a critique of the adaptationist programme. *Proceedings of the Royal Society London B*. 205:581–598.

Stephen Jay Gould and Richard C. Lewontin (both of whom spent much of their careers at Harvard University) were among the most provocative thinkers and articulate writers in evolutionary biology during the second half of the twentieth century. In this memorable paper, they joined forces in a fashion that showcased their gift for presenting evolutionary arguments, using evocative language and imagery drawn from a wide spectrum of the social and natural sciences. We will let this entertaining and eloquent article speak for itself.

RELATED READING

Gould, S. J. 1977. Ontogeny and phylogeny. Belknap Press, Cambridge, Massachusetts, USA.
Lewontin, R. C. 1974. The genetic basis of evolutionary change. Columbia University Press, New York, New York, USA.

Proc. R. Soc. Lond. B **205**, 581–598 (1979)
Printed in Great Britain

The spandrels of San Marco and the Panglossian paradigm: a critique of the adaptationist programme

By S. J. Gould and R. C. Lewontin
Museum of Comparative Zoology, Harvard University, Cambridge, Massachusetts 02138, *U.S.A.*

An adaptationist programme has dominated evolutionary thought in England and the United States during the past 40 years. It is based on faith in the power of natural selection as an optimizing agent. It proceeds by breaking an organism into unitary 'traits' and proposing an adaptive story for each considered separately. Trade-offs among competing selective demands exert the only brake upon perfection; non-optimality is thereby rendered as a result of adaptation as well. We criticize this approach and attempt to reassert a competing notion (long popular in continental Europe) that organisms must be analysed as integrated wholes, with *Baupläne* so constrained by phyletic heritage, pathways of development and general architecture that the constraints themselves become more interesting and more important in delimiting pathways of change than the selective force that may mediate change when it occurs. We fault the adaptationist programme for its failure to distinguish current utility from reasons for origin (male tyrannosaurs may have used their diminutive front legs to titillate female partners, but this will not explain *why* they got so small); for its unwillingness to consider alternatives to adaptive stories; for its reliance upon plausibility alone as a criterion for accepting speculative tales; and for its failure to consider adequately such competing themes as random fixation of alleles, production of non-adaptive structures by developmental correlation with selected features (allometry, pleiotropy, material compensation, mechanically forced correlation), the separability of adaptation and selection, multiple adaptive peaks, and current utility as an epiphenomenon of non-adaptive structures. We support Darwin's own pluralistic approach to identifying the agents of evolutionary change.

1. Introduction

The great central dome of St Mark's Cathedral in Venice presents in its mosaic design a detailed iconography expressing the mainstays of Christian faith. Three circles of figures radiate out from a central image of Christ: angels, disciples, and virtues. Each circle is divided into quadrants, even though the dome itself is radially symmetrical in structure. Each quadrant meets one of the four spandrels in the arches below the dome. Spandrels – the tapering triangular spaces formed by the intersection of two rounded arches at right angles (figure 1) – are necessary architectural by-products of mounting a dome on rounded arches. Each spandrel contains a design admirably fitted into its tapering space. An evangelist sits in the upper part flanked by the heavenly cities. Below, a man representing one of the four Biblical rivers (Tigris, Euphrates, Indus and Nile) pours water from a pitcher into the narrowing space below his feet.

The design is so elaborate, harmonious and purposeful that we are tempted to view it as the starting point of any analysis, as the cause in some sense of the surrounding architecture. But this would invert the proper path of analysis. The

FIGURE 1. One of the four spandrels of St Mark's; seated evangelist above, personification of river below.

system begins with an architectural constraint: the necessary four spandrels and their tapering triangular form. They provide a space in which the mosaicists worked; they set the quadripartite symmetry of the dome above.

Such architectural constraints abound and we find them easy to understand because we do not impose our biological biases upon them. Every fan vaulted ceiling must have a series of open spaces along the mid-line of the vault, where the sides of the fans intersect between the pillars (figure 2). Since the spaces must exist, they are often used for ingenious ornamental effect. In King's College Chapel in Cambridge, for example, the spaces contain bosses alternately embellished with the Tudor rose and portcullis. In a sense, this design represents an 'adaptation', but the architectural constraint is clearly primary. The spaces arise as a necessary by-product of fan vaulting; their appropriate use is a secondary effect. Anyone who tried to argue that the structure exists because the alternation of rose and portcullis makes so much sense in a Tudor chapel would be inviting the same ridicule that Voltaire heaped on Dr Pangloss: 'Things cannot be other than they

FIGURE 2. The ceiling of King's College Chapel.

are...Everything is made for the best purpose. Our noses were made to carry spectacles, so we have spectacles. Legs were clearly intended for breeches, and we wear them.' Yet evolutionary biologists, in their tendency to focus exclusively on immediate adaptation to local conditions, do tend to ignore architectural constraints and perform just such an inversion of explanation.

As a closer example, recently featured in some important biological literature on adaptation, anthropologist Michael Harner has proposed (1977) that Aztec human sacrifice arose as a solution to chronic shortage of meat (limbs of victims were often consumed, but only by people of high status). E. O. Wilson (1978) has used this explanation as a primary illustration of an adaptive, genetic predisposition for carnivory in humans. Harner and Wilson ask us to view an elaborate social system and a complex set of explicit justifications involving myth, symbol, and tradition as mere epiphenomena generated by the Aztecs as an unconscious rationalization masking the 'real' reason for it all: need for protein. But Sahlins (1978) has argued that human sacrifice represented just one part of an elaborate cultural fabric that, in its entirety, not only represented the material expression of Aztec cosmology, but also performed such utilitarian functions as the maintenance of social ranks and systems of tribute among cities.

We strongly suspect that Aztec cannibalism was an 'adaptation' much like evangelists and rivers in spandrels, or ornamented bosses in ceiling spaces: a secondary epiphenomenon representing a fruitful use of available parts, not a cause of the entire system. To put it crudely: a system developed for other reasons generated an increasing number of fresh bodies; use might as well be made of them. Why invert the whole system in such a curious fashion and view an entire culture as the epiphenomenon of an unusual way to beef up the meat supply. Spandrels do not exist to house the evangelists. (Moreover, as Sahlins argues, it is not even clear that human sacrifice was an adaptation at all. Human cultural practices can be orthogenetic and drive towards extinction in ways that Darwinian processes, based on genetic selection, cannot. Since each new monarch had to outdo his predecessor in even more elaborate and copious sacrifice, the practice was beginning to stretch resources to the breaking point. It would not have been the first time that a human culture did itself in. And, finally, many experts doubt Harner's premise in the first place (Ortiz de Montellano 1978). They argue that other sources of protein were not in short supply, and that a practice awarding meat only to privileged people who had enough anyway, and who used bodies so inefficiently (only the limbs were consumed, and partially at that) represents a mighty poor way to run a butchery.)

We deliberately chose non-biological examples in a sequence running from remote to more familiar: architecture to anthropology. We did this because the primacy of architectural constraint and the epiphenomenal nature of adaptation are not obscured by our biological prejudices in these examples. But we trust that the message for biologists will not go unheeded: if these had been biological systems, would we not, by force of habit, have regarded the epiphenomenal adaptation as primary and tried to build the whole structural system from it?

2. The adaptationist programme

We wish to question a deeply engrained habit of thinking among students of evolution. We call it the adaptationist programme, or the Panglossian paradigm. It is rooted in a notion popularized by A. R. Wallace and A. Weismann (but not, as we shall see, by Darwin) towards the end of the nineteenth century: the near omnipotence of natural selection in forging organic design and fashioning the best among possible worlds. This programme regards natural selection as so powerful and the constraints upon it so few that direct production of adaptation through its operation becomes the primary cause of nearly all organic form, function, and behaviour. Constraints upon the pervasive power of natural selection are recognized of course (phyletic inertia primarily among them, although immediate architectural constraints, as discussed in the last section, are rarely acknowledged). But they are usually dimissed as unimportant or else, and more frustratingly, simply acknowledged and then not taken to heart and invoked.

Studies under the adaptationist programme generally proceed in two steps:

(1) An organism is atomized into 'traits' and these traits are explained as structures optimally designed by natural selection for their functions. For lack of space, we must omit an extended discussion of the vital issue: 'what is a trait?' Some evolutionists may regard this as a trivial, or merely a semantic problem. It is not. Organisms are integrated entities, not collections of discrete objects. Evolutionists have often been led astray by inappropriate atomization, as D'Arcy Thompson (1942) loved to point out. Our favourite example involves the human chin (Gould 1977, pp. 381–382; Lewontin 1978). If we regard the chin as a 'thing', rather than as a product of interaction between two growth fields (alveolar and mandibular), then we are led to an interpretation of its origin (recapitulatory) exactly opposite to the one now generally favoured (neotenic).

(2) After the failure of part-by-part optimization, interaction is acknowledged via the dictum that an organism cannot optimize each part without imposing expenses on others. The notion of 'trade-off' is introduced, and organisms are interpreted as best compromises among competing demands. Thus, interaction among parts is retained completely within the adaptationist programme. Any suboptimality of a part is explained as its contribution to the best possible design for the whole. The notion that suboptimality might represent anything other than the immediate work of natural selection is usually not entertained. As Dr Pangloss said in explaining to Candide why he suffered from venereal disease: 'It is indispensable in this best of worlds. For if Columbus, when visiting the West Indies, had not caught this disease, which poisons the source of generation, which frequently even hinders generation, and is clearly opposed to the great end of Nature, we should have neither chocolate nor cochineal.' The adaptationist programme is truly Panglossian. Our world may not be good in an abstract sense, but it is the very best we could have. Each trait plays its part and must be as it is.

At this point, some evolutionists will protest that we are caricaturing their view of adaptation. After all, do they not admit genetic drift, allometry, and a variety of reasons for non-adaptive evolution? They do, to be sure, but we make a different point. In natural history, all possible things happen sometimes; you generally do not support your favoured phenomenon by declaring rivals impossible in theory. Rather, you acknowledge the rival, but circumscribe its domain of action so narrowly that it cannot have any importance in the affairs of nature. Then, you often congratulate yourself for being such an undogmatic and ecumenical chap. We maintain that alternatives to selection for best overall design have generally been relegated to unimportance by this mode of argument. Have we not all heard the catechism about genetic drift: it can only be important in populations so small that they are likely to become extinct before playing any sustained evolutionary role (but see Lande 1976).

The admission of alternatives in principle does not imply their serious consideration in daily practice. We all say that not everything is adaptive; yet, faced with an organism, we tend to break it into parts and tell adaptive stories as if trade-offs among competing, well designed parts were the only constraint upon perfection for each trait. It is an old habit. As Romanes complained about A. R. Wallace in 1900: 'Mr. Wallace does not expressly maintain the abstract impossibility of laws and causes other than those of utility and natural selection... Nevertheless, as he nowhere recognizes any other law or cause..., he practically concludes that, on inductive or empirical grounds, there *is* no such other law or cause to be entertained.'

The adaptationist programme can be traced through common styles of argument. We illustrate just a few; we trust they will be recognized by all:

(1) If one adaptive argument fails, try another. Zig-zag commissures of clams and brachiopods, once widely regarded as devices for strengthening the shell, become sieves for restricting particles above a given size (Rudwick 1964). A suite of external structures (horns, antlers, tusks) once viewed as weapons against predators, become symbols of intraspecific competition among males (Davitashvili 1961). The eskimo face, once depicted as 'cold engineered' (Coon *et al.* 1950), becomes an adaptation to generate and withstand large masticatory forces (Shea 1977). We do not attack these newer interpretations; they may all be right. We do wonder, though, whether the failure of one adaptive explanation should always simply inspire a search for another of the same general form, rather than a consideration of alternatives to the proposition that each part is 'for' some specific purpose.

(2) If one adaptive argument fails, assume that another must exist; a weaker version of the first argument. Costa & Bisol (1978), for example, hoped to find a correlation between genetic polymorphism and stability of environment in the deep sea, but they failed. They conclude (1978, pp. 132, 133): 'The degree of genetic polymorphism found would seem to indicate absence of correlation with the particular environmental factors which characterize the sampled area. The results suggest that the adaptive strategies of organisms belonging to different phyla are different.'

(3) In the absence of a good adaptive argument in the first place, attribute failure to imperfect understanding of where an organism lives and what it does. This is again an old argument. Consider Wallace on why all details of colour and form in land snails must be adaptive, even if different animals seem to inhabit the same environment (1899, p. 148): 'The exact proportions of the various species of plants, the numbers of each kind of insect or of bird, the peculiarities of more or less exposure to sunshine or to wind at certain critical epochs, and other slight differences which to us are absolutely immaterial and unrecognizable, may be of

the highest significance to these humble creatures, and be quite sufficient to require some slight adjustments of size, form, or colour, which natural selection will bring about.'

(4) Emphasize immediate utility and exclude other attributes of form. Fully half the explanatory information accompanying the full-scale Fibreglass *Tyrannosaurus* at Boston's Museum of Science reads: 'Front legs a puzzle: how *Tyrannosaurus* used its tiny front legs is a scientific puzzle; they were too short even to reach the mouth. They may have been used to help the animal rise from a lying position.' (We purposely choose an example based on public impact of science to show how widely habits of the adaptationist programme extend. We are not using glass beasts as straw men; similar arguments and relative emphases, framed in different words, appear regularly in the professional literature.) We don't doubt that *Tyrannosaurus* used its diminutive front legs for something. If they had arisen *de novo*, we would encourage the search for some immediate adaptive reason. But they are, after all, the reduced product of conventionally functional homologues in ancestors (longer limbs of allosaurs, for example). As such, we do not need an explicitly adaptive explanation for the reduction itself. It is likely to be a developmental correlate of allometric fields for relative increase in head and hindlimb size. This non-adaptive hypothesis can be tested by conventional allometric methods (Gould (1974) in general; Lande (1978) on limb reduction) and seems to us both more interesting and fruitful than untestable speculations based on secondary utility in the best of possible worlds. One must not confuse the fact that a structure is used in some way (consider again the spandrels, ceiling spaces and Aztec bodies) with the primary evolutionary reason for its existence and conformation.

3. Telling stories

'All this is a manifestation of the rightness of things, since if there is a volcano at Lisbon it could not be anywhere else. For it is impossible for things not to be where they are, because everything is for the best' (Dr Pangloss on the great Lisbon earthquake of 1755 in which up to 50000 people lost their lives).

We would not object so strenuously to the adaptationist programme if its invocation, in any particular case, could lead in principle to its rejection for want of evidence. We might still view it as restrictive and object to its status as an argument of first choice. But if it could be dismissed after failing some explicit test, then alternatives would get their chance. Unfortunately, a common procedure among evolutionists does not allow such definable rejection for two reasons. First, the rejection of one adaptive story usually leads to its replacement by another, rather than to a suspicion that a different kind of explanation might be required. Since the range of adaptive stories is as wide as our minds are fertile, new stories can always be postulated. And if a story is not immediately available, one can always plead temporary ignorance and trust that it will be forthcoming, as did Costa & Bisol (1978), cited above. Secondly, the criteria for acceptance of a story are so loose that many pass without proper confirmation. Often, evolutionists use *consistency* with natural selection as the sole criterion and consider their work done when they concoct a plausible story. But plausible stories can always be told. The key to historical research lies in devising criteria to identify proper explanations among the substantial set of plausible pathways to any modern result.

We have, for example (Gould 1978) criticized Barash's (1976) work on aggression in mountain bluebirds for this reason. Barash mounted a stuffed male near the nests of two pairs of bluebirds while the male was out foraging. He did this at the same nests on three occasions at 10 day intervals: the first before eggs were laid, the last two afterwards. He then counted aggressive approaches of the returning male towards both the model and the female. At time one, aggression was high towards the model and lower towards females but substantial in both nests. Aggression towards the model declined steadily for times two and three and plummeted to near zero towards females. Barash reasoned that this made evolutionary sense since males would be more sensitive to intruders before eggs were laid than afterwards (when they can have some confidence that their genes are inside). Having devised this plausible story, he considered his work as completed (1976, pp. 1099, 1100):

> 'The results are consistent with the expectations of evolutionary theory. Thus aggression toward an intruding male (the model) would clearly be especially advantageous early in the breeding season, when territories and nests are normally defended...The initial aggressive response to the mated female is also adaptive in that, given a situation suggesting a high probability of adultery (i.e. the presence of the model near the female) and assuming that replacement females are available, obtaining a new mate would enhance the fitness of males... The decline in male–female aggressiveness during incubation and fledgling stages could be attributed to the impossibility of being cuckolded after the eggs have been laid...The results are consistent with an evolutionary interpretation.'

They are indeed consistent, but what about an obvious alternative, dismissed without test by Barash? Male returns at times two and three, approaches the model, tests it a bit, recognizes it as the same phoney he saw before, and doesn't bother his female. Why not at least perform the obvious test for this alternative to a conventional adaptive story: expose a male to the model for the *first* time after the eggs are laid.

Since we criticized Barash's work, Morton *et al.* (1978) repeated it, with some variations (including the introduction of a female model), in the closely related eastern bluebird *Sialia sialis*. 'We hoped to confirm', they wrote, that Barash's conclusions represent 'a widespread evolutionary reality, at least within the genus *Sialia*. Unfortunately, we were unable to do so.' They found no 'anticuckoldry' behaviour at all: males never approached their females aggressively after testing the model at any nesting stage. Instead, females often approached the male model and, in any case, attacked female models more than males attacked male models.

'This violent response resulted in the near destruction of the female model after presentations and its complete demise on the third, as a female flew off with the model's head early in the experiment to lose it for us in the brush' (1978, p. 969). Yet, instead of calling Barash's selected story into question, they merely devise one of their own to render both results in the adaptationist mode. Perhaps, they conjecture, replacement females are scarce in their species and abundant in Barash's. Since Barash's males can replace a potentially 'unfaithful' female, they can afford to be choosy and possessive. Eastern bluebird males are stuck with uncommon mates and had best be respectful. They conclude: 'If we did not support Barash's suggestion that male bluebirds show anticuckoldry adaptations, we suggest that both studies still had "results that are consistent with the expectations of evolutionary theory" (Barash 1976, p. 1099), as we presume any careful study would.' But what good is a theory that cannot fail in careful study (since by 'evolutionary theory', they clearly mean the action of natural selection applied to particular cases, rather than the fact of transmutation itself).

4. The master's voice re-examined

Since Darwin has attained sainthood (if not divinity) among evolutionary biologists, and since all sides invoke God's allegiance, Darwin has often been depicted as a radical selectionist at heart who invoked other mechanisms only in retreat, and only as a result of his age's own lamented ignorance about the mechanisms of heredity. This view is false. Although Darwin regarded selection as the most important of evolutionary mechanisms (as do we), no argument from opponents angered him more than the common attempt to caricature and trivialize his theory by stating that it relied exclusively upon natural selection. In the last edition of the *Origin*, he wrote (1872, p. 395):

> 'As my conclusions have lately been much misrepresented, and it has been stated that I attribute the modification of species exclusively to natural selection, I may be permitted to remark that in the first edition of this work, and subsequently, I placed in a most conspicuous position – namely at the close of the Introduction – the following words: "I am convinced that natural selection has been the main, but not the exclusive means of modification." This has been of no avail. Great is the power of steady misinterpretation.'

Romanes, whose once famous essay (1900) on Darwin's pluralism versus the panselectionism of Wallace and Weismann deserves a resurrection, noted of this passage (1900, p. 5): 'In the whole range of Darwin's writings there cannot be found a passage so strongly worded as this: it presents the only note of bitterness in all the thousands of pages which he has published.' Apparently, Romanes did not know the letter Darwin wrote to *Nature* in 1880, in which he castigated Sir Wyville Thomson for caricaturing his theory as panselectionist (1880, p. 32):

> 'I am sorry to find that Sir Wyville Thomson does not understand the principle of natural selection...If he had done so, he could not have written the following sentence in the Introduction to the Voyage of the Challenger: "The character of the abyssal fauna refuses to give the least support to the theory which refers the evolution of species to extreme variation guided only by natural selection." This is a standard of criticism not uncommonly reached by theologians and metaphysicians when they write on scientific subjects, but is something new as coming from a naturalist...Can Sir Wyville Thomson name any one who has said that the evolution of species depends only on natural selection? As far as concerns myself, I believe that no one has brought forward so many observations on the effects of the use and disuse of parts, as I have done in my "Variation of Animals and Plants under Domestication"; and these observations were made for this special object. I have likewise there adduced a considerable body of facts, showing the direct action of external conditions on organisms.'

We do not now regard all of Darwin's subsidiary mechanisms as significant or even valid, though many, including direct modification and correlation of growth, are very important. But we should cherish his consistent attitude of pluralism in attempting to explain Nature's complexity.

5. A partial typology of alternatives to the adaptationist programme

In Darwin's pluralistic spirit, we present an incomplete hierarchy of alternatives to immediate adaptation for the explanation of form, function, and behaviour.

(1) No adaptation and no selection at all. At present, population geneticists are sharply divided on the question of how much genetic polymorphism within populations and how much of the genetic differences between species is, in fact, the result of natural selection as opposed to purely random factors. Populations are finite in size and the isolated populations that form the first step in the speciation process are often founded by a very small number of individuals. As a result of this restriction in population size, frequencies of alleles change by *genetic drift*, a kind of random genetic sampling error. The stochastic process of change in gene frequency by random genetic drift, including the very strong sampling process that goes on when a new isolated population is formed from a few immigrants, has several important consequences. First, populations and species will become genetically differentiated, and even fixed for different alleles at a locus in the complete absence of any selective force at all.

Secondly, alleles can become fixed in a population *in spite of natural selection*. Even if an allele is favoured by natural selection, some proportion of population, depending upon the product of population size N and selection intensity s, will become homozygous for the less fit allele because of genetic drift. If Ns is large this random fixation for unfavourable alleles is a rare phenomenon, but if selection coefficients are on the order of the reciprocal of population size ($Ns = 1$) or smaller, fixation for deleterious alleles is common. If many genes are involved in influencing a metric character like shape, metabolism or behaviour, then the intensity of selection on each locus will be small and Ns per locus may be small. As a result, many of the loci may be fixed for non-optimal alleles.

Thirdly, new mutations have a small chance of being incorporated into a population, even when selectively favoured. Genetic drift causes the immediate loss of most new mutations after their introduction. With a selection intensity s, a new favourable mutation has a probability of only $2s$ of ever being incorporated. Thus, one cannot claim that, eventually, a new mutation of just the right sort for some adaptive argument will occur and spread. 'Eventually' becomes a very long time if only one in 1000 or one in 10000 of the 'right' mutations that do occur ever get incorporated in a population.

(2) No adaptation and no selection on the part at issue; form of the part is a correlated consequence of selection directed elsewhere. Under this important category, Darwin ranked his 'mysterious' laws of the 'correlation of growth'. Today, we speak of pleiotropy, allometry, 'material compensation' (Rensch 1959, pp. 179–187) and mechanically forced correlations in D'Arcy Thompson's sense (1942; Gould 1971). Here we come face to face with organisms as integrated wholes, fundamentally not decomposable into independent and separately optimized parts.

Although allometric patterns are as subject to selection as static morphology itself (Gould 1966), some regularities in relative growth are probably not under immediate adaptive control. For example, we do not doubt that the famous 0.66 interspecific allometry of brain size in all major vertebrate groups represents a selected 'design criterion,' though its significance remains elusive (Jerison 1973). It is too repeatable across too wide a taxonomic range to represent much else than a series of creatures similarly well designed for their different sizes. But another common allometry, the 0.2 to 0.4 intraspecific scaling among homeothermic adults differing in body size, or among races within a species, probably does not require a selectionist story though many, including one of us, have tried to provide one (Gould 1974). R. Lande (personal communication) has used the experiments of Falconer (1973) to show that selection upon *body size alone* yields a brain–body slope across generations of 0.35 in mice.

More compelling examples abound in the literature on selection for altering the timing of maturation (Gould 1977). At least three times in the evolution of arthropods (mites, flies and beetles), the same complex adaptation has evolved, apparently for rapid turnover of generations in strongly r-selected feeders on superabundant but ephemeral fungal resources: females reproduce as larvae and grow the next generation within their bodies. Offspring eat their mother from inside and emerge from her hollow shell, only to be devoured a few days later by their own progeny. It would be foolish to seek adaptive significance in paedomorphic morphology *per se*; it is primarily a by-product of selection for rapid cycling of generations. In more interesting cases, selection for small size (as in animals of the interstitial fauna) or rapid maturation (dwarf males of many crustaceans) has occurred by progenesis (Gould 1977, pp. 324–336), and descendant adults contain a mixture of ancestral juvenile and adult features. Many biologists have been tempted to find primary adaptive meaning for the mixture, but it probably arises as a by-product of truncated maturation, leaving some features 'behind' in the larval state, while allowing others, more strongly correlated with sexual maturation, to retain the adult configuration of ancestors.

(3) The decoupling of selection and adaptation.

(i) Selection without adaptation. Lewontin (1979) has presented the following hypothetical example: 'A mutation which doubles the fecundity of individuals will sweep through a population rapidly. If there has been no change in efficiency of resource utilization, the individuals will leave no more offspring than before, but simply lay twice as many eggs, the excess dying because of resource limitation. In what sense are the individuals or the population as a whole better adapted than before? Indeed, if a predator on immature stages is led to switch to the species now that immatures are more plentiful, the population size may actually decrease as a consequence, yet natural selection at all times will favour individuals with higher fecundity.'

(ii) Adaptation without selection. Many sedentary marine organisms, sponges and corals in particular, are well adapted to the flow régimes in which they live. A wide spectrum of 'good design' may be purely phenotypic in origin, largely induced by the current itself. (We may be sure of this in numerous cases, when genetically identical individuals of a colony assume different shapes in different microhabitats.) Larger patterns of geographic variation are often adaptive and purely phenotypic as well. Sweeney & Vannote (1978), for example, showed that many hemimetabolous aquatic insects reach smaller adult size with reduced fecundity when they grow at temperatures above and below their optima. Coherent, climatically correlated patterns in geographic distribution for these insects – so often taken as *a priori* signs of genetic adaptation – may simply reflect this phenotypic plasticity.

'Adaptation' – the good fit of organisms to their environment – can occur at three hierarchical levels with different causes. It is unfortunate that our language has focused on the common result and called all three phenomena 'adaptation': the differences in process have been obscured and evolutionists have often been misled to extend the Darwinian mode to the other two levels as well. First, we have what physiologists call 'adaptation': the phenotypic plasticity that permits organisms to mould their form to prevailing circumstances during ontogeny. Human 'adaptations' to high altitude fall into this category (while others, like resistance of sickling heterozygotes to malaria, are genetic and Darwinian). Physiological adaptations are not heritable, though the capacity to develop them presumably is. Secondly, we have a 'heritable' form of non-Darwinian adaptation in humans (and, in rudimentary ways, in a few other advanced social species):

cultural adaptation (with heritability imposed by learning). Much confused thinking in human sociobiology arises from a failure to distinguish this mode from Darwinian adaptation based on genetic variation. Finally, we have adaptation arising from the conventional Darwinian mechanism of selection upon genetic variation. The mere existence of a good fit between organism and environment is insufficient evidence for inferring the action of natural selection.

(4) Adaptation and selection but no selective basis for differences among adaptations. Species of related organisms, or subpopulations within a species, often develop different adaptations as solutions to the same problem. When 'multiple adaptive peaks' are occupied, we usually have no basis for asserting that one solution is better than another. The solution followed in any spot is a result of history; the first steps went in one direction, though others would have led to adequate prosperity as well. Every naturalist has his favourite illustration. In the West Indian land snail *Cerion,* for example, populations living on rocky and windy coasts almost always develop white, thick and relatively squat shells for conventional adaptive reasons. We can identify at least two different developmental pathways to whiteness from the mottling of early whorls in all *Cerion*, two paths to thickened shells and three styles of allometry leading to squat shells. All 12 combinations can be identified in Bahamian populations, but would it be fruitful to ask why – in the sense of optimal design rather than historical contingency – *Cerion* from eastern Long Island evolved one solution, and *Cerion* from Acklins Island another?

(5) Adaptation and selection, but the adaptation is a secondary utilization of parts present for reasons of architecture, development or history. We have already discussed this neglected subject in the first section on spandrels, spaces and cannibalism. If blushing turns out to be an adaptation affected by sexual selection in humans, it will not help us to understand why blood is red. The immediate utility of an organic structure often says nothing at all about the reason for its being.

6. Another, and unfairly maligned, approach to evolution

In continental Europe, evolutionists have never been much attracted to the Anglo-American penchant for atomizing organisms into parts and trying to explain each as a direct adaptation. Their general alternative exists in both a strong and a weak form. In the strong form, as advocated by such major theorists as Schindewolf (1950), Remane (1971), and Grassé (1977), natural selection under the adaptationist programme can explain superficial modifications of the *Bauplan* that fit structure to environment: why moles are blind, giraffes have long necks, and ducks webbed feet, for example. But the important steps of evolution, the construction of the *Bauplan* itself and the transition between *Baupläne*, must involve some other unknown, and perhaps 'internal', mechanism. We believe that English biologists have been right in rejecting this strong form as close to an appeal to mysticism.

But the argument has a weaker – and paradoxically powerful – form that has not been appreciated, but deserves to be. It also acknowledges conventional selection for superficial modifications of the *Bauplan*. It also denies that the adaptationist programme (atomization plus optimizing selection on parts) can do much to explain *Baupläne* and the transitions between them. But it does not therefore resort to a fundamentally unknown process. It holds instead that the basic body plans of organisms are so integrated and so replete with constraints upon adaptation (categories 2 and 5 of our typology) that conventional styles of selective arguments can explain little of interest about them. It does not deny that change, when it occurs, may be mediated by natural selection, but it holds that constraints restrict possible paths and modes of change so strongly that the constraints themselves become much the most interesting aspect of evolution.

Rupert Riedl, the Austrian zoologist who has tried to develop this thesis for English audiences (1977 and 1975, now being translated into English by R. Jefferies), writes:

> 'The living world happens to be crowded by universal patterns of organization which, most obviously, find no direct explanation through environmental conditions or adaptive radiation, but exist primarily through universal requirements which can only be expected under the systems conditions of complex organization itself. . . This is not self-evident, for the whole of the huge and profound thought collected in the field of morphology, from Goethe to Remane, has virtually been cut off from modern biology. It is not taught in most American universities. Even the teachers who could teach it have disappeared.'

Constraints upon evolutionary change may be ordered into at least two categories. All evolutionists are familiar with *phyletic* constraints, as embodied in Gregory's classic distinction (1936) between habitus and heritage. We acknowledge a kind of phyletic inertia in recognizing, for example, that humans are not optimally designed for upright posture because so much of our *Bauplan* evolved for quadrupedal life. We also invoke phyletic constraint in explaining why no molluscs fly in air and no insects are as large as elephants.

Developmental constraints, a subcategory of phyletic restrictions, may hold the most powerful rein of all over possible evolutionary pathways. In complex organisms, early stages of ontogeny are remarkably refractory to evolutionary change, presumably because the differentiation of organ systems and their integration into a functioning body is such a delicate process, so easily derailed by early errors with accumulating effects. Von Baer's fundamental embryological laws (1828) represent little more than a recognition that early stages are both highly conservative and strongly restrictive of later development. Haeckel's biogenetic law, the primary subject of late nineteenth century evolutionary biology, rested upon a misreading of the same data (Gould 1977). If development occurs in integrated packages, and cannot be pulled apart piece by piece in evolution, then the adaptationist programme cannot explain the alteration of developmental programmes underlying nearly all changes of *Bauplan*.

The German palaeontologist A. Seilacher, whose work deserves far more attention than it has received, has emphasized what he calls '*bautechnischer*', or *architectural*, constraints (Seilacher 1970). These arise not from former adaptations retained in a new ecological setting (phyletic constraints as usually understood), but as architectural restrictions that never were adaptations, but rather the necessary consequences of materials and designs selected to build basic *Baupläne*. We devoted

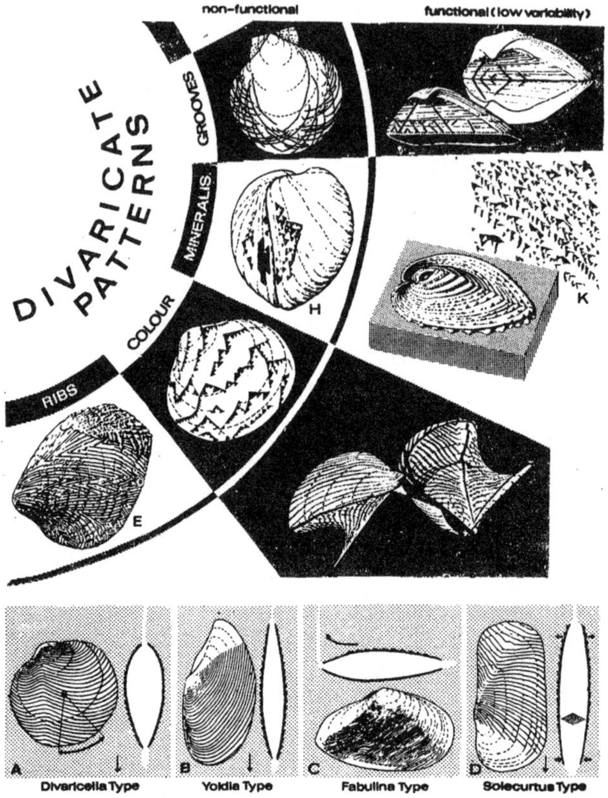

FIGURE 3. The range of divaricate patterns in molluscs. E, F, H, and L are non-functional in Seilacher's judgement. A–D are functional ribs (but these are far less common than non-functional ribs of the form E). G is the mimetic *Arca zebra*. K is *Corculum*. See text for details.

the first section of this paper to non-biological examples in this category. Spandrels must exist once a blueprint specifies that a dome shall rest on rounded arches. Architectural constraints can exert a far-ranging influence upon organisms as well. The subject is full of potential insight because it has rarely been acknowledged at all.

In a fascinating example, Seilacher (1972) has shown that the divaricate form of architecture (figure 3) occurs again and again in all groups of molluscs, and in brachiopods as well. This basic form expresses itself in a wide variety of structures: raised ornamental lines (not growth lines because they do not conform to the mantle margin at any time), patterns of coloration, internal structures in the mineralization of calcite, and incised grooves. He does not know what generates this pattern and feels that traditional and nearly exclusive focus on the adaptive value of each manifestation has diverted attention from questions of its genesis in growth and also prevented its recognition as a general phenomenon. It must arise from some characteristic pattern of inhomogeneity in the growing mantle, probably from the generation of interference patterns around regularly spaced centres; simple computer simulations can generate the form in this manner (Waddington & Cowe 1969). The general pattern may not be a direct adaptation at all.

Seilacher then argues that most manifestations of the pattern are probably non-adaptive. His reasons vary, but seem generally sound to us. Some are based on field observations: colour patterns that remain invisible because clams possessing them either live buried in sediments or remain covered with a periostracum so thick that the colours cannot be seen. Others rely on more general principles: presence only in odd and pathological individuals, rarity as a developmental anomaly, excessive variability compared with much reduced variability when the same general structure assumes a form judged functional on engineering grounds.

In a distinct minority of cases, the divaricate pattern becomes functional in each of the four categories (figure 3). Divaricate ribs may act as scoops and anchors in burrowing (Stanley 1970), but they are not properly arranged for such function in most clams. The colour chevrons are mimetic in one species (*Pteria zebra*) that lives on hydrozoan branches; here the variability is strongly reduced. The mineralization chevrons are probably adaptive in only one remarkable creature, the peculiar bivalve *Corculum cardissa* (in other species, they either appear in odd specimens or only as post-mortem products of shell erosion). This clam is uniquely flattened in an anterio-posterior direction. It lies on the substrate, posterior up. Distributed over its rear end are divaricate triangles of mineralization. They are translucent, while the rest of the shell is opaque. Under these windows dwell endosymbiotic algae!

All previous literature on divaricate structure has focused on its adaptive significance (and failed to find any in most cases). But Seilacher is probably right in representing this case as the spandrels, ceiling holes and sacrificed bodies of our first section. The divaricate pattern is a fundamental architectural constraint. Occasionally, since it is there, it is used to beneficial effect. But we cannot understand the pattern or its evolutionary meaning by viewing these infrequent and secondary adaptations as a reason for the pattern itself.

Galton (1909, p. 257) contrasted the adaptationist programme with a focus on constraints and modes of development by citing a telling anecdote about Herbert Spencer's fingerprints:

> 'Much has been written, but the last word has not been said, on the rationale of these curious papillary ridges; why in one man and in one finger they form whorls and in another loops. I may mention a characteristic anecdote of Herbert

Spencer in connection with this. He asked me to show him my Laboratory and to take his prints, which I did. Then I spoke of the failure to discover the origin of these patterns, and how the fingers of unborn children had been dissected to ascertain their earliest stages, and so forth. Spencer remarked that this was beginning in the wrong way; that I ought to consider the purpose the ridges had to fulfil, and to work backwards. Here, he said, it was obvious that the delicate mouths of the sudorific glands required the protection given to them by the ridges on either side of them, and therefrom he elaborated a consistent and ingenious hypothesis at great length. I replied that his arguments were beautiful and deserved to be true, but it happened that the mouths of the ducts did not run in the valleys between the crests, but along the crests of the ridges themselves.

We feel that the potential rewards of abandoning exclusive focus on the adaptationist programme are very great indeed. We do not offer a council of despair, as adaptationists have charged; for non-adaptive does not mean non-intelligible. We welcome the richness that a pluralistic approach, so akin to Darwin's spirit, can provide. Under the adaptationist programme, the great historic themes of developmental morphology and *Bauplan* were largely abandoned; for if selection can break any correlation and optimize parts separately, then an organism's integration counts for little. Too often, the adaptationist programme gave us an evolutionary biology of parts and genes, but not of organisms. It assumed that all transitions could occur step by step and underrated the importance of integrated developmental blocks and pervasive constraints of history and architecture. A pluralistic view could put organisms, with all their recalcitrant, yet intelligible, complexity, back into evolutionary theory.

References (Gould & Lewontin)

Baer, K. E. von 1828 *Entwicklungsgeschichte der Tiere*. Königsberg: Bornträger.

Barash, D. P. 1976 Male response to apparent female adultery in the mountain bluebird: an evolutionary interpretation. *Am. Nat.* **110**, 1097–1101.

Coon, C. S., Garn, S. M. & Birdsell, J. B. 1950 *Races*. Springfield, Ohio: C. Thomas.

Costa, R. & Bisol, P. M. 1978 Genetic variability in deep-sea organisms. *Biol. Bull.* **155**, 125–133.

Darwin, C. 1872 *The origin of species*. London: John Murray.

Darwin, C. 1880 Sir Wyville Thomson and natural selection. *Nature, Lond.* **23**, 32.

Davitashvili, L. S. 1961 *Teoriya polovogo otbora* [Theory of sexual selection]. Moscow: Akademii Nauk.

Falconer, D. S. 1973 Replicated selection for body weight in mice. *Genet. Res.* **22**, 291–321.

Galton, F. 1909 *Memories of my life*. London: Methuen.

Gould, S. J. 1966 Allometry and size in ontogeny and phylogeny. *Biol. Rev.* **41**, 587–640.

Gould, S. J. 1971 D'Arcy Thompson and the science of form. *New Literary Hist.* **2**(2), 229–258.

Gould, S. J. 1974 Allometry in primates, with emphasis on scaling and the evolution of the brain. In *Approaches to primate paleobiology. Contrib. Primatol.* **5**, 244–292.

Gould, S. J. 1977 *Ontogeny and phylogeny*. Cambridge, Mass.: Belknap Press.

Gould, S. J. 1978 Sociobiology: the art of storytelling. *New Scient.* **80**, 530–533.

Grassé, P.-P. 1977 *Evolution of living organisms.* New York: Academic Press.
Gregory, W. K. 1936 Habitus factors in the skeleton of fossil and recent mammals. *Proc. Am. phil. Soc.* **76**, 429–444.
Harner, M. 1977 The ecological basis for Aztec sacrifice. *Am. Ethnologist* **4**, 117–135.
Jerison, H. J. 1973 *Evolution of the brain and intelligence.* New York: Academic Press.
Lande, R. 1976 Natural selection and random genetic drift in phenotypic evolution. *Evolution* **30**, 314–334.
Lande, R. 1978 Evolutionary mechanisms of limb loss in tetrapods. *Evolution* **32**, 73–92.
Lewontin, R. C. 1978 Adaptation. *Scient. Am.* **239** (3), 156–169.
Lewontin, R. C. 1979 Sociobiology as an adaptationist program. *Behav. Sci.* (In the press.)
Morton, E. S., Geitgey, M. S. & McGrath, S. 1978 On bluebird 'responses to apparent female adultery'. *Am. Nat.* **112**, 968–971.
Ortiz de Montellano, B. R. 1978 Aztec cannibalism: an ecological necessity? *Science N.Y.* **200**, 611–617.
Remane, A. 1971 *Die Grundlagen des natürlichen Systems der vergleichenden Anatomie und der Phylogenetik.* Königstein-Taunus: Koeltz.
Rensch, B. 1959 *Evolution above the species level.* New York: Columbia University Press.
Riedl, R. 1975 *Die Ordnung des Lebendigen.* Hamburg: Paul Parey.
Riedl, R. 1977 A systems-analytical approach to macro-evolutionary phenomena. *Q. Rev. Biol.* **52**, 351–370.
Romanes, G. J. 1900 The Darwinism of Darwin and of the post-Darwinian schools. In *Darwin, and after Darwin*, vol. 2, new edn. London: Longmans, Green & Co.
Rudwick, M. J. S. 1964 The function of zig-zag deflections in the commissures of fossil brachiopods. *Palaeontology* **7**, 135–171.
Sahlins, M. 1978 Culture as protein and profit. *New York review of books*, 23 Nov., pp. 45–53.
Schindewolf, O. H. 1950 *Grundfragen der Paläontologie.* Stuttgart: Schweizerbart.
Seilacher, A. 1970 Arbeitskonzept zur Konstruktionsmorphologie. *Lethaia* **3**, 393–396.
Seilacher, A. 1972 Divaricate patterns in pelecypod shells. *Lethaia* **5**, 325–343.
Shea, B. T. 1977 Eskimo craniofacial morphology, cold stress and the maxillary sinus. *Am. J. phys. Anthrop.* **47**, 289–300.
Stanley, S. M. 1970 Relation of shell form to life habits in the Bivalvia (Mollusca). *Mem. geol. Soc. Am.* no. 125, 296 pp.
Sweeney, B. W. & Vannote, R. L. 1978 Size variation and the distribution of hemimetabolous aquatic insects: two thermal equilibrium hypotheses. *Science, N.Y.* **200**, 444–446.
Thompson, D. W. 1942 *Growth and form.* New York: Macmillan.
Waddington, C. H. & Cowe, J. R. 1969 Computer simulation of a molluscan pigmentation pattern. *J. theor. Biol.* **25**, 219–225.
Wallace, A. R. 1899 *Darwinism.* London: Macmillan.
Wilson, E. O. 1978 *On human nature.* Cambridge, Mass.: Harvard University Press.

Stebbins, G. L., and F. J. Ayala. 1981. Is a new evolutionary synthesis necessary? *Science* 213:967–971.

Two evolutionary controversies prevailed in the 1970s, both challenging the predominant role assigned to natural selection in the evolution of organisms. One derived from the neutrality theory of molecular evolution, which asserts that, at the molecular level, evolution is mostly governed by chance (see Kimura and Ohta 1971 in this volume). The second controversy emerged from the claim that at the large scale of the evolutionary process, called *macroevolution*, natural selection is much less significant than cladogenesis, the divergence of species. This divergence would be mostly a consequence of environmental variation, often drastic, in time and in space. It was argued that macroevolution was primarily characterized not by more-or-less steady gradual changes but rather by "punctuated equilibria," long periods of stasis, with little morphological change, alternating with periods of very rapid change (see Eldredge and Gould 1972, also in this volume).

The proponents of punctuated equilibrium as a dominant evolutionary process in macroevolution sought to augment the significance of paleontology in evolutionary theory by diminishing the role of natural selection and other microevolutionary processes and enhancing the significance of other, predominantly unspecified, macroevolutionary processes. But it is necessarily the case that natural selection, mutation, genetic drift, and migration play a role in macroevolution, since the evolutionary processes and outcomes of interest to paleontologists of necessity happened in past populations, in which the genetic processes of selection, mutation, drift, and migration were at work.

The proponents of punctuated equilibrium may legitimately claim that paleontology, as a discipline, is epistemologically autonomous, because there is no way in which macroevolutionary events could be predicted from the investigation of microevolutionary processes. This theoretical autonomy does not imply, however, that the component evolutionary processes occurring at any time in the populations of the past, even over long time scales, were other than the genetic processes of selection, mutation, drift and migration. Underlying the large-scale events that interest paleontologists are the component processes that interest geneticists. Nevertheless, the paleontological processes are theoretically autonomous, because they cannot be predicted from the constitutive genetic processes. Similarly, biological theories are autonomous relative to theories from physics and chemistry, even though all biological processes are made up of underlying physical and chemical processes.

RELATED READING

Avise, J. C. 1994. Molecular markers, natural history and evolution. Chapman and Hall, New York, New York, USA.

Conway Morris, S. 2003. Life's solution: inevitable humans in a lonely universe. Cambridge University Press, New York, New York, USA.

Erwin, D. M., and J. W. Valentine. 2013. The Cambrian explosion: the construction of animal diversity. Roberts, Greenwood, Colorado, USA.

Gould, S. J. 2002. The structure of evolutionary theory. Harvard University Press, Cambridge, Massachusetts, USA.

Kirschner, M. W., and J. C. Gerhart. 2005. The plausibility of life: resolving Darwin's dilemma. Yale University Press, New Haven, Connecticut, USA.

Knoll, A. H. 2003. Life on a young planet: the first three billion years of evolution on Earth. Princeton University Press, Princeton, New Jersey, USA.

Pigliucci, M., and G. B. Müller, editors. 2010. Evolution: the extended synthesis. MIT Press, Cambridge, Massachusetts, USA.

Simpson, G. G. 1944. Tempo and mode in evolution. Columbia University Press, New York New York, USA.

Is a New Evolutionary Synthesis Necessary?

G. Ledyard Stebbins and Francisco J. Ayala

Summary. The current (synthetic) theory of evolution has been criticized on the grounds that it implies that macroevolutionary processes (speciation and morphological diversification) are gradual. The extent to which macroevolution is gradual or punctuational remains to be ascertained. Macroevolutionary processes are underlain by microevolutionary phenomena and are compatible with the synthetic theory of evolution. But microevolutionary principles are compatible with both gradualism and punctualism; therefore, logically they entail neither. Thus, macroevolution and microevolution are decoupled in the important sense that macroevolutionary patterns cannot be deduced from microevolutionary principles.

The current theory of evolution, known as the "modern synthesis" (*1*), has been challenged by some scientists. Gould, for example, has written that "The modern synthesis, as an exclusive proposition, has broken down on *both of its fundamental claims*: extrapolationism (gradual allelic substitution as a model for all evolutionary change) and nearly exclusive reliance on selection leading to adaptation" (*2*; emphasis added). Gould goes on to voice the need for a "new and general evolutionary theory [that] will embody [the] notion of hierarchy and stress a variety of themes either ignored or explicitly rejected by the modern synthesis." Similar statements have been made by a few others (*3, 4*).

Many evolutionists would be surprised to see identified as the two "fundamental claims" of the modern synthesis those listed by Gould and most would not agree that the modern synthesis has "broken down." The impression that a "straw man" has been erected is confirmed when one discovers that the proposed new "themes" (*2–4*) are part and parcel of the modern synthesis (*1, 5–12*). However, the critics' appeal to the pluralistic structure of evolutionary theory, to the hierarchical nature of evolutionary processes, and to the distinctive contributions made by the study of macroevolutionary phenomena deserve attention.

Mutation and Selection

Genetic changes underlie the evolution of organisms; mutations are the ultimate source of the genetic variation that makes possible the evolutionary process. "Genetic mutations are changes in the hereditary materials. . . . They can be classified in one of two major categories: *gene* (or point) *mutations*, which affect only one or a few nucleotides within a gene; and *chromosomal mutations* (or aberrations), which affect the number of chromosomes, or the number or the arrangement of genes in a chromosome" (*1*, p. 57). Gene mutations are the source of allelic variation; chromosomal mutations make possible the evolution of the amount and organization of the hereditary material (DNA). It is misleading to criticize the modern synthesis on the alleged grounds of its exclusive reliance on "point mutations (micromutations)" (*2*). Chromosomal mutations (traditionally known as chromosomal abnormalities or aberrations) played an essential role in the development of the modern synthesis and remain one of its pivots (*6, 9–10*).

The frequencies of genes and gene arrangements change through the generations (evolve) owing to four processes: mutation, migration (gene flow), random drift, and natural selection. These four elementary processes of genetic change were already characterized by the early theorists who established the mathematical foundations of the modern synthesis (*5, 12, 13*). The highly organized character of organisms and their obvious adaptations are largely the result of natural selection operating under a variety of constraints, but responding to the demands of the environment. The constraints include the existing structure of organisms (and, hence, past history), the genetic variation available, as well as the particular circumstances of the physical and biotic components of the environment.

Without natural selection, populations of organisms would disintegrate over the generations because mutation and drift are random with respect to adaptation. But it is not correct that, for the synthetic theory, "All genetic change is adaptive"; or that "genetic drift certainly occurs—but only in populations so small and so near the brink that their rapid extinction will almost certainly ensue" (*2*, p. 120; *3*, pp. 20, 23–25). Controversy concerning the relative importance of random drift vis-à-vis natural selection has existed from the very beginning among the proponents of the modern synthesis. Fisher (*12*), for example, minimized the importance of random drift, but this is an important and decisive process in Wright's "shifting balance" version of the modern synthesis (*5*). Some evolutionists have relegated the importance of random drift to restricted—but by no means trivial—circumstances, such as "founder effects," which occur when a population is derived from only a few colonizers (*7*).

During the last decade no other issue has been more actively debated among evolutionists than the role of random drift. Molecular studies have shown that protein polymorphisms are pervasive in natural populations and that protein changes accompany the evolution of species (*14*). The neutrality theory of protein evolution proposes that evolution at the molecular level is largely due to random drift rather than being impelled by natural selection (*15*). But many evolutionists maintain that natural selection plays an essential role even at the molecular level (*14*). The "selectionist" and "neutralist" views of molecular evolution are competing hypotheses within the framework of the synthetic theory of evolution (*16*).

The Origin of Species

Living beings do not represent a continuum of all possible gene combinations generated at random, but are rather naturally grouped into species: arrays of populations between which intermediates are rare or absent. The distinctness of species is preserved by reproductive isolating mechanisms, that is, by biologically determined impediments to gene exchange, such as ethological or ecological

G. L. Stebbins is professor emeritus and F. J. Ayala is professor in the Department of Genetics, University of California, Davis 95616.

0036-8075/81/0828-0967$01.00/0

differentiation and hybrid sterility. In sexually reproducing organisms, species may be defined as "groups of interbreeding natural populations that are reproductively isolated from other such groups" (*6, 7*).

Interest in the process of speciation has recently burgeoned among paleontologists who sponsor the punctuated equilibria model (*3, 4, 17*), which is contrasted with the gradualistic model of macroevolution. Proponents of the punctualist model argue that, according to paleontological evidence, "species have tended to last for such long intervals of geological time that, once formed, they must have evolved very slowly. . . . This condition, when compared to the rapid pace of large-scale evolution, implies that most sizable evolutionary steps in the history of life must have occurred cryptically from a paleontological vantage point, during the rapid origination of certain species from small, localized populations of pre-existing species" (*3*, p. 3).

Whether macroevolution occurs according to the punctualist or the gradual model is something to be decided empirically. Certainly both modes have occurred in evolution, and the question then is their relative importance and the identification of factors that determine one or the other mode. Our primary concern here is, however, not this issue but rather whether any inconsistencies exist between the punctualist mode of evolutionary change and the synthetic theory's understanding of the speciation process.

We note, first, that the alleged relevance of punctuational evolution to speciation is based, at least in part, on two misunderstandings. The first one is a definitional artifact: paleontologists recognize species by their different morphologies as preserved in the fossil record (*18*). Thus, speciation events yielding little or no morphologically different products go totally unrecognized. Sibling (that is, morphologically indistinguishable) species are common in many groups of insects, in rodents, and in other well-studied organisms (*6, 7*). Speciation as seen by the paleontologist always involves substantial morphological change *because only when such change has occurred is the paleontologist able to recognize the presence of a new species*. The second misunderstanding concerns the time scale. When punctualists argue that paleontological evidence indicates that speciation is a rapid process (*3, 4, 17*), they are using a geological time scale. Instantaneous events in the paleontological scale, as in the transition between different geological strata, may involve thousands, at times many thousands, of years. In the microevolutionary scale of the population biologist, a thousand years is a long time, not an instant.

A more fundamental point is that rapid speciation, even in the microevolutionary scale, is not inconsistent with, and has been postulated by, the synthetic theory. Lewis's model of "saltational" speciation (*19*) and White's "stasipatric" speciation (*20*) are speciation models proposing that new species can arise in a few generations, as a result of the reproductive isolation produced by translocations and other chromosomal mutations. Polyploidy is the limiting case of rapid speciation—requiring only one or two generations—through chromosomal change (*6, 10*). Rapid speciation without chromosomal reorganization has not only been postulated by the proponents of the modern synthesis (*6, 7, 11*) but has been studied experimentally. A notable example—an incipient neospecies that arose in a *Drosophila paulistorum* culture, sometime between 1958 and 1963—was the subject of investigation by Dobzhansky for many years (*21*). In brief, the synthetic theory recognizes that there are a number of modes of speciation and that some of them, variously grouped under such terms as "saltational," "rapid," or "quantum" modes, require only a few generations and are effectively instantaneous in the geological time scale (*22*).

According to Gould, "The most exciting entry among punctuational models for speciation in ecological time is the emphasis, now coming from several quarters, on chromosomal alterations as isolating mechanisms" (*2*, p. 123). The role of chromosomal mutations in speciation is, like other important empirical questions, a subject of continued investigation and increased understanding. But the works to which Gould refers represent only the most recent accomplishments of a continuum that extends back to the 1930's (*6, 9, 10, 23*). Summarizing in 1950 the earlier work, Stebbins (*10*) concluded that (i) the most effective chromosome barriers of reproductive isolation come from the accumulation of small chromosomal changes; (ii) these changes may accumulate in a short time, such as 50 to 100 generations, to the point of resulting in reproductive isolation; (iii) these small changes occur largely independently of changes in the genes affecting external morphology; and, therefore, (iv) morphologically undifferentiated species may exhibit substantial chromosomal differences.

From Microevolution to Macroevolution

We come now to what has been called "the central question" posed by the proponents of punctualism, namely, "whether the mechanisms underlying microevolution can be extrapolated to explain macroevolution" (*24*). The argument has been succinctly expressed as follows: "if species originate in geological instants and then do not alter in major ways, then evolutionary trends cannot represent a simple extrapolation of allelic substitution within a population" (*2*, p. 125).

The question raised is the general issue of reduction as it applies to the different levels of the evolutionary process. Evolutionary trends are high-level phenomena predicated from events that encompass different species, as well as genera and higher taxa, and that extend over long periods of time. Microevolutionary studies are, on the contrary, concerned with evolutionary changes in populations that occur within "instants" of geological time. But, as so often happens with questions of reductionism, the issue of "whether the mechanisms underlying microevolution can be extrapolated" to macroevolution involves separate issues that must be distinguished in order to arrive at a satisfactory resolution.

Three separate questions, at least, are involved: (i) whether microevolutionary processes *operate* (and have operated in the past) throughout the different taxa in which macroevolutionary phenomena are observed; (ii) whether the microevolutionary processes identified by population geneticists (mutation, chromosomal change, random drift, natural selection) can account for the morphological changes and other macroevolutionary phenomena observed in higher taxa or, rather, whether additional kinds of genetic processes need to be postulated; and (iii) whether evolutionary trends and other macroevolutionary patterns can be predicted from knowledge of microevolutionary processes.

These distinctions may perhaps become clearer if we state them as they might be formulated by a biologist concerned with the question whether the laws of physics and chemistry can be extrapolated to biology. The first question would be whether the laws of physics and chemistry apply to the atoms and molecules present in living organisms. The second question would be whether biological phenomena can be accounted for as the result of interactions between atoms and molecules according to the laws known to physics and chemistry or whether the workings of organisms require additional kinds of interactions between atoms and molecules. The third question would be whether living phenomena can be predicted from the laws of physics and chemistry.

As to the first question, it is unlikely that any paleontologist would claim that mutation, chromosome change, drift, natural selection, and other microevolutionary processes do not apply to each of the populations of the higher taxa that are considered in macroevolution. There is, of course, an added dimension—macroevolutionists are largely concerned with phenomena of the past. Direct observation of microevolutionary processes in populations of long-extinct organisms is not possible. But there is no reason to believe that the processes of mutation, random drift, and natural selection, or the nature of the interactions between organisms and the environment would have been different in nature for, say, Paleozoic brachiopods and ostracoderms than for modern molluscs and fishes. Extinct and living populations—like different living populations—may have experienced quantitative differences in the relative importance of one or another process, but the processes could hardly have been different in kind.

The Origin of Differences Between Higher Taxa

The second question raised above has more substantive implications than the first. Can the microevolutionary processes studied by population geneticists account for macroevolutionary phenomena or do we need to postulate new kinds of genetic processes? The large morphological (phenotypic) changes observed in evolutionary history, and the rapidity with which they appear in the geological record, is one major matter of concern. Another issue is stasis—the apparent persistence of species, with little or no morphological change, for hundreds of thousands or millions of years. The apparent dilemma is that microevolutionary processes apparently yield small but continuous changes, while macroevolution as seen by punctualists occurs by large and rapid bursts of change followed by long periods without change.

Forty years ago Goldschmidt argued that the incompatibility is real: "The decisive step in evolution, the first step towards macroevolution, the step from one species to another, requires another evolutionary method than that of sheer accumulation of micromutations" (*25*). The specific solution postulated by Goldschmidt, that is, the occurrence of systemic mutations, yielding hopeful monsters, can be excluded in view of current genetic knowledge, but the issue raised by him deserves attention.

Single-gene or chromosome mutations may have large effects on the genotype because they act early in the embryo and their effects become magnified through development. Single-gene "macromutations" have been carefully analyzed, for example, in *Drosophila melanogaster*—mutations such as "bithorax" and the homeotic mutants that transform one body structure, for example, antennae, into another, such as legs. These large-effect mutations are not incompatible with the synthetic theory. Whether the kinds of morphological differences that characterize different taxa are due to such "macromutations" or to the accumulation of several mutations with small effect has been examined particularly in plants where fertile interspecific, and even intergeneric, hybrids can be obtained. The results of numerous studies do not support the hypothesis that the establishment of macromutations is necessary for divergence at the macroevolutionary level (*10*, *23*). In animals, even a familial character, the presence of three ocelli in drosophilids, can be changed by artificial selection, demonstrating that a family-distinctive trait can be produced by the accumulation of small mutations present in natural populations (*26*). Moreover, Lande has convincingly shown that major morphological changes, such as in the number of digits or limbs, can occur in a geologically rapid fashion through the accumulation of mutations each with a small effect (*27*). In general, the evidence from plants as well as from animals supports Fisher's (*12*) theoretical argument that the probability of incorporation of a mutation in a population is inversely proportional to the magnitude of the mutation's effect on the phenotype.

Nevertheless, rapid phenotypic evolution may be caused by relatively slight genetic changes that affect critical stages of development. Alberch (*28*) has described differences in the number and position of tarsal bones in salamanders of the genus *Plethodon*. It is not known at present whether only one mutation, or several with additive effects, is involved. But the important point is that only a few of the possible genetic changes can, in interaction with the rest of the genome, yield a functional phenotype; and, therefore, the organ can change in only one or very few directions. Phenotypic evolution is thus directed along certain channels that may be followed by separate lineages. To what extent canalization of development restricts the possible directions of morphological evolution is a question as yet unsolved.

How often mutations with large phenotypic effects are involved in the origin of new taxa is also an unsolved question. The punctualists' thesis that such mutations may have been largely responsible for macroevolutionary change is based on the rapidity with which morphological discontinuities appear in the fossil record (*2*, *3*). But the alleged evidence they present does not necessarily support the proposition. Microevolutionists and macroevolutionists use different time scales. The "geological instants" during which speciation and morphological shifts occur may involve intervals of the order of 50,000 years. There is little doubt that the gradual accumulation of small-effect mutations may yield sizable morphological changes during periods of that length. Anderson's study of body size in *Drosophila pseudoobscura* may serve as an example (*29*). Large populations, derived from a single set of parents, were set up at different temperatures and allowed to evolve on their own. A gradual, genetically determined, change in body size ensued, with flies kept at lower temperature becoming larger than those kept at higher temperatures. After 12 years, the mean size of the flies from the population kept at 16°C had become, when tested under standard conditions, approximately 10 percent greater than the size of the flies from the populations at 27°C; the change of mean value being greater than the standard deviation in size at the time when the tests were made. Assuming ten generations per year, the populations diverged at an average rate of 8×10^{-4} of the mean value per generation.

Paleontologists have emphasized the "extraordinary high net rate of evolution that is the hallmark of human phylogeny" (*3*). Interpreted in terms of the punctualist hypothesis, human phylogeny would have occurred as a succession of jumps, or geologically instantaneous saltations, interspersed by long periods without morphological change. Could these bursts of phenotypic evolution be due to the gradual accumulation of small changes? Consider cranial capacity, the character undergoing the greatest relative amount of change. The fastest rate of net change occurred between 500,000 years ago, when our ancestors were represented by *Homo erectus*, and 75,000 years ago, when Neanderthal man had acquired a cranial capacity similar to that of modern humans. In the intervening 425,000 years, cranial capacity evolved from about 900 cubic centimeters in Peking man to about 1400 cubic centimeters in Neanderthal people. Let us assume that the increase in brain size occurred in a single burst at the rate observed in *Drosophila pseudoobscura* of 8×10^{-4} of the mean value per generation. The change from 900 to 1400 cubic centimeters could have taken place in 540 generations or, if we assume 25 years per generation, in 13,500 years. Thirteen thousand years are, of course, a geological instant. Yet, this evolutionary "burst" could have taken place by gradual accumulation of small-effect mutations at rates compatible with those observed in microevolutionary studies (*30*).

We now raise the question of "stasis," the long-term persistence of species without morphological change. According to the model of punctuated equilibria, most phenotypic macroevolutionary change occurs in rapid bursts followed by long periods of stasis, during which little if any morphological change takes place. Phenotypic stability is compatible with microevolutionary processes; it ensues from stabilizing selection (*1, 6, 8*). Stebbins (*10*) in 1950 pointed out the morphological similarity, in forest trees and some herbs, between populations that have been separated from each other for millions of years. According to Dobzhansky (*31*), a successful morphology may persist unchanged for extremely long periods of time, even through speciation events. Some sibling species in *Drosophila* diverged from each other millions of years ago, yet their morphologies have remained identical to each other and to their ancestral species (*32*). Dobzhansky postulated that evolution in such cases continues, however, at the physiological or biochemical level; a prediction confirmed by recent molecular studies (*33*, p. 587).

Whether the phenomenon of paleontological stasis is as common as claimed by the punctualists needs to be carefully examined (*34*). As indicated by Levinton and Simon (*18*), paleontological taxonomy at the species level "requires the identification of species-specific characters which are invariant with time." Stasis may often be only apparent, as another artifact of the definition of species used.

Reduction, Hierarchy, and Macroevolution

We have just argued that the macroevolutionary patterns proposed by the model of punctuated equilibria—short periods of rapid phenotypic change followed by long spans of morphological stasis—are compatible with the theory of population genetics concerning microevolutionary processes. But does the theory predict that macroevolution will be punctuational? This is the third question formulated above, and the answer can only be no. The theory of population genetics is compatible with both punctualism and gradualism. Logically, therefore, it does not entail either. Whether macroevolution occurs predominantly according to the model of punctuated equilibria or to the model of phyletic gradualism is a question to be decided by studying macroevolutionary patterns, not by inference from our knowledge of microevolutionary processes (*35*).

Levinton and Simon (*18*) have written that "the implications of [the species-selection model proposed by the punctualists] should be of immediate concern to population biologists studying microevolutionary phenomena because it claims to negate the importance of population level phenomena in long term evolution," and they have gone on to "question the [punctualists'] belief that microevolution is decoupled from macroevolution." Statements of this kind need to be clarified. We have established above that at least three different issues are at stake, and have stated our solution to the first two issues. We may reiterate our points paraphrasing the terminology used by Levinton and Simon. Population level phenomena *are* important to long-term evolution because the populations in which macroevolutionary patterns are observed are the same populations that evolve at the microevolutionary level. Moreover, the study of microevolutionary phenomena *is* important to macroevolution because any theory of macroevolution that is correct must be compatible with well-established microevolutionary principles and theories; and indeed we have argued that the model of punctuated equilibria is compatible with the theory of population genetics. In these two senses—identity at the level of events and compatibility of theories—macroevolution cannot be decoupled from microevolution. But there is one sense (which epistemologically is most important) in which macroevolution and microevolution are decoupled, namely, in the sense that macroevolution is an autonomous field of study that must develop and test its own theories. In other words, macroevolutionary theories are not reducible (at least at the present state of knowledge and probably in principle) to microevolutionary theories.

Gould (2, p. 121) has pointed out that the study of evolution embodies "a concept of hierarchy—a world constructed not as a smooth and seamless continuum, permitting simple extrapolation from the lowest level to the highest, but as a series of ascending levels, each bound to the one below it in some ways and independent in others . . . 'emergent' features not implicit in the operation of processes at lower levels, may control events at higher levels."

The world of life is hierarchically structured. There is a hierarchy of levels: from atoms, through molecules, organelles, cells, tissues, organs, multicellular individuals and populations, to communities. Time adds another dimension of the hierarchy, with the interesting consequence that transitions from one level to another occur such that as time proceeds the descendants of a single species may include several species, genera, families, and so forth. Hierarchical organization often is such that the phenomena at a given level cannot be inferred from knowledge of the phenomena at a lower level of the hierarchy. Statements about "emergent" features imply this inability to predict from one level of organization to another. Consider, for example, the question whether water has emergent properties relative to its components, hydrogen and oxygen. One could argue that among the properties of hydrogen and oxygen one must include their ability to combine according to the formula H_2O and to exhibit the properties attributed to water. Proceeding accordingly, one could claim that the properties of oxygen and hydrogen include those of hemoglobin and other proteins as well as human speech and abstract thought, because oxygen and hydrogen have these properties when combined with other given atoms in certain ways. But this is a definitional maneuver that contributes little to the understanding of the relationships between complex systems and their constituent parts.

The consideration that is important is whether the properties of a complex object or system can be inferred from the study of component parts in isolation. It is for this reason that we do not usually include among the properties of hydrogen those of water, ethyl alcohol, proteins, or human beings (*36*).

The question of whether macroevolution is an autonomous field of knowledge is more appropriately posed in terms of the relationships between macroevolutionary and microevolutionary theories, rather than in terms of hierarchy of levels or emergent properties (*37*). The study of macroevolution is autonomous with respect to microevolutionary studies if the theories, hypotheses, and models of macroevolution cannot be "reduced" to the microevolutionary theories, hypotheses, or models. Two conditions are jointly necessary and sufficient for the reduction of one branch of science to another: derivability and connectability (*38*). The condition of derivability requires that the laws and theories of the branch of science to be reduced be derived as logical consequences from the laws and theories of some other branch of science. The condition of connectability requires that the distinctive terms of the secondary branch of science be redefined in the language of the branch of science to which it is reduced—this redefinition of terms is, of course, necessary in order to analyze the logical connections between the theories of the two branches of science.

Microevolutionary processes, as now known, are compatible with the two models of macroevolution—punctualism and gradualism. From microevolutionary knowledge, we cannot infer which one of those two macroevolutionary patterns prevails. Hence, the condition of derivability is not satisfied. Needless to say, the conflict between punctualism and gradualism is not the only macroevolutionary issue that cannot be decided by logical inference from microevolutionary principles. Consider, for example, the question of rates of morphological evolution. Three groups of crossopterygian fishes flourished during the Devonian. The lungfishes (Dipnoi) changed little for hundreds of millions of years and remain as relics. The coelacanths became highly successful in the open ocean until the Cretaceous, then declined and stagnated, leaving only the relictual *Latimeria*. The rhipidistians, in contrast, evolved into the amphibians, reptiles, and, finally, birds and mammals (*39*). Models to explain divergent rates of morphological evolution must incorporate factors other than microevolutionary principles, including rates of speciation and the environmental and biotic conditions that may account for successions of morphological change in some but not other lineages.

Distinctive macroevolutionary theories and models have been advanced concerning such issues as rates of morphological evolution, patterns of species extinctions, and historical factors regulating taxonomic diversity. As long as these theories and models are compatible with the theories and laws of population biology, the decision as to which one among alternative hypotheses is correct cannot be reached by recourse to microevolutionary principles. Such a decision must rather be based on appropriate tests with the use of macroevolutionary evidence (*3, 4, 18*). Thus, macroevolution is an autonomous field of evolutionary study and, in this epistemologically very important sense, macroevolution is decoupled from microevolution (*40*).

References and Notes

1. A recent summary of the theory of evolution can be found in Th. Dobzhansky *et al.*, *Evolution* (Freeman, San Francisco, 1977).
2. S. J. Gould, *Paleobiology* **6**, 119 (1980).
3. S. M. Stanley, *Macroevolution, Pattern and Process* (Freeman, San Francisco, 1979).
4. E. S. Vrba, *S. Afr. J. Sci.* **76**, 61 (1980).
5. S. Wright, *Genetics* **16**, 97 (1931); *Evolution and the Genetics of Populations* (Univ. of Chicago Press, Chicago, 1968–1978), vols. 1 to 4.
6. Th. Dobzhansky, *Genetics and the Origin of Species* (Columbia Univ. Press, New York, 1937); *Genetics of the Evolutionary Process* (Columbia Univ. Press, New York, 1970).
7. E. Mayr, *Systematics and the Origin of Species* (Columbia Univ. Press, New York, 1942); *Animal Species and Evolution* (Harvard Univ. Press, Cambridge, Mass., 1963).
8. G. G. Simpson, *Tempo and Mode in Evolution* (Columbia Univ. Press, New York, 1944).
9. M. J. D. White, *Animal Cytology and Evolution* (Cambridge Univ. Press, London, 1945); *Modes of Speciation* (Freeman, San Francisco, 1978).
10. G. L. Stebbins, *Variation and Evolution in Plants* (Columbia Univ. Press, New York, 1950).
11. H. L. Carson, *Am. Nat.* **109**, 83 (1975); G. L. Bush, *Annu. Rev. Ecol. Syst.* **6**, 339 (1975).
12. R. A. Fisher, *The Genetic Theory of Natural Selection* (Clarendon, Oxford, 1930).
13. J. B. S. Haldane, *The Causes of Evolution* (Harper, New York, 1932).
14. F. J. Ayala, Ed., *Molecular Evolution* (Sinauer, Sunderland, Mass., 1976).
15. M. Kimura, *Nature (London)* **217**, 624 (1968); J. L. King and T. H. Jukes, *Science* **164**, 788 (1969); M. Kimura and T. Ohta, *Nature (London)* **229**, 467 (1971).
16. F. J. Ayala, in *Molecular Evolution and Polymorphism*, M. Kimura, Ed., (National Institute of Genetics, Mishima, Japan, 1977), p. 73.
17. N. Eldredge, *Evolution* **25**, 156 (1971); S. J. Gould and N. Eldredge, *Paleobiology* **3**, 115 (1977).
18. J. S. Levinton and C. M. Simon, *Syst. Zool.* **29**, 130 (1980).
19. H. Lewis, *Science* **152**, 167 (1966).
20. M. J. D. White, *ibid.* **159**, 1065 (1968).
21. Th. Dobzhansky and O. Pavlovsky, *Genetics* **55**, 141 (1967); Th. Dobzhansky, *Science* **177**, 664 (1972).
22. We have characterized quantum speciation as "rapid, requiring only a few generations. The ancestor of new species . . . may consist of only one or a few individuals. . . . Quantum speciation usually and perhaps always includes one or more stochastic or chance events" (*1*, pp. 198–199).
23. J. Clausen, *Stages in the Evolution of Plant Species* (Cornell Univ. Press, Syracuse, N.Y., 1951); V. Grant, *Plant Speciation* (Columbia Univ. Press, New York, 1971).
24. R. Lewin, *Science* **210**, 883 (1980).
25. R. Goldschmidt, *The Material Basis of Evolution* (Yale Univ. Press, New Haven, Conn., 1940), p. 183.
26. J. M. Smith and K. C. Sondhi, *Genetics* **45**, 1039 (1960).
27. R. Lande, *Evolution* **30**, 314 (1976); *ibid.* **32**, 73 (1978).
28. P. Alberch, *Am. Zool.* **20**, 653 (1980).
29. W. W. Anderson, *Evolution* **27**, 278 (1973).
30. The mean cranial capacity of modern man is 1342 cm^3, with the standard deviation, based on several thousand individuals, being between 169 and 200 cm^3 [P. V. Tobias, *The Brain in Hominid Evolution* (Columbia Univ. Press, New York, 1971) p. 43]. Thus the standard deviation for cranial capacity in modern humans is greater than 10 percent of the mean—giving a coefficient of variation greater than that observed for body size in *Drosophila pseudoobscura* [see (*29*)].
31. Th. Dobzhansky, *Am. Nat.* **90**, 337 (1956).
32. The common occurrence of sibling species provides little support to the punctualists' postulated association between speciation events and morphological evolution.
33. F. J. Ayala, M. L. Tracey, D. Hedgecock, R. C. Richmond, *Evolution* **28**, 576 (1974).
34. T. J. M. Schopf, *Paleobiology* **6**, 380 (1980).
35. T. Hallam, *ibid.* **4**, 16 (1978); R. Lande, *ibid.* **6**, 233 (1980). We do not imply that all models of macroevolution are compatible with the theory of population genetics, but simply that the two models under consideration are, in the present state of knowledge, compatible with the theory. Some models of macroevolution, such as orthogenesis, are not compatible with our knowledge of microevolution. The conceptual field of macroevolutionary models compatible with population genetics is less than the universe of possible models, but is not reduced to only one such model.
36. Notice that, under the proposal herein made, whether or not a property is considered emergent depends on the state of knowledge; what appeared at one point as an emergent property might later be predictable from knowledge of the component parts. Notice also that the question of emergent properties is not limited to the world of life.
37. The reason is that the hierarchical differentiation of the subject matter is neither necessary nor sufficient for the autonomy of scientific disciplines. It is not necessary, because entities at a single hierarchical level can be the subject of diversified disciplines: cells are appropriate subject of study for cytology, genetics, immunology, and the like. In fact, identical events can be the subject of concern of different disciplines. The writing of this note can be studied by a physiologist interested in the muscular and nerve processes involved, by a philosopher interested in the epistemological question raised, by a psychologist concerned with thought processes, and so on. Moreover, hierarchical differentiation of subject matter is not a sufficient condition for the autonomy of the concerned disciplines; relativity theory applies all the way from subatomic particles to planetary motions, and genetic laws apply to multicellular organisms as well as to cellular and even subcellular entities.
38. E. Nagel, *The Structure of Science* (Harcourt, Brace, & World, New York, 1961); F. J. Ayala, *Am. Sci.* **56**, 207 (1968).
39. E. H. Colbert, *Evolution of the Vertebrates* (Wiley-Interscience, New York, ed. 3, 1980).
40. This does not imply, however, that macroevolutionary studies cannot be incorporated into the synthetic theory of evolution. Quite to the contrary, the modern theory of evolution is called "synthetic" because it incorporates knowledge from diverse autonomous disciplines, such as genetics, ecology, systematics, and paleontology.

41. Supported by NIH grant 1-P01-GM-2221.

Mayr, E. 1981. Biological classification: toward a synthesis of opposing methodologies. *Science* 214:510–516.

A solid knowledge of evolutionary facts and processes does not translate automatically into a sound scheme for biological classification, because taxonomy and systematics are to some extent arbitrary human endeavors directed toward categorization (in this case of biological objects). Thus, Linnaeus (1759) invented a method for biological classification without any inkling that evolution had occurred. As Darwin himself noted, a knowledge of genealogy or phylogeny by itself does not yield a solid classification (although it certainly can help). In the century following Darwin, a consensus gradually emerged that any biological classification should reflect, at least partially, the known or suspected phylogenetic relationships among the organisms being classified. There still remained much room for argument about the "best" stratagem for the classification of extant and extinct species, as evidenced by the emergence of three strong but conflicting schools of thought: numerical phenetics, cladistics, and evolutionary classification. Ernst Mayr reviews the merits and demerits of each of these three opposed ideologies and weighs in on his own preferred eclectic resolution of the heated debate between proponents of the different methods.

RELATED READING

Hennig, W. 1966. Phylogenetic systematics. University of Illinois Press, Urbana, Illinois, USA.
Linnaeus, C. 1759 (Reprinted 1964). Systema naturae. Wheldon & Wesley Ltd, New York, New York, USA.
Mayr, E., and P. D. Ashlock. 1991. Principles of systematic zoology. McGraw-Hill, New York, New York, USA.
Sokal, R. R., and P. H. A. Sneath. 1963. Principles of numerical taxonomy. Freeman, San Francisco, California, USA.

Biological Classification: Toward a Synthesis of Opposing Methodologies

Ernst Mayr

For nearly a century after the publication of Darwin's *Origin* (*1*) no well-defined schools of classifiers were recognizable. There were no competing methodologies. Taxonomists were unanimous in their endeavor to establish classifications that would reflect "degree of relationship." What differences there were among competing classifications concerned the number and kinds of characters that were used, whether or not an author accepted the principle of recapitulation, whether he attempted to "base his classification on phylogeny," and to what extent he used the fossil record (*2*). As a result of a lack of methodology, radically different classifications were sometimes proposed for the same group of organisms; also new classifications were introduced without any adequate justification except for the claim that they were "better." Dissatisfaction with such arbitrariness and seeming absence of any carefully thought out methodology, led in the 1950's and 1960's to the establishment of two new schools of taxonomy, numerical phenetics and cladistics, and to a more explicit articulation of Darwin's methodology, now referred to as evolutionary classification.

Summary. Currently a controversy is raging as to which of three competing methodologies of biological classification is the best: phenetics, cladistics, or evolutionary classification. The merits and seeming deficiencies of the three approaches are analyzed. Since classifying is a multiple-step procedure, it is suggested that the best components of the three methods be used at each step. By such a synthetic approach, classifications can be constructed that are equally suited as the basis of generalizations and as an index to information storage and retrieval systems.

The Major Schools of Taxonomy

Numerical phenetics. From the earliest preliterary days, organisms were grouped into classes by their outward appearance, into grasses, birds, butterflies, snails, and others. Such grouping "by inspection" is the expressly stated or unspoken starting point of virtually all systems of classification. Any classification incorporating the method of grouping taxa by similarity is, to that extent, phenetic.

In the 1950's to 1960's several investigators went one step further and suggested that classifications be based exclusively on "overall similarity." They also proposed, in order to make the method more objective, that every character be given equal weight, even though this would require the use of large numbers of characters (preferably well over a hundred). In order to reduce the values of so many characters to a single measure of "overall similarity," each character is to be recorded in numerical form. Finally, the clustering of species and their taxonomic distance from each other is to be calculated by the use of algorithms that operationally manipulate characters in certain ways, usually with the help of computers. The resulting diagram of relationship is called a phenogram. The calculated phenetic distances can be converted directly into a classification.

The fullest statement of this methodology and its underlying conceptualization was provided by Sokal and Sneath (*3*). They called their approach "numerical taxonomy," a somewhat misleading designation, since numerical methods, including numerical weighting, can be and have been applied to entirely different approaches to classification. The term numerical phenetics is now usually applied to this school. This has introduced some ambiguity since some authors have used the term phenetic broadly, applying it to any approach making use of the "similarity" of species and other taxa, while to the strict numerical pheneticists the term phenetic means the "theory-free" use of unweighted characters.

Cladistics (or cladism). This method of classification (*4*), the first comprehensive statement of which was published in 1950 by Hennig (*5*), bases classifications exclusively on genealogy, that is, on the branching pattern of phylogeny. For the cladist phylogeny consists of a sequence of dichotomies (*6*), each representing the splitting of a parental species into two daughter species; the ancestral species ceases to exist at the time of the dichotomy; sister groups must be given the same categorical rank; and the ancestral species together with all of its descendants must be included in a single "holophyletic" taxon.

Evolutionary classification. Phenetics and cladistics were proposed in the endeavor to replace the methodology of classification that had prevailed ever since Darwin and that was variously designated as the "traditional" or the "evolutionary" school, which bases its classifications on observed similarities and differences among groups of organisms, evaluated in the light of their inferred evolutionary history (*7*). The evolutionary school includes in the analysis all available attributes of these organisms, their correlations, ecological stations, and patterns of distributions and attempts to reflect both of the major evolutionary processes, branching and the subsequent diverging of the branches (clades). This school follows Darwin (and agrees in this point with the cladists) that classification must be based on genealogy and also agrees with Darwin (in contrast to the cladists) "that genealogy by itself does not give classification" (*8*).

The results of the evolutionary analysis are incorporated in a diagram, called a phylogram, which records both the branching points and the degrees of subsequent divergence. The method of inferring genealogical relationship with the help of taxonomic characters, as it was first carried out by Darwin, is an application of the hypothetico-deductive approach. Presumed relationships have to be tested again and again with the help of new characters, and the new evidence frequently leads to a revision of the inferences on relationship. This method is not circular (*9*) as has sometimes been suggested.

The author is Alexander Agassiz Professor Emeritus, Museum of Comparative Zoology, Harvard University, Cambridge, Massachusetts 02138.

Is There a Best Way to Classify?

Each of the three approaches to classification—phenetics, cladistics, and evolutionary classification—has virtues and weaknesses. The ideal classification would be one that would meet best as many as possible of the generally acknowledged objectives of a classification.

A biological classification, like any other, must serve as the basis of a convenient information storage and retrieval system. Since all three theories produce hierarchical systems, containing nested sets of subordinated taxa, they permit the following of information up and down the phyletic tree. But this is where the agreement among the three methods ends. Purely phenetic systems, derived from a single set of arbitrarily chosen characters, sometimes provide only low retrieval capacity as soon as other sets of characters are used. The effectiveness of the phenetic method could be improved by careful choice of selected characters. However, the method would then no longer be "automatic," because any selection of characters amounts to weighting.

Cladists use only as much information for the construction of the classification as is contained in the cladogram. They convert cladograms, quite unaltered, into classifications, only when the cladograms are strictly dichotomous. Even though cladists lose much information by this simplistic approach, the information on lines of descent can be read off their classifications directly. However, a neglect of all ancestral-descendant information reduces the heuristic value of their classifications. By contrast, since evolutionary taxonomists incorporate a great deal more information in their classifications than do the cladists, they cannot express all of it directly in the names and ranking of the taxa in their classifications. Therefore, they consider a classification simply to be an ordered index that refers them to the information that is stored elsewhere (in the detailed taxonomic treatments).

A far more important function of a classification, even though largely compatible with the informational one, is that it establishes groupings about which generalizations can be made. To the extent that classifications are explicitly based on the theory of common descent with modification, they postulate that members of a taxon share a common heritage and thus will have many characteristics in common. Such classifications, therefore, have great heuristic value in all comparative studies. The validity of specific observations can be generalized by testing them against other taxa in the system or against other kinds of characters (*10–12*).

Pheneticists, as well as cladists, have claimed that their methods of constructing classifications are nonarbitrary, automatic, and repeatable. The criticisms of these methods over the last 15 years (*13*) have shown, however, that these claims cannot be substantiated. It is becoming increasingly evident that a one-sided methodology cannot achieve all the above-listed objectives of a good classification.

The silent assumption in the methodologies of phenetics and cladistics is that classification is essentially a single-step procedure: clustering by similarity in phenetics, and establishment of branching patterns in cladistics. Actually a classification follows a sequence of steps, and different methods and concepts are pertinent at each of the consecutive steps. It seems to me that we might arrive at a less vulnerable methodology by developing the best method for each step consecutively. Perhaps the steps could eventually be combined in a single algorithm. In the meantime, their separate discussion contributes to the clarification of the various aspects of the classifying process.

Establishment of Similarity Classes

The first step is the grouping of species and genera by "inspection," that is, by a phenetic procedure. (I use phenetic in the broadest sense, not in the narrow one of numerical phenetics.) All of classifying consists of, or at least begins with, the establishment of similarity classes, such as a preliminary grouping of plants into trees, shrubs, herbs, and grasses. The reason why the method is so often successful is simply that—other things being equal—descendants of a common ancestor tend to be more similar to each other than they are to species that do not share immediate common descent. The method is thus excellent in principle. Numerical phenetics has nevertheless proved to be largely unsuccessful because (i) claims, such as "results objective and strictly repeatable," were not always justifiable since in practice different results are obtained when different characters are chosen or different programs of computation are used; (ii) the method was inconsistent in its claim of objectivity since subjective biological criteria were used in the assigning of variants (for example, sexes, age classes, and morphs) to "operational taxonomic units" (OTU's); and, most importantly, the method insisted on the equal weighting of all characters.

It is now evident that no computing method exists that can determine "true similarity" from a set of arbitrarily chosen characters. So-called similarity is a complex phenomenon that is not necessarily closely correlated with common descent, since similarity is often due to convergence. Most major improvements in plant and animal classifications have been due to the discovery of such convergence (*14*).

Different types of characters—morphological characters, chromosomal differences, enzyme genes, regulatory genes, and DNA matching—may lead to rather different grouping. Different stages in the life cycle may result in different groupings.

The ideal of phenetics has always been to discover a measure of total (overall) similarity. Since it is now evident that this cannot be achieved on the basis of a set of arbitrarily chosen characters, the question has been asked whether there is not a method to measure degrees of difference of the genotype as a whole. Improvements in the method of DNA hybridization offer hope that this method might give realistic classifications on a phenetic basis, at least up to the level of orders (*15*). The larger the fraction of the nonhybridizing DNA, the less reliable this method is, because it cannot be determined whether the nonmatching DNA is only slightly or drastically different.

Testing the Naturalness of Taxa

In the first step of the classifying procedure clusters of species were assembled that seemed to be more similar to each other than to species in other clusters. These clusters are the taxa we recognize tentatively (*16*). In order to make these clusters conform to evolutionary theory, two, operationally more or less inseparable tests, must be made: (i) determine for all species of a cluster (taxon) whether they are descendants of the nearest common ancestor and (ii) connect the taxa by a branching tree of common descent, that is, construct a cladogram. An indispensable preliminary of this testing is an analysis of the characters used to establish the similarity clusters.

Character analysis. A careful analysis shows almost invariably that some characters are better clues to relationship (have greater weight) than others. The fewer the number of available characters, the more carefully the weighting must be done. This weighting is one of the most controversial aspects of the classifying procedure. Investigators who come to systematics from the outside, say from mathematics, or who are beginners tend to demand objective or quantitative methods of weighting. There are such methods, principally ones based on the covariation of characters, but they are not nearly as informative as methods based on the biological evaluation of characters (*17*). But such an evaluation requires an understanding of many aspects of the to-be-classified group (that is, its life history, the inferred selection pressures to which it is exposed, and its evolutionary history) that may not be available to an outsider. This creates a genuine dilemma. If strictly taxometric methods were available that would produce satisfactory weighting, everyone would surely prefer them to weighting based on experience and biological knowledge. But so far such methods are still in their infancy.

The greatest difficulty for a purely phenetic method, indeed for any method of classification, is the discordance (noncongruence) of different sets of characters. Entirely different classifications may result from the use of characters of different stages of the life cycle as, for instance, larval versus adult characters. In a study of species of bees, Michener (*18*) obtained four different classifications when he sorted them into similarity classes on the basis of the characters of (i) larvae, (ii) pupae, (iii) the external morphology of the adults, and (iv) male genitalic structures. Phenetic delimitation of taxa unavoidably necessitates a great deal of decision-making on the use and weighting of characters. Often, when new sets of characters become available, their use may lead to a new delimitation of taxa or to a change in ranking.

Determination of the genealogy. Each group (taxon) tentatively established by the phenetic method is, so to speak, a hypothesis as to common descent, the validity of which must be tested. Is the delimited taxon truly monophyletic (*19*)? Are the species included in this taxon nearest relatives (descendants of the nearest common ancestor)? Have all species been excluded that are only superficially or convergently similar?

Methods to answer these questions have been in use since the days of Darwin, particularly the testing of the homology of critical characteristics of the included species. However, Hennig (*5*) was the first to articulate such methods explicitly, and these have been modified by some of his followers. These methods can be designated as the cladistic analysis.

Such an analysis involves first the partitioning of the joint characters of a group into ancestral ("plesiomorph" in Hennig's terminology) characters and derived ("apomorph") characters, that is, characters restricted to the descendants of the putative nearest common ancestor (*20*). The joint possession of homologous derived characters proves the common ancestry of a given set of species. A character is derived in relation to the ancestral condition of the character. The end product of such a cladistic character analysis is a cladogram, that is, a diagram (dendrogram) of the branching points of the phylogeny.

Although this procedure sounds simple, numerous practical difficulties have been pointed out (*21, 22*). Very often the branching points are inferred by way of single or very few characters and are affected by all the weaknesses of single character classifications. More serious are two other difficulties.

1) *Polarity*. A derived character is often simpler or less specialized than the ancestral condition. For this reason it can be difficult to determine polarity in a transformation series of characters, that is, to determine which end of the series is ancestral. Tattersall and Eldredge (*23*) stressed that "in practice it is hard, even impossible, to marshall a strong, logical argument for a given polarity for many characters in a given group." Are they primitive (ancestral) or derived? Much of the controversy concerning the phylogeny of the invertebrates, for instance, is due to differences of opinion concerning polarity. Hennig tried to elaborate methods for determining polarity but, as others (*24, 25*) have shown, with rather indifferent success. Since characters come and go in phyletic lines and since there is much convergence, the problem of polarity can rarely be solved unequivocally. There are three best types of evidence for polarity reconstruction. First is the fossil record. Although primitiveness and apparent ancientness are not correlated in every case, nevertheless as Simpson (*26*) stressed, "for any group with even a fair fossil record there is seldom any doubt that characters usual or shared by older members are almost always more primitive than those of later members." Second is sequential constraints. Consecutive chromosomal inversions (as in *Drosophila*) or sets of amino acid replacements (and presumably certain other molecular events) form definite sequences. Which end of the sequence is the beginning can usually not be read off from the sequence itself, but additional information (polarity of other character chains, geographical distribution, and the like) often permits an unequivocal determination of the polarity. Third is the reconstruction of the presumed evolutionary pathway. This can sometimes be done by studying evidence for adaptive shifts, the invasion of new competitors or the extinction of old ones, the behavior of correlated characters, and other biological evidence (*11*, pp. 886–887; *24*). Particular difficulties are posed when the polarity is reversed in the course of evolution, as documented in the fossil record.

2) *Kinds of derived characters*. Two taxa may resemble each other in a given character for one of three reasons: because the character existed already in the ancestry of the two groups before the evolution of the nearest common ancestor (symplesiomorphy in Hennig's terminology), because it originated in the common ancestor and is shared by all of his descendants (homologous apomorphy or synapomorphy), or because it originated independently by convergence in several descendant groups (nonhomologous or convergent apomorphy) (*27*). Since, according to the cladistic method, sister groups are recognized by the possession of synapomorphies, convergence poses a major problem. How are we to distinguish between homologous and convergent apomorphies? Hennig was fully aware of the critical importance of this problem, but it has been quietly ignored by many of his followers. Both grebes and loons, two orders of diving birds, have a prominent spur on the knee and were therefore called sister groups by one cladist. However, other anatomical and biochemical differences between the two taxa indicate that the shared derived feature was acquired by convergence. The reliability of the determination of monophyly of a group depends to a large extent on the care that is taken in discriminating between these two classes of shared apomorphy (*11*, pp. 880–890).

There is a third class of derived characters, so-called autapomorphies, which are characters that were acquired by and are restricted to a phyletic line after it branched off from its sister group.

The pheneticists do not undertake a character analysis. Cladists and evolutionary taxonomists agree with each other in principle on the importance of a careful character analysis. They disagree, however, fundamentally in how to use the findings of the character analysis in the construction of classifications, particularly the ranking procedure.

The Construction of a Classification

Cladistic classification. Cladists convert the cladogram directly into a cladistic classification. In such a classification taxa are delimited exclusively by holophyly, that is, by the possession of a common ancestor, rather than by a combination of genealogy and degree of divergence (*19*). This results in such incongruous combinations as a taxon containing only crocodiles and birds, or one containing only lice and one family of Mallophaga.

Taxa based exclusively on genealogy are of limited use in most biological comparisons. Since, as Hull (*28*) pointed out, cladists really classify characters rather than organisms, they have to make the arbitrary assumption that new apomorph characters originate whenever a line branches from its sister line. This is unlikely in most cases. Surely the reptilian species that originated the avian lineage lacked any of the flight specializations characteristic of modern birds, except perhaps the feathers (*29*).

Two principles govern the conversion of a cladogram into a cladistic classification: (i) all branchings are bifurcations that give rise to two sister groups, and (ii) branchings are usually connected with a change in categorical rank. Cladistic classifications are only representations of branching patterns, with complete disregard of evolutionary divergence, ancestor-descendant relationships, and the information content of autapomorph characters. Because these aspects of evolutionary change are neglected, the cladistic method of classification "either results in lumping very similar forms (parasites and their relatives) or in recognizing a multitude of taxa (perhaps also of other categories) regardless of the extreme similarity of some of them. Such simplistic procedures do violence to most biological attributes other than the pattern of the cladistic branching system, as well as to the function of a classification for convenient information transmittal and storage," as Michener remarked (*18*).

These objections show that the methodology of cladistic classification is not satisfactory. Anyone familiar with the history of taxonomy is strangely reminded of the principles of Aristotelian logical division when encountering cladistic classifications with their rigid dichotomies, the mandate that every taxon must have a sister group, and the principle of a straight-line hierarchy.

There has been much argument over the relationship between classification and phylogeny (*30*). Both cladists and evolutionary taxonomists agree that all members of a taxon must have a common ancestor. A phylogenetic analysis, and in particular a clear separation of homologous apomorphies from convergences, is a necessary component of the classifying procedure. Classificatory analysis often leads to new inferences on phylogeny, and new insights on phylogeny may necessitate changes in classification. These interactions are not in the least circular (*9*).

It is quite unnecessary in most cases to know the exact species that was the common ancestor of two diverging phyletic lines. An inability to specify such an ancestral species has rarely impeded paleontological research (*31, 32*). For instance, it is of little importance whether *Archaeopteryx* was the first real ancestor of modern birds or some other similar species or genus. What is important to know is whether birds evolved from lizard-like, crocodile-like, or dinosaur-like ancestors. If a reasonably good fossil record is available, it is usually possible, by the backward tracing of evolutionary trends and by the backward projection of divergent phyletic lines, to reconstruct a reasonably convincing facsimile of the representative of a phyletic line at an earlier time.

Simpson (*32*) has provided us with cogent arguments about why it is not permissible to reject information from the fossil record under the pretext that it fails to give the phylogenetic connections between fossil and recent taxa with absolute certainty. Hence, there is no merit in the suggestion to construct separate classifications for recent and for fossil organisms. After all, fossil species belong to the same tree of descent as living species. Indeed, enough evidence usually becomes available through a careful character analysis to permit relatively robust inferences on the most probable phylogeny. A number of recent endeavors have been made to develop a cladistic methodology that is quantitative and automatic. New methods in this area are published in rapid succession and it would seem too early to determine which is most successful and freest of possible flaws (*33*).

Evolutionary classification. The taxonomic task of the cladist is completed with the cladistic character analysis. The genealogy gives him the classification directly, since for him classification is nothing but genealogy. The evolutionary taxonomist carries the analysis one step further. He is interested not only in branching, but, like Darwin, also in the subsequent fate of each branch. In particular, he undertakes a comparative study of the phyletic divergence of all evolutionary lineages, since the evolutionary history of sister groups is often strikingly different. Among two related groups, derived from the same nearest common ancestor, one may hardly differ from the ancestral group, while the other may have entered a new adaptive zone and evolved into a novel type. Even though they are sister groups in the terminology of cladistics, they may deserve different categorical rank, because their biological characteristics differ to such an extent as to affect any comparative study. The importance of this consideration was stated by Darwin (*1*, p. 420): "I believe that the *arrangement* of the groups within each class, in due subordination and relation to the other groups, must be strictly genealogical in order to be natural, but that the *amount* of difference in the several branches or groups, though allied in the same degree in blood to their common progenitor, may differ greatly, being due to the different degrees of modification which they have undergone, and this is expressed by the forms being ranked under different genera, families, sections or orders." Darwin refers then to a diagram of three Silurian genera that have modern descendants; one has not even changed generically, but the other two have become distinct orders, one with three and the other with two families.

The question as to what extent an analysis of degrees of divergence is possible, is still debated. The cladist makes only "horizontal" comparisons, cataloging the synapomorphies of sister groups. The evolutionary taxonomist, however, also makes use of derived characters that are restricted to a single line of descent, so-called autapomorph characters (Fig. 1), which are apomorph characters restricted to a single sister group. The importance of autapomorphy is well illustrated by a comparison of birds with their sister group (*34*). Birds originated from that branch of the reptiles, the Archosauria, which also gave rise to the pterodactyls, dinosaurs, and crocodilians. The crocodilians are the sister group of the birds among living organisms; a stem group of archosaurians represents the common ancestry of birds and crocodilians. Although birds and crocodilians share a number of synapomorphies that originated after the archosaurian line had branched off from the other reptilian lines, nevertheless crocodilians are on the whole very similiar to other reptiles, that is, they have developed relatively few autapomorph characters. They represent the reptilian "grade," as many morphologists call it.

Birds, by contrast, have acquired a vast array of new autapomorph characters in connection with their shift to aerial living. Whenever a clade (phyletic lineage) enters a new adaptive zone that leads to a drastic reorganization of the clade, greater taxonomic weight may have to be assigned to the resulting transformation than to the proximity of joint ancestry. The cladist virtually ignores this ecological component of evolution.

The main difference between cladists and evolutionary taxonomists, thus, is in the treatment of autapomorph characters. Instead of automatically giving sister groups the same rank, the evolutionary taxonomist ranks them by considering the relative weight of their autapomorphies as compared to their synapomorphies (Fig. 1). For instance, one of the striking autapomorphies of man (in comparison to his sister group, the chimpanzee) is the possession of Broca's center in the brain, a character that is closely correlated with man's speaking ability. This single character is for most taxonomists of greater weight than various synapomorphous similarities or even identities in man and the apes in certain macromolecules such as hemoglobins and cytochrome *c*. The particular importance of autapomorphies is that they reflect the occupation of new niches and new adaptive zones that may have greater biological significance than synapomorphies in some of the standard macromolecules.

I agree with Szalay (*35*) when he says: "The loss of biological knowledge when not using a scheme of ancestor-descendant relationship, I believe, is great. In fact, whereas a sister group relationship may . . . tell us little, a postulated and investigated ancestor-descendant relationship may help explain a previously inexplicable character in terms of its origin and transformation, and subsequently its functional (mechanical) significance." In other words, the analysis of the ancestor-descendant relationships adds a great deal of information that cannot be supplied by the analysis of sister group relationships.

It is sometimes claimed that the analysis of ancestor-descendant relationships lacks the precision of cladistic sister group comparisons. However, as was shown above and as is also emphasized by Hull (*36*), the cladistic analysis is actually full of uncertainties. The slight possible loss of precision, caused by the use of autapomorphies, is a minor disadvantage in comparison with the advantage of the large amount of additional information thus made available.

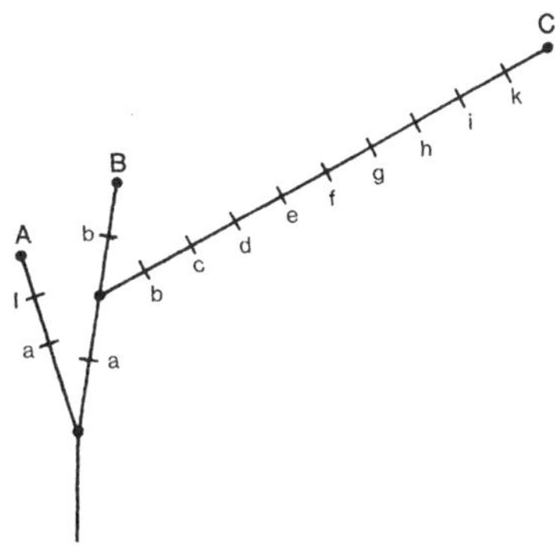

Fig. 1. Cladogram of taxa A, B, and C. Cladists combine B and C into a single taxon because B and C share the synapomorph character b. Evolutionary taxonomists separate C from A and B, which they combine, because C differs by many (c through k) autapomorph characters from A and B and shares only one (b) synapomorph character with B.

The information on autapomorphies permits the conversion of the cladogram into a phylogram. The phylogram differs from the cladogram by the placement of sister groups at different distances from the joint common ancestry (branching point) and by the expression of degree of divergence by different angles. Both of these topological devices can be translated into the respective categorical ranking of sister groups. These methods (*37*) generally attempt to discover the shortest possible "tree" that is compatible with the data. Yet, anyone familiar with the frequency of evolutionary reversals and of evolutionary opportunism, realizes the improbability of the assumption that the tree constructed by this so-called "parsimony method" corresponds to the actual phylogenetic tree. "To regard [the shortest tree method] as parsimonious completely misconceives the intent and use of parsimony in science" (*38*).

It is not always immediately evident whether a tree construction algorithm is based on cladists principles or on the methods of evolutionary classification. If the "special similarity" on which the trees are based are strictly synapomorphies, then the method is cladistic. If autapomorphies are also given strong weight, then the method falls under evolutionary classification.

The particular aspect of the method of evolutionary taxonomy found most unacceptable to cladists is the recognition of "paraphyletic" taxa. A paraphyletic taxon is a holophyletic group from which certain strikingly divergent members have been removed. For instance, the class Reptilia of the standard zoological literature is paraphyletic, because birds and mammals, two strikingly divergent descendants of the same common ancestor of all the Reptilia, are not included. Nevertheless, the traditional class Reptilia is monophyletic, because it consists exclusively of descendants from the common ancestor, even though it excludes birds and mammals owing to the high number of autapomorphies of these classes. The recognition of paraphyletic taxa is particularly useful whenever the recognition of definite grades of evolutionary change is important.

The Ranking of Taxa

Once species have been grouped into taxa the next step in the process of biological classification is the construction of a hierarchy of these taxa, the so-called Linnaean hierarchy. The hierarchy is constructed by assigning a definite rank such as family or order to each taxon, subordinating the lower categories to the higher ones. It is a basic weakness of cladistics that it lacks a sensitive method of ranking and simply gives a new rank after each branching point. The evolutionary taxonomist, following Darwin, ranks taxa by the degree of divergence from the common ancestor, often assigning a different rank to sister groups. Rank determination is one of the most difficult and subjective decision processes in classification. One aspect of evolution that causes difficulties is mosaic evolution (*39*). Rates of divergence of different characters are often drastically different. Conventionally taxa, such as those of vertebrates, are described and delimited on the basis of external morphology and of the skeleton, particularly the locomotory system. When other sets of morphological characters are used (for example, sense organs, reproductive system, central nervous system, or chromosomes), the evidence they provide is sometimes conflicting. The situation can become worse, if molecular characters are also used. The anthropoid genus *Pan* (chimpanzee), for instance, is very similar to *Homo* in molecular characters, but man differs so much from the anthropoid apes in traditional characters (central nervous system and its capacities) and occupation of a highly distinct adaptive zone that Julian Huxley even proposed to raise him to the rank of a separate kingdom—Psychozoa.

It has been suggested that different classifications should be constructed for each kind of character, or at least for morphological and molecular characters. Yet there is already much evidence that the acceptance of several classifications based on different characters would lead to insurmountable complications. By taking all available data into consideration simultaneously, a classification can usually be constructed that can serve conveniently as an all-purpose classification or, as Hennig (*5*) called it, "a general reference system."

It is usually possible to derive more than one classification from a phylogram, because higher taxa are usually composed of several end points of the phylogram, and different investigators differ by the degree to which they lump such terminal branches into a single higher taxon (*40*). An example is the phylogram of the higher ferns on which, as Wagner (*41*) has shown, six different classifications have been founded (Fig. 2) and many more are possible. The extent to which investigators "split" or "lump" higher taxa, thus, is of considerable influence on the classifications they produce.

Comparison of the Three Major Schools

Each school believes that its classification is the "best." Pheneticists as well as cladists claim that their respective methods have also the great merit of giving automatically nonarbitrary results. These claims cannot be substantiated. To be sure grouping by phenetic characters and determination of holophyly by cladistic analysis are valuable components of the procedure of biological classification. The great deficiency of both phenetics and cladistics is the failure to reflect adequately the past evolutionary history of taxa.

What needs to be emphasized once more is the fact that groups of organisms are the product of evolution and that no classification can hope to be satisfactory that does not take this fact fully into consideration. Both pheneticists and cladists are ambiguous in their attitude toward the evolutionary theory. The pheneticists claim that their approach is completely theory-free, but they nevertheless assume that their method will produce a hierarchy of taxa that corresponds to descent with modification. On the basis of this assumption, they also claim to be "evolutionary taxonomists" (*42*), but the fact that different phenetic procedures may produce very different classifications and that their procedure is not influenced by evolutionary considerations refutes this assertion. The cladists exclude most of evolutionary theory (for example, inferences on selection pressures, shifts of adaptive zones, evolutionary rates, and rates of evolutionary divergence) from their consideration (*43*) and tend increasingly not to classify species and taxa, but only taxonomic characters (*28*) and their origin. The connection with evolutionary principles is exceedingly tenuous in many recent cladistic writings.

By contrast, the evolutionary taxonomists, as indicated by the name of their school and by well-articulated statements of some of its major representatives (*7*), expressly base their classifications on evolutionary theory. They aim to construct classifications that reflect both of the two major evolutionary processes, branching and divergence (cladogenesis and anagenesis). They make full use of information on shifts into new adaptive zones and rates of evolutionary change and believe that the resulting classifications are a key to a far richer information content.

Although the three schools still seem rather fundamentally in disagreement, as far as the basic principles of classification are concerned, the more moderate representatives have quietly incorporated some of the criteria of the opposing schools, so that the differences among them have been partially obliterated. For instance, Farris' (*44*) clustering of special similarities is a phenetic method based on the weighting of characters. The evolutionary school uses phenetic criteria to establish similarity classes and to construct a classification, and cladistic criteria to test the naturalness of taxa. Comparing what McNeill (*45*) says in favor of phenetics (appropriately modified) and Farris (*44*) against it, we find that the gap has narrowed. I have no doubt that moderates will be able to develop an eclectic methodology, one that contains a proper balance of phenetics and cladistics that will produce far more "natural classifications" (*16*) than any one-sided approach that relies exclusively on a single criterion, whether it be overall similarity, parsimony of branching pattern, or what not. Evolutionary taxonomy, from Darwin on, has been characterized by the adoption of an eclectic approach that makes use of similarity, branching pattern, and degree of evolutionary divergence.

Classification and Information Retrieval

Biological classifications have two major objectives: to serve as the basis of biological generalizations in all sorts of comparative studies and to serve as the key to an information storage system. Up to this point, I have concentrated on

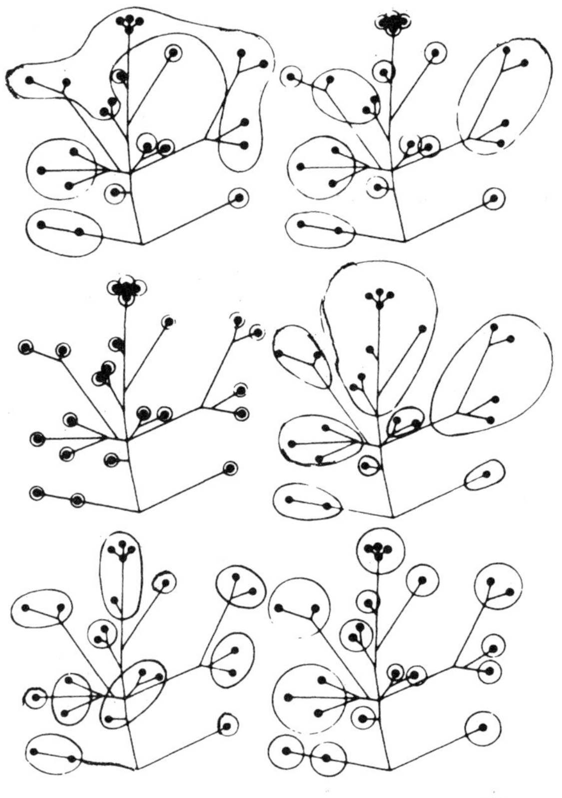

Fig. 2. Six different possible classifications of ferns, based on the same dendrogram. Each filled circle is a genus, and each open circle is a family. The differences are due to which and how many genera are combined to make up the families. [From W. H. Wagner (*41*, figure 7)].

those aspects of classifying that help to secure a sound basis for generalizations. This leaves unanswered the question of whether achievement of this first objective is, or is not, reconcilable with achievement of the second objective. Is the classification that is soundest as a basis of generalizations also most convenient for information retrieval? This, indeed, seems to have been true in most cases I have encountered. However, we can also look at this problem from another side.

It is possible at nearly each of the three major steps in the making of a classification to make a choice between several alternatives. These choices may be scientifically equivalent, but some may be more convenient in aiding information retrieval than others. If we choose one of them, it is not necessarily because the alternatives were "falsified," but rather because the chosen method is "more practical." In this respect, biological classifications are not unique. Scientific theories are nearly always judged by criteria additional to truth or falsity, for instance, by their simplicity or, in mathematics, by their "elegance." Therefore, it can be asserted that convenience in the use of a classification, including its function as key to information retrieval, is not necessarily in conflict with its more purely scientific objectives (*46–48*).

References and Notes

1. C. Darwin, *On the Origin of Species* (Murray, London, 1859).
2. For an illuminating survey of the thinking of that period see F. A. Bather [*Proc. Geol. Soc. London* **83**, LXII (1927)].
3. R. R. Sokal and P. H. A. Sneath, *Principles of Numerical Taxonomy* (Freeman, San Francisco, 1963). A drastically revised second edition was published in 1973.
4. The method was first published under the misleading name phylogenetic systematics, but since it is based on only a single one (branching) of the various processes of phylogeny, the terms cladism or cladistics have been substituted and are now widely accepted.
5. W. Hennig's original statement is *Grundzüge einer Theorie der Phylogenetischen Systematik* (Deutscher Zentralverlag, Berlin, 1950). A greatly revised second edition (reprinted in 1979) is *Phylogenetic Systematics*, D. D. Davis and R. Zangerl, Eds. (Univ. of Illinois Press, Urbana, 1966); see also W. Hennig (*47*). An independent phylogenetic analysis of characters was made by T. P. Maslin [*Syst. Zool.* **1**, 49 (1952)]. For an overview of the more significant recent literature see D. Hull (*36*) and J. S. Farris [*Syst. Zool.* **28**, 483 (1979)].
6. Some cladists in recent years have relaxed the requirements of strict dichotomy and have permitted tri- and polyfurcations or have quietly abandoned dichotomy by admitting empty internodes in the cladogram. Polyfurcations can be translated into several alternate bifurcations [see J. Felsenstein, *Syst. Zool.* **27**, 27 (1978)], and this makes the automatic conversion of the cladogram into a classification of sister groups impossible.
7. The classical statement of this theory is to be found in C. Darwin (*1*, pp. 411–434). G. G. Simpson [*Principles of Animal Taxonomy* (Columbia Univ. Press, New York, 1961)] and E. Mayr (*48*) provide comprehensive modern presentations of this theory. Several critical recent analyses are: W. Bock (*11*); C. D. Michener (*18*); *Syst. Zool.* **27**, 112 (1978); P. D. Ashlock (*12*).
8. F. Darwin, *Life and Letters of Charles Darwin* (Murray, London, 1887), vol. 2, p. 247.
9. D. Hull, *Evolution* **21**, 174 (1967); see also W. Bock (*11*).
10. F. E. Warburton, *Syst. Zool.* **16**, 241 (1967); W. Bock, *ibid.* **22**, 375 (1973).
11. W. Bock, in *Major Patterns in Vertebrate Evolution*, M. K. Hecht, P. C. Goody, B. M. Hecht, Eds. (NATO Advanced Study Institute Series, Plenum, New York, 1977), vol. 14, pp. 851–895.
12. P. D. Ashlock, *Syst. Zool.* **28**, 441 (1979).
13. I shall not, at this time, recount the almost interminable controversies among the three schools. For critiques of phenetics see E. Mayr (*48*, pp. 203–211), L. A. S. Johnson [*Syst. Zool.* **19**, 203 (1970)], and D. Hull [*Annu. Rev. Ecol. Syst.* **1**, 19 (1970)]. Some of the weaknesses pointed out by these early critics have been corrected in the 1973 edition of Sokal and Sneath (*3*) and by J. S. Farris (*44*). For critiques of cladistics see E. Mayr (*21*), R. R. Sokal (*22*), G. G. Simpson (*32*), D. Hull (*36*), P. D. Ashlock [*Annu. Rev. Ecol. Syst.* **5**, 81 (1974)]; and L. van Valen (*49*).
14. A particularly illuminating example is the breaking up of the plant group Amentiferae, which has been shown to consist of taxa secondarily adapted for wind pollination [R. F. Thorne, *Brittonia* **25**, 395 (1973)]. Examples among animals of radical reclassifications are the Rodentia, parasitic bees, certain beetle families, and the turbellarians.
15. C. G. Sibley, in preparation.
16. There have been arguments since before the days of Linnaeus about how to determine whether or not a system, a classification, is "natural." William Whewell, at a time before Darwin, had proclaimed his theory of common descent, expressed the then prevailing pragmatic consensus, "The maxim by which all systems professing to be natural must be tested is this: that the arrangement obtained from one set of characters coincides with the arrangement obtained from another set" [W. Whewell, *Philos. Inductive Sci.* **1**, 521 (1840)]. Interestingly, the covariance of characters is still perhaps the best practical test of the goodness of a classification. Since Darwin, of course, that classification is considered most natural that best reflects the inferred evolutionary history of the organisms involved.
17. For a tabulation and analysis of such qualitative methods of weighting, see E. Mayr (*48*, pp. 220–228).
18. C. D. Michener, *Syst. Zool.* **26**, 32 (1977).
19. I use the word monophyletic in its traditional sense, as a qualifying adjective of a taxon. Various definitions of monophyletic have been proposed but all of them for the same concept, a qualifying statement concerning a taxon. A taxon is monophyletic if all of its members are derived from the nearest common ancestor [E. Haeckel, *Natürliche Schöpfungsgeschichte* (Reimer, Berlin, 1868)]. Cladists have attempted to turn the situation upside down by placing all descendants of an ancestor into a taxon. Monophyletic thus becomes a qualifying adjective for descent, and a taxon is not recognized by its characteristics but only by its descent. The transfer of such a well-established term as monophyletic to an entirely different concept is as unscientific and unacceptable as if someone were to "redefine" mass, energy, or gravity by attaching these terms to entirely new concepts. P. D. Ashlock [*Syst. Zool.* **20**, 63 (1971)] has proposed the term holophyletic for the assemblage of descendants of a common ancestor. See also P. D. Ashlock (*12*, p. 443).
20. Terms like apomorph, synapomorph, derived, ancestral, and so forth always refer to characters of taxa at all levels. A genus may have synapomorphies with another genus, and so may an order with another order. It is this applicability of the same criteria for taxa of all ranks that permits the construction of the Linnaean hierarchy.
21. E. Mayr, *Z. Zool. Syst. Evolutionsforsch.* **12**, 94 (1974); reprinted in, E. Mayr, *Evolution and the Diversity of Life* (Harvard Univ. Press, Cambridge, Mass., 1976), pp. 433–478.
22. R. R. Sokal, *Syst. Zool.* **24**, 257 (1975).
23. I. Tattersall and N. Eldredge, *Am. Sci.* **65**, 204 (1977).
24. D. S. Peters and W. Gutmann, *Z. Zool. Syst. Evolutionsforsch.* **9**, 237 (1971).
25. O. Schindewolf, *Acta Biotheor.* **18**, 273 (1968); H. K. Erben, *Verh. Dtsch. Zool. Ges.* **79**, 116 (1979).
26. G. G. Simpson (*32*); see also L. van Valen (*49*).
27. For a diagram of these three categories of morphological resemblance see figure 1 in W. Hennig (*47*).
28. "Cladistic classifications do not represent the order of branching of sister groups, but the order of emergence of unique derived characters" [see D. Hull (*36*)].
29. G. G. Simpson [*The Major Features of Evolution* (Columbia Univ. Press, New York, 1953), p. 348] discusses the fallacy of the cladist assumption.
30. Phylogeny is equated by cladists with cladogenesis (branching), while the evolutionary taxonomist subsumes both branching and evolutionary divergence (anagenesis) under phylogeny.
31. C. W. Harper, *J. Paleontol.* **50**, 180 (1976).
32. G. G. Simpson, in *Phylogeny of the Primates*, W. Pluckett and F. S. Szalay, Eds. (Plenum, New York, 1975), pp. 3–19.
33. J. H. Camin and R. R. Sokal, *Evolution* **19**, 311 (1965); W. M. Fitch and E. Margoliash, *Science* **155**, 279 (1967); W. M. Fitch, in *Major Patterns of Vertebrate Evolution*, M. K. Hecht, P. C. Goody, B. M. Hecht, Eds. (NATO Advanced Study Institute Series, Plenum, New York, 1977), vol. 14, pp. 169–204.
34. There are literally hundreds of cases to illustrate this situation. I use again the classical case of birds and crocodilians because even a nonbiologist will understand the situation if such familiar animals are used. The holophyletic classification of the lice (Anoplura) derived from one of the suborders of the Mallophaga is another particularly instructive example [K. C. Kim and H. W. Ludwig, *Ann. Entomol. Soc. Am.* **71**, 910 (1978)].
35. F. S. Szalay, *Syst. Zool.* **26**, 12 (1977).
36. D. Hull, *ibid.* **28**, 416 (1979).
37. J. W. Hardin, *Brittonia* **9**, 145 (1957); W. H. Wagner, in *Plant Taxonomy: Methods and Principles*, L. Benson, Ed. (Ronald, New York, 1962), pp. 415–417; A. G. Kluge and J. S. Farris, *Syst. Zool.* **18**, 1 (1969); J. S. Farris, *Am. Nat.* **106**, 645 (1972).
38. L. H. Throckmorton, in *Biosystematics in Agriculture*, J. A. Romberger, Ed. (Wiley, New York, 1978), p. 237. Others who have questioned the validity of the so-called parsimony principle are M. Ghiselin [*Syst. Zool.* **15**, 214 (1966)] and W. Bock (*11*).
39. Unequal rates of evolution for different structures or for any other components of phenotypes or genotypes are designated mosaic evolution.
40. See E. Mayr (*48*, pp. 238–241) on the differences between splitters and lumpers.
41. W. H. Wagner, "The construction of a classification," in *Systematic Biology* (Publication 1692, National Academy of Sciences, Washington, D.C., 1969), pp. 67–90.
42. R. R. Sokal in (*22*), "I have yet to meet a nonevolutionary taxonomist."
43. Several leading cladists have recently published antiselectionist statements.
44. J. S. Farris, in *Major Patterns in Vertebrate Evolution*, M. K. Hecht, P. C. Goody, B. M. Hecht, Eds. (NATO Advanced Study Institute Series, Plenum, New York, 1977), vol. 14, pp. 823–850.
45. J. McNeill, *Syst. Zool.* **28**, 468 (1979).
46. For criteria by which to judge the practical usefulness of biological classifications, see E. Mayr (*48*, pp. 229–242).
47. W. Hennig, *Annu. Rev. Entomol.* **10**, 97 (1965).
48. E. Mayr, *Principles of Systematic Zoology* (McGraw-Hill, New York, 1969).
49. L. van Valen, *Evol. Theory* **3**, 285 (1978).
50. Drafts were read by P. Ashlock, J. Beatty, W. Bock, W. Fink, C. G. Hempel, and D. Hull, to all of whom I am indebted for valuable suggestions and critical comments, not all of which was I able to accept.

West-Eberhard, M. J. 1983. Sexual selection, social competition, and speciation. *The Quarterly Review of Biology* 58:155–183.

Darwin, in *The Origin*'s first paragraph of chapter 4, famously asserts: "can we doubt . . . that individuals having any advantage, however slight, over others, would have the best chance of surviving and of procreating their kind? On the other hand, we may feel sure that any variation in the least degree injurious would be rigidly destroyed. This preservation of favourable variations and the rejection of injurious variations, I call natural selection." In *The Origin*, as well as in most discussions of natural selection ever since, natural selection is explored in terms of "ecological" selection; that is, competition for environmental resources, such as food, nesting places and the like, and in reference to predators or prey, or to the "conditions of life." Darwin would later consider sexual selection, namely, in *The Descent of Man, and Selection in Relation to Sex*, published in 1871, surely because he saw that sexual selection had played a very important role in human evolution. In *The Origin*, Darwin dedicated a section of chapter 4 to "Sexual Selection" (pp. 87–90). In the final chapter, he asserts the significance of sexual selection in animal species: "With animals having separated sexes there will in most cases be a struggle between the males for possession of the females. The most vigorous individuals, or those that have most successfully struggled with their conditions of life, will generally leave most progeny. But success will often depend on having special weapons or means of defense, or on the charms of the males; and the slightest advantage will lead to victory" (p. 468).

West-Eberhard suitably points out that in sexual selection the resource for which animals compete is the other sex and sexual selection is a particular case of social selection, such as competition for social status. Indeed, "socially dominant individuals may monopolize essential resources, severely limiting the survival or reproductive success of others" (p. 158). Social competition may be an important agency leading to the origin of new species: "extreme and rapid divergence of the signals and weapons used in social competition can occur with or without ecological differences between isolated populations" (p. 163). Social competition often leads to exaggerated traits, such as the gigantic antlers of the Irish elk or the hypertrophied brain of *Homo sapiens*.

RELATED READING

Campbell, B., editor. 1972. Sexual selection and the descent of man, 1871–1971. Aldine, Chicago, Illinois, USA.

Frank, S. A., and B. J. Crespi. 2011. Pathology from evolutionary conflict with a theory of X chromosome versus autosome conflict over sexually antagonistic traits. Proceedings of the National Academy of Sciences, USA 108:10886–10893.

Jones, A. G., and N. L. Ratterman. 2009. Mate choice and sexual selection: what have we learned since Darwin? Pages 169–190, *in* Avise, J. C., and F. J. Ayala, In the light of evolution, vol. 3: Two centuries of Darwin. National Academies Press, Washington, D.C., USA.

Queller, D. C. 2011. Expanded social fitness and Hamilton's rule for kin, kith, and kind. Pages 5–25 *in* Strassmann, J. E., D. C. Queller, J. C. Avise, and F. J. Ayala, editors, In the light of evolution, vol. 5: Cooperation and conflict. National Academies Press, Washington, D.C., USA.

West-Eberhard, M. J. 2003. Developmental plasticity and evolution. Oxford University Press, Oxford, England.

VOLUME 58, No. 2 THE QUARTERLY REVIEW OF BIOLOGY JUNE 1983

THE QUARTERLY REVIEW of BIOLOGY

SEXUAL SELECTION, SOCIAL COMPETITION, AND SPECIATION

MARY JANE WEST-EBERHARD*
Smithsonian Tropical Research Institute
Balboa, Panama

ABSTRACT

Rapid divergence and speciation can occur between populations with or without ecological differences under selection for success in intraspecific social competition—competition in which an individual must win in contests or comparisons with conspecific rivals in order to gain access to some resource, including (under sexual selection) mates. Sexual selection theory is extended to encompass social competition for resources other than mates. Characters used in social competition can undergo particularly rapid and divergent evolution owing to (1) their great importance in determining access to critical resources, (2) the absence of a limit to change (except by selection in other contexts), (3) the generation-to-generation relentlessness of selection on these traits, (4) the potential for mutually accelerating evolution of preference and attractiveness in contests involving "choice," and (5) the very large number of factors that can initiate trends, including mutation and drift leading to use of different physiological or behavioral characteristics as signals, the role of novelty per se in the evolution of combat and display, ecological or habitat differences influencing the form of combat and of signals, and (in species capable of learning) imitation of idiosyncratic characteristics of successful individuals. Many species-specific signals heretofore attributed to selection for species recognition ("isolating mechanisms") are probably instead products of social selection. This may help explain the rarity of reproductive character displacement and other phenomena predicted by the species recognition hypothesis. Examples from a wide variety of organisms illustrate patterns predicted by social selection theory, including (1) exaggeration and rapid divergence of traits (e.g., weapons, pheromones, plumage, flowers, and song) used in social competition, (2) a correlation between type of social system (intensity of social selection) and distinctiveness and exaggeration of social traits, (3) sexually monomorphic extreme development of socially selected traits when both sexes compete socially, (4) occurrence of distinctive signals in allopatric populations lacking sympatric congeners, and (5) more rapid divergence (less phylogenetic conservatism) of socially competitive compared to non-competitive signals. Rapid divergence under social selection may accelerate speciation due to effects on pre-mating interactions, as well as on critical social determinants of survival and reproductive success which would put hybrids at a disadvantage. Maintenance of parapatric boundaries (extensive contact with little or no geographic overlap) between socially selected species may sometimes be due to competitive exclusion in sympatry between populations whose primary divergence has been social rather than ecological. Patterns of variation in socially selected characters demonstrate the wisdom of Darwin's distinction between natural and sexual selection, and the applicability of sexual selection theory to social competition in general.

* Address to mail: Escuela de Biología, Universidad de Costa Rica, Ciudad Universitaria, Costa Rica.

INTRODUCTION

EXPLAINING THE diversity of life is among the most ancient and persistent puzzles of philosophy and biology, and the one that inspired Darwin's *The Origin of Species.* A scientific understanding of organic diversity means understanding the multiplication of forms during the long history of life on earth. And this requires understanding the process by which new species originate—the process of "speciation."

Present-day discussions of speciation emphasize the importance of ecology in the origin of species. According to the most widely accepted theory of speciation (Mayr, 1963; see Lewontin, 1974), allopatric or geographically isolated populations initially diverge largely under selection for adaptation to the different environments in which they are found. Then if ecological divergence is sufficiently great, hybrids are at a disadvantage where the populations come into contact. Pre-mating "isolating mechanisms"—species-specific signals and courtship behavior that prevent mating between individuals of incompatible, or differently adapted, lineages—evolve either as an incidental by-product of the genetic divergence and reorganization of isolated populations (see Mayr, 1963, p. 551), or by selection against hybrids in zones of sympatry (e.g., see Fisher, 1930; Dobzhansky, 1940). Thus isolating mechanisms have been seen largely as, ultimately, products of ecological divergence. Ecological divergence plays a key role in all current theories of speciation. Theories differ over whether divergence occurs in geographic isolation, at different positions along a cline (Endler, 1977), or in different subdivisions (habitats) of the same locality (Bush, 1975). But they agree in describing the critical divergence as ecological: "The geographic variation of species is the inevitable consequence of the geographic variation of the environment" (Mayr, 1963, p. 311; see also Endler, 1977, p. 7, and his citations of Huxley, Dobzhansky, and Grant).

An outstanding exception to the rule of describing virtually all divergence in ecological terms is provided by Darwin's writings on speciation. Darwin (1859, 1871) showed how non-ecological selection, or "sexual selection"—competition for mates involving male combat or female choice—could be an important cause of the divergence characterizing new varieties and species. He argued extensively and explicitly that variation distinguishing subspecies and "allied" (closely related, recently diverged) species very commonly involves variation in sexually selected characters. Indeed, he devoted a book (Darwin, 1871) to showing that racial (subspecific) divergence in man has occurred primarily under sexual selection, and in order to substantiate this he cited massive evidence that the same was true in a wide variety of other organisms, including crustaceans, insects, fish, birds, reptiles, and mammals.

For various reasons (discussed by Ghiselin, 1974; O'Donald, 1977, 1980; Otte, 1979; Mayr, 1982a, p. 595 ff.), this prominent aspect of Darwin's theory of speciation was subsequently ignored. The Forgotten Era of sexual selection theory began in the 1930s when mathematical geneticists redefined fitness as a change in gene frequency, and thus emphasized the similarities of, rather than the differences between, natural and sexual selection (see Mayr, 1972); and when Huxley (1938) suggested that the term sexual selection be eliminated and the phenomena viewed as similar by Darwin be given diverse other explanations, including some based on "general" (that is, species) benefit (Huxley, 1938, p. 431). It ended in the 1970s (see especially Campbell, 1972) with the resurgence of interest in the evolutionary consequences of intraspecific competition (Williams, 1966). Ironically, the Forgotten Era of sexual selection theory coincided with the Golden Era of neoDarwinian speciation theory (highlighted by the contributions of Dobzhansky, 1937; Mayr, 1942, 1963; Simpson, 1944; Stebbins, 1950; and others), with the result that sexual selection theory, although occasionally mentioned (e.g., Mayr, 1963, pp. 199–201), played no important role in the formation of modern ideas about speciation. This is patricularly ironic since the principles most emphasized, and most solidly established, by this era of thought about speciation included the importance of geographic isolation in promoting divergence (Mayr, 1963), and the importance of courtship and pre-mating signal divergence in effecting reproductive isolation (Alexander, 1962). Sexually selected characters provide particularly dramatic evidence of both phenomena: as Darwin (1871) showed, divergence of sexually selected traits "useless" in the struggle for existence explains much of the geographic variation associated with race formation and the origin of species. And the signals and morphology involved in courtship are among the principal kinds of traits he so extensively cited.

Recent advances in evolutionary theory and the study of behavior have led to a renewed awareness that sexual selection is real and important in the lives of animals (e.g., see Campbell, 1972; Blum and Blum, 1979; Lloyd, 1979–81; Mayr, 1982a). Many authors (e.g., Hoenigsberg, de Navas, and Chejne, 1964; Spieth, 1974; Ringo, 1977; Carson, 1978; Alexander and Borgia, 1979; Lloyd, 1979; Thornhill, 1980; Lande, 1981, 1982; Thornhill and Alcock, in press) have mentioned the likelihood that sexual selection affects or accelerates speciation, or has led to unusual species diversity in particular groups. Nevertheless, there has been no new attempt at a general synthesis broadly relating sexual selection theory to speciation theory, and exploring the degree to which its predictions regarding divergence are upheld in nature (but see Thornhill and Alcock, in press). Probably many biologists would echo the questions raised by a recent critique (Templeton, 1979) of a sexual-selection explanation of speciation in Hawaiian *Drosophila:* Why would intrasexual selection favor intraspecific *variants* in courtship display? How can deviations from the norm be selected if accurate species recognition is at a premium. Is the special role of sexual selection in speciation confined to its effects on the evolution of species recognition and mate preference? And if not, why are the interspecific variants observed apparently "arbitrary" with respect to the environment? "Why, for instance, in one isolated population of bellbirds should the males develop three long bare wattles, . . . in another population a single wattle studded with small feathers, in another a beard consisting of a mass of stringy wattles on the throat, and in another a bare throat of colored skin?" (Snow, 1976, p. 88). And why the seemingly erratic and complex variation in horns of beetles with similar life histories and habitats (Howden, pers. commun.; see Howden, 1979)?

This paper begins to answer such questions by outlining the special features of sexual selection (and social competition in general) expected to accelerate divergence between isolated populations. It then discusses evidence that striking "non-ecological" divergence has occurred in a variety of organisms that compete socially for mates and other resources, and discusses the relevance of this to current ideas about speciation and the evolution of species-specific signals.

SOCIAL COMPETITION AND DIVERGENCE: THEORY

When Darwin (1871) distinguished between natural selection and sexual selection he referred to the difference between characters involved in the "struggle for existence" in the environment, and those concerned with competition among conspecific individuals for mates. He illustrated this distinction by comparing the natural-selection and sexual-selection functions of male prehensile organs. If, as in the case of some oceanic crustaceans, males have such organs in order to maintain a grasp on a female while being washed about by the waves of the sea, then their development has been the result of "ordinary or natural selection." But "if the chief service rendered to the male by his prehensile organs is to prevent the escape of the female before the arrival of other males, or when assaulted by them, these organs will have been perfected through sexual selection, that is, by the advantage acquired by certain individuals over their rivals" (Darwin, 1871, p. 569). Sexually selected characters, then, are molded to confront or surpass conspecific rivals rather than to deal with other aspects of the environment.

The special characteristics of sexual selection discussed by Darwin apply as well to social competition for resources other than mates. For this reason several authors (Wynne-Edwards, 1962; Crook, 1972; West-Eberhard, 1979) have attempted to generalize regarding selection under "social competition"—competition in which an individual must win in interactions or comparisons with conspecific rivals in order to gain access to some resource. The contested resources may include food, hibernation space, nesting material, mates, or a place to spend the night. Seen in this broader perspective, *sexual selection* refers to the subset of social competition in which the resource at stake is mates. And *social selection* is differential reproductive success (ultimately, differential gene replication) due to differential success in social competition, whatever the resource at stake.

In solitary species, or in the solitary stages of a life cycle, success depends primarily on the adequacy of characteristics enabling an individual to deal with environmental contingencies—success under what Darwin called natural selection. The expected result is evolutionary progress in non-social behavior and morphology—traits associated with feeding, hunting, defense against parasites and predators, and battling the elements. In certain circumstances, however (e.g., under ecological circumstances favoring aggregation or group life—see Alexander, 1974; Emlen and Oring, 1977), conspecific competitors may stand between the individual and one or more essential resource. Then social interactions can act as a screening process determining access to vital commodities. This can involve (1) a race (see Ghiselin, 1974; Parker, 1978; Lloyd, 1979); (2) the testing of a series of competitors in different places, as by a female Indigobird (Payne and Payne, 1977) or hammer-headed bat (Bradbury, 1977) visiting a number of displaying males in succession; or (3) a contest within a group. Whatever the forms of social competition, the morphology and behavior involved in battles, threats, and attractive or stimulatory display are expected to often show (1) rapid and extended evolution leading to exaggerated forms, and (2) a diversity of forms in different populations (Darwin, 1871). That is, there should be evidence of rapid divergence of socially selected traits when related populations (e.g., subspecies and species) are compared. The accelerating and diversifying causes of rapid divergence are discussed separately below.

Causes of Rapid and Prolonged Evolution under Social Selection

Strength of Social Selection

In species and life stages in which group living is obligatory or highly advantageous, one or a few socially dominant individuals may monopolize essential resources, severely limiting the survival or reproductive success of others, in some extreme cases even permanently sterilizing them (West-Eberhard, 1981). In such high-stakes social competition, social characters are critical determinants of survival and reproductive success: an individual superior in other respects may have its reproduction severely curtailed if unable to win in social contests. When this is true, social characters—the weapons used in fighting, and the signals used in ritualized combat and competitive display—should evolve rapidly, for they are under especially strong selection (slight variations in these characters are associated with great variance in reproductive success—see Warner, Robertson, and Leigh, 1975, for a well-analyzed example).

Unending Nature of Change

Another factor contributing to the exaggeration of social traits is the absence of a ceiling or limit to change, except by selection in other contexts. This property of sexual selection was noted by Darwin (1871), and it applies to the evolution of all socially competitive traits as well as to certain interspecific interactions (e.g., coevolved interactions of hosts and parasites, or predators and prey). As long as the interacting elements (e.g., competitive behavior or morphology of conspecifics) are mutually capable of evolutionary change or improvement, such change will continue in what has been called an unending evolutionary race. New traits keep leading to further competitive innovations until exaggerated characters are finally checked by some disadvantageous consequence (e.g., antlers grow so large as to hinder movement excessively, or the cost of their production outweighs their advantage). By contrast, change in ordinary or ecological characters—those responding to unchanging aspects of the physical environment, or organic aspects either not evolving or evolving very slowly in response to the adaptations in question—can approach a ceiling of perfection (optimum). Divergence in such characters in closely related species is therefore expected to be more limited than divergence in social traits.

Constancy of Social Selection

The generation-to-generation constancy of social competition makes social evolution an unusually relentless coevolutionary race. A particular parasite or predator may attack only a limited percentage of the individuals of a host or prey species, and may be able to switch to an alternative species if a primary host or food organism becomes rare or evolves an effective defense. But under intraspecific social competition every reproducing individual of every generation is involved in the same increasingly specialized unending contest, as long as the framework conducive to such competition (e.g., life in groups) persists. This constancy in the action of social selection should augment the amount of evolutionary change accumulated over time, even in species under weak social selection (species in which there is little variance in reproductive success associated with winning and losing—see Wade and Arnold, 1980; or species in which there is only low or transitory heritable variation in competitive ability).

Accelerating Effect of Novelty

The very nature of the functions—attraction, and combat—served by socially selected characters may sometimes put a premium on novelty per se (Darwin, 1871; Moynihan, 1975), and this may considerably increase the rate of evolution of weapons and display. In the evolution of weapons, a small change in behavior or morphology could prove decisive, owing in part to the lack of a countermeasure in opponents; and the very distinctiveness and conspicuousness of a novel signal might be advantageous in display (Armstrong, 1965, p. 251, 305). Hinde (1970) has discussed the role of novelty in the evolution of display and described its possible physiological basis. Perhaps the best demonstration of the advantage of sheer novelty under sexual selection is the rare-male phenomenon in *Drosophila* (see Ehrman, 1972, and references therein). The fact that flooding experimental mating chambers with odors and other cues produced by rare-type males eliminates the females' preference for them indicates that rareness per se is indeed selected (cf. an alternative interpretation involving female polymorphism for constant preferences—O'Donald, 1977). Although this frequency-dependent advantage would decline with the evolution of countermeasures (in the case of weapons) or with its increasing commonness (in the case of displays), its effect on the initial spread of innovations might be an important factor accelerating the overall rate at which the coevolutionary race proceeds, as well as in changing the focus of selection and, hence, the direction of evolution.

The Potential for Runaway Change

The only major insight regarding the special nature of social competition not mentioned by Darwin himself was contributed by Fisher (1930). Fisher pointed out that selection under female choice differs from that involving real or ritualized male combat in that there is the potential for a "runaway" process—rapid evolutionary exaggeration of competitive signals not constrained to reflect true superiority of their bearers (other than in the ability to signal per se). This would occur because any true indicator of superiority that is used by females in mate selection soon becomes an advantage in itself, due to the increased attractiveness of its bearers. Males showing the most developed expression of such traits are more successful at obtaining mates, and females mating with them gain an advantage through the greater attractiveness and mating success of their sons. [In male-male combat, on the other hand, selection on contestants to call the bluff of dishonest signallers should eventually limit the evolution of signals not indicating a true underlying ability or willingness to fight (see Zahavi, 1977; West-Eberhard, 1979). The latter paper errs in not making this distinction.] Fisher reasoned that runaway selection could lead to striking genetically correlated increases in the discriminatory powers of females and the exaggeration of signal characters of males, which would change "geometrically" in proportion to the development already achieved, until limited by (natural) selection in some other context (cf. O'Donald, 1980).

There is a potential for mutually accelerating selection for attractiveness and choice whenever one class of individuals is in a position to choose the winners among those competing. For example, in some social insects (e.g., honeybees and certain social wasps and ants) workers systematically persecute or kill all but one of several queens, or choose which of several to join in founding a new colony (Michener, 1974; West-Eberhard, 1978). Thus, there could be mutually reinforcing selection on the characters used to discriminate winner from loser queens, as well as on the workers' ability to distinguish them. Extremely attractive (to workers) queens would produce attractive daughter queens and thus yield a larger indirect genetic payoff to workers investing in their colonies (social traits of workers can be regarded as selected indirectly in relation to their effects on the reproductive success of genetically related queens). In this case a genetic correlation of genes for attractiveness and preference could develop not as a result of assortative mating, as in the case of sexual selection (see Lande, 1981), but as a result of the close genetic relatedness of colony members (West-Eberhard, 1973, 1978): male and female sexual offspring of attractive queens attended by discriminating workers are likely to carry genes for both attractiveness and choice. Such selection might affect characters like the "piping" and pheromone signals of honeybee queens (Michener, 1974) and the ritualized (and probably also pheromonal) dominance behavior of certain wasps (West-Eberhard, 1978, 1979, in press; Forsyth, 1980). The characters chosen as signals could initially be traits indicating reproductive superiority (robustness or egg-laying capacity), and then become elaborated under selection for signal effectiveness per se. Worker choice of queens might be importantly limited by the great significance of testing for the phenotypic quality of queens: there may be a great deal of variation in quality (especially, egg-laying capacity) among potential queens at the time of worker choice. Thus worker choice, like mate choice in species with large post-mating male investment in reproduction, should give relatively great weight to true indicators of *phenotypically* superior individuals, which would retard the evolution of signals (e.g., products of runaway selection) not truly indicative of quality. Worker choice might also sometimes be limited by the advantage to workers of favoring the queens most closely related to them—a not inconceivable possibility in light of recent research on social insects (Greenberg, 1979; the references in West-Eberhard, in press). This tendency, however, would also increase the genetic correlation of preference and attractiveness.

Parents are also frequently in a position to exercise favoritism, or parental choice, in treatment of their offspring; and offspring compete strenuously for parental attention, sometimes employing highly specialized and exaggerated signals, like the huge (and sometimes elaborately marked) gaping mouths of nestling birds (see Welty, 1962; Skutch, 1976), or the "hunger signals" of larval hornets (Ishay and Brown, 1975). Such characters could originate as releasers or guides of parental feeding behavior (Skutch, 1976), or even indicators of true superiority, then become elaborated under mutually reinforcing selection; since offspring of successful signallers would in turn be parentally favored, there would be a premium on the ability of parents to distinguish and favor them. Again, a genetic correlation of characters enhancing attractiveness and preference would develop due to genetic relatedness of interactants. And such a process would be checked eventually by natural selection against bearers of extreme characters, and by the advantage of parental genes contributing to the detection of phenotypically superior offspring (those most likely to be superior propagators of genes like the parent's).

The importance, in nature, of runaway selection like that visualized by Fisher (1930; see also Lande, 1981, 1982) is currently controversial (see Thornhill and Alcock, in press; Bradbury and Gibson, in press). Runaway selection can occur in only a limited subset of the characters under social selection, namely, some (but not all, see below) characters subject to choice. It can be arrested or slowed by two kinds of disadvantage: that to the survival ability of individuals producing and bearing an extreme trait; and that to individuals choosing superior signallers if the most desirable (genotypically or phenotypically, if the male aids the female) mates in the population cannot be identified by this means. That is, if male superiority in some other context is greater than that accruing to producers of a particularly extreme signal, then female attention to additional indicators of quality may reduce the consistency of a preference for the extreme signal. Recent studies indicate that this latter limit to runaway selection may be more important than heretofore appreciated. Theoretical consideration of the possible importance of resistance to parasites and disease as a basis for mate choice (Hamilton and Zuk, 1982) shows how true indicators of genetic quality in traits other than signalling ability can be important, even in polygynous species in which the male contributes nothing to the female other than genetic material. And experiments on female choice in frogs (Ryan, 1980) and crickets (Forrest, 1982) indicate use of a cue (call frequency) closely associated with a complexly determined phenotypic trait (size) that could serve as an index of general (genetic) superiority without being easily subject to runaway change. [Perhaps selection favors use of such generalized indicators of male superiority between what Fisher predicted would be relatively short-lived episodes of (eventually disadvantageous) runaway evolution.] Furthermore, female choice may sometimes prove to be indirect, as when females show a mating preference for the males who win in contests with other males (see Payne, in press), with females sometimes even inciting competition among males (see Thornhill and Alcock, in press). Under indirect female choice (called "passive selection" by Lloyd, 1979) males may produce spectacular male-male competitive signals, and females would exercise preference, but a runaway process would not occur, for the male characters (and the female preference) would be subject to the checks on bluff thought to characterize the evolution of male-male combat (see Borgia, 1979; West-Eberhard, 1979). Thornhill and Alcock (in press) provide examples demonstrating the importance of caution in attributing even greatly exaggerated sexually selected characters to Fisherian runaway selection.

Diversifying Factors

All of the factors discussed so far would contribute to rapid or continuing evolution of socially selected traits once a particular trend had started. But what determines the *direction* of evolution of social traits, and hence their diversity? Why should one expect not only exaggeration but also *variety* in the kinds of beetle horns and in the plumage and competitive displays of birds?

Unending Nature of Change

The lack of an optimum solution, or limit to change, under social selection means not only long-continued change, but also that a large variety of directions are possible. In the evolution of combat and display, a great diversity of novelties can serve as the basis for a whole new line of development (can be a strategic breakthrough). This applies not only to progress in the evolution of weapons, countermeasures, and enhanced display, but also to improvements in the sensing, or monitoring, ability of choosing and (in combat) contesting individuals. Sensory innovations are another source of new directions in the evolution of social behavior and communication. A great variety of characteristics constitute potential signals, given the ability to recognize them. As ethologists have long realized, social signals are often derived from movements or changes in color, posture, or odor, indicating motivational state or intention (see Hinde, 1970). And a large variety of cues—size, color, activity level, and numerous, more specific attributes—can serve to indicate the quality of competitors. For example, Hamilton and Zuk (1982) list detectable signs of infection and corresponding male displays that may be used by females to judge the level of resistence achieved by prospective mates; and courting and fighting males frequently perform complex and difficult feats which might be used to evaluate their quality (see Thornhill and Alcock, in press; and discussion below). A theoretically unlimited number of such characteristics may become signals under social selection, which would always favor improved ability to recognize them and respond appropriately.

The Role of Mutation and Drift

As in any evolutionary sequence, which among a large number of potential cues actually evolve as signals must depend in part upon random processes (mutation and drift) determining, respectively, (1) the initial ability to react to a trait as a signal, and (2) the commonness of particular potential signals in the sampled (interacting) population. Both processes would be expected to vary from one population to another.

The coevolutionary nature of change in weapons and displays may confer a special importance on the role of drift, or sampling of traits, in local populations. The success of a particular tactic depends on what others (or the majority of others) are doing. This may help explain why it is reasonable to conclude that random drift or founder effects are frequently important initiators of divergence and speciation in sexually selected groups such as Hawaiian *Drosophila* (Carson, 1978; see also Kirkpatrick, 1982). A dramatic illustration of the effect of local population composition on competitive tactics is provided by the highly sexually selected labrid fish, *Thalassoma bifasciatum,* which (facultatively) changes color and sometimes sex at different sizes, depending on the competitive situation on individual coral reefs (Warner, Robertson, and Leigh, 1975).

Ecological Factors

While I have argued above that divergence in social traits can proceed even without ecological differences between isolated populations, this is not to say that such evolution is independent of ecology. Indeed, environmental differences, when they exist, make the divergence of social traits even more likely. Smith (1977, pp. 348–52, 364–88) has reviewed some of the environmental factors that could initiate divergence in the characteristics of social communication (see also Gorman, 1968; Morton, 1975; Lloyd, 1979, 1983; Brenowitz, 1982; and references in Payne, in press). They include amount and location of sunlight available for visual displays; availability of leaves (or other substrates) suitable as sounding boards for particular kinds of acoustical displays; amount of background noise (e.g., visual and/or sonic) interfering with particular kinds of signals; density of vegetation obstructing particular signals; intensity of predation, which may limit the exaggeration of certain signals or the circumstances in which they are performed; and patterns of resource distribution, which may influence individual spacing and intensity of display. Spieth (1981) has described environmental factors possibly affecting the mode and morphology of male-male combat in *Drosophila.* The location of combat must often affect its form. For example, different species of male beetles fight in tunnels, at the entrances to cavities, or while clinging to grass stems, and have correspondingly different fighting behavior and morphology (see Eberhard, 1977a, b, 1979, 1980, 1981).

Different ecological circumstances may lead to the evolution of different patterns of distribution and parental care (Emlen and Oring, 1977), in turn affecting the nature of social signals (Alexander, 1975). And several authors have pointed out that the signal repertoire of a species is itself an aspect of its environment that can affect the further evolution of social signals. For example, social signals may evolve to increase contrast with signals used in different contexts (Hinde, 1970), or to achieve deceptive effects (mimicry) (West-Eberhard, 1975, p. 10; Lloyd, 1979). In dendrobatid frogs, species differences in the territorial aggressiveness and parental behavior of females are related to species differences in the role of distance signalling and appeasement in the courtship behavior of males (Wells, 1980). See Hinde and Tinbergen (1958) for a general discussion of social context as a source of signal diversity.

Learning

In animals capable of learning, evolutionary divergence of social signals may be influenced by initially fortuitous associations of individual differences and social success. Payne (1982, in press) suggests that locally distinctive birdcalls, or "dialects," sometimes originate via "song matching" when younger males imitate the distinctive call of a particular dominant or older, established male and thereby gain a competitive advantage. And Lloyd (1980) has suggested that comparable phenomena may occur in insects. Similarly, Darwin (1871) argued plausibly that the (inherited) physical differences among the races of man evolved under social selection in isolated populations having long histories of different learned "tastes" or culturally determined preferences affecting status and mating success. Learning could accelerate (genetic) divergence whenever a distinctive phenotype happens to be consistently associated (whether genetically or otherwise, e.g., hormonally or traditionally) with superior status. Then any genes contributing to the production of the successful phenotype would be favored.

In summary, extreme and rapid divergence of the signals and weapons used in social competition can occur with or without ecological differences between isolated populations. This is expected because of (1) the great importance of these characters in determining access to resources critical to survival and reproduction, (2) the potential for unending evolutionary change in socially competitive traits, (3) the generation-to-generation relentlessness of selection on these traits, (4) the effect of novelty in accelerating the initial spread of traits, and (5) the potential for mutually accelerating, genetically correlated evolution of preference and attractiveness in contests involving choice. A large number of factors can initiate divergent evolutionary trends, since a very large and theoretically unlimited array of physiological or behavioral characteristics may be used advantageously as signals, and ecological or habitat differences as well as mutation and (to an unusual degree) drift can produce local variants in those used. In species capable of learning, idiosyncratic traits of successful individuals may be advantageously imitated by others, and this may additionally influence the direction of evolution. Whatever the source of a new signal or weapon, it would be subject to strong selection for elaboration and improvement, the course of which would further vary under the influence of all of these accelerating and diversifying factors.

Alternative Hypotheses: Social Selection and Species Recognition

It has long been realized that sexually dimorphic characters like those used by Darwin (1871) to illustrate sexual selection can function in contexts other than competition for mates (see Wallace, 1878; Huxley, 1938; Mayr, 1963; Selander, 1972; Baker and Parker, 1979). They may function in identification of sex, species, and intention; in physiological synchrony or location of mates ("epigamic" displays); as adaptations to different ecological niches; or in defense against predators.

"Epigamic" courtship displays of males are believed to promote cooperation between the sexes by facilitating the location or stimulation of females. Such displays would likewise be subject to sexual selection, however, since superior performances by some males would lead to increased mating success in competition with others. As concluded by Mayr (1972, p. 97), ". . . sexual selection is presumably superimposed in all cases in which a male may gain a reproductive advantage owing to an extreme development of an epigamic character" (see also O'Donald, 1977). I therefore consider elaboration of so-called epigamic courtship displays an aspect of sexual selection in species where the female interacts with more than one male prior to copulation. In general, if there is direct evidence that a particular character is exposed or wielded in intraspecific competitive display or combat in a way illuminating its precise form or variability, and there is no comparable evidence for its use in other contexts (e.g., to frighten a predator), I consider this positive evidence that it has likely evolved primarily under social selection. (Such traits may of course have additional, secondary, functions.)

Species recognition has been the most influential alternative hypothesis explaining the diversity and species specificity of social signals. The species-recognition hypothesis holds that species-specific signals and morphology originate or persist in a particular form because they function as *isolating mecha-*

nisms (Dobzhansky, 1937; see also Fisher, 1930)—barriers to wasteful interaction or pair formation between members of differently adapted populations. Some authors (e.g., Mayr, 1963; see also discussion in Blair, 1960) have emphasized that isolating mechanisms can originate in isolated populations as incidental byproducts of genetic divergence under selection in other contexts. This latter view is compatible with the ideas presented here regarding the role of social selection in producing species-specific signals independent of or prior to a species-recognition function. It provides no explanation, however, for complex, coordinated divergence in the production and reception of species-specific signals, other than as an "incidental byproduct" or pleiotropic effect of general divergence (Mayr, 1963, pp. 551, 311). Perhaps for this reason, the species-recognition hypothesis seemed for a time the only sufficiently explicit explanation for such elaborate diversity, leading some authors to conclude that the "only obvious contexts of evolutionary change" in communication systems are "(1) perfection of intraspecific compatibility" (increased efficiency of interaction), "and (2) perfection of interspecific incompatibility (reproductive isolation)" (Alexander and Otte, 1967, p. 6). It is now clear, as I shall explain below, that "what had previously been regarded as species isolating mechanisms are to a large degree evolved instead in the context of sexual selection and competition within the species" (Alexander and Borgia, 1979, p. 437; see also Alexander, 1975). But during the Forgotten Era of sexual selection theory it would have been difficult to devise a hypothesis more perfectly suited than species recognition to displace sexual selection theory, and to distract biologists from its rediscovery. Many of the predictions of the two hypotheses are the same (see Payne, in press), or at least not contradictory. Both hypotheses predict species specificity of social signals. And several other phenomena cited in support of the species-recognition hypothesis (Alexander, 1962) can be explained as well by sexual selection theory. For example: (1) Some closely related species hybridize readily when distinguishing courtship or calling interactions are bypassed in the laboratory (e.g., by forcing non-conspecifics together). This could represent elimination of the species-recognition step in rapprochement; or in sexually selected species it could represent elimination of the step where the female would exercise (intraspecific) choice, with males from alien populations producing such inappropriate signals that they are normally discriminated against. (2) When related species overlap geographically they often have distinctive signals. This could evolve either as a mechanism of species recognition preventing wasteful interaction and hybridization; or it could represent divergence under sexual selection in reproductively isolated populations, either before or after sympatry. (3) Allopatrically (geographically) or allochronically (temporally) isolated populations sometimes lack signal distinctiveness. The species-recognition hypothesis can explain this as absence of selection for species recognition in the absence of overlap; or it could be due to absence or weakness of social selection, if the signals in question do not importantly affect access to critical resources. (4) Stereotypy (lack of individual variability of performance, at least of certain elements) could function to promote certain identification of conspecifics; or it could be the result of strong social selection having driven the character to fixation throughout the population (for a discussion of the selective basis of stereotypy in competitive signals see Zahavi, 1980).

Understanding the interaction of species recognition and social selection (intraspecific competition) in the evolution of species-specific communication is a crucial question raised by the revival of sexual selection theory. Fisher (1930) pointed out that species recognition signals would be subject to elaboration under sexual selection (female choice); and signals that originate under social selection must sometimes be used in, and may be channeled or maintained by, selection for species recognition (e.g., in sympatric vs. isolated populations; and see section on *Anolis,* below). The following discussion emphasizes examples (e.g., of non-sexual social selection and male-male competition) where it is possible to separate the two functions in order to establish social selection as an important cause of divergence. It must ultimately be considered, however, to be just one of several interacting causes.

There are several purely theoretical reasons for expecting social selection to be more often important than species recognition as a cause of divergence in socially competitive traits. Species in which social competition is important are subject to social selection in every generation and in every population, whether sympatric or allopatric with related species, with or without hybridization, and whether in the presence or absence of confusingly signalling neighbors. Furthermore, every reproducing individual is involved, not just those who happen to interact with inappropriate mates or respond to their signals in a zone of overlap, and at a time and site (habitat) when both are disposed to produce or react to signals. The species-recognition hypothesis is much more restrictive: it requires sympatry (or a history of sympatry) and disadvantageous hybridization or courtship interaction between populations that are genetically distinct but do not yet possess mechanisms for avoiding such interactions. It implies that distinctiveness evolves following and because of costly interaction between diverged populations. It is likely, however, that selection for species recognition would often favor discrimination of distinctive traits evolved in other contexts whenever such traits are available; in such cases species recognition could not be considered responsible for their divergence. By this reasoning, divergence under social selection may greatly reduce the number of situations in which divergence occurs as a result of selection for species recognition per se, since it can rapidly pre-adapt populations for species recognition by producing distinctive signals prior to contact and without special selection in the species-recognition context. On the other hand, signal divergence in the species-recognition context (if it occurs), would not restrict the scope of action of social selection (except to require maintenance of signal distinctiveness), and may even extend it by establishing characters subject to runaway change under female choice (Fisher, 1930). In sum, the effects of social selection are expected to predominate over, and often preclude, the effects of species recognition in the evolution of species-distinctive social signals.

It is therefore not surprising that the predictions of the species-recognition hypothesis are seldom borne out in socially selected groups, except when they coincide with those of the social selection hypothesis (above). For example, the species-recognition hypothesis predicts reproductive character displacement, or accentuated divergence of social signals, in areas of overlap with closely related species. This has seldom been demonstrated, even in the groups (e.g., singing Orthoptera and anurans, birds, and lizards) where it has most diligently been sought (see Walker, 1974; Blair, 1974; Payne, in press; Crews and Williams, 1977; and Ferguson, 1971, 1977, respectively). The failure to find reproductive character displacement common in these groups may be in part due to the difficulty of documenting its occurrence (Walker, 1974; Waage, 1979), and to the fact that courtship signals in these groups are known to be subject to sexual selection and are therefore likely to diverge *independent* of (and preclude) selection for species recognition (see Walker, 1974; Blair, 1974; and Williams and Rand, 1977 for examples). Reproductive character displacement may turn out to be most common in species under relatively weak social selection, since their signals are less likely to diverge in isolation prior to contact.

The species-recognition hypothesis also predicts that signal distinctiveness should be reduced on islands and in isolated (allopatric) populations. The plumage dullness (and increased sexual monomorphism) of male birds on remote, congener-free islands has long been considered decisive evidence for the importance of the species-isolation function in maintaining bright plumage in mainland populations having sympatric relatives (Sibley, 1957; Mayr, 1963, 1972). It may have, instead or in addition, a sexual-selection explanation (see Ghiselin, 1974; Selander, 1972). Recent studies of island waterfowl (Waller, 1980) indicate that pair bonds are maintained longer in island than in mainland (migrant) forms, a situation that would reduce sexual selection on males relative to females. And males of some species dedicate more time to brood care (Waller,

1980), a factor that might favor dull or cryptic plumage in males independent of selection (or relaxed selection) in the species-recognition context. Furthermore, loss of display distinctiveness in island populations is not a general phenomenon (see Ghiselin, 1974, p. 183 ff.). Gorman (1968) found the displays of isolated island *Anolis* species to be as distinctive and stereotyped as those of species having many sympatric congeners (e.g., those of Greater Antillean islands—see Ruibal, 1967), and cited similar diversity of displays in lizards (*Tropidurus*) isolated on different Galapagos Islands (Carpenter, 1966). The displays in question—agonistic and courtship movements—and their striking associated morphology (dewlaps and body coloration) are well known to function in intraspecific social competition (see below), which would account for their divergence in both sympatric and isolated populations. Other examples are given in a discussion of "superspecies," below.

Pre-mating isolating mechanisms are thought to evolve because of the advantage of early species recognition, prior to copulation, zygote formation, and other costly investment in disadvangaged (hybrid) offspring (Mayr, 1963; Alexander and Otte, 1967). Extending this argument, one would expect species recognition to occur early in courtship, and to involve brief interaction. Long, complex interactions, while predicted by the sexual selection hypothesis if females examine males using complex, repeated, or subtle comparisons, should usually be selected against in the species-recognition context. If species recognition occurs early in a courtship sequence, diverting mismatched pairs from further interaction, complexity and diversity in subsequent behavior cannot be regarded as being evolved or maintained under selection for species recognition. For example, in some *Drosophila* species, species recognition occurs prior to courtship, and involves different characters (Manning, 1966). And in Western grebes (*Aechmophorus occidentalis*) there is strong positive assortative mating between two color phases having virtually identical complex courtship displays, with recognition occurring as a result of phase-specific simple advertising displays given early during intersexual interactions and (unlike the courtship displays) showing greater phase specificity (or learned "response displacement") in areas where both phases are present (Neuchterlein, 1981a,b). In a review of song playback experiments involving various bird taxa, Emlen (1972; see also Neuchterlein, 1981b) concluded that in most species only a small fraction of available song features are essential for species recognition.

Failure to document the major predictions of the species-recognition hypothesis has given rise to several remedial hypotheses: the signal redundancy hypothesis (see Rand and Williams, 1970 on *Anolis*; discussed critically in relation to birds by Krebs and Kroodsma, 1980), which explains continued divergence as redundancy presumably improving the efficiency and certainty of species recognition; the lost neighbor hypothesis (the failed invasion hypothesis of Williams and Rand, 1977), which explains signal distinctiveness in allopatric isolates in terms of (hypothetical) former overlap with now extinct or allopatric populations of related species (Schodde, 1976; Fine, Winn, and Olle, 1977, p. 497); and the semi-assortative mating (or sub-species recognition) hypothesis (Crews and Williams, 1977), which envisions geographic variation in signals as having evolved to reduce possibly disadvantageous mating between forms adapted to somewhat different local ecologies. These hypotheses have the effect of salvaging the species-recognition idea when its more conventional interpretations are not supported by the available data. When cited without evidence to explain divergence of characters known to be under social selection, they should be regarded as explanations of last resort, since social selection can explain such divergence without special assumptions.

DIVERGENCE UNDER SOCIAL SELECTION: PREDICTIONS

Social selection theory makes the following predictions which distinguish it from the species-recognition hypothesis (see also Payne, in press; and Thornhill and Alcock, in press):

(1) Character exaggeration, and geographic variation suggesting relatively rapid divergence, should often occur in all kinds of

traits used in social competition—not only those used prior to or during courtship, but also weapons, threat signals used in male-male contests over mates, and other signals (e.g., of juveniles or females within groups) specialized to function in competition for resources other than mates.

(2) There should often be a correlation between social (or mating) system, and distinctiveness and exaggeration of social traits (modified after Payne, in press): the stronger the social selection in a particular social situation (the greater the variance in reproductive success due to social competition), the more complex and exaggerated the weapons or signals used, and the more rapid and greater the expected divergence between isolated populations. There should thus often be a positive association between degree of exaggeration of a social trait and its degree of geographic variability and species specificity.

(3) When social competition involves both sexes, the socially selected traits (weapons or signals) should be similarly or monomorphically extreme and species specific (or geographically variable) in both sexes. When only one sex is involved (or the sexes are involved unequally) sexual dimorphism in the socially selected characters should occur.

(4) Distinctive signals should often evolve even in allopatric isolated populations (in the absence of sympatric closely related species or others producing similar signals).

DIVERGENCE UNDER SOCIAL SELECTION: EXAMPLES

Predictions (1) and (2), above, were made with regard to sexual selection by Darwin (1871), who listed many illustrative examples of both invertebrates and vertebrates, especially insects, birds, and mammals. The following examples, while far from an exhaustive review, indicate that these predictions hold in a wide variety of organisms, and illustrate ways in which they can be tested. For information on divergence I have referred to taxonomic monographs and comparative behavior studies. If socially selected characters are used as "key" characters for the identification and description of species or subspecies this indicates that they have been among the fastest visible or measureable characters to diverge. Although many important behavioral and physiological characters are not taken into account by taxonomists, this at least gives a rough indication of the degree of divergence of social characters relative to non-social aspects of morphology. Non-social traits may, of course, also undergo rapid evolution under strong selection. Conversely, if a particular social character happens to be only weakly socially selected (e.g., it is rarely expressed in nature, or has little effect on survival or reproductive success), it would not follow the predictions suggested here.

I have relied mainly on descriptions of behavior as evidence for the competitive social function of the characters discussed below. They are considered basically competitive in nature if they fit Hinde's (1970) definition of "agonistic" behavior: "behavior directed towards another individual which could lead to physical injury to the latter and [or] often results in settling status, precedence, or access to some object or space between the two." Agonistic interactions sometimes lead to cooperation and group-beneficial social integration (see West-Eberhard, 1979, 1981), but there are good reasons for regarding their evolved nature as fundamentally competitive (see Williams, 1966).

Sexually Selected Displays

Among the most famous likely products of the sexual-selection subcategory of social selection are the spectacular plumes and markings of certain polygynous birds, which use their extravagant morphology in elaborate intraspecific male-male (threat) and male-female (nuptial) displays. Examples (with references describing displays) include those of grouse (Wing, 1946; Wiley, 1978); birds of paradise (Gilliard, 1969); manakins (Snow, 1976); cotingas and bellbirds (Armstrong, 1965; Snow, 1976); ducks (Armstrong, 1965); and pheasants (Armstrong, 1965). That these traits diverge relatively rapidly is reflected in the characters used to distinguish species and subspecies in these groups (see, respectively, Robbins, Bruun, and Zim, 1966; Gilliard, 1969; Davis, 1972, and Meyer de Schauensee and Phelps, 1978;

and Delacour, 1970). Although the females of related species are often very similar, males (the more strongly sexually selected sex) are easily distinguished by their plumage, indicating that it evolves more rapidly than that of females (the less strongly sexually selected sex). Furthermore, when behavior not involving plumage (e.g., bower construction or song) predominates in the competitive displays of males, plumage is relatively conservative and the competitive morphology or behavior is taxonomically and geographically distinctive (Sibley, 1957; Gilliard, 1969; Krebs and Kroodsma, 1980).

The species-specificity of bird plumage and song have often been considered products of selection for pre-mating isolating mechanisms, or species identification (e.g., see Marler, 1960; Mayr, 1963; and Smith, 1977). This hypothesis has been most explicitly tested and related to sexual selection as an alternative explanation in the case of vocalizations. There is evidence that bird song functions in both contexts: song repertoire affects the mating success of individual males (Yasukawa, Blank, and Patterson, 1980; Krebs and Kroodsma, 1980, pp. 147–8; Payne, in press); and females are preferentially attracted by songs of their own species and even local "dialects" (Baker, Spitler-Nabors, and Bradley, 1981; Payne, in press). Both theory and observations, however, support sexual selection as the more consistent cause of divergence (Payne, in prep.), without denying that species recognition could be an important (perhaps often secondary) effect. The fact that song affects male mating success, by means of effects on both male-male interactions and female choice (Payne, in press), means that it is subject to all of the divergence-accelerating factors discussed above; and there is no evidence for character displacement in song (Payne, in press). Payne (in press) has also tested the prediction that mating system and intensity of sexual selection (variance in male mating success) should correlate positively with the exaggeration of sexually selected traits, and finds that in general it does: sexual dimorphism in size and male ornamentation is greater in the lekking species in several families of birds, though not in all (lek-forming species were shown to have greater variance in male mating success than non-lek-forming species).

The major cheliped of male fiddler crabs (*Uca* species) is likewise a weapon used primarily in highly ritualized male-male displays (Crane, 1975). Crane (1975, p. 457) has called the giant claw of the male ". . . one of the most highly and variously specialized organs known to zoology, and certainly unsurpassed in the number of adaptations for ritualized combat." Cheliped displays are sometimes directed at females. But "regardless of the additional uses in the acoustics of courtship, the entire complex armature of both merus and claw forms a vast system reserved for intermale behavior"—both occasional real combat, and ritualized tapping displays. "Antler-bearing mammals, including even moose and Irish elk, show in contrast minor specializations. In fiddler crabs the claw alone, at its maximum relative weight, reached almost half the total weight of the crab" (Crane, 1975, p. 456). In fiddler crabs, morphology and behavior concerned with ecological adaptation and maintenance are "as a whole conservative" (p. 526), showing little variation within and among the 62 species recognized by Crane. In contrast, characters concerned with reproduction—the male gonopods (genitalia) and the major cheliped—show striking variation, with the claw being the most consistently used for distinguishing species and subspecies. Among the species-specific details of claw morphology are small knobs known to be used in male-male interactions (tapping displays). Since these aspects of morphology are unlikely to ever be perceived by females it is doubtful that they function in species recognition.

The dewlap of male (and in some species, female) *Anolis* lizards is a reptilian equivalent of the plumage and song of birds. It is a large, usually brightly colored, and variously patterned flap of skin that is extended during courtship and aggressive displays (see Gorman, 1968; Rand and Williams, 1970; Trivers, 1976). Field and laboratory observations (Greenberg and Noble, 1944; Ruibal, 1967; Gorman, 1968; Trivers, 1976) leave no doubt that dewlaps function in competitive social (territorial and courtship) displays and, hence, that they are subject to social selection.

As in the case of other socially selected traits, dewlaps and other lateral head and body markings are key taxonomic characters at the species level (Lazell and Williams, 1962; Williams, 1963; Gorman, 1968); and they show marked geographic divergence (Schwartz, 1968; Webster and Burns, 1973), often even in the absence of sympatric congeners (e.g., in the "solitary" island populations of *Anolis conspersus* and *A. lineatopus*—Williams and Rand, 1977; in island subspecies of *A. distichus* and *A. marmoratus*—Crews and Williams, 1977; and in island populations of the *A. roquet* group—Gorman, 1968). That is, dewlaps and other display characters probably do diverge independent of selection for reproductive isolation. This conclusion is reinforced by the fact that there is so far no unequivocal example of reproductive character displacement in *Anolis*. In one carefully studied case (the *brevirostris* species complex in Haiti), clinal variation in dewlap coloration within one species was in a direction that would maximize contrast with the closely related species contacted at opposite ends of the cline (Webster and Burns, 1973). This has been cited as the only good example of character displacement in reptiles (Ferguson, 1977). The geographic variation observed, however, could as well be a product of social selection.

Whereas there is thus good reason to believe that social selection is an important cause of display divergence in *Anolis,* there is also evidence that dewlap displays function in species recognition and reproductive isolation: experiments demonstrate an effect of dewlap color on mate selection (Ferguson, 1977), and studies of multi-species sympatric associations show that each species has a distinctive dewlap color (Rand and Williams, 1970; Williams and Rand, 1977), with the species that are most similar in size and general appearance differing most strikingly in dewlap color. In some populations, however, the displays seem to have diverged far beyond the degree necessary to effect species recognition: in a community containing eight species, Rand and Williams (1970) estimated the redundant information content of displays to be sufficient to separate up to 502 species!

I suggest the following interpretation of these facts: divergence under social selection is the primary source of intraspecific and interspecific variation in dewlaps and displays, and is a continuing source of distinctiveness whether species contact occurs or not. This must sometimes contribute to speciation by producing local differences in pre-mating signals, preadapting diverging populations for reproductive isolation. Such differences would be reinforced in the species-recognition context if, at species boundaries, there were selection against hybridization. In that case selection for species recognition may limit or direct the variants possible under social selection. But social selection would continue to increase the complexity of displays in sympatric populations. This could explain the apparent excess of diversity in display characters compared to that believed necessary for species recognition.

The genus *Anolis* thus serves to illustrate interaction of social selection and species recognition during speciation. It may prove a key genus in elucidating the roles of social competition, ecology (references in Jenssen, 1977, p. 204), and species recognition in the origin of species. Other well-studied and taxonomically useful ethological isolating mechanisms, such as cricket calls (Alexander, 1962), firefly flashes (Lloyd, 1966), frog calls (Blair, 1962), and bird songs (Lanyon, 1969), have undoubtedly diverged importantly under sexual selection (see, respectively, Ehrman, 1972, and Alexander, 1975; Lloyd, 1979; Ryan, in press; Payne, in press), and should therefore also serve to illuminate this interaction of factors. For a concise summary of an already well-analyzed example, see Thornhill and Alcock (in press) on *Drosophila.* Silberglied (in press) provides evidence that male intrasexual display is responsible for the diversity of brilliant male coloration in butterflies.

Competitive Displays of Plants

There is a clear (but not perfect) analogy between the evolution of competitive displays in animals, and that of comparable displays in plants—the colors and fragrances of flowers and fruits (see Stiles, 1982) of animal-pollinated and animal-dispersed species. When pollination is a limiting resource for plants, and nectar and pollen for pollinators, there is strong selection on the attractive displays of the plants as well as on the powers of discrimination and exploitation efficiency of pollinators. This has led to adaptations in animal-pollinated flowers that have been compared in flamboyancy and extravagance to the sexual displays of animals (Rothschild, 1975). The analogy with sexual selection for displays in animals is strongest in species such as some orchids that have one or a few specialized pollinators (Dodson, 1975; Dressler, 1981). And in such plants a spectacular diversity and specialization of pollinator attractants and pollination devices have sometimes evolved (see Darwin, 1862, on orchids; Meeuse and Schnieder, 1979, on water lilies). Flower structure is of great importance in the taxonomy of most groups (Ordnuff, 1978; Dressler, 1981), especially at the species level. This indicates that, as in the case of socially selected characters in animals, the competitive displays of plants diverge relatively rapidly. The intervention of a second species (the pollinator) in the case of plants means that a runaway process of the kind visualized by Fisher is impossible. But the other factors discussed above as favoring rapid signal divergence and speciation would apply, and pollinators introduce an additional important source of signal diversity: interspecific competition for pollination. The resource at stake (the effective services of pollinators) is often shared, and strongly contested, in geographically diverse mixtures of flowering species. Pollen of heterospecific competitors can seriously interfere with fertility and may lead to local divergence (character displacement in floral characteristics) (see review in Waser, in press); and the presence of highly attractive species can lead to local convergences (floral mimicry) (Heinrich and Raven, 1972). Both tendencies would contribute to the species and subspecies distinctiveness of animal-pollinated flowers. In addition, geographic variation in the relative abundance of different pollinators can lead to changes in flowering times associated with changes in pollinator (for possible examples see Frankie, 1975, p. 205). Such changes must sometimes also lead to changes in flower morphology and odor (Grant and Grant, 1965). The idea that competition for pollination contributes importantly to floral diversity is supported by the observation (Anderson, 1979) that remarkable conservatism in the pollinator-attracting features of flowers in the neotropical Malpighiaceae is associated with lack of nectar production and, hence, a limited clientele of potential pollinators (restricted avenues of local divergence). This interpretation would seem to be contradicted by the "endless" floral diversity of orchids (Darwin, 1862, p. 284), which likewise have extremely specialized pollinators. The correct resolution, however, may be that suggested by Darwin (1862), who believed the floral diversity of orchids to be driven by a low pollen:seed ratio, which, he argued, places a special premium on efficient pollen transfer. The population pollen:seed ratio, like the operational sex ratio in animals (see Emlen and Oring, 1977), must be an important determinant of the strength of social selection on competitive displays and, hence, mating (pollination) systems. In other words, *intra*-specific competition for efficient pollination may be unusually strong in the orchids; and this would increase the accuracy of the analogy with sexual selection and the applicability of generalizations regarding rapid evolution under intraspecific social selection. If Darwin's suggestion is correct, one would expect floral conservatism (low floral diversity) to be associated not only with low pollinator diversity but also sometimes with high pollen:seed ratios. As in animals, signal (floral characteristics) divergence in plants can affect reproductive isolation and speciation when animal pollinators are involved (see Grant, 1971, on "ethological isolating mechanisms" in plants).

Sexually Selected Weapons

The elaborate thoracic and head horns of male beetles offer a well-documented example of pattern in the evolution of weapons generally confined to one sex (the males). Beetle horns are used in male combat in all of a taxonomically wide variety of species in which the behavior of living males has been studied (Eberhard, 1977a–1981, and cited references). In each case the peculiar and sometimes outlandish shape of the horns proved appropriate for some distinctive method of fighting, such as prying, holding, flipping, ramming, pinching, twisting, or lifting conspecific male opponents of a particular (species characteristic) size or shape and behavior. There is no evidence for use of beetle horns in displays to females (Eberhard, 1979). And females do not fight. A recent, detailed taxonomic study of horned beetles of the genus *Blackburnium* (Scarabaeidae: Geotrupinae: Bolboceratini) (Howden, 1979) illustrates both intraspecific geographic variation in horns (e.g., in *B. angulicorne*) and interspecific differences in horns useful in the species-level classification of males. Females, on the other hand, show very little variation, as do non-sexually dimorphic characters of males other than genitalia (which may also be affected by sexual selection—Eberhard, in prep).

Similarly, the weapons and combat rituals of horned and antlered mammals show species-specific forms (see Mayr, 1974) and striking (clinal) geographic variation (Geist, 1971).

Non-Sexual Social Selection

Non-sexual social selection is selection involving competition for resources other than mates. A given character may be employed in both sexual and non-sexual social communication. For example, the "head-toss" of gulls occurs during food solicitation, courtship displays, and aggressive encounters (Hinde, 1970, p. 681). The purpose of this section is to provide evidence that non-sexual social selection is one important cause of rapid signal divergence, by citing examples in which species-specific or geographically variable characters are used in non-sexual social competition.

In mantis shrimps (stomatopod crustaceans) both males and females fight, using formidable weapons in the form of raptorial appendages specialized primarily for different modes of predation on other marine animals (Caldwell and Dingle, 1976). With these appendages stomatopods can break open the shells of clams, and (in some species) deliver lethal blows to conspecific competitors during territorial disputes. Among the structures specialized primarily (but not exclusively—see Schmitt, 1965) for intraspecific contests are the armor of the telson, or tail shell, which is modified as a defensive weapon to receive blows; and the meral spots, variously colored areas on the raptorial appendages that are conspicuous during threat displays. In accord with the predictions of social selection theory, these characters are (monomorphically) well developed in both sexes, and useful in the identification of closely related species (Manning, 1969; Caldwell and Dingle, 1976). The distinctiveness of the meral spot in sympatric species has been assumed to serve for species recognition (Caldwell and Dingle, 1976). But a striking positive correlation between the frequency of aggressive interactions (threat displays per minute) of a species and the prominence of its meral spots in six congeneric species (Caldwell and Dingle, 1976) indicates that meral spots have evolved importantly under social selection.

The same pattern holds in a number of groups of birds: species are sexually monomorphic for striking species-specific social-display morphology when both sexes participate in aggressive or territorial displays (Huxley, 1938, p. 426). Examples include toucans (*Ramphastos* and *Pteroglossus* species), whose enormous bills and bright facial markings may be used in ritualized agonistic behavior (see Skutch, 1958; Bourne, 1974); hummingbirds (*Amazilia* spp.) in which both sexes defend feeding territories (Wolf, 1969; Stiles and Wolf, 1970); colonial species of parrots (*Agapornis*) (Dilger, 1960); and jays (Brown, 1964). (For color plates documenting the predicted species-specificity in bright markings see Haffer, 1974, and Meyer de Schauensee and Phelps, 1978, on toucans and on hummingbirds; Dilger, 1960, and Forshaw, 1973, on parrots; and Davis, 1972, on jays.) The markings of monomorphically bright birds are unlikely to have evolved under sexual selection on males (with pleiotropic effect on females), as suggested by Darwin (1871). Most of these species (e.g., toucans, parrots, and jays) are monogamous for life, and pair formation may involve little or no display (see Dilger, 1960). The parrots studied by Dilger represent a series from solitary to highly social (group-living) species. Females most closely resemble males in the more social species, in which the females aggressively defend nest sites against intruding conspecifics regardless of sex. In the sexually dichromatic species, on the other hand, females confine their aggressive behavior mainly to attacks on other females, and males participate less in defense of the nest (Dilger, 1960, pp. 668–9). Similarly, toucans (*Selenidera*) that live as solitary pairs rather than in flocks are sexually dimorphic and have less exaggerated beaks than do more highly social toucans (*Aulachorhynchus, Rhamphastos*) (Meyer de Schauensee and Phelps, 1978). The trend toward monomorphic brightness in the more highly social species is as predicted in plumage signals if those species are more subject to selection (on both sexes) for successful social interaction within groups, and less strongly selected (primarily in males) for effective courtship. The species-recognition hypothesis is thrown in doubt in the case of parrots by the very large number of strikingly distinctive allopatric and island populations in this family of birds (Moreau, 1948; Forshaw, 1973; see discussion of "superspecies," below).

There is also a correlation between monomorphically bright signal coloration and territoriality by both sexes in some lizards (Ferguson, 1971), mammals (e.g., lemurs—Wilson, 1975, p. 530), and fish (Baylis, 1974). Jolly (1972, p. 155) describes lemurs as among the "showiest of mammals," comparing their species-specific signal pelage to the plumage of visually communicating birds. Some lemurs have undergone "extreme subspecific radiation," showing remarkable geographic variation in fur color and markings (Jolly, 1966, pp. 144–147), suggesting rapid evolution of these display characters. Although the functions of coloration in some reef fish (e.g., Chaetodontidae) are controversial (for a concise summary see Reese, 1975, pp. 38–39), there is no doubt that in some groups they have evolved under social selection (see Warner, Robertson, and Leigh, 1975). I suspect that much of the signal diversity in brightly colored and highly territorial reef species will prove explicable in terms of sexual and non-sexual social selection.

Severe non-sexual social competition occurs among temperate-zone wintering birds. In some species, individuals unable to hold territories generally perish (references in Rohwer, 1977). Experimental dying of the head and crown plumage has shown that the extent of black coloration communicated dominance status in winter hierarchies of Harris' sparrows (*Zonotrichia quereula*) (Rohwer, 1977; Rohwer and Rohwer, 1978). And this aspect of the plumage (also displayed in extreme form during the breeding season) is a key species-specific taxonomic character in North American species of the genus *Zonotrichia* (Robbins, Bruun, and Zim, 1966). It thus follows the prediction of rapid divergence under social selection. In the Harris' sparrow, individuals of both sexes and all age classes form flocks where they compete for resources within a single dominance hierarchy (Rohwer and Rohwer, 1978).

Completely non-sexual social selection occurs in the social insects. Within colonies of the social Hymenoptera (wasps, ants, and bees) only females compete for reproductive status, in contests involving both direct and ritualized dominance and sometimes mediated by pheromonal signals (see references in West-Eberhard, 1977, 1981). These contests result in very large differences in reproductive success between winners (egg-layers) and losers (non-egg-laying workers) (West-Eberhard, 1981). As predicted by social selection theory the aggressive rituals associated with this strong competition, like sexually selected displays, are complex and species specific in the few groups (*Polistes* wasps and stingless bees) where comparative data are available (see Sakagami, 1982; West-Eberhard, in press).

Additional Evidence

Both the degree of exaggeration of socially selected characters and their rate of evolution are expected to be functions of the intensity of social selection (Prediction 2, above). If this is so, there should be a positive association between degree of social specialization of a taxon and the rate of change in its socially selected characters.

Crane (1975, p. 534) recognized different degrees of social specialization among fiddler crabs: socially specialized species have more highly ritualized displays, longer and more complex or intense waving displays, more complex courtship, more time dedicated to territorial maintenance, and more striking male color change during display. As predicted, the subgenus *Celuca*, whose social behavior is the most specialized of the nine subgenera of fiddler crabs (Crane, 1975, p. 531), has undergone "explosive evolution" in the eastern Pacific, where it shows a far greater diversity of species and subspecies (recognized primarily on the basis of socially selected characters) than do the more primitively social subgenera. Of course such comparisons suffer from the difficulty of ascertaining how much of this divergence is due to different numbers and times of speciation-causing (isolating) events in the different subgenera, or to their age. A better test of the hypothesis that rate of divergence and speciation is accelerated by social selection would be to compare the amounts of divergence undergone by populations of different degrees of sociality after being split into allopatric isolates for a given known amount of time. Just such an experiment occurred with the last closing of a seaway through Central America near the end of the Pliocene (3 to 4 million years ago). This simultaneously divided eight species of *Uca* into pairs of populations isolated on the Pacific and Atlantic sides of the Isthmus of Panama (Crane found no evidence of fiddler migration through the Panama Canal). Two of them belong to the socially specialized subgenus *Celuca*; one to a "conservative" subgenus (*Boboruca*) resembling the most primitive social group (*Deltuca*); and five represent subgenera (*Uca, Minuca*) of intermediate and variable social specilization. As Crane pointed out, the classification of allopatric populations as "species" or "subspecies" is necessarily somewhat arbitrary. The important thing for the present comparison is the consistent application of the same criteria to the entire genus, so that the distinctions reflect consistent differences in degrees of divergence (especially, in socially selected features of the major cheliped and waving displays). Crane's taxonomic designations support the hypothesis that more strongly socially selected (specialized) populations evolve more rapidly: both of the most highly social species diverged in allopatry to the level of "species" (*batuenta/cumulanta*, and *limicola/leptodactyla*, on the Pacific/Atlantic sides, respectively), whereas the socially relatively unspecialized *U. thayeri* diverged only to the "subspecies" level (*t. umbratila/t. thayeri*). (Of the species of intermediate social specialization—subgenera *Uca* and *Minuca*—two diverged to the subspecies and three to the species level.)

Rohwer and Niles (1979) made a pioneer attempt to document the rapid evolution of a socially selected trait by studying plumage changes in male purple martins (*Progne subis*). During the period 1840 through the 1970s populations east of the Great Plains have become increasingly colonial due to the increased availability of favorable foraging areas (open fields) and artificial nest holes (martin houses). Rohwer and Niles hypothesized that this might produce evolutionary increase in female-mimicry (plumage dullness) of sub-adult males, which functions in social competition for mates. Although their measurements of museum specimens showed a trend toward increased dullness just short of statistical significance ($p \sim 0.15$), taken together their comparisons of eastern and western U.S. populations indicated a significant ($p = 0.011$) evolutionary effect of density-related geographic factors on subadult male plumage during a period of less than 150 years.

Both these analyses—of fiddler crabs, and of purple martins—would require further research to be considered definitive. But they suggest ways of documenting relative rates of change in socially selected characters using comparative study, in addition to Darwin's (1871) classical methods of (1) showing greater development of social versus non-social characters of the same species, and (2) showing greater variation in social versus non-social characters among different closely related populations (races or subspecies, and congeneric species).

The social-selection (but not the species-recognition) hypothesis is further supported by the occurrence of divergence in social signals within superspecies in a variety of taxa.

A superspecies is ". . . a monophyletic group of entirely or essentially allopatric species that are morphologically too different to be included in a single species" (Mayr, 1963, p. 499). That is, the populations of a superspecies have diverged beyond the subspecies to the species level as judged by taxonomists, with no or very little secondary contact. As such, they are "a particularly convincing illustration of the geographical nature of speciation" (Mayr, 1963, p. 501)—that is, of divergence occurring in isolation, and (by implication) without character displacement due to sympatry with related populations. Therefore, insofar as superspecies fit Mayr's (above) definition, they can be taken to illustrate divergence *not* likely to have originated in the species-recognition context.

Superspecies occur in many socially selected groups, including toucans (Haffer, 1974), birds of paradise (Mayr, 1963; Schodde and McKean, 1973), parrots (Keast, 1961), and fiddler crabs (Crane, 1975). In all of these groups the component species are distinguished primarily by divergence in socially selected characters, as pointed out above. Behavioral observations indicate that this divergence is certain to have been affected significantly by social selection. It cannot be attributed to selection for species recognition without giving more weight to hypothesized "lost neighbors" and selection in a very limited area of geographical contact, than to observed social interactions occurring throughout the species range and likely to be subject to strong selection.

Non-competitive signals (such as alarm call or anti-predator displays) often diverge less rapidly than competitive social signals. Some of them may be strongly selected to remain constant (Mayr, 1974). In one of the few available discussions of conservatism and lability in the evolution of different kinds of displays, Moynihan (1975) showed that three of the four kinds of cephalopod displays categorized as "conservative" (showing little variation among living species and even orders of cephalopods) are non-competitive startle reactions or alarm signals. By contrast, courtship displays are varied, even among closely related species. Moynihan hypothesized that conservative patterns are stable because they are adapted to influence a diversity of receivers (different age, size, and sex classes of the same species, or individuals of other species such as predators), with consequently severe limits on the kinds of signals that would be suitable to all. The contrasting rapid evolution of the competitive displays may be due to social selection. Similarly, in some birds with seasonal plumage change, drab winter or juvenile (camouflage) plumage is often less species-specific than the breeding (socially selected) plumage (e.g., in sandpipers, phalaropes, wood warblers, loons, and grebes—see Robbins, Bruun, and Zim, 1966).

In all of the above examples, the key characters for taxonomy at the species and subspecies levels have served to indicate the relative rapidity with which social characters evolve. Examples are therefore primarily limited to groups in which social competition happens to involve visible morphological characters or coloration. There must be many cases in which divergence important in social selection and speciation involves cryptic characters such as odor or behavior not usually observed by taxonomists (e.g., see Blaustein, 1981; Bornemissza, 1966; Bergstrom, Svensson, Appelgren, and Groth, 1981; and Blum, 1981, on the sex pheromones of mammals, scorpionflies, bumblebees, and ants, respectively). In such groups, taxonomic separation of biological species is often difficult until these characters are analyzed (see Alexander, 1962), and, in accord with the theory presented here, close study of living specimens should reveal divergence in the socially selected characters even when little or none is evident in morphology.

SPECIATION

The theory and examples given so far show a connection between social selection and rapid character divergence in isolated populations. Speciation is complete only when divergence is sufficient to lead to reproductive isolation. The question of how much genetic or phenotypic divergence is necessary to favor reproductive isolation between overlapping populations has never been resolved (Lewontin, 1974). (Electrophoretic studies of Hawaiian *Drosophila*—Carson, 1978—have exposed the inadequacy of using genetic distance to assign taxonomic status to different sexually selected populations.) The critical parameter is the degree to which the two sexes are incompatible or hybrids are at a disadvantage when male and female are from different populations, not simply the number of genes involved. Divergence in a critical signal could conceivably involve a small number of genes, yet have a disproportionately large effect on the sexual compatibility or competitive ability of hybrid offspring. For example, two species of Australian scorpionflies (*Harpobittacus*) that are scarcely distinguishable morphologically fail to pair in the laboratory due to divergent male sex pheromones (Bornemissza, 1966). The pheromonal divergence is almost certainly due to sexual selection rather than to selection for species recognition, since the two populations are known to have been completely allopatric since the Middle Cretaceous (Bornemissza, 1966). This seems to be a clear example of allopatric signal divergence under sexual selection leading to reproductive isolation and speciation without secondary contact and reproductive character displacement.

In general it seems reasonable to argue that if divergence can lead to speciation (reproduction isolation), then anything that accelerates divergence should tend to accelerate speciation—especially if characters critical to survival or reproductive success are concerned. It follows that social selection must often accelerate speciation in socially interacting organisms like those discussed in this article.

Spieth (1974) has remarked on the rapidity with which divergence and intra-island speciation is attained by lekking Hawaiian *Drosophila* compared to non-lek-forming subgroups, and has noted that in the lek species a single mountaintop has often served as a sufficient area to allow the evolution of a species. It is not surprising that many of the groups cited by Mayr (in press) to illustrate peripatric speciation (rapid acquisition of isolating mechanisms in small isolated populations) are groups often showing strong social or sexual selection (Hawaiian *Drosophila*, tropical birds, and lizards). As pointed out by Carson (1978), the distinctive sample of characters that happens to characterize such a small isolate (the "founder effect") could be enough to start the population on a new evolutionary direction under sexual selection.

When learning affects divergence it is difficult to generalize about the role of divergence on rates of speciation. In some species [e.g., white-crowned sparrows (*Zonotrichia leucophrys*), indigobirds (*Vidua* species), and Indigo buntings (*Passerina cyanea*)], individuals are so extremely flexible at mimicking local dialects and even songs of different species that it raises the possibility that learning retards, rather than promotes, the evolutionary (genetic) divergence of song. Instead, it suggests what Payne (in press) terms the "appealing" (but unproven) possibility that selection for flexibility under social competition in such species has led to the evolution of greater song learning ability, or intelligence. Indeed, it is probably no accident that the animals we regard as especially intelligent—e.g., crows and jays, dogs, porpoises, parrots, and primates—are also especially social (and hence subject to strong intragroup competition in which signal flexibility, mimicry, and other clever social manipulations may become highly advantageous) (see references in Alexander, 1979, p. 213).

But learning in some species seems to contribute to narrowness and rigidity (rather than flexibility) of breeding preferences. In white-crowned sparrows, for example, males learn songs in an early auditory-sensitive period (Baker, Spitler-Nabors, and Bradley, 1981), which would ordinarily mean that they could learn only the songs heard in their natal region. And there is a widespread tendency among social organisms to exclude outsiders, or at least to prefer to associate with members of their own flock or region on the basis of learned cues (Wilson, 1975). The extent to which this influences *mating* patterns is controversial (Bush, Case, Wilson, and Patton, 1977; Daly, 1981). Hardy (1966) hypothesized that the xenophobia, or clannishness of parrots—learned recognition cues, and their strong preference for pairing (in laboratory experiments) with members of the same flock, region, and species—may have contributed importantly to the rapid and superficial divergence of these birds, which never hybridize in the wild, yet like many other socially selected organisms (see below), when forced together in captivity can produce hybrids between species, genera, and even sub-families (Hardy, 1966, p. 80).

Sexually selected characters are often referred to as "superficial" (Sibley, 1957; Gilliard, 1969; Carson, 1978) in that conspicuous phenotypic differences can involve just a few genes (Carson, 1978). Two very distinctive varieties of the golden pheasant (*Chrysolophus pictus* mutants *obscurus* and *lutens*) are the results of single mutations (Gerrits, 1961). And interspecific and even intergeneric hybrids are common in sexually selected organisms in nature (Sibley, 1957; Mayr, 1963; Pace, 1974; Blackwell and Bull, 1978) and in captivity (Gerrits, 1961; Hardy, 1966). The great but superficial species diversity of orchids (indicated by ease of hybridization) may be due to a coevolutionary process like social selection involving flowers and pollinators. This kind of superficiality may render socially selected characters relatively useless in designating higher categories; few would deny the undesirability of the huge members of monotypic genera formerly recognized on the basis of socially selected characters in such groups as hummingbirds and birds of paradise (Sibley, 1957). At the species level, however, there is some justification for giving extra weight to socially selected characters because of their likely importance in effecting reproductive isolation.

The most notable result of social selection for the taxonomic studies of speciation is confusing variation. Darwin (1871) was impressed with the variability of sexually selected characters, and recognized it as being of several kinds: variation between the sexes (sexual dimorphism), which often makes it difficult to associate males and females of the same population in collections; geographic variation, making it difficult to ascertain the status (subspecific, specific, and generic) of allopatric populations; intrasexual and caste polymorphisms (e.g., high-low, or size-related dimorphisms, and immature vs. adult plumage of males), now seen to be associated with different mating or reproductive strategies (e.g., see Selander, 1972; West-Eberhard, 1979; Rohwer, Fretwell, and Niles, 1980; Eberhard, 1982); and individual variability, due to continuing selection (transient polymorphism of characters undergoing change, called "generative variability" by Darwin, 1859, p. 114). Taxa under sexual or social selection are thus commonly described as different groups. This problem is even more intractable for the systematist when variations are cryptic: for example, when there is biologically important but invisible divergence of behavioral, acoustical, or pheromonal characters without obvious morphological correlates. Conspicuous variation (e.g., in the acoutrements of visual displays) sometimes leads to inordinate "splitting" (elevation of geographic varieties to species rank) and to large numbers of allopatric and monotypic genera, whereas non-morphological variation leads to the opposite problem: lumping of variants, and unrecognized cryptic species.

One of the remaining controversies in speciation theory involves the significance of parapatric distributions, in which the borders of closely related species touch, often extensively and for long periods of time, without major overlap or massive hybridization (Mayr, 1982b). Parapatric distributions are reported in diverse organisms, including several of the socially selected taxa discussed in this article, such as birds of paradise (Gilliard, 1969), toucans (Haffer, 1974), parrots (Keast, 1961), lemurs (White, 1978, p. 89), and fiddler crabs (Crane, 1975). They must often have some ecological basis, with one species superior on each side of the boundary (Mayr, 1969), even though clear habitat or topographic barriers are not always visible to the human eye. As pointed out by Pace (1974, p. 73), however, maintenance of extensive contiguous distributions (e.g., see Pace, 1974, on leopard frogs) means that ". . . the limits of distribution of one member of any species pair are more clearly related to the presence of the other species than to any other feature of the physical or biotic environment, suggesting that significant, special interspecific interactions are involved in the biology of these species." What might the nature of these special interactions be? A combination of social-selection theory and classical ecological theory may hold an answer. When socially selected characters take the lead in speciation-related divergence, as suggested here, this implies that ecological divergence may sometimes lag behind. Niche similarity of allied species is expected to lead to competitive exclusion—elimination of one of the species from a region where it could exist alone, due to the competitive superiority of another species (see review by Hutchinson, 1975). Thus the low degree of ecological divergence of socially selected sibling species may particularly dispose them to parapatric distributions. And the aggressiveness and territoriality of socially competitive individuals, as well as the tendency to aggregate preferentially with conspecifics (e.g., in mating or nesting sites), may sometimes contribute to the maintenance of clearcut boundaries between species (e.g., see Diamond, 1973; Caldwell and Dingle, 1976). Gilliard (1969, p. 4) concludes that such a combination of ecological and behaviorial exclusion has affected the (parapatric) distributions and mating systems of closely related species of birds of paradise: "I found that the rapidly evolving polygynous 'genera' . . . usually behave somewhat like a semispecies in that they exclude each other ecologically; but the primary pressures of exclusion apparently involve the breeding grounds (the arenas), which they defend vigorously. . . ." Gilliard believes that these "exclusion pressures" may lead to the occurrence of generic tiering or layering of arenas at different heights within highland New Guinea forests, as well as to the (secondary) evolution of monogyny in some species and the evolution of bower construction behavior in some of those confined to the forest floor.

The evolution of divergent signals potentially serving as pre-mating isolating mechanisms can be an early rather than a late event in the speciation process in these groups; and divergence of social traits may be the basis of breeding incompatibility (hybrid disadvantage, or failure to interact as normal conspecifics) between populations that have diverged very little ecologically. Rapid divergence and speciation can thus conceivably occur between populations even when there is very little difference in their non-social environments.

CONCLUSION

The discovery in a taxon of species-specific social signals does not necessarily imply a primary species-recognition function. Social signals diverge in isolation, and can be elaborated independently of species recognition, even in sympatry with closely related species. Populations whose signals or appearance have diverged under social selection are preadapted for species recognition by the prior acquisition of species-specific markers, and need only be selected to distinguish them. If the diverged traits are so distinctive as to be severely disruptive to normal interaction, hybridization may not even occur upon recontact, and speciation (reproductive isolation) can be regarded as already complete. Any *assumption*, however, that even extravagant divergence under social selection would automatically have this result would be as invalid as the common but erroneous assumption that species specificity implies selection for species recognition.

The species-recognition hypothesis seems to have persisted as an explanation for divergence in social signals in many groups not because it was strongly supported by data, but because it was consistent with ethological and speciation theory at a time when sexual selection theory was largely forgotten. During that period biologists proved remarkably more creative at forcing the species-recognition interpretation upon contradictory data than they did at considering alternative explanations. Occasional attempts (e.g., Sibley, 1957) to revive Darwin's powerful arguments in support of sexual selection were not widely appreciated. The reasons for this would make an interesting study in the history of science.

Socially competitive characters, whether employed in a sexual or a non-sexual context, follow the same pattern of rapid divergence. This demonstrates the utility and wisdom of Darwin's (1871) painstaking insistence on the importance of the distinction between *natural* and *sexual* selection, as well as the importance of extending the latter category to include selection on non-sexual social characters. Only by making this distinction and applying Darwin's original generalizations regarding the special nature of sexual selection broadly (to all social competition) could one discover the common pattern in the evolution of beetles' horns, crabs' claws, pheasants' tails, toucans' beaks, birds' songs, and the dominance behavior of bees and wasps. The ideas presented here should eventually prove applicable to a very wide variety of organisms having behaviors and structures (e.g., complex genitalia and pheromonal signals) whose competitive functions are only beginning to be understood.

ACKNOWLEDGMENTS

The stimulus for writing this paper came from reading a manuscript on courtship and speciation in insects by R. Thornhill, to whom I am indebted for generous sharing of ideas and for having provided many key references. The antepenultimate version (February, 1982) was read and criticized in a seminar at the University of Michigan Museum of Zoology. Of the 22 able participants I especially thank L. Blumer, S. Dobson, W. Dominey, P. Ewald, G. Nuechterlein, R. Smuts, and M. Zuk for numerous suggestions and for access to manuscripts in preparation. For information and critical discussion of particular sections I thank J. R. Baylis, J. Bradbury, J. Bronstein, J. Crane, R. Dressler, P. Feinsinger, W. D. Hamilton, J. W. Hardy, G. Helfman, H. F. Howden, R. B. Manning, and M. Moynihan. R. D. Alexander, W. G. Eberhard, H. E. Evans, C. Haskins, E. G. Leigh, H. Lessios, J. Lloyd, R. Longair, E. Mayr, C. D. Michener, R. Payne, A. S. Rand, M. Ryan, A. Skutch, N. G. Smith, G. Stiles, and R. Thornhill read and criticized the entire manuscript in at least one stage. R. Lande developed similar arguments regarding sexual selection and speciation following a conversation (February, 1980) in which I outlined many of the ideas presented here. I benefitted from reading at that time his manuscript on runaway selection (later modified to include ideas on speciation—Lande, 1981), and from his comments on an early (October, 1980) draft of the present paper.

LIST OF LITERATURE

ALEXANDER, R. D. 1962. The role of behavioral study in cricket classification. *Syst. Zool.*, 11(2): 53–72.

——. 1974 The evolution of social behavior. *Annu. Rev. Ecol. Syst.*, 4: 325–383.

——. 1975 Natural selection and specialized chorusing behavior in acoustical insects. In D. Pimentel (ed.), *Insects, Science and Society*, p. 35–77. Academic Press, New York.

——. 1979. *Darwinism and Human Affairs*. Univ. Washington Press, Seattle.

ALEXANDER, R. D., and G. BORGIA. 1979. On the origin and basis of the male-female phenomenon. In M. S. Blum and N. A. Blum (eds.), *Sexual Selection and Reproductive Competition in Insects*, p. 417–440. Academic Press, New York.

ALEXANDER, R. D., and D. OTTE. 1967. The evolution of genitalia and mating behavior in crickets (Gryllidae) and other Orthoptera. *Misc. Publ. Mus. Zool. Univ. Mich.*, 133: 1–63.

ANDERSON, W. R. 1979. Floral conservatism in neotropical Malpighiaceae. *Biotropica*, 11: 219–223.

ARMSTRONG, E. A. 1965. *The Ethology of Bird Display and Behavior.* Dover Publications, New York.

BAKER, M. C., K. J. SPITLER-NABORS, and D. C. BRADLEY,. 1981. Early experience determines song dialect responsiveness of female sparrows. *Science*, 214: 819–821.

BAKER, R. R., and G. A. PARKER. 1979. The evolution of bird colouration. *Phil. Trans. Roy. Soc.*, B., 287: 63–130.

BAYLIS, J. R. 1974. The behavior and ecology of *Herotilapia multispinosa* (Teleostei, Cichlidae). *Z. Tierpsychol.*, 34: 115–146.

BERGSTROM, G., B. G. SVENSSON, M. APPLEGREN, and I. GROTH. 1981. Complexity of bumblebee marking pheromones: biochemical, ecological and systematical interpretations. In P. E. Howse and J.-L. Clement (eds.), *Biosystematics of Social Insects*, p. 175–184. Academic Press, New York.

BLACKWELL, J. H. and C. M. BULL. 1978. A narrow hybird zone between two Western Austrialian frog species *Ranidella insignifera* and *R. pseudoinsignifera:* The extent of introgression. *Heredity*, 40(1): 13–25.

BLAIR, W. F. (ed.). 1960. *Vertebrate Speciation.* Univ. Texas Press, Austin.

——. 1962. Non-morphological data in anuran classification. *Syst. Zool.*, 11: 72–84.

——. 1974. Character displacement in frogs. *Am. Zool.*, 14: 1119–1125.

BLAUSTEIN, A. R. 1981. Sexual selection and mammalian olfaction. *Am. Nat.*, 117: 1006–

1010.

BLUM, M. S. 1981. Sex pheromones in social insects: chemotaxonomic potential. In P. E. Howse and J.-L. Clement (eds.), *Biosystematics of Social Insects,* p. 163-174. Academic Press, New York.

BLUM, M. S., and N. A. BLUM (eds.). 1979. *Sexual Selection and Reproductive Competition in Insects.* Academic Press, New York.

BORGIA, G. 1979. Sexual selection and the evolution of mating systems. In M. S. Blum and N. A. Blum (eds.), *Sexual Selecton and Reproductive Competition in Insects*, p. 19-80. Academic Press, New York.

BORNEMISSZA, G. F. 1966. Observations on the hunting and mating behavior of two species of scorpionflies (Bittacidae: Mecoptera). *Aust. J. Zool.,* 14: 371-382.

BOURNE, G. R. 1974. The red-billed toucan in Guayana. *Living Bird*, 1974: 99-126.

BRADBURY, J. W. 1977. Lek mating behavior in the hammer-headed bat. *Z. Tierpsychol.,* 45: 225-255.

BRADBURY, J. W., and R. M. GIBSON. In press. In P. Bateson (ed.), *Mate Choice*, p. 109-138. Cambridge University Press, Cambridge.

BRENOWITZ, E. A. 1982. Long-range communication of species identity by song in the red-winged blackbird. *Behav. Ecol. Sociobiol.*, 10: 29-38.

BROWN, J. L. 1964. The integration of agonistic behavior in the Steller's jay *Cyanocitta stelleri* (Gmelin). *Univ. Calif. Pulb. Zool.,* 60: 223-328.

BUSH, G. L., 1975. Modes of animal speciation. *Annu. Rev. Ecol. Syst.,* 6: 339-364.

BUSH, G. L., S. M. CASE, A. C. WILSON, and J. L. PATTON. 1977. Rapid speciation and chromosomal evolution in mammals. *Proc. Nat. Acad. Sci. USA,* 74(9): 3942-3946.

CALDWELL, R. L., and H. DINGLE. 1976. Stomatopods. *Sci. Am.,* 234(1):81-89.

CAMPBELL, B. (ed.). 1972. *Sexual Selection and the Descent of Man, 1871-1971.* Aldine, Chicago.

CARPENTER, C. C. 1966. Comparative behavior of the Galapagos lava lizards (*Tropidurus*). In R. I. Bowman (ed.), *The Galapagos*, p. 269-273. Univ. California Press, Berkeley.

CARSON, H. L. 1978. Speciation and sexual selection in Hawaiian *Drosophilia.* In P. F. Brussard (ed.), *Ecological Genetics: The Interface,* p. 93-107. Springer-Verlag, New York.

CRANE, J. 1975. *Fiddler Crabs of the World.* Princeton Univ. Press, Princeton.

CREWS, D., and E. E. WILLIAMS. 1977. Hormones, behavior, and speciation. *Am. Zool.,* 17: 271-286.

CROOK, J. H. 1972. Sexual selection, dimorphism, and social organization in the primates. In B. Campbell (ed.), *Sexual Selection and the Descent of Man, 1871-1971,* p. 231-281. Aldine, Chicago.

DALY, J. C. 1981. Effects of social organization and environmental diversity on determining the genetic structure of a population of the wild rabbit, *Oryctologus cuniculus. Evolution,* 35(4): 689-706.

DARWIN, C. 1859 (1936). *The Origin of Species.* Modern Library, New York.

——. 1862 (1904). *The Various Contrivances By Which Orchids Are Fertilized By Insects.* John Murray, London.

——.1871 (1936). *The Descent of Man and Selection in Relation to Sex.* Modern Library, New York.

DAVIS, L. I. 1972. *Birds of Mexico and Central America.* Univ. Texas Press, Austin.

DELACOUR, J. 1970. *Pheasant Breeding and Care.* T. F. H. Publications, Jersey City.

DIAMOND, J. M. 1973. Distributional ecology of New Guinea birds. *Science,* 179: 759-769.

DILGER, W. C. 1960. The comparative ethology of the African parrot genus *Agapornis. Z. Tierpscychol.*, 17: 649-685.

DOBZHANSKY, T. 1937. *Genetics and the Origin of Species.* Columbia Univ. Press, New York.

——. 1940. Speciation as a stage in evolutionary divergence. *Am. Nat.,* 74: 312-321.

DODSON, C. H. 1975. Coevolution of orchids and bees. In L. E. Gilbert and P. H. Raven (eds.), *Coevolution of Animals and Plants.* p. 91-99. Univ. Texas Press, Austin.

DRESSLER, R. L. 1981. *The Orchids.* Harvard Univ. Press, Cambridge.

EBERHARD, W. G. 1977a. Fighting behavior of male *Golopha porteri* beetles (Scarabeidae: Dynastinae). *Psyche,* 83(3-4): 292-298

——. 1977b. La ecología y comportamiento del cucarron de la caña *Podischnus agenor* (Scarabeidae: Dynastinae). *Rev. Col. Entomol.*, 3(1-2): 17-21.

——. 1979. The function of horns in *Podischnus agenor* (Dynastinae) and other beetles. In M. S. Blum and N. Blum (eds.), *Sexual Selection and Reproduction Competition in Insects,* p. 231-258. Academic Press, New York.

——. 1980. Horned beetles. *Sci. Am.,* 242(3): 166-182.

——. 1981. The natural history of *Doryphora* sp. (Coleoptera, Chrysomelidae) and the function of its sternal horn. *Ann. Entomol. Soc. Am.*, 74: 445-448.

——. 1982. Beetle horn dimorphism: making the best of a bad lot. *Am. Nat.*, 119: 420-426.

EHRMAN, L. 1972. Genetics and sexual selection. In B. Campbell (ed.), *Sexual Selection and the*

Descent of Man, 1871-1971, p. 105-135. Aldine, Chicago.

EMLEN, S. T. 1972. An experimental analysis of the parameters of song eliciting species recognition. *Behaviour*, 41: 130-171.

EMLEN, S. T., and L. W. ORING. 1977. Ecology, sexual selection, and the evolution of mating systems. *Science*, 197: 215-223.

ENDLER, J. A. 1977. *Geographic Variation, Speciation, and Clines*. Princeton Univ. Press, Princeton.

FERGUSON, G. 1971. Variation and evolution of the push-up displays of the side-blotched lizard genus *Uta* (Iguanidae). *Syst. Zool.*, 20(1): 79-101.

——. 1977. Display and communications in reptiles: an historical perspective. *Am Zool.*, 17: 167-176.

FINE, M. C., H. E. WINN, and B. L. OLLE. 1977. Communication in fishes. In T. A. Sebeok (ed.), *How Animals Communicate*, p. 472-518. Indiana Univ. Press, Bloomington.

FISHER, R. A. 1930 (1958). *The Genetical Theory of Natural Selection*. Dover Publications, New York.

FORREST, T. G. 1982. Calling songs and mate choice in mole crickets. In D. Gwynne and G. K. Morris (eds.), *Orthopteran Mating Systems*, p. 185-204. Westview Press, Boulder.

FORSHAW, J. M. 1973. *Parrots of the World*. Lansdowne Press, Melbourne.

FORSYTH, A. B. 1980. Worker control of queen density in Hymenoptera. *Am. Nat.*, 116: 895-898.

FRANKIE, G. W. 1975. Tropical forest phenology and pollinator plant coevolution. In L. E. Gilbert and P. H. Raven (eds.), *Coevolution of Animals and Plants*, p. 192-209. Univ. Texas Press, Austin.

GEIST, V. 1971. *Mountain Sheep*. Univ. Chicago Press, Chicago.

GERRITS, H. A. 1961. *Pheasants*. Blanford Press, London.

GHISELIN, M. T. 1974. *The Economy of Nature and the Evolution of Sex*. Univ. California Press, Berkeley.

GILLIARD, E. T. 1969. *Birds of Paradise and Bower Birds*. Natural History Press, Garden City.

GORMAN, G. 1968. The relationships of *Anolis* and the *roquet* species group (Sauria: Iguanidae) III. Comparative study of display behavior. *Breviora*, 284: 1-31.

GRANT, V. 1971. *Plant Speciation*. Columbia Univ. Press, New York.

GRANT, V., and K. A. GRANT. 1965. *Flower Pollination in the Phlox Family*. Columbia Univ. Press, New York.

GREENBERG, B., and G. K. NOBLE. 1944. Social behavior of the American chameleon (*Anolis carolinensis*). *Physiol. Zool.*, 17: 392-439.

GREENBERG, L. 1979. Genetic component of odor in kin recognition. *Science*, 206:1095-1097.

HAFFER, J. 1974. Avian speciation in tropical South America. *Publ. Nuttal Ornithol. Club*, 14: 1-390.

HAMILTON, W. D., and M. ZUK. 1982. Heritable true fitness and bright birds: a role for parasites? *Science*, 218: 384-386.

HARDY, J. W. 1966. Physical and behavioral factors in sociality and evolution of certain parrots (*Aratinga*). *Auk*, 83: 66-83.

HEINRICH, B., and P. H. RAVEN. 1972. Energetics and pollination ecology. *Science*, 176: 597-602.

HINDE, R. A. 1970. *Animal Behaviour*. McGraw Hill, New York.

HINDE, R. A., and N. TINBERGEN. 1958. The comparative study of species-specific behavior. In A. Roe and G. G. Simpson (eds.)., *Behavior and Evolution*, p. 251-268. Yale Univ. Press, New Haven.

HOENIGSBERG, H. F., Y. G. DE NAVAS, and A. J. CHEJNE. 1964. Sexual selection in captive mutants of *Drosophila melanogaster*. *Z. Tierpsychol.*, 21(7): 786-793.

HOWDEN, H. F. 1979. A revision of the Australian genus *Blackburnium* Boucomont (Coleoptera: Scarabaeidae: Geotrupinae). *Aust. J. Zool. Suppl. Ser.*, 72: 1-88.

HUTCHINSON, G. E. 1975. Variations on a theme by Robert MacArthur. In M. L. Cody and J. M. Diamond (eds.), *Ecology and Evolution of Communities*, p. 492-521. Belknap Press, Cambridge.

HUXLEY, J. S. 1938. Darwin's theory of sexual selection and the data subsumed by it, in the light of recent research. *Am. Nat.*, 72: 416-433.

ISHAY, J., and M. B. BROWN. 1975. Pattern in the hunger signal of hornet larvae. *Experientia*, 31:1044-1046.

JENSSEN, T. A. 1977. Evolution of anoline lizard display behavior. *Am. Zool.*, 17:203-216.

JOLLY, A. 1966. *Lemur Behavior*. Univ. Chicago Press, Chicago.

——. 1972. *The Evolution of Primate Behavior*. Macmillan, New York.

KEAST, A. 1961. Bird speciation on the Australian continent. *Bull. Mus. Comp. Zool. Harv.*, 123: 305-495.

KIRKPATRICK, M. 1982. Sexual selection and the evolution of female choice. *Evolution*, 36: 1-12.

KREBS, J. R., and D. E. KROODSMA. 1980. Repertoires and geographical variation in bird song. *Adv. Stud. Behav.* 11: 143-177.

LANDE, R. 1981. Models of speciation by sexual

selection on polygenic traits. *Proc. Natl. Acad. Sci. USA*, 78: 3721-3725.
——. 1982. Rapid origin of sexual isolation and character divergence in a cline. *Evolution*, 36: 213-223.
LANYON, W. E. 1969. Vocal characters and avian systematics. In R. A. Hinde (ed.), *Bird Vocalizations*, p. 291-310. Cambridge Univ. Press, Cambridge.
LAZELL, J. D., JR., and E. E. WILLIAMS. 1962. The anoles of the Eastern Caribbean (Sauria, Iguanidae). *Bull. Mus. Comp. Zool. Harv.*, 127(9): 466-478.
LEWONTIN, R. C. 1974. *The Genetic Basis of Evolutionary Change*. Columbia Univ. Press, New York.
LLOYD, J. E. 1966. Studies on the flash communication system of *Photinus* fireflies. *Misc. Publ. Mus. Zool. Univ. Mich.*, 130: 1-95.
——. 1979. Sexual selection in luminescent beetles. In M. S. Blum and N. A. Blum (eds.), *Sexual Selection and Reproductive Competition in Insects*, p. 293-342. Academic Press, New York.
—— (ed.). 1979-1981. Insect behavioral ecology. *Fla. Entomol.*, 63(1): 1-111; 64(1): 1-118; 65(1): 1-104.
——. 1980. Sexual selection: individuality, identification, and recognition in a bumblebee and other insects. *Fla. Entomol.*, 64: 89-118.
——. 1983. Bioluminescence and communication in insects. *Annu. Rev. Entomol.*, 28: 131-160.
MANNING, A. 1966. Sexual behavior. *Symp. R. Entomol. Soc. Lond.*, 3: 59-68.
MANNING, R. B. 1969. Stomatopod crustacea of the Western Atlantic. *Stud. Trop. Oceanogr.*, 8: viii + 1-380.
MARLER, P. 1960. Bird songs and mate selection. In W. E. Lanyon and W. N. Tavolga (eds.), *Animal Sounds and Communication*, Vol. 7, p. 348-367. AIBS, Washington, D.C.
MAYR, E. 1942. *Systematics and the Origin of Species*. Columbia Univ. Press, New York.
——. 1963. *Animal Species and Evolution*. Harvard Univ. Press, Cambridge.
——. 1969. Bird speciation in the tropics. *Biol. J. Linn. Soc.*, 1: 1-17.
——. 1972. Sexual selection and natural selection. In B. Campbell (ed.), *Sexual Selection and the Descent of Man, 1871-1971*, p. 87-104. Aldine, Chicago.
——. 1974. Behavior programs and evolutionary strategies. *Am. Sci.* 62(6): 650-659.
——. 1982a. *The Growth of Biological Thought*. Harvard Univ. Press, Cambridge.
——. 1982b. Processes of speciation in animals. In C. Barigozzi (ed.), *Mechanisms of Speciation*, p. 1-15. Alan R. Liss, New York.
MEEUSE, B. J. D., and E. L. SCHNEIDER. 1979. *Nymphaea* revisited: a preliminary communication. *Isr. J. Bot.*, 28: 65-79.
MEYER DE SCHAUENSEE, R., and W. H. PHELPS. 1978. *Birds of Venezuela*. Princeton Univ. Press, Princeton.
MICHENER, C. D. 1974. *The Social Behavior of the Bees*. Harvard Univ. Press, Cambridge.
MOREAU, R. E. 1948. Aspects of evolution in the parrot genus *Agapornis*. *Ibis*, 90: 206-239.
MORTON, E. S. 1975. Ecological sources of selection on avian sounds. *Am. Nat.*, 109: 17-34.
MOYNIHAN, M. 1975. Conservatism of displays and comparable stereotyped patterns among cephalopods. In G. Baerends, C. Beer, and A. Manning (eds.), *Function and Evolution in Behaviour*, p. 276-291. Clarendon Press, Oxford.
NEUCHTERLEIN, G. L. 1981a. Courtship behavior and reproductive isolation between Western grebe color morphs. *Auk*, 98: 335-349.
——. 1981b. Variations and multiple functions of the advertising display of Western grebes. *Behaviour*, 76(3-4): 289-317.
O'DONALD, P. 1977. Theoretical aspects of sexual selection. *Theor. Popul. Biol.*, 12: 298-334.
——. 1980. *Genetic Models of Sexual Selection*. Cambridge Univ. Press, Cambridge.
ORDNUFF, R. 1978. Reproductive characters and taxonomy. *Syst. Bot.*, 3(4): 420-427.
OTTE, D. 1979. Historical development of sexual selection theory. In M. S. Blum and N. A. Blum (eds.), *Sexual Selection and Reproductive Competition in Insects*, p. 1-18. Academic Press, New York.
PACE, A. E. 1974. Systematic and biological studies of the leopard frogs (*Rana pipiens* complex) of the United States. *Misc. Publ. Mus. Zool. Univ. Mich.*, 148: 1-140.
PARKER, G. A. 1978. Evolution of competitive mate searching. *Annu. Rev. Entomol.*, 23: 173-196.
PAYNE, R. B. 1982. Ecological consequences of song matching: breeding success and intraspecific song mimicry in Indigo buntings. *Ecology*, 63(2): 401-411.
——. In press. Bird songs, sexual selection and female mating strategies. In S. Wasser (ed.), *Female Social Strategies*. Academic Press, New York.
PAYNE, R. B., and K. PAYNE. 1977. Social organization and mating success in local song populations of Village Indigobirds, *Vidua chalybeata*. *Z. Tierpsychol.*, 45: 113-173.
RAND, A. S., and E. E. WILLIAMS. 1970. An estimation of redundancy and information content of anole dewlaps. *Am. Nat.*, 104: 99-103.
REESE, E. S. 1975. A comparative field study of the social behavior and related ecology of reef

fishes of the family Chaetodontidae. *Z. Tierpsychol.*, 37: 37-61.
RINGO, J. M. 1977. Why 300 species of Hawaiian *Drosophila? Evolution*, 31: 695-754.
ROBBINS, C. S., B. BRUUN, and H. S. ZIM. 1966. *Birds of North America*. Golden Press, New York.
ROHWER, S. 1977. Status signalling in Harris sparrows: some experiments in deception. *Behaviour*, 61(1-2): 107-129.
ROHWER, S., S. D. FRETWELL, and D. M. NILES. 1980. Delayed maturation in passerine plumages and the deceptive acquisition of resources. *Am. Nat.*, 115: 400-437.
ROHWER, S., and D. M. NILES. 1979. The subadult plumage of male purple martins: variability, female mimicry and recent evolution. *Z. Tierpsychol.*, 51: 282-300.
ROHWER, S., and F. C. ROHWER. 1978. Status signalling in Harris sparrows: experimental deceptions achieved. *Anim. Behav.*, 26: 1012-1022.
ROTHSCHILD, M. 1975. Remarks on carotenoids in the evolution of signals. In L. E. Gilbert and P. H. Raven (eds.), *Coevolution of Animals and Plants*, p. 20-47. Univ. Texas Press, Austin.
RUIBAL, R. 1967. Evolution and behavior in West Indian anoles. In W. W. Milstead (ed.), *Lizard Ecology, A Symposium*, p. 116-140. Univ. Missouri Press, Columbia.
RYAN, M. J. 1980. Female mate choice in a neotropical frog. *Science*, 209: 523-525.
——. In press. *The Tungara Frog: A Study of Sexual Selection and Communication*. Univ. Chicago Press, Chicago.
SAKAGAMI, S. F. 1982. Stingless bees. In H. Hermann (ed.), *Social Insects*, Vol. III, p. 362-423. Academic Press, New York.
SCHMITT, W. L. 1965. *Crustaceans*. Univ. Michigan Press, Ann Arbor.
SCHODDE, R. 1976. Evolution in the birds of paradise and bowerbirds, a resynthesis. *Proc. 16th Int. Congr. Ornithology, Canberra:* 137-149.
SCHODDE, R., and J. L. MCKEAN. 1973. The species of the genus *Parotia* (Paradisaeidae) and their relationships. *Emu*, 73(4): 145-156.
SCHWARTZ, A. 1968. Geographic variation in *Anolis distichus* Cope (Lacertilia, Iguanidae) in the Bahama Islands and Hispaniola. *Bull. Mus. Comp. Zool.*, 137(2): 255-310.
SELANDER, R. K. 1972. Sexual selection and dimorphism in birds. In B. Campbell (ed.), *Sexual Selection and the Descent of Man, 1871-1971*, p. 180-230. Aldine, Chicago.
SIBLEY, C. G. 1957. The evolutionary and taxonomic significance of sexual dimorphism and hybridization in birds. *Condor*, 59: 166-191.
SILBERGLIED, R. In press. Visual communication and sexual selection among butterflies. In R. I. Vane-Wright and P. R. Ackery (eds.), *The Biology of Butterflies* (Roy. Entomol. Soc. Lond. Symposium No. 11). Academic Press, New York.
SIMPSON, G. G. 1944. *Tempo and Mode in Evolution*. Columbia Univ. Press, New York.
SKUTCH, A. 1958. The roosting and nesting of Aracara toucans. *Condor*, 60(4): 201-219.
——. 1976. *Parent Birds and Their Young*. Univ. Texas Press, Austin.
SMITH, W. J. 1977. *The Behavior of Communicating*. Harvard Univ. Press, Cambridge.
SNOW, D. W. 1976. *The Web of Adaptation*. Quadrangle, New York.
SPIETH, H. T. 1974. Mating behavior and evolution of the Hawaiian *Drosophila*. In M. J. D. White (ed.), *Genetic Mechanisms of Speciation in Insects*, p. 94-101. Australia and New Zealand Book Co., Sydney.
——. 1981. *Drosophila heteroneura* and *Drosophila silvestris:* head shapes, behavior and evolution. *Evolution*, 35: 921-930.
STEBBINS, G. L., JR. 1950. *Variation and Evolution in Plants*. Columbia Univ. Press, New York.
STILES, E. W. 1982. Fruit flags: two hypotheses. *Am. Nat.*, 120(4): 500-509.
STILES, F. G., and L. L. WOLF. 1970. Hummingbird territoriality at a tropical flowering tree. *Auk*, 87: 467-491.
TEMPLETON, A. R. 1979. Once again, why 300 species of Hawaiian *Drosophila? Evolution*, 33(1): 513-517.
THORNHILL, R. 1980. Competitive, charming males and choosy females: was Darwin correct? *Fla. Entomol.*, 63(1): 5-29.
THORNHILL, R., and J. ALCOCK. In press. *The Evolution of Insect Mating Systems*. Harvard Univ. Press, Cambridge.
TRIVERS, R. L. 1976. Sexual selection and resource-accruing abilities in *Anolis garmani*. *Evolution*, 30: 253-269.
WAAGE, J. K. 1979. Reproductive character displacement in *Calopteryx* (Odonata: Calopterygidae). *Evolution*, 33(1): 104-116.
WADE, M. J., and S. J. ARNOLD. 1980. The intensity of sexual selection in relation to male sexual behaviour, female choice, and sperm precedence. *Anim. Behav.*, 28:446-461.
WALKER, T. J. 1974. Character displacement and acoustical insects. *Am. Zool.*, 14: 1137-1150.
WALLACE, A. R. 1878. *Tropical Nature and Other Essays*. Macmillan and Co., London.
WALLER, M. W. 1980. *The Island Waterfowl*. Iowa State Univ. Press, Ames.
WARNER, R. R., D. R. ROBERTSON, and E. G. LEIGH, JR. 1975. Sex change and sexual selec-

tion. *Science*, 190: 633-638.

WASER, N. M. In press. Competition for pollination and floral character differences among sympatric plant species: a review of evidence. In C. E. Jones and R. J. Little (eds.), *Handbook of Experimental Pollination Ecology*. VanNostrand Reinhold, New York.

WEBSTER, T. P., and J. M. BURNS. 1973. Dewlap color variation and electrophoretically detected sibling species in a Haitian lizard, *Anolis brevirostris*. *Evolution*, 27: 368-377.

WELLS, K. D. 1980. Social behavior and communication of a dendrobatid frog (*Colostethus trunitatus*). *Herpetologica*, 36(2): 189-199.

WELTY, J. C. 1962. *The Life of Birds*. W. B. Saunders Co., Philadelphia.

West-Eberhard, M. J. 1973. Monogyny in "polygynous" social wasps. *Proc. Cong. Intl. Union for Study of Social Insects, London:* 396-403.

——. 1975. The evolution of social behavior by kin selection. *Q. Rev. Biol.*, 50: 1-33.

——. 1977. The establishment of dominance in social wasps. *Proc. Congr. Intl. Union for Study of Social Insects, Wageningen:* 223-227.

——. 1978. Temporary queens in *Metapolybia* wasps: Non-reproductive helpers without altruism? *Science*, 200: 441-443.

——. 1979. Sexual selection, social competition, and evolution. *Proc. Am. Philos. Soc.*, 123: 222-234.

——. 1981. Intragroup selection and the evolution of insect societies. In R. D. Alexander and D. W. Tinkle (eds.), *Natural Selection and Social Behavior*, p. 3-17. Chiron, New York.

——. In press. Communication in social wasps: predicted and observed patterns, with a note on the significance of behavioral and ontogenetic flexibility for theories of worker "altruism." In A. de Haro (ed.), *Communication chez les Insectes Sociaux*, Barcelona.

WHITE, M. J. D. 1978. *Modes of Speciation*. Freeman and Co., San Francisco.

WILEY, R. H., JR. 1978. The lek mating system of the sage grouse. *Sci. Am.*, 238: 114-125.

WILLIAMS, E. E. 1963. Studies on South American anoles, No. 9: Description of *Anolis mirus*, new species, from Rio San Juan, Colombia, with comment on digital dilation and dewlap as generic and specific characters in the anole. *Bull. Mus. Comp. Zool. Harv.*, 129(9): 463-480.

WILLIAMS, E. E., and A. S. RAND. 1977. Species recognition, dewlap function and faunal size. *Am. Zool.*, 17: 261-270.

WILLIAMS, G. C. 1966. *Adaptation and Natural Selection*. Princeton Univ. Press, Princeton.

WILSON, E. O. 1975. *Socibiology*. Harvard Univ. Press, Cambridge.

WING, L. 1946. Drumming flight in the blue grouse and courtship characters of the Tetraonidae. *Condor*, 48: 154-157.

WOLF, L. L. 1969. Female territoriality in a tropical hummingbird. *Auk*, 86: 490-504.

WYNNE-EDWARDS, V. C. 1962. *Animal Dispersion in Relation to Social Behaviour*. Oliver and Boyd, Edinburgh.

YASUKAWA, K., J. L. BLANK, and C. B. PATTERSON. 1980. Song repertoires and sexual selection in the red-winged Blackbird (*Agelaius phoenicus*). *Behav. Ecol. Sociobiol.*, 7: 233-241.

ZAHAVI, A. 1977. Reliability in communication systems and the evolution of altruism. In B. Stonehouse and C. M. Perrins (eds.), *Evolutionary Ecology*, p. 253-260. Macmillan Press, London.

——. 1980. Ritualization and the evolution of movement signals. *Behaviour*, 72(1-2): 77-81.

McClintock, B. 1984. The significance of responses of the genome to challenge. *Science* 226:792–801.

The publication at the dawn of the twenty-first century of an early draft of the human genome—the more than three billion base pairs that make up the DNA that each of the two parents contributes to their child—brought in an unexpected discovery to the understanding of the genetics of humans and, by implication, many other organisms. It turns out that a human genome consists of only about 24,000 protein-coding genes, amounting to about 1 to 2 percent of the genome. However, perhaps as much as 75 percent consists of mobile elements, DNA segments that move from place to place in the genome. Mobile elements may be either retrotransposons, which travel by means of an RNA intermediary, or transposons, which move by means of direct genomic excision and insertion. Retrotransposons of various sorts are the most abundant, accounting jointly for at least 42 percent of the human genome, while transposons are estimated to account for about 2.8 percent.

Barbara McClintock (1902–1992) is credited with the discovery of mobile elements. She published an important paper in 1950, in which she summarized her discovery of transposons in maize. The net effects of transposons are thought to be deleterious (Lynch 2007, 189), although they have incidental favorable consequences (Kidwell 2002; Wessler 2006; see Kidwell and Lisch 1997, this volume), a view advanced by McClintock herself, in the paper here reproduced, propounding that transposon activation during times of stress allows genomes to become modified in potentially beneficial ways.

Barbara McClintock, who spent most of her research career at the Cold Spring Harbor Laboratory in New York, received the Nobel Prize for Physiology or Medicine in 1983 for her discovery of transposons and additional related research. The paper that follows is the lecture she delivered on that occasion.

RELATED READING

Keller, E. F. 1983. A feeling for the organism: the life and work of Barbara McClintock. Freeman, San Francisco, California, USA.

Kidwell, M. G. 2002. Transposable elements and the evolution of genome size in eukaryotes. Genetica 115:49–63.

Lander, E. S., L. M. Linton, B. Birren, et al. 2001. Initial sequencing and analysis of the human genome. Nature 409:860–921.

Lynch, M. 2007. The origins of genome architecture. Sinauer, Sunderland, Massachusetts, USA (see particularly chapters 3 and 7).

McClintock, B. 1950. The origin and behavior of mutable loci in maize. Proceedings of the National Academy of Sciences, USA 36:344–355.

Venter, J. C., M. D. Adams, E. W. Myers, et al. 2001. The sequence of the human genome. Science 291:1304–1351.

Wessler, S. R. 2006. Transposable elements and the evolution of eukaryotic genomes. Proceedings of the National Academy of Sciences, USA 103:17600–17601.

The Significance of Responses of the Genome to Challenge

Barbara McClintock

There are "shocks" that a genome must face repeatedly, and for which it is prepared to respond in a programmed manner. Examples are the "heat shock" responses in eukaryotic organisms and the "SOS" responses in bacteria. Each of these initiates a highly programmed sequence of events within the cell that serves to cushion the effects of the shock. Some sensing mechanism must be present in these instances to alert the cell to imminent danger, and to set in motion the orderly sequence of events that will mitigate this danger. But there are also responses of genomes to unanticipated challenges that are not so precisely programmed. The genome is unprepared for these shocks. Nevertheless, they are sensed, and the genome responds in a discernible but initially unforseen manner.

An experiment conducted in the mid-1940's prepared me to expect unusual responses of a genome to challenges that the genome is unprepared to meet in an orderly, programmed manner. In most known instances of such challenges, the types of response are not predictable in advance of initial observations of them. Moreover, it is necessary to subject the genome repeatedly to the same challenge in order to observe and appreciate the nature of the changes it induces. Familiar examples are the production of mutation by x-rays and by some mutagenic agents.

It is the purpose of this discussion to consider some observations from my early studies that revealed programmed responses to threats that are initiated within the genome itself, as well as others similarly initiated, that lead to new and irreversible genomic modifications. These latter responses, now known to occur in many organisms, are significant for appreciating how a genome may reorganize itself when faced with a difficulty for which it is unprepared. Conditions known to provoke such responses are many. A few of these will be considered, along with several examples from nature implying that rapid reorganizations of genomes may underlie some species formations. Our present knowledge would suggest that these reorganizations originated from some "shock" that forced the genome to restructure itself in order to overcome a threat to its survival.

Because I became actively involved in the subject of genetics only 21 years after the rediscovery, in 1900, of Mendel's principles of heredity, and at a stage when acceptance of these principles was not general among biologists, I have had the pleasure of witnessing and experiencing the excitement created by revolutionary changes in genetic concepts that have occurred over the past sixty-odd years. I believe we are again experiencing such a revolution. It is altering our concepts of the genome: its component parts, their organizations, mobilities, and their modes of operation. Also, we are now better able to integrate activities of nuclear genomes with those of other components of a cell. Unquestionably, we will emerge from this revolutionary period with modified views of components of cells and how they operate, but only however, to await the emergence of the next revolutionary phase that again will bring startling changes in concepts.

Experiment with *Zea mays* in the Summer of 1944 and Its Consequences

The experiment that alerted me to the mobility of specific components of genomes involved the entrance of a newly ruptured end of a chromosome into a telophase nucleus. This experiment commenced with the growing of approximately 450 plants in the summer of 1944, each of which had started its development with a zygote that had received from each parent a chromosome with a newly ruptured end of one of its arms. The design of the experiment required that each plant be self-pollinated. This was in order to isolate from the self-pollinated progeny new mutants that were expected to appear, and to be confined to locations within the arm of a chromosome whose end had been ruptured. Each mutant was expected to reveal the phenotype produced by a minute homozygous deficiency and to segregate in a manner resembling that of a recessive in an F_2 progeny. Their modes of origin could be projected from the known behavior of broken ends of chromosomes in successive mitoses. In order to observe those mutants that might express an altered seedling character, 40 kernels from each self-pollinated ear were sown in a seedling bench in the greenhouse during the winter of 1944–1945.

Some seedling mutants of the type expected did segregate, but they were overshadowed by totally unexpected segregants exhibiting bizarre phenotypes. These segregants were variegated for type and degree of expression of a gene. Those variegated expressions given by genes associated with chlorophyll development were startlingly conspicuous. Within any one progeny chlorophyll intensities, and their pattern of distribution in the seedling leaves, were alike. Between progenies, however, both the type and the pattern differed widely. Variegated seedlings from the different progenies were transferred to pots in order to observe the variegated phenomenon in the later developing, larger leaves.

It soon became apparent that modified patterns of gene expression were being produced, and that these were confined to sharply defined sectors in a leaf. Thus, the modified expression appeared to relate to an event that had occurred in the ancestor cell that gave rise to the sector. It was this event that was responsible for altering the pattern or type (or both) of gene expression in descendant cells, often many cell generations removed from the event. It was soon evident that the event was related to some cell component that had been unequally segregated at a mitosis. Twin sectors appeared in which the patterns of gene expression in the two side-by-side sectors were reciprocals of each other. For example, one sector might have a reduced number of uniformly distributed fine green streaks in a white background in comparison to the number and distribution of such streaks initially appearing in the seedling and showing elsewhere on the same leaf. The twin, on the other hand, had a much increased number of such streaks. Because these twin sectors were side-by-side they were assumed to have arisen from daughter cells following a mitosis in which each daughter had been modified in a manner that would differentially regulate the pattern of gene expression in their progeny cells.

After observing many such twin sectors, I concluded that regulation of pattern of gene expression in these instances was associated with an event occurring at a mitosis in which one daughter cell had gained something that the other daughter cell had lost. Believing that I was viewing a basic genetic phenomenon, all attention was given, thereafter, to determining just what it was that one cell had gained that the other cell had lost. These proved to be transposable elements that could regulate gene expressions in precise ways. Because of this I called them "controlling elements." Their origins and their actions were a focus of my research for many years thereafter. It is their origin that is important for this discussion, and it is extraordinary. I doubt if this could have been anticipated before the 1944 experiment. It had to be discovered accidentally.

Early Observations of the Effect of X-rays on Chromosomes

The 1944 experiment took place 13 years after I had begun to examine the behavior of broken ends of chromosomes. It was knowledge gained in these years that led me to conceive this experiment. Initial studies of broken ends of chromosomes began in the summer of 1931. At that time our knowledge of chromosomes and genes was limited. In retrospect we might call it primitive. Genes were "beads" arranged in linear order on the chromosome "string." By 1931, however, means of studying the "string" in some detail was provided by newly developed methods of examining the ten chromosomes of the maize complement in microsporocytes at the pachytene stage of meiosis. At this stage the ten bivalent chromosomes are much elongated in comparison to their metaphase lengths. Each chromosome is identifiable by its relative length, by the location of its centromere, which is readily observed at the pachytene stage, and by the individuality of the chromomeres strung along the length of each chromosome. At that time maize provided the best material for locating known

The author is Distinguished Service Member of the Carnegie Institution of Washington, Cold Spring Harbor Laboratory, Cold Spring Harbor, New York 11724. This article is the lecture she delivered in Stockholm, Sweden, 8 December 1983, when she received the Nobel Prize in Physiology or Medicine. Minor corrections have been made by the author. It is published here with the permission of the Nobel Foundation and will also be included in the complete volume of *Les Prix Nobel en 1984* as well as in the series *Nobel Lectures* (in English) published by the Elsevier Publishing Company, Amsterdam and New York.

genes along a chromosome arm, and also for precisely determining the break points in chromosomes that had undergone various types of rearrangement, such as translocations and inversions. The usefulness of the salivary gland chromosomes of *Drosophila* for such purposes had not yet been recognized. This came several years later. In the interim, maize chromosomes were revealing, for the first time, some distinctive aspects of chromosome organization and behavior. One of these was the extraordinary effect of x-rays on chromosomes.

The reports of Muller in 1927 and 1928 (*1, 2*) and of Hanson in 1928 (*3*) on the use of x-rays for obtaining mutations in *Drosophila*, and similarly that of Stadler in 1928 (*4*) with the barley plant, had a profound effect on geneticists. Here was a way of obtaining mutations at will. One did not need to await their spontaneous appearances. Many persons over many years continued to use x-rays for such purposes. But x-rays did not fulfill initial expectations of their usefulness. For other purposes, however, they have been most valuable, particularly for obtaining various types of structural reorganizations of the genome, from minute deficiencies to multiple rearrangements of chromosomes. It was to observe the effects of x-rays on chromosomes of maize that brought me to the University of Missouri at Columbia in the summer of 1931.

Prior to 1931 Stadler had been using x-rays to obtain mutations in maize. He had developed techniques for isolating those mutations that occur at selected gene loci. One method was to irradiate pollen grains. Pollen grains carry the haploid male gametes. The irradiated male gametes in Stadler's experiments carried wild-type alleles of known recessive mutants. Irradiated pollen was placed on the silks of ears of plants that were homozygous for one or more recessive alleles located in known linkage groups. An x-ray–induced mutation altering the expression of the wild-type allele of one of these recessives should be identifiable in an individual plant derived from such a cross. By the summer of 1931, Stadler had many plants in his field at Columbia, Missouri, from which one could choose those that exhibited one or another of these recessive phenotypes. Stadler had asked me if I would be willing to examine such plants at the meiotic stages to determine what types of events might be responsible for these recessive expressions. I was delighted to do so as this would be a very new experience. After my arrival at Columbia in June 1931 plants were selected whose chromosomes were to be examined. The knowledge gained from these observations was new and impressive. Descriptions and photographs summarizing these observations appeared in a bulletin published by the University of Missouri Agricultural College and Experiment Station (*5*).

None of the recessive phenotypes in the examined plants arose from "gene mutation." Each reflected loss of a segment of a chromosome that carried the wild-type allele, and x-rays were responsible for inducing these deficiencies. They also were responsible for producing other types of chromosome rearrangements, some of them unexpectedly complex. A conclusion of basic significance could be drawn from these observations: broken ends of chromosomes will fuse, two-by-two, and any broken end with any other broken end. This principle has been amply proved in a series of experiments conducted over the years. In all such instances the break must sever both strands of the DNA double helix. This is a "double-strand break" in modern terminology. That two such broken ends entering a telophase nucleus will find each other and fuse, regardless of the initial distance that separates them, soon became apparent.

After returning to Cornell University in the fall of 1931, I received a reprint from geneticists located at the University of California, Berkeley. The authors described a pattern of variegation in *Nicotiana* plants that was produced by loss of a fragment chromosome during plant development. The fragment carried the dominant allele of a known recessive present in the normal homologues. Loss of the dominant allele allowed the recessive phenotype to be expressed in the descendants of those cells that had lost this fragment.

It occurred to me that the fragment could be a ring chromosome, and that losses of the fragment were caused by an exchange between sister chromatids after replication of the ring. This would produce a double-sized ring with two centromeres. In the following anaphase, passage of the centromeres to opposite poles would produce two chromatid bridges. This, I thought, could prevent the chromosome from being included in either telophase nucleus. I sent my suggestion to the geneticists at Berkeley who then sent me an amused reply. My suggestion, however, was not without logical support.

During the summer of 1931 I had seen plants in the maize field that showed variegation patterns resembling the one described for *Nicotiana*. The chromosomes in these plants had not been examined. I then wrote to Stadler asking if he would be willing to grow more of the same material in the summer of 1932 that had been grown in the summer of 1931. If so, I would like to select the variegated plants to determine the presence of a ring chromosome in each. Thus, in the summer of 1932 with Stadler's generous cooperation, I had the opportunity to examine such plants. Each plant did have a ring chromosome. It was the behavior of this ring that proved to be significant. It revealed several basic phenomena.

1) In most mitoses, replication of the ring chromosome produced two chromatids that were completely free from each other, and thus could separate without difficulty in the following anaphase.

2) Sister strand exchanges do occur between replicated or replicating chromatids, and the frequency of such events increases with increase in the size of the ring. These exchanges produce a double-size ring with two centromeres.

3) Mechanical rupture occurs in each of the two chromatid bridges formed at anaphase by passage of the two centromeres on the double-size ring to opposite poles of the mitotic spindle.

4) The location of a break could be at any one position along any one bridge.

5) The two broken ends entering a telophase nucleus then fuse.

6) The size and content of each newly constructed ring depend on the position of the rupture that had occurred in each bridge (*6–8*).

The conclusion seems inescapable that cells are able to sense the presence in their nuclei of ruptured ends of chromosomes and then to activate a mechanism that will bring together and then unite these ends, one with another. And this will occur, regardless of the initial distance in a telophase nucleus that separated the ruptured ends. The ability of a cell to sense these broken ends, to direct them toward each other, and then to unite them so that the union of the two DNA strands is correctly oriented, is a particularly revealing example of the sensitivity of cells to all that is going on within them.

Evidence from x-rays, ring chromosomes, and that obtained in later experiments (*9–12*), gives unequivocal support for the conclusion that broken ends will find each other and fuse. The challenge is met by a programmed response. This may be necessary, as both accidental breaks and programmed breaks may be frequent. If not repaired, such breaks could lead to genomic deficiencies having serious consequences.

Entrance into a Telophase Nucleus of a Broken End of a Chromosome

In the mid-1930's another event inducing chromosome rupture was discovered. It revealed why crossing over should be suppressed between the centromere and the nucleolus organizer in organisms in which chiasmata terminalize (that is, move from the initial location of a crossover to the end of the arm of the chromosome). In maize, terminalization occurs at the diplotene stage of meiosis. This is before the nucleolus breaks up, which it does at a later stage in the first meiotic prophase. It is known that the force responsible for terminalization is strong enough to induce chromosome breakage should the terminalization process be blocked before the terminalizing chiasma reaches the end of the arm of a chromosome. In maize the centromere and the nucleolus organizer on the nucleolus chromosome are relatively close together. No crossovers have been noted to occur between them. However, if a plant is homozygous for a translocation that places the centromere on the nucleolus chromosome some distance from its nucleolus organizer, crossing over does occur in the interval between them (*10*). A chiasma so located starts its terminalization process to reach the end of the arm. It is stopped, however, at the nucleolus border. The terminalizing chromatid strands cannot pass through the nucleolus. Instead, the two chromatids are ruptured at this border. Fusions then occur between the ruptured ends, establishing, thereby, a dicentric chromosome deficient for all of the chromatin that runs through the nucleolus and continues beyond to the end of the arm. At the meiotic anaphase, passage of the two centromeres of the dicentric chromosome to opposite poles of the spindle produces a bridge. This bridge is ruptured and again the rupture can occur at any one location along the bridge. Now a *single* ruptured end of a chromosome enters the telophase nucleus. How, then, does the cell deal with this novel situation?

In order to determine how a cell responds to the presence of a single ruptured end of a chromosome in its nucleus, tests were conducted with plants that were heterozygous for a relatively long inversion in the long arm of chromosome 4 of maize. It had been known for some time that a crossover within the inverted segment in plants that are heterozygous for an inversion in one arm of a chromosome would result in a dicentric chromosome, and also an acentric fragment composed of all the chromatin from the distal breakpoint of the inversion to the end of the arm. A chromatin bridge would form at the meiotic anaphase by passage of the two centromeres on the dicentric chromosome to opposite poles of the spindle. Mechanical rupture of this bridge as the spindle elongated would introduce a single broken end into the telophase nucleus, as illustrated in Fig. 1, a to d.

The intent of this experiment was to observe this chromosome in the following mitotic division in order to determine the fate of its ruptured end. This could be accomplished readily by observing the first mitotic division in the microspore. Meiosis on the male side gives rise to four haploid spores, termed microspores. Each spore enlarges. Its nucleus and nucleolus also enlarge. Approximately 7 days after completion of meiosis this very enlarged cell prepares for the first postmeiotic mitosis. This mitosis produces two cells, a very large cell with a large, active nucleus and nucleolus, and a small cell with compact chromatin in a small nucleus, surrounded by

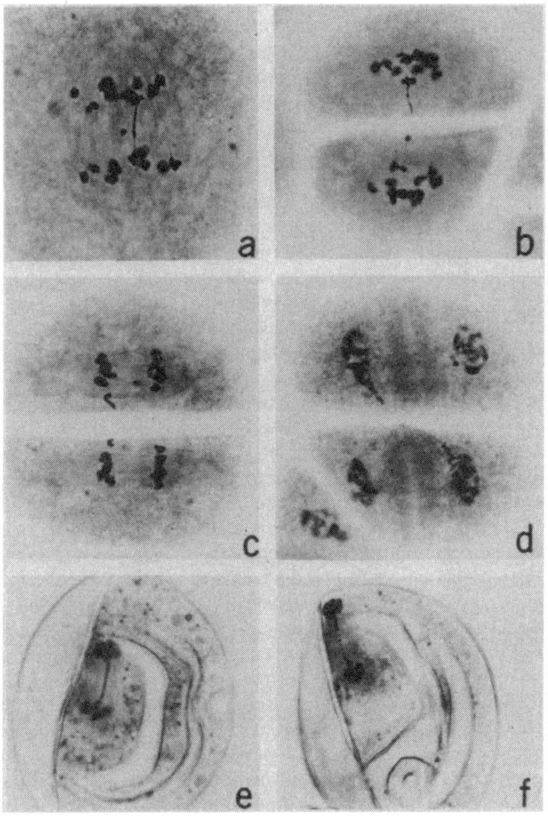

Fig. 1. Photographs illustrating the behavior of a newly ruptured end of a chromosome at the meiotic mitoses in microsporocytes and in the postmeiotic mitosis in the microspore. (a) Chromatin bridge at the first meiotic anaphase, with accompanying acentric fragment. Note the thin region in the bridge where rupture probably would have occurred at a slightly later stage. (b) Two sister cells at a very late prophase of the second meiotic mitosis. The rupture of the chromatid bridge that occurred at the previous anaphase severed the bridge at a nonmedian position. The larger segment so produced appears in the upper cell and opposite to the shorter segment in the lower cell. Their locations away from the other divalent chromatids relate to late entrances into the previous telophase nuclei caused by tension on the bridge before its rupture. Their placements show the positions they occupied in these telophase nuclei. The acentric fragment is in the lower cell, close to the cell membrane that was formed at the end of the first meiotic mitosis. (c) Anaphase of the second meiotic mitosis. The chromatid with a ruptured end is located closer to the wall in each cell that was formed after the first meiotic mitoses than are the other chromatids, for reasons given in (b). Note dissimilar lengths of the arms with ruptured ends. The acentric fragment is near the middle of the spindle in the upper cell. (d) Telophase of the second mitotic mitosis with extensions in two of the four nuclei pointing toward each other, one in the upper left nucleus and one in the lower right nucleus. The shapes of these nuclei reflect the off-positioning of chromatids having a newly ruptured end. Such off-positioning starts with the first meiotic telophase, continues throughout interphase and into the prophase, metaphase, and anaphase of the second meiotic mitosis and then into telophase, as shown here. The acentric fragment is adjacent to but not within the nucleus to the upper right. Note formation of the cell plate in each cell that anticipates the four spores that are products of meiosis. (e) Mitotic anaphase in the microspore showing a chromatid bridge produced by "fusion" of the replicated broken end. (f) Same as (e) but a slightly later stage showing rupture of the bridge. [Photographs adapted from *Missouri Agricultural Experiment Station Research Bulletin 290* (1938)]

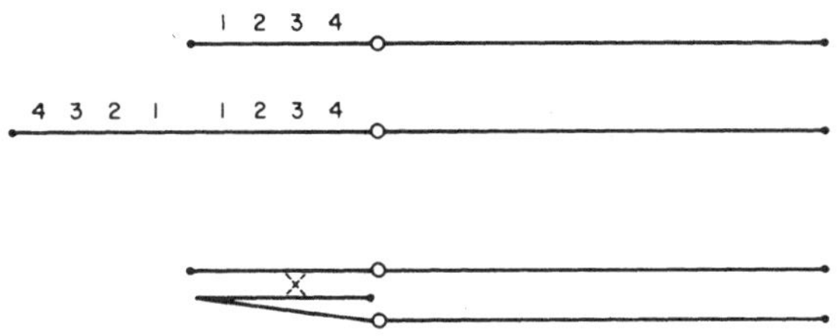

Fig. 2. Stylized representation of a crossover between a chromosome 9 with a normal short arm, upper line, and one with a duplication of this arm in reverse orientation, line below. In the lower two lines an exchange between two homologously associated arms is indicated by the cross. Such an exchange would give rise to a dicentric chromosome that simulates two normal chromosomes 9 attached together at the ends of their short arms, plus a small acentric fragment composed of the short arm of chromosome 9. The open circles represent centromeres. Telomeres are depicted as small knobs at the ends of chromosomes.

a thin layer of cytoplasm. This is the generative cell. Sometime later it undergoes a mitosis that will produce two condensed sperm cells. With completion of this division the pollen grain is nearly ready to function. The first division in the microspore may be observed readily merely by using a squash technique. The division of the generative cell, on the other hand, is obscured by the densely packed starch grains that accumulate during the interval between the two mitoses.

Examination of the first mitotic division in the microspore revealed a strange behavior of the single broken end that had entered a meiotic telophase nucleus. The replicated chromosome again was dicentric. The two chromatids produced by the replication process appeared to be fused at the location of the break that had occurred at the previous meiotic anaphase. In the spore, passage to opposite poles of the two centromeres of this newly created dicentric chromosome again produced a chromatid bridge that again was ruptured (Fig. 1, e and f). Thus, a newly ruptured end of the chromosome again entered each telophase nucleus. How would this newly broken end behave in subsequent mitoses? To determine this requires that the pollen grain with nuclei having such a ruptured end of a chromosome be functional. In the described instance, pollen grains whose nuclei had such a ruptured end would be deficient for a large terminal segment of the long arm of chromosome 4. Pollen grains whose nuclei have such a deficiency are unable to function.

The problem was resolved by obtaining plants having one chromosome of the maize complement with a duplication of all of its short arm in reverse orientation; its homologue had either a normal organization of its short arm, or better in the test to be performed, a short terminal deficiency of this arm that will not allow pollen grains receiving this chromosome to function. A crossover at the meiotic prophase, as shown in Fig. 2, produces a dicentric chromosome that simulates two normal chromosomes attached together at the ends of their short arms, and a fragment chromosome with telomeres at both ends. The dicentric chromosome produced by the crossover initiates the chromatid type of breakage-fusion-bridge cycle. This cycle, initially detected at the first mitosis in a microspore, could now be followed in subsequent mitoses. This is because the location of breaks in some of the anaphase bridges gave rise to chromosomes with at least a full complement of genes necessary for pollen functioning. Such functional pollen grains falling on the silks of ears will deliver their two sperm cells to the embryo sac inside a kernel-to-be. One sperm will contribute to the development of the embryo and the other will contribute to the development of the endosperm.

On the female side only a single cell in the kernel-to-be undergoes meiosis, and the embryo sac arises from only one of the four spores produced by the two meiotic mitoses. The other three spores degenerate. This one haploid cell, the megaspore, then undergoes three successive mitoses to form the embryo sac, or female gametophyte. Of the cells in the embryo sac only the egg cell and the much enlarged central cell need be considered here. The very large central cell has two haploid nuclei positioned close to each other and near the egg cell. After delivery of the two sperms to the embryo sac, one sperm nucleus fuses with the egg cell nucleus to form the diploid zygote. The other sperm nucleus and the two nuclei in the central cell fuse to form the primary endosperm nucleus, which is triploid. (The term "double fertilization" is commonly used in referring to these events.) Thus, the embryo and endosperm are formed separately, although both share the same genes, one set from each parent for the embryo, and two sets from the female parent and one set from the male parent for the endosperm. Although developing separately, the two structures are placed side-by-side in the mature kernel (Fig. 3).

It was soon learned that the chromatid type of breakage-fusion-bridge cycle, initiated at a meiotic anaphase, will continue during development of the pollen grain and the embryo sac. Whenever a sperm nucleus contributes a chromosome with a newly broken end to the primary endosperm nucleus, this cycle will continue throughout mitoses in the developing endosperm. Similarly, if the two nuclei in the central cell each have such a ruptured end of a chromosome, either the chromosome or chromatid type of breakage-fusion-bridge will occur throughout endosperm development. When, however, a single ruptured end of a chromosome is delivered to the zygote nucleus by either the egg or the sperm nucleus, the ruptured end will "heal" subsequently; the cycle ceases in the developing embryo. Although not yet proven at the molecular level, it is altogether likely that the healing process represents the formation of a new telomere at the ruptured end. This assures that the special requirement for DNA replication at free ends of chromosomes will be satisfied. This new telomere functions normally thereafter. It is as stable in this regard as any other telomere of the maize complement, and tests of this cover many cell and plant generations.

A cell capable of repairing a ruptured end of a chromosome must sense the presence of this end in its nucleus. This sensing activates a mechanism that is required for replacing the ruptured end with a functional telomere. That such a mechanism must exist was revealed by a mutant that arose in my stocks. When homozygous, this mutant would not allow the repair mechanism to operate in the cells of the plant. Entrance of a newly ruptured end of a chromosome into the zygote is followed by the chromatid type of breakage-fusion-bridge cycle throughout mitoses in the developing plant. This suggests that the repair mechanism in the maize strains I had been using is repressed in cells producing the male and female gametophytes and also in the endosperm, but is activated in the embryo. Although all of this was known before the 1944 experiment was conducted, the extent of trauma perceived by cells whose nuclei receive a *single* newly ruptured end of a chromosome that the cell cannot repair, and the speed with which this trauma is registered, was not appreciated until the winter of 1944–1945.

Initiation of Activations of Previously Silent Genomic Elements

By 1947 it was learned that the bizarre variegated phenotypes that segregated in many of the self-pollinated progenies grown on the seedling bench in the fall and winter of 1944–1945 were due to the action of transposable elements. It seemed clear that these elements must have been present in the genome, and in a silent state previous to an event that activated one or another of them. To my knowledge, no progenies derived from self-pollination of plants of the same strain, or related strains, had ever been reported to have produced so many distinctly different variegated expressions of different genes as had appeared in the progenies of these closely related plants that were grown in the summer of 1944. It was concluded that some traumatic event was responsible for these activations. The unique event in the history of these plants relates to their origin. Both parents of the plants, grown in the summer of 1944, had contributed a chromosome with a newly ruptured end to the zygote that gave rise to each of these plants. The rupture occurred, in the first instance, at a meiotic anaphase in each parent, and the ruptured end then underwent the succession of mitotic anaphase breaks associated with the chromatid type of breakage-fusion-bridge cycle during development of the male and female gametophytes—the pollen grain and the embryo sac. I suspected that an activating event could occur at some time during this phase of the life history of the parent plants. I decided, then, to test if this might be so.

The newly activated elements, isolated from the initial experiment, were observed to regulate gene expression after insertion of an element, or one of its derivatives, at a gene locus (*13–15*). In some instances the general mode of regulation resembled that produced by the *Dotted* gene on the standard recessive allele, *a*, of the *A* gene. This *a* allele represents the first recognized gene among a number of others whose action is required for production of anthocyanin pigment, either red or purple, in plant tissues and in several tissues of the kernel (*16*). In the mid-1930's Rhoades discovered this *Dotted* (*Dt*) element in a strain of Black Mexican sweet corn (*17, 18*). It behaved as a dominant gene that caused the otherwise very stable but nonfunctional *a* allele to mutate to new alleles that allowed anthocyanin pigment to be formed in both plant and kernel. The name *Dotted* given to it refers to the pattern of mutations that is expressed in plants and kernels homozygous for the *a* allele on chromosome 3 and having a *Dt* element located elsewhere in the chromosome complement. Small streaks of red or purple pigment appear in plants; the kernels have dots of this pigment distributed over the aleurone layer. (The aleurone layer is the outermost layer of the endosperm.)

In that this *Dt* may have originated from activation of a previously silent element in the maize genome, and in that such silent elements should be present in all maize genomes, it was decided to test whether the breakage-fusion-bridge cycle would activate one such silent *Dt* element. My stocks that were homozygous for the *a* allele had never given any indication of *Dt* action. Therefore, these stocks were used to test if a presumed silent *Dt* element could be activated by the chromatid type of breakage-fusion-bridge cycle. Plants homozygous for *a* and having a chromosome 9 constitution similar to that described for Fig. 2 were used as pollen parents in crosses to plants that also were homozygous for *a*. These pollen parents had the duplication of the short arm as shown in Fig. 2, but its homologue was deficient for a terminal segment of this arm that would not allow pollen grains having it to function. It was determined that 70 to 95 percent of the *functional* pollen grains produced by these plants carried sperms having a chromosome 9 with a newly ruptured end of its short arm, the initial rupture having occurred at the previous meiotic anaphase. Thus, most of the embryos and endosperms in the kernels on ears produced by the described cross started development with a newly ruptured end of the short arm of chromosome 9 in both embryo and endosperm. These kernels were searched for dots of pigment in their aleurone layer. A number of kernels had such dots. Many of these dots were confined to a restricted area of the aleurone layer, suggesting that this area represented a sector derived from a single cell in which a silent *Dt* element had been activated. One kernel had dots distributed over all of the aleurone layer, suggesting that the sperm nucleus contributing to the primary endosperm nucleus already had an activated *Dt* element in it. Tests of the plant arising from this kernel indicated that the sister sperm nucleus that had fused with the egg nucleus did not have such an activated element. Apparently, the activating event had occurred in the nucleus of only one of the two sperms. Significantly, this is only two mitoses removed from initiation of the chromatid type of breakage-fusion-bridge cycle. As was mentioned earlier, this cycle continues during successive mitoses in the development of the endosperm. This continuing cycle could explain the presence in some kernels of sectors with pigmented dots, and this, in turn, would imply that activations of silent elements could occur at any time that this cycle remained in operation (*19–21*).

A similar test was conducted some years later by Doerschug (*22*), using the same constitution of the pollen parent as that just described. He obtained similar results. In his tests, however, two kernels with spots of pigment distributed

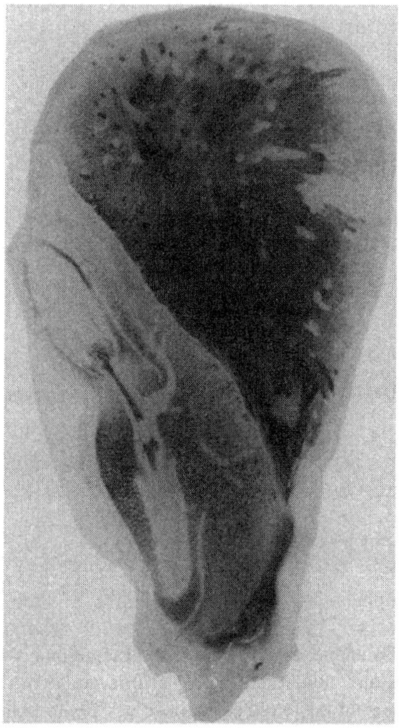

Fig. 3. Longitudinal section through a mature maize kernel to show its parts. The cut surface was treated with an iodine-potassium iodide solution to stain amylose in the starch granules of individual cells, and to indicate by density of stain the relative amounts of amylose in these cells. The narrow outer layer of the kernel is the pericarp, a maternal tissue. The embryo and endosperm are side-by-side but clearly delimited from each other. The endosperm is above and to the right of the embryo. In this photograph, four parts of the embryo may be noted. To the left and adjacent to the endosperm is the scutellum with its canals. The shoot, to the upper left, and the primary root below it, are connected to each other by the scutellar node. The different staining intensities in the endosperm cells reflect different amounts of amylose in them. These differences relate to the presence and action of a transposable *Ac* element at the *Wx* locus (*23*). The *Wx* gene is responsible for conversion of amylopectin to amylose, but only in the endosperm, not in the embryo.

over the entire aleurone layer proved to have an activated *Dt* element in the plant grown from each of these kernels. The behavior of these two newly activated *Dt* elements was extensively studied by Doerschug. The two elements differed from each other not only in their location in the chromosome complement, but also in their mode of control of the time and place of change in *a* gene action. We now know that such differences in performance of these elements are expected.

Doerschug's two *Dt* isolates are most significant for appreciating the speed of response of a genome to entrance of a newly ruptured end of a chromosome into a telophase nucleus. Each *Dt* element must have been activated in the microspore nucleus or not later than the generative nucleus produced by division of the microspore nucleus. The unexpected event probably is sensed and acted upon from the initial entrance of a single ruptured end of a chromosome into a telophase nucleus, and in each subsequent nucleus that receives such a newly ruptured end. It is recognized that *Dt* is only one among a number of silent but potentially transposable elements that are present in maize genomes. Most probably some of these other silent elements were activated during the described test, but they were not able to be recognized as were activations of *Dt* elements.

A similar approach could be used to detect such activations if a proper indicator stock were chosen for the test. Detection of silent elements is now made possible with the aid of DNA cloning techniques. Silent *Ac* (*Activator*) elements, as well as modified derivatives of it, have already been detected in several strains of maize (*23*). When other transposable elements are cloned it will be possible to compare their structural and numerical differences among various strains of maize. Present evidence suggests that wide differences may be found in this regard, as they have been found for middle and highly repetitious DNA sequences (*24*). In any one strain of maize the number of silent but potentially transposable elements, as well as other repetitious DNA's, may change, and most probably in response to challenges not yet recognized.

There are clear distinctions in the comportment of ends of chromosomes on entering telophase nuclei. These relate to (i) all chromosomes having normal ends, (ii) two chromosomes, each with a single ruptured end, or one chromosome with both ends ruptured, and (iii) one chromosome with a single broken end. Both ends of normal, unbroken chromosomes have a normal telomere, and no difficulties are experienced. Two ruptured ends, neither with a telomere, will find each other and fuse. In these instances there is no immediate telomere problem. A single broken end has no telomere, and no other broken end with which to fuse. If the cell cannot make a new telomere, which is the case in the maize gametophytes and the endosperm, the evidence indicates that a special trauma is experienced. Telomeres are especially adapted to replicate the free ends of chromosomes. When no telomere is present, attempts to replicate this uncapped end may be responsible for the apparent "fusions" of the replicated chromatids at the position of the previous break, and be responsible for perpetuating the chromatid type of breakage-fusion-bridge cycle in successive mitoses. Activation of potentially transposable elements, as well as other structural modifications of the chromosomes not considered here, are recognizable consequences of the cell's response to the continuing trauma.

Further Examples of Response of Genomes to Stress

Cells must be prepared to respond to many sources of stress. Mishaps that affect the operation of a cell must be occurring continuously. Sensing these and instigating repair systems are essential. We are aware of some of the mishaps affecting DNA, and also of their repair mechanisms, but many others could be difficult to recognize. Homeostatic adjustments to various accidents would be required if these accidents occur frequently. Many such mishaps and their adjustments would not be detected unless some event or observation directed attention to them. Some, however, are so conspicuous that they cannot fail to be noted. For example, in *Drosophila*, some sensing device recognizes when the amount of ribosomal DNA is above or below the standard amount, and then sets in motion the system that will make the proper adjustment. Similarly, amitotic divisions of macronuclei in ciliates may result in unequal distributions of DNA to daughter nuclei. These deviations are sensed in each daughter cell. To make adjustments, one cell may respond by increasing its DNA content to reach the standard amount. The other cell may discard the excess DNA. There must be numerous homeostatic adjustments required of cells. The sensing devices and the signals that initiate these adjustments are beyond our present ability to fathom. A goal for the future would be to determine the extent of knowledge the cell has of itself and how it utilizes this knowledge in a "thoughtful" manner when challenged.

One class of programmed responses to stress has received very little attention by biologists. Here a stress signal induces the cells of a plant to make a wholly new plant structure, and this to house and feed a developing insect, from egg to the emerging adult. A single *Vitis* plant, for example, may have on its leaves three or more distinctly different galls, each housing a different insect species. The stimulus associated with placement of the insect egg into the leaf will initiate reprogramming of the plant's genome, forcing it to make a unique structure adapted to the needs of the developing insect. The precise structural organization of a gall that gives it individuality must start with an initial stimulus, and each species provides its own specific stimulus. For each insect species the same distinctive reprogramming of the plant genome is seen to occur year-after-year. Some of the most interesting and elaborate plant galls house developing wasps. Each wasp species selects its own responding oak species, and the gall structure that is produced is special for each wasp to oak combination. All of these galls are precisely structured, externally and internally, as a rapid examination of them will show.

The galls on roots of legumes that are associated with nitrogen-fixing bacteria are readily available for examination. They illustrate in their own way an example of reprogramming of the plant genome by a stimulus received from a foreign organism. The induction of such reprogrammings by insects, bacteria, fungi, and other organisms that are not a required response of the plant genome at some stage in its life history is quite astounding. But this is no more astounding, it would seem, than the sharing of a single genome by two brilliantly designed organisms, the caterpillar and the moth. It is becoming increasingly apparent that we know little of the potentials of a genome. Nevertheless, much evidence tells us that they must be vast.

Many known and explored responses of genomes to stress are not so precisely programmed. Activation of potentially transposable elements in maize is one of these. We do not know when any particular element will be activated. Some responses to stress are especially significant for illustrating how a genome may modify itself when confronted with unfamiliar conditions. Changes induced in genomes when cells are removed from their normal locations and placed in tissue culture surroundings are outstanding examples of this.

The establishment of a successful tissue culture from animal cells, such as those of rat or mouse, is accompanied by readily observed genomic restructuring. None of these animal tissue cultures has given rise to a new animal. Thus, the significance of these changes for the organism as a whole is not yet directly testable. The ability to determine this significance is a distinct advantage of plant tissue cultures. Many plant tissue cultures have developed new plants and, in some instances, many plants from a single initial cell or tissue isolate. A reason for this difference in behavior of animal and plant tissue cultures is not difficult to find. In many animals the germline is set aside early in cleavage, allowing the soma—a dead-end structure—to develop by any means, including genome restructuring and nonreversible programming. In higher plants, each fertile flower has the equivalent of a "germline" in it. The flower makes the gametes and initiates embryo formation. In this regard, consider the many flowers that may be produced by a bush or a tree. Some system must operate to reprogram the genome in those cells of the flower that will produce the gametes and establish the zygote. This implies that the specific programming sequences, earlier initiated and required for flower production, must be "erased" in order to return the genome to its very early state. If this occurs in so many places in a bush or a tree, then it is not surprising that it may occur in a plant cell or a cluster of cells not within a flower. Also, in many plants such resettings are a common means of initiating new individuals from somatic cells. In these instances, however, the process of resetting is regulated, and the genome is not permanently restructured. This is not true for plants arising from many tissue cultures. The treatment, from isolation of the cell or cells of a plant, to callus formation, and then to production of new plants from the cells of these calluses, must inflict on the cells a succession of traumatic experiences. Resetting of the genome, in these instances, may not follow the same orderly sequence that occurs under natural conditions. Instead, the genome is abnormally reprogrammed or decidedly restructured. These restructurings can give rise to a wide range of altered phenotypic expressions. Some of the altered phenotypes are readily observed in the newly produced plants themselves. Others appear in their progeny. Some initially displayed altered phenotypes do not reappear in the progeny. Their association with genomic change remains problematic. Other altered phenotypes clearly reflect genomic restructuring, and various levels of this have been observed. It may be safe to state that no two of the callus derived plants are exactly alike, and none is just like the plant that donated the cell or cells for the tissue culture. The many levels of genomic modification that already are known and expressed as changed genotypes and phenotypes could be potent sources for selection by the plant breeder, and incidentally, for theoretical ponderings by the biologist.

Modifications in gene expression may be induced in plants when they are infected with an RNA virus. Instances of this may be detected merely by viewing infected plants in the field. For example, patterns of anthocyanin pigment distribution, normally highly regulated and prominently displayed in the flowers of a plant, may appear grossly distorted in those parts of a plant that clearly reveal the virus infection. Recently, it was learned that infection of maize plants with barley stripe mosaic virus, an RNA virus, may traumatize cells to respond by activating potentially transposable elements. These, in turn, may then enter a gene locus and modify its expression (*25*). Such changes in expression of known genes may be exhibited in the self-pollinated progeny of infected plants. More often they are detected in later generations. Yet, no virus genome has been detected in the immediate progeny of infected plants or in those plants shown to have a transposable element newly inserted at a known gene locus.

Species crosses are another potent source of genomic modification. Plants have provided many excellent examples of this. The advantage of plants is the ease of making crosses to obtain hybrids, the simplicity of growing them, the ready availability of their chromosomes, and the ability to obtain progeny in quantities, if necessary. The alterations produced when the genomes of two species are combined reflect their basic incompatibilities. Evidence of this is the appearance of the same types of genome change whenever the same two species are crossed. Expressions of incompatibilities do differ, but their nature is always in accordance with the particular two species whose genomes are combined. The genus *Nicotiana* has a large number of species that differ from each other in chromosome number, chromosome organization, and phenotypic expressions. Genome incompatibilities have been observed in a large number of two-by-two combinations of species. An illustration is the behavior of chromosomes in the hybrid plant produced by the cross of *N. tabacum* by *N. plumbaginifolia*. The chromosomes of *plumbaginifolia* are lost during development of this hybrid plant. Although whole chromosome losses appear to be common, other irregularities in chromosome behavior also occur. These are chromosome fragments, chromosome bridges in somatic anaphases, and the appearance in an occasional metaphase plate of a single, very much elongated chromosome, termed a "megachromosome." The presence of one or two such hugely elongated chromosomes, in some somatic metaphase plates, characterizes the hybrid derived from the cross of *N. tabacum* × *N. otophora*. In this instance it is known that heterochromatic segments in chromosomes of the *otophora* set contribute to these linear amplifications (*26, 27*). Hybrids produced by crosses of distantly related *Nicotiana* species are known to give rise to tumors, some of which resemble teratomas. In one instance it was shown that tumor production relates to a single heritable modification which was initiated in the hybrid (*28, 29*).

Major restructuring of chromosome components may arise in a hybrid plant and continue to arise in its progeny, sometimes over successive plant generations. The restructuring may range from apparently simple to obviously complex. These are associated with translocations, inversions, deficiencies, duplications, and the like, which are simple in some instances or variously intercalated in others. New stable or relatively stable "species" or "genera" have been derived from such initial hybrids. The commercially useful plant, *Triticale*, is an example. Wheat (*Triticum*) and rye (*Secale*) were crossed and the combined set of chromosomes was doubled to provide reproductive stability. Nevertheless, this genome was not altogether stable. Selections continued in later generations for better performances with considerable success, even though instabilities were not eliminated altogether. Some species of *Triticum* undoubtedly arose by a comparable mechanism as that outlined for *Triticale*, and different related genera made their contribution to some of these

Triticum species. Evidence for this is exceptionally clear (*30*).

Undoubtedly, new species can arise quite suddenly as the aftermath of accidental hybridizations between two species belonging to different genera. All evidence suggests that genomic modifications of some type would accompany formation of such new species. Some modifications may be slight and involve little more than reassortments of repetitious DNA's, about which we know so little. (The adjective "slight" refers to the apparent simplicity of the restructuring mechanism rather than the significance of its consequences.) Major genome restructuring most certainly accompanied formation of some species. Studies of genomes of many different species and genera indicate this. Appreciation of the various degrees of reassortment of components of a genome, that appear during and after various types of genome shock, allows degrees of freedom in considering such origins. It is difficult to resist concluding that some specific "genome shock" was responsible for origins of new species in the two instances described below.

The organization of chromosomes in many closely related species may resemble one another when viewed by light microscopy. Only genetic and molecular analyses would detect those differences in their genomes that could distinguish them as species. In some instances of this type distinctions relate to the assortment of repetitious DNA's over the genome, as if a response to shock had initiated mobilities of these elements. In other instances, distinctions between related species are readily observed at the light microscope level, such as polyploidizations that are common in plants, or amplifications of DNA that do not alter chromosome number or basic metaphase morphologies. Others relate to chromosome fusions or fragmentations, or readily observed differences in the placement of specific DNA segments. There are many descriptions of differences in chromosome organization among the species of a genus. Two instances of these latter differences warrant special consideration here, because the observed differences in chromosome organization suggest origins from a response to a single event. One response gave rise to extensive fusions of chromosomes. The other placed heterochromatic segments at new locations within the chromosomes of the set.

That such multiple chromosome changes may relate to some initial event, occurring in a cell of the germline, is proposed and defended in a review by King (*31*). An example that would fit his proposal is the organization of chromosomes of the Indian muntjac deer (*Muntiacus muntjak*) (*32*) when compared with its closely related species, *M. reevesi*, the Chinese muntjac. The latter species has 46 chromosomes as the diploid number, whereas the Indian muntjac has 6 chromosomes in the female and 7 chromosomes in the male, and these chromosomes are huge in comparison with those of the Chinese muntjac. Observations of the chromosomes in the hybrid between these two species strongly supports chromosome fusion as the mechanism of origin of the reduced number and huge size of the Indian muntjac chromosomes (*33*). In general, evidence of fusion of chromosomes is plentiful. When two or three chromosomes of a set appear to have arisen by fusion, the question of simultaneous or sequential events responsible for these fusions cannot be determined with certainty. In the case of the Indian muntjac, it is difficult to avoid the conclusion that the fusions of so many chromosomes resulted from some initial shocking event that activated a fusion mechanism already known to apply to fusions of individual chromosomes in many other organisms (Robertsonian translocations). Whatever the cause, the changed chromosome organization is stunning.

Another stunning example of differences in chromosome organization between species is reported by Beermann in an extraordinarily thorough and fascinating account (*34*). This report describes the chromosome organization in three species of the copepod genus *Cyclops*. The main differences among them to be considered here relate to distributions of conspicuous heterochromatic blocks in the chromosomes of each species. In one species, these blocks are confined to the ends of chromosomes. In another species, blocks of heterochromatin are at the ends of chromosomes, but also positioned to each side of the centromere. In the third species, blocks of heterochromatin are distributed all along the chromosomes. An additional feature of this heterochromatin is its unchanged presence in cells of the germline in contrast to its elimination at cleavages specific for each species and in cells destined to produce the soma. The elimination process is associated with formation of rings of DNA cut from the heterochromatin (*35*). Again it is difficult to avoid concluding that these distinctive distributions of heterochromatin relate to unusual and disturbing events, and that these events have activated mechanisms that can redistribute heterochromatin to specific sites.

Concluding Statement

The purpose of this discussion has been to outline several simple experiments conducted in my laboratory that revealed how a genome may react to conditions for which it is unprepared, but to which it responds in a totally unexpected manner. Among these is the extraordinary response of the maize genome to entrance of a single ruptured end of a chromosome into a telophase nucleus. It was this event that, basically, was responsible for activations of potentially transposable elements that are carried in a silent state in the maize genome. The mobility of these activated elements allows them to enter different gene loci and to take over control of action of the gene wherever one may enter. Because the broken end of a chromosome entering a telophase nucleus can initiate activations of a number of different potentially transposable elements, the modifications these elements induce in the genome may be explored readily. In addition to modifying gene action, these elements can restructure the genome at various levels, from small changes involving a few nucleotides, to gross modifications involving large segments of chromosomes, such as duplications, deficiencies, inversions, and other more complex reorganizations.

It was these various effects of an initial traumatic event that alerted me to anticipate unusual responses of a genome to various shocks it might receive, either produced by accidents occurring within the cell itself, or imposed from without, such as virus infections, species crosses, poisons of various sorts, or even altered surroundings such as those imposed by tissue culture. Time does not allow even a modest listing of known responses of genomes to stress that could or should be included in a discussion aimed at the significance of responses of genomes to challenge. The examples chosen illustrate the importance of stress in instigating genome modifications by mobilizing available cell mechanisms that can restructure genomes, and do so in quite different ways. A few illustrations from nature are included because they support the conclusion that stress, and the genome's reactions to it, may underlie many formations of new species.

In the future, attention undoubtedly will be centered on the genome, with greater appreciation of its significance as a highly sensitive organ of the cell that monitors genomic activities and corrects common errors, senses unusual and unexpected events, and responds to them, often by restructuring the genome. We know about the components of genomes that could be made available for such restructuring. We know nothing, however, about how the cell senses danger and instigates responses to it that often are truly remarkable.

References

1. H. J. Muller, *Science* **66**, 84 (1927).
2. ———, *Proc. Natl. Acad. Sci. U.S.A.* **14**, 714 (1928).
3. F. B. Hanson, *Science* **67**, 562 (1928).
4. L. J. Stadler, *ibid.* **68**, 186 (1928).
5. B. McClintock, *Mo. Agric. Exp. Sta. Res. Bull.* **163**, 1 (1931).
6. ———, *Proc. Natl. Acad. Sci. U.S.A.* **18**, 677 (1932).
7. ———, *Genetics* **23**, 315 (1938).
8. ———, *ibid.* **26**, 542 (1941).
9. ———, *ibid.*, p. 234.
10. ———, *Cold Spring Harbor Symp. Quant. Biol.* **9**, 72 (1941).
11. ———, *Proc. Natl. Acad. Sci. U.S.A.* **28**, 458 (1942).
12. ———, *Stadler Genet. Symp.* **10**, 25 (1978).
13. ———, *Carnegie Inst. Wash. Yearb.* **47**, 155 (1948).
14. ———, *ibid.* **48**, 142 (1949).
15. ———, *Proc. Natl. Acad. Sci. U.S.A.* **36**, 344 (1950).
16. R. A. Emerson, *Cornell Univ. Agr. Exp. Sta. Mem.* **16**, 231 (1918).
17. M. M. Rhoades, *J. Genet.* **33**, 347 (1936).
18. ———, *Genetics* **23**, 377 (1938).
19. B. McClintock, *Carnegie Inst. Wash. Yearb.* **49**, 157 (1950).
20. ———, *ibid.* **50**, 174 (1951).
21. ———, *Brookhaven Symp. Biol.* **18**, 162 (1965).
22. E. B. Doerschug, *Theor. Appl. Genet.* **43**, 182 (1973).
23. N. Fedoroff, S. Wessler, M. Shure, *Cell* **35**, 235 (1983).
24. S. Hake and V. Walbot, *Chromosoma* **79**, 251–270 (1980).
25. J. P. Mottinger, S. L. Dellaporta, P. B. Keller, *Genetics* **106**, 751 (1984).
26. D. U. Gerstel and J. A. Burns, *Chromosomes Today* **1**, 41 (1966).
27. ———, *Genetica* **46**, 139 (1976).
28. H. H. Smith, *Brookhaven Symp. Biol.* **6**, 55 (1954).
29. ———, *Brookhaven Lect. Ser.* **52**, 1–8 (1965).
30. M. Feldman, in *Evolution of Crop Plants*, N. W. Simmonds, Ed. (Longman, New York, 1976) p. 120.
31. M. King, *Genetica* **59**, 53 (1982).
32. D. H. Wurster and K. Benirschke, *Science* **168**, 1364 (1970).
33. Shi Liming, Ye Yingying, Duan Xingsheng, *Cytogenet. Cell Genet.* **26**, 22 (1980).
34. S. Berman, *Chromosoma* **60**, 297 (1977).
35. ——— and G. F. Meyer, *ibid.* **77**, 277 (1980).

Ayala, F. J. 1986. On the virtues and pitfalls of the molecular evolutionary clock. *Journal of Heredity* 77:226–235.

DNA, RNA, and proteins have been called "informational macromolecules," because they convey genetic information in the linear sequence of their nucleotide or amino acid components, similar to the way the sequence of letters in an English text conveys semantic information. Zuckerkandl and Pauling (1965) noticed that the number of component differences between homologous molecules from different organisms seemed to be proportional to the time elapsed since the evolutionary divergence of the species, or, in the case of paralogous genes, since the time when the gene duplication occurred. Investigating different hemoglobin polypeptides from the same or different species, Zuckerkandl and Pauling postulated that most nucleotide or amino acid replacements would likely be adaptively neutral and "*there may thus exist a molecular evolutionary clock*" (their emphasis).

Two years later, W. M. Fitch and E. Margoliash published "Construction of phylogenetic trees" (1967; in this volume), showing how cytochrome *c*, a small protein involved in cell respiration, could be used to reconstruct the evolutionary history of 20 extremely diverse species that had shared their last common ancestor more than one billion years ago. A rush of related publications followed, asserting that molecular evolution might be used as a "molecular clock of evolution" method to determine the time of past evolutionary events, just as radioactive decay is used to measure geological time.

Paleobiology, comparative anatomy, biogeography, and other traditional disciplines continue to be used in the reconstruction and timing of phylogeny, and they have distinctive advantages for certain purposes, such as paleontology for the investigation of the morphology, distribution, and other characteristics of extinct species. But molecular evolutionary studies have three notable advantages over the classical disciplines. One is *precision*, because the molecular information is readily quantifiable. The number of nucleotides or amino acids that are different in different organisms is easily established for a given macromolecule. It is simply a matter of aligning the nucleotides or amino acids between two or more species and counting the differences. The second advantage is *universality*, because comparisons can be made between organisms that are quite different from one another. There is very little that paleontology or comparative anatomy can say when organisms as diverse as yeasts, pine trees, and human beings are compared, but numerous DNA and protein sequences can be compared in all three. The third advantage is *multiplicity*. Each organism possesses thousands of genes and proteins, every one of which reflects the same evolutionary history. If the investigation of one particular gene or protein does not satisfactorily resolve the evolutionary relationships of a set of species, additional genes and proteins can be investigated until the matter has been settled.

RELATED READING

Kimura, M. 1969. The rate of molecular evolution considered from the standpoint of population genetics. Proceedings of the National Academy of Sciences USA 63:1181–1188.

Rodríguez-Trelles, F., R. Tarrío, and F. J. Ayala. 2001. Erratic overdispersion of three molecular clocks; GPDH, SOD, and XDH. Proceedings of the National Academy of Sciences USA 98:11405–11410.

Vogel, H. J., V. Bryson, and J. O. Lampen, editors. 1966. Informational macromolecules: a symposium. Academic Press, New York, New York, USA.

Zuckerkandl, E., and L. Pauling. 1965. Evolutionary divergence and convergence in proteins. Pages 97–166 *in* Bryson, V., and H. J. Vogel, editors, Evolving genes and proteins. Academic Press, New York, New York, USA.

On the virtues and pitfalls of the molecular evolutionary clock

Francisco J. Ayala

Abstract: "Informational" macromolecules—i.e., proteins and nucleic acids—have in their sequences a register of evolutionary history. Zuckerkandl and Pauling suggested in 1965 that these molecules might provide a "molecular clock" of evolution. The molecular clock would time evolutionary events and make it possible to reconstruct phylogenetic history—the branching relationships among lineages leading to modern species. Kimura's neutrality theory postulates that rates of molecular evolution are stochastically constant and, hence, that there is a molecular clock. A variety of tests have shown that molecular evolution does not behave like a stochastic clock. The variance in evolutionary rates is much too large and thus inconsistent with the neutrality theory. This, however, does not invalidate the clock, but rather leaves it without a theoretical foundation to anticipate its properties. Sequence comparisons show that molecular evolution is sufficiently regular to serve in many situations as a clock, but uncertainty concerning the properties of the clock (for example, about the circumstances that may yield large oscillations in substitution rates from time to time or from lineage to lineage) demands that it be used with caution. Few DNA or protein sequences are known from organisms that range from closely related, e.g., different mammals, to very remote, e.g., mammals and fungi. One example is cytochrome *c*, which has an acceptable clockwise behavior over the whole span, in spite of some irregularities. Another example is the copper-zinc superoxide dismutase (SOD), which behaves like a very erratic clock. The SOD average rate of amino acid substitution per 100 residues per 100 million years (MY) is 5.5 when fungi and animals are compared, 9.1 when comparisons are made between insects and mammals, and 27.8 when mammals are compared with each other. The question is which mode is more common over broad evolutionary spans: the regularity of cytochrome *c* or the capriciousness of SOD? Additional data sets will be required in order to obtain the answer and to develop expectations about the accuracy of the clock in particular instances. Until such data exist, conclusions solely based on the molecular clock are potentially fraught with error.

Dr. Ayala is professor of genetics, University of California, Davis, CA 95616. This lecture was delivered at the annual meeting of the Genetics Society of America, in Boston, on August 13, 1985. It is the twentieth in the series of Key lectures that was established by the American Genetic Association through funds bequeathed to the Association by Dr. Wilhelmine E. Key for the support of lectures in genetics. The author is very thankful to Drs. William A. Clemens, Walter M. Fitch, Richard Grantham, and Richard Holmquist for advice and discussions; to Dr. Alvaro Puga for providing his unpublished sequence of the cDNA for rat SOD; to Drs. Holmquist and Wilfried W. de Jong for copies of their manuscripts in press; and to Dr. Fitch for the amino acid differences between cytochrome *c* sequences. The efforts of Dr. Young Moo Lee who, with the collaboration of Mr. David J. Friedman, obtained in my laboratory the *Drosophila melanogaster* SOD sequence, deserve special mention as well as my appreciation.

Zuckerkandl and Pauling[32] were the first to propose that the rate of amino acid and nucleotide substitutions in the evolution of organisms might be constant over time. "Informational macromolecules" (i.e., proteins and nucleic acids) might then provide a molecular clock of evolution. This remarkable hypothesis has revolutionized the reconstruction of evolutionary history and the timing of evolutionary events.

The rationale advanced by Zuckerkandl and Pauling is that many amino acid substitutions may often be of relatively little functional consequence, so that the observed number of changes in the amino acid sequence of a protein may be approximately proportional to the evolutionary time elapsed. This insightful proposition was encased with other conjectures that would be dominant themes in molecular evolution studies during the ensuing two decades: 1) that the rate of amino acid substitutions of a given protein may be directly proportional to the number of sites that can be changed without radical alteration of function; 2) that rates of morphological evolution may be largely due to changes in gene regulation ("in the control of gene activity") and would not be reflected "in the rate of evolution of most polypeptide chains;" and 3) that because functionally highly significant amino acid substitutions are relatively rare, "the rates of evolutionarily effective amino acid substitutions during periods of rapid [morphological] evolution may not substantially differ from the rates that obtain during periods of slow evolution" (Zuckerkandl and Pauling[32] pp. 147–149).

Clocks and Clades

Table I gives, above the diagonal, the number of amino acid replacements observed in the cytochrome *c* of three species. The genetic code makes it possible to calculate the minimum number of nucleotide-pair substitutions required to change a codon for one amino acid to a codon for another. Some amino acid replacements are possible with the change of a single nucleotide, but other replacements require a minimum of two or even three nucleotide substitutions. The minimum number of nucleotide-pair substitutions required to account for the amino acid differences between the cytochrome *c* molecules of the three species are shown, below the diagonal, in Table I. The minimum number of nucleotide substitutions contains somewhat more information than the number of amino acid replacements.

The phylogeny (or cladogenetic history, i.e., the order of splitting of the different lineages or "clades") of humans, rhesus monkeys, and horses is known to be as represented in Figure 1. The data from Table I make it possible to calculate the number of changes that have occurred in each segment of the phylogeny; for example, the one amino acid difference between humans and rhesus monkeys must have been due to a change in the human lineage whereas none occurred in the rhesus lineage. Comparison with a fourth species that diverged from the previous three species before they diverged from each other, say the rattlesnake, would make it possible to dissociate in Figure 1 the number of changes that occurred from A to B and from A to E, which we know total 11 amino acid replacements (and no less than 16 nucleotide substitutions). When the number of species in a phylogeny is greater than three, the calculation of the number of changes in each segment of a phylogeny may be "overdetermined;" there may be more than one solution in each case. There are various ways for attempting to find the "best" solution but they need not detain us now.

The exercise just illustrated may serve the molecular clock hypothesis in either one of two ways: 1) if we know the times when the lineage-splitting events occurred, then we may test whether the molecular clock hypothesis is correct, i.e., whether the number of substitutions assigned to a particular segment of the phylogeny is proportional to the time elapsed during that segment of evolutionary history; 2) if the time of a splitting event is not known, or is controversial, the amino acid or nucleotide substitutions may be used to estimate the time elapsed. The epistemological status of these two activities is quite disparate. In (1) we are testing whether the molecular clock hypothesis is correct (and perhaps gaining information about the accuracy of the clock); in (2) we assume that the hypothesis is correct. The two activities, however, need not be considered incompatible. It is often the case in science that we use a hypothesis not yet well established in order to decide other issues. As we become satisfied with the solutions to these other problems (by their eventual corroboration by independent sources of evidence, for example), we gain greater confidence in the validity of the hypothesis. The process is not logically circular but helical: it gets somewhere, it advances knowledge.

There is yet another important use of data sets like the one in Table I: the reconstruction of phylogenetic history. Assume that we did not know the topological configuration of the phylogeny of the three species listed in Table I. If the rate of amino acid replacements in cytochrome *c* proceeds according to the molecular clock hypothesis, the configuration shown in Figure 1 would be the most likely one. The two other possible topologies require a highly disproportionate number of changes for comparable segment lengths (Figure 2). This application of the clock hypothesis has proven quite fertile in instances where a phylogeny is controversial either because the various sources of evidence are inconsistent or because no reliable evidence exists (as is the case for events in the pre-Cam-

FRANCISCO JOSÉ AYALA is a professor of genetics at the University of California, Davis, where he formerly has been Director of the Institute of Ecology. He is a member of the U.S. National Academy of Sciences, the American Academy of Arts and Sciences, and the American Philosophical Society. He is chairman of the Board of Basic Biology of the U.S. National Research Council and has been President of the Society for the Study of Evolution. He has received honorary degrees from several universities, including the Universidad Complutense of Madrid, Spain, where he began his undergraduate studies. He came to the United States in 1961 to study under Theodosius Dobzhansky and became a naturalized American in 1971. His research extends over the fields of population genetics, ecology, and evolutionary biology. He is the author of several textbooks and has written numerous articles on philosophical matters concerning evolution.

Table I. Number of amino acid replacements (above the diagonal) and minimum number of nucleotide substitutions (below the diagonal) between the cytochrome *c* molecules of humans, rhesus monkeys, and horses; the cytochrome *c* in these organisms has 104 amino acids

	Human	Rhesus monkey	Horse
Human	—	1	12
Rhesus monkey	1	—	11
Horse	17	16	—

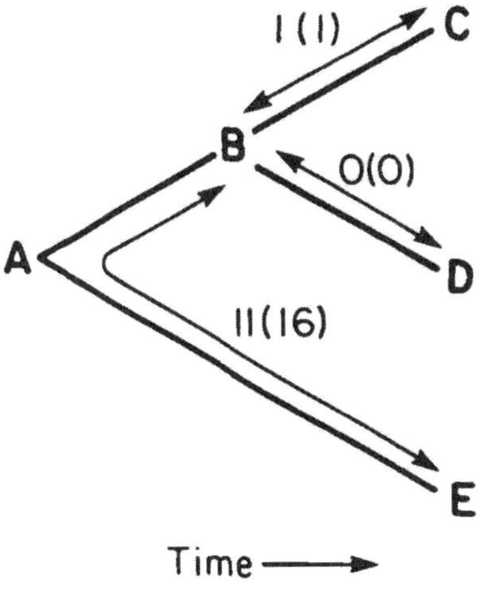

FIGURE 1 Number of cytochrome *c* amino acid replacements (and, in parentheses, minimum number of nucleotide substitutions) in the phylogeny of three contemporary species: humans (C), rhesus monkeys (D), and horses (E).

brian or very remote past). The most extensive use of the molecular clock hypothesis has been, no doubt, in the reconstruction of phylogenetic history. This use provides in turn one additional way for testing the clock hypothesis. If a topology inferred from molecular data is known to be incorrect, this may be because molecular evolution has not occurred in a clockwise fashion. We need not now be concerned with the various methods proposed to infer the most likely phylogenetic configuration from amino acid or nucleotide differences between pairs of species.

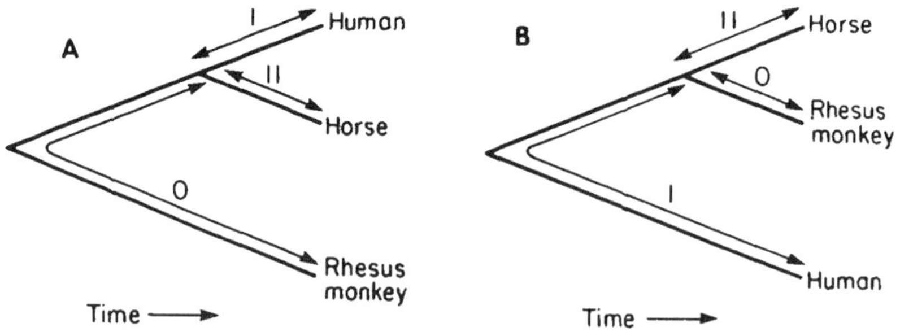

FIGURE 2 Two theoretically possible phylogenies of humans, rhesus monkeys, and horses. The numbers of amino acid substitutions required in each branch to account for the cytochrome *c* sequences indicate that neither of these two phylogenies is likely to be correct.

One of the most dramatic early applications of macromolecular information to the reconstruction of phylogeny was made by Walter M. Fitch and Emanuel Margoliash[4]. Table II gives the minimum number of nucleotide differences necessary to account for the amino differences in the cytochrome *c* molecules of 20 organisms. A phylogeny based on that data matrix, as well as the minimum number of nucleotide substitutions required in each branch, is shown in Figure 3. The phylogenetic relationships displayed in this figure correspond well, on the whole, with the phylogeny of the organisms as determined from the fossil record and other sources. There are three conspicuous disagreements, however. Chickens appear more closely related to penguins than to ducks and pigeons; the turtle, a reptile, is closer to the birds than to the rattlesnake; men and monkeys diverge from the other mammals before the marsupial kangaroo separates from the nonprimate placentals. Despite these erroneous relationships, it is remarkable that the study of a single protein (and a small one at that) should yield such an accurate representation of the phylogeny of 20 organisms as diverse as those in Figure 3. The amino acid sequences of proteins store considerable evolutionary information, as Zuckerkandl and Pauling had argued.

The evidence contained in Table II and Figure 3 indicates that amino acid substitutions occur in a gradual manner, so that lineages that diverged earlier exhibit greater numbers of amino acid differences than those of more recent separation. But how should we evaluate the discrepancies just noted? And how should we ascertain whether the regularity of change is due to a molecular clock that would maintain the same rate of change in different lineages or at different times? A specific proposal about the nature of the molecular clock was advanced by the neutrality theory of protein (molecular) evolution[14,16,18].

A Theoretical Foundation for the Clock: Neutral Evolution

The neutrality theory of protein evolution proposes that the rates of amino acid replacement in proteins and of nucleotide substitution in DNA may be approximately constant because the vast majority of such changes are selectively neutral. New alleles appear in a population by mutation. If alternative alleles have identical fitness, changes in allelic frequencies through time will occur only by accidental sampling errors from generation to

Table II. Minimum number of nucleotide substitutions in the gene coding for cytochrome *c* molecules in 20 organisms (from Fitch and Margoliash[4])

Organism	1	2	3	4	5	6	7	8	9	10	11	12	13	14	15	16	17	18	19	20
1. Human	—	1	13	17	16	13	12	12	17	16	18	18	19	20	31	33	36	63	56	66
2. Monkey		—	12	16	15	12	11	13	16	15	17	17	18	21	32	32	35	62	57	65
3. Dog			—	10	8	4	6	7	12	12	14	14	13	30	29	24	28	64	61	66
4. Horse				—	1	5	11	11	16	16	16	17	16	32	27	24	33	64	60	68
5. Donkey					—	4	10	12	15	15	15	16	15	31	26	25	32	64	59	67
6. Pig						—	6	7	13	13	13	14	13	30	25	26	31	64	59	67
7. Rabbit							—	7	10	8	11	11	11	25	26	23	29	62	59	67
8. Kangaroo								—	14	14	15	13	14	30	27	26	31	66	58	68
9. Duck									—	3	3	3	7	24	26	25	29	61	62	66
10. Pigeon										—	4	4	8	24	27	26	30	59	62	66
11. Chicken											—	2	8	28	26	26	31	61	62	66
12. Penguin												—	8	28	27	28	30	62	61	65
13. Turtle													—	30	27	30	33	65	64	67
14. Rattlesnake														—	38	40	41	61	61	69
15. Tuna															—	34	41	72	66	69
16. Screwworm fly																—	16	58	63	65
17. Moth																	—	59	60	61
18. *Neurospora*																		—	57	61
19. *Saccharomyces*																			—	41
20. *Candida*																				—

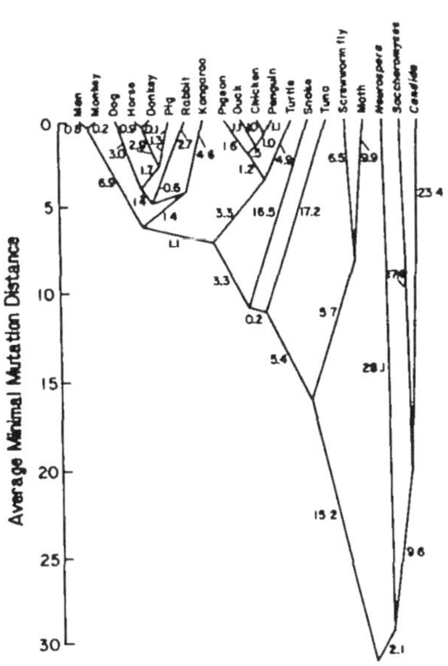

FIGURE 3 Phylogeny of 20 organisms, based on the cytochrome *c* matrix in Table II. The phylogeny agrees fairly well with phylogenetic relationships inferred from the fossil record and other sources. This good agreement is remarkable since it is based on the study of a single protein and encompasses organisms ranging from yeast, through insects, fish, reptiles, amphibians, birds, and mammals, to man. The numbers on the branches are the estimated number of nucleotide substitutions that have taken place in evolution (after Fitch and Margoliash[4]).

generation, that is, by genetic drift. Rates of allelic substitution would be stochastically constant—they would occur with a constant probability for a given protein. That probability can be shown to be simply the mutation rate for neutral alleles.

The neutrality theory of molecular evolution recognizes that, for any gene, a large proportion of all possible mutants are harmful to their carriers; these mutants are eliminated or kept at very low frequency by natural selection. The evolution of morphological, behavioral, and ecological traits is governed largely by natural selection, because it is determined by the selection of favorable mutants against deleterious ones. It is assumed, however, that a number of favorable mutants, adaptively equivalent to one another, can occur at each locus. These mutants are not subject to selection relative to one another because they do not affect the fitness of their carriers (nor do they modify their morphological, physiological, or behavioral properties). According to the neutrality theory, evolution at the molecular level consists for the most part of the gradual, random replacement of one neutral allele by another that is functionally equivalent to the first. The theory assumes that although favorable mutations occur, they are so rare that they have little effect on the overall evolutionary rate of nucleotide and amino acid substitutions.

Table III. **Statistical tests of the constancy of evolutionary rates of seven proteins in 17 species of mammals (from Fitch[3])**

Rates tested	χ^2	*df*	*P*
Overall rates (comparisons among branches over all seven proteins)	82.4	31	4×10^{-6}
Relative rates (comparisons among proteins within branches)	166.3	123	6×10^{-2}
Total	248.7	154	6×10^{-6}

Neutral alleles are not defined as having fitnesses that are identical in the mathematical sense. Operationally, neutral alleles are those the differential contribution of which to fitness is so small that their frequencies change more owing to drift than to natural selection. Assume that two alleles, A_1 and A_2, have fitnesses 1 and $1 - s$ (where s is a positive number smaller than 1). The two alleles are effectively neutral if, and only if,

$$4N_e s \ll 1$$

where N_e is the effective size of the population (approximately, the number of breeding individuals). In a random mating population with N diploid individuals, the rate of substitution of the neutral alleles, k, per unit time is

$$k = 2Nux$$

where u is the neutral mutation rate per gamete per unit time (time measured in the same units as for k) and x is the probability of ultimate fixation of a neutral mutant. The derivation of this equation is straightforward: there are $2Nu$ mutants per unit time, each with a probability x of becoming fixed.

A population of N individuals has $2N$ genes at each autosomal locus. If the alleles are neutral, all genes have the identical probability of becoming fixed, which is simply

$$x = \frac{1}{2N}$$

Substituting this value of x in the previous equation, we obtain

$$k = 2Nu\frac{1}{2N} = u$$

That is, the rate of substitution of neutral alleles is precisely the rate at which the neutral alleles arise by mutation, independently of the size of the population and any other parameters.

This is not only a remarkably simple result, but also one with momentous implication if it indeed applies to molecular evolution. If the neutral mutation rate is constant over evolutionary time, there would, indeed, be a molecular clock; for each protein, the number of amino acid differences between organisms would be expected to be directly proportional to the time since their divergence from a common ancestor. Different proteins (or DNA sequences) would evolve at different rates, namely their rate of neutral mutation. But all proteins and DNA sequences would be timing the same evolutionary events and thus each would provide an independent clock of phylogenetic history.

The molecular clock postulated by the neutrality theory is, of course, not a metronomic clock, like timepieces in ordinary life that measure time exactly. The neutrality theory predicts, instead, that molecular evolution is a "stochastic clock," like radioactive decay. The *probability* of change is constant, although some variation occurs. Over fairly long periods, a stochastic clock is nevertheless quite accurate, and the joint results of several proteins or DNA sequences would, in any case, provide a fairly accurate clock.

The neutrality theory provides a well-defined theoretical basis to the molecular clock hypothesis. It also is subject to the possibility of empirical testing, because precise predictions can be derived from the hypothesis. A great variety of tests indeed have been performed. The results of a particularly meaningful one are given in Table III[19].

The test uses seven proteins sequenced in 17 mammals, and starts by adding up the proteins one after another and treating them as if they were one single sequence. The minimum number of nucleotide substitutions that accounts for the descent of the amino acid sequences from a common ancestor is found; the numbers of substitutions are then assigned to the various branches in the phylogeny. Two independent tests are made. First, the total number of substitutions per unit time is examined for different times; the hypothesis tested is whether the *overall* rate of change is uniform over time. The probability that the variation observed is due to chance is 4×10^{-6}, which is statistically highly significant. The conclusion follows that the proteins have not evolved at a constant rate with a Poisson variance. It is possible, however, that the proteins have all changed their rates *proportionately*. This possibility is examined by testing whether the rates of evolution of one protein *relative* to those of another are uniform through time. There is a marginally significant deviation from expectation ($P \approx 0.06$). The probability that all the variation observed (*total*) is due to chance is extremely small, 6×10^{-6}. (Gillespie and Langley[7] have pointed out that the variance expected from neutral evolution could be greater than the Poisson variance expected from a stochastic process like radioactive decay, due to the existence of protein polymorphisms in the common ancestral populations. However, for the example they use—the data of Langley and Fitch[19] for β hemoglobins—the variance is 2.65 times greater than expected, whereas the levels of protein polymorphism typically observed would make the variance less than one percent greater than expected. See also Golding[8].)

The test of Langley and Fitch[19] is particularly valid because it makes no use of palentological dates. The phylogeny is constructed using the protein data alone; this maximizes the probability of agreement between the data and the hypothesis of statistically constant rates of molecular evolution. Other tests use molecular data from "star phylogenies," groups of species whose lineages branch all from each other at about the same time. The variance is consistently greater than expected under the assumption of a constant probability of amino acid (or nucleotide) substitution. The variance observed for a given protein is on the average more than twice as large as expected[6]. It has been suggested that the large variances observed might be due to variation in evolutionary rates between lineages rather than within a given lineage[27]. But variation between lineages *implies* variation within lineages. Consider two lineages, *A* and *B*, assumed to have evolved at different rates since *they* diverged from each other at time *t*. This necessarily implies that the rate of evolution of at least one of the lineages was different before *t* and after *t*.

Empirical Evidence of the Clock and its Limitations

The variation observed in molecular evolutionary rates has been shown to be consistent with particular models of evolution by natural selection[6]. But the demise of the neutrality theory as the explanation for the molecular clock leaves us without a theoretical foundation for estimating the probable error of evolutionary dates and of phylogenies determined from molecular data. Because the observed variance is typically no more than two or three times larger than the Poisson variance, it would seem that any desired degree of accuracy can be achieved by making two or three times as many observations as would be required for achieving the same accuracy if rates of molecular evolution were stochastically constant[3]. But until such a time as we do have an acceptable theory for the molecular clock, we must depend on empirical data as the only guide to the reliability of the clock in any particular case, and use it with caution when it contradicts other sources of evidence or no reliable evidence exists.

Figure 4 plots the paleontological dates of divergence against the total minimum number of nucleotide substitutions required to account for the amino acid replacements observed in seven proteins from 17 mammalian species. This figure suggests that the molecular clock may be fairly reliable when two conditions are met: 1) data for several proteins (or DNA sequences) are combined; 2) the time elapsed is large (on the evolutionary scale). This is simply a consequence of the "law" of large numbers. As time increases, periods of rapid evolution and periods of slow evolution in any one lineage will tend to cancel each other out; and as the number of proteins examined becomes large, rate variations from lineage to lineage in any one protein will often be compensated by variations in other proteins.

The average rate of evolution of the seven proteins used in Figure 4 would give largely erroneous estimates of divergence time for the lineages that separated within the last 40 million years (MY). These are mostly primate lineages, in which these proteins have evolved at a lower than average rate (although cytochrome *c*, one of the seven proteins, has evolved faster in the primates than in other mammalian lineages). When data for a single protein are considered, persistent variations in evolutionary rates either for different periods or in different lineages may give grossly erroneous estimates of divergence even when long lapses of time are involved. An example of gross departures from clocklike regularity is provided by the copper-zinc superoxide dismutase, an enzyme known to exist in most eukaryotic organisms.

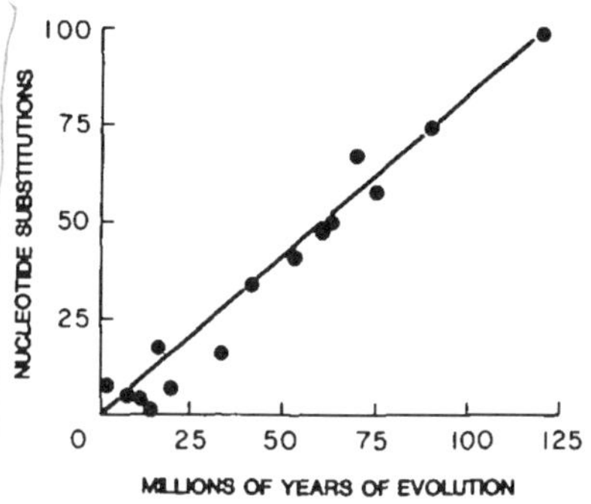

FIGURE 4 Nucleotide substitutions versus paleontological time. The minimum numbers of nucleotide substitutions for seven proteins (cytochome *c*, fibrinopeptides A and B, hemoglobins α and β, myoglobin, and insulin C-peptide), sequenced *in 17 species of mammals*, have been calculated for comparisons between pairs of species whose ancestors diverged at the time indicated on the abscissa. The line has been drawn from the origin to the outermost point. Most points fall near the line, except for some representing comparisons between primates (points below the line at lower left), in which protein evolution seems to have occurred at a lower than average rate (after Fitch[3]).

An Erratic Clock: Superoxide Dismutase

The superoxide dismutases are essential in the defense of organisms against the toxicity of oxygen. The Cu-Zn superoxide dismutase (SOD) of *Drosophila melanogaster* is a dimer molecule made up of two identical subunits associated with two Cu^{++} and two Zn^{++} per molecule. Each subunit has a molecular weight of 15 750 and consists of 151 amino acids[20]. The complete amino acid sequence of the *D. melanogaster* SOD subunit, obtained in our laboratory[21] is shown in Figure 5. This figure also gives the amino acid sequence of SOD in seven other eukaryotes.

The SOD subunit consists of 153 amino acids in human, horse, yeast, and mold, but of 151 amino acids in rat, cow, swordfish, and *Drosophila*. In order to maximize the congruence between the animal SOD sequences, two amino acids have been deleted in four animal sequences in the following positions: rat at sites 1 and 2, cow at 24 and 25, swordfish at 103 and 154, *Drosophila* at 14 and 15. The swordfish deletions are at the sites suggested by Rocha et al.[28] who obtained the sequence. The two-residue gap in *Drosophila* has been decoupled from that in cow and moved 10 positions to the left relative to its location by Lee et al.[21], which increases the similarity between *Drosophila* and the other animals. The cow gap could be moved up to two positions to the left or three to the right with little consequence. The two-residue gap placed at the beginning of the rat sequence is the only sensible choice and is suggested by Puga (pers. comm.) who provided the sequence. The two fungi sequences have been shifted one frame to the right relative to the animal sequences so that they start at site 2; this shift is compensated by placing a one-residue gap in the six animals at position 38.

Three regions of functional importance appear highly conserved among the eight

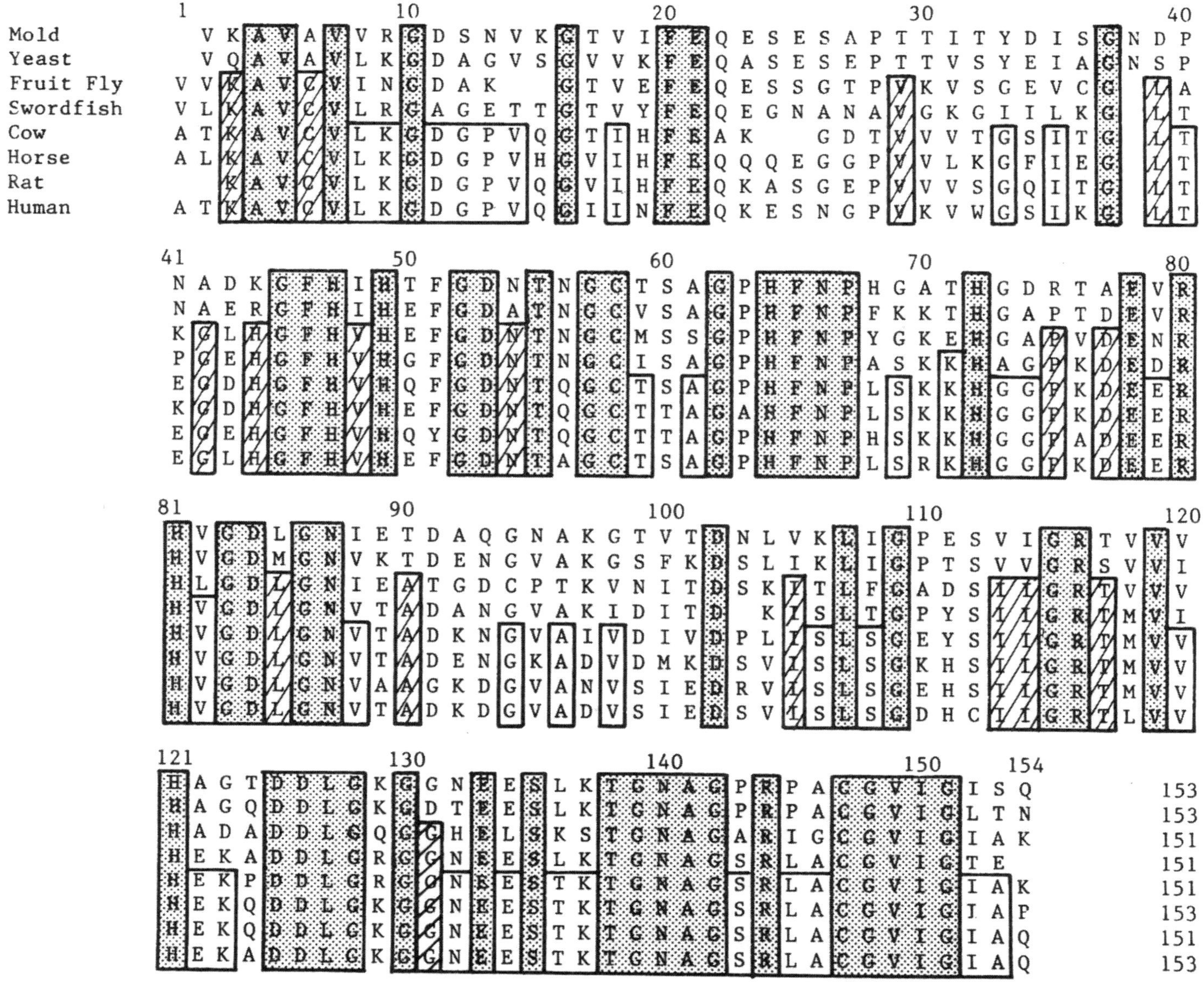

FIGURE 5 The amino acid sequences of Cu-Zn superoxide dismutase in eight organisms, aligned to maximize the number of homologous sites. The amino acids shared by all eight organisms are darkly shaded; those shared only by the six animals are cross-hatched; and those shared only by the four mammals are enclosed in boxes. The sequences are obtained from the following sources: mold, *Neurospora crassa* (Lerch and Schenk[23]); yeast, *Saccharomyces cerevisiae* (Johansen et al.[13]); fruit fly, *Drosophila melanogaster* (Lee et al.[21]); swordfish liver (Rocha et al.[28]); cow erythrocyte (Steinman et al.[30]); horse liver (Lerch and Ammer[22]); rat, inferred from the cDNA nucleotide sequence (Alvaro Puga, pers. comm.); human, inferred from the cDNA nucleotide sequence (Sherman et al.[29]).

SOD enzymes[21]. First is the sequence from residue 49 to 81, which forms the exposed long loop containing the disulfide and zinc ligand regions; second is the sequence from 121 to 142, which forms the second exposed loop (active-site lid loop); third is the COOH-terminal region from residue 144 to 154. The six histidine residues at positions 47, 49, 64, 72, 81, and 121, as well as Asp-84, which are involved in metal liganding are all conserved. The glycine residues, essential to the formation of the characteristic β-strand barrel structure of Cu-Zn SOD, also are remarkably conserved in their location as well as in abundance: there are from 22 to 26 glycine residues in a given sequence, 17 of them have identical positions in all eight organisms and three more in seven of the eight sequences. Two regions are, on the other hand, quite variable: one from residue 12 to 43, the other turn 88 to 101 as previously noted[21]. The second variable region includes residue 99, which is polymorphic in natural populations of *D. melanogaster*[20].

The organisms listed in Figure 5 belong to two different kingdoms, fungi and animals, that diverged from each other probably more than one billion years ago. The fungi include the two remotely related yeast and mold; the animals include an insect, a fish, and four mammals. The considerable degree of sequence conservation recorded in the previous paragraph would seem to suggest that SOD is a slowly evolving protein. Indeed all eight organisms share the same residue at 55 (38.2 percent) of the 144 sites common to all the sequences. This conjecture of slow evolution is, however, not confirmed when it is noticed that the four mammals, whose ancestral lineages diverged from each other within the last 75 MY, share the same amino acid in only 108 (72.5 percent) of their 149 common sites. This is a degree of homology characteristic of very fast evolving proteins.

The phylogenetic history of the eight organisms is shown in Figure 6. There can be little doubt that the topology of the figure is correct; that is, that the order of branching of the various lineages is as indicated in the figure. The times of divergence shown are less certain. The divergence between the two ungulate lineages or between the rodent and the primate lineages is given as 63 MY ago, a figure accepted by paleontologists with a likely error of less than five MY. The most recent ancestor to the four mammal lineages is placed at 75 MY, a value that could be in error also by a few, but probably no more than 10, million years. The divergence between the fish and mammal lineages is placed at 450 MY, near the beginning of the Silurian; the ray-fin fishes ancestral to modern fishes and the lobe-fin fishes from which the terrestrial vertebrates derived are already present in the Silurian and become common in the Devonian fossil record. The arthropod and vertebrate lineages had diverged by the beginning of the Cambrian, 570 MY ago. The divergence time given, 600 MY, may be an underestimate, but it is unlikely that it would be by more than 100 MY. More tentative yet are the divergence times between the yeast and the mold, given at 600 MY on the basis of protein sequence data that assume the molecular clock; and between the two kingdoms, animals and fungi, given as 1200 MY, a figure that could be off by as much as 200 MY.

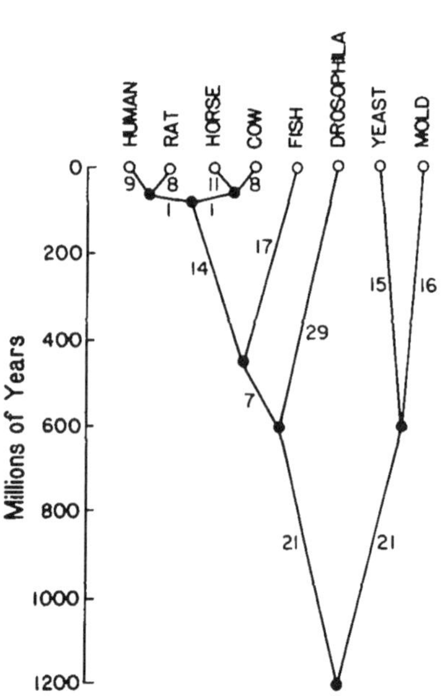

FIGURE 6 Phylogenetic relationships among eight organisms with the approximate time scale. The numbers along the branches of the phylogeny represent the inferred number of amino acid substitutions (PAM values) that have taken place per 100 residues in the evolution of SOD. It is assumed, when no other information is available, that the rate of change per unit is constant along the two branches originating from a node, i.e., from that node until the next node or organism in the branch.

Table IV. Number of amino acid replacements (above the diagonal) and minimum number of nucleotide substitutions (below the diagonal) between the SOD sequences of various organisms; the total number of residues compared is given in parentheses

	Human	Rat	Horse	Cow	Fish	Fly	Yeast	Mold
Human	—	25 (151)	30 (153)	26 (151)	48 (151)	58 (151)	69 (152)	69 (152)
Rat	33 (453)	—	28 (151)	23 (149)	49 (149)	59 (149)	63 (151)	69 (151)
Horse	41 (459)	37 (453)	—	27 (151)	48 (151)	61 (151)	65 (152)	69 (152)
Cow	34 (453)	28 (447)	35 (453)	—	41 (147)	60 (147)	68 (150)	65 (150)
Fish	66 (453)	68 (447)	67 (453)	59 (441)	—	61 (147)	67 (150)	65 (150)
Fly	83 (453)	77 (447)	86 (453)	81 (441)	81 (441)	—	69 (150)	65 (150)
Yeast	93 (456)	83 (453)	91 (456)	92 (450)	92 (450)	94 (450)	—	47 (153)
Mold	86 (456)	95 (453)	96 (456)	92 (450)	94 (450)	87 (450)	61 (459)	—

Table IV gives, above the diagonal, the number of amino acid residues that are different between any two SOD sequences. The minimum numbers of nucleotide substitutions required to effect these amino acid replacements are given below the diagonal. Both sets of values are for the overlapping residues only; where one of the two sequences compared has a gap at a certain position, this is treated as no difference.

The topology of the phylogeny shown in Figure 6 makes it possible to group comparisons between pairs of species that are separated by equal lengths of evolutionary time. This has been done in Table V, where the amino acid and nucleotide differences are given not for the whole sequence but per hundred residues, which corrects for differences in sequence length.

It is apparent in Table V that there are large discrepancies between the observed numbers of amino acid (or nucleotide) substitutions and the times since the divergence of the lineages (given in the last column of Table V). The number of substitutions are

2.6 (or 2.7) times greater between the fungi and the animals than between the two pairs of closely related mammals, whereas the time elapsed is 19 times longer.

Corrections for Hidden Substitutions

However, the amino acid differences observed between two living species are only lower-bound estimates of the number of substitutions that have occurred in phylogentic history: amino acid differences now observed may have been mediated by others during the evolution of the lineages, and substitutions may be reversed. The longer the time since the divergence of two lineages, the more likely it is that "hidden" substitutions may have taken place. A commonly used procedure to correct for superimposed and reversed substitutions is to transfrom the observed residue differences into "accepted point mutations" or PAMs (Dayhoff[2], Tables 36 and 37). The PAM values for both, amino acids and nucleotides, are given in Table V. These are the average values obtained after calculating the PAMs for each one of the appropriate pairwise comparisons. The PAM values reduce the discrepancy between the times elapsed and the molecular changes, but only very little.

PAMs may sometimes give erroneous estimates of the proportion of hidden substitutions[31]. I have, therefore, used other methods of evaluating the changes that have occurred in the evolutionary history of lineages on the basis of protein sequences. Some results are included in Table V, where all figures are averages of the values obtained when the appropriate pairwise comparisons are made.

D_E is a measure proposed by Kimura and Ohta[17], which estimates the average number of nucleotide substitutions per codon that separate the DNA sequences encoding two proteins on the assumption that all codon and nucleotide sites are free to fix mutations. *REH* ("random evolutionary hits") is a measure of evolutionary change that does *not* assume that all sites can change (Holmquist 1978, *in press*). REH_e estimates the total number of accepted nucleotide replacements separating the two encoding DNA sequences taking into account the number of variable codon sites that also is estimated from the data[11]. REH_o differs from REH_e primarily in that it does not assume that all substitutions are equally likely (and in that it is made computationally simple). REH_o is particularly designed for understanding the dynamic behavior of experimentally observable changes in homologous proteins (or genes) as the comparisons move from sequences with few changes to distantly related sequences with many changes[12], which is precisely the situation for the SOD sequences herein considered.

Table V shows that every one of the various measures indicates that the evolution of SOD grossly departs from the regularity expected from a molecular clock. (The calculations in this table ignore sites where one of the two sequences has a deletion. When such sites are treated as one amino acid replacement or three nucleotide substitutions per site, the results do not differ in any material way from those given in the table). Among all the statistics used, *REH* gives relatively the greatest amount of evolutionary change for the lineages furthest removed in time. But the discrepancy between the times elapsed and the amount of change remains enormous. The estimated number of substitutions between the two kingdoms is less than four times, and between the two animal phyla less than three times, the number of substitutions between the two ungulates or between the rat and man, although the times elapsed are about 19 and 10 times greater, respectively.

PAM, D_E, REH_e, and REH_o imply models that make various assumptions as to the dynamics of the amino acid and nucleotide substitutions during evolution. The assumptions of any one or all the models might be challenged, although these models have been used to exhibit agreement between other sets of data and the expectations of a molecular clock. Table V gives the amino acid substitutions observed, and the PAM values, for cytochrome *c*, a protein that like SOD exists in all eukaryotes and that has been studied in many. It can be seen that the PAM values are in fair agreement with the times elapsed in the evolution of the various lineages. (Cytochrome *c* is known to have evolved particularly fast in primates, which is apparent in line 2 of Table V. The PAM value for the cytochrome *c* comparison between yeast and mold is disproportionately greater than expected if the 600 MY given in the table for their divergence is nearly correct. But this is a date based on molecular data and for which we have little confidence.) The amino acid PAM values for cytochrome *c* and SOD are fairly similar when we compare the fungi and the animals (58 and 66, respectively); for the comparison between the ungulates, the SOD value is nearly seven times larger than the cytochrome *c* value (19.5 vs. 2.9). Clearly,

Table V. Differentiation between the SOD sequences of organisms with increasingly remote common ancestors. The various statistics are described in the text; the differences for amino acids and nucleotides are per hundred residues; the values normalized to those for the most closely related organisms are given in parentheses

Comparison	Amino acids observed	Amino acids PAM	Nucleotides observed	Nucleotides PAM	D_E	REH_e	REH_o	Cytochrome *c* amino acids* observed	Cytochrome *c* amino acids* PAM	Million years since divergence
1. Horse to cow and rat to human	17.2 (1.0)	19.5 (1.0)	7.5 (1.0)	7.8 (1.0)	0.237 (1.0)	86.0 (1.0)	202.3 (1.0)	2.9 (1.0)	2.9 (1.0)	63 (1.0)
2. Horse or cow to rat or human	18.7 (1.1)	20.1 (1.0)	7.7 (1.0)	8.1 (1.0)	0.247 (1.0)	84.9 (1.0)	211.8 (1.0)	10.6 (3.6)	11.4 (3.9)	75 (1.2)
3. Fish to mammal	31.0 (1.8)	40.0 (2.1)	14.4 (1.9)	15.8 (2.0)	0.483 (2.0)	217.7 (2.5)	559.4 (2.8)	18.3 (6.3)	20.9 (7.2)	450 (7.1)
4. Insect to vertebrate	39.9 (2.3)	55.9 (2.9)	18.1 (2.4)	20.6 (2.6)	0.628 (2.6)	244.8 (2.8)	592.6 (2.9)	20.0 (6.9)	23.3 (8.1)	600 (9.5)
5. Yeast to mold	30.7 (1.8)	39.4 (2.0)	13.3 (1.8)	14.4 (1.8)	0.439 (1.9)	171.6 (2.0)	384.9 (1.9)	37.5 (13.0)	52.5 (18.2)	600 (9.5)
6. Fungus to animal	44.6 (2.6)	66.2 (3.4)	20.3 (2.7)	23.4 (3.0)	0.705 (3.0)	297.0 (3.5)	789.2 (3.9)	40.9 (14.2)	58.0 (20.0)	1200 (19.0)

* The rat is not included in the cytochrome *c* comparisons, which are calculated using data kindly provided by Dr. Walter Fitch; the fish sequence used is for the dogfish rather than the swordfish

SOD is not evolving with the kind of regularity seen in cytochrome *c*. (In any case, one is compelled to admit that SOD and cytochrome *c* cannot *both* have evolved in a clockwise fashion.)

The numbers of amino acid substitutions (PAM values per hundred sites) that have occurred in each branch of the phylogeny of the eight organisms are given in Figure 6. These have been calculated to minimize their total difference with the matrix of PAM distances between all species pairs. When no other information is available, it has been assumed that the rate of change per unit time has been the same in the two branches originating from a common node, that is, from that node until the next node or organism. The apparent rate of amino acid replacements is least in the fungi lineages and during the first 600 MY of evolutionary history. The mammals exhibit a much faster rate of SOD evolution than any of the other lineages.

The neutrality theory of molecular evolution implies that the rate of evolutionary substitutions should be the same as the neutral mutation rate. Mutation rates, however are usually a function of the number of generations, not of absolute time. Yet the observed rates of molecular evolution approximate a time constancy (see Kimura[15] for a discussion of this issue). The SOD data aggravate this difficulty, because the rate of evolution is greater precisely in the mammal lineages, where fewer generations have elapsed given that the mammals have on the whole much longer generation times than the other organisms.

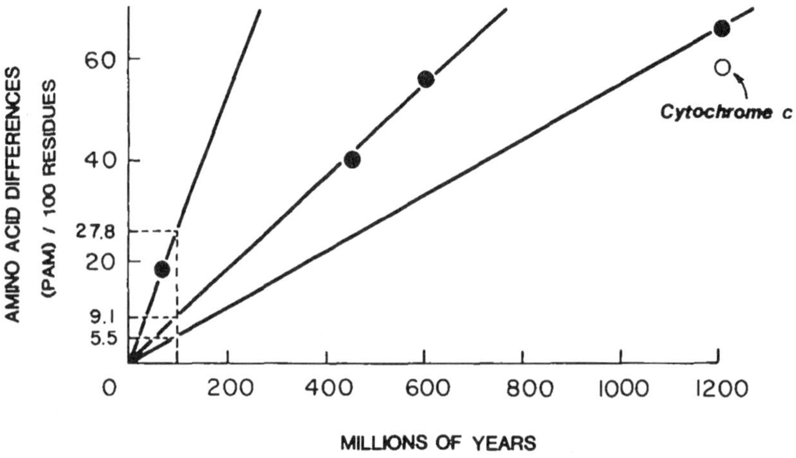

FIGURE 7 Rate of amino acid substitutions (PAM values) per 100 residues in the evolution of SOD. From left to right, the solid circles represent, respectively, the average differentiation between four mammals (70 MY), between the swordfish and the mammals (450 MY), between the fruit fly and the five vertebrates (600 MY), and between the two fungi and the six animals (1200 MY). It is not possible to draw a single straight line through the four points. The two points for the comparisons between the fruit fly and the swordfish with each other and with the mammals can be encompassed within a single straight line. The three straight lines give three different rates of evolution: 5.5, 9.1, and 27.8 amino acid substitutions per hundred residues per 100 MY. (Notice that these rates are for the amino acid *differences* between two species; the rates of substitution within a single lineage would be half as large—if the two lineages are evolving at the same rate.) The open circle represents the average differentiation between the two fungi and the animals for cytochrome *c*, a protein known to evolve at a fairly constant rate.

Rates of SOD Evolution

The disparity between the rates of evolution in different lineages and/or different times is further displayed in Figure 7. When the average differentiation between all four mammals is examined, the rate of amino acid substitution (in PAMs) per 100 residues per 100 MY is 27.8, a fairly fast rate of evolution. The PAM distances between the swordfish and the mammals or between any of them and *Drosophila* give a rate of evolution of 9.1 amino acid substitutions per 100 residues per 100 MY, which is fairly slow. A very slow rate of evolution is obtained from the PAM distances between the animals and the fungi: 5.5 amino acids per 100 residues per 100 MY, similar to the rate obtained when the same comparisons are made for cytochrome *c* (see open circle in the figure), which is a very slowly evolving protein.

The conclusion seems warranted that SOD is a very erratic evolutionary clock. Other instances of rather erratic rates of molecular evolution are known[9,10,26], but few if any exhibit such large variations over so long an evolutionary span as SOD. What circumstances might account for these large variations?

One conceivable explanation is that a number of sites in the protein are fixed: a certain amino acid and no other must be present at a given site if function is to be preserved. The number of amino acid replacements would presumably increase with evolutionary time, approaching asymptotically the maximum allowed. Once the maximum number of mutable sites would have been changed, the number of amino acid differences would not further increase with time.

I have pointed out above that there are three highly conserved regions and other sites in SOD, particularly those involved in metal liganding. It seems, however, unlikely that the number of SOD mutable sites has been fulfilled with the eight organisms studied. The four mammals give a total of 45 variable sites; the addition of a fish, an insect, and two fungi increases to 99 the number of variable sites. It seems highly probable that SOD sequences from other organisms, including invertebrates and plants, will increase the number of mutable sites.

Three other considerations are relevant. First is that the largest number of amino acid substitutions between any two species is 69, whereas there are 99 mutable sites; hence the maximum possible number of substitutions has not been reached by far in any one sequence. It can be argued, against this observation, that not all 20 amino acids can occur at all sites, but rather that at certain sites only a few amino acids are compatible with a functional enzyme. This would reduce the apparent rate of amino acid substitution over long evolutionary spans, because the same amino acid would reappear at a certain site of a sequence after several substitutions had taken place. Nevertheless, only if very few amino acids would be tolerated at most mutable sites would this phenomenon have a significant effect. Such an extreme restriction is unlikely and is contradicted by the data that exist, and would not be sufficient in any case to account for the large discrepancies in the SOD evolutionary rates.

The second consideration is that various measures used in Table V do try to correct for superimposed and reversed mutations. REH_e and REH_o in particular assume that not all sites are mutable, and utilize the data for estimating the fraction of fixed sites in order to make the appropriate corrections.

The third consideration is that other proteins do not exhibit the irregularities of rate noticed in SOD. Cytochrome *c* has no fewer constraints than SOD with respect to the tolerance of amino acid substitutions; yet it conforms fairly well to the expectations of a molecular clock.

The existence of a large number of variable sites when evolutionary distant organisms are compared does not imply that all the variable sites could change at any time. Fitch and Margoliash[5] put forward the concept of concominantly variable codons ("covarions") which refers to the fraction of all amino acid positions that can accept replacements in any one species at any one point time. They concluded for cytochrome *c* that the number of covarions was only about 10 percent, although more than two-thirds of all sites had substitutions when the many species sequenced, ranging from fungi to humans, are examined. A biologically plausible explanation of this phenomenon is that the ability of a site to accept a replacement depends on the interaction of that amino acid with those in other positions: when a replacement takes place, the interactions change and hence the sites at which further replacements can occur also change. But the existence of a limited number of covarions at any one time does not imply that a protein will evolve with different rates at different times or in different lineages, so long as the number of covarions remains the same.

A drastic possibility to account for the irregularity of SOD evolution is "horizontal" gene transfer between lineages: the SOD gene might have been transmitted from one to another contemporaneous species (horizontally, in contrast to the usual vertical transmission of genes from parents to their descendants). Martin and Fridovich[25] have suggested that the similarity in amino acid composition between the Cu-Zn SOD of the ponyfish, *Leiognathus splendens*, and its bacterial symbiont *Photobacterium leiognathi* may be due to horizontal transfer of the gene from the fish to the bacterium in recent evolutionary time. Bannister and Parker[1] have supported this suggestion but the detailed analysis of Leunissen and de Jong[24] convincingly shows that horizontal transfer did not occur in this case. Moreover, horizontal transfer would seem a most unlikely possibility for accounting for the apparent irregular rates of evolution seen in the eukaryotic SODs, because several transfers would need to be postulated at different times between different lineages, with no justification for postulating the transfers other than an attempt to salvage the molecular clock hypothesis.

A Tame Conclusion

Regardless of the causes responsible for the apparent discrepancy in the superoxide dismutase evolutionary rates, the conclusion remains that this protein is not an acceptable evolutionary clock. This observation, in turn, demands that caution be used before making evolutionary inferences (about the timing of evolutionary events and even about the branching order of lineages in a phylogeny) on the basis of the amino acid and/or nucleotide sequence of a single protein or gene. We lack a valid theory that would allow us to calculate the probable error of estimates based on the molecular clock. There is empirical evidence of a certain degree of regularity in protein and DNA evolution, but we have no reason to believe that the evolutionary rate of a certain gene or protein is the same in different lineages or persists in the same lineage for long periods of time. Before we learn about the reliability of the clock, we need extensive data sets of a kind scarcely available at present. Homologous DNA or protein sequences must be obtained from organisms covering extensive degrees of relatedness, from those from the same genus or family to those from different phyla or kingdoms. The molecular clock is sloppy, but we do not yet know how sloppy it is.

References

1. Bannister, J. V. and M. W. Parker. The presence of a copper/zinc superoxide dismutase in the bacterium *Photobacterium leiognathi*: A likely case of gene transfer from eukaryotes to prokaryotes. *PNAS* 82:149–152. 1985.
2. Dayhoff, M. D. Atlas of Protein Sequence and Structure. Natl. Biomed. Res. Found., Washington, D.C. 1978.
3. Fitch, W. M. Molecular evolutionary clocks. *In* Molecular Evolution, F. J. Ayala, Ed. Sinuaer, Sunderland, MA. p. 160–178. 1976.
4. —— and E. Margoliash. Construction of phylogenetic trees. *Science* 155:279–284. 1967.
5. —— and ——. The usefulness of amino acid and nucleotide sequences in evolutionary studies. *In* Evolutionary Biology, vol. 4, p. 67–109. 1970.
6. Gillespie, J. H. Natural selection and the molecular clock. *Mol. Biol. Evol.* 3:138–155. 1986.
7. —— and C. H. Langley. Are evolutionary rates really variable? *J. Mol. Evol.* 13:27–34. 1979.
8. Golding, G. B. Estimates of DNA and protein sequence divergence: An examination of some assumptions. *Mol. Biol. Evol.* 1:125–142. 1983.
9. Goodman, M., G. W. Moore, J. Barnabas, and G. Matsuda. The phylogeny of human globin genes investigated by the maximum parsimony method. *J. Mol. Evol.* 3:1–48. 1974.
10. ——, ——, and G. Matsuda. Darwinian evolution in the genealogy of hemoglobin. *Nature* 253:603–608. 1975.
11. Holmquist, R. Evaluation of compositional nonrandomness in proteins. *J. Mol. Evol.* 11:349–360. 1978.
12. ——. Construction of a 4-class model for nonrandom genetic divergence. *J. Mol. Evol.* In press. 1986.
13. Johansen, J. T., C. Over Balle-Petersen, B. Martin, V. Hasemann, and I. Svendsen. The complete amino acid sequence of copper, zinc superoxide dismutase from *Saccharomyces cerevisiae. Carlsberg Res. Commun.* 44:201–217. 1979.
14. Kimura, M. Evolutionary rate at the molecular level. *Nature* 217:624–626. 1968.
15. ——. The Neutral Theory of Molecular Evolution. Cambridge Univ. Press, Cambridge, MA. 1983.
16. —— and T. Ohta. Protein polymorphism as a phase of molecular evolution. *Nature* 229:467–469. 1971.
17. —— and ——. On the stochastic model for estimation of mutational distance between homologous proteins. *J. Mol. Evol.* 2:87–90. 1972.
18. King, J. L. and T. H. Jukes. Non-Darwinian evolution. *Science* 164:788–798. 1969.
19. Langley, C. H. and W. M. Fitch. An examination of the constancy of the rate of molecular evolution. *J. Mol. Evol.* 3:161–177. 1974.
20. Lee, Y. M., M. P. Misra, and F. J. Ayala. Superoxide dismutase in *Drosophila melanogaster*: Biochemical and structural characterization of allozyme variants. *PNAS* 78:7052–7055. 1981.
21. ——, D. J. Friedman, and F. J. Ayala. Superoxide dismutase: An evolutionary puzzle. *PNAS* 82:824–828. 1985.
22. Lerch, K. and D. Ammer. Amino acid sequence of copper-zinc superoxide dismutase from horse liver. *J. Biol. Chem.* 256:11545–11551. 1981.
23. —— and E. Schenk. Primary structure of copper-zinc superoxide dismutase from *Neurospora crassa. J. Biol. Chem.* 260:9559–9566. 1985.
24. Leunissen, J. A. M. and W. W. de Jong. Copper/zinc superoxide dismutase: How likely is gene transfer from ponyfish to *Photobacterium leiognathi? J. Mol. Evol.* In press. 1986.
25. Martin, J. P. and I. Fridovich. Evidence for a natural gene transfer from the ponyfish to its bioluminescent bacterial symbiont *Photobacter leiognathi. J. Biol. Chem.* 256:6080–6089. 1981.
26. Matsubara, H., T. Hase, S. Wakabayashi, and K. Wada. Gene duplications during evolution of chloroplast-type ferrodoxins. *In* Evolution of Protein Molecules, H. Matsubara and T. Yamanaka, Eds. Japan Scientific Soc. Press, Tokyo. p. 209–220. 1978.
27. Ohta, T. and M. Kimura. On the constancy of the evolutionary rate of cistrons. *J. Mol. Evol.* 1:18–25. 1971.
28. Rocha, H., W. H. Bannister, and J. V. Bannister. The amino-acid sequence of copper/zinc superoxide dismutase sequences. *Eur. J. Biochem.* 145:477–484. 1984.
29. Sherman, L., N. Dafni, J. Lieman-Hurwitz, and Y. Groner. Nucleotide sequence and expression of human chromosome 21-encoded superoxide dismutase mRNA. *PNAS* 80:5465–5469. 1983.
30. Steinman, H. M., V. R. Naik, J. L. Abernethy, and R. L. Hill. Bovine erythrocyte superoxide dismutase. *J. Biol. Chem.* 249:7326–7338. 1974.
31. Wilbur, W. J. On the PAM matrix model of protein evolution. *Mol. Biol. Evol.* 2:434–447. 1985.
32. Zuckerkandl, E. and L. Pauling. Evolutionary divergence and convergence in proteins. *In* Evolving Genes and Proteins, V. Bryson and H. J. Vogel, Eds. Academic Press, NY. p. 97–166. 1965.

Avise, J. C., J. Arnold, R. M. Ball, E. Bermingham, T. Lamb, J. E. Neigel, C. A. Reeb, and N. C. Saunders. 1987. Intraspecific phylogeography: the mitochondrial DNA bridge between population genetics and systematics. *Annual Review of Ecology and Systematics* 18:489–522.

The field of evolutionary genetics traditionally had two separate foci: microevolution within species (the realm of population genetics) and macroevolution above the species level (the realm of phylogenetics). Given that evolution inevitably encompasses both of these foci, it was ironic that these two disciplines had limited contact, as evidenced by the fact that population geneticists and phylogeneticists typically had different educational backgrounds and spoke quite different evolutionary "dialects." Evolution to a population geneticist meant shifts in population allelic or genotypic frequencies due to microevolutionary processes such as mutation, genetic recombination, natural selection, and genetic drift, whereas evolution to a systematist was described in terms of supraspecific lineages, phylogenetic trees, clades, and histories of speciation and extinction. This separation between micro- and macroevolutionary perspectives was largely artificial, but it plagued and stymied efforts toward unification or synthesis of these two major evolutionary realms.

This state of affairs began to change rather abruptly, in 1987, following the introduction of the novel concept of *intraspecific phylogeography*, which itself had been motivated by unconventional evolutionary perspectives originating from studies of animal mitochondrial (mt) DNA. The paper included here introduced the term *phylogeography*, inaugurated phylogeographic perspectives on evolution, and set the stage for what would become a blossoming of the field of phylogeography throughout the ensuing decades (e.g., Avise 2000) and continuing today. As the title of the paper indicates, the seminal contribution of phylogeography has been to build empirical and conceptual bridges and thereby help to unify population genetics and phylogenetics.

RELATED READING

Avise, J. C. 2000. Phylogeography: the history and formation of species. Harvard University Press, Cambridge, Massachusetts, USA.

Ann. Rev. Ecol. Syst. 1987. 18:489–522

INTRASPECIFIC PHYLOGEOGRAPHY: The Mitochondrial DNA Bridge Between Population Genetics and Systematics

John C. Avise[1], *Jonathan Arnold*[1], *R. Martin Ball*[1], *Eldredge Bermingham*[1,2], *Trip Lamb*[1,3], *Joseph E. Neigel*[1,4], *Carol A. Reeb*[1], *and Nancy C. Saunders*[1,5]

[1]Department of Genetics, University of Georgia, Athens, Georgia 30602; [2]NMFS/CZES, Genetics, 2725 Montlake Boulevard East, Seattle, Washington 98112; [3]Savannah River Ecology Laboratory, Drawer E, Aiken, South Carolina 29801; [4]Department of Microbiology and Immunology, School of Medicine, University of California, Los Angeles, California 90024; [5]School of Veterinary Medicine, Virginia Tech University, Blacksburg, Virginia 24046

INTRODUCTION

A recurring debate in evolutionary biology is over the extent to which microevolutionary processes operating within species can be extrapolated to explain macroevolutionary differences among species and higher taxa (36, 38, 45, 46, 53, 67, 68, 80). As discussed by Stebbins & Ayala (83), several issues involved must be carefully distinguished, such as (*a*) whether microevolutionary processes (e.g. mutation, chromosomal change, genetic drift, natural selection) have operated throughout the history of life (presumably they have); (*b*) whether such known processes can by themselves *account* for macroevolutionary phenomena; and (*c*) whether these processes can *predict* macroevolutionary trends and patterns. In another, phylogenetic sense, macroevolution is ineluctably an extrapolation of microevolution: Organisms have parents, who in turn had parents, and so on back through time. Thus, the branches in macroevolutionary trees have a substructure that consists of smaller branches and twigs, ultimately resolved as generation-to-generation pedigrees (Figure 1). It is through these pedigrees that genes have been transmitted, tracing the stream of heredity that is phylogeny.

It would seem that considerations of phylogeny and heredity should provide a logical starting point for attempts to understand any connections of macroevolution to microevolution. Yet amazingly, the discipline traditionally associated with heredity and microevolutionary process (population genetics) developed and has remained largely separate from those fields associated with phylogeny and macroevolution (systematics and paleontology). Thus, several classic textbooks in population genetics (35, 39, 64) do not so much as index "phylogeny," "systematics," or "speciation," while the equally important textbooks in systematics (55, 81, 96) can be read and understood with only the most rudimentary knowledge of Mendelian and population genetics. Notwithstanding some evidence for recent increased communication between these disciplines (40, 71), too many systematists and population geneticists continue to operate in largely separate realms, employing different languages and concepts to address issues that should be of importance to all.

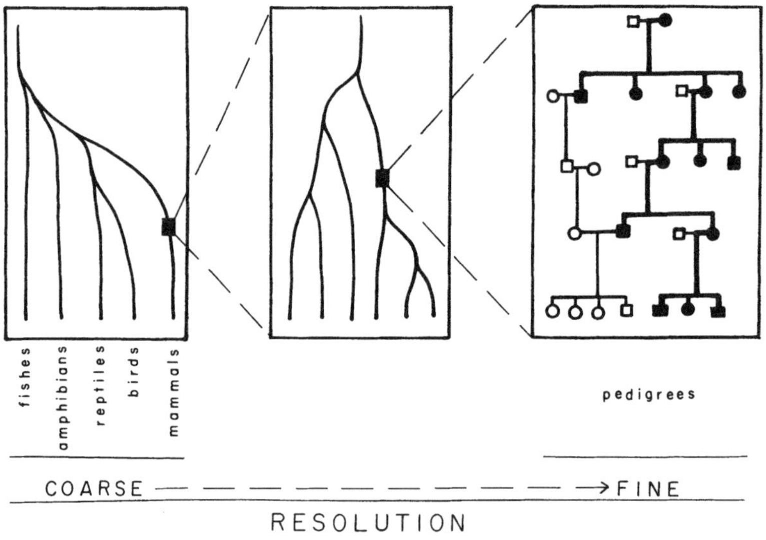

Figure 1 At closer levels of examination, macroevolutionary trees (such as the one on the left summarizing relationships among some of the vertebrate classes) must in principle have a substructure consisting of smaller and smaller branches, ultimately resolvable as family pedigrees through which genes have been transmitted. Some branches in the pedigree on the right have been darkened to indicate the transmission path of mtDNA from the earliest pictured female.

It might also be supposed that the newer field of molecular evolution, with its obvious grounding in genetics and yet a concern with phylogeny, would have facilitated a firmer linkage between micro- and macroevolutionary study. And to some extent it has, by allowing evaluations of large-scale evolutionary trees in terms of DNA and protein characters with a known genetic basis. But movement in the opposite direction—extending phylogenetic principles and reasoning to the microevolutionary level—has been negligible. Thus, when molecular evolutionists work at the intraspecific level, they tend to adopt the terms and concepts of population genetics, such as "variances in allele frequencies," "genetic drift," "mutation-selection balance," "fitness," and so forth, but not the terms and concepts of systematics, such as "monophyletic groups," "parsimony networks," "clades," or "synapomorphic character states." Conversely, when systematists work at the intraspecific level, for example to describe subspecies, it is usually with morphological or behavioral traits whose genetic basis (or even control) is poorly known. Worse yet, geographic locale per se is too often the primary basis for assigning newly collected specimens to "subspecies," so that attempts to understand any relationship between phylogenetic differentiation and spatial separation flirt with circular reasoning.

The reasons that the field of molecular evolution has not contributed greatly to the incorporation of phylogenetic principles into population genetics are, we suspect, primarily twofold. First, the inception and early development of molecular evolution (reviewed in 18, 63, 70) largely coincided with (and stimulated) the rise of the neutralist school of thought (58, 59), which challenged a common view that genetic variability was molded primarily by natural selection. Thus, molecular evolutionists were (justifiably) preoccupied with the selectionist-neutralist debate and never gained very close contact with various schools of systematic thought that were also growing actively at that time (55, 81). Second, in the early years protein electrophoresis was the only molecular technique readily applicable to comparisons at the intraspecific level. Yet allozymes of a particular locus are qualitative, multistate traits the phylogenetic order of which cannot be safely inferred from the observable property, electrophoretic mobility. Furthermore, allozymes are encoded by nuclear genes that segregate and recombine during each generation of sexual reproduction. These attributes of allozymes understandably directed methods of data analysis toward concerns with allele frequencies and heterozygosities, which in turn channeled thinking back into the traditional framework and language of population genetics and away from phylogeny.

The purpose of this report is to make a case that animal mitochondrial DNA (mtDNA) (by virtue of its maternal, nonrecombining mode of inheritance, rapid pace of evolution, and extensive intraspecific polymorphism) permits and even demands an extension of phylogenetic thinking to the microevolutionary level. As such, data from mtDNA can provide a liaison service for expanding communication between systematists and population geneticists. With empirical and conceptual channels opened, it might then be possible to reconsider various connections between micro- and macroevolutionary change as interpreted against a continuous genealogical backdrop. This review will be a success if it stimulates further dialogue in these areas.

mtDNA—NOT "JUST ANOTHER" MOLECULAR MARKER

If one were to specify the properties desired of an ideal molecular system for phylogenetic analysis, the wish list might include the following. The molecule should: (*a*) be distinctive, yet ubiquitously distributed, so that secure homologous comparions could be made among a wide variety of organisms; (*b*) be easy to isolate and assay; (*c*) have a simple genetic structure lacking complicating features such as repetitive DNA, transposable elements, pseudogenes, and introns; (*d*) exhibit a straightforward mode of genetic transmission, without recombination or other genetic rearrangements; (*e*) provide suites of qualitative character states whose phylogenetic interrelationships could be inferred by reasonable parsimony criteria; and, for purposes of microevolutionary analysis, (*f*) evolve at a rapid pace such that new character states commonly arise within the lifespan of a species. To a remarkable degree, the mitochondrial DNA of higher animals meets all of the above criteria.

Molecular properties of animal mtDNA have been reviewed previously (10, 25, 26), so only a brief synopsis sufficient for current discussion is given here. The reader is directed to the earlier papers for details and qualifications. In higher animals, mtDNA is a small, covalently closed circular molecule, about 16–20 kilobases long. It is tightly packed with genes for 13 messenger RNA's, 2 ribosomal RNA's, and 22 transfer RNA's. In addition to these 37 genes, an area known as the "D-loop" (in vertebrates and echinoderms) or "A + T-rich" region (in *Drosophila*), roughly 0.8 kilobases long, appears to exercise control over mtDNA replication and RNA transcription. Introns, repetitive DNA, pseudogenes, and even sizeable spacer sequences between genes, are all absent. Gene arrangement appears very stable, at least within a taxonomic class or phylum. For example, gene order is identical in assayed mammals and frogs but differs from that in *Drosophila*. Nonetheless, evolution at the nucleotide sequence level is rapid, perhaps 1–10 times faster than typical single-copy nuclear DNA (28, 92). Most of the genetic changes are simple base substitutions; some are small addition/deletions (one or a few nucleotides); and fewer still involve large length differences (up to several hundred nucleotides). The size differences are usually (though not exclusively; 69) confined to the control region of the molecule, which in general is evolving especially rapidly. The final and perhaps most important point is that, to the best of current knowledge (50, 60), inheritance of animal mtDNA is strictly maternal. Thus, unlike the situation for nuclear DNA, the mtDNA mutations arising in different individuals are not recombined during sexual reproduction.

No molecular system is likely to be perfect for phylogenetic analysis, and mtDNA does have some potential and real limitations that need to be recognized:

Heteroplasmy

Most somatic cells (and mature oocytes) contain hundreds or thousands of mtDNA molecules, so that at its inception a new mutation will either generate or add to a heteroplasmic condition in which two or more genotypes coexist within an individual. On theoretical grounds, it was originally feared that heteroplasmy might be extensive and hopelessly complicate mtDNA study, but empirical experience proved this worry to be unjustified. Cases of heteroplasmy have been discovered (20 and references therein) but are unusual and therefore of little impact in routine surveys of animal mtDNA. Current thinking is that mutations within a cell line (as opposed to paternal leakage of mtDNA via sperm) generate most instances of heteroplasmy, and that the heteroplasmic state is quite transitory, due to rapid sorting of mtDNA molecules in germcell lineages (34, 52, 76, 86). Thus, as phrased by Wilson et al (97), "The vast majority of individuals tested seem effectively haploid as regards the number of types of mtDNA transmitted to the next generation (although polyploid as regards the number of mtDNA copies per cell)."

Homoplasy

An ideal phylogenetic marker would be free from reversals as well as parallel or convergent evolutionary change (homoplasy). In one respect, mtDNA falls short of this standard—many restriction sites have been observed to "blink" on and off repeatedly during evolution (e.g. 43, 61). This phenomenon is presumably most often attributable to recurrent transitional base substitutions (3) at some nucleotide sites. If particular positions in the mtDNA genome are considered "characters," and if evolutionary change at these positions has been especially rapid with respect to the time since separation of assayed lineages, then the small number of alternative character states assumable insures that some homoplasious changes will have occurred. Nonetheless, because mtDNA genomes are nonrecombining, the entire molecule can justifiably be considered the "character," in which case the number of possible character states becomes astronomical.

Typical empirical surveys of mtDNA (see beyond) effectively involve assay of at least several hundred base-pairs of information per individual. When viewed this way, any widespread and intricate similarities present in mtDNA are most unlikely to have arisen by convergent evolution, and so they must primarily reflect phylogenetic descent (or, conversely stated, any widespread and intricate differences observed among mtDNA molecules could not be overcome by wholesale convergent mutation). The effects of homoplasious change in mtDNA are thus probably limited to introduction of circumscribed ambiguity in tree or network placements of mtDNA genotypes. Furthermore, approaches for recognizing and treating homoplasy in mtDNA have been suggested (10, 89, 90).

Scale

Some nucleotide positions in mtDNA are far more labile evolutionarily than are others, presumably due to relaxed selective constraints (4). The initial rapid pace of mtDNA differentiation (estimated at about 2% sequence divergence per million years in mammals; 28) is attributable primarily to changes at these sites, after which further mtDNA differences accumulate much more slowly. The overall effect is that beyond perhaps about 8–10 million yr, a plot of mtDNA nucleotide sequence divergence (p) against time (t) becomes curvilinear, eventually reaching a plateau where estimation of t from p is pointless (28). For this reason, unless special precautions are taken to work only with more slowly evolving portions of the molecule, meaningful phylogenetic comparisons from conventional mtDNA surveys will normally be confined to conspecific populations and closely related species whose separations date to within the last few million years.

At the other end of the scale, for very recently disjoined populations or species, it is likely that a substantial fraction of observed mtDNA sequence differences arose prior to population separation (i.e. they represent retention of polymorphisms originally present in ancestral parental stock). There are at least two ways to deal with this potential complication. First, from a population genetic perspective, statistical corrections can be applied (72, 97). For example, let δ_X, δ_Y, δ_A, and δ_{XY} represent the mean pairwise mtDNA divergence values between individuals of population X, of population Y, of the ancestral population, and between individuals in population X versus Y, respectively. Although δ_A cannot be observed directly, it can be estimated by assuming that

$$\delta_A = 0.5\ (\delta_X + \delta_Y).$$

Then the corrected distance estimate between populations X and Y becomes

$$\delta = \delta_{XY} - \delta_A.$$

A second way to deal with the predicament involves a shift to a phylogenetic perspective. Since mtDNA genotypes in different lines do not recombine, individual organisms (rather than populations or species) can justifiably be considered as the basic operational taxonomic units (OTU's) in a phylogenetic reconstruction (62). This straightforward approach, in which individuals (or, more precisely, their mtDNA genotypes) constitute the tips of hypothesized evolutionary trees (or nodes of evolutionary networks), can be especially informative. For example, with respect to matriarchal phylogeny, it is biologically quite plausible that some individuals may truly be more closely related to members of another species than they are to conspecifics, owing solely to particular patterns of maternal lineage survival and extinction accompanying the speciation process (73, 85; also see below).

Selection Versus Neutrality

The longstanding debate about whether the dynamics of genetic variation are governed primarily by natural selection or by genetic drift of neutral mutations, can also be extended to mtDNA. In our view, the phylogenetic value of mtDNA does not, however, completely hinge on the outcome. Thus, even if mtDNA genotypes prove commonly to differ with respect to fitness, properly identified synapomorphic (shared-derived) character states should still permit recognition of monophyletic assemblages (clades) of molecules. Nonetheless, in some kinds of data analyses involving genetic distance estimates and molecular clock concepts to date separation events, it would be especially important to know whether mtDNA variability is neutral (although, particularly when longer spans of time are involved and much genetic information is assayed, the magnitude of genetic differentiation under some models of natural selection should also be well-correlated with time; 9, 44).

Two senses in which mtDNA variability might be deemed neutral need to be carefully distinguished. First, in a mechanistic sense, we already "know" that most of the particular mtDNA genotypic variants segregating in populations probably have, by themselves, absolutely no differential effect on organismal fitness. These include, for example, base substitutions in silent positions of protein-coding genes, and some substitutions and small addition/deletions in the nontranscribed D-loop region. These changes are disproportionately common in mtDNA (25) and are ones for which only the most ardent selectionist would argue a direct link to organismal fitness. On the other hand, mtDNA contains genes whose products (usually in collaboration with those of the nuclear genome) are crucial to production of energy necessary for animal survival and reproduction (49, 74). Some mtDNA mutations must, then, be highly visible to selection. When they arise, each such mutation will by chance be associated with a particular array of mechanistically neutral variants elsewhere in the molecule. Since mtDNA is maternally inherited, these associations will not be dissoluted by recombination (95). In this second, dynamic sense, mechanistically neutral mtDNA variants may, through linkage to selected mtDNA mutations, have evolutionary histories that are at times influenced or even dominated by effects of natural selection. A deeper understanding of such possibilities poses a stiff challenge for future study. Not only will knowledge be required of the periodicity and intensity of selection on fitness-related mtDNA mutations, but historical accidents of association with neutral markers will have to be taken into account. Furthermore, all this action must be understood within the context of ever-changing nuclear gene backgrounds whose epistatic interactions with mtDNA are likely to be of great importance (5, 25, 49, 78).

Lineage Sampling Bias

The phylogenies inferred from mtDNA comparisons represent the presumed historical sequences of mutational events accompanying the differentiation of maternal lines. An mtDNA phylogeny is thus an example of a *molecular genealogy*—a record of evolutionary changes in a piece of DNA, in this case one that has a history of maternal transmission. In general, any *organismal phylogeny* must in some sense represent a composite attribute of many molecular genealogies, including those for all nuclear genes, each of which in any generation could have been transmitted through male or female parents. As phrased by Wainscoat (93), "We inherit our mitochondrial DNA from just one of our sixteen great-great grandparents, yet this maternal ancestor has only contributed one-sixteenth of our nuclear DNA." The asexual, maternal transmission of mtDNA is thus a double-edged sword. Although the information recorded in mtDNA represents only one of many molecular tracings in the evolutionary histories of organisms, it is nonetheless a specified genealogical history (female → female → female), and one whose molecular record has not been complicated by the effects of recombination.

INTRASPECIFIC PHYLOGENY AND GEOGRAPHIC POPULATION STRUCTURE

Most mtDNA surveys of natural populations have involved the technically expedient restriction enzyme approach. MtDNA is isolated from individual animals, digested with particular endonucleases, and the resulting digestion products separated by molecular weight through gels. The "raw" data then consist of restriction fragment digestion profiles on gels, or with some additional effort, restriction site maps. The evolutionary changes in restriction sites underlying the differing digestion profiles or site maps can often be inferred simply, and a parsimony network summarizing the presumed history of genotypic interconversions can be generated. A straightforward example involving *Bst*EII sites observed in the mtDNA from *Peromyscus maniculatus* (61) is presented in Figure 2. A typical survey now often includes data from ten or more enzymes and involves, on average, 40–100 or more restriction sites per individual. The recognition sequence of each employed endonuclease is either four, five, or six base-pairs in length, so a routine survey of mtDNA would effectively screen individuals for genetic differences at several hundred nucleotide positions.

To exemplify more fully the kinds of phylogenetic implications inherent in such data, we briefly summarize results from a typical natural population survey. Bermingham & Avise (19) used 13 restriction endonucleases to score an average of 54 sites per individual in the mtDNA of 75 bowfin fish *(Amia calva)* collected from river drainages from South Carolina to Mississippi. A total of 13 distinct mtDNA genotypes (which for simplicity can be called "clones") were observed. Figure 3A shows a hand-drawn parsimony network (constructed by an extension of the approach exemplified in Figure 2) interconnecting these clonal genotypes, and in Figure 3B this network is superimposed over the geographic sources of the collections. Two major genetic (and geographic) assemblages of mtDNA clones are apparent—an eastern assemblage of nine related clones observed in bowfin from South Carolina, Georgia, and Florida; and a western assemblage of four related

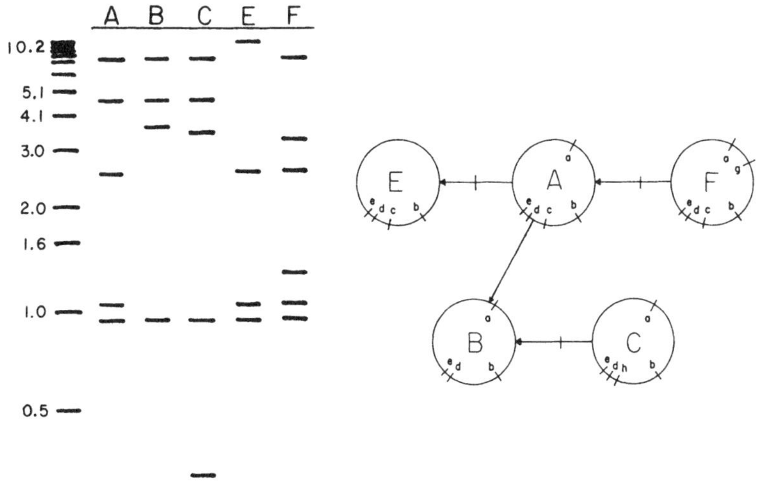

Figure 2 (left) Diagrammatic representation of the five *Bst*EII digestion profiles observed among mtDNA's isolated from samples of *Peromyscus maniculatus* (61). The leftmost lane shows selected sizes (in kilobases) of fragments in a molecular weight standard. (right) Restriction site maps (obtained from double-digestion procedures) corresponding to the fragment profiles on the left. These site maps have been interconnected into a parsimony network reflecting probable evolutionary relationships among the *Bst*EII patterns. Arrows indicate direction of site loss and not necessarily direction of evolution.

clones in bowfin from Alabama and Mississippi. *At least* four assayed restriction site changes distinguish any eastern from any western genotype.

As drawn, these parsimony networks are unrooted, but additional hypotheses about phylogenetic orientation can be advanced. By several criteria, mtDNA clone 1 is a likely candidate for the ancestral genotype within the eastern assemblage of *Amia calva:* (*a*) It is by far the most common eastern genotype, occurring in 30 of 59 assayed specimens; (*b*) it is geographically the most widespread, observed in nine of the ten eastern river drainages surveyed; and (*c*) in the parsimony analysis, it forms the hub of a network whose spokes connect separately to seven other mtDNA genotypes (Figure 3A). Clone 1 is also at least one mutation step closer to the western mtDNA

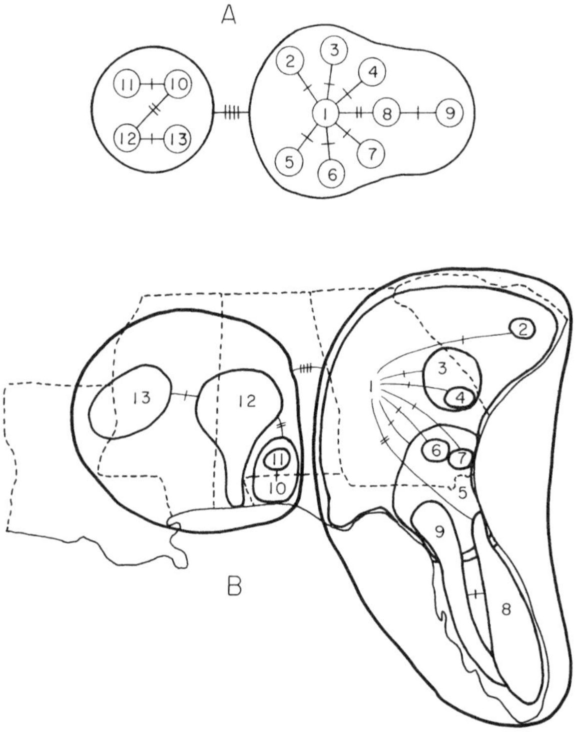

Figure 3 Phylogenetic networks and phenograms summarizing evolutionary relationships among 13 mtDNA genotypes observed in a sample of 75 bowfin fish, *Amia calva* (19). (A) Hand-drawn parsimony network. Slashes crossing branches indicate restriction site changes along a path; heavier lines encompass 2 major arrays of mtDNA genotypes distinguishable by at least 4 restriction site changes. (B) The parsimony network in A superimposed over the geographic sources of collections. (C) Wagner parsimony network computer generated from a presence-absence site matrix. Inferred restriction site changes are indicated, and numbers in the network represent levels of statistical support (by bootstrapping) for various clades. (D) UPGMA phenogram, where p is estimated nucleotide sequence divergence.

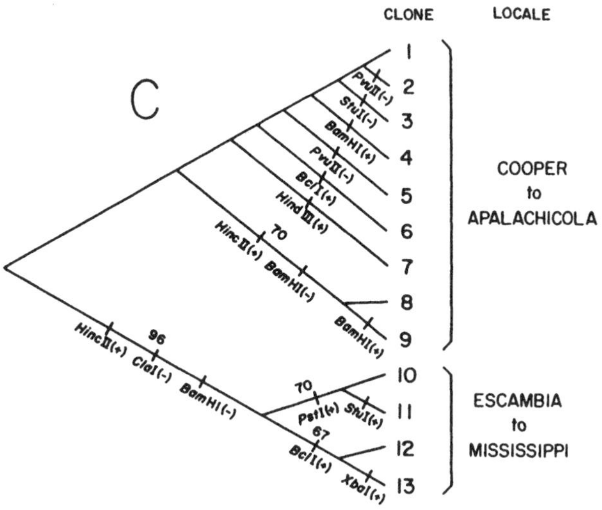

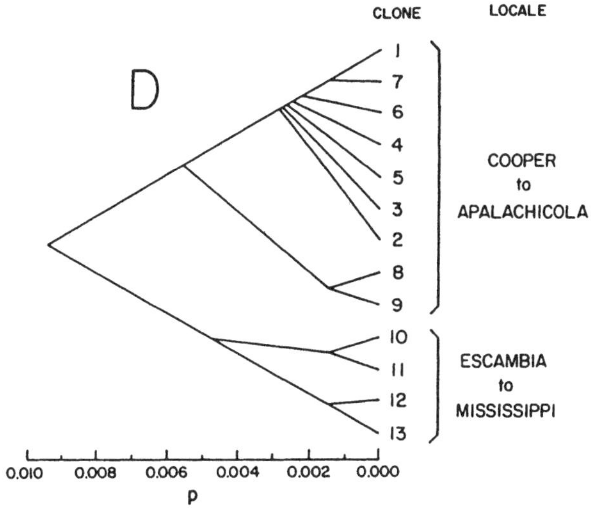

clade than are any other mtDNA genotypes in the east. In the western genotypic array, clones 10 and 12, which occur in the drainages most proximate to those in the east (Figure 3B), are genetically closest to clone 1 (each differs by four assayed mutation steps); and fish in the most westerly drainage show further distinction from these clones (Figures 3A and 3B).

Some data sets are far too large for such easy analysis by hand, and computer assistance is required. Several tree-building software packages are available (42); we routinely employ various algorithms in the PHYLIP package distributed by Joe Felsenstein. For example, Figure 3C shows a Wagner parsimony network (from the METRO annealing algorithm in PHYLIP) of mtDNA genotypes in bowfin fish generated from a matrix consisting of presence-absence information for each restriction site in each mtDNA clone. Particular site changes along various branches of the network are shown, and numbers indicate the levels of statistical support (the proportion of times that a group was distinguished in a bootstrap analysis; 41) for a given hypothesized mtDNA clade.

It is also possible to convert mtDNA fragment or site data into estimates of nucleotide sequence divergence (p) between genotypes (e.g. 72, 91), and the resulting distance matrixes can provide the basis for tree or phenogram construction. Figure 3D shows a UPGMA phenogram (81) for the mtDNA clones in *Amia calva*. The eastern versus western clonal assemblages are again apparent and differ in nucleotide sequence by an average of about 1%. In general then, many qualitative and quantitative methods of tree construction can be applied to mtDNA data. It is beyond the scope of this review to address the ongoing debate about "best" methods for phylogeny reconstruction (and Avise's views have been presented elsewhere; 6). Suffice it to say that in our experience, tree-constructing algorithms involving philosophically distinct methodologies usually produce very similar outcomes when applied to a given set of mtDNA data. The pictured networks and phenograms for *Amia calva* (Figure 3) are merely a case in point.

In our laboratory, similar surveys of geographic variation in mtDNA have now been completed or are in progress for about 20 species, including mammals, birds, reptiles, amphibians, marine and freshwater fishes, and an invertebrate (the horseshoe crab). The remainder of this section summarizes major features of these data (Table 1) in the context of qualitative patterns of geographic population structure. The original papers should be consulted for details.

In principle, intraspecific phylogenies overlaid on geographic maps could yield many kinds of outcomes. Five major categories of possibilities and their provisional interpretations are summarized in Figure 4. For example, an mtDNA phylogeny itself could show discontinuities (or genetic "breaks") in which arrays of related genotypes differ from other such arrays by many mutational steps. Such genetically distinct mtDNA assemblages might occupy separate geographic regions within the range of a species (category I, Figure 4), or they could co-occur geographically (category II). Alternatively, mtDNA phylogenies themselves might be more or less continuous genetically, and spatially either disjunct (category III), totally overlapping (category IV), or nested (category V). We have empirical examples approximating almost all of these theoretical outcomes.

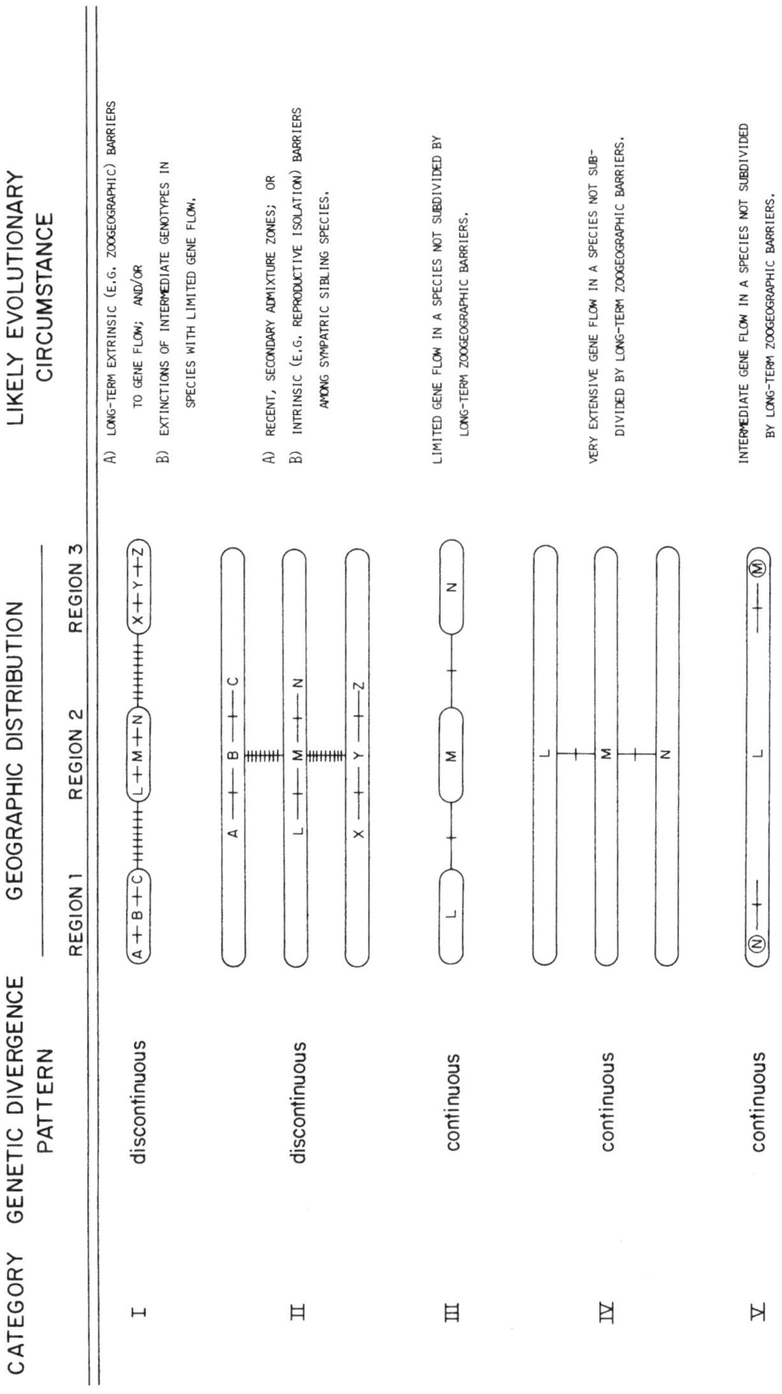

CATEGORY
GENETIC DIVERGENCE PATTERN
GEOGRAPHIC DISTRIBUTION
REGION 1
REGION 2
REGION 3
LIKELY EVOLUTIONARY CIRCUMSTANCE
I
discontinuous
A B C
L M N
X Y Z
A) LONG-TERM EXTRINSIC (E.G. ZOOGEOGRAPHIC) BARRIERS TO GENE FLOW; AND/OR
B) EXTINCTIONS OF INTERMEDIATE GENOTYPES IN SPECIES WITH LIMITED GENE FLOW.
II
discontinuous
A B C
L M N
X Y Z
A) RECENT, SECONDARY ADMIXTURE ZONES; OR
B) INTRINSIC (E.G. REPRODUCTIVE ISOLATION) BARRIERS AMONG SYMPATRIC SIBLING SPECIES.
III
continuous
L
M
N
LIMITED GENE FLOW IN A SPECIES NOT SUBDIVIDED BY LONG-TERM ZOOGEOGRAPHIC BARRIERS.
IV
continuous
L
M
N
VERY EXTENSIVE GENE FLOW IN A SPECIES NOT SUB-DIVIDED BY LONG-TERM ZOOGEOGRAPHIC BARRIERS.
V
continuous
N
L
M
INTERMEDIATE GENE FLOW IN A SPECIES NOT SUBDIVIDED BY LONG-TERM ZOOGEOGRAPHIC BARRIERS.

Figure 4 General phylogeographic patterns (relationships between phylogeny and geography) theoretically observable in mtDNA surveys (see text).

Category I—Phylogenetic Discontinuities, Spatial Separation

In our experience, this is the most commonly encountered situation. It applies for example, to the *Amia calva* case history already detailed above, in which arrays of related mtDNA genotypes occurred in eastern versus western river drainages in the southeastern United States, and the two arrays were differentiable by at least four assayed mutation steps (and $\bar{p} \cong 0.01$). The magnitudes of genetic breaks distinguishing populations from different geographic regions are in fact often considerably greater than that observed in *Amia calva*. For example, in the redear sunfish *Lepomis microlophus*, which was sampled from the same river drainages as *Amia*, eastern versus western arrays of mtDNA genotypes differed by 17 or more assayed mutation steps ($\bar{p} \cong 0.09$), yet *maximum* differentiation within either the eastern or western mtDNA assemblages was always less than $p = 0.007$ (19). Other species in which we have observed discontinuous intraspecific mtDNA phylogenetic networks, with a strong geographic orientation, include the pocket gopher *(Geomys pinetis)*, deer mouse *(Peromyscus maniculatus)*, bluegill sunfish *(Lepomis macrochirus)*, spotted sunfish *(L. punctatus)*, warmouth sunfish *(L. gulosus)*, mudpuppy salamanders *(Necturus alabamensis* and relatives), desert tortoise *(Scaptochelys agassizii)*, and horseshoe crab *(Limulus polyphemus)*. References and relevant data from these studies are summarized in Table 1.

The most likely explanation for major genetic discontinuities that display geographic orientation involves long-term, extrinsic (i.e. zoogeographic) barriers to gene flow, such that conspecific populations occupy easily recognizable branches on an intraspecific evolutionary tree. Another related possibility, not mutually exclusive, is extinction of intermediate genotypes in widely distributed species with limited dispersal and gene flow capabilities.

Apart from the mtDNA *phylogeographic* patterns per se, is there additional support for the significance of historical zoogeography in shaping intraspecific genetic architectures? At least two empirical lines of evidence can be advanced. First, populations separated for long times by zoogeographic barriers should also accumulate differences in the nuclear genome. Few studies have assayed nuclear genes (or their products) in concert with mtDNA, but two that have done so—involving the pocket gopher, *Geomys pinetis* (13) and the bluegill sunfish, *Lepomis macrochirus* (12)—found dramatic allozyme frequency distinctions between major mtDNA phylogenetic groups (Figure 5). Second, strong biogeographic barriers should mould the genetic structures of independently evolving species in concordant fashion. Five species of freshwater fishes have been assayed for mtDNA differentiation across river drainages in the southeastern United States—and, remarkably, all showed strong patterns of congruence in the geographic placements of the major mtDNA phylogenetic breaks (Figures 3, 5, and 6). To account for these

Table 1 Summary information from the larger phylogeographic surveys of mtDNA conducted in our laboratory. Original studies should be consulted for details.

Species	Number of			Geographic scale of survey	Mean number restriction sites scored	Phylogeographic category (Figure 4)	Mean number base substitutions per nucleotide	Reference
	Individuals	Locales	Different mtDNA genotypes					
Invertebrates								
Limulus polyphemus (horseshoe crab)	99	15	10	New Hampshire to Florida	41	I	0.020	77
Freshwater fishes								
Amia calva (bowfin)	78	20	13	5 southeastern states	54	I	0.009	20
Lepomis punctatus (spotted sunfish)	79	16	17	6 southeastern states	39	I	0.062	20
Lepomis microlophus (redear sunfish)	77	17	7	6 southeastern states	48	I	0.087	20
Lepomis gulosus (warmouth sunfish)	74	17	32	6 southeastern states	50	I	0.063	20
Lepomis macrochirus (bluegill sunfish)	189	9	—	4 southeastern states	37	I	0.085	12
Marine or catadromous fishes								
Anguilla rostrata (American eel)	109	7	23	Maine to Louisiana	78	IV	0.001	14

Table 1 Summary information from the larger phylogeographic surveys of mtDNA conducted in our laboratory. Original studies should be consulted for details.

Species	Number of: Individuals	Number of: Locales	Number of: Different mtDNA genotypes	Geographic scale of survey	Mean number restriction sites scored	Phylogeographic category (Figure 4)	Mean number base substitutions per nucleotide	Reference
Marine or catadromous fishes								
Arius felis (hardhead catfish)	60	10	11	North Carolina to Louisiana	57	IV	0.005	16
Barge marinus (gafftopsail catfish)	12	5	4	Georgia to Florida panhandle	49	IV	0.005	16
Opsanus tau (oyster toadfish)	43	9	5	Massachusetts to Georgia	50	III	0.005	16
Opsanus beta (gulf toadfish)	17	4	8	Florida to Louisiana	54	III or V	0.004	16
Amphibians								
Necturus sp.[b,c] (mud puppies)	50	24	>22	North Carolina to Louisiana	38	I	≅0.060	Bermingham et al., in prep.
Bufo terrestris[c] (southern toad)	117	24	45	North Carolina to Louisiana	47	III	0.010	Bermingham and Avise, in prep.
Reptiles								
Malaclemys terrapin (diamondback terrapin)[c]	41	7	4	Massachusetts to Florida	54	III or V	.002	Lamb and Avise, in prep.

Trachemys scripta (yellow-bellied turtle)[c]	55	20	2	North Carolina to Texas	46	I or III	.007	Bermingham and Avise in prep.
Scaptochelys agassizii (desert tortoise)[c]	44	19	6	southwestern states and Mexico	40	I	.051	Lamb and Avise, in prep.
Birds								
Agelaius phoeniceus (red-winged blackbird)[c]	130	21	33	North American continent	71	IV or V	.004	Ball et al., in prep.
Mammals								
Geomys pinetis (pocket gopher)	87	24	23	3 southeastern states	28	I	0.034	13
Peromyscus maniculatus (deer mouse)	135	40	61	North American continent	39	I	0.040	61
Peromyscus polionotus (old-field mouse)	68	15	22	3 southeastern states	37	III or V	0.011	17
Peromyscus leucopus (white-footed mouse)	14	9	12	eastern and central United States	43	I	0.020	17

[a] Between major phylogenetic arrays of mtDNA clones (if the outcome fell into phylogeographic category I), or among mtDNA clones within the entire species (if the outcome fell into phylogeographic categories III, IV, or V).
[b] *N. alabamensis* and related forms and species whose taxonomic status is in dispute.
[c] Study still in progress, and hence entries in this table are provisional and subject to possible revision.

results, we advanced a detailed biogeographic reconstruction—one that implicates historical patterns of river drainage isolation and coalescence associated with Pliocene and Pleistocene changes in sea level (19).

Preliminary evidence suggests that intraspecific phylogeographic discontinuities in mtDNA may commonly align with the boundaries of zoogeographic provinces as identified by more conventional biogeographic data. For example, from lists of distributional limits of freshwater fish species, Swift et al (84) identified two major zoogeographic provinces (east versus west of the Apalachicola River), plus additional subprovinces, in the southeastern United States. Boundary zones between these regions agreed quite well with the concentrations of intraspecific phylogenetic breaks in

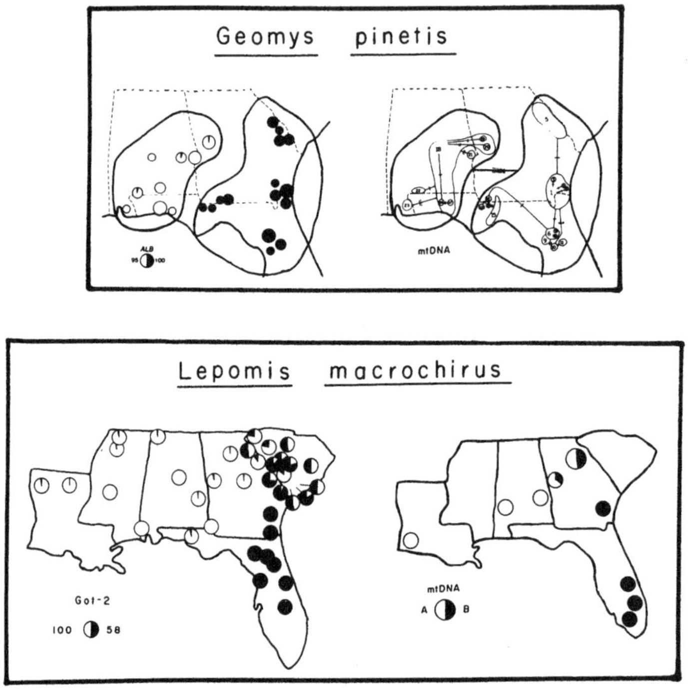

Figure 5 Empirical examples in which highly divergent mtDNA phylogeographic groupings also proved distinct in allozyme frequencies. Above: Data for southeastern pocket gopher, *Geomys pinetis* (13). On the left are pie diagrams summarizing geographic distributions (in three southern states) of the two electromorphs (labeled "95" and "100") of the albumin locus. On the right is an mtDNA phylogenetic network, the most dramatic feature of which is the large genetic gap ($\bar{p}$ 0.034) distinguishing the same arrays of eastern versus western samples. Below: Data for the bluegill sunfish, *Lepomis macrochirus*. On the left are pie diagrams summarizing geographic distributions of electromorphs (labeled "100" and "58") of the *Got-2* nuclear locus (from 11). On the right are pie diagrams of frequencies of two highly distinct ($\bar{p}$ = 0.085) mtDNA genotypes (from 12).

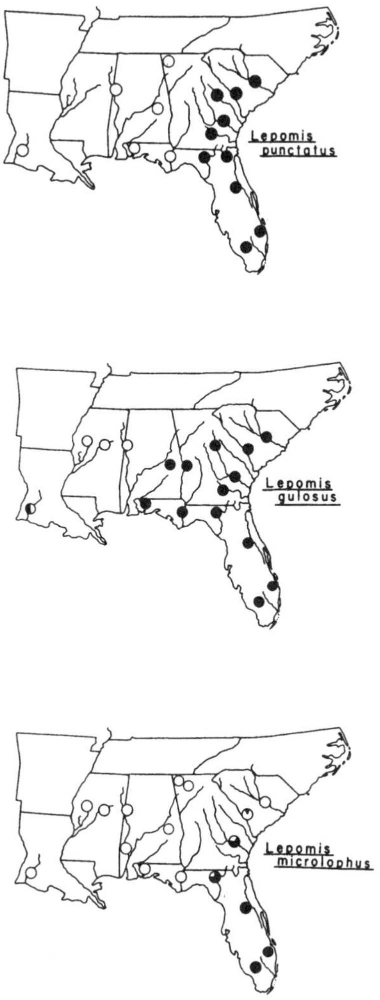

Figure 6 Geographic distributions of major mtDNA clades in three additional species of sunfish, *Lepomis* (from 19). Within each species, major mtDNA phylogenetic breaks ($\bar{p}$ = 0.062, 0.063, and 0.087 for *L. punctatus, L. gulosus,* and *L. microlophus,* respectively) distinguish fish in eastern rivers from those in drainages further to the west.

mtDNA for the five widely distributed fish species assayed by Bermingham & Avise (19). In another example, a genetic discontinuity between two phylogenetic assemblages of mtDNA's in the coastal horseshoe crab (77) occurred near Cape Canaveral, Florida, a region long-recognized as transitional between warm-temperate and tropical marine faunas (1, 22). Recognizable biogeographic provinces presumably exist because of environmental impediments (ecological and/or physical; historical as well as contemporary) to dispersal and gene flow. These impediments are conventionally recognized as reflected in concentrations of distributional limits for many species; perhaps they may also be reflected in concentrations of intraspecific phylogenetic discontinuities within species that have geographic distributions extending across zoogeographic provinces.

Category II—Phylogenetic Discontinuities, Lack of Spatial Separation

In surveys from our laboratory, we have no good empirical examples of the situation diagrammed in category II, Figure 4—mtDNA phylogenetic discontinuities not associated with spatial separation. Indeed, it has even been rare to observe large mtDNA differences (i.e. greater than about 1–2% nucleotide sequence divergence) between conspecific individuals collected at any given geographic site. One example involved the deer mouse *Peromyscus maniculatus* (61), in which, for unknown reasons, two or more moderately divergent mtDNA clones were occasionally found within particular localities in the eastern United States. For example, two mtDNA clones collected in Giles County, Virginia, differed by five restriction sites changes (in assays with eight endonucleases) and an estimated sequence divergence of $p \cong 0.013$. Even in *P. maniculatus,* however, the largest genetic differences in mtDNA (p values greater than about 0.03) were invariably between mice from different regions of North America (61), so that the overall pattern is more consistent with category I in Figure 4.

In a large sample of bluegill sunfish *(Lepomis macrochirus)* collected from Lake Oglethorpe in north Georgia, two grossly different mtDNA genotypes ($p \cong 0.085$) did co-occur in roughly equal frequency (12). Further analysis of this situation, involving more extensive geographic sampling as well as comparisons with nuclear genotypes (11), revealed that the Lake Oglethorpe population is probably a random-mating, hybrid swarm arising from secondary contact between allopatrically evolved races of bluegill. In such secondary admixture zones (as well as in cases where reproductively isolated sibling species are inadvertently assayed as if belonging to a single species), mtDNA phylogenetic discontinuities in the absence of current spatial separation are of course to be expected.

Category III—Phylogenetic Continuity, Spatial Separation

Not all assayed species have exhibited the large mtDNA phylogeographic "breaks" characteristic of category I. Another commonly encountered situation is one in which mtDNA parsimony networks are more or less continuous, with consistently small numbers of mutational steps (and fairly low p values) between phylogenetically adjacent clones, each of which is nonetheless confined to a subset of the geographic range of the species (III, Figure 4). Such a situation is approximated in the marine oyster toadfish *Opsanus tau* (16).

Among 43 individuals sampled from Massachusetts to Georgia, five closely related but geographically localized mtDNA genotypes were observed, each differing from its apparent closest relative by only one or two assayed restriction sites (and associated $p < 0.008$). The two most common genotypes were respectively confined to collections north versus south of the Cape Hatteras area in North Carolina (another boundary region between zoogeographic provinces (22). Yet mean sequence divergence between the five mtDNA genotypes in *O. tau* was only $\bar{p} \cong 0.005$. Other species in which we have observed limited differentiation yet geographic localization of mtDNA clones include the gulf toadfish *(Opsanus beta)*, diamondback terrapin *(Malaclemys terrapin)*, and old-field mouse *(Peromyscus polionotus)* (Table 1). This phylogeographic pattern is also characteristic of the differentiation observed *within* particular regions for most of the category-I species previously listed (in other words, category III is similar to the within-region pattern of category I—Figure 4).

The most likely explanation for geographic localization of mtDNA clones and clades, in the absence of major phylogenetic breaks, involves historically limited gene flow between populations in species not subdivided by firm long-term zoogeographic barriers to dispersal. Thus, recently arisen mutations are confined to subsets of the species' range, and the overall population structure may conform more or less to either the "island" or "stepping stone" models in traditional population genetics (54).

Category IV—Phylogenetic Continuity, Lack of Spatial Separation

Within a few species, closely related mtDNA genotypes appear not to be geographically localized. Perhaps the best example involves the American eel, *Anguilla rostrata* (14). In 109 eels taken from seven locales between Maine and Louisiana, numerous related mtDNA genotypes were detected, yet each (when present in two or more individuals) was geographically widespread. Similarly, in the hardhead sea catfish *(Arius felis)*, two related clades of mtDNA genotypes ($\bar{p} \cong 0.006$) were both widely distributed along the South Atlantic and Gulf of Mexico coastlines (16). Other assayed species exhibiting limited mtDNA phylogenetic diversity and relatively little geographic structure include the marine gafftopsail catfish *(Bagre marinus)*, the red-winged blackbird *(Agelaius phoeniceus)* (Table 1), and, to an argued extent, humans (24, 94).

We propose that geographic populations of species exhibiting this category of intraspecific phylogeography have had relatively extensive and recent historical interconnections through gene flow. This would require the absence of firm and longstanding zoogeographic barriers to movement, as well as life histories conducive to dispersal. All of the above examples can at least provisionally be understood in these terms. American eels have a catadromous life history—mass spawning takes place in the western tropical mid-Atlantic Ocean, and larvae are transported (perhaps passively) to coastal regions by ocean currents. Young eels mature in freshwater before completing the life cycle by migrating back to the mid-Atlantic for spawning. Thus, any freshwater population in the Americas may contain a nearly random draw of genotypes from what is effectively a single mating pool. The eel life-history pattern is highly unusual, but other marine fishes, such as the marine catfish which are active swimmers as adults, may also prove to exhibit the "category-IV" phylogeographic pattern (8, 16, 48). Marine fish occupy a realm *relatively* free of solid geographic barriers to dispersal (at least over major portions of their ranges and in comparison to those of their freshwater counterparts that are necessarily confined to specific drainages for moderate lengths of evolutionary time). Many marine fishes also possess great dispersal capabilities, either as pelagic larvae, juveniles, and/or adults.

Birds constitute another group of potentially highly mobile animals for which the "category-IV" phylogeographic pattern may prove to be common. For example, populations of the red-winged blackbird, *Agelaius phoeniceus*, (the only avian species extensively assayed at the time of this writing) exhibited very limited mtDNA phylogeographic structure across all of North America (Table 1). Red-winged blackbirds are known to be moderately nest-site philopatric (average distances between banding and recovery at nesting sites in successive years are generally less than 50 km; 37), so the documentation of a "category-IV" pattern clearly cannot be taken to imply panmixia or even long-distance gene flow on a generation-to-generation scale. Rather, we suspect that these blackbirds (and other species in phylogeographic category IV) have had a relative fluidity of movement (in birds, obviously facilitated by the capacity for flight) over a recent evolutionary time scale such that populations have been in solid genetic contact within, perhaps, the last few tens of thousands of generations (see next section).

For humans, assays of mtDNA from individuals of diverse racial and geographic origin revealed only a weak tendency for phylogenetic structuring of groups, according to Cann et al (33; see also 24, 27, 30, 31, 32). Based on a conventional mtDNA clock calibration of 2% nucleotide sequence divergence per million years, Cann et al (31) proposed a mean interracial divergence time in humans of about 50,000 years, and Brown (24) and Cann et al (33) hypothesized a common (female) ancestor for all humans about 200,000 years ago. Cann et al (33) also argue from the mtDNA data that this female ancestor lived in Africa. For any species whose numbers and ranges have expanded dramatically from a single refugium or place of origin in recent evolutionary times, mtDNA phylogeographic differentiation should similarly be quite limited.

Using independently obtained data, Johnson et al (57) report a greater degree of geographic and racial clustering of human mtDNA genotypes than did the Cann et al (31, 32, 33) research group. MtDNA genotypes thought to be ancestral for humans were geographically and racially widespread, but genotypes presumed derived were often race specific. Thus, according to the Johnson et al data (57; see also 94), humans are better characterized as exhibiting the category-V phylogeographic pattern (see below).

Category V—Phylogenetic Continuity, Partial Spatial Separation

The four phylogeographic categories listed above are of course somewhat arbitrarily selected though distinct points from a wide field of possibilities. We include category V here and in Figure 4 only to provide an example of one type of intermediate situation. In this category, some mtDNA genotypes are geographically widespread, while allied genotypes are localized, such that the overall pattern is one of a nested series of phylogeographic relationships. Besides humans, we have already mentioned one other example. In the eastern mtDNA clonal assemblage of the bowfin fish *Amia calva* (Figure 3), mtDNA genotype 1 was present in nine of ten surveyed river drainages, while each of eight other genotypes was apparently confined to one or a few adjacent drainages within the range of genotype 1. A reasonable hypothesis for this eastern assemblage is that genotype 1 is plesiomorphic (ancestral), while the other genotypes are apomorphic (derived). Individual fish sharing the derived states (i.e. possessing synapomorphic mtDNA traits) form various monophyletic groups (with respect to maternal ancestry). But because of the possibility of joint retention of the ancestral condition, individuals sharing genotype 1 do not necessarily form a clade within the eastern assemblage of bowfin (although compared to bowfin in the western drainages, they may still form a broader clade including all eastern genotypes).

Phylogeographic category V might be anticipated in species or subsets of species with historically intermediate levels of gene flow between geographic populations. Thus, unlike category III, presumed ancestral genotypes occur over a broad area; while unlike category IV, newly arisen mutations have not yet spread throughout the range of a species. Nonetheless, in practice this intermediate situation may normally be difficult to distinguish clearly from categories III or IV, respectively (Table 1).

mtDNA EVOLUTIONARY TREES ARE SELF-PRUNING

mtDNA transmission is the female analogue of "male surname transmission" in many human societies: Progeny of both sexes inherit mitochondria from their mothers, but only daughters subsequently transmit mtDNA to future generations. Thus, mtDNA (and surnames) are examples of asexually transmitted traits within otherwise sexually reproducing species. Realistic statistical models of mtDNA evolution must accommodate this mode of inheritance;

they must also somehow account for the empirical rarity of major mtDNA phylogenetic gaps within local populations (category II, Figure 4), and the common occurrence of monophyletic groupings among allopatric populations (categories I and III, Figure 4). The method of applying generating functions to the distributions of family size in a branching process (51) is a relevant probabilistic approach that has been used to study the dynamics of surnames (65, 66, 82) as well as mtDNA lineages (15, 73).

Assume, for example, that adult females within a population produce daughters according to a Poisson distribution with mean μ. The probability of loss of a given female lineage after one generation (or the proportion of such lineages lost from the population) is then $e^{-\mu}$, and the probability of loss after G generations is given by the generating function $p_G = e^{\mu(x-1)}$, where x equals the probability of loss in the previous generation (82). (Generating functions are also available for other parametric family size distributions, such as the binomial.) In the Poisson case, if mothers leave on average one surviving daughter, about 37% of the maternal lineages will by chance go extinct in the first generation, and less than 2% of the original mothers will likely have successfully contributed mtDNA molecules to the population 100 generations later.

Avise et al (15) used an extension of this approach to estimate probabilities (π) of survival of *two or more* independent mtDNA lineages through time. In the Poisson situation, with μ (and hence also the variance, v) equal to 1.0, within about $4n$ generations all individuals within a stable-sized population begun with n females will with high probability trace maternal ancestries to a single foundress (Figure 7); and the times in generations to intermediate levels of π are roughly n to $2n$. That is, all mtDNA sequence differences would almost certainly have arisen less than $4n$ generations earlier and more probably within n to $2n$ generations. Lineage sorting can be much more rapid than this when the variance in progeny numbers across females is greater. For example, in computer simulations where females produced daughters according to a negative binomial distribution with $\mu = 1.0$ and $v = 5.0$, individuals *invariably* stemmed from a single female ancestor less than $2n$ generations earlier (Figure 7). The variances in progeny survival between families are probably large in many species.

In general then, stochastic mtDNA lineage extinction within a population is expected to occur at a (counterintuitively) rapid pace, with the net effect of continually truncating the frequency spectrum of times to common mtDNA ancestry. In other words, due simply to the stochastic lineage turnover associated with the vagaries of reproduction, mtDNA evolutionary trees are continually "self-pruning." This line of reasoning may largely account for limited mtDNA sequence divergence values (e.g. $\bar{p}$ usually much less then about 0.01) observed within local populations, or within entire species char-

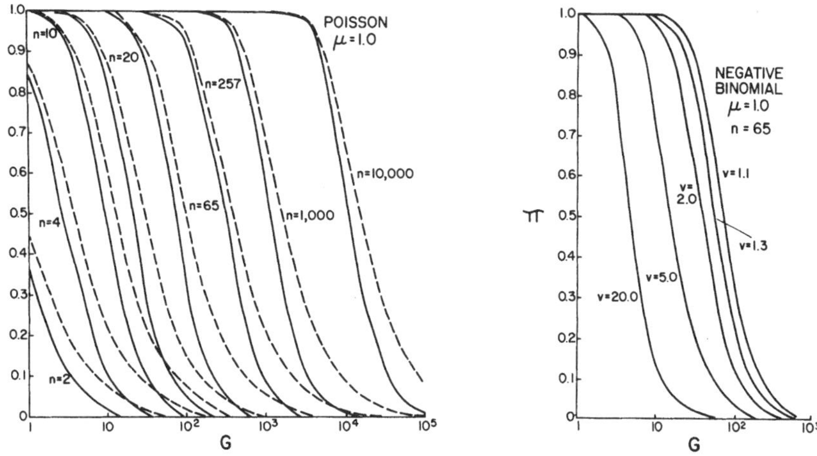

Figure 7 left: Solid lines are theoretical probabilities (π) of survival of two or more mtDNA lineages through G generations within populations founded by n females producing daughters according to a Poisson distribution with mean 1.0. Dashed lines are conditional probabilities (π_c) that two or more lineages survive, given that the population remains extant. Right: π values within populations founded by n = 65 females producing daughters according to a negative binomial distribution with mean 1.0 and variances (v) ranging from 1.1 to 20.0 (from 15).

acterized by historically high levels of gene flow and/or recent expansion from a single refugium.

On the other hand, the empirical existence of major mtDNA phylogenetic gaps within so many species (Table 1) also implies the operation of processes acting dramatically to inhibit extinction of some mtDNA clades, such that much larger genetic differences than those normally observed *within* populations have had time to accumulate. Since such phylogenetic gaps are almost invariably observed between allopatric populations, it seems reasonable that long-term population isolation is responsible. Suppose a particular species has been subdivided historically into two or more spatially isolated populations. Although the genetic distances between lineages within each population will be limited by the balance between the rate of novel mtDNA mutations and female lineage extinction, at least one mtDNA lineage per extant population will be retained indefinitely, and the distinction between these mtDNA lineages could be no less than that which had accumulated since the time of the original population separation.

Neigel & Avise (73) have, by computer simulation of branching processes, formalized these latter ideas and couched them in the language of systematics. (Our models were developed for "species differences" in mtDNA, but they apply equally well to expected relationships between spatially isolated conspecific populations). Suppose that from ancestral stock, populations A and B have separated recently (less than about n generations earlier, where n is the carrying capacity of females in each daughter population). Then because of stochastic mtDNA lineage sorting at and subsequent to population separation, it is likely that some individuals within A are in reality more closely related (i.e. have shared a female ancestor more recently) to some individuals of B than they are to other members of A; and conversely, some members of B are phylogenetically closer to some A's than to some B's. *Populations* A and B could thus be said to be polyphyletic in matriarchal ancestry. However, through time mtDNA lineage extinction continues inexorably within populations, such that after about $3n$ to $4n$ generations, populations A and B would with high probability appear monophyletic with respect to one another. At intermediate times of separation from ancestral stock, populations A and B could reasonably be expected to exhibit a paraphyletic relationship. Figure 8 plots results of a typical computer simulation summarizing probabilities of poly-, para-, and monophyly of isolated populations as a function of time since population separation. These times were observed under the Poisson distribution of family size. For larger variances in progeny numbers (or when founder effect is severe), expected times to reciprocal monophyly of populations would be even lower. Thus, particularly for species composed of many small demes (and limited gene flow between them), allopatric populations should often be monophyletic in matriarchal ancestry. Whether or not they would appear to be so in an mtDNA survey might well depend on the level of laboratory effort expended (i.e. the number of restriction sites assayed) in the search for synapomorphic character states. The larger phylogenetic gaps (category I, Figure 4) should occur between demic arrays

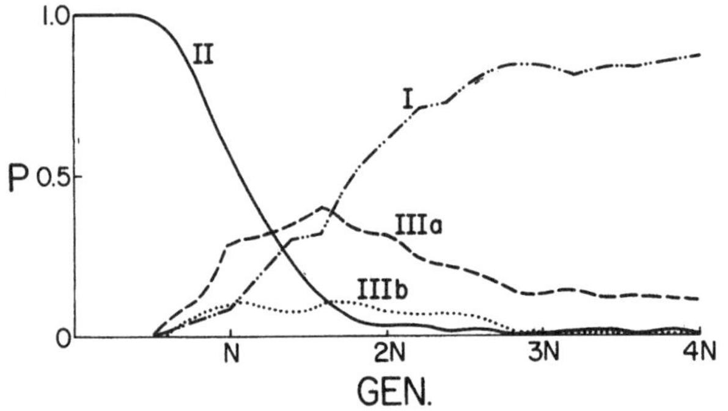

Figure 8 Example of results from computer simulations monitoring probabilities (P) of monophyly (curve I), polyphyly (II) and paraphyly (III a,b) of two isolated populations G generations after their separation from ancestral stock (from 73). In each of 400 replicate simulations, the daughter populations were founded by 300 and 200 individuals, respectively, drawn at random from an ancestral stock. The daughter populations were subsequently density regulated at carrying capacities k = 300 and k = 200 by constraining the mean number of female progeny per mother as follows: $\mu = e^{(k-n)/k}$. N is equal to 300.

isolated for especially long periods of time and should be easiest to detect (as well as most strongly supported statistically by procedures such as bootstrapping; Figure 3C).

Can these models be related to observed rates of mtDNA evolution and census population sizes in more concrete fashion? From empirical experience, individuals in localized geographic areas, and in entire species in phylogeographic category IV, usually show estimates of mtDNA nucleotide sequence divergence much less than about $p = 0.008$. Using Brown et al's (28) conventional mtDNA clock calibration, this implies an *upper bound* on times since common female ancestry of about 400,000 years. This would be roughly compatible with expectations for a population (or set of populations well inteconnected by gene flow) of perhaps $n \cong 400{,}000$ females, provided the population has a generation length of 1 year, has been fairly stable in size, and has μ and v near 1.0. But because branching process theory yields only probabilistic outcomes, and because mtDNA lineage survivorship is likely to have a large stochastic component, it would not at all be surprising for that same population to have been of size anywhere from, say, $n = 200{,}000$ to 1,000,000 or more (Figure 7). And some very different demographic scenarios would not be ruled out. For example, absolute population size could have been vastly larger throughout much of the evolution of the species, but by chance, two mtDNA lineages dating to 400,000 years ago happened to squeeze through more recent bottlenecks in population size. Thus, evolutionary reconstructions regarding population size and times of ancestry should be presented with due caution; and for most populations, we may never have direct and detailed knowledge of historical demography against which to evaluate possible inferences from present-day mtDNA diversities.

Nonetheless, unless the rate of mtDNA evolution is anomalously high in particular populations, the major phylogeographic gaps observed in many species strongly suggest long times since common female ancestry for some conspecifics—much longer than is observed empirically *within* local populations. For example, from the mtDNA data for the eastern versus western monophyletic groupings of the redear sunfish *Lepomis microlophus* (Figure 6), mean population separation (corrected for within-region divergence) occurred about 4 million years ago, and some mtDNA lineages within the species may date to as much as 5 million years B P (20). Similar values apply to allopatric clades within several other species in Table 1.

ECOGEOGRAPHY AND PHYLOGEOGRAPHY

Data on within-species variability in mtDNA thus lend themselves to examination from two vantages: (*a*) phylogenetic interrelationships among the mtDNA molecules themselves and (*b*) geographic distributions of the phylogenetic groupings. Jointly, these elements constitute concerns of a discipline that might be termed *intraspecific phylogeography*.

Notwithstanding occasional examples of concern with the influence of historical population subdivision in shaping genetic architecture at the intraspecific level (e.g. 2, 75), attention seems more conventionally to have been focused on possible adaptive explanations (the "adaptationist paradigm") for geographic differences in attributes such as morphology or behavior (47). One line of evidence for this preoccupation has been the formulation of several "ecogeographic rules" summarizing recognizable trends in presumed adaptive responses to geographically varying environmental conditions (23). For example, Bergmann's rule notes a tendency in homeotherms for larger body sizes at higher latitudes (presumably a surface/volume adaptation for heat conservation in colder climates); Allen's rule notes a latitudinal trend in lengths of limbs (shorter extremities may similarly conserve heat in cold climates); and Gloger's rule notes a tendency for populations in humid areas to be more heavily pigmented (probably a manifestation of selection for background-matching related to predation and competition). While these and other ecogeographic rules at best represent general trends with many exceptions (98), they have been provocative and informative constructs. In this same spirit, we want to suggest several *phylogeographic hypotheses* that may serve as a stimulus for further considerations of geographic trends in intraspecific phylogeny.

We take it as axiomatic that the extended pedigree within any species constitutes its intraspecific phylogeny and that genes transmitted through this pedigree can in principle provide genealogical tracings of hereditary history. As emphasized, data from mtDNA allow estimation of one specified component of the pedigree—the matriarchal phylogeny. Thus, the historical picture recorded in mtDNA is far from a complete characterization of intraspecific phylogeny, and that picture may be especially distorted if males and females differ in phylogeographically relevant characteristics, such as variances in progeny numbers or levels of dispersal (87, 88). Yet techniques of mtDNA assay have provided the first extensive and readily accessible data in the form of "gene genealogies" at the intraspecific level. The following phylogeographic hypotheses are motivated by the mtDNA data and theory currently available and are offered within that context.

Phylogeographic Hypotheses

(*a*) *Most species are composed of geographic populations whose members occupy different branches of an intraspecific, phylogenetic tree*. Such geographic partitioning of phylogenetic branches can be termed phylogeographic population structure. The magnitude of genetic distance between branches can range from small to great, but not uncommonly, geographic clades are distinguished by large phylogenetic gaps or breaks.

(*b*) *Species with limited phylogeographic population structure have life histories conducive to dispersal and have occupied ranges free of firm impediments to gene flow.* Such species have had a relative fluidity of geographic movement over recent evolutionary time and may be especially common in certain groups such as flying insects, birds, and marine fishes, or in species such as the human that have expanded recently from a single refugium. Genealogical distances within such species are constrained because of the inevitable extinction of lineages expected within populations behaving as a single demographic unit in evolution.

(*c*) *Monophyletic groups distinguished by large phylogenetic gaps usually arise from long-term extrinsic (zoogeographic) barriers to gene flow.* Since reproduction leads to a continual turnover of lineages, isolated populations should evolve through time to a condition of reciprocal monophyly, and the time of isolation should be positively correlated (all else being equal) with the magnitude of genealogical differentiation. This hypothesis has a series of corollaries that also serve as predictions for further empirical tests of the expectation:

(i) *As time since isolation increases, the degree of phylogeographic concordance across separate gene genealogies increases.* That is, phylogenetic differentiation between long-isolated populations (either in refugia or in situ) should be reflected in appropriate assays of numerous nuclear as well as cytoplasmic genes.

(ii) *The geographic placements of phylogenetic gaps are concordant across species.* That is, long-term barriers to gene flow should tend to mold the intraspecific genetic architectures of species with similar life histories in geographically concordant fashion.

(iii) *Phylogenetic gaps within species are geographically concordant with boundaries between traditionally recognized zoogeographic provinces.* That is, to the extent that biogeographic provinces reflected in species' distributional limits exist because of environmental barriers to gene flow, such barriers may also tend to result in geographic concentrations of boundaries between well-differentiated clades within species.

Whether or not these hypotheses are confirmed with additional data, we feel that concern with intraspecific phylogeography should assume a place in evolutionary study at least commensurate with ecogeography. Indeed, ecogeography will also benefit from this new enterprise. Let us give two empirical examples. In the deer mouse *Peromyscus maniculatus*, mammalogists have recognized two distinct morphotypes—a long-tailed, long-eared form typically associated with forest environments, and a short-tailed, short-eared form more characteristic of grasslands (21). Data from mtDNA clearly indicate that at least with respect to matriarchal ancestry, these morphotypes do not constitute separate evolutionary clades (61). The extensive mtDNA phylogenetic structure in *P. maniculatus* across North America is strongly oriented to geography and bears no consistent relationship to these morphological distinctions. Such findings add support to earlier suggestions that the ear and tail length differences represent selection-driven responses to ecological challenges posed by forests and grassland and have arisen more than once in separate evolutionary lines. For a counterexample, in the bluegill sunfish *Lepomis macrochirus,* ichthyologists have also recognized two distinct morphological and physiological forms (56). In this case the morphological "races" proved to belong to highly divergent branches in an intraspecific evolutionary tree (Figure 5; 11, 12). This of course in no way excludes natural selection as a possible factor influencing the evolution of these racial differences.

In a recent review of geographic variation in allozymes, Selander & Whittam (79) concluded: "studies of protein polymorphisms indicate that a great variety of organisms, ranging from bacteria to humans . . ., are strongly structured genetically and that their evolution cannot be understood without reference to this structure." Data from mtDNA have revealed an even greater degree of population structure for many species. But more importantly, the nature of assayable mtDNA differences has allowed relatively unambiguous documentation of a strong phylogenetic component to geographic differentiation. Most species have a rich phylogeographic diversity characterized by localized clades and, not infrequently, important phylogenetic gaps between allopatric populations. Many mtDNA lineages within species date to common ancestors several million years BP. Thus, no longer will it be defensible to consider species as phylogenetically monolithic entities in scenarios of speciation or macroevolution. Phylogenetic differences within species are qualitatively of the same kind as, though often smaller in magnitude than, those normally pictured in higher-order phylogeny reconstructions. To paraphrase and update the statement by Selander & Whittam quoted above: Studies of mtDNA polymorphisms indicate that a great many species are strongly structured phylogenetically and that their evolution cannot be fully understood without references to this intraspecific phylogeographic structure.

SUMMARY

Mitochondrial DNA has provided the first extensive and readily accessible data available to evolutionists in a form suitable for strong genealogical inference at the intraspecific level. The rapid pace of mtDNA nucleotide substitution, coupled with the special mode of maternal nonrecombining mtDNA inheritance, offers advantages for phylogenetic analysis at the microevolutionary level that will not be matched easily by any nuclear gene system. These peculiarities of mtDNA data have literally forced the addition of a phylogenetic perspective to studies of intraspecific evolutionary process and as such have provided an empirical and conceptual bridge between the nominally rather separate disciplines of systematics and population genetics.

MtDNA has also served to clarify thinking about the distinction between (yet relevance of) gene genealogies to organismal phylogeny.

Many species have proved to exhibit a deep and geographically structured mtDNA phylogenetic history. Study of the relationship between genealogy and geography constitutes a discipline that can be termed intraspecific phylogeography. We present several phylogeographic hypotheses that were motivated by available data and that represent possible trends whose broader generality remains to be tested. Study of intraspecific phylogeography should assume a place in evolutionary biology at least commensurate with that of ecogeography, with mutual benefit resulting to both disciplines. Theories of speciation and macroevolution must now recognize and accommodate the reality of phylogeographic differentiation at the intraspecific level.

ACKNOWLEDGMENTS

We wish to thank Dr. Bob Lansman for introducing us to mitochondrial DNA. John C. Avise's laboratory has been supported by grants from NSF. Publication costs were funded by contract DE-AC09-76SR00-819 between the US Department of Energy and the University of Georgia Institute of Ecology.

Literature Cited

1. Abbott, R. 1957. The tropical western Atlantic province. *Proc. Phila. Shell Club* 1:7–11
2. Ammerman, A. J., Cavalli-Sforza, L. L. 1984. *The Neolithic Transition and the Genetics of Populations in Europe*. Princeton, NJ: Princeton Univ. Press
3. Aquadro, C. F., Greenberg, B. D. 1983. Human mitochondrial DNA variation and evolution: Analysis of nucleotide sequences from seven individuals. *Genetics* 103:287–312
4. Aquadro, C. F., Kaplan, N., Risko, K. J. 1984. An analysis of the dynamics of mammalian mitochondrial DNA sequence evolution. *Molec. Biol. Evol.* 1:423–34
5. Asmussen, M. A., Arnold, J., Avise, J. C. 1987. Definition and properties of disequilibrium statistics for associations between nuclear and cytoplasmic genotypes. *Genetics*. 115:755–68
6. Avise, J. C. 1983. Protein variation and phylogenetic reconstruction. In *Protein Variation: Adaptive and Taxonomic Significance*, ed. G. Oxford, D. Rollinson, pp. 103–30. London: Syst. Assoc. Publ. Br. Mus. Nat. History
7. Avise, J. C. 1986. Mitochondrial DNA and the evolutionary genetics of higher animals. *Phil. Trans. R. Soc. Lond.* B312:325–42
8. Avise, J. C. 1987. Identification and interpretation of mitochondrial DNA stocks in marine species. In *Proc. Stock Identification Workshop*, ed. H. Kumpf, E. L. Nakamura, pp. Panama City, FL: Publ. Natl. Oceanogr. Atmos. Admin. In press
9. Avise, J. C., Ayala, F. J. 1975. Genetic change and rates of cladogenesis. *Genetics* 81:757–73
10. Avise, J. C., Lansman, R. A. 1983. Polymorphism of mitochondrial DNA in populations of higher animals. In *Evolution of Genes and Proteins*, ed. M. Nei, R. K. Koehn, pp. 147–64. Sunderland, Mass: Sinauer
11. Avise, J. C., Smith, M. H. 1974. Biochemical genetics of sunfish. I. Geographic variation and subspecific intergradation in the bluegill, *Lepomis macrochirus*. *Evolution* 28:42–56
12. Avise, J. C., Bermingham, E., Kessler, L. G., Saunders, N. C. 1984. Characterization of mitochondrial DNA variability in a hybrid swarm between subspecies of bluegill sunfish *(Lepomis macrochirus)*. *Evolution* 38:931–41
13. Avise, J. C., Giblin-Davidson, G.,

Laerm, J., Patton, J. C., Lansman, R. A. 1979. Mitochondrial DNA clones and matriarchal phylogeny within and among geographic populations of the pocket gopher, *Geomys pinetis*. *Proc. Natl. Acad. Aci. USA* 76:6694–98
14. Avise, J. C., Helfman, G. S., Saunders, N. C., Hales, L. S. 1986. Mitochondrial DNA differentiation in North Atlantic eels: Population genetic consequences of an unusual life history pattern. *Proc. Natl. Acad. Sci. USA* 83:4350–54
15. Avise, J. C., Neigel, J. E., Arnold, J. 1984. Demographic influences on mitochondrial DNA lineage survivorship in animal populations. *J. Mol. Evol.* 20:99–105
16. Avise, J. C., Reeb, C. A., Saunders, N. C. 1987. Geographic population structure and species differences in mitochondrial DNA of mouthbrooding marine catfishes (Ariidae) and demersal spawning toadfishes (Batrachoididae). *Evolution*. In press
17. Avise, J. C., Shapira, J. F., Daniel, S. W., Aquadro, C. F., Lansman, R. A. 1983. Mitochondrial DNA differentiation during the speciation process in *Peromyscus*. *Mol. Biol. Evol.* 1:38–56
18. Ayala, F. J., ed. 1976. *Molecular Evolution*. Sunderland, Mass: Sinauer
19. Bermingham, E., Avise, J. C. 1986. Molecular zoogeography of freshwater fishes in the southeastern United States. *Genetics* 113:939–65
20. Bermingham, E., Lamb, T., Avise, J. C. 1986. Size polymorphism and heteroplasmy in the mitochondrial DNA of lower vertebrates. *J. Heredity* 77:249–52
21. Blair, W. F. 1950. Ecological factors in speciation of *Peromyscus*. *Evolution* 4:253–75
22. Briggs, J. C. 1974. *Marine Zoogeography*. New York: McGraw-Hill
23. Brown, J. H., Gibson, A. C. 1983. *Biogeography*. St. Louis, Mo: C. V. Mosby
24. Brown, W. M. 1980. Polymorphism in mitochondrial DNA of humans as revealed by restriction endonuclease analysis. *Proc. Natl. Acad. Sci. USA* 77:3605–9
25. Brown, W. M. 1983. Evolution of animal mitochondrial DNA. In *Evolution of Genes and Proteins*, ed. M. Nei, R. K. Koehn, pp. 62–88. Sunderland, Mass: Sinauer
26. Brown, W. M. 1985. The mitochondrial genome of animals. In *Molecular Evolutionary Genetics*, ed. R. J. MacIntyre pp. 95–130. New York: Plenum
27. Brown, W. M., Goodman, H. M. 1979. Quantitation of intrapopulation variation by restriction analysis of human mitochondrial DNA. In *Extrachromosomal DNA*, ed. D. J. Cummings, P. Borst, I. B. Dawid, S. M. Weissman, C. F. Fox, pp. 485–500. New York: Academic
28. Brown, W. M., George, M. Jr., Wilson, A. C. 1979. Rapid evolution of animal mitochondrial DNA. *Proc. Natl. Acad. Sci. USA* 76:1967–71
29. Brown, W. M., Prager, E. M., Wang, A., Wilson, A. C. 1982. Mitochondrial DNA sequences of primates: Tempo and mode of evolution. *J. Mol. Evol.* 18:225–39
30. Cann, R. L., Wilson, A. C. 1983. Length mutations in human mitochondrial DNA. *Genetics* 104:699–711
31. Cann, R. L., Brown, W. M., Wilson, A. C. 1982. Evolution of human mitochondrial DNA: A preliminary report. In *Human Genetics, Part A, The unfolding genome*, ed. B. Bonné-Tamir, P. Cohen, and R. N. Goodman, pp. 157–65. New York: Liss
32. Cann, R. L., Brown, W. M., Wilson, A. C. 1984. Polymorphic sites and the mechanism of evolution of human mitochondrial DNA. *Genetics* 106:479–99
33. Cann, R. L., Stoneking, M., Wilson, A. C. 1987. Mitochondrial DNA and human evolution. *Nature* 325:31–36
34. Chapman, R. W., Stephens, J. C., Lansman, R. A., Avise, J. C. 1982. Models of mitochondrial DNA transmission genetics and evolution in higher eucaryotes. *Genet. Res.* 40:41–57
35. Crow, J. F., Kimura, M. 1970. *An Introduction to Population Genetic Theory*. New York: Harper & Row
36. Dobzhansky, T. 1937. *Genetics and the Origin of Species*. New York: Columbia Univ. Press
37. Dolbeer, R. A. 1978. Movement and migration patterns of red-winged blackbirds: A continental overview. *Bird Banding* 49:17–34
38. Eldrege, N., Gould, S. J. 1972. Punctuated equilibria: An alternative to phyletic gradualism. In *Models in Paleobiology*, ed. T. J. M. Schopf, pp. 82–115. San Francisco, Calif: Freeman, Cooper
39. Falconer, D. S. 1981. *Introduction to Quantitative Genetics* New York: Longman. 2nd ed.
40. Felsenstein, J. 1982. Numerical methods for inferring evolutionary trees. *Q. Rev. Biol.* 57:379–404
41. Felsenstein, J. 1985. Confidence limits on phylogenies: An approach utilizing the bootstrap. *Evolution* 39:783–91

42. Fink, W. L. 1986. Microcomputers and phylogenetic analysis. *Science* 234: 1135–39
43. George, M. Jr, Ryder, O. A. 1986. Mitochondrial DNA evolution in the genus *Equus*. *Mol. Biol. Evol.* 3:535–46
44. Gillespie, J. H. 1986. Rates of molecular evolution. *Ann. Rev. Ecol. Syst.* 17:637–65
45. Goldschmidt, R. 1940. *The Material Basis of Evolution*. New Haven, Conn: Yale Univ. Press
46. Gould, S. J. 1980. Is a new and general theory of evolution emerging? *Paleobiology* 6:119–30
47. Gould, S. J., Lewontin, R. C. 1979. The spandrels of San Marco and the Panglossian paradigm: A critique of the adaptationist programme. *Proc. Royal Soc. London Ser. B* 205:581–98
48. Graves, J. E., Ferris, S. D., Dizon, A. E. 1984. Close genetic similarity of Atlantic and Pacific skipjack tuna *(Katsuwonus pelamis)* demonstrated with restriction endonuclease analysis of mitochondrial DNA. *Mar. Biol.* 79:315–19
49. Grivell, L. A. 1983. Mitochondrial DNA. *Sci. Am.* 248:78–89
50. Gyllensten, U., Wharton, D., Wilson, A. C. 1985. Material inheritance of mitochondrial DNA during backcrossing of two species of mice. *J. Hered.* 76:321–24
51. Harris, T. 1963. *The Theory of Branching Processes*. Berlin: Springer-Verlag
52. Hauswirth, W. W., Laipis, P. J. 1985. Transmission genetics of mammalian mitochondria: A molecular model and experimental evidence. In *Achievements and Perspectives of Mitochondrial Research,* Vol. II: *Biogenesis,* ed. E. Quagliariello, E. C. Slater, F. Palmieri, C. Saccone, A. M. Kroon, pp. 49–60. New York: Elsevier
53. Hecht, M. K., Hoffman, A. 1986. Why not neo-Darwinism? A critique of paleobiological challenges. *Oxford Surv. Evol. Biol.* 3:1–47
54. Hedrick, P. W. 1983. *Genetics of Populations*. New York: Van Nostrand Reinhold
55. Hennig, W. 1966. *Phylogenetic Systematics*. Urbana: Univ. Ill. Press
56. Hubbs, C. L., Allen, E. R. 1944. Fishes of Silver Springs, Florida. *Proc. Fla. Acad. Sci.* 6:110–30
57. Johnson, M. J., Wallace, D. C., Ferris, S. D., Rattazzi, M. C., Cavalli-Sforza, L. L. 1983. Radiation of human mitochondrial DNA types analyzed by restriction endonuclease cleavage patterns. *J. Mol. Evol.* 19:255–71
58. Kimura, M., Ohta, T. 1971. *Theoretical Aspects of Population Genetics*. Princeton, NJ: Princeton Univ. Press
59. King, J. L., Jukes, T. H. 1969. Non-Darwinian evolution: Random fixation of selectively neutral mutations. *Science* 164:788–98
60. Lansman, R. A., Avise, J. C., Huettel, M. D. 1983. Critical experimental test of the possibility of "paternal leakage" of mitochondrial DNA. *Proc. Natl. Acad. Sci. USA* 80:1969–71
61. Lansman, R. A., Avise, J. C., Aquadro, C. F., Shapira, J. F., Daniel, S. W. 1983. Extensive genetic variation in mitochondrial DNAs among geographic populations of the deer mouse, *Peromyscus maniculatus*. *Evolution* 37:1–16
62. Lansman, R. A., Shade, R. O., Shapira, J. F., Avise, J. C. 1981. The use of restriction endonucleases to measure mitochondrial DNA sequence relatedness in natural populations. III. Techniques and potential applications. *J. Mol. Evol.* 17:214–26
63. Lewontin, R. C. 1974. *The Genetic Basis of Evolutionary Change*. Columbia Univ. Press, NY.
64. Li, C. C. 1955. *Population Genetics*. Chicago: Univ. Chicago Press
65. Lotka, A. J. 1931. Population analysis—the extinction of families. I. *J. Wash. Acad. Sci.* 21:453–59
66. Lotka, A. J. 1931. Population analysis—the extinction of families. II. *J. Wash. Acad. Sci.* 21:453–59
67. Mayr, E. 1963. *Animal Species and Evolution*. Cambridge, Mass: Belknap Press, Harvard
68. Mayr, E. 1982. Speciation and macroevolution. *Evolution* 36:1119–32
69. Moritz, C., Brown, W. M. 1986. Tandem duplication of D-loop and ribosomal RNA sequences in lizard mitochondrial DNA. *Science* 233:1425–27
70. Nei, M. 1975. *Molecular Population Genetics and Evolution*. New York: Elsevier
71. Nei, M. 1987. *Molecular Evolutionary Genetics*. New York: Columbia Univ. Press
72. Nei, M., Li, W.-H. 1979. Mathematical model for studying genetic variation in terms of restriction endonucleases. *Proc. Natl. Acad. Sci. USA* 76:5269–73
73. Neigel, J. E., Avise, J. C. 1986. Phylogenetic relationships of mitochondrial DNA under various demographic models of speciation. In *Evolutionary Processes and Theory,* ed. E. Nevo, S. Karlin, pp. 515–34. New York: Academic

74. O'Brien, T. W., Denslow, N. D., Harville, T. O., Hessler, R. A., Matthews, D. E. 1980. Functional and structural roles of proteins in mammalian mitochondrial ribosomes. In *The Organization and Expression of the Mitochondrial Genome,* ed. A. M. Kroon, C. Saccone, pp. 301–305. New York: Elsevier
75. Ochman, H., Jones, J. S., Selander, R. K. 1983. Molecular area effects in *Cepaea. Proc. Natl. Acad. Sci. USA* 80:4189–93
76. Rand, D. M., Harrison, R. G. 1986. Mitochondrial DNA transmission genetics in crickets. *Genetics* 114:955–70
77. Saunders, N. C., Kessler, L. G., Avise, J. C. 1986. Genetic variation and geographic differentiation in mitochondrial DNA of the horseshoe crab, *Limulus polyphemus. Genetics* 112:613–27
78. Schweyen, R. J., Wolf, K., Kaudewitz, F. eds. 1983. *Mitochondria 1983, Nucleo-Mitochondrial Interactions.* New York: de Gruyter Publ.
79. Selander, R. K., Whittam, T. S. 1983. Protein polymorphism and the genetic structure of populations. In *Evolution of Genes and Proteins,* ed. M. Nei, R. K. Koehn, pp. 89–114. Sunderland, Mass: Sinauer
80. Simpson, G. G. 1944. *Tempo and Mode in Evolution.* New York: Columbia Univ. Press
81. Sneath, P. H. A., Sokal, R. R. 1973. *Numerical Taxonomy.* San Francisco: W. H. Freeman
82. Spiess, E. B. 1977. *Genes in Populations.* New York: Wiley
83. Stebbins, G. L., Ayala, F. J. 1981. Is a new evolutionary synthesis necessary? *Science* 213:967–71
84. Swift, C. C., Gilbert, C. R., Bortone, S. A., Burgess, G. H., Yerger, R. W. 1986. Zoogeography of the freshwater fishes of the southeastern United States: Savannah River to Lake Ponchartrain. In *Zoogeography of North American Freshwater Fishes,* ed. C. H. Hocutt, E. O. Wiley, pp. 213–65. New York: Wiley
85. Tajima, F. 1983. Evolutionary relationship of DNA sequences in finite populations. *Genetics* 105:437–60
86. Takahata, N. 1985. Population genetics of extranuclear genomes: A model and review. In *Population Genetics and Molecular Evolution,* ed. T. Ohta, K. Aoki, pp. 195–212. Berlin: Springer-Verlag
87. Takahata, N., Palumbi, S. R. 1985. Extranuclear differentiation and gene flow in the finite island model. *Genetics* 109:441–57
88. Takahata, N., Slatkin, M. 1984. Mitochondrial gene flow. *Proc. Natl. Acad. Sci. USA* 81:1764–67
89. Templeton, A. R. 1983. Phylogenetic inference from restriction endonuclease cleavage site maps with particular reference to the evolution of man and the apes. *Evolution* 37:221–44
90. Templeton, A. R. 1983. Convergent evolution and nonparametric inferences from restriction data and DNA sequences. In *Statistical Analysis of DNA Sequence Data,* ed. B. S. Weir, pp. 151–79. New York: Marcel Dekker
91. Uphold, W. B. 1977. Estimation of DNA sequence divergence from comparison of restriction endonuclease digests. *Nucleic Acids Res.* 4:1257–65
92. Vawter, L., Brown, W. M. 1986. Nuclear and mitochondrial DNA comparisons reveal extreme rate variation in the molecular clock. *Science* 234:194–96
93. Wainscoat, J. 1987. Out of the garden of Eden. *Nature* 325:13
94. Wallace, D. G. 1983. Structure and evolution of organelle DNAs. In *Endocytobiology,* Vol. II, ed. H. E. A. Schenk, W. Schwemmler, pp. 87–100. New York: de Gruyter
95. Whittam, T. S., Clark, A. G., Stoneking, M., Cann, R. L., Wilson, A. C. 1986. Allelic variation in human mitochondrial genes based on patterns of restriction site polymorphism. *Proc. Natl. Acad. Sci. USA* 83:9611–15
96. Wiley, E. O. 1981. *Phylogenetics.* New York: Wiley
97. Wilson, A. C., Cann, R. L., Carr, S. M., George, M. Jr., Gyllensten, U. B., Helm-Bychowski, K. M., Higuchi, R. G., Palumbi, S. R., Prager, E. M., Sage, R. D., Stoneking, M. 1985. Mitochondrial DNA and two perspectives on evolutionary genetics. *Biol. J. Linn. Soc.* 26:375–400
98. Zink, R. M., Remsen, J. V. Jr. 1986. Evolutionary processes and patterns of geographic variation in birds. In *Current Ornithology,* Vol. 4, ed. R. F. Johnston, pp. 1–69. New York: Plenum

Charlesworth, B. 1989. The evolution of sex and recombination. *Trends in Ecology and Evolution* 4:264–267.

Consider two populations of a given species in which population A is parthenogentic and population B reproduces sexually. In population A, all will be females; in population B, only half will be females (assuming a 50–50 split between females and males). If both populations consist of equal number of individuals and if the females are equally fertile in both populations, population A will produce twice as many daughters as population B, which will have also produced an equal number of males. Assuming that there are not environmental limitations, in the next generation population A will produce twice as many progeny because it consists of twice as many females as population B, all of which will be females, that is, four times as many females as population B. The number of progeny will double again in each successive generation in population A relative to population B. The same disadvantage in rate of reproduction will occur when the sexually reproducing and the parthenogenetic females are members of the same population. The sexually reproducing females produce only half as many females as the parthenogenetic females, because half of their progeny are males. One would expect that the sexually reproducing females will rapidly decrease in frequency relative to the parthenogenetic females.

Yet, sexual reproduction is common in multicellular and in many protozoan organisms. How does sexual reproduction persist and, indeed, how is it that it predominates in eukaryotes in spite of its 50 percent disadvantage? A large number of possible answers to this question have been proposed over the past century, some as speculative possibilities, others formulated as hypotheses subject to the possibility of empirical testing. Typically, these hypotheses are difficult to test, because of the generality of both the biological processes involved and the different sorts of organisms. The hypotheses are often differentiated into ecological (or environmental) and genetic (better called "mutational," because even in the case of ecological hypotheses, the processes of adaptation are genetic). In this article, evolutionary geneticist Brian Charlesworth reviews these ideas about the significance of sex and recombination and concludes that a variety of evolutionary factors are probably involved.

RELATED READING

Hartfield, M., and P. D. Keightley. 2012. Current hypotheses for the evolution of sex and recombination. Integrative Zoology 7:192–209.

Maynard Smith, J. 1978. The evolution of sex. Cambridge University Press, Cambridge, Massachusetts, USA.

Michod, R. E., and B. R. Levin, editors. 1988. The evolution of sex: an examination of current ideas. Sinauer, Sunderland, Massachusetts, USA.

West, S. A., C. M. Lively, and A. F. Read. 1999. A pluralistic approach to sex and recombination. Journal of Evolutionary Biology 12:1003–1012.

Williams, G. C. 1975. Sex and evolution. Princeton University Press, Princeton, New Jersey, USA.

TREE vol. 4, no. 9, September 1989

The Evolution of Sex and Recombination

Brian Charlesworth

The evolution and maintenance of sexual reproduction, and the associated process of genetic recombination, are still controversial issues. Two recent books have provided overviews of the ideas and observations in this field. This article reviews some of the major ideas that have been proposed to account for sex and recombination, and comments on the results of attempts at empirical tests. While there is now an impressive body of well-formulated evolutionary models, it has proved hard to discriminate between them, either experimentally or by means of comparative data. It may well be that there is no unitary selective advantage to sex and recombination, but that a variety of forces are involved.

In eukaryotes, both unicellular and multicellular, the majority of species are sexually reproducing, with individuals produced by the fusion of haploid gametes to give a diploid zygote. New gametes are produced by reduction division (meiosis), during which genetic recombination occurs between genes on different chromosomes, as a result of independent segregation, or between genes on the same chromosome, as a result of crossing over. Asexual species in eukaryotes are usually sporadic in their taxonomic distribution, and have clearly identifiable sexual relatives, suggesting that they have originated relatively recently in evolutionary time[1–3]. The most primitive unicellular eukaryotes have a life cycle in which meiosis follows zygote formation, so that individuals are haploid. Advanced eukaryotes defer meiosis, and individuals are multicellular and diploid.

In bacteria, exchange of genetic material may occur between separate individuals as a result of conjugation, transduction and transformation[4]. In DNA viruses, crossing over may take place between two viral genomes that have simultaneously infected the same host cell[5]. Some RNA viruses have substitutes for recombination, e.g. splitting of the genome into a number of separate RNA molecules enclosed in the same protein coat[6]. Thus, even in groups that do not possess the elaborate mechanisms of sexual reproduction characteristic of higher organisms, the possibility of exchange of genetic information between parents exists, with the result that an individual may combine characteristics that were originally present in its two different parents. In this sense, sexuality is an almost universal feature of the living world.

In the absence of recombination, no new combinations of alleles at separate loci can occur in a population, and so the number of different genetic types present in a population is much more limited than when sexual reproduction and recombination take place. The structure of genetic variation in many bacterial populations is such that it is clear that rather little recombination can be taking place naturally: populations usually contain only a few different clonal genotypes with respect to electrophoretic loci[4]. A similar genetic structure (except that individuals are diploid instead of haploid) is found in asexual populations of multicellular organisms, e.g. *Daphnia*[7].

Advantages and disadvantages of sex

Before considering how these genetic consequences of sexual reproduction may be involved in the evolution of sex and recombination, we have to note that asexual reproduction in species where gametes are differentiated into sperm and eggs carries an intrinsic advantage, generating what is known as the 'cost of sex' or 'cost of meiosis'[1,8]. In a population with a 1:1 sex ratio, a line of asexual females with the same fertility as sexual females, but which produces only asexual daughters, will have a two-fold reproductive advantage over sexual females[1]. A sexually reproducing population is thus vulnerable to invasion by mutations causing females to reproduce asexually. Why, then, do sexual species predominate among the eukaryotes?

Similarly, why is genetic recombination maintained in sexual species? Although there is no cost to recombination analogous to that for sex, there is evidence that recombination may often lower fitness, due to the breakdown of favourable gene combinations[1]. Theoretical analysis shows that, under these circumstances, selection will favor reduced crossing over in randomly mating populations[1,9].

There are three possible answers to these questions:

(1) Constraints of development prevent meiosis being effectively replaced by mitosis or its genetic equivalent[10]. In mammals, for instance, there is strong evidence that normal development of the embryo cannot proceed without chromosomes that have been passaged through sperm, so that viable individuals cannot be produced without a paternal contribution[11]. Even if asexuality can be achieved, it is possible that the mechanics of the replacement of meiosis by an essentially mitotic process are so inefficient that the fertility of asexuals is drastically reduced, and they cannot outcompete the sexuals[3]. However, there are many examples in other groups in which successful asexual reproduction has been achieved[1–3], so this cannot be a universal explanation. It also fails to explain crossing over, since there is abundant evidence for genetic variation in crossing over rates in a variety of species[9].

(2) Asexual mutants do arise from time to time and take over the species, causing an evolutionary transition from sexual to asexual reproduction, but the long-term fitness of asexual populations is lower than that of related sexual populations[1], so that they become extinct with higher probability. This species-level advantage explains the fact that asexual species are usually of relatively recent origin[12].

(3) There are individual-level advantages to sex, such that an asexual variant ends up with a lower net fitness than the sexual individuals with which it is competing, and cannot establish itself.

Since the rate of response to

Brian Charlesworth is at the Dept of Ecology and Evolution, University of Chicago, 1103 E. 57th St, Chicago, IL 60637, USA.

selection (natural or artificial) on a character is known to depend on the amount of genetic variation in the population, the fact that sexual reproduction promotes a wider variety of genotypes than asexual reproduction suggests that an advantage to sex may accrue from a greater ability of sexual populations to respond to selection. This was one of the earliest species-level explanations for sex[1,13].

In a recent refinement of this principle, sex and recombination have been shown to have an advantage when the population is exposed to a changing environment, so that new mean values of adaptively significant characters are favoured by selection[14]. Under many circumstances, directional selection acting on a quantitative character leads to lower genetic variance among the individuals surviving the action of selection, compared with the population before selection (see Fig. 1a). This reduction causes negative linkage disequilibrium between the alleles at the different loci affecting the character[15], i.e. an allele at one locus with a favorable effect on the character tends to be associated with an allele with an opposite effect at another locus. If reproduction is asexual, or if there is no genetic recombination, then this reduction in variance is transmitted to the next generation, since genotypes are handed on intact. But if there is sex and recombination, allelic combinations are partially randomized, and the genetic variance increases towards its previous value.

Thus, a higher level of genetic variance is maintained in the presence of recombination, and the population can respond faster to selection. Furthermore, genes that increase the amount of recombination between the loci concerned can increase in frequency, when the environment fluctuates with an appropriate period[14]. The nature of the environmental changes that cause shifts in the selective optimum is left open under this model; they could be either physical or biotic.

A similar effect occurs when deleterious variants are introduced each generation into the population by mutations at a large number of different loci, since there is a perpetual pressure of selection to reduce the number of deleterious genes; the number of mutant alleles per individual can be treated as a quantitative character under selection, and the above argument applied to this case[16,17]. This mechanism, strongly advocated by Kondrashov[18], requires that fitness falls off faster than log-linearly with number of mutations carried by an individual, since this is necessary for the variance in numbers of mutations per individual to be reduced by selection (Fig. 1b). An asexual population always has a higher equilibrium fitness than a sexual one under these conditions[16,17], and a gene that increases the amount of recombination is always selected if the mutation rate per genome is sufficiently high[18].

Other mechanisms have been proposed for the evolutionary advantage of sex, but perhaps seem less plausible than the above two as a general means for promoting recombination throughout the genome, since they require more specialized conditions. One is the occurrence of temporal changes in the environment, such that different combinations of genes are favored in different generations[1,19]. A version of this theory that has attracted especial interest in recent years is the possibility that the coevolution of hosts and parasites may generate fluctuations in the frequencies of gene combinations conferring resistance and pathogenicity, in a way that favors sex and recombination[20]. Detailed modelling of this process has yielded somewhat equivocal results: recombination may be favored under some kinds of model but not under others[21].

Another possibility is the occurrence of spatial variation with respect to which genotypes are favored by selection, accompanied by sibling competition[8]. If different genotypes of offspring produced by the same parent are competing for limited resources in the same patch, then genotypic diversity among the offspring may be advantageous, since it ensures a higher probability of production of a 'winning' genotype[1,8,22]. Alternatively, diversity may be favored if the environment contains different niches for which related individuals are competing[22]. Other models for the evolution of sex, which are more or less closely related to these, are reviewed in Refs 23 and 24.

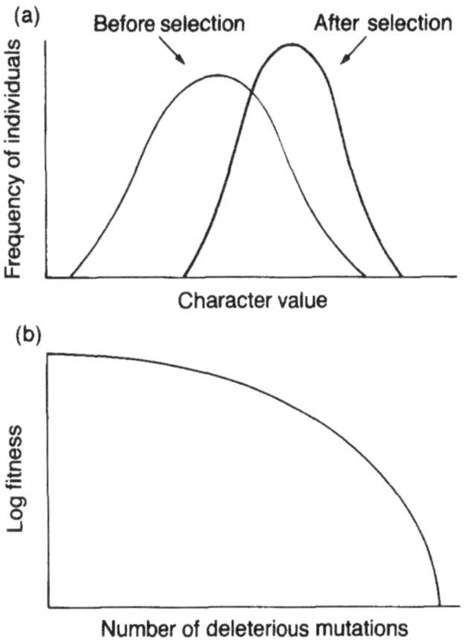

Fig. 1. (a) The effect of directional selection for a larger value of a character, showing the frequency distributions of individuals before and after selection in a single generation. (b) The form of the relationship between the logarithm of fitness and the number of deleterious mutations per individual that generates a decreased variance in number of mutations.

These advantages can be viewed in the context of both species- and individual-level selection. However, calculations show that such advantages are usually smaller than the cost of sex, except possibly for the mechanism involving mutation when the genome is sufficiently large that there is a high rate of occurrence of deleterious mutations per gamete[18]. This does not present a problem for the maintenance of sex by species-level selection, or for the maintenance of genetic recombination, but suggests that sex may not normally be maintained by individual selection, unless asexual variants usually suffer from fertility losses. An exception may be the case of cyclical parthenogenesis, in which there is a life cycle in which several generations of asexual reproduction alternate with a single generation of sexual reproduction, and where the cost of sex can be shown to be small[25].

Testing the hypotheses

What evidence do we have to discriminate between these possible models? The answer is: very little. We know that the environment changes over time, and that the biometrical analysis of fossil lineages reveals irregular fluctuations in the means of quantitative traits that are consistent with selective responses to fluctuating optima[26]. Thus, there is good reason to believe that directional selection on quantitative traits occurs in nature, and so must contribute to selection for increased recombination. Selection for DDT resistance in *Drosophila* has resulted in an increase in the frequency of crossing over[27], in accordance with this idea. Similarly, all loci of functional significance must mutate to deleterious alleles, albeit at a low frequency per locus per generation, and so there is a substantial mutation rate per genome in higher organisms[18]. It is less clear that fitness generally falls off as a function of number of mutant alleles in the fashion required by Kondrashov's theory, although studies of the effects of mutation in *Drosophila* support the postulated faster-than-linear decline[16].

Comparative data on mammalian chiasma frequencies suggest that taxa with late reproduction tend to have higher frequencies of crossing over than those with shorter generation times[28,29]. This has been taken as evidence in support of models of temporal fluctuations in favored gene combinations, due to host–parasite coevolution[28], but the validity of this inference has been questioned[30]. For example, in an environment where the optimum for a quantitative character is changing, species with a long generation time may experience a higher net selection pressure per generation than species with short generation times. A similar problem of interpretation arises in relation to the interesting observation that the internal partners in mutualistic associations tend to abandon sex more often than the external partners[31]. This could be due to the internal partners being exposed to fewer biotic interactions with other species than external partners, or it could be due to their being exposed to less environmental change generally. Finally, experiments on plants to test the sib-competition model have succeeded in demonstrating a substantial fitness advantage to sexually produced offspring over asexuals, but this advantage appears to have little to do with sib competition[32,33].

Repair and recombination

The fact that bacteria have well-developed mechanisms for promoting crossing over between homologous DNA molecules in the same cell, but that sexual reproduction in natural populations of bacteria is infrequent[4], suggests that crossing over did not originally evolve in order to promote recombination[34]. There is very strong genetic and molecular evidence that the enzymes involved in crossing over are also responsible for the repair of damage to DNA[5]. The recA protein of *E. coli* promotes the invasion of intact double-stranded DNA by single strands, an essential step in both repair and crossing over. Synthesis of the recA protein is increased more than ten times following DNA damage. Molecular models of crossing over assume that the process is initiated by breaks in DNA[5]. Since the fossil evidence shows that bacteria were the only type of organism present on earth for more than a billion years, it seems likely that the molecular mechanism of crossing over originated in them, and was related to their need to repair damage to their DNA.

Some have suggested that repair provides a sufficient explanation for sex and recombination in all species[34]. However, it is hard to see why the processes of gamete fusion and meiosis should have evolved if this is all there is to it[35]. Since there can be shown to be little or no cost of sex in a primitive eukaryote species like *Chlamydomonas*, in which gametes are of equal size[1], a modest advantage to sex due to one or more of the processes discussed above could have promoted the evolution of regular sexual reproduction of the eukaryote kind.

Conclusions

Once a system of regular sexual reproduction of the kind found in *Chlamydomonas* was established in evolution, probably about 2 billion years ago, the stage was set for the subsequent elaboration of the machinery of sexual reproduction. This elaboration involves many phenomena of great biological interest, e.g. the differentiation of gametes into male (numerous, small, and motile) and female (few, large, and stationary); the separation of sexes into male and female individuals (as opposed to hermaphrodites that produce both male and female gametes), the evolution of a diploid, multicellular life history phase; the evolution of systems of chromosomal sex determination, with Y chromosomes becoming largely devoid of genetic information.

These processes are rather better understood than the evolution of sexual reproduction itself[1,23], and it may be that there is no single force promoting sex and recombination. This would make it hard to establish the nature of the factors that are involved by the use of the comparative method, especially as the relative advantages and disadvantages of sex may be modulated by ecological factors that have no relation to the forces actively promoting sex and recombination. One such factor is 'reproductive assurance'; in environments where the density of individuals in the species is low, the reproductive success of asexuals may be higher than that of sexual individuals, who have to be fertilized by other individuals in order to breed[3,30,36]. This may explain many of the well-established correlations between asexuality and biotically sparse environments[2,3,36]. Experiments designed to test the validity of the assumptions enshrined in some of the models (e.g. the synergistic effects of mutations on fitness postulated by Kondrashov[18]) may be a more fruitful way of making progress.

References

1 Maynard Smith, J. (1978) *The Evolution of Sex*, Cambridge University Press
2 Bell, G. (1982) *The Masterpiece of Nature*, Croom Helm
3 Bierzychudek, P. (1987) in *The Evolution of Sex and its Consequences* (Stearns, S.C., ed.), pp. 163–174 and 197–217, Birkhaüser
4 Levin, B.R. (1988) in *The Evolution of Sex* (Michod, R.E. and Levin, B.R., eds), pp. 194–211, Sinauer
5 Devoret, R. (1988) in *The Evolution of Sex* (Michod, R.E. and Levin, B.R., eds), pp. 24–44, Sinauer
6 Pressing, J. and Reanney, D.C. (1984) *J. Mol. Evol.* 20, 135–146
7 Hebert, P.D.N., Ward, R.D. and Weider, L.J. (1988) *Evolution* 42, 147–159
8 Williams, G.C. (1975) *Sex and Evolution*, Princeton University Press

9 Brooks, L.D. (1988) in *The Evolution of Sex* (Michod, R.E. and Levin, B.R., eds), pp. 87–105, Sinauer
10 Margulis, L. and Sagan, D. (1986) *Origins of Sex*, Yale University Press
11 Surani, M.A., Reik, W. and Allen, N.D. (1988) *Trends Genet.* 4, 59–62
12 Van Valen, L. (1975) *Evolution* 29, 87–98
13 Mather, K. (1943) *Biol. Rev.* 18, 32–64
14 Maynard Smith, J. (1988) *Genet. Res.* 51, 59–63
15 Bulmer, M.G. (1980) *The Mathematical Theory of Quantitative Genetics*, Oxford University Press
16 Crow, J.F. (1970) in *Mathematical Topics in Population Genetics* (Kojima, K., ed.), pp. 128–177, Springer-Verlag
17 Kondrashov, A.S. (1982) *Genet. Res.* 44, 325–332
18 Kondrashov, A.S. (1988) *Nature* 336, 435–440
19 Sturtevant, A.H. and Mather, K. (1938) *Am. Nat.* 72, 447–452
20 Seger, J. and Hamilton, W.D. (1988) in *The Evolution of Sex* (Michod, R.E. and Levin, B.R., eds), pp. 176–193, Sinauer
21 Bell, G. and Maynard Smith, J. (1987) *Nature* 328, 66–68
22 Young, J.W.P. (1982) *J. Theor. Biol.* 88, 755–756
23 Stearns, S.C., ed. (1987) *The Evolution of Sex and its Consequences*, Birkhaüser
24 Michod, R.E. and Levin, B.R., eds (1988) *The Evolution of Sex*, Sinauer
25 Charlesworth, B. (1980) *J. Theor. Biol.* 87, 517–528
26 Charlesworth, B. (1984) *Paleobiology* 10, 308–318
27 Flexon, P.B. and Rodell, C.F. (1982) *Nature* 298, 672–674
28 Burt, A. and Bell, G. (1987) *Nature* 326, 803–805
29 Sharp, P.J. and Hayman, D.L. (1988) *Heredity* 60, 77–85
30 Charlesworth, B., Greenwood, J.J.D. and Koella, J.C. (1987) *Nature* 330, 116–118
31 Law, R. and Lewis, D.H. (1983) *Biol. J. Linn. Soc.* 20, 249–276
32 Bierzychudek, P. (1987) in *The Evolution of Sex and its Consequences* (Stearns, S.C., ed.), pp. 163–174, Birkhaüser
33 Kelley, S.E., Antonovics, J.A. and Schmitt, J.A. (1988) *Nature* 331, 714–716
34 Bernstein, H., Hopf, F.A. and Michod, R.E. (1988) in *The Evolution of Sex* (Michod, R.E. and Levin, B.R., eds), pp. 139–160, Sinauer
35 Maynard Smith, J. (1988) in *The Evolution of Sex* (Michod, R.E. and Levin, B.R., eds), pp. 106–125, Sinauer
36 Lloyd, D.G. (1980) *Evol. Biol.* 13, 69–110

Hillis, D. M., J. J. Bull, W. E. White, M. R. Badgett, and I. J. Molineux. 1992. Experimental phylogenetics: generation of a known phylogeny. *Science* 255:589–592.

Evolutionary biology is sometimes castigated as being a descriptive rather than an experimental science, and this might seem to be especially true in the evolutionary subdiscipline of phylogenetics. Phylogeneticists attempt to recover the genealogical history of life on Earth using comparative assessments of molecular, morphological, or other traits from extant and extinct species. The success of this endeavor is measured by the extent to which branches in the Tree of Life are discovered and assembled into a coherent and accurate picture of life's diversification. But how can one be sure that a phylogenetic depiction is valid? The traditional approach centers on the degree of concordance across characters. When many independent traits (molecular, morphological, or other) paint the same picture of phylogenetic relationships for a given set of taxa, biologists can be reasonably assured that they are on the correct historical track.

This compelling paper by David Hillis and colleagues is included in this volume for two reasons: (1) it showcases the burgeoning field of molecular phylogenetics, which is a substantial component of modern evolutionary biology, and (2) it demonstrates that even the most descriptive areas of evolutionary biology can be highly critical in an experimental sense, at least on occasion. Hillis et al. use phylogenies reconstructed by means of molecular sequences in order to compare them to previously established phylogenies. At their best, the reconstructed molecular phylogenies were 98.6 percent correct. Not too bad for a descriptive science!

RELATED READING

Hall, B. G. 2001. Phylogenetic trees made easy: a how-to manual for molecular biologists. Sinauer, Sunderland, Massachusetts, USA.

Hillis, D. M., C. Moritz, and B. K. Mable. 1996. Molecular systematics, 2nd ed. Sinauer, Sunderland, Massachusetts, USA.

Nei, M., and S. Kumar. 2000. Molecular evolution and phylogenetics. Oxford University Press, Oxford, England.

Page, R. D. M., and E. C. Holmes. 1998. Molecular evolution: a phylogenetic approach. Blackwell, Oxford, England.

Experimental Phylogenetics: Generation of a Known Phylogeny

DAVID M. HILLIS, JAMES J. BULL, MARY E. WHITE,
MARTY R. BADGETT, IAN J. MOLINEUX

Although methods of phylogenetic estimation are used routinely in comparative biology, direct tests of these methods are hampered by the lack of known phylogenies. Here a system based on serial propagation of bacteriophage T7 in the presence of a mutagen was used to create the first completely known phylogeny. Restriction-site maps of the terminal lineages were used to infer the evolutionary history of the experimental lines for comparison to the known history and actual ancestors. The five methods used to reconstruct branching pattern all predicted the correct topology but varied in their predictions of branch lengths; one method also predicts ancestral restriction maps and was found to be greater than 98 percent accurate.

THE DEVELOPMENT OVER THE PAST four decades of explicit methods for phylogenetic inference (*1*) has permitted biologists to reconstruct the broad outlines of evolutionary history and to interpret comparative biological studies within an evolutionary framework (*2*). However, evolutionary history usually cannot be observed directly, at least over the course of relevant magnitudes of change, so that assessment of phylogenetic methods has relied on numerical simulations. Although simulations have provided considerable insight into the effectiveness of various

Departments of Zoology and Microbiology, University of Texas, Austin, TX 78712.

phylogenetic algorithms, they are limited by an incomplete knowledge of biology: all models incorporate untested assumptions about evolutionary processes.

Direct tests of organismal phylogenetic histories are limited to a few studies of strains of laboratory animals (*3*) and plant cultivars (*4*). Even these cases have shortcomings: the organisms underwent little genetic differentiation, phylogenies were produced over the course of decades or centuries, and the histories are incompletely known. In contrast, viruses can be manipulated in the laboratory through thousands of generations per year, and mutation rates of viruses can easily be elevated through the use of mutagens, so that experimental studies of phylogenies with viruses should be feasible (*5, 6*). We report the creation of a known phylogeny of lineages derived from bacteriophage T7 and provide fine-scale restriction maps of the entire genomes of the experimental lineages, including the ancestors. Our purpose was to test the effectiveness of methods for inferring phylogeny and ancestral genetic character states by comparing the inferred evolutionary history against a known phylogeny and the true ancestors.

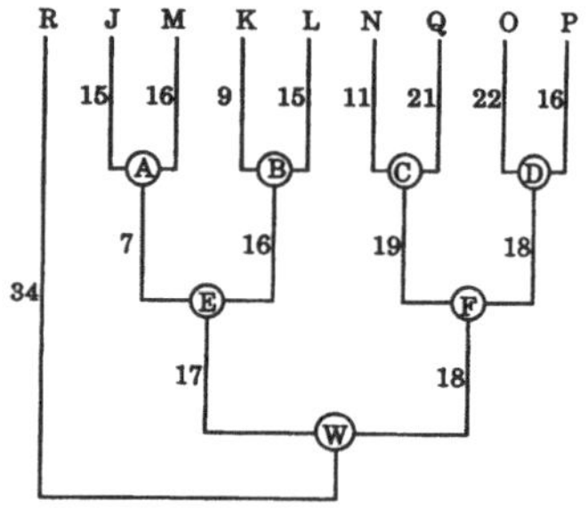

Fig. 1. True phylogeny for the experimental lineages of bacteriophage T7. The ancestors at each node are labeled with letters A through F and W (the latter represents wild-type T7). The numbers represent the number of restriction-site differences scored between the phages at each node of the phylogeny.

We chose to construct a symmetric phylogeny with equal distances among nodes (Fig. 1) (*7*). This topology (tree shape) has proven especially amenable to accurate reconstruction in theoretical studies (*8–10*) and may thus be regarded as a "null" model against which other topologies may be compared. We created a phylogeny of nine taxa (eight ingroup lineages and one outgroup lineage to root the tree) by serially propagating T7 phage in the presence of a mutagen and dividing the lineages at predetermined intervals (*7*); a clonal stock of wild-type T7 phage was the common ancestor of all lineages. There are 135,135 possible bifurcating trees for this many taxa, so the likelihood of inferring the correct phylogeny by chance alone is minimal. We compared the actual phylogeny (Fig. 1) to estimated phylogenies from five reconstruction methods; estimates were based on restriction-site maps produced for 34 restriction endonucleases in all terminal lineages (Figs. 2 and 3). To avoid bias, the actual phylogeny was unknown to the person mapping the restriction sites. We also produced restriction maps for the ancestral phage at each of the nodes of the true phylogeny (Fig. 3). Three aspects of the inferred phylogeny were compared to the actual phylogeny: branching topology, branch lengths, and ancestral states. The five methods of phylogenetic inference evaluated were parsimony (*12*), the Fitch-Margoliash method (*13*), the Cavalli-Sforza method (*14*), neighbor-joining (*15*), and the unweighted pair-group method of arithmetic averages (UPGMA) (*16*).

All methods predicted the correct branching order of the known phylogeny, but no method predicted the actual branch lengths for every branch (*17*). To compare the five methods for their ability to predict branch lengths, the correlation between observed and predicted branch lengths was calculated for each method. These five correlations were significantly heterogeneous, with parsimony yielding the highest value and UPGMA yielding the lowest value (*18*). The UPGMA method is known to be sensitive to unequal rates of change (*1*), and the number of changes per branch was quite variable in the true phylogeny [although the number of changes per ingroup branch is not significantly heterogeneous from the expectation under a Poisson distribution; test from (*19*)].

The experimental system also enabled us to determine ancestral states directly. Of the methods tested, only parsimony makes predictions about ancestral character states (parsimony may be used to optimize states onto phylogenies inferred by other methods, but, for these data, all methods estimated the same branching pattern). In comparing inferred ancestral states to the actual ancestral states, three outcomes are possible: the ancestral states may be (i) correctly inferred, (ii) incorrectly inferred, or (iii) ambiguous (when more than one character optimization is possible). For 202 variable sites assayed in each of seven ancestors, parsimony correctly inferred 1369 states (97.3%), incorrectly inferred 18 states (1.3%), and was ambiguous about 20 states (1.4%). Seven states (of four sites) could not be observed in some ancestors because they fell under deletion mutations (Fig. 3). If the 91 wild-type states that were invariant in all lineages are included, the inferred restriction maps are an average of 98.6% identical to the actual maps (with either delayed or accelerated

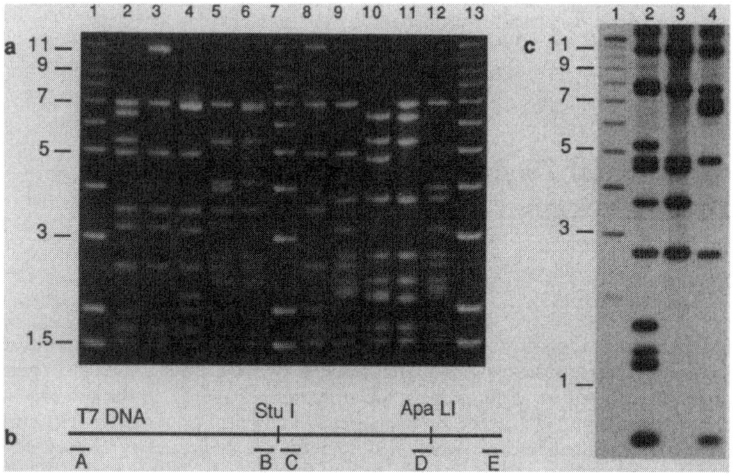

Fig. 2. (**a**) DNA from each terminal lineage cleaved with the restriction enzyme Xmn I, separated by electrophoresis on 0.8% wedge-shaped agarose gels and stained with ethidium bromide. Lanes 1, 7, and 13 contain size standards (sizes shown in kilobase pairs); lane 2 contains wild-type T7 DNA; lanes 3 through 6 contain DNA from lineages R, Q, P, and O, respectively; lanes 8 through 12 contain DNA from lineages N, M, L, K, and J, respectively. (**b**) Strategy for mapping restriction site variation. Viral DNA was cleaved with the restriction enzyme Stu I (which has a single recognition site in T7 that remained invariant in all lineages) and then partially digested with the restriction enzyme to be mapped. After electrophoresis on 0.8% wedge-shaped agarose gels, Southern blots were successively hybridized with four oligonucleotides that matched sequences located at each end of T7 DNA and on either side of the Stu I fragment (A through C and E). To improve resolution of the map for several restriction sites, we cleaved with the restriction enzyme Apa LI (which also has an invariant recognition site in all lineages), partially digested with the enzyme to be mapped, and hybridized the Southern blot with an oligonucleotide located at position D. (**c**) An example of an autoradiogram produced as described in (b), probed with oligonucleotide B. Lane 1 contains size standards (indicated in kilobase pairs); lanes 2 through 4 contain DNA from lineages O, P, and Q partially cleaved with Sau 3A1 (an isoschizomer of Mbo I.

Enzyme/site		J	K	L	M	N	O	P	Q	R	A	B	C	D	E	F	W
BglII*	0.4	-	-	-	-	-	+	-	-	-	-	-	-	(+)	-	-	-
SpeI	0.8	-	-	-	-	-	-	-	-	+	-	-	-	-	-	-	-
MboI	0.8	-	-	-	-	+	+	+	+	-	-	-	+	+	-	(-)	-
MboI	1.0	+	-	-	-	-	-	-	-	-	-	-	-	-	-	-	-
MboI	1.1	-	-	-	-	-	+	-	-	-	-	-	-	(+)	-	-	-
MboI	1.3	-	-	-	-	-	+	+	-	-	-	-	-	+	-	(+)	-
BclI*	1.3	-	-	-	-	-	+	+	-	-	-	-	-	+	-	-	-
HpaI	1.4	-	-	-	-	+	-	-	+	-	-	-	+	-	-	-	-
Del: 1.5–2.7		-	-	-	-	+	-	-	+	-	-	-	+	-	-	-	-
Del: 1.7–2.6		+	+	+	+	-	-	-	-	-	+	+	-	-	+	-	-
Del: 2.0–2.9		-	-	-	-	-	+	+	-	-	-	-	-	+	-	-	-
BglII*	1.5	-	+	-	-	D	-	-	D	+	-	-	D	-	-	-	-
MboI	1.6	-	-	-	-	D	+	-	D	-	-	-	D	-	-	-	-
SpeI	1.9	D	D	D	D	D	+	+	D	-	D	D	D	+	D	[+]	[-]
ClaI*	2.7	-	-	-	-	-	D	D	-	+	-	-	-	D	-	-	-
MboI	3.0	-	-	-	-	-	-	+	-	-	-	-	-	-	-	-	-
MboI	3.1	-	-	-	-	-	-	-	-	+	-	-	-	-	-	-	-
MboI	3.2	-	-	-	+	-	-	-	-	-	-	-	-	-	-	-	-
BclI*	3.3	-	-	-	-	-	-	-	-	-	-	-	-	(+)	-	-	-
PstI	3.7	-	-	-	-	-	-	+	-	-	-	-	-	-	-	-	-
MboI	3.9	-	-	-	-	-	-	-	-	+	-	-	-	-	-	-	-
MboI	4.0	-	-	-	-	-	-	-	+	-	-	-	-	-	-	-	-
MboI	4.3	+	-	-	-	-	-	-	-	-	-	-	-	-	-	-	-
ClaI	4.6	-	-	-	-	-	-	-	-	+	-	-	-	-	-	-	-
MboI	5.0	-	-	-	-	-	-	-	-	+	-	-	-	-	-	-	-
BamHI*	5.2	-	-	-	-	-	-	-	-	+	-	-	-	-	-	-	-
EcoNI	5.5	+	-	-	+	-	-	-	-	-	+	-	-	-	-	-	-
Ase I	5.5	-	+	+	-	-	-	-	-	-	-	+	-	-	-	-	-
BclI*	5.7	-	-	-	-	-	-	-	-	+	-	-	-	-	-	-	-
MboI	5.8	-	-	+	+	-	-	-	-	-	-	-	-	-	-	-	-
EcoRV	5.9	-	-	-	+	-	-	-	-	-	-	-	-	-	-	-	-
MboI	6.0	+	+	+	+	-	-	-	-	-	+	+	-	-	+	-	-
BclI*	6.2	+	+	+	+	-	-	-	-	-	+	+	-	-	+	-	-
NsiI	6.2	-	-	-	-	-	-	-	-	+	-	-	-	-	-	-	-
MboI	6.3	-	-	-	-	-	-	-	+	+	-	-	-	-	-	-	-
BglII*	6.4	-	-	-	-	-	+	-	-	-	-	-	-	-	-	-	-
SacI	6.7	-	-	-	-	-	-	-	-	+	-	-	-	-	-	-	-
MboI	6.9	+	+	+	+	-	-	-	-	-	+	+	-	-	+	-	-
EcoRI	7.0	-	-	-	-	+	+	+	+	-	-	-	+	+	-	+	-
SspI	7.0	-	-	-	-	+	-	-	+	-	-	-	+	-	-	-	-
XmnI	7.1	+	+	+	+	-	-	-	-	-	+	+	-	-	+	-	-
SspI	7.2	-	+	+	-	-	-	-	-	-	-	+	-	-	-	-	-
SspI	7.3	+	+	+	+	-	-	-	-	-	+	+	-	-	+	-	-
HpaI	7752	+	+	+	+	-	-	-	-	+	+	+	-	-	+	-	+
BstNI	8188	+	-	-	+	+	+	+	+	+	+	-	+	+	+	+	+
DraI	8.6	-	-	-	-	-	-	+	-	-	-	-	-	-	-	-	-
SalI	8.8	-	-	-	-	-	-	-	-	+	-	-	-	-	-	-	-
PvuII	8.9	-	-	-	-	-	-	-	+	-	-	-	-	-	-	-	-
MboI	9.0	-	-	-	-	+	-	-	-	-	-	-	-	-	-	-	-
HpaI	9.0	+	+	+	-	-	-	-	-	-	[+]	+	-	-	[+]	-	-
HindIII	9.4	-	-	-	+	-	-	-	-	-	-	-	-	-	-	-	-
XmnI	9569	+	+	+	+	-	-	-	-	+	+	+	-	-	+	-	+
SspI	10.2	-	-	-	-	-	-	-	-	-	-	(+)	-	-	-	-	-
SacI	10.4	-	-	-	-	-	-	-	-	+	-	-	-	-	-	-	-
EcoRI	10.5	-	-	-	-	+	-	-	-	-	-	-	-	-	-	-	-
MboI	11.2	+	+	+	+	-	+	-	-	+	+	+	-	[-]	+	[-]	-
PvuII	11481	+	+	+	+	+	+	+	+	-	+	+	+	+	+	+	+
HpaI	11804	+	+	+	-	+	+	+	+	+	+	+	+	+	+	+	+
EcoRI	11.8	+	+	+	+	-	-	-	-	-	+	+	-	-	+	-	-
SspI	12.1	-	-	-	-	-	+	-	-	-	-	-	-	-	-	-	-
BclI*	12.2	-	-	-	-	+	-	-	+	-	-	-	+	-	-	-	-
PvuI*	12.3	-	+	+	-	-	-	-	-	-	-	+	-	-	-	-	-
EcoRV	12.4	-	-	-	-	-	-	-	-	+	-	-	-	-	-	-	-
EcoRI	12.5	-	+	-	-	-	-	-	-	-	-	(+)	-	-	-	-	-
BclI*	12.6	-	-	-	-	+	-	-	+	-	-	-	+	-	-	-	-
MboI	12.8	-	-	-	+	-	-	-	-	+	-	-	-	-	-	-	-
EcoRI	12.9	-	-	-	-	+	+	+	+	-	-	-	+	+	-	+	-

Enzyme/site		J	K	L	M	N	O	P	Q	R	A	B	C	D	E	F	W
MboI	13.0	-	-	-	-	-	-	-	-	-	-	-	-	-	-	(+)	-
XbaI	13.0	-	-	-	-	-	-	-	+	-	-	-	-	-	-	-	-
XmnI	13.2	+	+	+	+	-	+	+	-	-	+	+	-	+	+	[-]	[-]
MluI	13.3	-	-	-	-	-	+	-	-	-	-	-	-	-	-	-	-
BclI*	13.6	-	-	-	-	+	-	-	-	-	-	-	-	-	-	-	-
MboI	13.8	-	-	-	-	-	+	-	-	-	-	-	-	-	-	-	-
SspI	14.0	-	-	-	+	-	-	-	-	-	-	-	-	-	-	-	-
AseI	14.2	-	-	-	-	-	-	-	-	+	-	-	-	-	-	-	-
NdeI	14306	+	+	+	+	-	-	-	-	+	+	+	-	-	+	-	+
BclI*	14.4	-	-	-	-	-	+	+	-	-	-	-	-	+	-	-	-
ClaI	14.7	-	-	-	-	-	-	+	-	-	-	-	-	-	-	-	-
BamHI*	15.3	-	-	-	-	-	+	+	-	-	-	-	-	+	-	(+)	-
MboI	15.4	-	-	-	-	-	-	-	-	+	-	-	-	-	-	-	-
ScaI	15515	+	+	+	+	-	+	+	+	+	+	+	+	+	+	+	+
XmnI	16008	+	+	+	+	-	+	+	+	-	+	+	(-)	+	+	+	+
MboI	16.1	+	-	-	-	-	-	-	-	-	-	-	-	-	-	-	-
MboI	16.3	+	+	+	+	+	+	+	+	-	+	+	+	+	+	+	(-)
SpeI	16.7	-	-	-	-	-	+	-	-	-	-	-	-	-	-	-	-
SalI	16.8	-	-	-	-	-	+	-	-	+	-	-	-	-	-	-	-
MboI	17.2	-	-	-	-	-	+	-	-	-	-	-	-	(+)	-	-	-
SpeI	17279	+	+	+	+	+	-	-	+	+	+	+	+	-	+	+	+
AseI	17.4	-	-	-	+	-	-	-	-	-	-	-	-	-	-	-	-
MboI	17.5	+	-	-	-	-	-	-	-	-	-	-	-	-	-	-	-
KpnI	17.6	-	-	-	-	-	-	-	-	+	-	-	-	-	-	-	-
BclI*	17.6	-	-	-	-	-	+	-	-	-	-	-	-	-	-	-	-
ClaI	17.7	-	-	-	-	-	+	+	-	-	-	-	-	+	-	-	-
MboI	17.8	-	-	-	-	-	+	-	-	-	-	-	-	(+)	-	-	-
EcoRI	17.9	+	-	+	-	-	-	-	-	-	-	-	-	-	-	-	-
MboI	18.2	-	+	+	-	-	-	-	-	-	-	+	-	-	-	-	-
MboI	18.3	-	-	+	-	-	+	-	+	-	-	-	-	-	-	-	-
EcoRI	18.4	+	+	+	+	-	-	-	+	-	+	+	-	-	+	-	-
AseI	18.6	+	-	-	+	-	-	-	-	-	+	-	-	-	-	-	-
BstBI	19213	+	+	+	+	+	+	+	+	-	+	+	+	+	+	+	+
XmnI	19.3	+	-	-	+	-	-	-	-	-	+	-	-	-	-	-	-
SacI	19.6	-	-	+	-	-	-	-	-	-	-	-	-	-	-	-	-
MboI	19.7	+	+	-	+	-	-	-	-	-	+	[-]	-	-	[+]	-	-
MboI	19.9	-	+	+	-	-	-	-	+	-	-	+	-	-	-	-	-
BstEII	20065	+	+	+	+	+	-	+	+	+	+	+	+	(-)	+	+	+
MboI	20.6	-	-	-	-	-	-	-	+	-	-	-	-	-	-	-	-
XbaI	20.8	-	-	-	-	-	-	-	+	-	-	-	-	-	-	-	-
MboI	20.9	+	-	-	+	-	-	-	-	-	+	-	-	-	-	-	-
SalI	21.2	-	-	-	-	-	-	-	+	-	-	-	-	-	-	-	-
MboI	21.6	-	-	-	-	+	-	-	+	-	-	-	+	-	-	-	-
MboI	21.7	-	-	+	-	-	-	-	-	-	-	-	-	-	-	-	-
MboI	21.9	-	-	-	-	-	-	-	+	-	-	-	-	-	-	-	-
MboI	22.1	-	-	-	-	+	+	-	+	-	-	-	+	[+]	-	[-]	-
HindIII	22.5	-	-	-	-	+	+	+	+	-	-	-	+	+	-	+	-
NsiI	22.6	-	-	-	-	-	-	-	+	-	-	-	-	-	-	-	-
MboI	22.7	-	-	-	-	-	-	-	+	-	-	-	-	-	-	-	-
MboI	22.8	-	+	+	-	-	-	-	-	-	-	+	-	-	-	-	-
EcoRV	22.9	+	-	-	-	-	-	-	-	-	-	-	-	-	-	-	-
MboI	22.9	-	-	-	-	+	-	-	-	-	-	-	-	-	-	-	-
NheI	2296	-	+	+	-	-	+	+	-	+	-	+	-	+	+	+	+
MboI	23.4	-	-	-	-	+	+	+	+	-	-	-	+	+	-	+	-
MboI	23.6	-	+	+	-	-	-	-	-	-	-	(-)	-	-	-	-	-
KpnI	23740	+	-	-	+	+	+	+	-	+	+	-	+	+	+	+	+
HpaI	24.1	+	-	-	-	-	-	-	-	-	-	-	-	-	-	-	-
EcoRV	24.5	-	-	-	-	+	-	-	-	-	-	-	-	-	-	-	-
MboI	24.6	+	-	-	-	-	-	-	-	-	-	-	-	-	-	-	-
NsiI	24.8	+	-	-	-	-	-	-	-	-	-	-	-	-	-	-	-
EcoRI	25.0	-	-	-	-	+	-	-	+	-	-	-	+	-	-	-	-
MboI	25.2	-	-	-	-	-	-	-	+	-	-	-	-	-	-	-	-
AseI	25.3	-	-	-	-	+	-	-	-	-	-	-	-	-	-	-	-
MboI	25.6	-	-	-	-	-	+	+	-	-	-	-	-	(-)	-	-	-
SalI	25.8	-	-	-	-	-	-	-	+	-	-	-	-	-	-	-	-
MboI	25.8	-	-	+	-	-	-	-	-	-	-	-	-	-	-	-	-
MboI	26.1	-	-	-	+	-	-	-	-	-	-	-	-	-	-	-	-

Enzyme/site		J	K	L	M	N	O	P	Q	R	A	B	C	D	E	F	W
ScaI	26.3	-	-	-	-	-	-	-	-	+	-	-	-	-	-	-	-
MboI	26.4	-	-	-	-	-	-	-	+	-	-	-	-	-	-	-	-
MboI	27.0	-	-	-	-	+	+	+	+	-	-	-	+	+	-	+	-
HpaI	27.1	-	-	-	-	+	+	+	+	-	-	-	+	+	-	+	-
XmnI	27.3	-	-	+	-	-	-	-	-	-	-	-	-	-	-	-	-
MboI	27.7	-	-	-	-	-	-	-	-	+	-	-	-	-	-	-	-
NdeI	27724	+	+	+	+	+	+	-	+	+	+	+	+	+	+	+	+
EcoRV	28.0	-	-	-	+	-	-	-	-	-	-	-	-	-	-	-	-
BamHI*	28.2	-	-	-	-	-	+	-	-	-	-	-	-	-	-	-	-
MboI	28.2	-	-	+	-	-	+	-	-	-	-	-	-	-	-	-	-
EcoNI	28749	+	-	-	+	+	+	+	+	+	+	-	+	+	+	+	+
HpaI	29.1	-	-	-	-	-	-	-	-	+	-	-	-	-	-	-	-
AseI	29.2	+	+	+	+	-	-	-	-	-	+	+	-	-	+	-	-
BclI*	29.4	-	-	-	-	+	-	-	+	-	-	-	+	-	-	-	-
NsiI	29.5	-	+	-	-	-	-	-	-	-	-	-	-	-	-	-	-
HpaI	29587	+	+	+	+	-	-	-	-	+	+	+	-	-	+	-	+
MboI	29.6	-	-	+	-	-	-	-	-	-	-	-	-	-	-	-	-
NcoI	29717	+	+	-	+	-	-	+	-	+	+	+	-	[+]	+	[+]	+
MboI	29.8	-	-	-	-	+	-	-	+	-	-	-	+	-	-	-	-
EcoRI	29.8	-	+	+	-	-	-	-	-	+	-	+	-	-	[-]	-	[-]
MboI	30.0	-	-	-	-	-	+	-	+	-	-	-	(+)	-	-	-	-
PvuI*	30.0	-	-	-	-	-	+	-	-	-	-	-	-	-	-	-	-
BstBI	30.0	+	-	-	-	-	-	-	-	-	-	-	-	-	-	-	-
HindIII	30.1	+	+	+	-	-	-	-	-	-	[+]	+	-	-	[+]	-	-
BstBI	30.3	+	-	-	-	+	+	+	+	-	-	-	+	+	-	+	-
HpaI	30.8	-	-	-	-	-	-	-	-	+	-	-	-	-	-	-	-
EcoRV	31.2	-	-	-	-	-	-	-	-	+	-	-	-	-	-	-	-
EcoRI	31.2	-	-	-	-	-	-	+	-	-	-	-	-	-	-	-	-
XmnI	31202	+	+	+	+	+	-	+	+	+	+	+	+	+	+	+	+
MboI	31.4	-	-	-	-	+	+	+	+	-	-	-	+	+	-	+	-
DraI	31.6	-	-	+	-	-	-	-	-	-	-	-	-	-	-	-	-
AseI	31.7	-	-	+	-	-	-	-	-	-	-	-	-	-	-	-	-
BglII*	32.0	+	-	-	-	-	-	-	-	-	-	-	-	-	-	-	-
ClaI*	32.1	-	-	-	-	+	-	-	-	-	-	-	-	-	-	-	-
DraI	32.2	-	-	-	-	+	-	-	+	-	-	-	+	-	-	-	-
MboI	32.2	+	-	-	+	-	-	-	-	-	+	-	-	-	-	-	-
XmnI	32.2	-	+	+	-	-	-	-	-	-	-	+	-	-	-	-	-
MboI	32.4	+	+	+	+	-	-	-	-	-	+	+	-	-	+	-	-
ScaI	32.6	+	-	-	+	-	-	-	-	+	+	-	-	-	[-]	-	[-]
MboI	32.8	-	-	-	-	+	-	-	-	-	-	-	-	-	-	-	-
HindIII	32.8	-	-	-	-	-	-	-	-	+	-	-	-	-	-	-	-
MboI	33.0	-	-	-	-	-	+	+	-	-	-	-	-	+	-	-	-
XmnI	33.0	-	-	-	-	-	-	-	+	-	-	-	-	-	-	-	-
MboI	33.4	-	-	-	-	-	-	-	+	-	-	-	-	-	-	-	-
EcoRV	33.5	-	-	-	-	-	+	-	-	-	-	-	-	-	-	-	-
MboI	33.5	-	-	-	-	+	+	+	+	-	-	-	+	+	-	+	-
XmnI	33.8	-	+	-	-	-	-	-	-	-	-	-	-	-	-	-	-
NsiI	33.9	-	-	-	-	-	+	-	-	-	-	-	-	-	-	-	-
ScaI	34099	+	+	+	-	+	+	+	+	+	+	+	+	+	+	+	+
ClaI*	34.9	-	+	-	-	-	-	-	-	-	-	-	-	-	-	-	-
PvuI*	34.9	-	+	-	-	-	-	-	-	-	-	-	-	-	-	-	-
XmnI	35.1	+	-	-	-	-	-	+	-	-	-	-	-	-	-	-	-
MboI	35.4	-	-	-	-	-	+	+	-	-	-	-	-	+	-	-	-
MboI	35684	+	+	+	+	-	+	+	-	+	+	+	-	+	+	+	+
BstEII	35.8	+	+	+	+	-	-	-	-	-	+	+	-	-	+	-	-
XmnI	36134	+	-	-	+	+	+	+	+	+	+	-	+	+	+	+	+
MboI	36.7	-	+	-	-	-	-	-	-	-	-	-	-	-	-	-	-
ClaI	37078	+	+	+	+	+	-	-	+	+	+	+	+	-	+	+	+
EcoRI	37.2	-	-	-	+	-	-	-	-	+	-	-	-	-	-	-	-
NsiI	37.2	-	-	-	-	-	+	-	-	-	-	-	-	-	-	-	-
MboI	37.8	-	-	-	+	-	-	-	-	-	-	-	-	-	-	-	-
MboI	37.9	-	-	-	-	+	-	-	-	-	-	-	-	-	-	-	-
MboI	38.2	-	-	-	-	-	+	-	-	-	-	-	-	(+)	-	-	-
MboI	38.4	-	-	-	-	-	+	-	-	-	-	-	-	-	-	-	-
MboI	38.5	-	+	+	-	-	-	-	+	-	-	+	-	-	-	-	-
ClaI	38.9	+	-	-	-	-	-	-	-	-	-	-	-	-	-	-	-
NsiI	39133	-	+	+	+	+	+	+	+	+	+	+	+	+	+	+	+
MboI	39.3	-	-	-	+	-	-	-	-	-	-	-	-	-	-	-	-

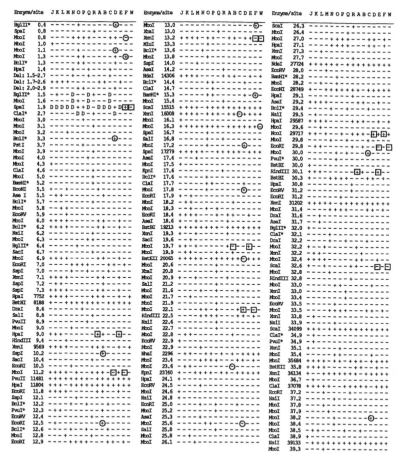

Fig. 3. Variable restriction enzyme cleavage sites and deletions of the terminal lineages (J through R) and ancestral nodes (A through F) compared to the wild-type T7 genome (W). Wild-type sites for these enzymes that are not listed are found among all the lineages. Sites with positions shown without decimals are wild-type sites that have been lost in some of the lineages; their positions were determined from the complete sequence of T7 DNA (*11*). The remaining site locations were mapped as shown in Fig. 2; these positions are indicated in kilobases. Sites marked with an asterisk are also Mbo I sites; they are listed only once unless multiple changes produced differences between Mbo I and the larger recognition sequence in some of the taxa (for example, the Mbo I and Bam HI sites at position 5.2). Sites in ancestors that were inferred incorrectly by parsimony analysis of the terminal lineages are circled; those for which inference was ambiguous are surrounded by squares. A letter "D" indicates that the site occurs in a region that has been deleted in the respective lineage; these sites were coded as unknown in the analyses.

character optimization used) (*1*). Three of the incorrectly inferred states involved sites that were found only in the ancestral lineages, hence would not have been detected as variable characters if ancestors were unavailable (as is typically the case in phylogenetic studies).

The results of this study directly support the legitimacy of methods for phylogenetic estimation, not only with regard to reconstructing branching relationships, but also branch lengths and ancestral genotypes. Perhaps more importantly, they point the direction to a field of research in which methods of reconstruction can be tested against various known phylogenies of real organisms differing in topological and other evolutionary characteristics in the same fashion that tests have been conducted with simulated, theoretical phylogenies. Experimental phylogenetics is not a substitute for numerical studies, nor is it likely that laboratory phylogenies will ever display the full complexity of phylogenies produced over long-term evolution, but such studies will fill an important void in the science of phylogenetic reconstruction.

REFERENCES AND NOTES

1. D. L. Swofford and G. J. Olsen, in *Molecular Systematics*, D. M. Hillis and C. Moritz, Eds. (Sinauer, Sunderland, MA, 1990), pp. 411–501.
2. N. Eldredge and J. Cracraft, *Phylogenetic Patterns and the Evolutionary Process* (Columbia Univ. Press, New York, 1980); J. Felsenstein, *Am. Nat.* **125**, 1 (1985); D. R. Brooks and D. A. McLennan, *Phylogeny, Ecology, and Behavior* (Univ. of Chicago Press, Chicago, 1991); P. H. Harvey and M. D. Pagel, *The Comparative Method in Evolutionary Biology* (Oxford Univ. Press, Oxford, 1991).
3. W. M. Fitch and W. R. Atchley, *Science* **228**, 1169 (1985); in *Molecules and Morphology in Evolution: Conflict or Compromise?*, C. Patterson, Ed. (Cambridge Univ. Press, Cambridge, 1987), pp. 203–216; W. R. Atchley and W. M. Fitch, *Science* **254**, 554 (1991).
4. B. R. Baum, in *Cladistics: Perspectives on the Reconstruction of Evolutionary History*, T. Duncan and T. F. Stuessy, Eds. (Columbia Univ. Press, New York, 1984), pp. 192–220.
5. F. W. Studier, in *Genes, Cells, and Behavior: A View of Biology Fifty Years Later*, N. H. Horowitz and E. Hutchings, Jr., Eds. (Freeman, San Francisco, 1980), pp. 72–78.
6. M. E. White, J. J. Bull, I. J. Molineux, D. M. Hillis, in *Proceedings of the Fourth International Congress of Systematic and Evolutionary Biology*, E. Dudley, Ed. (Dioscorides Press, Portland, OR, 1991), pp. 935–943.
7. T7 phage was grown in 1-ml cultures of *Escherichia coli* strain W3110 in the presence of the mutagen *N*-methyl-*N'*-nitro-*N'*-nitrosoguanidine (20 μg/ml). After lysis, a new culture was infected with 10 μl of the lysate. After every five serial lysates, phage were plated on agar and a single plaque was randomly selected for further propagation. The distance between each node of the ingroup phylogeny (lineages J through P and their ancestors) was equal to 40 serial lysates; each lysate is the consequence of two to three bursts (generations) of phage. The outgroup lineage (R) was derived from wild-type T7 by a similar protocol (*6*). To guard against contamination among lineages, we assayed DNA from each isolated plaque with restriction enzymes whose restriction patterns evolve rapidly in this system (either Mbo I or a combination of Hae II, Hpa I, Kpn I, and Spe I) and looked for massive convergence of the restriction patterns. Two cases of contamination were detected by this procedure; after verifying the contamination with additional restriction digests, we regrew the lineages from the last contamination-free ancestor.
8. J. Felsenstein, *Syst. Zool.* **27**, 401 (1978).
9. J. P. Huelsenbeck, *ibid.* **40**, 257 (1991).
10. J. A. Lake, *Mol. Biol. Evol.* **4**, 167 (1987).
11. J. J. Dunn and F. W. Studier, *J. Mol. Biol.* **166**, 477 (1983).
12. D. L. Swofford, *Phylogenetic Analysis Using Parsimony* (Illinois Natural History Survey, Champaign, 1990).
13. W. M. Fitch and E. Margoliash, *Science* **155**, 279 (1967).
14. L. L. Cavalli-Sforza and A. W. F. Edwards, *Am. J. Hum. Genet.* **19**, 233 (1967).
15. N. Saitou and M. Nei, *Mol. Biol. Evol.* **4**, 406 (1987).
16. R. R. Sokal and C. D. Michener, *Univ. Kansas Sci. Bull.* **38**, 1409 (1958).
17. For the parsimony reconstruction, bootstrap confidence intervals were 100% for all nodes except A (92%; refer to Fig. 1 for this node), based on 1000 replications; the consistency index for informative characters was 0.750.
18. Correlations between the actual and predicted branch lengths were 0.91 for parsimony (average of accelerated and delayed character transformation), 0.89 for the Fitch-Margoliash method, 0.88 for the Cavalli-Sforza method, 0.88 for the neighbor-joining method, and 0.82 for the UPGMA method. As the estimates of branch lengths by these methods may violate the assumptions underlying parametric statistics, we tested the possibility of statistically significant differences among the five correlations in the following fashion. For each branch *i*, each of the five estimates of that branch was assigned at random and without replacement to one of the five reconstruction methods; the resulting matrix was thus a shuffling of the original matrix, with the constraint that an estimate of branch *i* was always assigned as an estimate of branch *i*, even though the method to which it was assigned varied. Correlations between actual and estimated branch lengths were computed for each of the methods in this randomized matrix, and these data were recorded; the process was repeated 10,000 times. The heterogeneity of corre-

lations from the actual estimates was found to be significantly greater than that of the randomized data ($P = 0.017$), with the correlation for UPGMA being significantly less than the randomized value minimum ($P < 0.011$). The heterogeneity of correlations among the set of four methods with UPGMA removed is no longer significantly large, but the correlation for parsimony is larger than the maximum of randomized values at $P = 0.05$. By these criteria, UPGMA appears to be significantly worse than the other methods, and there is some evidence that parsimony is superior.
19. G. W. Snedecor and W. G. Cochran, *Statistical Methods* (Iowa State Univ. Press, Ames, ed. 7, 1980).
20. Supported by the NSF (to D.M.H.) and the NIH (to I.J.M.). We thank F. W. Studier, I. Tessman, and T. Kunkel for advice on mutagenesis in phage.

8 August 1991; accepted 22 November 1991

Williams, G. C. 1992. *Gaia*, nature worship and biocentric fallacies. *Quarterly Review of Biology* 67:479–486.

George Williams, a highly influential American biologist, had a special talent for piercing to the intellectual core of an evolutionary argument and offering searingly insightful commentary. He was a career-long champion of natural selection, following the topic to its logical conclusions, wherever they might lead. He wrote about diverse and eclectic topics, ranging from the evolution of sex to the nature of adaptation and from teleology to evolutionary medicine.

In this article, Williams is at his eloquent best as he critiques what he perceives to be naïve invocations of group selection under the Gaia concept. In ancient Greece, Gaia was an Earth goddess, so it is perhaps appropriate that her name was adopted by James Lovelock (1979, 1988) when he proposed that the Earth's biota is like a superorganism devoted to the management of the environment in the biota's collective interests. Although the Gaia hypothesis has been heartily applauded by some ecologists and others, Williams finds serious evolutionary fault with its central thesis. Read the article to see why.

A view of life on Earth completely antagonistic to Gaia is Peter Ward's *The Medea Hypothesis: Is Life on Earth Ultimately Self-Destructive?* (2009).

RELATED READING

Lovelock, J. 1979. Gaia: a new look at life on Earth. Oxford University Press, Oxford, England.

Lovelock, J. 1988. The ages of Gaia. Norton, New York, New York, USA.

Neese, R. M., and G. C. Williams. 1994. Why we get sick: the new science of Darwinian medicine. Times Books, New York, New York, USA.

Ward, P. 2009. The Medea hypothesis: is life on Earth ultimately self-destructive? Princeton University Press, Princeton, New Jersey, USA.

Williams, G. C. 1966. Adaptation and natural selection. Princeton University Press, Princeton, New Jersey, USA.

Williams, G. C. 1975. Sex and evolution. Princeton University Press, Princeton, New Jersey, USA.

Williams, G. C. 1992. Natural selection: domains, levels, and challenges. Oxford University Press, Oxford, England.

VOLUME 67, NO. 4 THE QUARTERLY REVIEW OF BIOLOGY DECEMBER 1992

COMMENTARY

GAIA, NATURE WORSHIP AND BIOCENTRIC FALLACIES

GEORGE C. WILLIAMS

Department of Ecology & Evolution, State University of New York
Stony Brook, New York 11794-5245 USA

Let us understand, once for all, that the ethical progress of society depends, not on imitating the cosmic process, still less in running away from it, but in combating it (T. H. Huxley, 1894:83).

I AGREE with Huxley's statement, but others do not. Romantic traditions persist in finding practical lessons and moral directives in natural phenomena and, at least by implication, in urging back-to-nature sentiments or in somehow imitating the cosmic process. Others prefer to run away, or at least turn their backs on nature's hostility and pretend it is not there. Their main method is to use the verbal camouflage of nice names for adverse conditions or ethically unacceptable actions. Here I will attempt to support Huxley's position, that the universe is hostile to human life and values, and to counter prevalent romanticism and what Lillie (1913) called the "biocentric" view of the universe. I am motivated by recent statements in *Science* (Kerr, 1988; Mann, 1991) on the *Gaia* concept, in opposition to Huxley's and my own view of biological principles and the human condition.

The idea that the universe is especially designed to be a suitable abode for life in general and for human life in particular is, of course, an old one. It had to be abandoned in its early forms with the triumph of Copernican astronomy in the Renaissance, but some scholars still find it possible to argue that the Earth, at least, can be regarded as especially suited for human life. This idea found eloquent expression in Henderson's (1913) *The Fitness of the Environment*. Its main modern manifestation is in the Gaia concept of Lovelock and Margulis (1974) and Lovelock (1979, 1988).

THE UNFITNESS OF THE COSMIC ENVIRONMENT

I will be concerned mainly with the Earth and with material properties shown under physical conditions found on the surface of this planet, but it may foster a balanced perspective to begin at a more macroscopic level. A clear message may be read in the fact that our galaxy is an infinitesimal part of the universe, the solar system a minuscule part of the galaxy, the Earth only about a hundred thousandth part of the solar system, and the biologically relevant part of the Earth only a thin film on its surface. There is nothing in this deployment of materials to suggest any cosmic importance for what happens on Earth. It is also noteworthy that what we regard as normal environmental conditions, conducive to animal and plant life as we know it, have prevailed for only about a quarter of the history of our planet.

Consider just one detail of traditional cosmology, that the Sun was put there to heat and illuminate the surface of the Earth. The

The Quarterly Review of Biology, December 1992, Vol. 67, No. 4

0033-5770/92/6704-0003$1.00

Sun is about 150,000,000 km away, and the Earth's diameter about 13,000 km. A bit of geometry with these figures shows that the Earth intercepts about a billionth of the solar output, the rest radiating out in other directions. Obviously no consideration of energy efficiency went into the design of the Earth's power system.

With a more parochial focus on our current and immediate environment, it may appear that conditions are eminently suitable for ourselves and other organisms. This impression stems from failure to appreciate how completely one-sided adaptation is, and what it can be expected to accomplish. Living organisms are elaborately adapted to their particular ways of life in the environments in which they evolved. There is no evidence for any other kind of adaptation.

OUR PHYSICAL ENVIRONMENT

Life, at least on this planet, takes place in an aqueous medium, and it is closely dependent in many ways on the special properties of water. This led Henderson (1913) to argue that water was designed as a medium ideally suitable for life. His argument was not a traditionally theological one, but as I read him, he was a vitalist and a teleologist in his belief in a predetermined course of evolution. Not only was organic evolution proceeding toward a goal but, prior to organic evolution, the properties of matter were formed by a process that adapted it to its biological role. This "biocentric" cosmology was quickly criticized by Lillie (1913), and other aspects of Henderson's work have been criticized by others, most recently and cogently by Craik (1989). Here I will merely deal with a few additional criticisms, related mainly to the properties of water, which formed an essential part of Henderson's argument.

In a list of the attributes of common substances, water is found to have extremely high values for specific heat and for heats of fusion and vaporization. As Henderson and many others have argued, these properties undoubtedly make a planet largely covered by a deep ocean much less variable in temperature than it would be if these values were lower. A more thermally variable planet might be considered less suitable for life, but this is gratuitous. The biosphere of the Earth in fact has a range of commonly encountered temperatures, and organisms are observed to adapt to them (see the discussion of Gaia below for more on the thermal adaptability of organisms). I presume that if a broader temperature range were experienced over long periods of geologic time, organisms would adapt to that too.

The thermal inertia of water not only makes the environment difficult to heat or cool, but also has the same effect on organisms themselves, and Henderson cites this stability as an adaptive aspect of water. If it were not for its high specific heat, we would be heated by a given exertion more than we are. This, as Henderson emphasizes, would be maladaptive for a mammal in hot weather. The greater heating, for the same mammal, would be adaptive in cold weather. The high specific heat of water is clearly disadvantageous for any reptile or insect or other animal that depends on basking to reach a favorable temperature on a cold day.

I suspect that, with moderate effort, additional adverse effects from the properties of water and other common substances could be found and emphasized for the thesis that our physical environment is diabolically perverse. Many such arguments could be based on the same features that Henderson cites as especially favorable to life, such as the expansion of water when it freezes. This means that ice floats and that a body of water in contact with sufficiently cold air freezes at the surface. It is thereby insulated against further thermal exchanges with the air, so that the moderation of climate by the ocean is largely frustrated when it might be regarded as most needed.

An additional problem, if thermal stability is thought to be biologically desirable, is that snow has the ideally wrong color. It greatly exacerbates winter chill by reflecting solar radiation into space, an effect emphasized in the white-earth models of climatologists (Wetherald and Manabe, 1975:2057). If ever an ice age advances beyond a critical threshold, it must continue advancing with no possibility of retreat, and the Earth will stay frozen forever. There would be no such threat if snow had been given a darker color by Henderson's biocentric designer.

There is a more directly biological effect of the expansion of water on forming ice. Any freezing organism will be killed by the mechanically disruptive effects of expanding ice crystals, unless it has special adaptations to prevent such effects. I imagine that an extraterrestrial observer, from a planet that either lacks water or never experiences freezing temperatures, would conclude that terrestrial life of middle or high latitudes must be destroyed every winter with the first freeze. The observer would probably confine any search for perennial organisms to the tropics or deep water.

Or perhaps only deep water. Surfaces exposed to the air or with only a shallow covering of water would be vulnerable to sunlight, which an educated visitor might well assume would rule out all vital activity. The rapidly lethal nature of both ultraviolet and visible wave lengths can be convincingly shown by earthworms and trout embryos and other organisms that lack elaborate defenses against this environmental threat (Perlmutter, 1961). The extreme transparency of water makes it extraordinarily ineffective as a shield against radiation.

An even greater discouragement to an extraterrestrial visitor's search for life would be the high concentration of oxygen in the Earth's atmosphere, which might plausibly be attributed to the photolysis of water. It is most unlikely that the visitor would speculate that this process might be used by illuminated organisms to power the recovery of carbon from the minute trace of CO_2 in the atmosphere. The abundance of oxygen would force the conclusion that all materials at the Earth's surface must be oxidized (i.e., dead). Only in locally anoxic aquatic environments would there be any hope of finding living organisms, but the capacity of water to dissolve oxygen would be another of its unfortunate attributes.

THE GAIA CONCEPT

The modern form of the idea that adaptation is symmetrical, with organisms adapted to their environments and the environments adapted to organisms, is the Gaia concept (Lovelock and Margulis, 1974; Lovelock, 1979, 1988). *Gaia* was an Earth goddess of the ancient Greeks. It is an appropriate name for the collective biota of the Earth if that collective is, as the advocates claim, a superorganism devoted to the management of the environment in her own collective interest. Overt mysticism is not an essential element of the Gaia concept. Its logical structure is not entirely clear to me, but it seems rather similar to the *invisible hand* model in economics: If each organism freely pursues its own interests, the average result over the ecosystem will be favorable to organisms.

Discussions in support of Gaia are often unclear as to what sort of interests organisms are supposed to be pursuing. The advocates claim that the idea is fully in accord with mainstream opinion in evolutionary biology, but examples are often overtly group-selectionist (e.g., Lovelock, 1988:39, 143–149). The characters that a species evolves, according to Lovelock, are those that benefit the species, rather than those that enable its members to compete better with each other. There is no illogic or inconsistency in Gaia advocates seeking support in group selection theory (e.g., from Van Valen, 1975 or Wilson, 1980), because no one doubts the reality of group selection. What currently prevails in biology is a great doubt about its strength relative to individual selection (Futuyma, 1986: 258–266; Krebs and Davies, 1987:14–21). If the Gaia concept depends heavily on group selection, it should not be represented as being in harmony with modern Darwinism.

Its conflict with current evolutionary theory forms the basis of much recent criticism of Gaia (e.g., Dawkins, 1982:234–237). I think the criticisms are valid, but not decisive. If Gaia is real and has the properties claimed, the proper reaction is to modify evolutionary theory, not deny the facts. My criticism will focus on two factual complaints. The first is that there is no evidence that the metaphorical invisible hand really manipulates the evolutionary process so as to produce results favorable to organisms. The second is that, even if this invisible-hand effect were valid, it would not justify the claim that Gaia is "a total planetary being," which practices "homeostasis" (Lovelock, 1988:19). It would not even justify Kerr's (1988) weaker claim for planetary homeostasis.

WHAT DOES GAIA DO?

I do not know whether Adam Smith's invisible hand manipulates economies the way he proposed. I do know that the biological equivalent often manipulates ecologies in seriously maladaptive ways, and I will give two of many possible examples. The first is the widespread habit of trees of producing, in the pursuit of their individual interests, conditions that favor forest fires. Conceivably this need not have been so. On a more arid Earth, the vegetation might nowhere be more abundant than widely spaced clumps of shrubs. If so, I would expect the Gaia advocates to seize upon this circumstance for a favorable argument: The vegetation is careful not to produce continuous high-density concentrations of combustible materials, which could be ruinous to the whole community.

Any recent visitor to Yellowstone Park will see dramatic evidence that this is not so. Each individual tree tries to grow out of the shade of its neighbors so as to intercept enough light to produce an abundance of pollen and seeds. This can be done only by supporting its photosynthetic tissues on a massive infrastructure of cellulose. The collective result, over enormous regions, is the accumulation of a high concentration of combustible material. Sooner or later there will be disastrous forest fires, like those that recently struck Yellowstone. Many kinds of forests can be destroyed by fire, even normally moist rainforests (Mutch, 1970).

It may be pointed out that some trees and other plants are benefited by occasional fires, and may even go extinct if fires are absent for 30 to 100 years (Noble and Slatyer, 1980). This fire dependence is especially prevalent in Australia, where the dominant eucalypts may be termed *pyrophytes* (Pyne, 1991:20–21). They would be replaced by acacias or other forest trees if it were not for fire. This fire dependence does not mean that fire itself is in any general sense favorable to life. It merely gives an example of the *helpful-stress effect*: In adapting to even a severe sort of stress, an organism may come to depend on it indirectly, in competition with others not so adapted, or even absolutely, so that it cannot survive at all without the stress (see examples in Williams, 1992:116–118).

My second example affects a far larger part of the Earth's biota, the marine epipelagic and all dependent communities. Some marine productivity is contributed by benthic plants, such as the grasses and seaweeds of rocky shallows, but this is globally trivial compared to the phytoplankton contribution over about seven tenths of the surface of the planet. Planktonic algae, like the plants of a forest or corn field, use sunlight to convert CO_2 and dissolved minerals into organic matter. This matter forms the base of the food chain upon which almost all the animal life depends.

Phytoplankton productivity depends on the strength of illumination and the concentration of dissolved nutrients, of which nitrates and related compounds are most critical. Unfortunately for the marine ecosystem, the efforts of planktonic organisms to survive and reproduce give rise to a severe depression of collective productivity. The invisible hand imposes a perverse dilemma: Where there is light enough for abundant photosynthesis there is a severe shortage of nutrients, where nutrients are abundant the light is grossly inadequate (Fig. 1).

The sparse algal community is confined to near-surface layers where (at certain latitudes and times of day) there is light to power its growth and maintenance. Growth depends on the necessarily slow capture of nutrients from water already severely depleted. Any success in this effort further depresses nutrient levels and productivity. Nutrients are returned to the water when the cells die, but dead cells sink to lower levels, and other biological processes also conspire to rob the surface layers of nutrients. Dead animals and animal products, such as feces and molted exoskeletons, sink to lower levels as they decompose and release nutrients. Many animals feed on plankton near the surface at night, but then descend and do much of their metabolizing and releasing of plant nutrients in poorly illuminated depths by day.

Replenishment of surface nutrients depends on the nonbiological processes of turbulent mixing by wind, or on seasonally or geographically restricted upwelling. Gaia's destructive influence is more pervasive and influential than these inorganic processes. She depresses marine productivity to levels far be-

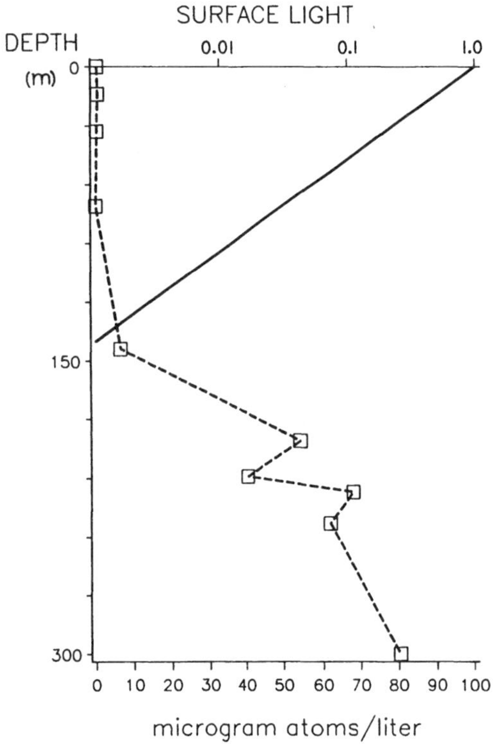

Fig. 1. Subversion of Productivity in a Marine Pelagic Community.

Intensity of blue light (most penetrating wave length) as a proportion of surface illumination is shown by the continuous line. Nitrate concentration is shown by the dashed line. Values for light are hypothetical, but realistic for extremely clear water. Nitrate values were measured in the western Gulf of Mexico in June 1951 (from Collier, 1958). The conditions shown can be considered typical for most of the ocean most of the time. The highest nitrate values shown would be low for terrestrial soils. Light of less than a thousandth of full sunlight (left edge of graph) is inadequate for phytoplankton maintenance.

low what would prevail if there were some countervailing biological process that would transport plant nutrients from their gigantic accumulation in deep water up to levels where they can do some good.

This argument is valid, of course, only if I have properly identified the sort of good that Gaia is supposed to accomplish. Is it really Gaia's role to maximize marine productivity? Unfortunately there are no adequately explicit Gaia-theory axioms from which to deduce such a proposition, but it is clearly Lovelock's (1988:135) position that Gaia prefers more productive to less productive, and he proposes that "the health of Gaia is measured by the abundance of life." If so, Gaia must be sick indeed over most of the Earth, where the biological community is a desert and life's abundance is at a paltry level.

Mere competition would be bad enough. The success of one organism will often mean failure for another, but the real mischief in the pelagic ecosystem is in the relentless downward transport of nitrogen and phosphorus and other nutrients. Other communities may be poisoned by waste products. The accumulation of combustibles causing a much later catastrophe, for successes and failures alike, is a more remote and indirect effect, but surely a real one. One can also, with the Gaia advocates, identify ways in which one organism may have a directly or indirectly beneficial effect on some other organism or on a widely scattered group of organisms. Should the question of the reality and importance of Gaia be decided on the basis of which side can muster the more impressive list of examples?

Surely not, because it will often be difficult to distinguish Gaia's benefits from Gaia's harm to which organisms have adapted. Since the late Precambrian, an effect of Lovelock's (1988: 19) *total planetary being* has been to keep a high concentration of oxygen in the atmosphere. This free oxygen is now a vital necessity to most of the organisms we think of as important, but this could not have been true originally. In the early stages of oxygen accumulation, at least some organisms evolved ways to avoid getting oxidized. From some of them, modern obligate aerobes such as ourselves evolved. To organisms with aerobic metabolism, oxygen is a resource that allows much higher levels of energy use than is possible for any anaerobe. This does not mean that the first appearance of free oxygen was the production of a beneficial substance. It means that some organisms were able to evolve a beneficial usage. Oxygen may be the outstanding example of the phenomenon of helpful stress, mentioned above.

The likelihood of confusing beneficial environmental factors with their secondarily evolved beneficial uses was recognized by Lillie (1913), who observed that "this world may be the best possible environment for the organisms that have come to exist in it, but it might not be so for the living beings of another and quite different cosmos" (p. 339). Very much the same sentiment was independently expressed by Craik (1989).

THE FALLACY OF PLANETARY HOMEOSTASIS

No matter how many general benefits may arise from the activities of organisms, they can never provide a valid parallel between the collective effects and an organism's homeostatic controls. The analogy confuses mass-action equilibria or input-output balances with homeostasis. Homeostasis requires a feedback loop from a sensor of the controlled variable to the machinery by which the control is achieved. The thermostatic control of house temperature is an excellent if trite example of homeostasis, which is found only in living organisms or, like that persistently comfortable house, in their products.

Homeostasis should be recognized only if special sensory and control machinery can be shown. Stability of some condition, even if the condition is rapidly restored after disturbance, should not be uncritically attributed to homeostatic machinery. For instance, if I could magically cause all atmospheric moisture to condense and fall to the surface, the resulting zero humidity everywhere on Earth would be an extremely temporary condition. Evaporation from the oceans and other wet surfaces would, perhaps in a few days, restore the atmospheric humidity to something in the normal range. This would not be homeostasis. The long-term maintenance of the normal range of total atmospheric moisture is a direct result of a balance between evaporation and condensation. There is no sensor that monitors the humidity and no control machinery activated by abnormality.

The regularity of periodic geyser eruptions is Falk's (1981) well-chosen example of stability without homeostasis. A nearly constant inflow of shallow subterranean water with nearly constant heating far below sooner or later results in an explosive release of superheated water. The only operative factors are the hole, the heat, the water, and its eruptions, which Falk calls the "main effects." No feedback loop activates the eruption machinery as a result of monitoring temperature or pressure or the passage of time. Falk concludes that Old Faithful is not an organism because it is not homeostatic. I would say the same about Gaia for the same reason.

WHAT WOULD A REALLY FIT ENVIRONMENT BE LIKE?

Adaptation is always asymmetrical; organisms adapt to their environments, never vice versa. If the environments at the surface of the Earth seem well suited to living organisms, it is simply because those are the environments to which organisms have adapted. It must certainly be true that life, however broadly conceived, can exist only within a certain range of physical conditions. Terrestrial environments fall within that otherwise unknown range. Lunar and martian ones apparently do not.

I have trouble imagining any kind of life that is not based on chemical reactions between surfaces and diffusible substrates in a fluid medium. Perhaps the medium must be liquid, with no adequately controlled metabolism possible in a gas. It may even be, as Henderson (1913) argued, that all other candidates (e.g., hydrocarbons and liquid ammonia) can be ruled out, and that only liquid water is a suitable medium. If so, the cosmos is indeed hostile to life, because liquid water can exist only under special conditions.

But perhaps not as special as they seem at first glance. The boiling point is not a property of water, but a relationship between water and pressure. Liquid water can exist in a household pressure cooker more than 20°C hotter than in the open air. The freezing point of pure water is also a function of pressure, so that subterranean waters can stay liquid under permafrost at many degrees below 0°C. Equally important is the fact that both boiling and freezing points are sensitive to solute concentrations. Seawater freezes at about −2°C, and boils at about 103°C at a pressure of one atmosphere.

Lovelock (1988) believes that Gaia keeps most habitats between 0° and 30°C because that is the most favorable range for organisms. He also believes that aquatic organisms are favored by moderate solute concentrations, and that the salt content of seawater is near the upper end of the favorable range. It is certainly true that the greatest diversity of organisms lives in these ranges of temperature and salinity, and that life is less abundant and diverse outside this range.

My explanation (and that of Craik, 1989) of these observations is that organisms will be most abundant under the most common and persistent conditions, where they will have had the greatest opportunity to adapt. Diversity decreases with increasing latitude, not because low temperature is biologically unfavorable, but because the far north and far south have included seasonally liquid water for only a few thousand years. Diversity is likewise low in hot springs and hypersaline desert sumps, not because the conditions in such places are extreme, but because they are local and new on an evolutionary time scale. Coral reef and rainforest communities are more diverse because they have been there for many millions of years, perhaps 70 million for the southeast Asiatic rainforest (Thórhallsdóttir, 1989).

Salinity gives a good test of extremity-versus-brevity interpretations. Oceans and fresh waters have been present as large continuous habitats throughout the Phanerozoic. Medium-salinity intermediates have been local and ephemeral. Today's brackish estuaries, such as the Chesapeake and the lower Shannon, are products of Recent sea-level changes, and are only a few thousand years old. Biotic diversity in these new and scattered habitats is severely depressed compared to nearby fully marine or fully freshwater habitats (Giere, 1968). Suppose it were the intermediate salinities, like intermediate temperatures, that were most common and had the most diverse communities. I would expect Gaia proponents to argue that brackish water is most favorable to organisms, and that this is the reason that Gaia maintains mostly intermediate salinities.

Craik (1989) and Pool (1990) have provided handy summaries of what is known of some organisms' tolerance of extreme conditions. Some bacteria thrive in deep-sea hydrothermal vents at temperatures at least up to 110°C and perhaps considerably higher. Other bacteria thrive in pH ranges from 1.0 to 11.5. Most processes studied by biochemists are severely disturbed by pressures of 100 atmospheres, but deep-sea bacteria and members of all major animal phyla, including vertebrates, may live out their lives at pressures of more than a thousand atmospheres.

Some marine algae show a tolerance of extreme conditions not considered by Craik or Pool. They are trapped in sea ice as the surface freezes and must spend the winter there. The return of springtime light enables them to photosynthesize, and then the summer sun melts the ice and disperses them (McConnville and Wetherbee, 1983; Palmisano and Sullivan, 1983). The formation of sea ice separates ice crystals from trapped water, which must therefore reach ever higher salinities as temperatures fall and more ice forms. Measurement of physical conditions in the microscopic channels of ice-bound water is difficult, but the resident algal cells obviously experience salinities far above that of normal seawater, and temperatures far below the seawater freezing point of −2°C. It is possible that they live in saturated brine near Fahrenheit's theoretical minimum of −18°C. Not only do they survive this imprisonment in sea ice, they can photosynthesize and grow at an appreciable rate under those extreme conditions.

It would seem that evolution has been capable of producing organisms that can be active and thrive at temperatures from perhaps as low as −18°C to at least 110°C, in the entire range of naturally occurring salinities, at pressures from a fraction of an atmosphere to more than a thousand, and over most of the possible pH spectrum. Preoccupation by Lovelock (1988) and other Gaia proponents with the maintenance of the narrow range of conditions normally experienced by human life seems a bit parochial. Until some real exobiological data are available, we can only speculate about what physical and chemical conditions are most favorable for the origin and evolution of life. Perhaps such data will come from exploring the dense hydroatmospheres of the major planets, where chemically complex conditions are stable for immense periods of time.

EVOLUTION AND ETHICS

Huxley (1894) was concerned with several issues of immense philosophical importance: the hostility of the cosmos to human life and aspirations, a moral evaluation of the evolutionary process and its products, and the evolutionary origin of the human moral impulse. My discussions above dealt with the first of these topics. I have already had much to say about the second and third (Williams, 1988; 1989a,b). Here I will merely point out that these works abundantly support Huxley's (1894:73–74) view that "immeasurable mischief" has been done by "the moralizing of sentimentalists" and their belief that nature can provide "an exemplar for human conduct." I should also point out that Richards (1987:Appendix 2) has provided a literate and well-informed defense of a contrary opinion.

REFERENCES

Collier, A. 1958. Gulf of Mexico physical and chemical data from *Alaska* cruises. *U.S. Fish and Wildlife Service, Special Scientific Report—Fisheries*, 249:1–417.

Craik, J. C. A. 1989. The Gaia hypothesis—fact or fancy. *J. Mar. Biol. Assoc. U. K.*, 69:759–768.

Dawkins, R. 1982. *The Extended Phenotype.* W. H. Freeman, Oxford.

Falk, A. E. 1981. Purpose, feedback, and evolution. *Philos. Sci.*, 48:198–217.

Futuyma, D. J. 1986. *Evolutionary Biology*, 2nd ed. Sinauer Associates, Sunderland.

Giere, O. 1968. Die Fluctuationen des marinen Zooplankton im Elbe-Aestuar. *Arch. Hydrobiol.*, 31(Suppl.):379–546.

Henderson, L. J. 1913. *The Fitness of the Environment*, 1958 (reprint). Beacon Press, Boston.

Huxley, T. H. 1894. *Evolution and Ethics.* In J. Paradis and G. C. Williams (eds.), *T. H. Huxley's* Evolution and Ethics, *With New Essays on its Victorian and Sociobiological Context*, 1989 (reprint). Princeton University Press, Princeton.

Kerr, R. A. 1988. No longer willful, Gaia becomes respectable. *Science*, 240:393–395.

Krebs, J. R., and N. B. Davies. 1987. *An Introduction to Behavioral Ecology.* Blackwell, Oxford.

Lillie, R. S. 1913. The fitness of the environment. *Science*, 38:337–342.

Lovelock, J. 1979. *Gaia: A New Look at Life on Earth.* Oxford University Press, Oxford.

———. 1988. *The Ages of Gaia.* W. W. Norton & Co., New York.

Lovelock, J., and L. Margulis. 1974. Biological modulation of the Earth's atmosphere. *Icarus*, 21:471–489.

McConnville, M. J., and R. Wetherbee. 1983. The bottom-ice microalgal community from annual ice in the inshore waters of east Antarctica. *J. Phycol.*, 19:431–439.

Mann, C. 1991. Lynn Margulis: Science's unruly earth mother. *Science*, 252:378–381.

Mutch, R. W. 1970. Wildfires and ecosystems: a hypothesis. *Ecology*, 51:1046–1051.

Noble, I. R., and R. O. Slatyer. 1980. The use of vital attributes to predict successional changes in plant communities subject to recurrent disturbances. *Vegetatio*, 43:5–21.

Palmisano, A. C., and C. W. Sullivan. 1983. Sea-ice microbial communities (SIMCO). 1. Distribution, abundance, and primary production of ice microalgae in McMurdo Sound, Antarctica, in 1980. *Polar Biol.*, 2:171–177.

Perlmutter, A. 1961. Possible effect of lethal visible light on year-class fluctuations of aquatic animals. *Science*, 133:1081–1082.

Pool, R. 1990. Pushing the envelope of life. *Science*, 247:158–160.

Pyne, S. J. 1991. *Burning Bush: A Fire History of Australia.* Henry Holt and Co., New York.

Richards, R. J. 1987. *Darwinism and the Emergence of Evolutionary Theories of Mind and Behavior.* The University of Chicago Press, Chicago.

Thórhallsdóttir, T. E. 1989. Regnskógar hitabeltisins [rainforests of the tropical zone]. *Náttúrufraedingurinn*, 59:9–37.

Van Valen, L. M. 1975. Group selection, sex, and fossils. *Evolution*, 29:87–94.

Wetherald, R. T., and S. Manabe. 1975. The effects of changing the solar constant on the climate of a general circulation model. *J. Atmos. Sci.*, 32:2044–2059.

Williams, G. C. 1988. Huxley's *Evolution and Ethics* in sociobiological perspective. *Zygon*, 23:383–407.

———. 1989a. A sociobiological expansion of *Evolution and Ethics*, pp. 179–214, 228–236. In J. Paradis and G. C. Williams (eds.), *T. H. Huxley's* Evolution and Ethics, *With New Essays on its Victorian and Sociobiological Context.* Princeton University Press, Princeton.

———. 1989b. Review of Richards, R. J., *Darwinism and the Emergence of Evolutionary Theories of Mind and Behavior. J. Evol. Biol.*, 2:385–387.

———. 1992. *Natural Selection: Domains, Levels, and Challenges.* Oxford University Press, New York.

Wilson, D. S. 1980. *The Natural Selection of Populations and Communities.* Benjamin/Cummings, Menlo Park.

Carroll, S. B. 1995. Homeotic genes and the evolution of arthropods and chordates. *Nature* 376:479–485.

Biology in the twenty-first century faces a great research frontier: ontogenetic decoding, or the egg-to-adult transformation, and the problem of how the unidimensional genetic information encoded in the DNA of a single cell becomes transformed into a four-dimensional being, the individual that lives in time to grow, mature, reproduce, and die. Cancer, disease, and aging are epiphenomena of ontogenetic decoding.

The instructions that guide the ontogenetic process are carried in the hereditary material. The theory of biological heredity was formulated in 1865 by the Augustinian monk Gregor Mendel, but it became generally known by biologists only in 1900. The first important step toward understanding how the genetic information is decoded came when G. W. Beadle and E. L. Tatum (1941; found in this volume) demonstrated that genes determine the synthesis of enzymes, which are the catalysts that control chemical reactions in living beings. Chemical reactions must occur in an orderly manner. Organisms must have ways of switching genes on and off, since different sets of genes are active in different cells and at different times. The first control mechanism was discovered in 1961 by François Jacob and Jacques Monod for a gene that encodes enzymes that digest the sugar lactose in the bacterium *Escherichia coli*.

The investigation of gene control systems in mammals and other complex organisms became possible in the mid-1970s with the development of recombinant DNA techniques. In addition to the alternating sequence of coding and noncoding DNA segments, mammalian genes contain short control sequences, like those in bacteria but typically more numerous and complex, that act as control switches and signal where the coding sequence begins and ends. In mammals, insects, and other complex organisms, there are control circuits that operate at higher levels than the control mechanisms that activate and deactivate individual genes. Some of these higher-level circuits act on sets rather than individual genes. For example, *homeotic* genes encode factors that regulate gene transcription in such a way as to help govern somatic differentiation along the primary body axes and also play regulatory roles in the construction of secondary body axes such as limbs. In this article, Sean Carroll explains and explores many questions that must be resolved in order to elucidate the egg-to-adult transformation in multicellular animals.

RELATED READING

Carroll, S. B. 2006. Endless forms most beautiful: the new science of evo devo and the making of the animal kingdom. Norton, New York, New York, USA.

Frankel, N., S. Wang, and D. L. Stern. 2012. Conserved regulatory architecture underlies parallel genetic changes and convergent phenotypic evolution. Proceedings of the National Academy of Sciences USA 109:20975–20979.

Gehring, W. J. 2002. The genetic control of eye development and its implications for the evolution of various eye types. International Journal of Developmental Biology 46:65–73.

Hoffer, A., J. Xiang, and L. Pick. 2013. Variation and constraint in *Hox* gene evolution. Proceedings of the National Academy of Sciences USA 110:2211–2216.

Jacob, F., and J. Monod. 1961. Genetic regulatory mechanisms in the synthesis of proteins. Journal of Molecular Biology 3:318–356.

Lewis, E. B. 1978. A gene complex controlling segmentation in *Drosophila*. Nature 276:565–570.

Nüslein-Volhard, C., and E. Wieschaus. 1980. Mutations affecting segment number and polarity in *Drosophila*. Nature 287:795–801.

Shubin, N. 2008. Your inner fish: a journey into the 3.5-billion-year history of the human body. Pantheon, New York, New York, USA.

Homeotic genes and the evolution of arthropods and chordates

Sean B. Carroll

Clusters of homeotic genes sculpt the morphology of animal body plans and body parts. Different body patterns may evolve through changes in homeotic gene number, regulation or function. Recent evidence suggests that homeotic gene clusters were duplicated early in vertebrate evolution, but the generation of arthropod and tetrapod diversity has largely involved regulatory changes in the expression of conserved arrays of homeotic genes and the evolution of interactions between homeotic proteins and the genes they regulate.

ARTHROPODS, in terms of their diversity and sheer numbers, and chordates, because of their anatomical and behavioural complexity, are two of the most successful animal phyla. The evolution of both groups has been marked by numerous innovations and modifications to their respective body plans, the archetypes of which arose more than 500 million years ago[1]. Although arthropod and vertebrate phylogeny have long been approached through systematics and palaeontology, the genetic basis for the morphological diversity of these or any other animals has, until recently, been beyond the reach of biology.

How do body plans and body parts evolve? Differences in morphology are, of course, the developmental products of genetic differences between animals, but it has not been clear how many or what kind of differences underlie changes in body patterns. Do gross changes in morphology, such as the evolution of fish, snakes and mammals, or the invention of the insect wing, involve distinct genetic mechanisms from the evolution of different kinds of butterflies?

Early investigations into the nature of genetic evolution identified two potential mechanisms underlying the origin of new features. Ohno[2], for example, stressed the role of gene duplication and divergence in evolution, whereas Wilson[3] and Jacob[4] emphasized the power of regulatory changes in gene expression. But in order to determine how new genes or regulatory innovations might affect morphological evolution, it is essential to know which genes control morphology[3,5,6].

One of the most important biological discoveries of the past decade is that arthropods and chordates, and indeed most or all other animals, share a special family of genes, the homeotic (or *Hox*) genes, which are important for determining body pattern. The diversity of *Hox*-regulated features in arthropods (segment morphology, appendage number and pattern) and vertebrates (vertebral morphology, limb and central nervous system pattern) suggests that the *Hox* genes are implicated at some level in the morphological evolution of these animals.

After a brief review of *Hox* gene organization and function in development, I will describe several examples of morphological differences among arthropods and chordates involving evolution at the *Hox* gene level that have recently been discovered. These include changes in the regulation of *Hox* genes and the evolution of interactions between *Hox* gene products and the genes they regulate. The evidence indicates that the duplication of *Hox* clusters and other developmental genes between primitive chordates and early vertebrates enabled the evolution of the anatomical complexity of vertebrates. However, the diversity of arthropods, insects and vertebrates has arisen primarily through regulatory evolution.

Homeotic genes in development

Early studies of homeotic genes were almost entirely restricted to *Drosophila melanogaster*, where these genes were known to control the identity (that is, the unique appearance) of different body segments along the anteroposterior (A–P) axis of the embryo and adult fly. Subsequently, *Hox* genes have been found in all sorts of animals, including hydra[7], nematodes[8], and all arthropods[9] and chordates[10]. Three remarkable conserved features unite the *Hox* genes of higher animals: (1) their organization in gene complexes; (2) their expression in discrete regions in the same relative order along the main (A–P) body axis[11]; and (3) their possession of a sequence of 180 base pairs (the homeobox) encoding a DNA-binding motif (the homeodomain).

***Hox* gene organization and expression.** In *Drosophila* there are eight *Hox* genes among two complexes, the Antennapedia Complex (ANT-C)[12] and Bithorax Complex (BX-C)[13], which act in different regions of the animal (other homeobox genes of the ANT-C that are not homeotic genes will not be considered here; see ref. 9 for discussion). There is a striking correlation between the order of these genes on the chromosome and the position of their expression in the developing animal (Fig. 1, top). In beetles, one gene complex contains the same set of genes, and they control the unique appearance of different beetle segments in an anterior to posterior order[14]. Vertebrate *Hox* genes are also organized in clusters which are deployed in an anterior to posterior order according to their position in the complex[15]. Vertebrates have four clusters of 9 to 11 *Hox* genes which can be aligned into 13 sets (called paralogous groups) based upon their sequence, organization and thus homology to their *Drosophila* relatives (Fig. 1). Not all genes are represented in each vertebrate cluster, and some genes within clusters appear to have been duplicated since the invertebrate/vertebrate divergence. For example, the *Hox* gene most similar to *Abdominal-B* has been duplicated several times and exists in multiple copies (*Hox-9–13*) in three of the four clusters (A, C and D), suggesting that its representation was expanded before the clusters were duplicated.

The conservation of *Hox* genes may at first seem paradoxical. How can animals that have the same array of *Hox* genes appear so different? For example, flies and beetles have the same complement of *Hox* genes. Moreover, mammals and teleost fish[16] (R. Krumlauf, personal communication; D. Duboule, personal communication) each have four *Hox* clusters (and at least the *Hox B* clusters contain the same set of genes (R. Krumlauf, personal communication)), and these genes are expressed in a conserved relative order along the main body axis[11]. For insight into this puzzle, we must appreciate that *Hox* genes act only to demarcate relative positions in animals rather than to specify any particular structure[11]. Within one species, different *Hox* genes control the morphology of different body regions (without *Hox* genes all insect segments appear alike[13,17]). Between species, the same *Hox* gene can regulate the homologous segment or body region in different ways. The key to understanding how *Hox*

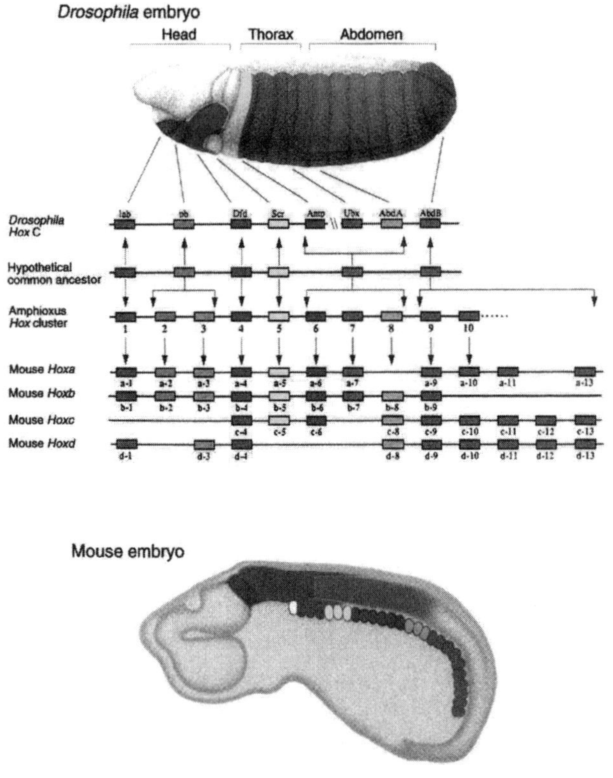

FIG. 1 *Hox* gene organization and expression. Top, the A–P domains of *Drosophila Hox* gene expression correspond to the order of the genes within the Hox complex. Middle, the evolutionary relationship between the *Drosophila*, Amphioxus and mouse *Hox* clusters, and the deduced complement of *Hox* genes in the presumed common ancestor of arthropods and chordates. Bottom, the A–P domains of mouse *Hox* genes within the developing mouse also correspond to gene order in the Hox complexes. Adapted from refs 50, 52 and 75.

genes control morphology and diversity is based on their action as regulatory proteins and the wide range of target genes regulated by different *Hox* genes in one animal, and by the same *Hox* gene in different animals.

Hox proteins and the genes they regulate. The large effects of *Hox* genes on morphology suggest that they regulate, directly and indirectly, large numbers of genes. The promiscuity of *Hox* protein function may stem from the relatively simple sequences recognized by this class of DNA-binding proteins. The *Hox* proteins bind to a four-base core sequence[18,19], and there is a lot of flexibility concerning the sequence surrounding this core. This makes it easy to imagine how *Hox* binding sites can be gained or lost by target genes. Indeed, the regulation of several *in vivo* targets of the yeast $\alpha2$ homeodomain protein has been examined[20], and a notable tolerance for different bases within $\alpha2$ recognition sites was found. Most mutations in target sequences produced only small effects on homeodomain binding *in vitro* and target expression *in vivo*, suggesting that the 'relaxed' DNA-binding specificity of homeodomains may allow new regulatory interactions to evolve readily amaong *Hox* proteins and potential target genes.

The identity of genes regulated by the *Hox* proteins is a crucial issue because these genes must mediate the cellular processes involved in morphogenesis. In *Drosophila*, proven *Hox* targets include genes encoding other transcriptional regulatory proteins, secreted signalling proteins, and structural proteins, but these few examples may be just the tip of the iceberg[21]. It has recently been estimated that the *Drosophila* genome contains 85 to 170 genes that are regulated by the product of the *Ultrabithorax* gene[22]. This estimation underscores the monumental challenge

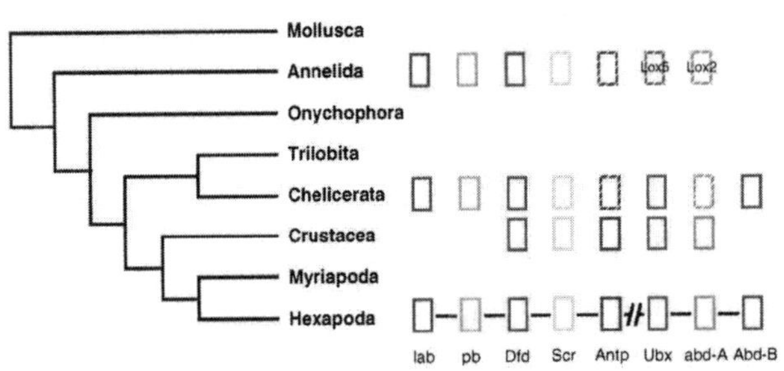

FIG. 2 *Hox* genes and arthropod phylogeny. The array of *Hox* genes detected in arthropods and other invertebrates is displayed in comparison to the *Drosophila Hox* genes. By comparing annelids with arthropods and insects, it appears that most *Hox* genes pre-date the annelid/arthropod divergence, and all insect *Hox* genes pre-date the insect/crustacean divergence. Solid boxes represent confirmed *Hox* genes; stippled boxes are *Hox* genes that are too similar to another to be distinguished on the basis of available data; missing boxes represent the absence of information. Data are from surveys of the *Hox* genes of annelids[29], chelicerates[59] and crustacea[28]; the arthropod phylogeny on the left is from ref. 76. Other arthropod phylogenies are possible but do not affect the conclusions regarding *Hox* gene origins.

ahead in deciphering how *Ubx* regulation of these genes in *Drosophila* distinguishes the morphology of body segments and appendages. It also helps to explain why the homeodomain protein sequences are so constrained in evolution (because mutations in the *Ubx* protein would alter the regulation, directly and indirectly, of ptentially more than 100 genes). Finally, it suggests that divergence among these large sets of target genes may distinguish animal patterns.

Homeotic genes in evolution

There are six potential genetic mechanisms through which *Hox* genes could influence morphological evolution: (1) an expansion in the structural diversity of *Hox* genes within a *Hox* complex (for example, when did the different *Hox* genes such as *lab*, *pb*, *Dfd*, *Antp* and *abd-B* arise?); (2) an expansion in the number of *Hox* genes of a given class (for example, the multiple *abd-B*-like genes (paralogues 9–13) found in vertebrates); (3) an expansion in the number of *Hox* complexes; (4) the loss of one or more *Hox* genes; (5) a change in the position, timing or level of *Hox* gene expression; and (6) changes in the regulatory interactions between *Hox* proteins and their targets. Several of these mechanisms are now correlated with the diversification of arthropods and chordates.

The Lewis hypothesis: new genes for new animals? In his landmark 1978 review, years before any *Hox* genes had been cloned, Lewis proposed one of the first explicit hypotheses concerning the genetic basis of morphological evolution, in this case the evolution of insects and flies[13]. Founded upon his pioneering genetic studies of homeotic genes in *Drosophila*, this 'Lewis hypothesis' consisted of three central ideas. First, during the evolution of insects, 'leg-suppressing genes' evolved which removed legs from abdominal segments of millipede-like ancestors. Second, 'haltere-promoting genes' evolved to suppress the second pair of wings of a four-winged ancestor. Third, a tandem array of redundant genes presumably diversified by mutation to produce the BX-C. It was this third idea that prompted the initial structural comparisons of homeotic genes a decade ago, and led directly to the identification of the homeobox[23-25]. The discovery of the homeobox-encoded DNA-binding motif in each *Drosophila* homeotic gene tied together both the history and the biochemical function of homeotic genes, and provided the crucial tool for the study of *Hox* genes in the development and evolution of animals.

The first two ideas suggest that genes of the BX-C, which control the morphology of the posterior thorax and abdomen of *Drosophila*, arose in the course of insect and fly evolution. A survey of annelid[26,27] and arthropod[9,28,29] *Hox* genes demonstrates that this is not the case. All of the arthropod *Hox* genes evolved before the origin of insects, as demonstrated by the occurrence of these genes in crustacea and chelicerates (Fig. 2), two arthropod classes that pre-date the insects. In addition, the unexpectedly diverse array of *Hox* genes that exists in annelids (Fig. 2) suggests that the diversification of the *Hox* genes must have occurred before the annelid/arthropod divergence. Similarly, the array of vertebrate *Hox* genes demonstrates that 6 or 7 *Hox* genes were present in the common ancestor of all three groups[30,31] (Fig. 1, middle), although it is not known what that creature would have looked like[1].

Evolution of *Hox* gene regulation and arthropod body plans. If the array of *Hox* genes is conserved among insects, crustacea and chelicerates, how do we explain the different body plans characteristic of these classes? The major differences between arthropods involves the number, type and organization of body appendages, such as antennae, claws, mouthparts and legs, all of which evolved from a common, ancestral arthropod limb[32]. In insects the homeotic genes regulate where body appendages may form and the type of appendage found on a particular segment. For example, in *Drosophila*, the products of the BX-C suppress limb formation on the abdomen by repressng the *Distal-less* gene which is required for proximodistal axis formation[33], whereas the *Antp* protein promotes leg formation and suppresses antennal development in the second thoracic segment[34]. The homeotic genes are not actualy required to make a limb[17,35], as that potential appears to exist in all segments; rather, the homeotic genes either suppress limb development or modify it to create unique appendage morphologies.

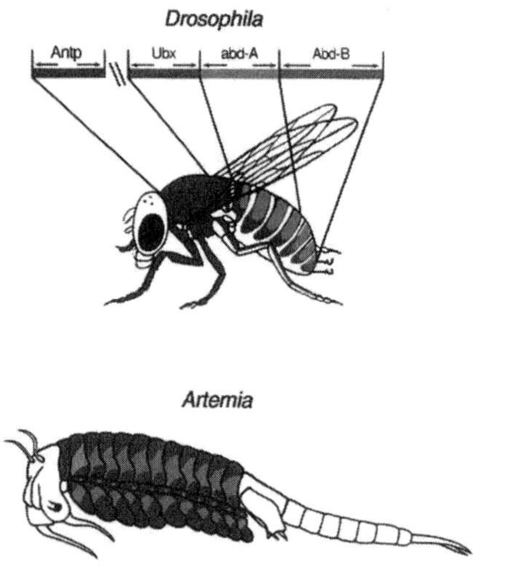

FIG. 3 Evolution of *Hox* gene regulation and the crustacean and insect body plans. The domains of BX-C gene expression have shifted with respect to each other in the thorax and abdomen of insects and crustacea. Top, the *Antp* (purple), *Ubx* (brown), *abd-A* (orange) and *Abd-B* (green) genes sculpt the morphology of the insect trunk. Bottom, *Ubx* and *abd-A* are expressed in relatively more anterior positions in crustacea[36].*

The insect body plan is divided into three main body regions (or tagma), with a head bearing numerous appendages, a thorax bearing three pairs of walking legs, and a limbless abdomen. This pattern is sculpted by the homeotic genes. By contrast, chelicerates have just two distinct tagma, and the number and type of appendages vary considerably within the three crustacean tagma[32]. To determine the extent to which the deployment of homeotic genes reflects the tagmatization of the arthropods, the expression of the three *Hox* proteins, Antp, Ubx and abd-A, has been examined in a crustacean[36]. Surprisingly, Ubx and abd-A are expressed in nearly all of the limb-bearing thoracic segments, far more anteriorly than in insects (Fig. 3). This demonstrates that the expression domains of *Hox* genes have shifted considerably between crustacea and insects, and that the interactions between *Hox* proteins and some key developmental programs, such as limb formation, have changed as well.

The scope of regulatory evolution among the *Hox* genes of arthropods is not yet known. The analysis of different crustacea (which exhibit a wide range of body plans[32]), chelicerates, myriapods and the putative sister group of the arthropods, the onychophora, will be crucial. It is of special interest to know where the same *Hox* genes are deployed in these last two groups, in which all trunk segments are alike, as this represents the probable ancestral arthropod condition. The *Hox* genes are likely to be present, but they might not regulate limb formation or even ectodermal patterns in these animals.

Modifications of the insect body plan

It now seems likely that all insect diversity has evolved from a body plan sculpted by the same set of homeotic genes. Yet the insect body plan has not been static; there have been numerous changes in the number and type of appendages, which are

* *This figure is reproduced in color following the index.*

reflected by extant and extinct orders. For example, differences in larval limb and adult wing number, as well as appendage morphology, are regulated by homeotic genes. The study of these characters in modern insects may allow us to gain fundamental insights into their origin and diversification.

Prolegs: a *Hox*-regulated atavism? Abdominal limbs known as prolegs are found on the larvae of various insect species belonging to several orders, but are ubiquitous in the Lepidoptera (moths and butterflies). These limbs were probably present in insect ancestors so it may be that prolegs have reappeared through the de-repression of an ancestral limb developmental program (that is, they may be atavistic). To investigate how homeotic genes influence proleg formation, the expression of several homeotic genes during butterfly development was examined. Proleg formation involves an obvious change in the regulation of *abd-A* expression during embryogenesis[37]. The initial expression of this gene is in the anterior abdomen, as in all insects examined thus far (where it represses limb formation), but it is selectively turned off in small patches of cells within four abdominal segments which then activate limb-patterning genes and begin to form a proleg. It is not known whether switching off *abd-A* underlies proleg formation in other insects, although the phylogenetic and segmental distribution of prolegs[38] suggests that they may arise by a regulatory switch operating on homeotic genes.

The origin and evolution of winged insects. The most important insect innovation was flight, which catalysed the radiation of the most successful subclass of animals, the pterygotes. What was the role of *Hox* genes in this crucial development? The first pterygote orders are long extinct, so we can only gain insight by studying modern insects. The answer is surprising: *Hox* genes were not involved in the development of wings[39]. This conclusion relies mainly on the developmental genetics of wing formation in *Drosophila*[34], and partly on the available fossil record[40,41]. In *Drosophila* and all pterygotes, wings form on the second thoracic segment. This is principally the domain of the *Antp* gene, which suggests that *Antp* would be part of a regulatory code for making a wing. However, this is not true because the wing primordia, the imaginal wing disc and the adult wing all appear normal when the *Antp* gene is removed[39]. Indeed, *Antp* is barely expressed at all in the developing wing[39,42]. The lack of homeotic input into wing formation makes even more sense when we consider the fossil record. When wings first evolved they appeared on all thoracic and abdominal segments[40,41]; this lack of segmental restriction suggests that no homeotic gene positively or negatively regulated wing formation.

In the subsequent course of pterygote evolution, however, homeotic genes did indeed become important, but as repressors and modifiers of wing formation. Orders appeared that lacked abdominal wings, and wings were also lost from the first thoracic segment. This suggests homeotic regulation and, in *Drosophila*, the BX-C and *Scr* genes do repress the wing primordia in these body segments[39]. In flies, which bear one pair of wings on the second thoracic segment and a miniature flight appendage (the haltere) on the third thoracic segment, the *Ubx* gene also represses the size of the flight appendage primordia[38] and modifies the morphology of the developing haltere[43].

The morphological evolution of homologous structures between species. Different *Hox* genes regulate the morphology of serially homologous structures within a species, but what about the homologous structures of different species? In insect terms, for example, can we explain how the hindwings of flies (the haltere), beetles and butterflies are distinct not only from their respective forewings but also from each other? We know in each of these orders that the *Ubx* gene is expressed in and

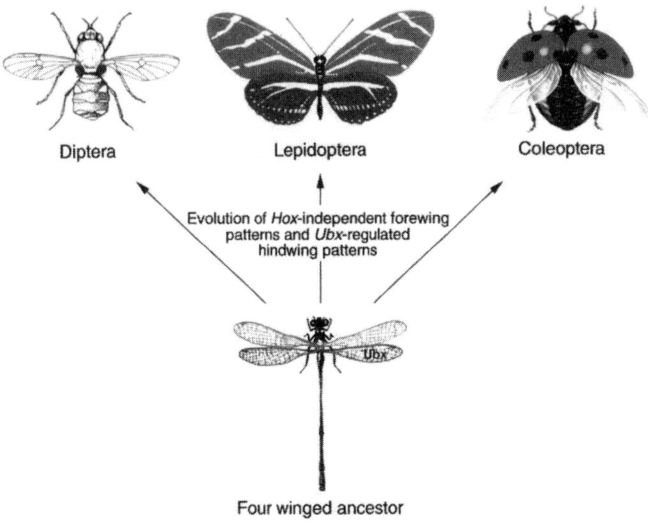

FIG. 4 *Hox* genes and the morphological divergence of homologous structures. The forewings and hindwings differ between insects, and usually from each other. Based on the developmental genetics of wing formation in *Drosophila* and comparisons of homeotic gene expression and function between insects, it appears that forewing patterns evolve independently of homeotic genes, whereas hindwing patterns are *Ubx*-regulated modifications of the respective forewing patterns. In primitive four-winged insects the forewings and hindwings appear similar but have evolved quite different morphologies in various insect orders. The different hindwing morphologies are probably due in part to the divergence of the potentially large sets of wing-patterning genes regulated by *Ubx* in different orders. This divergence is brought about by changes in *cis*-regulatory sequences in target genes that affect their regulation by *Ubx*.

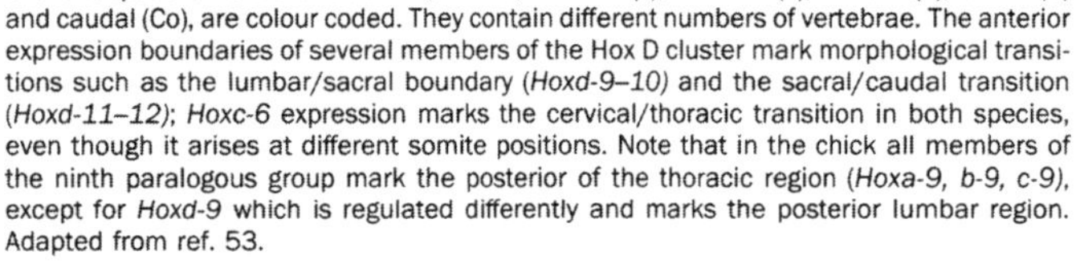

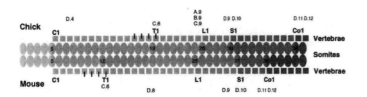

FIG. 5 *Hox* genes and the evolution of vertebrate axial morphology. The expression of various *Hox* genes along the chick and mouse body axes are depicted. The different regions of the respective vertebral columns, that is, cervical (c), thoracic (T), lumbar (L), sacral (S) and caudal (Co), are colour coded. They contain different numbers of vertebrae. The anterior expression boundaries of several members of the Hox D cluster mark morphological transitions such as the lumbar/sacral boundary (*Hoxd-9–10*) and the sacral/caudal transition (*Hoxd-11–12*); *Hoxc-6* expression marks the cervical/thoracic transition in both species, even though it arises at different somite positions. Note that in the chick all members of the ninth paralogous group mark the posterior of the thoracic region (*Hoxa-9, b-9, c-9*), except for *Hoxd-9* which is regulated differently and marks the posterior lumbar region. Adapted from ref. 53.

differentiates the hindwing from the forewing[14,37,44,45]. Because *Antp* is not involved in forewing patterning[39], the differences between insect forewings are all independent of homeotic genes. However, the considerable differences in the relative size and morphology of the hindwings must also be due to differences in the wide range of target genes regulated by *Ubx* (Fig. 4). If the set of potential target genes in *Drosophila* is as large as recent studies suggest[22], it is easy to imagine that many differences in *Ubx*-regulated target genes exist between species, and that such differences may evolve readily. For example, butterfly hindwing patterns are usually distinct from forewings and are species specific[46]. This diversity could be generated at least in part through the gain or loss of *Ubx* binding sites in the *cis*-regulatory sequences of a myriad of potential target genes.

Rephrasing the Lewis hypothesis. The general theme emerging from developmental studies of *Hox* genes[10,11,47], and echoed by the above examples, is that homeotic genes do not constitute an instructional code that says 'make a wing' or 'make a leg'. Rather, they modify developmental programs such that what would otherwise be a wing becomes a haltere, or what would develop as an antenna becomes a leg. In terms of homeotic genes and Lewis's original hypothesis, what has evolved in the course of insect and fly evolution are not new genes but new regulatory interactions between BX-C proteins and genes involved in limb formation and wing morphogenesis.

Hox genes and chordate evolution

The course of vertebrate evolution has been marked by many changes in the chordate body plan. The evolution of the head with paired eyes, jaws and teeth, the origin of the tetrapods and subsequent evolution of limbless forms, and the diverse axial morphologies of modern vertebrates are dramatic examples of large-scale morphological evolution. Developmental analyses of *Hox* gene expression and function implicate the homeotic genes in craniofacial development, hindbrain organization, axial morphology, limb patterning and many other facets of vertebrate development[15,47], which begs the question of their potential roles in the evolution of the chordates. The issues parallel those of the arthropods, discussed above: did the appearance of new genes enable the origin of new structures? How have these new structures been modified in the course of evolution?

The examination of these issues in the chordates is facilitated by the relative richness of vertebrate palaeontology, but made more difficult by the genetic and anatomical complexity of the vertebrates. The duplicated *Hox* complexes of modern vertebrates may have allowed the diversification of various homeotic genes but also increases the possibility of genetic redundancy, which conceals individual *Hox* functions in gene knockout experiments[48]. Furthermore, during the formation of a given structure such as a limb, the number of *Hox* genes involved, the spatial and temporal dynamics of gene expression, and the increasing cellular complexity of a growing three-dimensional structure makes individual mutant phenotypes difficult to interpret[49]. Nevertheless, the comparison of *Hox* gene organization and expression in primitive chordates and representative vertebrates has provided several examples of *Hox* gene involvement in chordate evolution.

The origin of the vertebrate body plan. Could the origin of new *Hox* genes underlie the origin of vertebrate characters? To address this question, the organization of *Hox* genes was examined[50] in amphioxus, a cephalochordate possessing a notochord, dorsal nerve cord, and paired somites, but lacking most vertebrate head structures, appendages and various other features. Remarkably, this primitive creature possesses a single *Hox* complex that appears to be the archetype of *Hox* clusters of modern vertebrates. Two features of the amphioxus cluster stand out. First, it contains homologues of at least the first ten groups of vertebrate *Hox* genes in a collinear array. Second, it contains at least two *Abd-B*-like genes, which indicates that the tandem duplication of these genes and their selective retention in the cephalochordates preceded, or was at least underway before, the evolution of features unique to vertebrates (such as the limbs in which these genes are important[49]). Thus the origin of new genes via the expansion and diversification of the *Hox* clusters may well have been important in enabling the evolution of more complex chordate body plans.

The number and organization of *Hox* genes in other primitive chordates and vertebrates are now being examined to determine whether there were intermediate stages of *Hox* cluster duplication in the evolution of the vertebrates. In jawless fish (such as lamprey and hagfish) there appear to be fewer *Hox* genes and perhaps as few as two *Hox* clusters (although there could be more)[51]. Based upon the available *Hox* data and several other gene families, it has been suggested[30] that there were two phases of cluster duplication, one close to the origin of the vertebrates and one close to the origin of jawed vertebrates.

***Hox* gene regulation and the evolution of vertebrate axial morphology.** The mesodermal segments of vertebrates, called somites, are serially homologous and give rise to the vertebrae. The number and morphology of vertebrae that contribute to each distinct region of the vertebral column (the cervical, thoracic, lumbar, sacral and caudal vertebrae) often differs between vertebrates (Fig. 5). For example, mammals (whether mice or giraffes) usually have seven cervical vertebrae, birds have between 13 and 25, and snakes have up to 454 pre-caudal vertebrae. Because *Hox* genes are expressed at distinct levels along the A–P axis of vertebrate embryos, and loss or ectopic expression of individual *Hox* genes can transform vertebrae from one type to another, it has been proposed that the combination of *Hox* genes expressed in each somite specifies the different vertebral morphologies[52]. This would imply that different axial morphologies could be created by shifting *Hox* gene expression domains up or down the A–P axis or by changing the response of *Hox*-regulated genes to a fixed pattern of *Hox* gene expression (for example, at a given somite number).

By comparing the expression of a large set of *Hox* genes between mice and birds, it has been demonstrated[53] that *Hox* gene domains shift in parallel with vertebral anatomy and are independent of the number of vertebrae within a given body region (summarized in Fig. 5). For example, *Hox* genes of the fifth paralogue group (for example, *Hoxc-5*) are expressed at the level of the forelimb which arises at somite level 17–18 in the chick and 10–11 in the mouse. Similarly, the anterior expression boundary of *Hoxc-6* falls at the boundary of the neck (cervical) and thorax in mice, chickens, geese (which have 17 cervical vertebrae, 3 more than chickens) and frogs (with a highly derived vertebral morphology and only 3 or 4 cervical vertebrae). The thoracic–lumbar transition appears to be associated with expression of the *Hoxa-9, Hoxb-9* and *Hoxc-9* genes, but the fourth member of this paralogue group, *Hoxd-9*, is expressed out of register. This may be significant because the thoracic–lumbar distinction is not general among tetrapods. It may be that shifts within the *Hox-9* group were important in the evolution of this transition from a more uniform trunk, perhaps even in the evolution of the tetrapods from fish. For *Hox* gene expression domains to shift according to anatomy and not somite level, either the upstream regulators of *Hox* gene expression are deployed differently in different vertebrates, and/or the response of *Hox* genes to these regulators has changed. The fact that individual members of paralogous groups have different sites of expression within a species demonstrates that the response to the same upstream regulators can differ, and suggests that the evolution of regulatory diversity within paralogue groups may have enabled major changes in axial morphology (as well as in limb patterns; see below).

Origin of the tetrapods. The earliest tetrapod-like fossils suggest that the vertebrate hindlimb evolved first from the pelvic fin of fish[54], with the forelimb evolving subsequently from the pectoral fin. It is difficult to explain how the fore- and hindlimbs of tetrapods are so similar to each other when they evolved from different fish fins. It has been suggested that a major change in *Hox* gene expression brought about the serial homology of the forelimb and hindlimb[55].

Hox genes are expressed both in the mesoderm adjacent to the limbs and in the developing limb buds of modern tetrapods and are required for anteroposterior and proximodistal patterning. Different Hox C genes are expressed in the fore- and hindlimb with their respective expression extending from the adjacent axial mesoderm. However, the Hox A and Hox D cluster genes expressed in the two limbs are those that are normally expressed in the posterior of the main vertebrate body axis, and they are deployed in very similar spatiotemporal patterns in each limb. Because the Hox A and Hox D genes are being expressed far anterior to their axial site of expression, it has been suggested[55] that, in the course of tetrapod evolution, the Hox A and Hox D cluster genes were ectopically activated in the pectoral fin which led to its transformation from fin to limb and accounts for the serial homology between fore- and hindlimbs of modern terrestrial vertebrates. The ectopic activation of *Hox* genes could most readily be accomplished by the expression of a new signalling centre in the nascent forelimb. The inspection of *Hox* gene expression patterns in living relics, such as the lungfish, coelacanths and sharks, could test the validity of this intriguing model for *Hox* gene function in tetrapod origins.

Hox genes and constraint: could snakes learn to walk (again)? Much of this review has focused on the evolutionary inferences drawn from studies of *Hox* gene organization, expression and function in a few animals. It may also be worthwhile to consider what might be evolutionarily possible but has not been observed (yet). For example, there is no clear case of *Hox* gene loss being implicated in the modification of the vertebrate body plan. Is there such constraint on *Hox* organization and function that even single members of paralogue groups are indispensable? Where might we test such an idea? Perhaps in snakes, which forfeited their limbs, much of their axial morphology, and elements of craniofacial architecture (for example, ears) in the course of their evolution as burrowing reptiles approximately 150 Myr ago[56]. (Their recent radiation as surface-dwelling mammal-hunters is linked to the abundance of prey and the evolution of prey-immobilizing toxins, not the acquisition of new structures.) Given that individual *Hox* genes were lost after the expansion to four *Hox* clusters in vertebtate ancestors[57], and that mice can survive without certain individual genes[15], it seems possible that snakes could possess fewer *Hox* genes than other vertebrates.

Snakes could reveal what sort of constraints exist on the *Hox* clusters. For example, if the *Hox* genes have all been preserved then regulatory changes can probably explain their unique axial morphology. This could also suggest that the potential for specifying limb position and morphology is still preserved (a possibility supported by the presence of hindlimb rudiments in pythons and boas). If so, is it conceivable that snakes or other limbless tetrapods might regain their limbs? Probably not, as the evolution of limblessness is strongly correlated with body elongation and increased vertebral number, and the constraints against a return towards the primitive tetrapod form may be prohibitive[58]. If *Hox* genes have been lost, however, then we might learn some valuable lessons about genetic redundancy from the identity of the lost genes.

The other side of the constraint issue is illustrated by the horseshoe crab, *Limulus polyphemus*. This chelicerate arthropod may possess up to four *Hox* clusters which include the genes found in the *Drosophila* cluster as well as additional genes present in vertebrate *Hox* clusters[59]. Although the multiple *Hox* clusters could be due to a genome-wide polyploidization event, this must have occurred at least 75 Myr ago. The static body pattern of horseshoe crabs suggests that *Hox* complexity and anatomical complexity are not strictly correlated. If *Hox* genes create possibilities, *Limulus* has apparently not found any creative uses for a lot of *Hox* genes (or other genes), even though selection has retained them. Might this reflect some constraint imposed by the chelicerate or arthropod body plan that prevents its modification[60]? After 600 Myr to expand their *Hox* clusters, it seems surprising that arthropods don't have more *Hox* genes; is this because they couldn't find ways to use them?

The primacy of regulatory evolution

The available evidence suggests that primitive arthropods and chordates each possessed a single *Hox* complex containing the diverse array of *Hox* genes found in their modern descendants. Although this cluster was duplicated in the chordates, presumably in one or two early phases in their evolution, it appears that the subsequent course of vertebrate evolution from primitive bony fishes to mammals, and the entire course of arthropod evolution, was founded upon conserved sets of *Hox* genes. The phylogeny of *Hox* genes and the many examples cited above of large-scale morphological changes associated with diversity in *Hox* gene regulation and target regulation suggest that the primary genetic mechanism enabling morphological diversity among arthropods and vertebrates is regulatory evolution.

The anatomical complexity of vertebrates, reflected by a larger relative number of different cell types[61], may be a consequence of a greater number of developmental genes. For example, important developmental gene families such as the *Hox*, *Wnt*, *TGF-β* and *Dll/Dlx* genes are generally several-fold larger in vertebrates than primitive chordates, arthropods or annelids (Table 1 and references cited therein). The structural and regulatory diversity among these genes (especially, for example, the *Wnt* and *TGF-β* families) demonstrates that gene duplication and diversification have probably been an important force in the evolution of the cellular and anatomical complexity of vertebrates. However, the complement of *Hox*[62,16], *Wnt*[63] and *Dlx*[64] genes is comparable in fish and mammals. This suggests that morphological diversity within vertebrates is the product of regulatory evolution within larger, but essentially fixed, families of developmental genes.

How can regulatory evolution be sufficient to explain the differences between trilobites and butterflies, or dinosaurs and sparrows? The creative potential of regulatory evolution lies in the hierarchical and combinatorial nature of the regulatory networks that guide the organization of body plans and the morphogenesis of body parts. We now know that *Hox* genes are regulated by many upstream factors, and that *Hox* proteins act as sculptors that modify the basic arthropod or chordate metamere by modulating the expression of potentially dozens of interacting genes, the products of which determine the cellular events of morphogenesis. Variation in the morphogenetic output of such a multigenic network can arise at many levels simply by tinkering with the relative timing of developmental gene expression (heterochrony) or the interactions between members of the regulatory network. Such regulatory changes are presumably the consequence of alterations in *cis*-regulatory sequences through simple mutations as well as duplications or deletions of regulatory elements. In this manner, one aspect of gene function can

TABLE 1 Different trends in chordates and arthropods

	Principal Hox genes	*Wnt* genes	*TGF-β*-related genes	*Dlx* genes
Annelids	⩾7	⩾2		
Insects	8	4	3	1
Cephalochordates	⩾10	⩾2		
Bony fish	~38	14		⩾5
Mammals	38	14	>25	6

Gene family data are based on the following references: for Hox, refs 15, 27, 50, and D. Duboule and G. Wagner, personal communication; for *Wnt*, refs 30, 63, 70; for *TGF-β*, refs 71, 72; for *Dlx*, refs 64, 73, 74 and, in the zebrafish, M. Westerfield, personal communication.

evolve without altering others. Single genes are often regulated by arrays of discrete regulatory elements which control the pattern, position, timing and level of gene expression, and these features can differ between species[65]. Key body-patterning genes such as the homeotic[66,67] or *Distal-less*[33] genes contain many such elements, which suggests that the evolution of regulatory elements is much more common, and therefore a more continuous source of variation, than the duplication of entire genes.

Conclusion

Our understanding of the role of *Hox* genes in evolution depends on both developmental and comparative studies. One of the most challenging problems is the elucidation of *Hox* protein function in regulating morphology in model experimental species. When the regulatory targets of individual *Hox* proteins are better known then the differential regulation of these genes by different *Hox* proteins in a single species, or by the same protein in different species, can be assessed. The evolutionary value of this information will rely upon the judicious choice of species selected for comparative study.

At a broader level, it is not known how *Hox* genes are organized or expressed in several major arthropod and vertebrate taxa. At the very least, we need to investigate the *Hox* genes of primitive arthropods and possible sister groups as well as all protochordate and vertebrate classes. Of course, arthropods and chordates are only 2 of the 35 or so metazoan phyla and, although studying then provides an understanding of general ways of modifying body plans, it does not explain the origins of *Hox* gene diversity or the evolution of other body plans. The exploration of other phyla, such as the cnidaria, echinoderms, molluscs and flatworms, might lead us to discover when this ancient regulatory 'toolbox' came into existence, and to understand its role in the genesis and modification of these diverse animal body plans.

Finally, it must be appreciated that the comparisons of differences between higher taxa only reveal what has changed, not how it changed. We do not know the rate at which these changes arose or the extent of variation in *Hox*-regulated characters in populations. The idea of macroevolution in a single step, the 'hopeful monster' so often insinuated in the discussion of homeotic genes[68], is widely discredited[69]. The new perspective emerging from the study of *Hox* genes in phylogeny and their regulatory roles in developmental needs to be integrated within the evolutionary frameworks of palaeontology and population biology.

Note added in proof: Sardino *et al.* (*Nature* **375**, 678–681; 1995) have recently shown that the tetrapod foot may be a new structure formed under the control of a destruct phase of *Hox* gene expression in the limb bud. □

Sean B. Carroll is at the Howard Hughes Medical Institute, University of Wisconsin at Madison, R. M. Bock Laboratories, 1525 Linden Drive, Madison, Wisconsin 53706, USA.

1. Conway Morris, S. *Nature* **361,** 219–225 (1993).
2. Ohno, S. *Evolution by Gene Duplication* (Springer, New York, 1970).
3. Wilson, A. C. *Scient. Am.* **253,** 164–173 (1985).
4. Jacob, F. *Science* **196,** 1161–1166 (1977).
5. Gould, S. *Ontogeny and Phylogeny* (Belknap, Cambridge, MA, 1977).
6. Raff, R. & Kaufman, T. *Embryos, Genes and Evolution* (Indiana Univ. Press, Bloomington, 1983).
7. Schummer, M., Scheurlen, I., Schaller, C. & Galliot, B. *EMBO J.* **11,** 1815–1823 (1992).
8. Wang, B. *et al. Cell* **74,** 29–42 (1993).
9. Akam, M. *et al. Development* (suppl.) 209–215 (1994).
10. Kenyon, C. *Cell* **78,** 175–180 (1994).
11. Slack, J., Holland, P. & Graham, C. *Nature* **361,** 490–492 (1993).
12. Kaufman, T., Seeger, M. & Olsen, G. in *Genetic Regulatory Hierarchies in Development* (ed. Wright, T.) (Academic, San Diego, 1990).
13. Lewis, E. B. *Nature* **276,** 565–570 (1978).
14. Beeman, R., Stuart, J., Brown, S. & Denell, R. *BioEssays* **15,** 439–444 (1993).
15. Krumlauf, R. *Cell* **78,** 191–201 (1994).
16. Misof, B. & Wagner, G. *CCE Tech. Report 24, Yale Univ.* (1995).
17. Stuart, J., Brown, S., Beeman, R. & Denell, R. *Nature* **350,** 72–74 (1991).
18. Ekker, S. C. *et al. EMBO J.* **13,** 3551–3560 (1994).
19. Gehring, W. *et al. Cell* **78,** 211–233 (1994).
20. Smith, D. L. & Johnson, A. D. *EMBO J.* **13,** 2378–2387 (1994).
21. Botas, J. *Curr. Biol.* **5,** 1015–1022 (1993).
22. Mastick, G., McKay, R., Oligino, T., Donovan, K. & Lopez, A. *Genetics* **139,** 349–363 (1995).
23. McGinnis, W., Levine, M., Hafen, E., Kuroiwa, A. & Gehring, W. *Nature* **308,** 428–433 (1984).
24. Scott, M. P. & Weiner, A. J. *Proc. natn. Acad. Sci. U.S.A.* **81,** 4115–4119 (1984).
25. McGinnis, B. *Genetics* **137,** 607–611 (1994).
26. Wysocka-Diller, J. W., Aisemberg, G. O., Baumgarten, M., Levine, M. & Macagno, E. R. *Nature* **341,** 760–763 (1989).
27. Shankland, M., Martindale, M., Nardelli-Haefliger, D., Baxter, E. & Price, D. *Development* (suppl.), 29–38 (1991).
28. Averof, M. & Akam, M. *Curr. Biol.* **3,** 73–78 (1993).
29. Dick, M. & Buss, L. *Molec. Phylog. Evol.* **3,** 146–158 (1994).
30. Holland, P. W. H., Garcia-Fernandez, J., Williams, N. A. & Sidow, A. *Development* (suppl.) 125–133 (1994).
31. Schubert, F., Nieselt-Struwe, K. & Gruss, P. *Proc. natn. Acad. Sci. U.S.A.* **90,** 143–147 (1993).
32. Brusca, R. & Brusca, G. *Invertebrates* (Sinauer, Sunderland, MA, 1990).
33. Vachon, G. *et al. Cell* **71,** 437–450 (1992).
34. Struhl, G. *Proc. natn. Acad. Sci. U.S.A.* **79,** 7380–7384 (1982).
35. Mann, R. *Development* **120,** 3205–3212 (1994).
36. Averof, M. & Akam, M. *Nature* **376,** 420–423 (1995).
37. Warren, R., Nagy, L., Selegue, J., Gates, J. & Carroll, S. *Nature* **372,** 458–461 (1994).
38. Birket-Smith, S. J. R. *Prolegs, Legs and Wings of Insects* (Scandinavian Science, Copenhagen, 1984).
39. Carroll, S., Weatherbee, S. & Langeland, J. *Nature* **375,** 58–61 (1995).
40. Kukalova-Peck, J. *J. Morph.* **156,** 53–126 (1978).
41. Kukalova-Peck, J. *Can. J. Zool.* **61,** 1618–1669 (1983).
42. Condie, J., Mustard, J. & Brower, D. *Drosoph. Inf. Serv.* **70,** 52–54 (1991).
43. Morata, G. & Garcia-Bellido, A. *Wilhelm Roux Arch. Entw. Mech. Org.* **179,** 125–143 (1976).
44. Beachy, P. A., Helfand, S. L. & Hogness, D. S. *Nature* **313,** 545–551 (1985).
45. White, R. A. H. & Wilcox, M. *Cell* **39,** 163–171 (1984).
46. Nijhout, H. F. *The Development and Evolution of Butterfly Wing Patterns* (Smithsonian Institution, Washington DC, 1991).
47. McGinnis, W. & Krumlauf, R. *Cell* **68,** 283–302 (1992).
48. Rancourt, D., Teruhisa, T. & Capecchi, M. *Genes Dev.* **9,** 108–122 (1995).
49. Graham, A. *Curr. Biol.* **4,** 1135–1137 (1994).
50. Garcia-Fernandez, J. & Holland, P. *Nature* **370,** 563–566 (1994).
51. Pendleton, J., Nagai, B., Murtha, M. & Ruddle, F. *Proc. natn. Acad. Sci. U.S.A.* **90,** 6300–6304 (1993).
52. Kessel, M. & Gruss, P. *Science* **249,** 374–379 (1990).
53. Burke, A., Nelson, C., Morgan, B. & Tabin, C. *Development* **121,** 333–346 (1995).
54. Ahlberg, P. *Nature* **354,** 298–301 (1991).
55. Tabin, C. & Laufer, E. *Nature* **361,** 692–693 (1993).
56. Carroll, R. *Vertebrate Paleontology and Evolution* (Freeman, New York, 1988).
57. Ruddle, F., Bentley, K., Murtha, M. & Risch, N. *Development* (suppl.) 155–161 (1994).
58. Gans, C. *Am. Zool* **15,** 455–467 (1975).
59. Cartwright, P., Dick, M. & Buss, L. *Molec. Phylog. Evol.* **2,** 185–192 (1993).
60. Jacobs, D. *Proc. natn. Acad. Sci. U.S.A.* **87,** 4406–4410 (1990).
61. Bonner, J. *The Evolution of Complexity* (Princeton Univ. Press, Princeton, NJ, 1988).
62. Duboule, D. *BioEssays* **14,** 375–384 (1992).
63. Sidow, A. *Proc. natn. Acad. Sci. U.S.A.* **89,** 5098–5102 (1992).
64. Simeone, A. *et al. Proc. natn. Acad. Sci. U.S.A.* **91,** 2250–2254 (1994).
65. Dickinson, W. J. in *Evolutionary Biology* (eds Dobzhansky, T., Hecht, M. & Steere, W. C.) 127–173 (Appleton-Century-Crofts, New York, 1991).
66. Galloni, M., Gyurkovics, P., Schedl, P. & Karch, F. *EMBO J.* **12,** 1087–1097 (1993).
67. Gindhart, J. Jr, King, A. & Kaufman, T. *Genetics* **139,** 781–795 (1995).
68. Goldschmidt, R. *The Material Basis of Evolution* (Yale Univ. Press, New Haven, CT, 1940).
69. Wallace, B. *Q. Rev. Biol.* **60,** 31–42 (1985).
70. Kostriken, R. & Weisblat, D. *Devl Biol.* **151,** 225–241 (1992).
71. Kingsley, D. *Genes Dev.* **8,** 133–146 (1994).
72. Hogan, B., Blessing, M., Winnier, G. & Suzuki, N. *Development* (suppl.) 53–60 (1994).
73. Cohen, S., Bronner, F., Kuttner, G., Jurgens, G. & Jäckle, H. *Nature* **338,** 432–434 (1989).
74. Panganiban, G., Nagy, L. & Carroll, S. *Curr. Biol.* **4,** 671–675 (1994).
75. McGinnis, W. & Kuziora, M. *Scient. Am.* **270,** 58–66 (1994).
76. Wheeler, W., Cartwright, P. & Hayashi, C. *Cladistics* **9,** 1–39 (1993).

ACKNOWLEDGEMENTS. We thank members of our laboratory and J. Langeland, A. Burke, P. Carroll and the referees for constructive comments; N. Patel, M. Averof, M. Akam, D. Duboule, G. Wagner, M. Westerfield, R. Krumlauf and P. Holland for communication of unpublished data; J. Wilson for preparation of the manuscript; and L. Olds for artwork. This work was supported by the NSF, the Shaw Scientists Program of the Milwaukee Foundation and the HHMI.

Kidwell, M. G., and D. Lisch. 1997. Transposable elements as sources of variation in animals and plants. *Proceedings of the National Academy of Sciences, USA* 94:7704–7711.

Perhaps the most startling discovery in evolutionary genomics in recent decades involves the ubiquity and abundance of transposable elements, also known informally as "jumping genes." Plants and animals, including humans, house legions of such elements that often have (or previously had) the capacity to colonize new chromosomal locations by replicatively hopping from one site to another within a cell lineage. Active or deceased mobile elements constitute at least 45 percent of the human genome, but the true fraction may be 75 percent or more, if the tally were to include DNA regions that probably originated as jumping genes but are no longer firmly identifiable as such because of their decay by means of postformational mutations. Barbara McClintock discovered mobile elements in her studies of maize beginning in the 1940s, and in 1983 she received a Nobel Prize for her efforts (see McClintock 1984 in this volume).

Prevailing wisdom holds that mobile elements exist in such abundance because they are selfish pieces of DNA that proliferate inside host genomes much like infectious viruses (which they closely resemble in structure and behavior). In a sexual species, any DNA sequence that gains a capacity to generate and disperse copies of itself across chromosomal sites almost inevitably enhances its prospects for survival to succeeding generations.

None of this is to imply that transposable elements never confer benefits on their hosts. To the contrary, in the coevolutionary dances (choreographed by natural selection) between jumping genes and host organisms, it is to be expected that the genomes of the host occasionally recruit or co-opt formerly parasitic mobile elements into host-beneficial functions. This article by Margaret Kidwell and Damon Lisch is a succinct yet comprehensive synopsis exploring the evolutionary ramifications of transposable elements, including their possible adaptive significance for the organisms in which they are housed.

RELATED READING

Avise, J. C. 2010. Inside the human genome: a case for non-intelligent design. Oxford University Press, Oxford, England.
Lynch, M. 2007. The origins of genome architecture. Sinauer, Sunderland, Massachusetts, USA.
McDonald, J. F., editor. 1993. Transposable elements and evolution. Springer-Science, Dordrecht, the Netherlands.
Saedler, H., and A. Gierl, editors. 1996. Transposable elements. Springer, Heidelberg, Germany.

Proc. Natl. Acad. Sci. USA
Vol. 94, pp. 7704–7711, July 1997
Colloquium Paper

This paper was presented at a colloquium entitled "Genetics and the Origin of Species" organized by Francisco J. Ayala (Co-chair) and Walter M. Fitch (Co-chair), held January 30 to February 1, 1997, at the National Academy of Sciences Beckman Center in Irvine, CA.

Transposable elements as sources of variation in animals and plants

MARGARET G. KIDWELL* AND DAMON LISCH

Department of Ecology and Evolutionary Biology and The Center for Insect Science, University of Arizona, Tucson, AZ 85721

ABSTRACT A tremendous wealth of data is accumulating on the variety and distribution of transposable elements (TEs) in natural populations. There is little doubt that TEs provide new genetic variation on a scale, and with a degree of sophistication, previously unimagined. There are many examples of mutations and other types of genetic variation associated with the activity of mobile elements. Mutant phenotypes range from subtle changes in tissue specificity to dramatic alterations in the development and organization of tissues and organs. Such changes can occur because of insertions in coding regions, but the more sophisticated TE-mediated changes are more often the result of insertions into 5′ flanking regions and introns. Here, TE-induced variation is viewed from three evolutionary perspectives that are not mutually exclusive. First, variation resulting from the intrinsic parasitic nature of TE activity is examined. Second, we describe possible coadaptations between elements and their hosts that appear to have evolved because of selection to reduce the deleterious effects of new insertions on host fitness. Finally, some possible cases are explored in which the capacity of TEs to generate variation has been exploited by their hosts. The number of well documented cases in which element sequences appear to confer useful traits on the host, although small, is growing rapidly.

The book whose publication we are celebrating in this colloquium indicates that Theodosius Dobzhansky had a very special interest in gene mutation and its causes. Dobzhansky recognized mutation as the "raw material" on which natural selection acts and as the first of three steps necessary for evolution to take place. However, the discovery of transposable elements (TEs) in the 1940s by Barbara McClintock occurred a decade later, and it was a further 30 years before the significance of her findings started to be fully appreciated. Sixty years ago, Dobzhansky was well aware of the mutagenic properties of ionizing radiation discovered in 1927 by H. J. Muller but acknowledged that much less than 1% of spontaneous mutations were attributable to this cause. He distinguished between spontaneous and induced mutations: "The former are those which arise in strains not consciously exposed to known or suspected mutation-producing agents." He also pointed out that "since the name spontaneous constitutes only a thinly-veiled [sic] admission of the ignorance of the phenomenon to which it is applied, the quest for the causes of mutation has always occupied the attention of geneticists." Although at that time no clues to its nature were yet available, Dobzhansky realized that a major piece of the mutation puzzle was still missing. We believe he would have been intrigued with the discoveries of TEs in natural populations that have taken place during the last 20 years and that he would have been an active participant in the continuing debate about their role in evolution.

Distribution and Classification

TEs are discrete segments of DNA that are distinguished by their ability to move and replicate within genomes. Since their discovery by Barbara McClintock ≈50 years ago (1), TEs have been found to be ubiquitous in most living organisms. They comprise a major component of the middle repetitive DNA of genomes of animals and plants. They are present in copy numbers ranging from just a few elements to tens, or hundreds, of thousands per genome. In the latter case, they can represent a major fraction of the genome, especially in some plants. For example, TEs recently have been estimated to make up >50% of the maize genome (2). In *Drosophila*, ≈10–15% of the genome is estimated to be made up of TEs, most of which are found in distinct regions of centric heterochromatin (3).

TEs are classified in families according to their sequence similarity. Two major classes are distinguished by their differing modes of transposition (4). Class I elements are retroelements that use reverse transcriptase to transpose by means of an RNA intermediate. They include long terminal repeat retrotransposons and long and short interspersed elements (*LINES* and *SINES*, respectively). Long terminal repeat retrotransposons are closely related to other retroelements of major interest, such as retroviruses (5). The *gypsy* element in *Drosophila* is an example of a rare type of retrotransposon that can sometimes also behave as a retrovirus (6).

Class II elements transpose directly from DNA to DNA and include transposons such as the *Activator-Dissociation* (*Ac-Ds*) family in maize, the *Tam* element in *Antirrhinum*, the *P* element in *Drosophila*, and the *Tc1* element in the worm, *Caenhorabditis elegans*. Recently. a category of TEs has been discovered (7) whose transposition mechanism is not yet known. These miniature inverted-repeat TE (*MITEs*) have some properties of both class I and II elements. They are short (100–400 bp in length), and none so far has been found to have any coding potential. They are present in high copy number (3,000–10,000) per genome and have target site preference for TAA or TA in plants. *MITEs* such as the *Tourist* element in maize and the *Stowaway* element in Sorghum (7) are found frequently in the 5′ and 3′ noncoding regions of genes and are frequently associated with the regulatory regions of genes of diverse flowering plants. TEs with similar properties also have been described in *Xenopus* (8), humans (9, 10), and the yellow fever mosquito, *Aedes aegypti* (11).

Most, but not all, TE families are made up of both autonomous and nonautonomous elements. Whereas autonomous

Abbreviations: TE, transposable elements; MITE, miniature inverted-repeat TE.

*To whom reprint requests should be addressed. e-mail: mkidwell@ccit.arizona.edu.

elements code for their own transposition, nonautonomous elements lack this ability and usually depend on autonomous elements from the same, or a different, family to provide a reverse transcriptase or transposase *in trans*.

This paper aims first to provide a brief, general description of the types of genetic variation caused by TEs in animals and plants and then to examine this variation within an evolutionary framework: (*i*) direct selection on TEs at the level of the DNA sequence (parasitic DNA); (*ii*) coevolution of TEs and their animal and plant hosts to avoid or mitigate the deleterious effects of insertion; and (*iii*) positive selection on elements that have evolved to provide some positive benefit to their hosts in addition to simply minimizing the harm they do.

Types of TE-Induced Genetic Variation

Like new mutations produced by any mutator mechanism, the majority of new TE-induced mutations are expected to be deleterious to their hosts. Those mutations that survive over long periods of evolutionary time are expected to be a small subsample of newly induced mutations. The property that distinguishes TE-induced mutations from those produced by other mutational mechanisms is their remarkable diversity and the degree to which their induction is regulated by both the host and the TE itself.

The genetic variability resulting from TEs ranges from changes in the size and arrangement of whole genomes to changes in single nucleotides. It may produce major effects on phenotypic traits or small silent changes detectable only at the DNA sequence level. It is important to note that TEs produce their mutagenic effects not simply on initial insertion into host DNA. TEs may also produce mutations when they excise, leaving either no identifying sequence or only small "footprints" of their previous presence. In addition, some TE-induced mutations that may be of evolutionary significance to their hosts, such as mutations in regulatory sequences (12), may take long periods of time to evolve new functions or these new functions may have been acquired a long time ago. Consequently, they may have lost their original identification as TEs. For these reasons, the reliance solely on the distribution of TE sequences in the genomes of contemporary species of animals and plants to deduce the long term evolutionary importance of TEs may produce a biased result that may not adequately reflect TE-associated events that occurred long in the past.

Deleterious effects of TEs can result not only from mutations caused by the insertion or excision of these elements at a single chromosomal site but also from genomic-level disruptive effects associated with TE transposition. For example, massive chromosome breakage in larval cells resulting from excision and transposition of genomic *P* elements has been implicated as the cause of temperature-dependent pupal lethality and sterility in hybrid dysgenesis in *Drosophila melanogaster* (13, 14).

A brief description of the types of genetic variability caused by TE activity follows, based largely on the types of host DNA involved. Some of the mutations described were generated in the laboratory and have been subjected to artificial selection under unnatural and noncompetitive conditions. Although these are generally not the class of mutations that are of interest from an evolutionary perspective, we include them here to provide some indication of the potentially wide spectrum of phenotypic changes associated with TE activity.

Insertions of TEs into Exons of Host Genes. On average, TEs that insert within the exons of genes are most likely to result in null mutations because of the sensitivity of these regions to frame shift mutations and the lack of tolerance of highly conserved regions to most mutations of any kind. However, those mutations that are not simply inviable can provide interesting and sometimes spectacular phenotypic variability. In *Drosophila*, a series of null alleles at the X-linked, white locus allowed the first identification of the *P* element in *D. melanogaster* as the causal agent of P–M hybrid dysgenesis (15). The insertion of both the *P* element and the *copia* element into exon sequences interrupted the coding sequences and the production of the red eye pigment by the wild-type gene. The result is a bleached white eye phenotype that reflects the lack of pigmentation. Such a null mutation can be maintained in the laboratory but is unlikely to survive in natural populations.

A good classic example is the insertion of an element of the *Ac-Ds* family into wx-*m9*, an allele of the *waxy* locus in maize first discovered by McClintock (16). The mutation is caused by the insertion of *Ds* (*Dissociator*) into the 10th exon of the *waxy* locus. This was the first element to be cloned from maize, and it is of continuing interest because it is spliced, resulting in partial revertant activity (17, 18). In this case, the effect of the insertion is attenuated by the loss through splicing of the TE after transcription.

Insertions into Regulatory Regions of Genes. An excellent example of this type is the insertion of *gypsy* into the 5′ upstream region of the *yellow* gene in *Drosophila*, which causes a loss of expression of the *yellow* gene in specific tissues (19). The loss of expression in some tissues and not others in this case is the result of the interaction of the element, tissue-specific enhancers upstream of the element, and specific host factors. In *Antirrhinum*, a *Tam3* element was observed to insert into a region 5′ of the *niv* gene, which is involved in the synthesis of anthocyanin pigments. The initial insertion was observed to down-regulate expression of the gene. However, a series of rearrangements mediated by this element resulted in a change in the level and tissue specificity of expression of *niv* (Fig. 1). The net effect is a new and novel distribution of anthocyanin pigment in the flower tube (20). This series of mutations exemplifies the potential for TE-mediated "rewiring" of regulatory networks, in this case by bringing new regulatory sequences in proximity to exonic sequences via an inversion, followed by an imprecise excision event.

TE activity can result in even more complex rearrangements that can have effects on gene regulation. In maize, the insertion of a *Mu* (*Mutator*) element into the TATA box of *Adh1* changes the tissue specificity of RNA expression (21). Of interest, excision of this element caused a complex series of duplications and inversions whose net effect was to cause additional changes in tissue specificity (22). Kloeckener–Gruissem and Freeling (22) suggest that this kind of "promoter scrambling" may represent a more general process by which transposons produce variants of a type not produced by other mechanisms. Furthermore, in this case, like that of the *Tam* element in the *niv* gene of *Antirrhinum*, the TE footprint left behind after the element has generated the new mutation is small enough to be invisible to TE probes.

Insertions in Introns. TEs that insert into introns generally have a greater chance to survive because these insertions are less visible to natural selection. Many of them are probably successfully spliced out during mRNA processing and have no obvious effect on the function of the gene. Even when spliced, however, introns are sometimes the site of regulatory sequences. In these cases, TE insertions into introns can affect gene regulation in surprising ways. For instance, the insertion of *Mu* elements into an intron of the *Knotted* locus in maize induces ectopic expression of the gene, suggesting that the intron carries sequences normally required to repress expression of the gene in certain tissues (23). Similarly, in *Antirrhinum*, complementary floral homeotic phenotypes result from opposite orientations of a *Tam3* transposon in an intron of the *ple* gene (24) .

Insertions in Heterochromatin. Middle repetitive DNA sequences, including TEs, are an important component of β heterochromatin in *Drosophila*, and retrotransposons constitute a considerable fraction of this DNA (25). Recent work by Pimpinelli and coworkers (3) has revealed that TE clustering into discrete regions of heterochromatin is a general property of elements in *Drosophila*. The cause of this distribution pattern is an open question. In some cases, a heterochromatic location probably reduces the probability of elimination of inserted se-

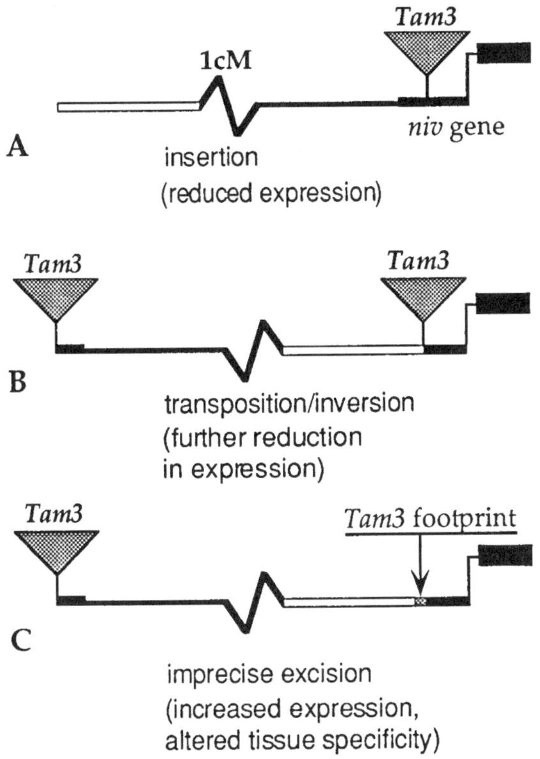

FIG. 1. Rearrangements associated with TE activity near a gene result in altered tissue specificity. (*A*) In the original isolate, a *Tam3* element had inserted 64 bp upstream of the start of transcription of the *niv* gene of *Antirrhinum*. The result of this insertion was a reduced level of expression. (*B*) A derivative of the initial insertion allele carries an inversion flanked by two copies of the transposon. This allele confers an additional reduction in the level of expression. (*C*) Excision of the element closest to the *niv* gene left a short (26 bp) "footprint." This rearrangement resulted in an increase in the level of *niv* gene expression as well as a novel pattern of expression, presumably due to the juxtaposition of a novel sequence with the *niv* gene TATA box and coding sequences [adapted from Lister *et al.* (20)].

quences from the genome by ectopic recombination, the mechanism believed to be largely responsible for controlling TE copy number in chromosome arms (26). However, some centric heterochromatic regions have been described as a graveyard for dead elements, rather than a safe haven for active elements, because the majority found there appear to be inactive and highly diverged sequences. For example, *LINE*-like *I* elements cloned from the *Charolles*-reactive strain of *D. melanogaster* contain no active euchromatic *I* factors, only defective copies that are embedded in clusters of defective copies of other retroelements (27). In contrast, elements such as *mdg1* have been found to have a nested arrangement within other retrotransposons located in euchromatic chromosome regions (28, 29). These sites appear to exhibit properties of intercalary heterochromatin (25) and may be responsible for the properties of ectopic pairing, susceptibility to breakage, and late replication that are characteristic of this type of chromatin.

Mediation of Recombination. TE-mediated increases in the rate of recombination have consequences not only for genetic variation at individual loci. This recombination activity can also result in more general changes in both fine and gross structural characteristics of chromosomes. For instance, analysis of mutations in the *mei-41* and *mus302* genes required for normal postreplication repair in *D. melanogaster* (30) revealed a striking stimulation of site-specific gene conversion and recombination mediated by *P* element transposition. As a consequence of the selection against the negative effects of ectopic recombination, this is postulated to be the mechanism chiefly responsible for the removal of certain subsets of TEs from genomes and a means for controlling copy number (26, 31). It is worth noting, however, that not all rearrangements caused by ectopic recombination are necessarily selected against. From an evolutionary perspective, rare surviving chromosomal rearrangements could be of significance. For example, Lyttle and Haymer (32) demonstrated the presence of the *hobo* element at the breakpoints of several endemic, but not cosmopolitan, inversions in *D. melanogaster* natural populations from Hawaii. This result is consistent with the recent introduction of *hobo* elements to *D. melanogaster* by horizontal transfer (33) and the subsequent production of these inversions as a consequence of the activity of these elements.

Effects on Quantitative Variability. The *P* element in *Drosophila* provides one of the most compelling demonstrations of TE-induced genetic variability in quantitative genetic traits. A series of experiments (34, 35) has shown that quantitative variability for bristle number is induced by P–M hybrid dysgenesis and is demonstrable by directional artificial selection. Furthermore, in well controlled experiments (36), a dramatic increase in new additive genetic variation in abdominal bristle number was observed that was 30 times greater than that expected from spontaneous mutation. In another *Drosophila* study (37), an excess of *P* element mutations having large effects on metabolic characters was observed relative to those expected. Significant among-line heterogeneity indicated that the mutational target site for enzyme activity is large and that most of the mutations must be regulatory. It was concluded that the large pleiotropic effects observed had important consequences for metabolic characters.

Evolutionary Considerations

The idea that TEs are primarily parasitic is not at all inconsistent with a role for these elements in the evolution of their hosts. Indeed, as documented below and elsewhere (12, 38–40), there is a growing body of evidence for coadaptation by both elements and their hosts to the long term presence in the genome of these parasitic sequences. In some cases, it appears that this coadaptation even may have lead to the use of TE sequences for essential and beneficial host functions. In this section, we explore all three perspectives.

TEs as Genomic Parasites

The intrinsically parasitic nature of active TEs (41–43) accounts for their undisputed ability to invade new species, increase in copy number, and survive over long periods of evolutionary time. The replicative advantage of TEs (44) is responsible for this ability, which is facilitated by their generally compact structure and inclusion of the coding capacity for transposition within their sequences. Natural selection acting on TEs at the level of the DNA sequence is responsible for maintaining their essentially parasitic properties. For example, *P* elements can rapidly invade a naive population of *D. melanogaster*, despite extremely strong negative selection at the host level in the form of high frequencies of temperature-dependent gonadal sterility (45).

A proclivity for horizontal transfer is consistent with the role of TEs as genomic parasites. The life cycle of TEs in any single phylogenetic lineage can apparently last for many thousands or millions of years and can be considered as a succession of three phases: dynamic replication, inactivation, and degradation (46, 47). The transposition of both major classes of elements is error-prone and produces nonautonomous elements that often repress the transposition rate of active elements. Over long periods of evolutionary time, there is a tendency for a family of elements to degrade in coding capacity, but horizontal transfer to another host lineage provides the opportunity for active TEs to move to another lineage and begin the cycle over again (46, 48, 49). There is evidence that TEs do transfer horizontally more frequently than nonmobile genes (46). Class II elements such as *mariner* and *P* elements provide good examples of TE horizontal transfer (50–52), but a major puzzle remains regarding the mechanism by which horizontal transfer is achieved (49).

Coadaptations to Mitigate Reduced Host Fitness

Like viruses, TEs are dependent on their host organisms for survival, but unlike viruses, most TEs do not have a phase in their life cycle in which they can survive independent of their hosts. Therefore, coevolution and coadaptation of TEs with host genomes is expected to play a particularly important role in the long term survival of these element families. Given that the majority of new insertions tend to be deleterious to hosts, it is in the interests of both parties to mitigate or remove such deleterious effects. TEs can insert in many locations other than exons. In these noncoding sequences, they are likely to increase their probability of survival because of less visibility to natural selection. Examples of some of the ways that coadaptations by both mobile elements and their hosts appear to have evolved are described below and are summarized in Table 1.

Insertion Bias for Noncoding Regions. A dramatic demonstration of the preference for new *P* element insertions into potentially regulatory regions of genes, rather than exons, is provided by *P* elements in *Drosophila* (53). Only a small minority of *P* elements was observed to have inserted in coding sequences, and these elements were in the 5′ portion. Thus, there is a strong bias in favor of insertion in the 5′ end of the gene and especially for the 5′ untranslated region. Because these insertions all caused an obvious mutant phenotype (they failed to compliment a deficiency), this result is actually an understatement of the true number of insertions into or near regions of genes often involved in regulation. This preference is supported by the remarkable observation that, in another experiment, >65% of the 500 independent *P* element enhancer trap insertions were expressed in a spatially and temporally restricted fashion (54) .

Another good example of preferential insertion of elements outside of coding regions is provided in yeast (55). Of over 100 new insertions of *Ty1* observed in chromosome III, nearly all were inserted into or near either tRNA genes or preexisting long terminal repeats; only 3% were found in ORFs. Distribution patterns favoring the 5′ noncoding sequences of genes were observed for *MITE* elements in plants (7) and animals (e.g., see ref. 11). However, it is not clear whether this pattern represents an insertion preference or whether it results from strong earlier selection against insertions into other regions. Similarly, in view of the large number of group II and group III introns present in the chloroplast genome of *E. gracilis*, the complete absence of introns in rRNA and tRNA genes is striking (56). One possibility is that secondary structure features of rRNA and intron RNA or tRNA and intron RNA (if they were present in the same pre-mRNA molecule) would interact in such a way as to prevent one or both from functioning.

Preference for Insertion into Preexisting Elements. In some cases, it is clear that unrestricted transposition would be absolutely disastrous for the host. As mentioned above, a remarkable number of retroelements, probably representing well over 50% of the genome, are found between genes in maize (2). The five most abundant families make up more than 25% of the genome, and one family alone, *Opie*, makes up 10–15% of the maize genome. Despite this ubiquity, few homologies in the database were found among maize genes; these elements appeared to have a pronounced preference for regions outside of genes. Indeed, fully half of the elements examined were found nested within other elements.

Splicing from Pre-mRNA Transcripts. It has been suggested that splicing of TEs from exonic sequences has evolved as a means by which TEs can minimize their deleterious host impact. We refer back to the *waxy* mutation described earlier (18). In that case, splicing of the *Ds* element inserted into the *waxy* gene results in partial reversion of the mutation caused by the original insertion. Presumably, even a partial amelioration of the mutant phenotype can provide some selective advantage to those TEs capable of providing it. Additional examples are found in *Drosophila*, *C. elegans*, and other plants (57–59). In addition, Marillonnet and Wessler have observed tissue-specific splicing of an element (S. Wessler, personal communication), suggesting that TEs could potentially play a role in the evolution of tissue-specific regulation of some genes. These examples may be illustrating an evolutionary spectrum, from purely parasitic behavior to functional significance for the host. The original capacity to be spliced may have arisen as a way to minimize the impact of insertions into coding sequences but on the road from poorly spliced variants that simply ameliorate mutant phenotypes to fully effective and selectively invisible splicing may have come the opportunity to develop new classes of regulation, such as tissue specificity.

Tissue Specificity of TE Activity. A good example of a likely adaptation of a TE to its host is the restriction of transposition of *P* and *I* elements to the germ line (14). It is of mutual benefit for an element to transpose in those tissues that will ensure transmission to the next host generation, but to curtail activity in somatic tissues is likely to result in loss of host fitness without providing any benefit to the transposon. Repression of *P* element transposition in somatic cells occurs on the level of RNA processing (60). The 2–3 intron is spliced only in the germ cells, resulting in the absence of transposase in somatic cells. Splicing of this intron is prevented in the somatic cells by an 87-kDa protein that binds to a site in exon 2 located 12–31 bases from the 5′ splice site (61). An existing host-splicing mechanism apparently has been coopted for this purpose (61, 62) that has been highly conserved during evolution.

Host Regulation of TE Copy Number. Good examples of host regulation of copy number are found in maize and

Table 1. Possible coevolved mechanisms to mitigate reduction in host fitness

Mechanism	Examples
Insertion bias for noncoding regions	Preferential insertion in regulatory regions (1); nested retrotransposons in maize (2); clustered *I* elements (27); *mdg1* in *Drosophila* (28, 29)
Pre-mRNA splicing	Splicing from the maize *wx* gene (18); various genes in *Drosophila* (57), *C. elegans* (59), and plants (58)
Tissue specificity of transposition	Repression of *P* element transposition in somatic cells at the level of RNA processing (60)
TE copy number regulation	Methylation of maize *Ac*, *Spm*, and *Mu* by host factors (63, 64); type I and type II *P* element-encoded repressors (13, 68, 70–72); *Ac* dosage effects (16); *Spm* repressor action (73–75)
Regulators of mutant phenotype expression	Transposase-dependent expression of mutant phenotype caused by insertion of *Spm* in maize (73); masking of mutant phenotypes by alleles of host suppressor genes in *Drosophila* (40)

Drosophila. Unknown host-encoded factors specifically methylate *Ac*, *Spm*, and *Mutator* elements in maize (63, 64). With *Mu* (64), the example is particularly striking because it represents the global methylation of dozens of previously active elements simultaneously in a single generation. The methylation is not simply related to structural features of insertion sites of *Mu* elements because they become specifically methylated even when inside of genes; modification is rarely detected in the flanking sequences within the genes (65) .

In *D. melanogaster* females, expression of the *gypsy* element *envelope* gene is strongly repressed by one copy of the nonpermissive allele of *flamenco* [reviewed by Bucheton (6)]. A less dramatic reduction in the accumulation of other transcripts and retrotranscripts also is observed. These effects correlate well with the inhibition of *gypsy* transposition in the progeny of these females and are therefore likely to be responsible for this phenomenon. The effects of *flamenco* on *gypsy* expression apparently are restricted to the somatic follicle cells that surround the maternal germline.

Self-Regulation of TE Copy Number. It has been shown theoretically that self-regulation of TEs cannot evolve if it is assumed that deleterious effects on host fitness are caused by increased copy number alone or are not caused by dominant lethals (66). However, if the deleterious effects are immediate and occur as a direct consequence of transposition itself, then there may be a selective advantage to elements with reduced transposition rates that still allow them to spread in the genome but at a reduced cost to their host [reviewed by Brookfield (67)].

The activity of the *P* elements in *D. melanogaster* is regulated by element-encoded repressor products. These repressors fall into two discrete categories, type I and type II. Type I repressors are responsible for a cellular condition known as P cytotype, which depends on a 66-kDa, *P* element-encoded, repressor of transposition and excision (68). The genomic position of repressor elements determines the maternal vs. zygotic inheritance of *P* cytotype (69). Type II repressors usually have large internal deletions, are sensitive to genomic location, but show no maternal inheritance (13, 70–72) .

In plants, the *Ac* element shows dosage effects; an increase in number of elements results in a decreased number of transpositions of the element (16). This could be interpreted as a response of the plant to increases in the level of *Ac* transposase or as an autoregulatory mechanism. Similarly, *Spm:tnpA* can protect *Spm* from methylation but may also act as a repressor of *Spm* (73, 74). Additionally, some deleted *Spm* elements can repress full length *Spm* elements in *trans* (75) .

Regulators of Expression of Mutant Phenotypes. Very early in the investigation of mutable alleles in maize, it was discovered that the expression of some alleles depended on the presence or absence of a second factor (76). In these cases, that factor was the source of the transposase. However, as is clear from the example of some *gypsy* insertions whose mutant phenotype is only manifested in the presence of both *Su(Hw)* and a second host encoded factor, *mod(mdg4)* (77), TE-induced mutant alleles can also become ameliorated by other factors as well.

The mutant phenotypes associated with many retrotransposon insertions are masked by alleles of host suppressor genes that act as trans-regulators of retrotransposon expression (40). The argument made is that such suppressor action may allow insertion mutations to partially, or completely, escape the action of purifying selection and allow them to persist or even increase in frequency in natural populations. There is evidence for the presence of host genes with suppressor function in natural populations of *Drosophila* (78).

TE-Induced Characters Having Benefit to the Host

There has been considerable debate whether, in addition to deleterious effects on fitness, TE-induced variability has any significance for host organisms over evolutionary time (40). The generally unpredictable nature of TE movements, coupled with the paucity of fixed insertion sites for TEs in species such as *Drosophila* (26), has lead some to reject the possibility of TEs having any significant evolutionary importance, other than as molecular parasites. However, there is a rapidly growing list of possible examples of TEs having evolved highly sophisticated functions, as shown by the examples briefly described below and summarized in Table 2.

Insertions with Host Gene Regulatory Functions. It has been speculated for some time that changes in cis-regulatory regions of duplicated genes may be more important for the evolution and divergence of functional and morphological characters than mutations in coding sequences (see, e.g., refs. 79 and 80). However, only recently has evidence started to accumulate to support this hypothesis. In *Drosophila,* for instance, the three homeotic genes *paired* (*prd*), *gooseberry* (*gsb*), and *gooseberry neuro* (*gsbn*) have evolved from a single ancestral gene, following gene duplication. They now have distinct developmental functions during embryogenesis. The three corresponding proteins PRD, GSB and GSBN are transcription factors. Li and Noll (81) demonstrated that the three proteins are interchangeable with respect to their regulatory functions and that their distinct developmental functions are a consequence of changes in the regulatory sequences rather than in the proteins themselves. Because they lend themselves so well to changes in the architecture of promoter regions, it is likely that TE mutations have been involved in this kind of regulatory evolution (82).

The potential importance of TEs as modifiers of the expression of normal plant genes has been highlighted by recent findings in plants. Long terminal repeat retrotransposons and *MITE*s have

Table 2. Examples of TEs having functions that benefit their hosts

New function	Examples
Insertions with regulatory functions	More than 20 examples of insertions into regulatory regions of genes (12, 38, 39)
"Molecular domestication"	*P* element tandem repeats in *D. obscura* group may provide a new host gene function (89)
Source of new introns	Introns and twintrons in *Euglena gracilis* plastids (24)
Replacement of normal host functions	Repair of damaged chromosome ends by *HET-A* and *TART* in *Drosophila* (92, 93)
A role in host cell repair mechanisms	Endogenous retroelements associated with repair of DSBs in yeast (94, 95)
Mediation of concerted evolution	*P* element-mediated changes in subtelomeric repeat numbers in *D. melanogaster* (96)
Possible functions of heterochromatic TE clusters	Developmentally programmed changes in DNA content; expression of heterochromatically embedded loci; genomic housekeeping functions (25)

DSB, double-strand chromosome break.

been found to be associated with the genes of many plants where some of these TEs contribute regulatory sequences (7). Furthermore, the *MITE* elements recently discovered in *Aedes aegypti* (11) also are associated closely with genes. In domesticated rice, *Oryza sativa*, a computer-based search revealed 32 common sequences belonging to nine putative mobile element families (83). Four of these families had been previously described, but five families were first discovered through this computer search, and four of these five had characteristics of *MITES*.

New Patterns of Tissue-Specific Expression. TEs can contribute to the functional diversification of genes by supplying cis-regulatory domains altering expression patterns. Earlier, we described the insertion of the gypsy element into a 5′ upstream region of the *yellow* gene in *Drosophila* causing a loss of expression of this gene in specific tissues (84). The tissue-specific alterations in expression (a kind of mutation that is more subtle than simply knocking out a gene) is due to the presence of a specific sequence of DNA that is bound by *Su(Hw),* which is thought to be a transcription factor. More interesting, the *Su(Hw)* binding sequence seems to act as a general "buffer" that helps to define structural domains in the chromatin (77). Thus, *gypsy* may serve to introduce domains of regulation into given regions of the chromosome. This may have arisen initially as a means by which the TE could buffer itself from its chromosomal environment, but this kind of domain alteration could certainly also result in interesting variations in gene regulation as well.

In addition to simply buffering chromosomal regions, many TEs are specifically expressed only in particular tissues at particular times. Based on recent findings, it appears that tissue specificity is a general feature of all retrotransposons in *Drosophila*. The expression patterns of 15 different families of long terminal repeat-containing retrotransposons were examined by Ding and Lipshitz (85) during normal development in different wild-type strains of *D. melanogaster.* Each family exhibited a pattern typical of spatial and temporal expression during embryogenesis, suggesting that each TE harbors cis-regulatory factors that interact specifically with host transcription factors. These mobile cis-regulatory factors could potentially act to modify the expression of any number of host genes.

Other Types of Insertions with Regulatory Functions. Some of the examples given thus far are anecdotal; they represent laboratory observations as to the kinds of changes that TEs can introduce into the host genome, rather than changes that have actually contributed to the evolution of the host. However, Britten (12, 38, 39) has used stringent criteria for the identification of strong cases of the involvement of TEs in the actual evolution of gene regulation. He maintains that a long term perspective is necessary in identifying and understanding mutations important for gene regulation. The number of cases he has identified is small, but growing. In addition to the plant *MITE* examples discussed above, he includes cases involving *Alu*-containing, T cell-specific enhancers in the human *CD8a* gene (86), the association of a retrovirus-related element with androgen regulation of the sex-limited protein (*Slp*) gene in mouse (87), and inverted repeats in the *CyIIIa* actin gene of sea urchin (88).

Tandem Repeats of *P*-Related Sequences in *Drosophila*. A number of tandem *P* element repeats in three closely related species of the *obscura* group provides a very interesting example of several unrelated TE sequences evolving together that may provide a type of host gene function, which Miller *et al.* have termed "molecular domestication" (89). In this case, the *P* elements have lost all of their terminal repeats and thus can no longer transpose. Remarkably, each cluster unit consists of a cis-regulating section composed of insertion sequences derived from unrelated TEs, followed by the first three exons that, in mobile *P* elements, code for a 66-kDa protein that represses *P* element transposition. In contrast to this normal repressor function, these stationary *P* element repeats are hypothesized to have evolved the function of transcription factors (90).

A Source of New Introns. Some retroelements are apparently fully adapted to their niche within exonic sequences. For example, the 143-kb *Euglena gracilis* plastid genome contains 155 group II and group III introns (56), nearly 10 times the number in any other known plastid DNA. The original introns were likely mobile, retrotransposable genetic elements that invaded the genome from another organism, relying in part on internally encoded enzyme activities for mobility. The group III introns appear to be streamlined versions of group II introns, sharing a common evolutionary ancestor with a group II intron. Among the *E. gracilis* introns are a number of introns-within-introns (twintrons), suggesting that these elements themselves have been targets of intron insertions. In one particularly interesting example (91), a group III intron is formed from domains of two individual group II introns. The authors suggest the possibility that "the introduction of one catalytic RNA into a functional domain of another catalytic RNA, through a process similar to twintron formation, can result in new combinations of sequences and structural domains that might lead to new RNA catalyzed reactions significant for RNA evolution."

Telomeres in *Drosophila*. An unusually finely tuned system between the host genome and mobile elements has evolved in *Drosophila* to take over a basic cellular function. Several retroelements, such as *HET-A* and *TART*, carry out the function of replacing damaged chromosome ends that is performed by telomerase in other insects (92, 93). The insertion frequency of the TEs involved has become adapted to match the average rate of telomere loss to maintain constant chromosome size. This is the best example to date of a TE providing a vital function to its host.

As a Repair Mechanism of DNA Double-Strand Breaks. Although *SINEs* and *LINEs* and pseudogenes are abundant in eukaryotic genomes, indicating that reverse transcriptase-mediated phenomena are important in genome evolution, the mechanisms responsible for their spread are largely unknown. The results of two recent experiments with the yeast *Saccharomyces cerevisiae* (94, 95) have linked reverse transcriptase-mediated events with double-strand chromosome breaks in the absence of normal repair. This suggests a possible role for endogenous retroelements in the repair of double-strand chromosome breaks under certain circumstances. Note that, in this case, as in others described here, coadaptation may have grown out of apparently parasitic element behavior; double-strand chromosome breaks may simply represent an especially good target for efficient TE insertion, and in turn, these insertions may sometimes be the most efficient repair pathway available to the host. The net effect, rapid insertional repair of breaks, is expected to benefit both host and TE.

Mediation of the Concerted Evolution of Repetitive Gene Families. Evidence for the ability of TEs to directly influence the constitution of repetitive DNA was provided by experiments using genetically marked *P* elements located in a subtelomeric repeat of *D. melanogaster* (96). After *P* element mobilization, the number of repeats frequently was observed to be altered, with decreases being more common than increases, due to unequal gene conversion events. Therefore, TEs may play an important role in the evolution of heterochromatin.

Changes in Genome Size. As described above, TEs may represent a variable and sometimes surprisingly high proportion of genomes, particularly in plants. By means of variation in sheer bulk, it is possible that TEs affect variability in life history traits and related characteristics because of the correlation between genome size, cell size, and various aspects of plant life form, such as growth rate and developmental time (97).

Other Possible Functions. The idea that heterochromatic clusters of nomadic elements are merely graveyards of dead transposons appears to be giving way to the idea that these regions may also be involved in a number of important cellular processes (25). These include developmentally programmed changes in DNA content, expression of heterochromatically embedded gene loci, and housekeeping functions such as chromosome pairing, sister chromatid adhesion, and centromere function. The TE content of these regions may be important for these processes in ways that are not yet understood.

Discussion

One of the most compelling questions that arise when considering the new data on the preponderance of TE-derived sequences in some plant genomes is how enormous numbers of TE copies can accumulate in a single genome. For example, how is it that a single TE can make up 10% of the maize genome? Obviously, the recombinogenic properties of TEs that are hypothesized to maintain a relatively low, constant copy number in other organisms are not relevant in these cases. It may be that, in the case of low copy elements transposing at relatively low frequencies, recombination is able to purge some elements from euchromatin, but the maize example suggests that there may be a vast number of elements interspersed between genes in many locations. Recombination, then, may not be a particularly effective mechanism for purging the vast majority of repetitive sequences, many of which are clearly not located in heterochromatin. It can be argued that, wherever there is a concentration of TE sequences, heterochromatin-like structural features begin to evolve to down-regulate their expression. In turn, this would also tend to reduce the frequency of removal by ectopic recombination.

These considerations motivate us to postulate that there are at least two types of elements that occupy two very different niches in the ecology of the genome: first, a type that preferentially inserts into regions distant from host gene sequences, such as heterochromatin or the regions between genes [e.g., the many retrotransposons found inserted between the genes on the third chromosome in maize (2)]; and second, a type that lives more dangerously by being more prone to insert into, or near, single copy sequences. We suggest that the first type escapes the "trap" of inactivation (via methylation or heterochromatinization) in regions outside of single copy host genes through the use of various buffer sequences; it has become specifically adapted to (or even makes up much of) these regions. As a strategy to minimize their potentially devastating effects on their hosts, these elements target regions in which recombination is minimal and where essential genes are scarce. The second type travels light and has evolved to take advantage of relatively accessible chromosomal architecture, a high concentration of transcription factors, host enhancer sequences, and horizontal transfer to maximize replication advantage. This type, represented by elements like *Mu* (which target single copy sequences) and *P* elements (at least 65% of insertions are located near enhancers) trades the disadvantage of an increased risk of negative selection for the advantages of occupying regions which are enriched for factors promoting efficient transcription and replication. This second type is postulated to be the one most likely to be discovered by geneticists (it is more likely to cause mutations) and also the one most likely to be lost through recombination (by targeting actively transcribed regions of the genome in which recombination is more frequent). We suggest that, when these elements insert in heterochromatin, they become inactive because they are not well adapted to that environment.

We therefore need to consider the possibility that there may be more than one strategy to being a transposon and that each strategy, although successful from an evolutionary perspective, has a very different dynamic. Each type would be expected to affect host evolution in a different way. Type 1 would affect the overall architecture of the host chromosome, rather than the specific expression characteristics of individual genes. In contrast, type 2 would participate more directly in changes in gene regulation, such as is observed at the *Adh1* locus in maize.

A second area of considerable interest from an evolutionary perspective is the stress-induced mutability that is characteristic of some TEs (98). A gradualist argument leveled against the idea that regulatory changes resulting from TE-induced mutations may be important in evolution is that such "macromutations," like Goldschmit's hopeful monsters, would be unlikely to arise at the precise time when a new ecological niche became available (40). However, there is increasing evidence that TE-induced mutation rates are far from constant. High frequencies of mutations are expected to appear in waves, such as those resulting from hybrid dysgenesis that accompany element invasions of new populations or species. TE-induced mutations have been recorded to occur in transpositional bursts (99) whose cause is not well understood but is likely related to inbreeding and other forms of genomic or environmental stress, possibly akin to the genomic stress referred to by McClintock (100). For example, it appears that plant retroelements are normally quiescent but can be activated by stress (98), such as cell culturing (101) or microbial infection (102). We suggest that the proximal, or adaptive, function in these cases is to increase element copy number during periods of stress to ensure a high probability of transmission by those host variants that happen to survive. With respect to the evolution of the host, however, the preadaptive, or exaptive function is to provide variation during periods of stress. In this case, as in the other cases outlined above, the transposon does not have to "know" that it is contributing to the evolution of its host nor has it evolved to do so, but out of its elemental parasitic behavior arises the potential for both dramatic and subtle changes in the genome of its host.

Conclusions

We are only just beginning to glimpse the complexity of possible interactions in the coevolution of TEs and their hosts. A full understanding of the population and evolutionary dynamics of these interactions, and the consequences to hosts, must await the results of further research. However some tentative conclusions can be made on the basis of current information.

The primary parasitic nature of these sequences during their invasion of host populations is beyond dispute, but we believe that this does not by any means represent the whole story. A number of features of both TEs and their hosts can be interpreted as coadaptations to mitigate or abolish the reduction of fitness due to unbridled transposition. Furthermore, the number of well documented cases in which TE sequences have been coopted successfully by the host to provide a useful function is small but is growing rapidly. We suggest that the process by which elements and their hosts coevolve mutually beneficial strategies may lend itself to the production of genetic variation that would not otherwise have arisen.

Although the role of TEs in evolution may not turn out to be precisely what McClintock had in mind when she first described controlling elements in maize, the importance of their role in the evolution of gene regulation and other host functions may yet surprise us. To paraphrase Dobzhansky's famous phrase, there is good reason to believe that "Nothing about mobile elements makes sense except in the light of evolution."

We thank Dr. Zhijian Tu for comments on the manuscript. This work was supported by National Science Foundation Grant DEB9119349 to M.K. D.L. was supported by National Institutes of Health Training Program in Insect Science 1T32 AI07475.

1. McClintock, B. (1948) *Carnegie Inst. Wash. Yearbook* **47,** 155–169.
2. SanMiguel, P., Tikhonov, A., Jin, Y. K., Motchoulskaia, N., Zakharov, D., Melake-Berhan, A., Springer, P. S., Edwards, K. J., Lee, M., Avramova, Z. & Bennetzen, J. L. (1996) *Science* **274,** 765–768.
3. Pimpinelli, S., Berloco, M., Fanti, L., Dimitri, P., Bonaccorsi, S., Marchetti, E., Caizzi, R., Caggesse, C. & Gatti, M. (1995) *Proc. Natl. Acad. Sci. USA* **92,** 3804–3808.
4. Finnegan, D. J. (1992) *Curr. Opin. Genet. Dev.* **2,** 861–867.
5. McClure, M. A. (1993) in *Reverse Transcriptase* (Cold Spring Harbor Lab. Press, Plainview, NY), pp. 425–443.
6. Bucheton, A. (1995) *Trends Genet.* **11,** 349–353.

7. Wessler, S., Bureau, T. E. & White, S. E. (1995) *Curr. Opin. Genet. Dev.* **5,** 814–821.
8. Unsal, K. & Morgan, G. T. (1995) *J. Mol. Biol.* **248,** 812–823.
9. Morgan, J. T. (1995) *J. Mol. Biol.* **254,** 1–5.
10. Smit, A. F. A. & Riggs, A. D. (1996) *Proc. Natl. Acad. Sci. USA* **93,** 1443–1448.
11. Tu, J. (1997) *Proc. Natl. Acad. Sci. USA,* in press.
12. Britten, R. J. (1996) *Proc. Natl. Acad. Sci. USA* **93,** 9374–9377.
13. Engels, W. R. (1996) in *Transposable Elements*, eds. Saedler, H. & Gierl, A. (Springer, Berlin), pp. 103–123.
14. Bregliano, J. C. & Kidwell, M. G. (1983) in *Mobile Genetic Elements*, ed. Shapiro, J. A. (Academic, New York), pp. 363–410.
15. Rubin, G. M., Kidwell, M. G. & Bingham, P. M. (1982) *Cell* **29,** 987–994.
16. McClintock, B. (1951) *Cold Spring Harbor Symp. Quant. Biol.* **16,** 13–47.
17. Wessler, S., Baran, G., Varagona, M. & Dellaporta, S. (1986) *EMBO J.* **5,** 2427–2432.
18. Wessler, S., Baran, G. & Varagona, M. (1987) *Science* **237,** 916–918.
19. Corces, V. G. & Geyer, P. K. (1991) *Trends Genet.* **7,** 86–90.
20. Lister, C., Jackson, D. & Martin, C. (1993) *Plant Cell* **5,** 1541–1553.
21. Kloeckener-Gruissem, B., Vogel, J. M. & Freeling, M. (1992) *EMBO J.* **11,** 157–166.
22. Kloeckener-Gruissem, B. & Freeling, M. (1995) *Proc. Natl. Acad. Sci. USA* **92,** 1836–1840.
23. Greene, B., Walko, R. & Hake, S. (1994) *Genetics* **138,** 1275–1285.
24. Bradley, D., Carpenter, R., Sommer, H., Hartley, N. & Coen, E. (1993) *Cell* **72,** 85–95.
25. Arkhipova, I. R., Lyubomirskaya, N. V. & Ilyin, Y. V. (1995) *Drosophila Retrotransposons* (Landes, Austin, TX).
26. Charlesworth, B. & Langley, C. H. (1989) *Annu. Rev. Genet.* **23,** 251–287.
27. Vaury, C., Bucheton, A. & Pelisson, A. (1989) *Chromosoma* **98,** 215–224.
28. Tchurikov, N. A., Zelentsova, E. S. & Georgiev, G. P. (1980) *Nucleic Acids Res.* **8,** 1243–1258.
29. Tchurikov, N. A., Ilyin, Y. V., Skryabin, K. G., Anan'ev, E. V., Bayev, A. A., Krayev, A. S., Zelentsova, E. S., Kulguskin, V. V., Lyubomirskaya, N. V. & Georgiev, G. P. (1981) *Cold Spring Harbor Symp. Quant. Biol.* **45,** 655–665.
30. Banga, S. S., Velazquez, A. & Boyd, J. B. (1991) *Mutat. Res.* **255,** 79–88.
31. Charlesworth, B., Sniegowski, P. & Stephan, W. (1994) *Nature (London)* **371,** 215–220.
32. Lyttle, T. W. & Haymer, D. S. (1993) in *Transposable Elements and Evolution*, ed. McDonald, J. F. (Kluwer, Dordrecht, The Netherlands).
33. Simmons, G. M. (1992) *Mol. Biol. Evol.* **9,** 1050–1060.
34. Mackay, T. F. C. (1987) *Genet. Res.* **49,** 225–233.
35. Mackay, T. F., Lyman, R. F. & Jackson, M. S. (1992) *Genetics* **130,** 315–332.
36. Torkamanzehi, A., Moran, C. & Nicholas, F. W. (1992) *Genetics* **131,** 73–78.
37. Clark, A. G., Wang, L. & Hulleberg, T. (1995) *Genetics* **139,** 337–348.
38. Britten, R. J. (1996) *Mol. Phylogenet. Evol.* **5,** 13–17.
39. Britten, R. J. (1997) *Gene*, in press.
40. McDonald, J. F. (1995) *Trends Ecol. Evol.* **10,** 123–126.
41. Doolittle, W. F. & Sapienza, C. (1980) *Nature (London)* **284,** 601–603.
42. Orgel, L. E. & Crick, F. H. C. (1980) *Nature (London)* **284.**
43. Hickey, D. A. (1982) *Genetics* **101,** 519–531.
44. Plasterk, R. A. (1995) in *Mobile Genetic Elements*, ed. Sherratt, D. J. (IRL, Oxford), pp. 18–37.
45. Kiyasu, P. K. & Kidwell, M. G. (1984) *Genet. Res.* **44,** 251–259.
46. Kidwell, M. G. (1993) *Ann. Rev. Genet.* **27,** 235–256.
47. Miller, W. J., Kruckenhauser, L. & Pinsker, W. (1996) in *Transgenic Organisms: Biological and Social Implications*, eds. Tomiuk, J., Woehrmann, K. & Sentker, A. (Birkhaeuser, Basel), pp. 21–35.
48. Hurst, G. D. D., Hurst, L. D. & Majerus, M. E. N. (1992) *Nature (London)* **356,** 659–660.
49. Kidwell, M. G. (1994) *J. Hered.* **85,** 339–346.
50. Robertson, H. M. (1993) *Nature (London)* **362,** 241–245.
51. Lohe, A. R., Moriyama, E. N., Lidholm, D. A. & Hartl, D. L. (1995) *Mol. Biol. Evol.* **12,** 62–72.
52. Clark, J. B., Maddison, W. P. & Kidwell, M. G. (1994) *Mol. Biol. Evol.* **11,** 40–50.
53. Spradling, A. C., Stern, D. M., Kiss, I., Roote, J., Laverty, T. & Rubin, G. M. (1995) *Proc. Natl. Acad. Sci. USA* **92,** 10824–10830.
54. Bellen, H., O'Kane, C. J., Wilson, C., Grossniklaus, U., Pearson, R. K. & Gehring, W. J. (1989) *Genes Dev.* **3,** 1288–1300.
55. Ji, H., Moore, D. P., Blomberg, M. A., Braiterman, L. T., Voytas, D. F., Natsoulis, G. & Boeke, J. D. (1993) *Cell* **73,** 1007–1018.
56. Hallick, R. B., Hong, L., Drager, R. G., Favreau, M. R., Montfort, A., Orsat, B., Spielmann, A. & Stutz, E. (1993) *Nucleic Acids Res.* **21,** 3537–3544.
57. Fridell, R. A., Pret, A. M. & Searles, L. L. (1990) *Genes Dev.* **4,** 559–566.
58. Purugganan, M. & Wessler, S. (1992) *Genetica* **86,** 295–303.
59. Rushforth, A. M. & Anderson, P. (1996) *Mol. Cell. Biol.* **16,** 422–429.
60. Laski, F. A., Rio, D. C. & Rubin, G. M. (1986) *Cell* **44,** 7–19.
61. Tseng, J. C., Zollman, S., Chain, A. C. & Laski, F. A. (1991) *Mech. Dev.* **35,** 65–72.
62. Bingham, P. M., Chou, T. B., I., M. & Zachar, Z. (1988) *Trends Genet.* **4,** 134–138.
63. Chomet, P. S., Wessler, S. & Dellaporta, S. L. (1987) *EMBO J.* **6,** 295–302.
64. Chandler, V. & Walbot, V. (1986) *Proc. Natl. Acad. Sci. USA* **83,** 1767–1771.
65. Bennetzen, J. L. (1996) in *Transposable Elements*, eds. Saedler, H. & Gierl, A. (Springer, Berlin), pp. 195–229.
66. Charlesworth, B. & Langley, C. H. (1986) *Genetics* **112,** 359–383.
67. Brookfield, J. F. Y. (1995) in *Mobile Genetic Elements*, ed. Sherratt, D. J. (IRL, Oxford), pp. 130–153.
68. Robertson, H. M. & Engels, W. R. (1989) *Genetics* **123,** 815–824.
69. Misra, S. & Rio, D. C. (1990) *Cell* **62,** 269–284.
70. Black, D. M., Jackson, M. S., Kidwell, M. G. & Dover, G. A. (1987) *EMBO J.* **6,** 4125–4135.
71. Jackson, M. S., D. M. Black & G. A. Dover. (1988) *Genetics* **120,** 1003–1013.
72. Rasmusson, K. E., Raymond, J. D. & Simmons, M. J. (1993) *Genetics* **133,** 605–622.
73. Fedoroff, N., Schlappi, M. & Raina, R. (1995) *Bioessays* **17,** 291–297.
74. Schlappi, M., Raina, R. & Fedoroff, N. (1994) *Cell* **77,** 427–437.
75. Cuypers, H., Dash, S., Peterson, P. A., Saedler, H. & Gierl, A. (1988) *EMBO J.* **7,** 2953–2960.
76. McClintock, B. (1958) *Carnegie Inst. Wash. Yearbook* **57,** 415–429.
77. Gdula, D. A., Gerasimova, T. I. & Corces, V. G. (1996) *Proc. Natl. Acad. Sci. USA* **93,** 9378–9383.
78. Csink, A. & McDonald, J. F. (1990) *Genetics* **126,** 375–385.
79. King, M. C. & Wilson, A. C. (1975) *Science* **188,** 107–116.
80. Britten, R. J. & Davidson, E. H. (1971) *Q. Rev. Biol.* **46,** 111–138.
81. Li, X. & Noll, M. (1994) *Nature (London)* **367,** 83–87.
82. Fincham, J. R. S. & Sastry, G. R. K. (1974) *Annu. Rev. Genet.* **8,** 15–50.
83. Bureau, T. E., Ronald, P. C. & Wessler, S. R. (1996) *Proc. Natl. Acad. Sci. USA* **93,** 8524–8529.
84. Pelisson, A., Song, S. U., Prud'homme, N., Smith, P. A., Bucheton, A. & Corces, V. G. (1994) *EMBO J.* **13,** 4401–4411.
85. Ding, D. & Lipshitz, H. D. (1994) *Genet. Res.* **64,** 167–181.
86. Hambor, J. E., Mennone, J., Coon, M. E., Hanke, J. H. & Kavathas, P. (1993) *Mol. Cell. Biol.* **13,** 7056–7070.
87. Stavenhagen, J. B. & Robins, D. M. (1988) *Cell* **55,** 247–254.
88. Anderson, R., Britten, R. J. & Davidson, E. H. (1994) *Dev. Biol.* **163,** 11–18.
89. Miller, W. J., McDonald, J. F. & Pinsker, W. (1997) *Genetica*, in press.
90. Miller, W. J., Paricio, N., Hagemann, S., Martinez-Sebastian, M. J., Pinsker, W. & de Frutos, R. (1995) *Gene* **156,** 167–174.
91. Hong, L. & Hallick, R. B. (1994) *Genes Dev.* **8,** 1589–1599.
92. Biessmann, H., Valgeirsdottir, K., Lofsky, A., Chin, C., Ginther, B., Levis, R. W. & Pardue, M. L. (1992) *Mol. Cell. Biol.* **12,** 3910–3918.
93. Pardue, M. L., Danilevskaya, O. N., Lowenhaupt, K., Slot, F. & Traverse, K. L. (1996) *Trends Genet.* **12,** 48–52.
94. Moore, J. K. & Haber, J. E. (1996) *Nature (London)* **383,** 644–646.
95. Teng, S. C., Kim, B. & Gabriel, A. (1996) *Nature (London)* **383,** 641–644.
96. Thompson-Stewart, D., Karpen, G. H. & Spradling, A. C. (1994) *Proc. Natl. Acad. Sci. USA* **91,** 9042–9046.
97. Smyth, D. R. (1993) in *Control of Plant Gene Expression*, ed. Verma, D. P. S. (CRC, Boca Raton, FL).
98. Wessler, S. R. (1996) *Curr. Biol.* **6,** 959–961.
99. Gerasimova, T. I., Matjunima, L. V., Mizrokhi, L. J. & Georgiev, G. P. (1985) *EMBO J.* **4,** 3773–3779.
100. McClintock, B. (1984) *Science* **226,** 792–801.
101. Pouteau, S., Huttner, E., Grandbastien, M. A. & Caboche, M. (1991) *EMBO J.* **10,** 1911–1918.
102. Pouteau, S., Grandbastien, M. A. & Boccara, M. (1994) *Plant J.* **5,** 535–542.

Margulis, L., M. F. Dolan, and R. Guerrero. 2000. The chimeric eukaryote: origin of the nucleus from the karyomastigont in amitochondriate protists. *Proceedings of the National Academy of Sciences, USA* 97:6954–6959.

In 1967, Lynn Margulis (who at that time was married to Carl Sagan) published an article whose short title belied the huge impact her paper would have in evolutionary biology: "On the origin of mitosing cells."

In that paper (which, according to her testimony, had been rejected by more than a dozen scientific journals), Margulis laid out at great length an extraordinary hypothesis for the evolutionary origin of organelles (such as mitochondria and chloroplasts) in the cytoplasm of eukaryotic cells. According to her endosymbiont theory, such organelles originated in the distant past through symbiotic mergers when free-living bacteria were engulfed by and became incorporated into protoeukaryotic host cells. This panoramic interpretation of the chimeric evolution of cellular structures was then a radical idea, but it has withstood the test of time to the extent that at least major outlines of the endosymbiont theory are now universally accepted by evolutionary biologists.

The original 1967 paper is much too long to reproduce in this volume, so instead we include a latter-day surrogate that updates and extends Margulis's broader worldview and showcases her predilection for seeing evolutionary footprints of cooperation and symbiosis throughout the biological world. As her career proceeded, Margulis became increasingly enamored (some critics would say too enamored) of the notion that interdependence and collaboration—more so than competition and conflict—underlie numerous aspects of organogenesis and are major hallmarks of evolution. Readers should not interpret the 2000 paper by Margulis and her colleagues as well substantiated in all details (for example, amitochondrial protists are now suspected to have lost their mitochondria secondarily), but readers should view it as an example of the depth and breadth of her career-long fascination with symbiosis and the chimeric origins of eukaryotic life-forms.

RELATED READING

Margulis, L. 1981. Symbiosis in cell evolution. Freeman, New York, New York, USA.

Margulis, L., and D. Sagan. 1997. Microcosmos: four billion years of microbial evolution. Summit Books, New York, New York, USA.

Margulis, L., and D. Sagan. 2000. What is life? University of California Press, Berkeley, California, USA.

Mereschkowski, K. S. 1909. Theory of two plasms as the basis of symbiogenesis: a new study on the origin of organisms (in Russian). Kazan, Russia.

Niklas, K. J. 2010. Book review: Boris M. Kozo-Polyansky, *Symbiogenesis: a new principle of evolution*. Symbiosis 52:49–50.

Sagan, L. 1967. On the origin of mitosing cells. Journal of Theoretical Biology 14:225–274.

Wallin, I. E. 1927. Symbionticism and the origin of species. Williams and Wilkins, Baltimore, Maryland, USA.

Wilson, E. B. 1925. The cell in development and heredity. Macmillan, New York, New York, USA.

Colloquium

The chimeric eukaryote: Origin of the nucleus from the karyomastigont in amitochondriate protists

Lynn Margulis*, Michael F. Dolan*†, and Ricardo Guerrero‡

*Department of Geosciences, Organismic and Evolutionary Biology Graduate Program, University of Massachusetts, Amherst, MA 01003; and ‡Department of Microbiology, and Special Research Center Complex Systems (Microbiology Group), University of Barcelona, 08028 Barcelona, Spain

We present a testable model for the origin of the nucleus, the membrane-bounded organelle that defines eukaryotes. A chimeric cell evolved via symbiogenesis by syntrophic merger between an archaebacterium and a eubacterium. The archaebacterium, a thermoacidophil resembling extant *Thermoplasma*, generated hydrogen sulfide to protect the eubacterium, a heterotrophic swimmer comparable to *Spirochaeta* or *Hollandina* that oxidized sulfide to sulfur. Selection pressure for speed swimming and oxygen avoidance led to an ancient analogue of the extant cosmopolitan bacterial consortium "Thiodendron latens." By eubacterial-archaebacterial genetic integration, the chimera, an amitochondriate heterotroph, evolved. This "earliest branching protist" that formed by permanent DNA recombination generated the nucleus as a component of the karyomastigont, an intracellular complex that assured genetic continuity of the former symbionts. The karyomastigont organellar system, common in extant amitochondriate protists as well as in presumed mitochondriate ancestors, minimally consists of a single nucleus, a single kinetosome and their protein connector. As predecessor of standard mitosis, the karyomastigont preceded free (unattached) nuclei. The nucleus evolved in karyomastigont ancestors by detachment at least five times (archamoebae, calonymphids, chlorophyte green algae, ciliates, foraminifera). This specific model of syntrophic chimeric fusion can be proved by sequence comparison of functional domains of motility proteins isolated from candidate taxa.

Archaeprotists | spirochetes | sulfur syntrophy | *Thiodendron* | trichomonad

Two Domains, Not Three

All living beings are composed of cells and are unambiguously classifiable into one of two categories: prokaryote (bacteria) or eukaryote (nucleated organisms). Here we outline the origin of the nucleus, the membrane-bounded organelle that defines eukaryotes. The common ancestor of all eukaryotes by genome fusion of two or more different prokaryotes became "chimeras" via symbiogenesis (1). Long term physical association between metabolically dependent consortia bacteria led, by genetic fusion, to this chimera. The chimera originated when an archaebacterium (a thermoacidophil) and a motile eubacterium emerged under selective pressure: oxygen threat and scarcity both of carbon compounds and electron acceptors. The nucleus evolved in the chimera. The earliest descendant of this momentous merger, if alive today, would be recognized as an amitochondriate protist. An advantage of our model includes its simultaneous consistency in the evolutionary scenario across fields of science: cell biology, developmental biology, ecology, genetics, microbiology, molecular evolution, paleontology, protistology. Environmentally plausible habitats and modern taxa are easily comprehensible as legacies of the fusion event. The scheme that generates predictions demonstrable by molecular biology, especially motile protein sequence comparisons (2), provides insight into the structure, physiology, and classification of microorganisms.

Our analysis requires the two- (Bacteria/Eukarya) not the three- (Archaea/Eubacteria/Eukarya) domain system (3). The prokaryote vs. eukaryote that replaced the animal vs. plant dichotomy so far has resisted every challenge. Microbiologist's molecular biology-based threat to the prokaryote vs. eukaryote evolutionary distinction seems idle (4). In a history of contradictory classifications of microorganisms since 1820, Scamardella (5) noted that Woese's entirely nonmorphological system ignores symbioses. But bacterial consortia and protist endosymbioses irreducibly underlie evolutionary transitions from prokaryotes to eukaryotes. Although some prokaryotes [certain Gram-positive bacteria (6)] are intermediate between eubacteria and archaebacteria, no organisms intermediate between prokaryotes and eukaryotes exist. These facts render the 16S rRNA and other nonmorphological taxonomies of Woese and others inadequate. Only all-inclusive taxonomy, based on the work of thousands of investigators over more than 200 years on live organisms (7), suffices for detailed evolutionary reconstruction (4).

When Woese (8) insists "there are actually three, not two, primary phylogenetic groupings of organisms on this planet" and claims that they, the "Archaebacteria" (or, in his term that tries to deny their bacterial nature, the "Archaea") and the "Eubacteria" are "each no more like the other than they are like eukaryotes," he denies intracellular motility, including that of the mitotic nucleus. He minimizes these and other cell biological data, sexual life histories including cyclical cell fusion, fossil record correlation (9), and protein-based molecular comparisons (10, 11). The tacit, uninformed assumption of Woese and other molecular biologists that all heredity resides in nuclear genes is patently contradicted by embryological, cytological, and cytoplasmic heredity literature (12). The tubulin-actin motility systems of feeding and sexual cell fusion facilitate frequent viable incorporation of heterologous nucleic acid. Many eukaryotes, but no prokaryotes, regularly ingest entire cells, including, of course, their genomes, in a single phagocytotic event. This invalidates any single measure alone, including ribosomal RNA gene sequences, to represent the evolutionary history of a lineage.

As chimeras, eukaryotes that evolved by integration of more than a single prokaryotic genome (6) differ qualitatively from prokaryotes. Because prokaryotes are not directly comparable to symbiotically generated eukaryotes, we must reject Woese's three-domain interpretation. Yet our model greatly appreciates his archaebacterial-eubacterial distinction: the very first anaerobic eukaryotes derived from both of these prokaryotic lineages. The enzymes of protein synthesis in eukaryotes come primarily

This paper was presented at the National Academy of Sciences colloquium "Variation and Evolution in Plants and Microorganisms: Toward a New Synthesis 50 Years After Stebbins," held January 27–29, 2000, at the Arnold and Mabel Beckman Center in Irvine, CA.

†To whom reprint requests should be addressed. E-mail: mdolan@geo.umass.edu.

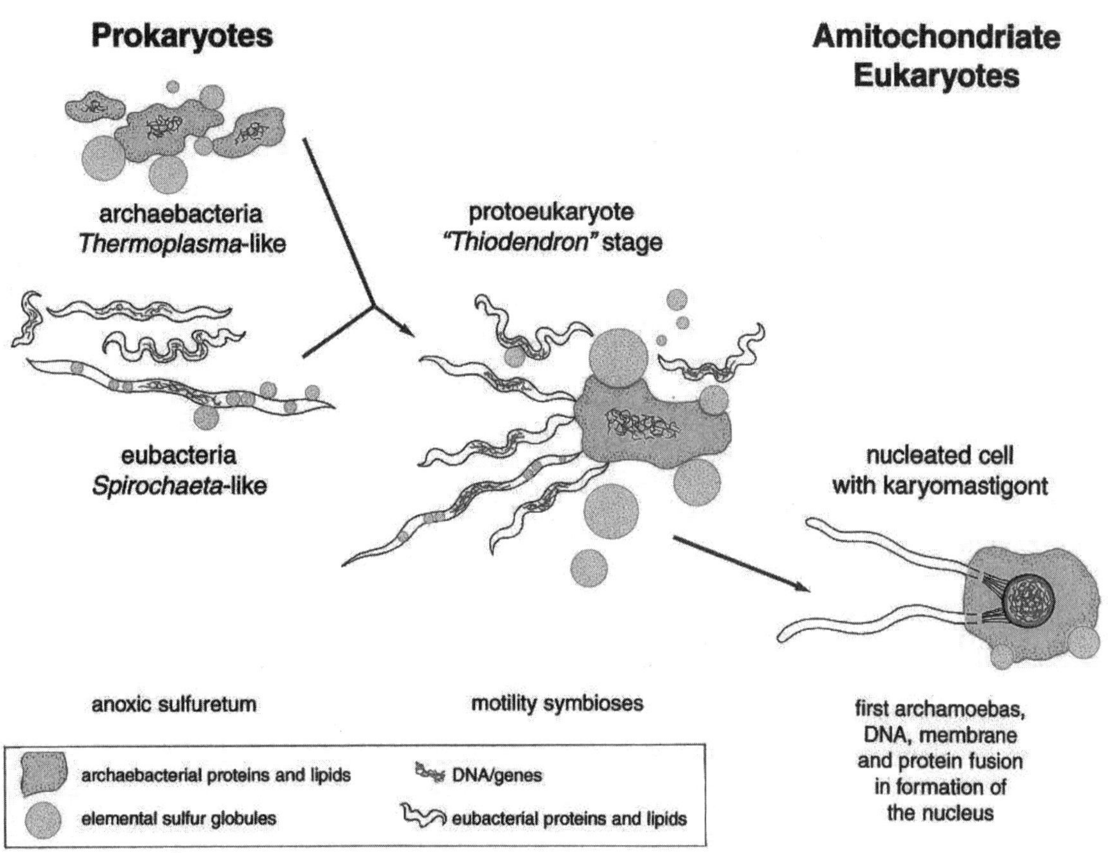

Fig. 1. Origin of the chimeric eukaryote with karyomastigonts from a motile sulfur-bacteria consortium.

from archaebacteria whereas in the motility system (microtubules and their organizing centers), many soluble heat-shock and other proteins originated from eubacteria (9). Here we apply Gupta's idea (from protein sequences) (10) to comparative protist data (13) to show how two kinds of prokaryotes made the first chimeric eukaryote. We reconstruct the fusion event that produced the nucleus.

The Chimera: Archaebacterium/Eubacterium Merger

Study of conserved protein sequences [a far larger data set than that used by Woese *et al.* (3)] led Gupta (10) to conclude "all eukaryotic cells, including amitochondriate and aplastidic cells received major genetic contributions to the nuclear genome from both an archaebacterium (very probably of the eocyte, i.e., thermoacidophil group and a Gram-negative bacterium . . . [t]he ancestral eukaryotic cell never directly descended from archaebacteria but instead was a chimera formed by fusion and integration of the genomes of an archaebacerium and a Gram-negative bacterium" (p. 1487). The eubacterium ancestor has yet to be identified; Gupta rejects our spirochete hypothesis. In answer to which microbe provided the eubacterial contribution, he claims: "the sequence data . . . . suggest that the archaebacteria are polyphyletic and are close relatives of the Gram-positive bacteria" (p. 1485). The archaebacterial sequences, we posit, following Searcy (14), come from a *Thermoplasma acidophilum*-like thermoacidophilic (eocyte) prokaryote. This archaebacterial ancestor lived in warm, acidic, and sporadically sulfurous waters, where it used either elemental sulfur (generating H_2S) or less than 5% oxygen (generating H_2O) as terminal electron acceptor. As does its extant descendant, the ancient archaebacterium survived acid-hydrolysis environmental conditions by nucleosome-style histone-like protein coating of its DNA (14) and actin-like stress-protein synthesis (15). The wall-less archaebacterium was remarkably pleiomorphic; it tended into tight physical association with globules of elemental sulfur by use of its rudimentary cytoskeletal system (16). The second member of the consortium, an obligate anaerobe, required for growth the highly reduced conditions provided by sulfur and sulfate reduction to hydrogen sulfide. Degradation of carbohydrate (e.g., starch, sugars such as cellobiose) and oxidation of the sulfide to elemental sulfur by the eubacterium generated carbon-rich fermentation products and electron acceptors for the archaebacterium. When swimming eubacteria attached to the archaebacterium, the likelihood that the consortium efficiently reached its carbon sources was enhanced. This hypothetical consortium, before the integration to form a chimera (Fig. 1), differs little from the widespread and geochemically important "*Thiodendron*" (17, 18).

The "*Thiodendron*" Stage

The "*Thiodendron*" stage refers to an extant bacterial consortium that models our idea of an archaebacteria-eubacteria sulfur syntrophic motility symbiosis. The partners in our view merged to become the chimeric predecessor to archaeprotists. The membrane-bounded nucleus, by hypothesis, is the morphological manifestation of the chimera genetic system that evolved from a *Thiodendron*-type consortium. Each phenomenon we suggest, from free-living bacteria to integrated association, enjoys extant natural analogues.

Study of marine microbial mats revealed relevant bacterial consortia in more than six geographically separate locations. Isolations from Staraya Russa mineral spring 8, mineral spring Serebryani, Lake Nizhnee, mud-baths; littoral zone at the White Sea strait near Veliky Island, Gulf of Nilma; Pacific Ocean hydrothermal habitats at the Kurile Islands and Kraternaya Bay; Matupi Harbor Bay, Papua New Guinea, etc. (17) all yielded "*Thiodendron latens*" or very similar bacteria. Samples were taken from just below oxygen-sulfide interface in anoxic waters (17, 18). Laboratory work showed it necessary to abolish the genus *Thiodendron* because it is a sulfur syntrophy. A stable ectosymbiotic association of two bacterial types grows as an anaerobic consortium between 4 and 32°C at marine pH values and salinities. Starch, cellobiose, and other carbohydrates (not cellulose, amino acids, organic acids, or alcohol) supplemented by heterotrophic CO_2 fixation provide it carbon. *Thiodendron* appears as bluish-white spherical gelatinous colonies, concentric in structure within a slimy matrix produced by the consortium bacteria. The dominant partner invariably is a distinctive strain of pleiomorphic spirochetes: they vary from the typical walled *Spirochaeta* 1:2:1 morphology to large membranous spheres, sulfur-studded threads, gliding or nonmotile cells of variable width (0.09–0.45 μm) and lengths to millimeters. The other partner, a small, morphologically stable vibrioid, *Desulfobacter* sp., requires organic carbon, primarily acetate, from spirochetal carbohydrate degradation. The spirochetal *Escherichia coli*-like formic acid fermentation generates energy and food. *Desulfobacter* sp. cells that reduce both sulfate and sulfur to sulfide are always present in the natural consortium but in far less abundance than the spirochetes. We envision the *Thiodendron* consortium of "free-living spirochetes in geochemical sulfur cycle" (ref. 18, p. 456) and spirochete motility symbioses (19) as preadaptations for chimera evolution. *Thiodendron* differs from the archaebacterium-eubacterium association we hypothesize; the marine *Desulfobacter* would have been replaced with a pleiomorphic wall-less, sulfuric-acid tolerant soil *Thermoplasma*-like archaebacterium. New thermoplasmas are under study. We predict strains that participate in spirochete consortia in less saline, more acidic, and higher temperature sulfurous habitats than *Thiodendron* will be found.

When "pure cultures" that survived low oxygen were first described [by B. V. Perfil'ev in 1969, in Russian (see refs. 17 and 18] a complex life history of vibrioids, spheroids, threads and helices was attributed to "*Thiodendron latens*". We now know these morphologies are artifacts of environmental selection pressure: Dubinina *et al.* (ref. 17, p. 435), reported that "the pattern of bacterial growth changes drastically when the redox potential of the medium is brought down by addition of 500 mg/l of sodium sulfide." The differential growth of the two tightly associated partners in the consortium imitates the purported *Thiodendron* bacterial developmental patterns. The syntrophy is maintained by lowering the level of oxygen enough for spirochete growth. The processes of sulfur oxidation-reduction and oxygen removal from oxygen-sensitive enzymes, we suggest, were internalized by the chimera and retained by their protist descendants as developmental cues.

Metabolic interaction, in particular syntrophy under anoxia, retained the integrated prokaryotes as emphasized by Martin and Müller (20). However, we reject their concept, for which no evidence exists, that the archaebacterial partner was a methanogen. Our sulfur syntrophy idea, by contrast, is bolstered by observations that hydrogen sulfide is still generated in amitochondriate, anucleate eukaryotic cells (mammalian erythrocytes) (21).

T. acidophilum in pure culture attach to suspended elemental sulfur. When sulfur is available, they generate hydrogen sulfide (16). Although severely hindered by ambient oxygen, they are

Table 1. Karyomastigont distribution in unicellular protoctists

	Archaeprotista*		
Class	Karyomastigont	Kinetosome	Nucleus
Pelobiontids	+†	−	+
Metamonads	+	+	+/−
Parabasalids	+	+	+
Trichomonads	+	+	+/−
Hypermastigids	−	+	−
	Chlorophyta		
Genus	Karyomastigont	Kinetosome	Nucleus
Chlamydomonas	+	+	−
Chlorella	−	−	−
Acetabularia	+	+	+
	Ciliophora		
Subphyla	Karyomastigont	Kinetosome	Nucleus
Postciliodesmatophora	−	+	+
Rabdophora	−	+	+
Cyrtophora	−	+	+
	Discomitochondria		
Class	Karyomastigont	Kinetosome	Nucleus
Amoebomastigotes	+	−	+/−
Kinetoplastids	−	−	−
Euglenids	−/?	−	−
Pseudociliates	−	+	−
	Granuloreticulosa		
Class	Karyomastigont	Kinetosome	Nucleus
Reticulomixids	−/?	−	+
Foraminiferans	+	−	+
	Hemimastigophora		
Genus	Karyomastigont	Kinetosome	Nucleus
Stereonema	−	+	+/−
Spironema	−	+	−
Hemimastix	−	+	−
	Zoomastigota		
Class	Karyomastigont	Kinetosome	Nucleus
Jakobids	?	−	−
Bicosoecids	+/?	+	−
Proteromonads	+	−	−
Opalinids	−	+	+
Choanomastigotes	+	−	−

*Bold entries are protoctist phyla. All species of Archaeprotists lack mitochondria. "Karyomastigont," "kinetosome," and "nucleus," refer to relative proliferation of these organelles. Members of the phylum Archaeprotista group into one of three classes: Pelobiontid giant amoebae; Metamonads, which include three subclasses: Diplomonads (*Giardia*), Retortamonads (*Retortamonas*), and Oxymonads (such as *Pyrsonympha* and *Saccinobaculus*); and Parabasalia. The Class Parabasalia unites trichomonads, devescovinids, calonymphids, and hypermastigotes such as *Trichonympha*. The phylum Discomitochondria includes amoebomastigotes, kinetoplastids (*Trypanosoma*), euglenids, and pseudociliates (*Stephanopogon*). The Hemimastigophora comprise a new southern-hemisphere phylum of free-living mitochondriate protists (28). Hemimastigophorans probably evolved from members of the kinetoplastid-euglenid taxon (29). If so, they represent a seventh example of release of the nucleus from the karyomastigont and subsequent kinetosome proliferation. The phylum Granuloreticulosa includes the shelled (Class Foraminifera) and unshelled (Class Reticulomyxa) foraminiferans. The phylum Zoomastigota includes five classes of single-celled, free-living and symbiotrophic mitochondriate protists: Jakobids, Bicosoecids, Proteromonads, Opalinids, and Choanomastigotes. Details of the biology are in the work by Margulis *et al.* (30). A current phylogeny is depicted in Fig. 2.

†Structure known but not demonstrated for all species at the electron microscopic level.

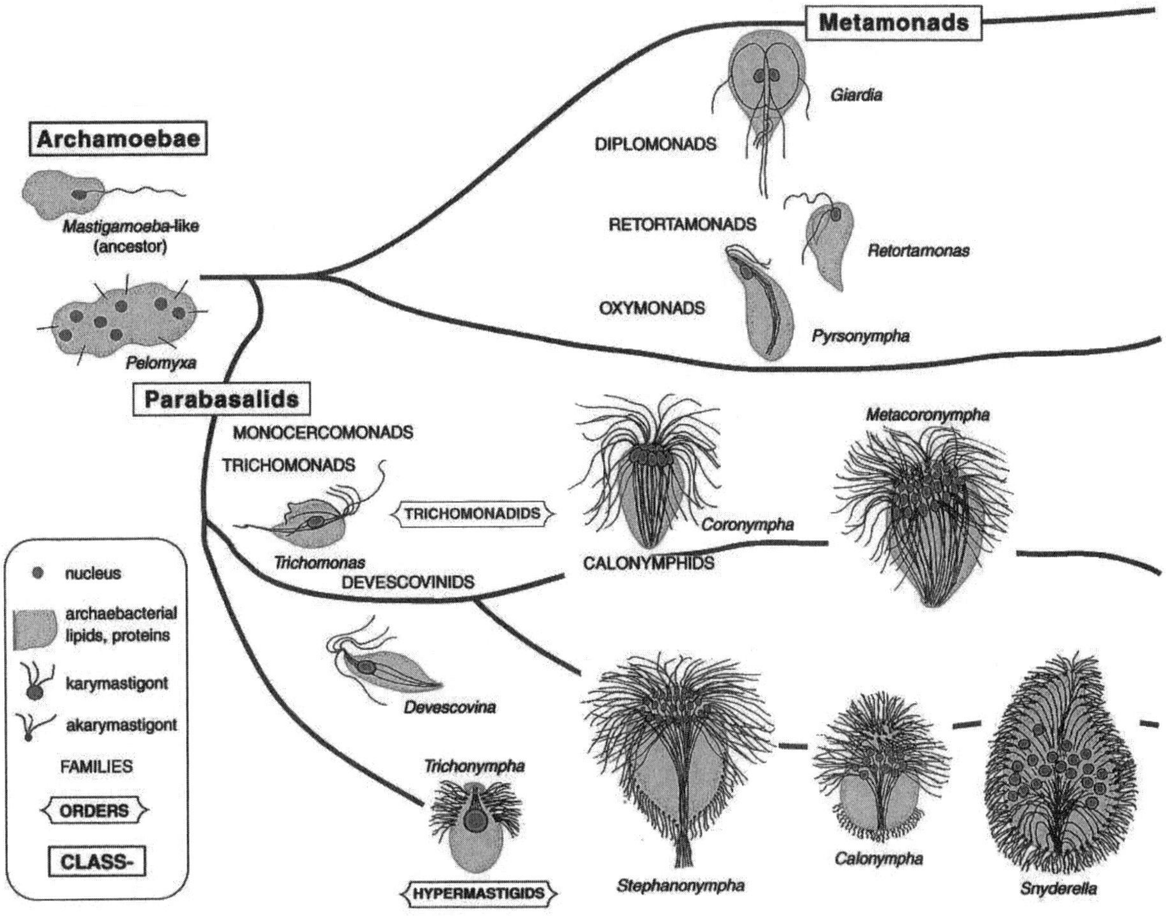

Fig. 2. Biological phylogeny of chimeric eukaryotes taken to be primitively amitochondriate.

microaerophilic in the presence of small quantities (<5%) of oxygen. The *Thermoplasma* partner thus would be expected to produce sulfide and scrub small quantities of oxygen to maintain low redox potential in the spirochete association. The syntrophic predecessors to the chimera is metabolically analogous to *Thiodendron* where *Desulfobacter* reduces sulfur and sulfate producing sulfide at levels that permit the spirochetes to grow. We simply suggest the replacement of the marine sulfidogen with *Thermoplasma*. In both the theoretical and actual case, the spirochetes would supply oxidized sulfur as terminal electron acceptor to the sulfidogen.

The DNA of the *Thermoplasma*-like archaebacterium permanently recombined with that of the eubacterial swimmer. A precedent exists for our suggestion that membrane hypertrophies around DNA to form a stable vesicle in some prokaryotes: the membrane-bounded nucleoid in the eubacterium *Gemmata obscuriglobus* (22). The joint *Thermoplasma*-like archaebacterial DNA package that began as the consortium nucleoid became the chimera's nucleus.

The two unlike prokaryotes together produced a persistent protein exudate package. This step in the origin of the nucleus—the genetic integration of the two-membered consortium to form the chimera—is traceable by its morphological legacy: the karyomastigont. The attached swimmer partner, precursor to mitotic microtubule system, belonged to genera like the nearly ubiquitous consortium-former *Spirochaeta* or the cytoplasmic tubule-maker *Hollandina* (19). The swimmer's attachment structures hypertrophied as typically they do in extant motility symbioses (19). The archaebacterium-eubacterium swimmer attachment system became the karyomastigont. The proteinaceous karyomastigont that united partner DNA in a membrane-bounded, jointly produced package, assured stability to the chimera. All of the DNA of the former prokaryotes recombined inside the membrane to become nuclear DNA while the protein-based motility system of the eubacterium, from the moment of fusion until the present, segregated the chimeric DNA. During the lower Proterozoic eon (2,500–1,800 million years ago), many interactions inside the chimera generated protists in which mitosis and eventually meiotic sexuality evolved. The key concept here is that the karyomastigont, retained by amitochondriate protists and later by their mitochondriate descendants, is the morphological manifestation of the original archaebacterial-eubacterial fused genetic system. Free (unattached) nuclei evolved many times by disassociation from the rest of the karyomastigont. The karyomastigont, therefore, was the first microtubule-organizing center.

Karyomastigonts Preceded Nuclei

The term "karyomastigont" was coined by Janicki (23) to refer to a conspicuous organellar system he observed in certain protists: the mastigont ("cell whip," eukaryotic flagellum, or undulipodium, the [9 (2) + (2)] microtubular axoneme underlain by its [9 (3) + 0)] kinetosome) attached by a "nuclear connector" or "rhizoplast" to a nucleus. The need for a term came from Janicki's work on highly motile trichomonad symbionts in the intestines of termites where karyomastigonts dominate the cells. When kinetosomes, nuclear connector, and other components were present but the nucleus was absent from its predictable position, Janicki called the organelle system an "akaryomastigont." In the Calonymphidae, one family of entirely multinucleate trichomonads, numerous karyomastigonts, and akaryomastigonts are simultaneously present in the same cell (e.g., *Calonympha grassii*) (24).

The karyomastigont, an ancestral feature of eukaryotes, is present in "early branching protists" (25–27). Archaeprotists, a large inclusive taxon (phylum of Kingdom Protoctista) (7) are heterotrophic unicells that inhabit anoxic environments. All lack mitochondria. At least 28 families are placed in the phylum Archaeprotista. Examples include archaemoebae (*Pelomyxa* and *Mastigamoeba*), metamonads (*Retortamonas*), diplomonads (*Giardia*), oxymonads (*Pyrsonympha*), and the two orders of Parabasalia: Trichomonadida [*Devescovina*, *Mixotricha*, *Monocercomonas*, *Trichomonas*, and calonymphids (*Coronympha*, *Snyderella*)] and Hypermastigida (*Lophomonas*, *Staurojoenina*, and *Trichonympha*). These cells either bear karyomastigonts or derive by differential organelle reproduction (simple morphological steps) from those that do (Table 1). When, during evolution of these protists, nuclei were severed from their karyomastigonts, akaryomastigonts were generated (31). Nuclei, unattached, at least temporarily, to undulipodia were freed to proliferate and occupy central positions in cells. Undulipodia, also freed to proliferate, generated larger, faster-swimming cells in the same evolutionary step.

The karyomastigont is the conspicuous central cytoskeleton in basal members of virtually all archaeprotist lineages [three classes: Archamoeba, Metamonads, and Parabasalia (32)] (Fig. 2). In trichomonads, the karyomastigont, which includes a parabasal body (Golgi complex), coordinates the placement of hydrogenosomes (membrane-bounded bacterial-sized cell inclusions that generate hydrogen). The karyomastigont reproduces as a unit structure. Typically, four attached kinetosomes with rolled sheets of microtubules (the axostyle and its extension the pelta) reproduce as their morphological relationships are retained. Kinetosomes reproduce first, the nucleus divides, and the two groups of kinetosomes separate at the poles of a thin microtubule spindle called the paradesmose. Kinetosomes and associated structures are partitioned to one of the two new karyomastigonts. The other produces components it lacks such as the Golgi complex and axostyle.

Nuclear α-proteobacterial genes were interpreted to have originated from lost or degenerate mitochondria in at least two archaeprotist species [*Giardia lamblia* (33); *Trichomonas vaginalis* (34, 35)] and in a microsporidian (36). Hydrogenosomes, at least some types, share common origin with mitochondria. In the hydrogen hypothesis (20), hydrogenosomes are claimed to be the source of eubacterial genes in amitochondriates. That mitochondria were never acquired in the ancestors we consider more likely than that they were lost in every species of these anaerobic protists. Eubacterial genes in the nucleus that are not from the original spirochete probably were acquired in amitochondriate protists from proteobacterial symbionts other than those of the mitochondrial lineage. Gram-negative bacteria, some of which may be related to ancestors of hydrogenosomes, are rampant as epibionts, endobionts, and even endonuclear symbionts—for example, in *Caduceia versatilis* (37).

Karyomastigonts freed (detached from) nuclei independently in many lineages both before and after the acquisition of mitochondria. Calonymphid ancestors of *Snyderella* released free nuclei before the mitochondrial symbiosis (13), and *Chlamydomonas*-like ancestors of other chlorophytes such as *Acetabularia* released the nuclei after the lineage was fully aerobic (38). In trophic forms of protists that lack mastigote stages, the karyomastigont is generally absent. An exception is *Histomonas*, an amoeboid trichomonad cell that lacks an axoneme but bears enough of the remnant karyomastigont structure to permit its classification with parabasalids rather than with rhizopod amoebae (39). This organellar system appears in the zoospores, motile trophic forms, or sperm of many organisms, suggesting the relative ease of karyomastigont development. The karyomastigont, apparently in some cells, is easily lost, suppressed, and regained. In many taxa of multinucleate or multicellular protists (foraminifera, green algae) and even in plants, the karyomastigont persists only in the zoospores or gametes.

In yeast, nematode, insect, and mammalian cells, nonkaryomastigont microtubule-organizing centers are "required to position nuclei at specific locations in the cytoplasm" (40). The link between the microtubule organizing center and the nuclei "is mysterious" (40). To us, the link is an evolutionary legacy, a remnant of the original archaebacterial-eubacterial connector. The modern organelles (i.e., centriole-kinetosomes, untethered nuclei, Golgi, and axostyles) derive from what first ensured genetic continuity of the chimera's components: the karyomastigont, a structure that would have been much more conspicuous to Proterozoic investigators than to us.

We thank our colleagues Ray Bradley, Michael Chapman, Floyd Craft, Kathryn Delisle (for figures), Ugo d'Ambrosio, Donna Reppard, Dennis Searcy, and Andrew Wier. We acknowledge research assistance from the University of Massachusetts Graduate School via Linda Slakey, Dean of Natural Science and Mathematics, from the Richard Lounsbery Foundation, and from the American Museum of Natural History Department of Invertebrates (New York). Our research is supported by National Aeronautics and Space Administration Space Sciences and Comision Interministerial de Ciencia y Tecnologia Project No. AMB98-0338 (to R.G.).

1. Golding, G. B. & Gupta, R. S. (1995) *Mol. Biol. Evol.* **12,** 1–6.
2. Chapman, M., Dolan, M. F. & Margulis, L. (2000) *Q. Rev. Biol.*, in press.
3. Woese, C. R., Kandler, O. & Wheelis, M. L. (1990) *Proc. Natl. Acad. Sci. USA* **87,** 4576–4579.
4. Mayr, E. (1998) *Proc. Natl. Acad. Sci. USA* **95,** 9720–9723.
5. Scamardella, J. M. (1999) *Int. Microbiol.* **2,** 207–216.
6. Gupta, R. S. (1998) *Mol. Microbiol.* **29,** 695–708.
7. Margulis, L. & Schwartz, K. V. (1998) *Five Kingdoms: An Illustrated Guide to the Phyla of Life on Earth* (Freeman, New York).
8. Woese, C. R. (1998) *Proc. Natl. Acad. Sci. USA* **95,** 11043–11046.
9. Margulis, L. (1996) *Proc. Natl. Acad. Sci USA* **93,** 1071–1076.
10. Gupta, R. S. (1998) *Microbiol. Mol. Biol. Rev.* **62,** 1435–1491.
11. Gupta, R. S. (1998) *Theor. Popul. Biol.* **54,** 91–104.
12. Sapp, J. (1999) *Hist. Philos. Life Sci.* **20,** 3–38.
13. Dolan, M. F., d'Ambrosio, U., Wier, A. & Margulis, L. (2000) *Acta Protozool.* **39,** 135–141.
14. Searcy, D. G. (1992) in *The Origin and Evolution of the Cell*, eds. Hartman, H. & Matsuno, K. (World Scientific, Singapore), pp. 47–78.
15. Searcy, D. G. & Delange, R. J. (1980) *Biochim. Biophys. Acta* **609,** 197–200.
16. Searcy, D. & Hixon, W. G. (1994) *BioSystems* **10,** 19–28.
17. Dubinina, G. A., Leshcheva, N. V. & Grabovich, M. Y. (1993) *Microbiology* **62,** 432–444.
18. Dubinina, G. A., Grabovich, M. Y. & Lesheva, N. V. (1993) *Microbiology* **62,** 450–456.
19. Margulis, L. (1993) *Symbiosis in Cell Evolution* (Freeman, New York).
20. Martin, W. & Müller, M. (1998) *Nature (London)* **392,** 37–41.
21. Searcy, D. & Lee, S. H. (1998) *J. Exp. Zool.* **282,** 310–322.
22. Fuerst, J. A. & Webb, R. I. (1991) *Proc. Natl. Acad. Sci. USA* **88,** 8184–8188.
23. Janicki, C. (1915) *Z. Wiss. Zool.* **112,** 573–691.
24. Kirby, H. & Margulis, L. (1994) *Symbiosis* **16,** 7–63.
25. Dacks, J. B. & Redfield, R. (1998) *J. Eukaryotic Microbiol.* **45,** 445–447.
26. Delgado-Viscogliosi, P., Viscogliosi, E., Gerbod, D., Juldo, J., Sogin, M. L. & Edgcomb, V. (2000) *J. Eukaryotic Microbiol.* **47,** 70–75.
27. Edgcomb, V., Viscogliosi, E., Simpson, A. G. B., Delgado-Viscogliosi, P., Roger, A. J. & Sogin, M. L. (1998) *Protist* **149,** 359–366.
28. Foissner, W., Blatterer, H. & Foissner, I. (1988) *Eur. J. Protistol.* **23,** 361–383.

29. Foissner, W. & Foissner, I. (1993) *J. Eukaryotic Microbiol.* **40,** 422–438.
30. Margulis, L., McKhann, H. I. & Olendzenski, L., eds. (1993) *Illustrated Glossary of the Protoctista* (Jones & Bartlett, Sudbury, MA).
31. Kirby, H. (1949) *Rev. Soc. Mex. Hist. Nat.* **10,** 57–79.
32. Brugerolle, G. (1991) *Protoplasma* **164,** 70–90.
33. Roger, A. J., Srard, S. G., Tovar, J., Clark, C. G., Smith, M. W., Gillin, F. D. & Sogin, M. L. (1998) *Proc. Natl. Acad. Sci. USA* **95,** 229–234.
34. Roger, A. J., Clark, C. G. & Doolittle, W. M. (1996) *Proc. Natl. Acad. Sci. USA* **93,** 14618–14622.
35. Germot, A., Philippe, H. & Le Guyader, H. (1996) *Proc. Natl. Acad. Sci. USA* **93,** 14614–14617.
36. Sogin, M. L. (1997) *Curr. Opin. Gen. Dev.* **7,** 792–799.
37. d'Ambrosio, U., Dolan, M., Wier, A. & Margulis, L. (1999) *Eur. J. Protistol.* **35,** 327–337.
38. Hall, J. & Luck, D. J. L. (1995) *Proc. Natl. Acad. Sci. USA* **92,** 5129–5133.
39. Dyer, B. (1990) in *Handbook of Protoctista*, eds. Margulis, L., Corliss, J. O., Melkonian, M. & Chapman, D. J. (Jones & Bartlett, Sudbury, MA), pp. 252–258.
40. Raff, J. W. (1999) *Curr. Biol.* **9,** R708–R710.

Lynch, M., and J. S. Conery. 2000. The evolutionary fate and consequences of duplicate genes. *Science* 290:1151–1155.

Evolution proceeds when natural selection, genetic drift, gene flow, and recombination act on standing population-genetic variation that originated in various lineages as mutational departures from more ancestral conditions. In addition, genomes change in size through processes of gene duplication and loss, and these too provide important fodder for natural selection and other evolutionary processes. In particular, gene duplication offers tremendous evolutionary potential, because, according to standard wisdom, one of the duplicate loci may retain its former function, thus in effect freeing the duplicate copy to evolve in new directions and perhaps to acquire entirely new or modified functional roles in a cell. Michael Lynch and John Conery trace the evolutionary trajectories of large numbers of duplicate genes and thereby elevate the study of gene duplications to the status of a sophisticated quantitative branch of evolutionary genomics.

RELATED READING

Li, W.-H. 1999. Molecular evolution. Sinauer, Sunderland, Massachusetts, USA.
Lynch, M. 2007. The origins of genome architecture. Sinauer, Sunderland, Massachusetts, USA.
Ohno, S. 1970. Evolution by gene duplication. Springer-Verlag, New York, New York, USA.
Ohta, T., and K. Aoki, editors. Population genetics and molecular evolution. Springer-Verlag, Berlin, Germany.

The Evolutionary Fate and Consequences of Duplicate Genes

Michael Lynch[1]* and John S. Conery[2]

Gene duplication has generally been viewed as a necessary source of material for the origin of evolutionary novelties, but it is unclear how often gene duplicates arise and how frequently they evolve new functions. Observations from the genomic databases for several eukaryotic species suggest that duplicate genes arise at a very high rate, on average 0.01 per gene per million years. Most duplicated genes experience a brief period of relaxed selection early in their history, with a moderate fraction of them evolving in an effectively neutral manner during this period. However, the vast majority of gene duplicates are silenced within a few million years, with the few survivors subsequently experiencing strong purifying selection. Although duplicate genes may only rarely evolve new functions, the stochastic silencing of such genes may play a significant role in the passive origin of new species.

Duplications of individual genes, chromosomal segments, or entire genomes have long been thought to be a primary source of material for the origin of evolutionary novelties, including new gene functions and expression patterns (*1–3*). However, it is unclear how duplicate genes successfully navigate an evolutionary trajectory from an initial state of complete redundancy, wherein one copy is likely to be expendable, to a stable situation in which both copies are maintained by natural selection. Nor is it clear how often these events occur.

Theory suggests three alternative outcomes in the evolution of duplicate genes: (i) one copy may simply become silenced by degenerative mutations (nonfunctionalization); (ii) one copy may acquire a novel, beneficial function and become preserved by natural selection, with the other copy retaining the original function (neofunctionalization); or (iii) both copies may become partially compromised by mutation accumulation to the point at which their total capacity is reduced to the level of the single-copy ancestral gene (subfunctionalization) (*1–12*). Because the vast majority of mutations affecting fitness are deleterious (*13*), and because gene duplicates are generally assumed to be functionally redundant at the time of origin, virtually all models predict that the usual fate of a duplicate-gene pair is the nonfunctionalization of one copy. The expected time that elapses before a gene is silenced is thought to be relatively short, on the order of the reciprocal of the null mutation rate per locus (a few million years or less), except in populations with enormous effective sizes (*11, 12*).

These theoretical expectations are only partially consistent with the limited data that we have on gene duplication. First, comparative studies of nucleotide sequences suggest that although both copies of a gene may often accumulate degenerative mutations at an accelerated rate following a duplication event, selection may not be relaxed completely (*14–16*). Second, the frequency of duplicate-gene preservation following ancient polyploidization events, often suggested to be in the neighborhood of 30 to 50% over periods of tens to hundreds of millions of years (*17–20*), is unexpectedly high.

Further insight into the rates of origin of duplicate genes and their evolutionary fates can now be acquired by using the genomic databases that have emerged for several species. We focused on nine taxa for which large numbers of protein-coding sequences are available through electronic databases: human (*Homo sapiens*), mouse (*Mus musculus*), chicken (*Gallus gallus*), nematode (*Caenorhabditis elegans*), fly (*Drosophila melanogaster*), the plants *Arabidopsis thaliana* and *Oryza sativa* (rice), and the yeast *Saccharomyces cerevisiae*. For each of these species, the complete set of available open reading frames was screened to eliminate se-

[1]Department of Biology, University of Oregon, Eugene, OR 97403, USA. [2]Department of Computer and Information Science, University of Oregon, Eugene, OR 97403, USA.

*To whom correspondence should be addressed. E-mail: mlynch@oregon.uoregon.edu quences that were unlikely to be functional proteins (*21*). Each sequence retained after this initial filtering was then compared against all other members of the intraspecific set to identify pairs of gene duplicates, which were then analyzed for the degree of nucleotide divergence (*21*). The analyses for *C. elegans*, *D. melanogaster*, and *S. cerevisiae* were based on the complete genomic sequences available for these species.

The traditional approach to inferring the magnitude of selective constraint on protein evolution focuses on codons, comparing the rates of nucleotide substitution at replacement and silent sites (*7, 15, 16*). With this sort of analysis, only the cumulative pattern of nucleotide substitution is identified, making it difficult to determine whether duplicate genes typically undergo different phases of evolutionary divergence, e.g., an early phase of near neutrality followed by a later phase of selective constraint. Some clarification of this issue can be achieved by considering the features of sets of gene duplicates separated by an array of divergence times.

Under the assumption that silent substitutions are largely immune from selection and accumulate at a stochastic rate that is proportional to time, we take the number of substitutions per silent site, S, separating two members of a pair of duplicates to be a measure of the relative age of the pair. Letting R denote the number of substitutions per replacement site, a net (cumulative) selective constraint since the time of origin of a pair of duplicates will be reflected in an R/S ratio < 1, whereas a net acceleration of protein evolution will be revealed by an R/S ratio > 1. Complete relaxation of selection will result in $R/S \approx 1$. For the duplicate genes that we have identified, there is often considerable scatter around the neutral expectation when $S < 0.05$ (Fig. 1), suggesting that early in their history, many gene duplicates experience a phase of relaxed selection or even accelerated evolution at replacement sites. The progressive decline of R/S beyond this point reflects a gradual increase in the magnitude of selective constraint. The vast majority of gene duplicates with $S > 0.1$ exhibits an R/S ratio $\ll 1$.

From the qualitative behavior of the cumulative R/S ratio, some insight into the temporal development of increasing selective constraint on duplicate-gene evolution can be obtained by considering a simple model in which R declines relative to S, according to the function

$$\frac{dR}{dS} = \frac{1}{a - be^{-mS}} \qquad (1)$$

Under this model, assuming positive m, the ratio of rates of replacement to silent substitutions initiates with an expected value of $1/(a - b)$ at $S = 0$ (reflecting the evolutionary properties of newly arisen duplicates) and declines to $1/a$ as $S \to \infty$ (reflecting ancient duplicates). Integrating this equation, the expected cumulative number of substitutions per replacement site (R) can be described as a function of the cumulative number of substitutions per silent site (S),

$$R = \frac{1}{am}\left[mS - \ln\left(\frac{a - b}{a - be^{-mS}}\right)\right] \qquad (2)$$

The parameters a, b, and m can then be estimated by performing least-squares analysis on the pairwise gene-specific estimates of R and S (*22*).

Given the inherently stochastic nature of molecular evolutionary processes, Eq. 2 describes the average rate of accumulation of amino acid–replacing substitutions fairly well, explaining more than 50% of the variance in the data in all cases (Fig. 1). Moreover, the pattern is quite similar across species. The estimates of dR/dS at low S are all < 1, with a narrow range of 0.37 to 0.46 and a mean value of 0.43 (SE = 0.01), and dR/dS gradually declines to asymptotic values in the range of 0.022 to 0.106 (mean = 0.053, SE = 0.009) (Table 1). These results imply that, early in their evolutionary history, duplicate genes tend to be under moderate selective constraints with the rate of amino acid substitution averaging about 43% of the neutral expectation. The efficiency of purifying selection subsequently increases approximately 10-fold, to the point at which only about 5% of amino acid–changing mutations are able to rise to fixation.

Some caveats in the interpretation of these results are in order. First, the nucleotide divergence statistics describe the average pattern of molecular evolution. Individual codons may, in many cases, deviate substan-

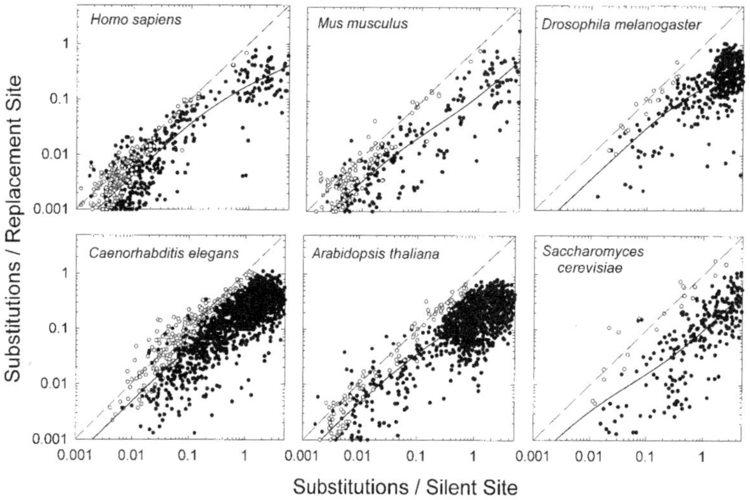

Fig. 1. Cumulative numbers of observed replacement substitutions per replacement site as a function of the number of silent substitutions per silent site. Each point represents a single pair of gene duplicates. The dashed line denotes the expectation under the neutral model, whereas the solid line is the least-squares fit of Eq. 2 to the data (*22*). Open points denote gene pairs for which the ratio *R/S* is not significantly different from the neutral expectation of 1.

Table 1. Fitted coefficients for the function describing cumulative replacement substitutions per replacement site versus silent substitutions per silent site, Eq. 2, and for the function describing the rate of loss of young duplicates, Eq. 3. The value r^2 gives the proportion of variance in the observed values described by the model; standard errors are in parentheses.

Species	Equation 2				Equation 3	
	m	$(dR/dS)_{S=0}$	$(dR/dS)_{S=\infty}$	r^2	d	r^2
H. sapiens	0.412	0.442	0.038	0.759	23.9 (2.0)	0.954
M. musculus	6.574	0.388	0.106	0.730	13.9 (3.2)	0.698
G. gallus	0.829	0.382	0.032	0.720	–	–
Danio rerio	0.857	0.450	0.022	0.677	–	–
D. melanogaster	0.564	0.372	0.050	0.533	8.2 (1.6)	0.766
C. elegans	0.547	0.500	0.062	0.647	7.0 (1.5)	0.735
A. thaliana	0.695	0.458	0.043	0.750	17.6 (5.0)	0.605
O. sativa	0.500	0.412	0.034	0.540	–	–
S. cerevisiae	20.357	0.433	0.090	0.531	7.5 (2.4)	0.538

tially from the norm. Second, for gene pairs with $S > 1$, potentially large inaccuracies in the estimates of nucleotide divergence are expected to result from multiple substitutions per site. Nevertheless, as can be seen in Fig. 1, the patterns that we describe are fully apparent within the subset of gene duplicates with $S < 1$. Third, although we have taken special precautions to avoid the inclusion of nonfunctional gene duplicates in our analyses (*21*), in the absence of actual expression pattern data, we cannot be certain that all of the genes we have included are functional. However, the fact that most of the pairs that we have identified have $R/S < 1$ and that many pairs with small S have $R/S \gg 1.0$ suggests that we have not inadvertently included many pseudogenes in our analyses.

Assuming that the number of silent substitutions increases approximately linearly with time, the relative age-distribution of gene duplicates within a genome can be inferred indirectly from the distribution of S (*23*). For all species, the highest density of duplicates is contained within the youngest age classes, with the density dropping off very rapidly with increasing S (Fig. 2). For *Arabidopsis*, there is a conspicuous secondary peak in the age distribution centered around $S = 0.8$, which is consistent with conclusions from comparative mapping data that the lineage containing this species experienced an ancient polyploidization event (*24*). Using an estimated rate of silent-site substitution of 6.1 per silent site per billion years (*25*), this event dates to approximately 65 million years ago. Unfortunately, this type of analysis cannot shed much light on the debate over whether complete genome duplications preceded the divergence of ray-finned fishes and tetrapods (*1–3, 26–28*). With a divergence time between these two lineages at approximately 430 million years ago (*29*), the average S for a pair of older duplicates would be expected to be in excess of 1.0. Levels of substitution of this magnitude are estimated with a great degree of inaccuracy, which would weaken the signature of ancient genome-duplication events.

For levels of divergence less than $S = 0.25$, problems with saturation effects in the estimation of substitutions per site should be minimal, and the time scale is short enough that it is reasonable to expect the rate of evolution at silent sites to be approximately constant. If the origin and loss of duplicates is then viewed as having been an essentially steady-state process over the time period $S = 0$ to 0.25, the rate of loss of gene duplicates can be estimated by using the survivorship function

$$N_S = N_0 e^{-dS} \qquad (3)$$

where N_S is the number of duplicates observed at divergence level S, and N_0 and d are fitted constants obtained by linear regression of the log-transformed data (Fig. 3) (*30*). For the species for which adequate data are available for analysis, the loss coefficients fall in the range of $d = 7$ to 24, with a mean value of 13.0 (SE = 2.8) (Table 1). For $d = 7$, 13, and 24, the half-life of a gene duplicate on the scale of S is 0.099, 0.053, and 0.029, respectively, and 95% loss is expected at 4.3 times these S values. Thus, assuming they are not nonfunctional at the time of origin, most gene duplicates are apparently nonfunctionalized by the time silent sites have diverged by only a few percent.

Some insight into the absolute time to duplicate-gene loss can be acquired for the groups in which estimated rates of nucleotide evolution at silent sites are available. The average estimate of d for mouse and human is 18.9, which, using an average rate of silent substitution in mammalian genes of 2.5 per silent site per billion years (*31*), translates to 7.3 million years. The estimates of d for the two invertebrates *Drosophila* and *Caenorhabditis* are very similar, averaging to 7.6. Although a direct estimate of the rate of silent substitution is not available for nematodes, indirect evidence suggests that the rate of molecular evolution in *C. elegans* is elevated relative to that in other invertebrates (*32*). Using the estimated rate of silent-site substitution in *Drosophila* of 15.6 per silent site per BY (*7*), we obtain a possibly upwardly biased estimate of 2.9 million years as the average half-life of duplicate genes in invertebrates. For *Arabidopsis*, $d = 17.6$, which translates into a half-life of 3.2 million years using the silent substitution rate cited above.

Finally, we note that for the three species for which the complete genomic sequence is available, the rate of origin of gene duplicates can be estimated from the abundance of the very youngest pairs. For *D. melanogaster*, there are 10 pairs of duplicates with $S < 0.01$, which translates to a rate of origin of approximately 31 new duplicates per genome per million years, or by using the estimated 13,601 genes per genome (*33*), to 0.0023 per gene per million years. There are 32 identifiable duplicates in yeast with $S < 0.01$. Although no direct estimates of the rate of nucleotide substitution exist for fungi, there is no evidence that the fungal rate is very different from that of animals or plants either. Using the average silent substitution rate for mammals, *Drosophila*, and vascular plants (8.1 per nucleotide site per BY), the crudely estimated number of new duplicates arising in the yeast genome per million years is 52; with a total genome of approximately 6241 open reading frames, this translates to 0.0083 per million years. The rate of origin of gene duplicates in *C. elegans* over the past few hundred thousand years appears to be substantially greater than that for *D. melanogaster* and *S. cerevisiae*. There are 164 pairs of gene dupli-

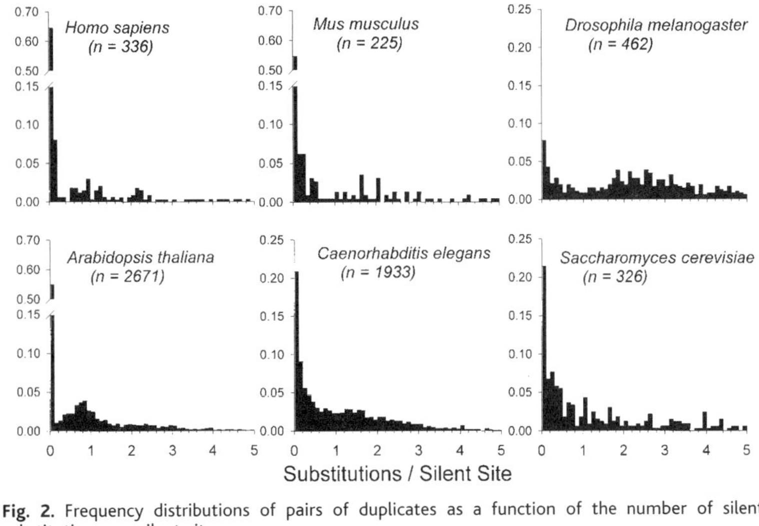

Fig. 2. Frequency distributions of pairs of duplicates as a function of the number of silent substitutions per silent site.

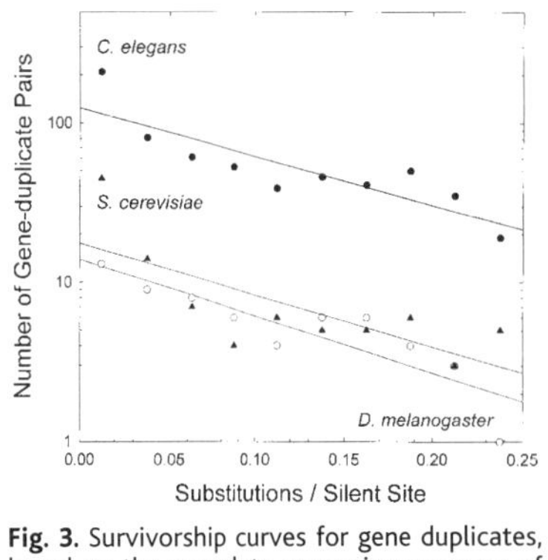

Fig. 3. Survivorship curves for gene duplicates, based on the complete genomic sequences of *C. elegans* (●), *D. melanogaster* (○), and *S. cerevisiae* (▲). The fitted parameters for these and other species are contained in Table 1.

cates with $S < 0.01$ in *C. elegans*. Again using the rate of silent-site substitution from *Drosophila*, the rate of origin of new duplicates in this species is at least 383 per genome per million years; with a genome size of approximately 18,424 open reading frames (*33*), this translates to a per-gene rate of duplication of 0.0208 per million years.

These estimated rates of origin of new gene duplicates could be inflated if gene conversion keeps substantial numbers of older duplicates appearing as if they were younger. Of the young duplicates identified in the previous paragraph, 100% of those in *Drosophila*, 56% of those in *Saccharomyces*, and 71% of those in *Caenorhabditis* are located on the same chromosome. However, although significant, the correlation between *S* and the physical distance between duplicates residing on the same chromosome tends to be quite weak, and many spatially contiguous gene duplicates are highly divergent (see figure at www.csi.uoregon.edu/projects/genetics/duplications). In addition, a genome-wide analysis of *C. elegans* suggests that gene-conversion events arise only rarely in duplicate genes and are largely concentrated in multigene families (*34*). Such multigene families have been excluded from our analyses (*21*).

These results suggest a conservative estimate of the average rate of origin of new gene duplicates on the order of 0.01 per gene per million years, with rates in different species ranging from about 0.02 down to 0.002. Given this range, 50% of all of the genes in a genome are expected to duplicate and increase to high frequency at least once on time scales of 35 to 350 million years. Thus, even in the absence of direct amplification of entire genomes (polyploidization), gene duplication has the potential to generate substantial molecular substrate for the origin of evolutionary novelties. The rate of duplication of a gene is of the same order of magnitude as the rate of mutation per nucleotide site (*7*).

However, the fate awaiting most gene duplicates appears to be silencing rather than preservation. For the species that we have examined, the average half-life of a gene duplicate is approximately 4 million years, consistent with the theoretical predictions mentioned above (*11, 12*). The contrast between the high rate of silencing observed in this study and the high level of duplicate-gene preservation that occurs in polyploid species (*17–20*) may be reconciled if dosage requirements play an important role in the selective environment of gene duplicates. Polyploidization preserves the necessary stoichiometric relationships between gene products, which may be subsequently maintained by stabilizing selection, whereas duplicates of single genes that are out of balance with their interacting partners may be actively opposed by purifying selection.

Despite the rather narrow window of opportunity for evolutionary exploration by gene duplicates, such genes may play a prominent role in the generation of biodiversity by promoting the origin of postmating reproductive barriers (*35, 36*). Consider a young pair of functionally redundant duplicate genes in an ancestral species. If a geographic isolating event occurs, a random copy will be silenced in the two sister taxa with very high probability within the next one to 2 million years. The probability that alternative copies will be silenced in the two sister taxa is 0.5, so if the copies are unlinked and the two taxa are then brought back together, there will be a 0.0625 probability that an F_2 derivative will be a double-null homozygote for the two loci. With tens to hundreds of young, unresolved gene duplicates present in most eukaryotic genomes, such genes may provide a common substrate for the passive origin of isolating barriers. Moreover, this process does not simply rely on gene duplicates in ancestral species. With rates of establishment of 0.002 to 0.02 duplicates per gene per million years and a moderate genome size of 15,000 genes, we can expect on the order of 60 to 600 duplicate genes to arise in a pair of sister taxa per million years, many of which will subsequently experience divergent resolution.

The passive build-up of reproductive isolation induced by gene duplicates, with no loss (and in most cases, no gain) of fitness in sister taxa, provides a simple mechanism for speciation that is consistent with the Bateson-Dobzhansky-Muller model (*37*), without requiring the presence of negative epistatic interactions between gene products derived from isolated genomes. The microchromosomal repatterning induced by recurrent gene duplication is also consistent with the chromosomal model for speciation (*38*), without requiring the large-scale rearrangements that are typically thought to be necessary (*39*). Finally, the time scale of the process is consistent with what we know about the average time to postreproductive isolation (*40, 41*).

References and Notes

1. S. Ohno, *Evolution by Gene Duplication* (Springer-Verlag, Heidelberg, Germany, 1970).
2. P. W. H. Holland, J. Garcia-Fernandez, N. A. Williams, A. Sidow, *Development (suppl.)*, p. 125 (1994).
3. A. Sidow, *Curr. Opin. Genet. Dev.* **6**, 715 (1996).
4. J. B. S. Haldane, *Am. Nat.* **67**, 5 (1933).
5. H. J. Muller, *Genetics* **17**, 237 (1935).
6. J. B. Walsh, *Genetics* **110**, 345 (1985).
7. W.-H. Li, *Molecular Evolution* (Sinauer, Sunderland, MA, 1999).
8. A. Force *et al.*, *Genetics* **151**, 1531 (1999).
9. A. L. Hughes, *Proc. R. Soc. London Ser. B* **256**, 119 (1994).
10. A. Stoltzfus, *J. Mol. Evol.* **49**, 169 (1999).
11. M. Lynch, A. Force, *Genetics* **154**, 459 (2000).
12. G. A. Watterson, *Genetics* **105**, 745 (1983).
13. M. Lynch, B. Walsh, *Genetics and Analysis of Quantitative Traits* (Sinauer, Sunderland, MA, 1998).
14. W.-H. Li, in *Population Genetics and Molecular Evolution*, T. Ohta, K. Aoki, Eds. (Springer-Verlag, Berlin, 1985), pp. 333–352.
15. M. K. Hughes, A. L. Hughes, *Mol. Biol. Evol.* **10**, 1360 (1993).
16. B. S. Gaut, J. F. Doebley, *Proc. Natl. Acad. Sci. U.S.A.* **94**, 6809 (1997).
17. S. D. Ferris, G. S. Whitt, *J. Mol. Evol.* **12**, 267 (1979).
18. J. H. Nadeau, D. Sankoff, *Genetics* **147**, 1259 (1997).
19. A. Amores *et al.*, *Science* **282**, 1711 (1998).
20. J. F. Wendel, *Plant Mol. Biol.* **42**, 225 (2000).
21. For each organism, the complete set of available putative amino acid sequences was downloaded from GenBank and stored in a local file. We first filtered out possible nonfunctional protein sequences by removing all those that did not start with methionine, and all sequences that were annotated as known or suspected pseudogenes or transposable elements were also discarded. We then used BLAST [S. F. Altschul *et al.*, *Nucleic Acids Res.* **25**, 3389 (1997)] to compare all pairs of protein sequences in the set, retaining those pairs for which the alignment score was below 10^{-10}. To avoid the inclusion of large multigene families (and as a secondary guard against the inclusion of transposable elements), we further excluded all genes that identified more than five matching sequences. All remaining sequences that were similar to known transposable elements were also removed at this point. As a final step in preparing the data set for analysis, we attempted to minimize the noise inherent in poorly aligned sequences by adopting a conservative strategy for retaining only unambiguously aligned sequences. Using the protein alignment generated by BLAST as a guide, at each alignment gap we scanned to the right and discarded all sequence until an anchor pair of identical amino acids was located, then retained all subsequent sequence (exclusive of the anchor site) provided another match was found within the next six amino acid sites. This procedure was iterated to the right and to the left of each alignment gap until acceptable anchor sites were encountered. We used a two-out-of-seven rule because the probability of a lower level of sequence identity arising by chance is >0.05, and we excluded anchor sites to minimize upward bias in estimated sequence similarities. Using the final set of amino acid alignments as guides, we then retrieved and aligned all of the necessary nucleotide sequences. The numbers of nucleotide substitutions per silent and replacement sites were estimated by using the maximum-likelihood procedure in the PAML package, version 2.0k [Z. Yang, *Comput. Appl. Biosci.* **13**, 555 (1997)]. Estimated rates of nucleotide substitution can be highly sensitive to the relative rates of occurrence of transitions and transversions when the amount of sequence divergence is high. Therefore, before our final analyses, we obtained species-specific estimates of the transition/transversion bias by considering the observed substitutions at all fourfold redundant sites in all pairs of sequences that were similar enough that multiple substitutions per site were unlikely to have occurred (we used only proteins for which the divergence at such sites is ≤15%, after verifying that the transition/transversion ratio is essentially constant below this point). The estimated transition/transversion ratios were then treated as constants in the maximum-likelihood analyses.
22. The parameter estimates for Eq. 2 are those that minimize the sum of squared deviations between predicted and observed values. In these analyses, the natural logarithms of observed and expected values were used, so as not to give undue weight to sequence pairs with large *R*. Pairs of sequences for which *S* or $R = 0$ were not included in this analysis. In addition, for human and mouse, we excluded gene pairs for which $S < 0.01$ in the curve-fitting procedure to avoid the accidental inclusion of alleles at the same locus; this problem should be negligible for the other large genome projects, where the species are either largely selfing (plants and *C. elegans*) or the final data exclude the possibility of allelic inclusion (*D. melanogaster*). The final data, including gene numbers, can be located at www.csi.uoregon.edu/projects/genetics/duplications.
23. Such analysis implicitly assumes an unbiased sample of gene duplicates, which should be valid for most

analysis herein, because they are based on either completely sequenced genomes or random sequencing projects.
24. D. Grant, P. Cregan, R. C. Showemaker, *Proc. Natl. Acad. Sci. U.S.A.* **97**, 4168 (2000).
25. This estimate is the average of two surveys based on analyses of multiple genes in vascular plants (*7, 42*).
26. L. Skrabanek, K. H. Wolfe, *Curr. Opin. Genet. Dev.* **8**, 694 (1998).
27. A. L. Hughes, *J. Mol. Evol.* **48**, 565 (1999).
28. A. P. Martin, *Am. Nat.* **154**, 111 (1999).
29. P. E. Ahlberg, A. R. Milner, *Nature* **368**, 507 (1994).
30. In large genomic databases like those analyzed here, some sequencing errors may inflate the apparent level of divergence, but this error should be independent of *S*, and in any event, is unlikely to add more than 0.01 to individual estimates of *S*. Thus, the impact of such error on our statistical analyses should be negligible.
31. This estimate is the average of the results obtained in three broadly compatible studies, all of which surveyed a large number of genes [S. Easteal, C. Collet, *Mol. Biol. Evol.* **11**, 643 (1994); T. Ohta, *J. Mol. Evol.* **40**, 56 (1995); (*7*)].
32. A. M. A. Aguinaldo *et al.*, *Nature* **387**, 489 (1997).
33. G. M. Rubin *et al.*, *Science* **287**, 2204 (2000).
34. C. Semple, K. H. Wolfe, *J. Mol. Evol.* **48**, 555 (1999).
35. C. R. Werth, M. D. Windham, *Am. Nat.* **137**, 515 (1991).
36. M. Lynch, A. Force, *Am. Nat.*, in press.
37. H. A. Orr, *Genetics* **144**, 1331 (1996).
38. M. J. D. White, *Modes of Speciation* (Freeman, San Francisco, CA, 1978).
39. G. Fischer, S. A. James, I. N. Roberts, S. G. Oliver, E. J. Louis, *Nature* **405**, 451 (2000).
40. H. R. Parker, D. P. Philipp, G. S. Whitt, *J. Exp. Zool.* **233**, 451 (1985).
41. J. A. Coyne, H. A. Orr, *Evolution* **51**, 295 (1997).
42. M. Lynch, *Mol. Biol. Evol.* **14**, 914 (1997).
43. This research was supported by NIH grant RO1-GM36827. We thank A. Force and A. Wagner for helpful comments.

22 June 2000; accepted 4 October 2000

Schluter, D. 2001. Ecology and the origin of species. *Trends in Ecology and Evolution* 16:372–380.

In everyday experience, we identify different kinds of organisms by their appearance. People belong to the human species and are different from cats and dogs, which in turn are different from each other. There are differences among people, as well as among cats and dogs; but individuals of the same species are generally more similar among themselves than they are to individuals of other species. External similarity is a commonsense basis for identifying individuals as members of the same species, but there is more to a species than appearance. A dalmatian, a greyhound, and a terrier look very different, but they are all dogs because they can interbreed. People can also interbreed with one another and so can cats with other cats, but people cannot interbreed with dogs or cats, nor can these with each other.

In organisms that reproduce sexually, species may be defined as groups of interbreeding natural populations that are reproductively isolated from other such groups. Sexually reproducing organisms include a majority of plants, animals, and fungi. Bacteria and archaea do not reproduce sexually but instead by fission. Organisms that lack sexual reproduction are classified into different species according to criteria such as external characteristics and morphology, chemical and physiological properties, and genetic makeup. These features are also used for identifying sexually reproducing species.

The ability to interbreed is of great evolutionary importance, because it defines species as independent evolutionary units. Genetic changes originate in single individuals; they can spread by natural selection to all members of the species but not to individuals of other species. Individuals of a species share a common gene pool that is not shared with individuals of other species. Different species have independently evolving gene pools because they are reproductively isolated.

There is no way to test whether or not today's humans could interbreed with those who lived thousands or millions of years ago. Biologists distinguish species of organisms that lived at different times by means of a commonsense morphological criterion. If two organisms differ from each other about as much as two living individuals belonging to two different species, they are classified in separate species and given different names. Species that lived at different times but are related by descent are often called *chronospecies*.

In cladistic taxonomy, species are identified as different branches in a phylogenetic tree. Species that come about as two branches of a line of descent are considered different from each other and different from the species from which the two branches emerged (see Simpson 1951 in this volume, in particular, panel D in figure 2).

Darwin expounded on the origin of new species by means of divergence driven by natural selection, a view that has resurfaced time and again in recent decades. Here, Dolph Schluter returns again to this Darwinian theme by emphasizing the ecological factors that may often underlie the speciation process.

RELATED READING

Coyne, J. A., and H. A. Orr. 2004. Speciation. Sinauer, Sunderland, Massachusetts, USA.

Fei, E. J., and B. G. Spratt. 2001. Recombination and the population structures of bacterial pathogens. Annual Review of Microbiology 55:561–590.

Hey, J., W. M. Fitch, and F. J. Ayala, editors. 2004. Systematics and the *Origin of Species*: on Ernst Mayr's 100th anniversary. National Academies Press, Washington, D.C., USA.

Howard, D. J., and S. H. Berlocher, eds. 1998. Endless forms: species and speciation. Oxford University Press, Oxford, England.

Ecology and the origin of species

Dolph Schluter

The ecological hypothesis of speciation is that reproductive isolation evolves ultimately as a consequence of divergent natural selection on traits between environments. Ecological speciation is general and might occur in allopatry or sympatry, involve many agents of natural selection, and result from a combination of adaptive processes. The main difficulty of the ecological hypothesis has been the scarcity of examples from nature, but several potential cases have recently emerged. I review the mechanisms that give rise to new species by divergent selection, compare ecological speciation with its alternatives, summarize recent tests in nature, and highlight areas requiring research.

ECOLOGICAL SPECIATION (see Glossary) occurs when DIVERGENT SELECTION on traits between populations or subpopulations in contrasting environments leads directly or indirectly to the evolution of REPRODUCTIVE ISOLATION. The concept of ecological speciation dates back to the 1940s, from the time the BIOLOGICAL SPECIES CONCEPT was developed. Dobzhansky[1] believed that 'SPECIATION in *Drosophila* proceeds mainly through evolving physiological complexes which are successful each in its environment'. Mayr[2] recognized that many of the accumulated genetic differences between populations 'particularly those affecting physiological and ecological characters, are potential isolating mechanisms'. Acceptance of this perspective by many evolutionists in the mid-20th century resulted from the inherent appeal and simple plausibility of ecological speciation. However, until recently, neither was there evidence to support ecological speciation, nor had tests been devised and applied to distinguish ecological speciation from other mechanisms that might also cause speciation in the wild, such as GENETIC DRIFT (Box 1).

Mechanisms of ecological speciation

'Ecological speciation' is a concept that unites speciation processes in which reproductive isolation evolves ultimately as a consequence of divergent (including DISRUPTIVE) selection on traits between environments. 'Environment' refers to biotic and abiotic elements of habitat (e.g. climate, resources and physical structure) as well as to interactions with other species (e.g. resource competition, predation, mutualism and various forms of interspecific interference). A diversity of evolutionary processes might be involved. Ecological speciation might occur in ALLOPATRY or in SYMPATRY. It might lead to mainly premating isolation, mainly postmating isolation, or a combination of both. It includes several (but not all) modes of speciation involving SEXUAL SELECTION. Ecological speciation might come about indirectly as a consequence of natural selection on morphological, physiological or behavioral traits, or it might include direct selection on premating isolation (REINFORCEMENT). Distinguishing the ways in which divergent selection has led to reproductive isolation is among the greatest challenges of the empirical study of ecological speciation.

In what is perhaps the classic scenario for ecological speciation (Fig. 1), reproductive isolation between two populations starts to build in allopatry as populations accumulate adaptations to unique aspects of their environments. Premating isolation then evolves to completion by reinforcement after sympatry is secondarily established. The timing of SECONDARY CONTACT is flexible, however, and the extremes of possibility lead to departures from this classic model. At one extreme is pure ALLOPATRIC SPECIATION, in which the sympatric phase is entirely absent. New species might eventually become sympatric, but this occurs after reproductive isolation is complete. The other extreme is full-blown SYMPATRIC SPECIATION, which completely lacks the allopatric stage. Although debate about the plausibility of these extremes has long polarized research on speciation, from an ecological perspective, the more fundamental issue concerns the mechanisms that drive the evolution of reproductive isolation, which is the focus of this article.

By-product mechanisms

In the simplest models of ecological speciation, reproductive isolation builds between populations incidentally as a by-product of adaptation to alternative selection regimes[2,3]. Reproductive isolation is not directly favored by selection, but is a secondary consequence of genetic differentiation driven by selection on other traits. This BY-PRODUCT MECHANISM could lead to premating isolation and to various forms of postmating isolation.

Several laboratory experiments with *Drosophila* have simulated the early stages of by-product speciation (reviewed in Ref. 4), and these hint at how the process might work in nature. Kilias *et al.*[5] raised different lines of *Drosophila melanogaster* for five years in either a cold–dry–dark or a warm–damp–light environment. Dodd[6] examined mating preferences in replicate lines of *Drosophila pseudoobscura* raised for one year on either starch-based or maltose-based larval medium. In both studies, some premating reproductive isolation evolved between lines that had experienced contrasting environments, but no premating isolation evolved between independent lines raised in the same environment (Fig. 2). The populations in these experiments were fully allopatric, as in the first part of the classic scenario (Fig. 1).

Dolph Schluter
Zoology Dept and The Center for Biodiversity Research, The University of British Columbia, Vancouver, BC, Canada V6T 1Z4.
e-mail: schluter@zoology.ubc.ca

0169-5347/01/$ – see front matter © 2001 Elsevier Science Ltd. All rights reserved. PII: S0169-5347(01)02198-X

Box 1. The major modes of speciation, according to initial causes

Speciation modes have been classified historically by the geographical arrangement of populations undergoing the process (allopatric, sympatric or parapatric), a classification that focuses on the inhibitory effects of gene flow on the evolution of reproductive isolation. Some have argued that an alternative classification centering on mechanisms that drive the evolution of reproductive isolation would be more productive[a–c] (Via[d], this issue). Table I breaks modes of speciation events into four categories according to initial cause, and compares some other principal features of those categories.

Natural selection is involved at an early stage of two different modes of speciation. Under ecological speciation, populations in different environments, or populations exploiting different resources, experience contrasting natural selection pressures on traits that directly or indirectly bring about the evolution of reproductive isolation. However, divergence might also occur under uniform selection, for example, if different advantageous (but incompatible) mutations arise in separate populations occupying similar environments (Turelli *et al.*[e], this issue). Reproductive isolation between populations brought about by conflict and coevolution between the sexes within populations[f–h] is an example of this second mode.

If populations are in contact before the evolution of reproductive isolation is completed, natural selection might be involved at a late stage of all four modes of speciation, during reinforcement of premating isolation (Turelli *et al.*[e] this issue)[i]. Similarly, sexual selection might be involved in every speciation mode, depending on the cause of divergence in mate preferences. For this reason, the presence of natural or sexual selection *per se* is probably not a good basis for classifying speciation events. Nevertheless, the classification in Table I is coarse, and finer subdivisions are necessary to highlight differences in the roles of natural and sexual selection within each mode, as well as different roles for genetic drift in speciation initiated by that mechanism.

References

a Schluter, D. (1996) Ecological speciation in postglacial fishes. *Philos. Trans. R. Soc. London Ser. B* 351, 807–814

b Schluter, D. (1998) Ecological causes of speciation. In *Endless Forms: Species and Speciation* (Howard, D. and Berlocher, S., eds), pp. 114–129, Oxford University Press

c Orr, M.R. and Smith, T.B. (1998) Ecology and speciation. *Trends Ecol. Evol.* 13, 502–506

d Via, S. (2001) Sympatric speciation in animals: the ugly duckling grows up. *Trends Ecol. Evol.* 16, 381–390

e Turelli, M. *et al.* (2001) Theory and speciation. *Trends Ecol. Evol.* 16, 330–343

f Palumbi, S.R. (1998) Species formation and the evolution of gamete recognition loci. In *Endless Forms: Species and Speciation* (Howard, D. and Berlocher, S., eds), pp. 271–278, Oxford University Press

g Rice, W.R. (1998) Intergenomic conflict, interlocus antagonistic coevolution, and the evolution of reproductive isolation. In *Endless Forms: Species and Speciation* (Howard, D. and Berlocher, S., eds), pp. 261–270, Oxford University Press

h Gavrilets, S. (2000) Rapid evolution of reproductive isolation driven by sexual conflict. *Nature* 403, 886–889

i Dobzhansky, T. (1951) *Genetics and the Origin of Species* (3rd edn), Columbia University Press

Table I. Modes of speciation

Mode of speciation	Mechanism of initial divergence	Initial form of reproductive isolation	Proximate basis of reduced hybrid fitness	Examples of the roles of natural selection	Example roles of sexual selection
Ecological speciation	Divergent natural selection	Prezygotic or postzygotic	Ecological selection, genetic incompatibility and sexual incompatibility	**Initial:** Drive divergence in phenotypic traits **Final:** Reinforcement	Amplify divergence of mate preferences initiated by natural selection Reinforcement
Speciation by divergence under uniform selection	Different advantageous mutations occur in separate populations experiencing similar selection pressures	Prezygotic or postzygotic	Genetic incompatibility and sexual incompatibility	**Initial:** Drive fixation of incompatible mutations in different populations **Final:** Reinforcement	Drive fixation of alternative incompatible mutations in different populations Reinforcement
Speciation by genetic drift	Genetic drift	Prezygotic or postzygotic	Genetic incompatibility and sexual incompatibility	**Initial:** None; or opposes divergence **Final:** Reinforcement caused by drift	Amplify differences in mate preferences Reinforcement
Polyploid speciation	Hybridization and polyploidy	Postzygotic	Genetic incompatibility	**Initial:** None; or promotes further genetic divergence **Final:** Reinforcement	Reinforcement

However, complete allopatry is not crucial for the by-product mechanism, an idea that extends the classic scenario in two ways[4]. First, selection can continue to strengthen reproductive isolation via the by-product mechanism even when the allopatric phase is over and the sympatric phase is in progress. Reinforcement is not the only way to complete the evolution of reproductive isolation after secondary contact. Second,

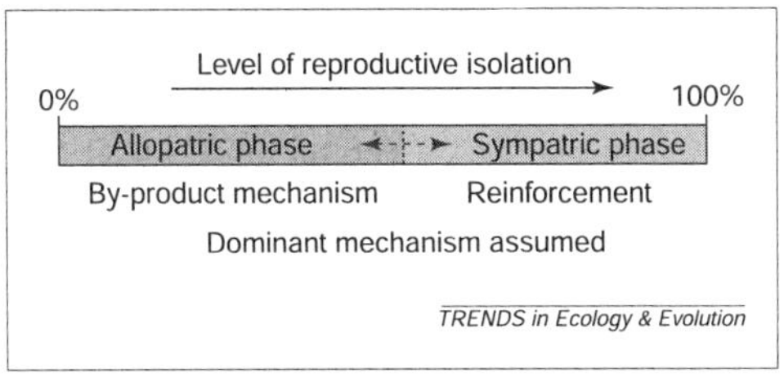

Fig. 1. The classic scenario of an ecological speciation event, from beginning to end. Reproductive isolation builds in allopatry (green) as an incidental by-product of adaptation to alternative environments (by-product mechanism). Reinforcement of premating isolation, driven by reduced hybrid fitness, completes the speciation process during the sympatric phase (blue). The timing of secondary contact is flexible (indicated by arrows at the boundary between the allopatric and sympatric phases).*

an initial allopatric phase is not always essential for the by-product mechanism to work, and can be dispensed with if selection is strong enough or if gene flow between subpopulations is not too high (see Turelli *et al.*[7], this issue, and Via[8], this issue).

Many different environmental agents of divergent selection could lead to the evolution of reproductive isolation as a by-product, and identifying these alternatives is of major interest. During the allopatric phase, such agents might include contrasting resources, predators, competitors, cytoplasmic symbionts, structural habitat features affecting locomotion or transmission of communication signals, and other biotic and abiotic factors. Allopatric populations might also confront distinct constellations of other closely related species, which could lead to divergence of mating and social signals if these signals evolve partly in response to interspecific interference. The last mechanism extends the by-product concept to divergence in a wider set of traits, such as color and song, than those immediately related to resource acquisition. For example, divergence of color in response to interspecific reproductive or aggressive interference with other species could explain the rapid divergence of sexual signals and mate preferences between spatially separated populations of African cichlid fish that are otherwise similar in food and habitat requirements[9,10] (see Barraclough and Nee[11], this issue).

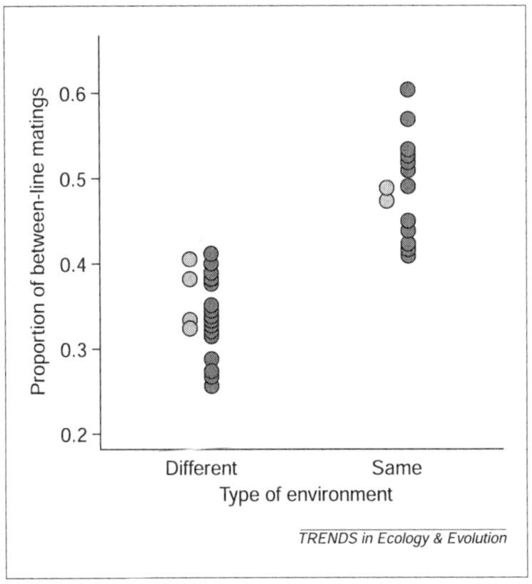

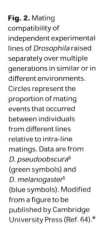

Fig. 2. Mating compatibility of independent experimental lines of *Drosophila* raised separately over multiple generations in similar or in different environments. Circles represent the proportion of mating events that occurred between individuals from different lines relative to intra-line matings. Data are from *D. pseudoobscura*[6] (green symbols) and *D. melanogaster*[6] (blue symbols). Modified from a figure to be published by Cambridge University Press (Ref. 64).*

Many agents of divergent selection that drive the evolution of reproductive isolation between two allopatric populations would continue to strengthen isolation following secondary contact. However, other agents of selection that arise from interactions between the two nascent species are added during the sympatric phase, and these too might drive reproductive isolation to completion. For example, competition for resources between sympatric populations could lead to exaggerated divergence in phenotype, and further enhance reproductive isolation as a by-product. Alternatively, divergent selection could generate body size differences that result in predation on the smaller species by individuals of the larger. Any evolution of behavioral defenses in the smaller species would probably reduce the frequency of crossbreeding as a by-product. This last mechanism was raised as a possible explanation for enhanced premating isolation between a pair of threespine stickleback *Gasterosteus* spp. in which females of the larger species prey upon eggs guarded by males of the smaller species[12]. The influence of such interactions on the evolution of reproductive isolation during the sympatric phase of speciation has received very little attention.

Reinforcement and sympatric speciation

Divergent natural selection in the allopatric phase might build up differences leading to reduced hybrid fitness after secondary contact that subsequently favor reinforcement[3,13]. Reinforcement is distinct from the by-product mechanism because selection directly favors enhanced premating isolation as a consequence of the inferiority of hybrid offspring. Dobzhansky[3] viewed reinforcement as dominating the completion of ecological speciation after secondary contact (Fig. 1), a claim that remains to be proven. Of course, reinforcement might occur even in non-ecological speciation (Box 1). Consequently, testing ecological speciation requires the examination of processes acting at an earlier stage (e.g. during the allopatric phase).

Most conceptions of sympatric speciation invoke a process similar to that of reinforcement, except that there is no previous build up of phenotypic and genetic differences leading to lowered hybrid fitness. Instead, the fitness of intermediates of a single population is reduced from the start by direct ecological selection pressures. For example, intermediate phenotypes (including hybrids) might exploit available resources less effectively than do extreme phenotypes[14,15], or intermediate phenotypes might suffer greater overall resource competition[16]. Doebeli and Dieckmann[17] have argued that selection against intermediate phenotypes is the expected outcome of many types of ecological interactions, including competition, shared predation and shared mutualism (see also Ref. 18) and, under certain conditions, these might favor the evolution of premating reproductive isolation in sympatry (see Via[8], this issue).

* *This figure is reproduced in color following the index.*

Table 1. Tests of ecological speciation in nature

Test	Taxon[a]	Result and implication: consistent (c) or inconsistent (i) with ecological speciation	Refs
Rate of evolution of reproductive isolation is correlated with strength of divergent selection	Hawaiian versus mainland *Drosophila* spp. pairs	Evolution of reproductive isolation is not faster in Hawaiian *Drosophila* than in mainland *Drosophila* (i)	34
	Lake whitefish in different postglacial lakes	Gene flow is inversely correlated with level of morphological divergence (c)	35
Ecological selection reduces hybrid fitness	Threespine sticklebacks in postglacial lakes	F1 hybrid growth rate is lower in the wild than in the lab; rank order of growth rates of backcross hybrids reverses between parental habitats (c)	15,[b]
	Darwin's ground finches[1]	F1 hybrid survival fluctuates with resource distribution (c)	45,46
	Butterfly races	F1, F2 and backcross hybrids have higher viability in the lab than in the wild, and predation is major agent of selection (c)	39
Trait under divergent selection influences reproductive compatibility	Threespine sticklebacks in postglacial lakes	Mating compatibility is influenced by body size and by nuptial coloration (c)	49,58
	Monkey flowers[1] on different soils	Postzygotic isolation is a pleiotropic effect of a gene encoding tolerance of copper-contaminated soil (c)	47
	Monkey flowers[2]	Divergent selection by pollinators is detected on floral traits and underlying quantitative trait loci (c)	48
	Darwin's ground finches[2]	Beak and body size, under divergent selection, influence vocal signals and are used as cues in interspecific mate discrimination (c)	50,51
	Pea aphids on different host plants	Divergent selection occurs on host-plant choice, which determines premating isolation (c)	41
	Host races of apple maggot fly on different host plants	Divergent selection on diapause influences timing of emergence (c)	40,61
Parallel evolution of mating incompatibilities over similar environmental gradients	Threespine sticklebacks in postglacial lakes	Independently evolved populations of the same ecotype show strong premating isolation, whereas populations of different ecotypes show little or no premating isolation (c)	55
	Freshwater amphipod from two types of environment	Mating compatibility between populations is high when body size and environment type are similar, but is low when they are different (c)	59
	Leaf beetles on different host plants	Premating isolation is high between populations from same host plants, but is low between those on different host plants (c)	60

[a]Latin names: Apple maggot fly, *Rhagoletis pomonella*; Butterfly, *Heliconius erato*; Darwin's ground finches, *Geospiza fuliginosa*[1] and *G. fortis*[1], and *Geospiza* spp.[2]; Freshwater amphipod, *Hyalella azteca*; Fruit flies, *Drosophila* spp.; Lake whitefish, *Coregonus* spp.; Leaf beetles, *Neochlamisus bebbianae*; Monkey flowers[1], *Mimulus guttatus*; Monkey flowers[2], *Mimulus cardinalis* and *M. lewisii*; Pea aphids, *Acyrthosiphon pisum*; Threespine sticklebacks, *Gasterosteus* spp.
[b]H.D. Rundle, unpublished.

Ecology and sexual selection
Sexual selection is regarded as a potent force driving the evolution of premating and/or postmating isolation between populations (see Panhuis *et al.*[19], this issue). From an ecological perspective, the key question is how divergent mate preferences become established in the first place. In most theoretical models of speciation by sexual selection, divergent natural selection plays a dominant role, as spatial variation in selection on SECONDARY SEXUAL TRAITS[20], as divergent selection on sensory systems[21], or as differences between environments in the most effective modes of mating signal transmission[22,23]. By contrast, genetic drift[24] or unique mutations favored by intersexual conflict[25,26] play the dominant role in the evolution of reproductive isolation in non-ecological models of speciation by sexual selection. Therefore, a demonstration that sexual selection is involved in the speciation process does not, by itself, restrict the range of speciation models under investigation (Box 1). The mechanisms ultimately driving divergence of mate preference, in particular the role of divergent natural selection, must still be identified.

Tests of ecological speciation
Tests of ecological speciation must consider the alternatives that need to be distinguished (Box 1): speciation by ordinary genetic drift or genetic drift during population BOTTLENECKS[27,28]; speciation by fixation of alternative advantageous genes in populations experiencing similar selection pressures[29]; and speciation by polyploidy[30,31]. Speciation by polyploidy can be readily diagnosed genetically, but although it is more common in plants than animals, polyploidy accounts for only 2–4% of plant speciation events[32]. Even in plants, the cause of most speciation events remains to be identified.

Several tests have been carried out in recent years that indicate a role for divergent selection in the origin of species in nature (Table 1). Demonstrating a role for divergent selection in speciation, however, is only the first step to detecting an ecological speciation event. The next step is to understand the process by which divergent selection has led to the evolution of reproductive isolation (e.g. by-product alone or with reinforcement, mechanisms of hybrid fitness, etc.), a step in which less progress has been made. I have not reviewed evidence from molecular studies that infer natural selection from an unusually high rate of sequence divergence between sister species[33] because that method does not distinguish divergent from uniform selection (Box 1).

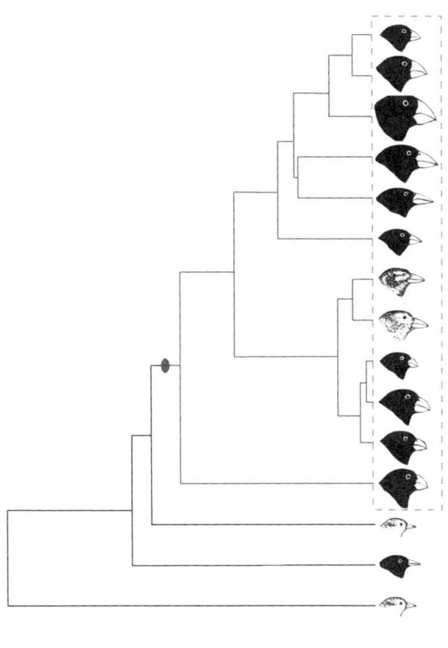

Fig. 3. A phylogenetic tree showing the possible correlation between ecological diversification and speciation rate in the Darwin's finches (*Geospiza* spp.). The high diversity of beak traits among species within the CLADE of tree and ground finches (outlined in red) contrasts with the lower diversity of beak traits among species of the three older LINEAGES. Speciation rates are also highly uneven, being significantly greater in the tree and ground finch clade than in the rest of the tree ($P = 0.011$, calculated using the Nee *et al.*[62] equal-rates test for multiple lineages). Data are taken from Ref. 63. Bird images are reproduced, with permission, from Ref. 65.*

Divergent selection and the rate of evolution of reproductive isolation

The study of ADAPTIVE RADIATION suggests that speciation rates are often elevated during periods of ecological and phenotypic differentiation[34] (see Barraclough and Nee[19], this issue). Darwin's finches appear to exemplify this pattern (Fig. 3): one explanation is that if the level of phenotypic differentiation is indicative of strength of divergent selection, then reproductive isolation evolves most quickly when divergent selection is strongest. A correlation between strength of divergent selection and rate of evolution of reproductive isolation is a prediction of ecological speciation (Table 1).

The correlation between strength of divergent selection and rate of evolution of reproductive isolation has not been measured, probably because divergent selection is difficult to quantify. Instead, researchers have used indirect measures (Fig. 3). Lu and Bernatchez[35] compared gene flow with the level of morphological differentiation between INCIPIENT SPECIES of dwarf and normal lake whitefish *Coregonus* in several postglacial lakes of eastern Canada and northern Maine (USA). Estimated levels of gene flow between sympatric dwarf and normal populations were lowest where morphological differences between the forms were greatest. This is consistent with the prediction of ecological speciation if the degree of morphological differentiation is a good index of strength of divergent selection. Alternatively, the strength of divergent selection is not different between lakes, and variation in morphological differentiation results instead from variation in gene flow. Preliminary measurements of the resources available in the lakes suggest that the strength of divergent selection is different between lakes[35].

In a second test of the prediction, I used Coyne and Orr's[36,37] survey of *Drosophila* to compare the strength of reproductive isolation in pairs of Hawaiian picture-winged *Drosophila* with the strength of isolation in pairs of populations and species of continental *Drosophila* of similar age[34]. The Hawaiian *Drosophila* are a diverse group having a high speciation rate and possibly a higher overall rate of phenotypic and ecological differentiation. However, no difference was detected between Hawaiian and continental *Drosophila* in the average strength of premating or postmating isolation between similar-aged populations, a result that does not support the prediction of ecological speciation. The reliability of the test is uncertain, however, because nothing is known about the strength of divergent selection in either Hawaiian or continental *Drosophila*.

Ecological selection against hybrids

Stronger evidence for ecological speciation is that divergent selection arising from features of the environment directly reduces fitness of hybrids (and other individuals of intermediate phenotype) between coexisting species (ECOLOGICAL MECHANISM OF REDUCED HYBRID FITNESS). Such postmating isolation can arise because an intermediate phenotype is less efficient at capturing prey in the wild, or because intermediate defenses leave the hybrid susceptible to predation and parasitism. This type of isolation is environment dependent and should vanish in a common laboratory setting[4,38]. By contrast, genetic mechanisms of reduced hybrid fitness arise from INTRINSIC INCOMPATIBILITIES between genes inherited from the parent species, and should be manifested in every environment. Whereas genetic mechanisms of reduced hybrid fitness could arise during ecological and non-ecological speciation, direct reduction of hybrid fitness by ecological selection pressures are a unique prediction of ecological speciation (Box 1).

Demonstrations of ecological selection against hybrids are still few (Table 1 lists only the strongest tests). F1 hybrids between the limnetic and benthic species of threespine sticklebacks have a high fitness in the laboratory but an intermediate phenotype that compromises their ability to acquire food from the two main habitats in their native lakes. The result is slower growth of F1 hybrids relative to either parent species when transplanted to the habitat of that parent[15]. Furthermore, relative growth rates of limnetic and benthic backcrosses are reversed between habitats, as would be expected if hybrid fitness is directly reduced by ecological selection pressures (H.D. Rundle, unpublished). A similar pattern of high viability of hybrid crosses in the laboratory, coupled with reduced viability in the wild, is seen in crosses between two PARAPATRIC RACES of the

* *This figure is reproduced in color following the index.*

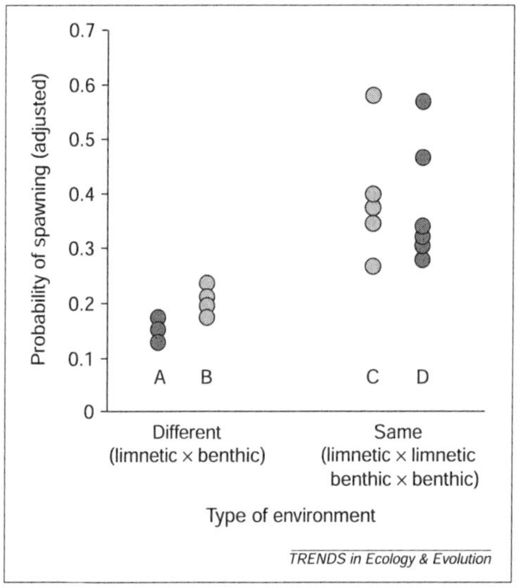

Fig. 4. Parallel evolution of premating isolation between benthic and limnetic threespine sticklebacks (*Gasterosteus* spp.) from three lakes. Each circle indicates frequency of matings between males and females from a pair of populations, measured in no-choice laboratory mating trials. Pairs of populations from the same lake are indicated in green; values in column D are mating frequencies between males and females from the same population. Pairs of populations from different lakes are indicated in blue. Pairs of populations occurring in the same type of environment (C and D) mate with higher frequency than do pairs of populations from different types of environments (A and B). Modified, with permission, from Ref. 55.*

butterfly *Heliconius erato* in Peru[39]. Selection against rare warning color phenotypes in nature by avian predators appears to be the cause.

Differences in timing and duration of diapause of host races of the apple maggot fly *Rhagoletis pomonella* represent adaptations to the timing of fruit production of different host plants in relation to the timing of winter[40]. Hybrid offspring of individuals that switch hosts should be heavily disadvantaged as a consequence, although this has not been demonstrated directly. Selection against F1 hybrids between host races of pea aphids *Acyrthosiphon pisum* on host plants of both parent species probably has an ecological basis, but this has not yet been confirmed[41]. Craig *et al.*[42] showed that F1 and F2 hybrids between two host races of the fly *Eurosta solidaginis* survived poorly on the host plants of their parents, although the pattern of fitnesses was complex. An extended season of leaf production by hybrids between poplar *Populus* spp. might explain the much higher levels of insect herbivory that they experience compared with those on the parent species[43]. Reciprocal transplants along an elevation gradient of two subspecies of sagebrush *Artemisia tridentata* and their hybrids indicated that each of the three populations has highest fitness in its own environment, with the hybrids being most fit at an intermediate elevation[44]. Hybrids in the last two studies were neither F1 nor F2 hybrids, but were individuals from populations of hybrid origin.

Ecologically based postmating isolation is expected to change as environments change. For example, Grant and Grant[45,46] recorded the fates of individual hybrids made of crosses between two species of ground finch (*Geospiza fuliginosa* and *G. fortis*) over 20 years on a Galápagos island. Hybrid survival was typically poor, owing to the low abundance of seeds that the hybrids could handle efficiently. However, seed abundances were changed significantly in the years following a dramatic El Niño event in which rainfall increased by an order of magnitude. Remarkably, the greater relative abundance of small seeds eliminated the difference in survival between hybrids and the dominant parent species, *G. fortis*.

Ecological traits underlying reproductive isolation

Evidence for ecological speciation is gained when specific genes or phenotypic traits known to be under divergent selection between environments are found to be the basis of reproductive isolation (or are genetically correlated with traits that are the basis of reproductive isolation) (Table 1). For example, in the monkey flower *Mimulus guttatus*, alleles conferring tolerance to soils contaminated with copper are lethal when combined in the offspring of crosses with plants from uncontaminated soils[47]. Reproductive isolation between two other monkey flowers (*M. cardinalis* and *M. lewisii*) is associated with differences in floral traits that attract different pollinators and therefore contribute to premating isolation. *Mimulus lewisii* has broad, flat, pink petals with yellow nectar guides, small nectar volume, and is pollinated primarily by bumblebees, whereas *M. cardinalis* has a narrow tubular corolla and large nectar rewards and is pollinated primarily by hummingbirds. These divergent adaptations to contrasting pollinators contribute to premating isolation, and pollinators are attracted to artificial F2 hybrids in proportion to the mixture of genes from the preferred parent[48].

Body size is strongly divergent between sympatric stickleback species, and several lines of evidence suggest that this difference is the result of contrasting natural selection between the main habitats that they exploit[49]. Body size was also found to affect strongly the probability of interspecific hybridization in no-choice laboratory trials: crossbreeding occurred only between the largest individuals of the smallest species and the smallest individuals of the largest species[49]. Similarly, in Darwin's finches, size and shape of the beak and body, which are strongly selected for efficient exploitation of different foods, are also used as cues in interspecific mate discrimination[50]. Some divergence in song, albeit incomplete, might accompany divergence of size and shape of the beak in Darwin's finches, and this could also influence premating isolation[51].

Adaptive life-history differentiation might lead to reproductive isolation in many insects[52]. For example, different timing of emergence between the apple and hawthorn races of the apple maggot fly is linked to changes in the timing and duration of diapause[40]. Development time in populations of *Drosophila mojavensis* has diverged between populations on different cactus hosts, and this trait is genetically correlated with behavioral traits that influence reproductive isolation[53].

** This figure is reproduced in color following the index.*

Parallel speciation

Evidence for ecological speciation is gained when traits determining mating compatibility evolve in parallel in different populations experiencing similar environments ('PARALLEL SPECIATION'[54,55]). Threespine sticklebacks provide the clearest case (Fig. 4). Sympatric limnetic and benthic species of threespine sticklebacks have arisen independently as many as four times in separate lakes[56,57]. The two species within a lake rarely (if ever) hybridize in the wild. Frequency of hybrid mating is raised to 10–15% in no-choice laboratory mating trials (Fig. 4, column A), which is significantly below the value for males and females from the same population (Fig. 4, column D). Remarkably, limnetics and benthics from different lakes also hybridize at low frequency in the laboratory (Fig. 4, column B), whereas populations of the same ECOTYPE from different lakes (i.e. both limnetic or both benthic) mate at high frequency (Fig. 4, column C). This pattern implies that traits influencing mating compatibility have evolved in parallel under similar environmental conditions, strongly implicating divergent natural selection in the origin of stickleback species[55]. Some nonecological processes of speciation might also yield parallel evolution of mating compatibility (e.g. polyploid speciation), but no consistent association between mating compatibility and environment is expected. The traits determining mating compatibility in sticklebacks are not known, but body size[49] and nuptial coloration[58] probably each play a role.

Freshwater amphipods *Hyalella azteca* occur in two types of lakes. Lakes with predatory sunfish contain a small-bodied ecotype, whereas lakes lacking fish predators contain a large-bodied ecotype. Body size differences between amphipod populations are genetically based and are not closely correlated with electrophoretic distance[59], suggesting that multiple transitions in size have occurred. Individuals from populations of the same ecotype (i.e. both large or both small) readily interbreed, whereas individuals from different ecotypes do not[59]. Again, environment and phenotype predict mating compatibility better than do genetic relationships.

Funk[60] examined levels of premating isolation between populations of the leaf beetle *Neochlamisus bebbianae*, a species that exploits different host plants in different parts of its range. Mating trials were carried out with two populations adapted to maple *Acer rubrum*, one adapted to birch *Betula nigra*, and one from willow *Salix bebbiana*. Leaves from the host plants were present in half the trials and absent from the other half, but this made no difference to the results. Reproductive isolation was strong between the birch population and both maple populations, and between the willow population and both maple populations (isolation between willow and birch populations was not tested), but was absent between the two maple populations. However, genetic divergence (based on mitochondrial DNA) was greater between the two maple populations than between one of the maple populations and the willow population, suggesting that reproductive isolation is better predicted by environment than by PHYLOGENY.

Discussion

Speciation is one of the least understood major features of evolution. The main obstacle to progress is the variety of mechanisms that might lead to the evolution of reproductive isolation (Box 1), any one of which can be difficult to rule out in a specific case. The upshot is that it is still difficult to point to even two species in nature and state with confidence the mechanism that produced them. The exceptions are speciation events resulting from polyploidy, because polyploidy leaves a clear genetic signature for a substantial period of time. However, speciation by polyploidy is relatively common only in plants, and, even in that taxon, probably accounts for only a minority of speciation events[32]. The vast majority of speciation events in nature must therefore be explained by other processes.

One of these other processes is ecological speciation, driven by divergent natural selection on traits and resulting from features of the environment. Ecological speciation is probably more easily tested than is speciation by genetic drift or speciation resulting from the accumulation of alternative incompatible mutations under uniform selection. This is because, similar to speciation by polyploidy, divergent natural selection often leaves a signature in the pattern of reproductive isolation, at least for a time (further genetic divergence after speciation is complete might eliminate the signature). For example, whereas genetic mechanisms of reproductive isolation are the expected outcome of every speciation process (Box 1), ecologically based selection against hybrid phenotypes is a unique prediction of ecological speciation[15] (H.D. Rundle, unpublished).

Similarly, parallel speciation is compelling evidence that divergent natural selection has ultimately brought about the evolution of reproductive isolation, as is the finding that traits under divergent natural selection are the basis of reproductive isolation (or are genetically correlated with traits that form the basis of reproductive isolation). A correlation between strength of divergent selection and the rate of evolution of reproductive isolation would be further evidence of ecological speciation, but strength of divergent selection is not easily measured. Finally, persistence of ecologically differentiated populations in the face of gene flow, and evidence of sympatric speciation coupled with strong ecological differentiation, also point to ecological speciation (Via[8], this issue), but alternative mechanisms can produce such patterns and must be tested.

Recent progress has been made in testing ecological speciation in nature (Table 1). Although few in number, candidate examples already indicate the multiple ways that divergent selection might lead to reproductive isolation. For example, at least one candidate for sympatric speciation is represented (*Rhagoletis*), although most species pairs appear to have histories that include an allopatric phase (e.g. threespine sticklebacks, Darwin's ground finches and *Coregonus*). The cases of ecological speciation include populations and species whose hybrids fail partly because of intrinsic genetic incompatibilities (e.g. *Mimulus guttatus* and *Coregonus*), and hybrids of other populations and species whose fitness depends on features of environment (e.g. threespine sticklebacks; Darwin's ground finches and *Heliconius erato*). Evidence for reinforcement is seen in one case (threespine sticklebacks) but even in this example, as in others (Table 1), considerable reproductive isolation appears to have evolved purely as a by-product of divergent selection. Our understanding of the range of processes involved in ecological speciation will increase as more examples accumulate.

Nevertheless, the evidence for ecological speciation is incomplete. More tests from nature are badly needed, not only on the systems already identified as good candidates (Table 1) but also in other systems. The generality of ecological speciation is a long way from being decided, but at least there are both tools available with which to address the problem, and some promising indications from a few studies of species in the wild.

The mechanisms driving ecological speciation also need to be more fully understood. What are the ecological agents of divergent selection? How important are species interactions to strengthening of reproductive isolation during the sympatric phase? Is the by-product mechanism effective by itself, and responsible for the bulk of the evolution of reproductive isolation in allopatry and sympatry, or does reinforcement in sympatry play a vital role in the final production of two coexisting species from a single ancestor? Is divergent sexual selection often the outcome of divergent natural selection, or does it arise independently from other processes?

The link between ecological speciation and adaptive radiation also needs to be assessed[34]. Divergent natural selection is an important process in phenotypic differentiation in adaptive radiation and, for this reason, it might be expected to contribute also to speciation, but this is less well understood. Is speciation in adaptive radiation chiefly ecological speciation? How do the agents of divergent selection, the traits that are its targets, and the consequences for reproductive isolation, change as an adaptive radiation proceeds and the numbers of species in the environment builds? How durable are 'ecological species' (the products of ecological speciation): given the importance of ecological context in determining hybrid fitness, at least when the species are young, are ecological species particularly prone to extinction when environments change? If so, how do the mechanisms of species origin affect the build up of species during adaptive radiation?

These questions emphasize the substantial challenges for research posed by Dobzhansky's[3] claim, one of the clearest early statements of the hypothesis of ecological speciation, that 'the genotype of a species is an integrated system adapted to the ecological niche in which the species lives. Gene recombination in the offspring of species hybrids may lead to formation of discordant gene patterns'. Here, Dobzhansky was referring to the build up of genetic mechanisms of postmating isolation, but the hypothesis is general and applies equally well to traits causing premating isolation – these too might be the product of adaptation to environments. Happily, we are unlikely to have to wait another 50 years before the challenge is taken up, and the hypothesis receives the full evaluation that it deserves.

Acknowledgements
I thank N. Barton, S. Via and other anonymous reviewers for their suggestions. My research is funded by the Natural Sciences and Engineering Research Council of Canada (NSERC).

References

1 Dobzhansky, T. (1946) Complete reproductive isolation between two morphologically similar species of *Drosophila*. *Ecology* 27, 205–211
2 Mayr, E. (1942) *Systematics and the Origin of Species*, Columbia University Press
3 Dobzhansky, T. (1951) *Genetics and the Origin of Species* (3rd edn), Columbia University Press
4 Rice, W.R. and Hostert, E.E. (1993) Laboratory experiments on speciation: what have we learned in 40 years? *Evolution* 47, 1637–1653
5 Kilias, G. *et al.* (1980) A multifactorial genetic investigation of speciation theory using *Drosophila melanogaster*. *Evolution* 34, 730–737
6 Dodd, D.M.B. (1989) Reproductive isolation as a consequence of adaptive divergence in *Drosophila pseudoobscura*. *Evolution* 43, 1308–1311
7 Turelli, M. *et al.* (2001) Theory and speciation. *Trends Ecol. Evol.* 16, 330–343
8 Via, S. (2001) Sympatric speciation in animals: the ugly duckling grows up. *Trends Ecol. Evol.* 16, 381–390
9 Seehausen, O. *et al.* (1997) Cichlid fish diversity threatened by eutrophication that curbs sexual selection. *Science* 277, 1808–1811
10 Albertson, R.C. *et al.* (1999) Phylogeny of a rapidly evolving clade: The cichlid fishes of Lake Malawi, East Africa. *Proc. Natl. Acad. Sci. U. S. A.* 96, 5107–5110
11 Barraclough, T. and Nee, S. (2001) Phylogenetics and speciation. *Trends Ecol. Evol.* 16, 391–399
12 Rundle, H.D. and Schluter, D. (1998) Reinforcement of stickleback mate preferences: sympatry breeds contempt. *Evolution* 52, 200–208
13 Butlin, R. (1989) Reinforcement of premating isolation. In *Speciation and its Consequences* (Otte, D. and Endler, J.A., eds), pp. 85–110, Sinauer
14 Wilson, D.S. and Turelli, M. (1986) Stable underdominance and the evolutionary invasion of empty niches. *Am. Nat.* 127, 835–850
15 Hatfield, T. and Schluter, D. (1999) Ecological speciation in sticklebacks: environment-dependent hybrid fitness. *Evolution* 53, 866–873
16 Dieckmann, U. and Doebeli, M. (1999) On the origin of species by sympatric speciation. *Nature* 400, 354–357
17 Doebeli, M. and Dieckmann, U. (2000) Evolutionary branching and sympatric speciation caused by different types of ecological interactions. *Am. Nat.* 156, S77–S101
18 Abrams, P.A. (2000) Character shifts of prey species that share predators. *Am. Nat.* 156, S45–S61
19 Panhuis, T. *et al.* (2001) Sexual selection and speciation. *Trends Ecol. Evol.* 16, 364–371
20 Lande, R. (1982) Rapid origin of sexual isolation and character divergence in a cline. *Evolution* 36, 213–223
21 Ryan, M.J. and Rand, A.S. (1993) Species recognition and sexual selection as a unitary problem in animal communication. *Evolution* 47, 647–657
22 Endler, J.A. (1992) Signals, signal conditions, and the direction of evolution. *Am. Nat.* 139, S125–S153

23 Schluter, D. and Price, T. (1993) Honesty, perception and population divergence in sexually selected traits. *Proc. R. Soc. London B Biol. Sci.* 253, 117–122
24 Lande, R. (1981) Models of speciation by sexual selection on polygenic traits. *Proc. Natl. Acad. Sci. U. S. A.* 78, 3721–3725
25 Rice, W.R. (1998) Intergenomic conflict, interlocus antagonistic coevolution, and the evolution of reproductive isolation. In *Endless Forms: Species and Speciation* (Howard, D. and Berlocher, S., eds), pp. 261–270, Oxford University Press
26 Gavrilets, S. (2000) Rapid evolution of reproductive isolation driven by sexual conflict. *Nature* 403, 886–889
27 Wright, S. (1940) The statistical consequences of mendelian heredity in relation to speciation. In *The New Systematics* (Huxley, J.S., ed.), pp. 161–183, Clarendon Press
28 Mayr, E. (1954) Change of genetic environment and evolution. In *Evolution as a Process* (Huxley, J. *et al.*, eds), pp. 157–180, Allen & Unwin
29 Muller, H.J. (1940) Bearings of the *Drosophila* work on systematics. In *The New Systematics* (Huxley, J.S., ed.), pp. 185–268, Clarendon Press
30 Stebbins, G.L. (1950) *Variation and Evolution in Plants*, Columbia University Press
31 Ramsey, J. and Schemske, D.W. (1998) Pathways, mechanisms, and rates of polyploid formation in flowering plants. *Annu. Rev. Ecol. Syst.* 29, 467–501
32 Otto, S.P. and Whitton, J. (2000) Polyploid incidence and evolution. *Annu. Rev. Genet.* 34, 401–437
33 Ting, C-T. *et al.* (1998) A rapidly evolving homeobox at the site of a hybrid sterility gene. *Science* 282, 1501–1504
34 Schluter, D. (2000) *The Ecology of Adaptive Radiation*, Oxford University Press
35 Lu, G. and Bernatchez, L. (1999) Correlated trophic specialization and genetic divergence in sympatric lake whitefish ecotypes (*Coregonus clupeaformis*): support for the ecological speciation hypothesis. *Evolution* 53, 1491–1505
36 Coyne, J.A. and Orr, H.A. (1989) Patterns of speciation in *Drosophila*. *Evolution* 43, 362–381
37 Coyne, J.A. and Orr, H.A. (1997) 'Patterns of speciation in *Drosophila*' revisited. *Evolution* 51, 295–303
38 Schluter, D. (1998) Ecological causes of speciation. In *Endless Forms: Species and Speciation* (Howard, D. and Berlocher, S., eds), pp. 114–129, Oxford University Press
39 Mallet, J. *et al.* (1998) Mimicry and warning color at the boundary between races and species. In *Endless Forms: Species and Speciation* (Howard, D. and Berlocher, S., eds), pp. 390–403, Oxford University Press
40 Feder, J.L. (1998) The apple maggot fly, *Rhagoletis pomonella*: flies in the face of conventional wisdom about speciation? In *Endless Forms: Species and Speciation* (Howard, D. and Berlocher, S., eds), pp. 130–144, Oxford University Press
41 Via, S. *et al.* (2000) Reproductive isolation between divergent races of pea aphids on two hosts: selection against migrants and hybrids in the parental environments. *Evolution* 54, 1626–1637
42 Craig, T.P. *et al.* (1997) Hybridization studies on the host races of *Eurosta solidaginis*: implications for sympatric speciation. *Evolution* 51, 552–560
43 Floate, K.D. *et al.* (1993) Elevated herbivory in plant hybrid zones: *Chrysomela confluens*, *Populus* and phenological sinks. *Ecology* 74, 2056–2065
44 Wang, H. *et al.* (1997) Narrow hybrid zone between two subspecies of big sagebrush (*Artemisia tridentata*: Asteraceae). IV. Reciprocal transplant experiment. *Evolution* 51, 95–102
45 Grant, P.R. and Grant, B.R. (1992) Hybridization of bird species. *Science* 256, 193–197
46 Grant, B.R. and Grant, P.R. (1993) Evolution of Darwin's finches caused by a rare climatic event. *Proc. R. Soc. London B Biol. Sci.* 251, 111–117
47 Macnair, M.R. and Christie, P. (1983) Reproductive isolation as a pleiotropic effect of copper tolerance in *Mimulus guttatus*? *Heredity* 50, 295–302
48 Schemske, D.W. and Bradshaw, H.D., Jr (1999) Pollinator preference and the evolution of floral traits in monkey flowers (*Mimulus*). *Proc. Natl. Acad. Sci. U. S. A.* 96, 11910–11915
49 Nagel, L. and Schluter, D. (1998) Body size, natural selection, and speciation in sticklebacks. *Evolution* 52, 209–218
50 Ratcliffe, L.M. and Grant, P.R. (1983) Species recognition in Darwin's finches (*Geospiza*, Gould). I. Discrimination by morphological cues. *Anim. Behav.* 31, 1139–1153
51 Podos, J. (2001) Correlated evolution of morphology and vocal signal structure in Darwin's finches. *Nature* 409, 185–188
52 Miyatake, T. and Shimizu, T. (1999) Genetic correlations between life-history and behavioral traits can cause reproductive isolation. *Evolution* 53, 201–208
53 Etges, W.J. (1998) Premating isolation is determined by larval rearing substrates in cactophilic *Drosophila mojavensis*. IV. Correlated responses in behavioral isolation to artificial selection on a life-history trait. *Am. Nat.* 152, 129–144
54 Schluter, D and Nagel, L.M. (1995) Parallel speciation by natural selection. *Am. Nat.* 146, 292–301
55 Rundle, H.D. *et al.* (2000) Natural selection and parallel speciation in sticklebacks. *Science* 287, 306–308
56 Taylor, E. B. *et al.* (1997) History of ecological selection in sticklebacks: uniting experimental and phylogenetic approaches. In *Molecular Evolution and Adaptive Radiation* (Givnish, T.J. and Sytsma, K.J., eds), pp. 511–534, Cambridge University Press
57 Taylor, E.B. and McPhail, J.D. (2000) Historical contingency and ecological determinism interact to prime speciation in sticklebacks, *Gasterosteus*. *Proc. R. Soc. London B Biol. Sci.* 267, 2375–2384
58 Boughman, J.W. Divergent sexual selection enhances reproductive isolation in sticklebacks. *Nature* (in press)
59 McPeek, M.A. and Wellborn, G.A. (1998) Genetic variation and reproductive isolation among phenotypically divergent amphipod populations. *Limnol. Oceanogr.* 43, 1162–1169
60 Funk, D.J. (1998) Isolating a role for natural selection in speciation: host adaptation and sexual isolation in *Neochlamisus bebbianae* leaf beetles. *Evolution* 52, 1744–1759
61 Filchak, K.E. *et al.* (2000) Natural selection and sympatric divergence in the apple maggot, *Rhagoletis pomonella*. *Nature* 407, 739–742
62 Nee, S. and Barraclough, T.G. (1996) Temporal changes in biodiversity detecting patterns and identifying causes. In *Biodiversity* (Gaston, K.J., ed.), pp. 230–252, Oxford University Press
63 Petren, K. *et al.* (1999) A phylogeny of Darwin's finches based on microsatellite DNA length variation. *Proc. R. Soc. London B Biol. Sci.* 266, 321–330
64 Rundle, H.D. and Schluter, D. (2001) Natural selection and speciation in sticklebacks from beginning to end. In *Adaptive Speciation* (Dieckmann, U. and Doebeli, M., ed.), pp. 000–000, Cambridge University Press
65 Grant, P.R. (1986) *Ecology and Evolution of Darwin's Finches*, Princeton University Press

Moran, N. A. 2002. Microbial minimalism: genome reduction in bacterial pathogens. *Cell* 108:583–586.

The hereditary molecule, deoxyribonucleic acid (DNA), has three attributes that are fundamental for life. First, DNA holds the genetic information that directs all life processes. The information is encased in sequences of the four nucleotides (A, C, G, and T). Genes are DNA segments that code for proteins. There are two types of proteins. Some are important structural components of organisms; for example, collagen is the main protein component of bone. Others are enzymes, catalysts that mediate chemical reactions in organisms. Enzymes may be seen as molecular machines that mediate all living processes inside cells.

Second, DNA accounts for the precision of biological heredity. The two strands in the DNA double helix are complementary; both carry the same genetic information, and either one of the two strands can serve as a template for the synthesis of a complementary strand, identical to the original . Each of the four nucleotides pairs with only one particular nucleotide in the complementary strand: A pairs only T, and C pairs with G. This complementarity accounts for the fidelity of biological heredity.

The third property is mutation, which makes possible the evolution of organisms. The information encoded in the nucleotide sequence is, as a rule, faithfully reproduced during replication, but occasionally mutations occur in the DNA molecule during replication, so that the daughter cells differ from the parental cell (and from each other) in the nucleotide sequence of the DNA. Mutations often involve one single nucleotide, but occasionally may encompass several or many nucleotides.

The genomes of bacteria, according to current information, range in size from 0.16 to 13 megabases (mb). The genomes of archaea range from 0.49 mb to 5.7 mb. Protozoan genomes range from 2.3 mb to more than 6×10^5 mb. Among multicellular eukaryotes, plants, arthropods, fish, and amphibians have the greatest ranges of variation, from somewhat more than 10^2 mb to nearly 10^6 mb. Reptiles, birds, and mammals exhibit much less variation, from about 5 to 20 x 10^3 mb. According to some recent estimates, the human genome consists of 2910 mb, about one fifth as much DNA as some salamanders and only 2 percent as much as the African lungfish. In eukaryotes, only a small fraction of the genome consists of coding DNA, about 2 to 10 percent of the total. Most of the DNA consists of introns (intervening sequences) and mobile elements known as transposons (see McClintock 1984 in this volume).

To appreciate the amount of genetic information encoded in the DNA, consider that if each nucleotide is represented by one letter (A, C, G, and T), as is usual, printing one human genome would take 1,000 volumes of 1,000 pages each with 3,000 letters per page. Scientists do not print the full genomes of humans or other organisms; the DNA information is stored in computers or other electronic devices.

Clearly, the sizes of genomes expand and contract across evolutionary time. In this article, Nancy Moran focuses on the causes and consequences of genomic reductions in pathogenic microbes.

RELATED READING

Bentley, S. D., and J. Parkhill. 2004. Comparative genomic structure of prokaryote. Annual Review of Genetics 38:7721–792.

Doolittle, W. F. 1998. You are what you eat: a gene transfer ratchet could account for bacterial genes in eukaryotic nuclear genomes. Trends in Genetics 14:307–311.

Lynch, M., and J. S. Conery. 2003. The origins of genome complexity. Science 302:1401–1404.

Margulis, L., and K. V. Swartz. 1998. Five kingdoms: an illustrated guide to the phyla of life on Earth. Freeman, New York, New York, USA.

Maynard Smith, J., and E. Szathmáry. 1995. The major transitions in evolution. Freeman, San Francisco, California, USA.

Cell, Vol. 108, 583–586, March 8, 2002, Copyright ©2002 by Cell Press

Microbial Minimalism: Genome Reduction in Bacterial Pathogens

Minireview

Nancy A. Moran[1]
Department of Ecology and Evolutionary Biology
University of Arizona
Tucson, Arizona 85721

When bacterial lineages make the transition from free-living or facultatively parasitic life cycles to permanent associations with hosts, they undergo a major loss of genes and DNA. Complete genome sequences are providing an understanding of how extreme genome reduction affects evolutionary directions and metabolic capabilities of obligate pathogens and symbionts.

Introduction

Known genome sizes of bacteria range from under 0.6 to ~10 megabases (mb). At the lowest extreme of this range are the mycoplasmas and related bacteria, with genome sizes reported as low as 530 kilobases. Among the many early revelations from molecular phylogenetic studies of bacteria (Woese, 1987) was the recognition that the mycoplasmas represented an evolutionarily derived condition rather than a primitive one, as once believed. Now that phylogenetic relationships and genome sizes are determined for a broader array of organisms, it is clear that the mycoplasmas are just one example of genome shrinkage that has occurred in a variety of obligately host-associated bacteria. Other prominent examples are *Rickettsia* and related pathogens within the α-proteobacteria; insect symbionts within the γ-proteobacteria, as exemplified by *Buchnera aphidicola* in aphids; the chlamydiae; and the parasitic spirochetes, such as *Borrelia burgdorferi* (the agent of Lyme disease).

Small genome size in these organisms is associated with other distinctive genetic features, including rapid evolution of polypeptide sequences and low genomic G+C content (Figure 1). The repeated evolution of these features in unrelated bacteria indicates that an obligate association with host tissues somehow promotes genome reduction. Understanding the causes of these genome level changes will help to reveal the processes that are important in pathogen and symbiont evolution.

Over 50 eubacterial genomes are now fully sequenced and annotated, with many more near completion. These sequences have corroborated a link between obligate host-associated lifestyles and a distinctive set of genomic features that include small size. Furthermore, they are yielding detailed information on the evolutionary basis for DNA loss and the functional implications of this loss.

Which Genes Disappear

Bacterial genomes are comprised mostly of coding genes: in almost all of the fully sequenced genomes, over 80% of the sequence consists of intact ORFs. This, combined with the fact that gene length is effectively constant (averaging ~1 kb per gene) across genomes, implies that small genomes have few genes and correspondingly limited metabolic capabilities. Whereas bacteria with free-living stages, such as *Escherichia coli*, *Salmonella* species, or *Bacillus* species, typically encode 1500 to 6000 proteins, obligately pathogenic bacteria often encode as few as 500 to 1000 proteins (Figure 1).

The simplest possibility would be that reduced genomes converge on a set of universal genes that underlie the core processes of cellular growth and replication, with each genome also containing some loci corresponding to that species' ecology or host-relationship. But this possibility is contradicted by the full genome sequences. The set of orthologs that are universal, or nearly so, among eubacteria constitutes only a small proportion (<15%) of each genome, totaling about 80 genes (Koonin, 2000). Thus, each lineage has taken a different evolutionary route to minimalism. Since universal cellular processes require many more than 80 genes, differences in gene inventories imply that the same functions can be achieved by retention of nonhomologous genes.

Use It or Lose It

Nevertheless, some intelligible patterns do emerge from comparing gene sets of fully sequenced genomes. One clear basis for genome reduction is that bacteria living continuously in hosts can obtain many compounds of intermediate metabolism from host cytoplasm or tissue; thus, they can discard the corresponding biosynthetic pathways and genes. Such elimination of unneeded pathways explains a substantial proportion of observed gene losses. For instance, many of the genes involved in energy metabolism are eliminated from *Rickettsia* species, *Mycoplasma* species, and *Buchnera*, which can rely on consistent availability of particular energy substrates from hosts (Figure 2).

Likewise, most small genomes have eliminated genes underlying biosynthesis of amino acids, which are taken up from host cells. A remarkable exception—of the type that proves the rule—occurs in *Buchnera*, an obligate maternally transmitted symbiont of aphids. A basis for the mutualism is the *provisioning* of essential amino acids to hosts, and *Buchnera* retains 54 genes (comprising ~10% of its genome) for biosynthesis of essential amino acids, but has lost pathways for amino acids that the host can produce itself (Shigenobu et al., 2000). Pathways for nucleotide biosynthesis, and vitamin biosynthesis, are also missing from many reduced genomes. Individual genomes retain unique combinations of anabolic pathways, probably relating to different environmental conditions.

Small genomes have lost many regulatory elements, including sigma factors. This aspect also may be partly attributable to a lack of need: living continuously within the host eliminates the extreme environmental fluctuations encountered by free-living bacteria.

Use It, but Lose It Anyway

The premise that useful genes are retained and useless ones eliminated oversimplifies the evolutionary processes that affect persistence of genes in genomes.

[1]Correspondence: nmoran@u.arizona.edu

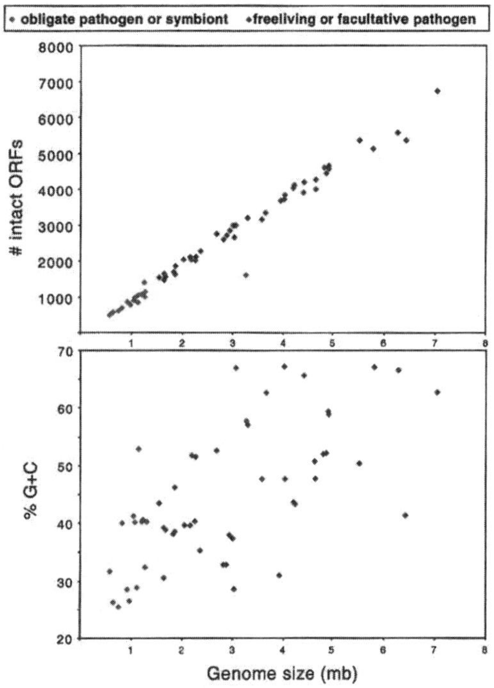

Figure 1. Size of Eubacterial Genomes in Relation to Number of intact ORFs (top) and % G+C Content in Genome (bottom)

The red spots correspond to obligate pathogens, belonging to a variety of unrelated lineages. *M. leprae* contains inactivated genes not included in the tally of intact ORFs. Included are all eubacterial genomes available January 2002 in NCBI Entrez Genomes (http://www.ncbi.nlm.nih.gov:80/PMGifs/Genomes/eub_g.html), with numbers of intact ORFS taken from this annotation.*

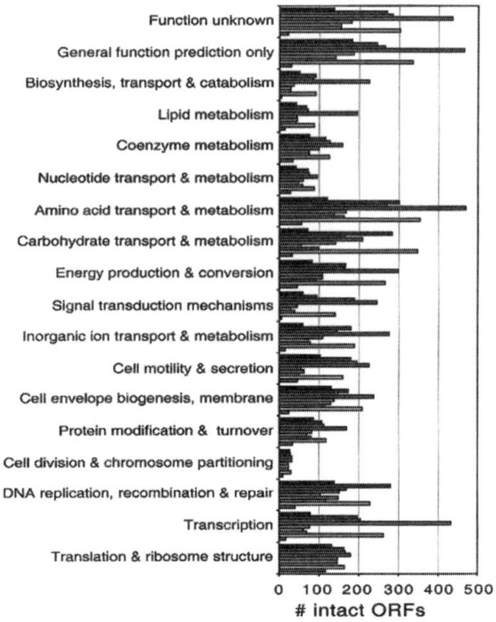

Figure 2. Numbers of ORFs Assigned to Different Functional Categories for Fully Sequenced Genomes of Members of the β- and γ-Proteobacteria

From top, they are *Xylella fastidiosa*, *Yersinia pestis*, *Vibrio cholerae*, *Pseudomonas aeruginosa*, *Pasteurella multocida*, *Neisseria meningitidus*, *Haemophilus influenzae*, *Escherichia coli* K12, and *Buchnera aphidicola*. *Buchnera*, the only organism in this group that is obligately associated with hosts and that has a highly reduced genome, shows reduced numbers of ORFs in all categories. Categories are those corresponding to the COGs database (Tatusov, R.L., Koonin, E.V., and Lipman, D.J., 1997, Science *278*, 631-637, http://www.ncbi.nlm.nih.gov/COG/xindex.html).*

And it cannot explain why many of the discarded genes encode products that would seem to be just as useful in obligate pathogens as in other organisms. Many eliminated genes encode accessory proteins or regulatory products involved in universal cellular processes, including replication, transcription, and translation (e.g., Andersson, et al. 1998; Moran and Wernegreen, 2000; Figure 2). Some genes underlying DNA recombination and repair pathways are eliminated from every small genome, although the precise set discarded varies. Also, small genomes contain fewer tRNAs, retaining only one for many amino acids. Thus, a single anticodon must pair with multiple codons, presumably resulting in less efficient translation machinery. It is not clear why obligate intracellular pathogens would benefit by retaining fewer tRNAs and fewer DNA repair enzymes.

Also important in the evolutionary processes determining patterns of gene persistence are the changes in population structure that accompany a shift to an obligately pathogenic lifestyle. Acquisition of an obligately host-associated lifestyle will often greatly diminish the genetic population size of a lineage, due to restricted habitat (hosts), and to bottlenecks in bacterial numbers at the time of infection (Andersson and Kurland, 1998; Moran and Wernegreen, 2000). The resulting genetic drift can lead to the fixation of mutations that inactivate genes that are useful (but obviously not essential), or they can decrease the efficiency of gene products. Thus, entirely useless genes, such as those not needed in a newly acquired niche, will lose functionality due to mutations that disrupt the coding region, but even beneficial genes may be lost or degraded if genetic drift precludes effective purifying selection.

How Many Genes Are Required?

Studies aiming to define the minimal gene inventory of a cellular organism received new focus with the publication of the complete sequence of *M. genitalium*, the smallest sequenced bacterial genome (Maniloff, 1996). As mentioned, the set of universally distributed genes is small and insufficient for independent cellular growth and replication, implying that small genome organisms accomplish the same set of cellular processes by retaining different sets of genes. This is achieved in part through nonorthologous gene displacement: the role of one gene is replaced by an unrelated gene accomplishing the same function (Koonin, 2000). Redundancy within the ancestral, large genome appears to be eliminated through different routes. The final gene set may depend on the gene content of chromosomal deletions that occur early in the course of genome reduction.

* *This figure is reproduced in color following the index.*

Even the tiny genome of *M. genitalium* harbors many genes that are dispensable, at least for growth in vitro. Based on a study in which single genes of *M. genitalium* were inactivated using transposon-mediated mutagenesis, at least 129 of that organism's 484 ORFs were unnecessary for growth. Thus, a substantially smaller genome is plausible. It must be remembered, however, that the bare minimum is not necessarily what we expect in naturally occurring organisms, in which selection will eliminate less competitive genotypes.

Every sequenced genome contains a set of ORFs with no assigned function, with the proportion varying among genomes according to the phylogenetic proximity to a laboratory model organism. One kind of insight to be gained from small genome sequences is the identification of unassigned genes that are good targets for further study. For example, almost every *Buchnera* gene has a clear ortholog in *E. coli*, indicating that *Buchnera* provides a good approximation of a minimal *E. coli* genome (in the context of an intracellular environment). For about 50 of these shared genes, no function is assigned. As more small genomes in the γ-proteobacteria are completed, the set of genes indispensable within this group will be defined; these can focus experimental studies directed at extending knowledge of gene function in this model clade of organisms.

Selective Advantage for Smallness?

A frequently proposed explanation for genome reduction is that selection has favored small genome size for the sake of growth efficiency or competitiveness within the host. The implication is that selection for efficient replication would suffice to eliminate DNA corresponding to some genes. But changes in DNA content, on the scale corresponding to individual genes, have not been shown to affect rate of bacterial cell replication. Also, genome sequence analyses contradict some predictions of the hypothesis that selection drives elimination of DNA. One line of evidence against significant selection for small genome size is the retention of nonfunctional DNA in the form of pseudogenes within small genomes, as in *Rickettsia* and *Buchnera*. If genome reduction were driven by a replication advantage of a minimal genome, gene inactivation would not be expected to precede loss of DNA. Finally, small genomes are no more tightly packed, based on overall length of spacer regions or on comparison of homologous chromosomal regions between the tiny *Buchnera* genome and the much larger genome of the related *E. coli* (Mira et al., 2001).

Mutation to Smallness (Deletional Bias)

In contrast to eukaryotes, in which nonfunctional DNA often persists, bacterial genomes are tightly packed with genes, implying elimination of nonfunctional DNA. Analyses of pseudogene sequences in *Rickettsia* and a wide range of other bacteria reveals widespread mutational bias toward DNA loss (deletional bias) (Andersson and Andersson, 2001; Mira et al., 2001). Thus, DNA is retained in the bacterial genome only if selection is acting effectively to preserve it. If gene functions are rendered useless, due to redundancy within the host environment, then mutations will inactivate the sequences, and the corresponding DNA will be eroded over time through mutational patterns favoring deletion.

Although small genomes present a clear progression from fragmentation or disruption of an ORF to elimination of the corresponding DNA, it is not yet obvious when inactivated genes cease to be transcribed and translated. In the genome of *Rickettsia conorii*, numerous interrupted ORFs were found to be (at least) transcribed (Ogata et al., 2001). Transcription of nonfunctional or unneeded genes might impose a selective cost, possibly favoring elimination of the corresponding DNA. Thus, even if nontranscribed, inactivated genes impose no selective cost; deletional bias may confer a benefit through elimination of DNA that is transcribed but not useful.

Mutational Pressure for A+T Enrichment

Many obligate pathogenic bacteria display A/T-enriched sequences that reflect mutational bias. In each A/T-biased small genome, the bias toward A/T is evident in all types of genes and positions but is strongest in neutral DNA positions, such as noncoding spacers and third codon positions of ORFs. However, the tendency to greater genomic A+T content affects even those nucleotides that effect amino acid replacements. As a result, polypeptides of small genome bacteria are enriched in amino acids, such as lysine, that contain more A or T in the codon family. One consequence is higher predicted isoelectric point (pI) of polypeptides; for example, the average pIs for polypeptides of *Buchnera* and of its relative *E. coli* are 9.6 and 7.2, respectively (Shigenobu et al., 2000). Among currently sequenced genomes, the most extreme A/T bias occurs in *Ureaplasma urealyticum* (25.5% GC), in which the most A/T biased ORFs correspond to genes shown to be expendable in gene knockout studies on the related *M. genitalium*. Thus, the A+T content of a particular set of nucleotide positions reflects the opposing pressures of mutational bias toward A+T versus purifying selection for preservation of gene function.

The basis of mutational pressure toward increased A+T may reflect the elimination of genes encoding DNA repair enzymes, or the decreased efficiency of these enzymes. In particular, the incorporation of uracil into DNA, due either to replication error or to C → U deamination, will result in mutational pressure toward A+T if not prevented or corrected, and the enzymes mediating these changes are sometimes missing or less efficient (Glass et al., 2000). Another possible explanation for the A/T mutational bias might involve nucleotide pools favoring A or T. However, this is unlikely to provide a general reason for the pattern as small genomes vary both in capability for nucleotide biosynthesis and in location within hosts.

Reconstructing Genome Reduction

In most cases, the small eubacterial genomes are only distantly related to any larger genome organisms, a situation that precludes a reliable reconstruction of how the genome reduction occurred. For example, *Rickettsia* species and relatives such as *Wolbachia pipientis* and *Ehrlichia* species comprise an α-proteobacterial clade characterized by consistently small genomes, and this clade is only distantly related to species with larger genomes. Likewise, the mycoplasmas and chlamydiae are embedded in ancient groups with uniformly small genome size. However, some of the symbiotic bacteria of insects and other arthropods fall are quite closely related to larger genome species within the Enterobacteriaceae such as *E. coli*, *Yersinia pestis*, and *Salmonella* species. These symbiotic bacteria, which include *Buchnera* in aphids and *Wigglesworthia* in tsetse flies, provide the opportunity to reconstruct the process of genome reduction. Such an attempt to reconstruct the pattern of gene deletions during the evolution of *Buchnera* suggested that, in addition to gradual erosion of some individual genes through small deletions, some deletions were large and spanned dozens of ancestral genes (Moran and Mira, 2001). One plausible scenario is that the initial transition to the obligately symbiotic (or pathogenic) lifestyle is accompanied by massive genomic changes with some large deletions being fixed within the lineage. These deletions might establish themselves due to a combination of reduced competition and selection in the newly invaded niche and of increased genetic drift arising from population bottlenecks that occur at the time of infection (with many hosts invaded by a single genotype). It is clear as well that some *Buchnera* genes were lost individually through a process of inactivation followed by decay (Moran and Mira, 2001, Silva et al., 2001), as documented in *Rickettsia* (Andersson and Andersson, 2001).

A snapshot of genome degradation in progress is provided by the complete sequence of the genome of *Mycobacterium leprae*, the infectious agent for leprosy. *M. leprae* stands out among the mycobacteria in having a genome that is both reduced and rearranged (Cole et al., 2001). Comparing *M. leprae* to its more typical relative, *Mycobacterium tuberculosis*, indicates that the *M. leprae* lineage has discarded more than 2000 genes. DNA corresponding to more than 1000 of these genes is still present as partial copies or as nonfunctional pseudogenes. Proteome analyses confirm that *M. leprae* does express a much reduced complement of proteins as compared to *M. tuberculosis*, indicating that the apparent pseudogenes have indeed been silenced. This organism has the largest proportion of noncoding DNA of any fully sequenced bacterial genome; only about half of its sequence encodes proteins, as contrasted with 90% in *M. tuberculosis*. In addition, *M. leprae* exhibits some very large deletions that span multiple loci, and its genome size is considerably reduced (3.3 mb as compared to 4.4 mb in other mycobacteria). It has also undergone a shift in base composition toward lower G+C%. The most plausible interpretation of this unusual bacterial genome is that the reductive evolution of *M. leprae* is recent, perhaps linked to its becoming an obligate pathogen during the last few million years.

Lineages that have acquired pathogenic lifestyles recently appear to have embarked on some of the same processes of gene decay and deletion that have progressed to extensive genome shrinkage in more ancient pathogenic groups. For example, large numbers of pseudogenes have been identified in both *Yersinia pestis*, the agent for plague, and *Salmonella enterica* serovar Typhi (Parkhill et al., 2001). The acquisition of a pathogenic life cycle may impose a relatively constant environment, rendering many genes useless, as well as population bottlenecks, resulting in greater levels of genetic drift and resulting gene inactivation. Support for the latter comes from recent studies of the population genetics of human pathogens *Y. pestis* and *M. tuberculosis*, which show very low levels of polymorphism and indicate high levels of genetic drift relative to related free-living bacteria (Achtman et al., 1999).

Outlook

From the extensive documentation of lateral transfer of pathogenicity islands, we know that gene acquisition often enables pathogenic life. Yet the evolutionarily ancient obligate pathogens possess genomes in which gene loss is far more extensive than gene acquisition. Analysis of the varying solutions to genome minimalism, as presented by different small genome organisms, promises to yield information about interdependencies of gene products. Such information is not evident from studies based on single gene knockouts. Understanding of the basis for the observed differences in gene inventories will depend in part on identifying the kinds of DNA deletions that occur at different evolutionary stages of genome reduction, and this will soon be possible, in view of the current rapid rate of publication of new genome sequences.

Selected Reading

Achtman, M., Zurth, K., Morelli, C., Torrea, G., Guiyoule, A., and Carniel, E. (1999). Proc. Natl. Acad. Sci. USA *96*, 14043–14048.

Andersson, J.O., and Andersson, S.G.E. (2001). Mol. Biol. Evol. *18*, 829–839.

Andersson, S.G.E., and Kurland, C.G. (1998). Trends Microbiol. *6*, 263–278.

Andersson, S.G.E., Zomorodipour, A., Andersson, J.O., Sicheritz-Ponten, T., Alsmark, U.C.M., Podowski, R.M., Naslund, A.K., Eriksson, A.S., Winkler, H.H., and Kurland, C.G. (1998). Nature *396*, 133–140.

Cole, S.T., Eiglmeier, K., Parkhill, J., James, K.D., Thomson, K.D., Wheeler, P.R., Honore, N., Garnier, T., Churcher, C., Harris, D., et al. (2001). Nature *409*, 1007–1011.

Goebel, W., and Gross, R. (2001). Trends Microbiol. *9*, 267–273.

Glass, J.L., Lefkowitz, E.J., Glass, J.S., Hlener, C.R., Chen, E.Y., and Cassell, G.H. (2000). Nature *407*, 757–762.

Himmelreich, R., Plagens, H., Hilbert, H., Reiner, B., and Herrmann, R. (1997). Nucleic Acids Res. *25*, 701–712.

Hutchison, C.A., Peterson, S.N., Gill, S.R., Cline, R.T., White, O., Fraser, C.M., Smith, H.O., and Venter, J.C. (1999). Science *286*, 2165–2169.

Koonin, E.V. (2000). Ann. Rev. Genom. Hum. Gen. *1*, 99–116.

Maniloff, J. (1996). Proc. Natl. Acad. Sci. USA *93*, 10004–10006.

Moran, N.A., and Mira, A. (2001). Genome Biol., in press.

Mira, A., Ochman, H., and Moran, N.A. (2001). Trends Genet. *17*, 589–596.

Moran, N.A., and Wernegreen, J.J. (2000). Trends Ecol. Evol. *15*, 321–326.

Mushegian, A.R., and Koonin, E.V. (1996). Proc. Natl. Acad. Sci. USA *93*, 10268–10273.

Ogata, H., Audic, S., Renesto-Audiffren, P., Fournier, P.E., Barbe, V., Samson, D., Roux, V., Cossart, P., Weissenbach, J., Claverie, J.M., and Raoult, D. (2001). Science *293*, 2093–2098.

Parkhill, J., Dougan, G., James, K.D., Thomson, N.R., Pickard, D., Wain, J., Churcher, C., Mungall, K.L., Bentley, S.D., Holden, M.T.G., et al. (2001). Nature *413*, 848–852.

Shigenobu, S., Watanabe, H., Hattori, M., Sakaki, Y., and Ishikawa, H. (2000). Nature *407*, 81–86.

Silva, F.J., Latorre, A., and Moya, A. (2001). Trends Genet. *17*, 615–618.

Woese, C.R. (1987). Microbiol. Rev. *51*, 221–271.

Grant, P. R., and B. R. Grant. 2002. Unpredictable evolution in a 30-year study of Darwin's finches. *Science* 296:707–711.

Evolution can be seen as a two-step process: one random, the other nonrandom. First, hereditary variation arises; second, natural selection occurs of those genetic variants that will be differentially passed to the following generations because those variants favor the adaptation of the organisms to the environments where they live more than alternative genetic variants do; that is, they increase the probability of survival and reproduction of their carriers.

Hereditary variation entails two processes that are random: the spontaneous mutation of one genetic variant to another and, in sexually reproducing organisms, the recombination of those variants through the sexual process. The transmission of the hereditary variants from one generation to another is largely governed by natural selection, a nonrandom process, which often is not predictable, because natural selection occurs in response to the vagaries of the environment. Environments change physically—in climate, configuration and so on—but also biologically, because the predators, parasites, competitors, and food sources with which an organism interacts are also evolving.

Some evolutionary changes can be predicted, at least in part; for example, when organisms migrate from one environment to another or when long-term environmental changes are occurring. One well known example is the evolution of what is called *industrial melanism* in the peppered moth, *Biston betularia*, in the humid forests of England and elsewhere, as well as in other moths and butterflies. Since the middle of the nineteenth century, the appearance and spread of darkly pigmented moth variants was recorded in industrial regions where the vegetation was blackened because of pollution from soot and other wastes. About 100 species have shown industrial melanisms, while in unpolluted areas the light ones persisted. After strict clean air laws were introduced in various regions, and soot was largely replaced by lichens and other natural covers on the bark of trees, the process was reversed and the original light-colored moths and butterflies replaced the melanic ones. Another example typically occurring over longer time periods is the atrophy of eyes in fishes and other animals living in caves.

The paper by Peter R. Grant and B. Rosemary Grant summarizes a long-term study of evolution in a natural population, where some trends could be ascertained but not predicted.

RELATED READING

Endler, J. A. 1986. Natural selection in the wild. Princeton University Press, Princeton, New Jersey, USA.

Grant, P. R., and B. R. Grant. 2008. How and why species multiply. Princeton University Press, Princeton, New Jersey, USA.

Grant, P. R., and B. R. Grant, editors. 2010. In search of the causes of evolution: from field observations to mechanisms. Princeton University Press, Princeton, New Jersey, USA.

Lenski, R. E., and M. Travisiano. 1994. Dynamics of adaptation and diversification: a 10,000-generation experiment with bacterial populations. Proceedings of the National Academy of Sciences USA 91:6808–6814.

Nachman, M. W., H. E. Hoekstra, and S. L. D'Argostino. 2003. The genetic basis of adaptive melanism in pocket mice. Proceedings of the National Academy of Sciences USA 100:5268–5273.

Reznick, D. N., M. Mateos, and M. Springer. 2002. Independent origins and rapid evolution of the placenta in the fish genus *Poeciliopsis*. Science 298:1018–1020.

Stern, D. L., and V. Orgogozo. 2009. Is genetic evolution predictable? Science 323:746–751.

Unpredictable Evolution in a 30-Year Study of Darwin's Finches

Peter R. Grant and B. Rosemary Grant

Evolution can be predicted in the short term from a knowledge of selection and inheritance. However, in the long term evolution is unpredictable because environments, which determine the directions and magnitudes of selection coefficients, fluctuate unpredictably. These two features of evolution, the predictable and unpredictable, are demonstrated in a study of two populations of Darwin's finches on the Galápagos island of Daphne Major. From 1972 to 2001, *Geospiza fortis* (medium ground finch) and *Geospiza scandens* (cactus finch) changed several times in body size and two beak traits. Natural selection occurred frequently in both species and varied from unidirectional to oscillating, episodic to gradual. Hybridization occurred repeatedly though rarely, resulting in elevated phenotypic variances in *G. scandens* and a change in beak shape. The phenotypic states of both species at the end of the 30-year study could not have been predicted at the beginning. Continuous, long-term studies are needed to detect and interpret rare but important events and nonuniform evolutionary change.

The value of long-term studies in ecology has become widely recognized among scientists and the media (*1–3*). Less widely appreciated is the similar value to be gained from long-term studies of evolution in nature. A classic study, spanning 49 years, was carried out by H. D. Ford and E. B. Ford (*4*) on phenotypic variation in a European butterfly, *Melitaea aurinia* (Marsh Fritillary). The most important discovery was made after about 40 years of monitoring that began with the collecting of specimens by amateur naturalists in 1881. In the early 1920s an outburst of phenotypic and presumed genetic variation occurred in association with a rapid increase in butterfly numbers from an extremely low density caused by parasitoids. Variation then declined to a lower and stable level, with phenotypes in the late 1920s being recognizably different from those in the same population at the beginning of the study. The inferred genetic reorganization helped to frame ideas about evolution in contemporary time, inspired other long-term studies of butterflies (*5*) and moths (*6*), and contributed to the development of at least one model of speciation (*7*).

Long-term studies of evolution involving annual or more frequent sampling have many potential benefits. These include documentation and understanding of slow and cryptic directional evolutionary change, perhaps in association with gradual global warming, reversals in the direction of evolution, rare events with strong effects such as genetic bottlenecks caused by population crashes, phenomena recurring at long intervals, and processes with high interannual variability such as erratic and intermittent gene flow. These benefits are beginning to be realized (*8–15*), but few studies have persisted long enough for us to be able to generalize about the temporal pattern and predictability of basic evolutionary processes in unconstrained natural populations.

Here, we report the results of a 30-year study of evolution of size and shape traits in two populations of Darwin's finches based on annual sampling and measurement. Distinctive features of the study are its length, continuity, entirely natural environmental setting, the availability of pedigree information to construct and interpret evolutionary change, and the macroevolutionary context of an adaptive radiation. The study reveals the irregular occurrence, frequency, and consequences of two evolutionary processes that are more often inferred than directly studied: natural selection and introgressive hybridization.

Natural selection and evolution. Populations of *Geospiza fortis* (medium ground finch) and *G. scandens* (cactus finch) have been studied on the Galápagos island of Daphne Major every year since 1973; adults that year were born (hatched) no later than 1972. Survival of marked and measured individuals has been recorded every year, and reproduction of most individuals has been recorded in most years (*16*). Six measured traits on adults whose growth has ceased have been reduced by principal components analyses to three interpretable synthetic traits: body size, beak size, and beak shape (*17–20*). The null expectation is that, subject to sampling error, means of these traits have remained constant across the period of study.

This expectation of no change is clearly not supported by the data (Fig. 1). Lack of independence of samples in successive years precludes year-by-year significance testing of the total samples. Nevertheless, comparisons across years show nonoverlapping 95% confidence estimates of the means at different times. Mean body size and beak shape were markedly different at the end of the period (2001) than at the beginning (1973) in both species (*21*). Between these two times mean body and beak size of *G. fortis* initially decreased, then increased sharply, and decreased again more slowly (Fig. 1, A and B). Beak shape abruptly became more pointed in the mid-1980s and remained so for the next 15 years (Fig. 1C). *G. scandens*, a larger species, displayed more gradual and uniform trends toward smaller size and blunter beaks (Fig. 1, D to F), thereby converging toward *G. fortis* in morphology.

Apart from random sampling effects, annual changes in morphological means are caused by selective losses, as a result of mortality and emigration, and selective gains, as a result of breeding and immigration (*22*). Previous work has demonstrated directional natural selection on beak and body size traits associated with survival, in *G. fortis* at three times and in *G. scandens* once, when a scarcity of rain caused a change in the composition of the seed supply that forms their dry-season diets (*23–25*). Evolutionary responses of *G. fortis* to the two strongest selection episodes occurred in the following generations (*26*), as expected from the high heritabilities of the morphological traits [$h^2 = 0.5$ to 0.9 after corrections for misidentified paternity arising from extrapair copulations (*18, 27*)].

Figure 2 provides the long-term perspective of repeated natural selection in both species (*28*). There are four main features of the figure. First, body and beak size traits were subject to selection more often than was beak shape. Setting α at 0.01, to allow for the lack of complete independence of the traits (*29*), we find that body size was subject to selection about once every 3 years in both species (*30*), that is, once each generation of 4.5 years (*G. fortis*) or 5.5 years (*G. scandens*) (*31*) on average. Second, considering only the statistically significant selection differentials, the species differed in the directions of net selection on size traits. *G. fortis* experienced selection in both directions with equal frequency (Fig. 2, A to C), whereas *G. scandens* experienced selection that repeatedly favored large body size and in no instance favored small beak size (Fig. 2, D and E). Third, unidirectional selection occurred in successive years, up to a maximum of 3 years in

Department of Ecology and Evolutionary Biology, Princeton University, Washington Road, Princeton, NJ 08544–1003, USA. E-mail: prgrantprinceton.edu both species (Fig. 2, A, D, and E). Fourth, selection events in the two species were usually not synchronous, except in the late 1970s, when large size was selectively favored in both species during a drought (*23*). The demonstration here of natural selection occurring repeatedly in the same populations over a long time complements the widespread detection of natural selection in many different species of plants and animals over much shorter times (*32*, *33*, *34*). As in these broad surveys, and in three studies of birds lasting for 11 to 18 years (*15*, *35*, *36*), the magnitude of selection on the finch populations was usually less than 0.15 SD and rarely more than 0.50 SD (*33*, *34*). Median values (0.03 to 0.06) are well within the normal range (0.00 to 0.30) of other studies (*34*).

Evolution followed as a consequence of selection in both species because all traits are highly heritable (*18*, *20*, *27*). We compared the mean of a trait before selection with the mean of the same trait in the next generation by one-tailed *t* tests ($P < 0.05$) (*26*). Significant evolutionary events occurred in *G. fortis* eight times (body size, four; beak size, three; beak shape, one) and in *G. scandens* seven times (body size, two; beak size, five). Evolution below the level of statistical detectability may have followed other instances of directional selection, may have been masked by annual variation in environmental effects on growth to final size (*37*), or may have been nullified by countervailing selection on correlated traits not included in the analyses (*32*). Magnitudes of evolution of the two independent beak traits (size and shape) are correlated with values predicted from the products of selection differentials and heritabilities (Fig. 3). Similar results were obtained in analyses of the direct effects of selection on the six measured traits of *G. fortis* at two times of intense selection, taking into account genetic correlations among them (*26*). Thus evolution, as an immediate response to selection, was predictable.

Introgressive hybridization. Annual changes in morphology (Fig. 1) are largely but not entirely accounted for by selective losses. The greatest discrepancy is in the 1990s when the single occurrence of natural selection on beak shape in *G. scandens* at the beginning (at $P < 0.05$; Fig. 2F) does not account for the continuing change in mean beak shape over the decade (Fig. 1F). Therefore, we next consider selective gains as a result of nonrandom recruitment to the adult population.

There are four potential contributors to nonrandom additions: conspecific and heterospecific residents and immigrants. Breeding immigrants are not known in *G. scandens* and are extremely rare in *G. fortis* (*16*). Biased conspecific breeding has minor effects on morphological trajectories. Prior analyses indicate some evidence for sexual selection on morphological traits (*38*), yet little influence of morphological variation on lifetime fitness as measured by the production of offspring that survive to breed (*39*). However, hybridization does occur rarely between resident *G. fortis* and *G. scandens*, and *G. fortis* also breeds with a rare immigrant species, *G. fuliginosa* (small ground finch) (*40*). In both cases there is generally little or no fitness loss (*41*, *42*). After the dry period of the late 1970s, and beginning in the extraordinarily prolonged wet season of 1983 (El Niño year), successful breeding of F_1 hybrids and backcrosses was documented (*40*–*43*). Effects of introgression on morphological means and variances have not been tested before, but are to be expected in view of the large additive genetic variation underlying the size and shape traits of both *G. fortis* and *G. scandens* (*18*, *20*, *27*).

A specific prediction of the introgression hypothesis is an increase in variance and skewness in the morphological distributions, beginning in 1983 in *G. fortis* and 1987 in *G. scandens* (*42*). Increases are expected to be greater in *G. scandens* than in *G. fortis*, despite bidirectional gene exchange, because F_1 hybrids and first-generation backcrosses made a proportionately greater numerical contribution to the *G. scandens* samples (Fig. 4, A and B) (*43*). The predicted increases in beak shape variance and skewness are observed in *G. scandens* (Fig. 5, E and F) and are scarcely noticeable in *G. fortis* (Fig. 5, B and C). Beak shape variance in *G. scandens* doubled from 0.430 in 1973 (95% confidence intervals 0.311, 0.636; $n = 62$) to 1.026 (0.674, 1.762; $n = 35$) in 2001, whereas the variance in *G. fortis* beak shape remained stationary: 0.627 (0.493, 0.824; $n = 173$) in 1973, and 0.887 (0.744, 1.316; $n = 114$) in

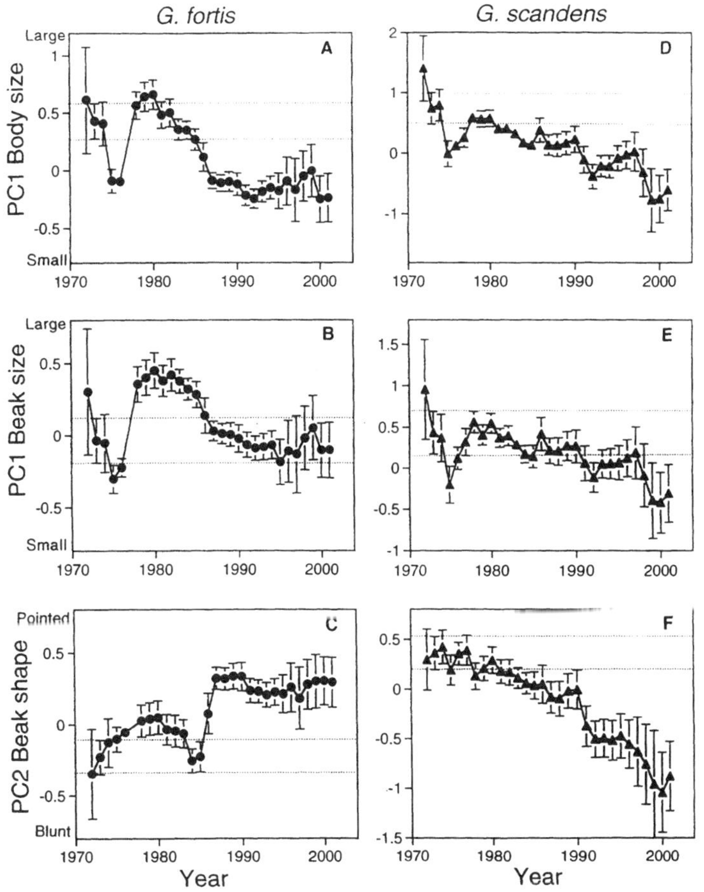

Fig. 1. Morphological trajectories of adult *Geospiza fortis* (**A** to **C**) and *G. scandens* (**D** to **F**). In the absence of change, mean trait values should have remained within the 95% confidence intervals (horizontal broken lines) of the estimates from the 1973 samples (body size: *G. fortis*, $n = 115$, *G. scandens*, $n = 37$; beak traits: *G. fortis*, $n = 173$, *G. scandens*, $n = 62$). Sample sizes varied from 45 (1997) to 976 (1991) for *G. fortis* and from 30 (1999) to 336 (1983) for *G. scandens*. The 1972 sample is composed of the adults (=1 year old) in 1973.

2001. Beak size and body size variances (not shown) are not as well estimated and show no significant variation across the study period (95% confidence intervals broadly overlap). Skewnesses in beak size and body size distributions vary in parallel with beak shape skewness: negatively in *G. scandens*, projecting toward *G. fortis*; and positively in *G. fortis*, projecting toward *G. scandens*.

A second prediction is that variance and skewness will decrease if F_1 hybrids and first-generation backcrosses are deleted from the total samples. Deletions of these two classes of birds do indeed effectively eliminate the increases in skewness and variance in *G. scandens* beak shape, and reduce the degree to which the mean changed (Fig. 5, D to F), because on average F_1 hybrids and first-generation backcrosses are smaller in body size and beak size and have less pointed beaks than the parental *G. scandens* species (*44*). We conclude that introgressive hybridization caused a change in means and other moments of the frequency distributions of *G. scandens* measurements. Selection may have contributed as well but to a minor extent (Fig. 2F). Deletions had no obvious effect on the distributions of *G. fortis* traits (Fig. 5, A to C) (*44*, *45*).

The proportionally greater gene flow from *G. fortis* to *G. scandens* than vice versa has an ecological explanation. Adult sex ratios of *G. scandens* became male biased after 1983 (Fig. 4C) as a result of heavy mortality of the socially subordinate females. High mortality was caused by the decline of their principal dry-season food, *Opuntia* cactus seeds and flowers; rampantly growing vines smothered the bushes (*16*). *G. fortis*, more dependent on small seeds of several other plant species, retained a sex ratio close to 1:1 (Fig. 4C). Thus, when breeding resumed in 1987 after 2 years of drought, competition among females for mates was greater in *G. fortis* than in *G. scandens*. All 23 *G. scandens* females paired with *G. scandens* males, but two of 115 *G. fortis* females paired interspecifically. All their F_1 offspring later bred with *G. scandens* (*43*) because choice of mates is largely determined by a sexual imprinting-like process on paternal song (*42*).

Conclusion. The long-term study of Darwin's finch populations illustrates evolutionary unpredictability on a scale of decades. Mean body size and beak shape of both species at the end of the study could not have been predicted at the beginning. Moreover, sampling at only the beginning and at the end would have missed beak size changes in *G. fortis* in the middle. The temporal pattern of change shows that reversals in the direction of selection do not necessarily return a population to its earlier phenotypic state. Evolution of a population is contingent upon environmental change, which may be highly

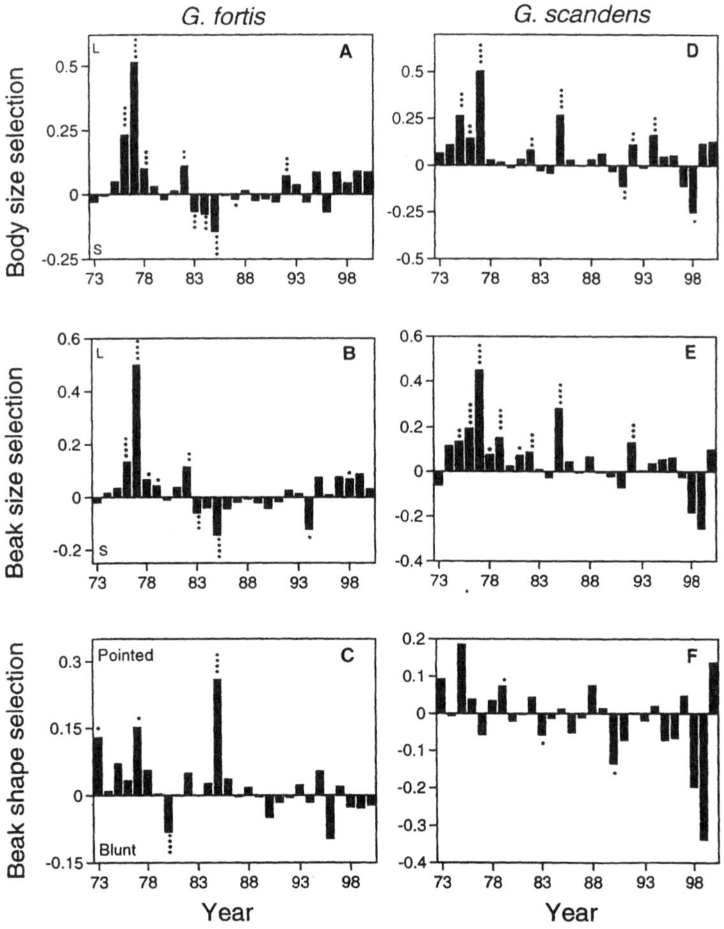

Fig. 2. (**A** to **F**) Standardized selection differentials, calculated for each sample surviving from year *x* to year *x*+*1*. Positive values indicate selection for large size or pointed beaks; L, large; S, small. Significance levels are shown without correction for multiple testing or lack of independence (*29*). Males and females were combined with adults of unknown sex because separate selection analyses of males and females give similar results (*23*). Differentials are temporally autocorrelated to varying degrees; autocorrelation coefficients are 0.416, 0.302, and 0.093 for *G. fortis* body size, beak size, and beak shape, respectively, and 0.171, 0.373, and 0.103 for the same traits in *G. scandens*. *$P < 0.05$; **$P < 0.01$; ***$P < 0.005$; ****$P < 0.001$.

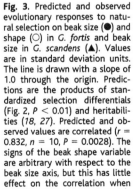

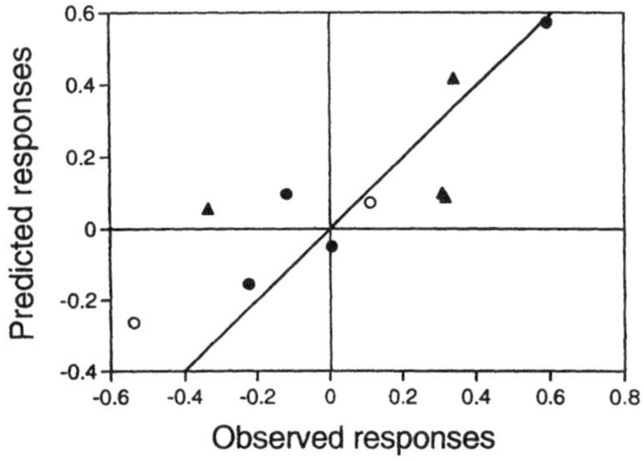

Fig. 3. Predicted and observed evolutionary responses to natural selection on beak size (●) and shape (○) in *G. fortis* and beak size in *G. scandens* (▲). Values are in standard deviation units. The line is drawn with a slope of 1.0 through the origin. Predictions are the products of standardized selection differentials (Fig. 2, $P < 0.01$) and heritabilities (*18*, *27*). Predicted and observed values are correlated ($r = 0.832$, $n = 10$, $P = 0.0028$). The signs of the beak shape variable are arbitrary with respect to the beak size axis, but this has little effect on the correlation when beak shape signs are reversed ($r = 0.781$, $n = 10$, $P = 0.0077$). Data for *G. scandens* after 1986 have not been included because of complications arising from introgressive hybridization (Figs. 4 and 5). Body size has not been included because it is not independent of beak size (*29*).

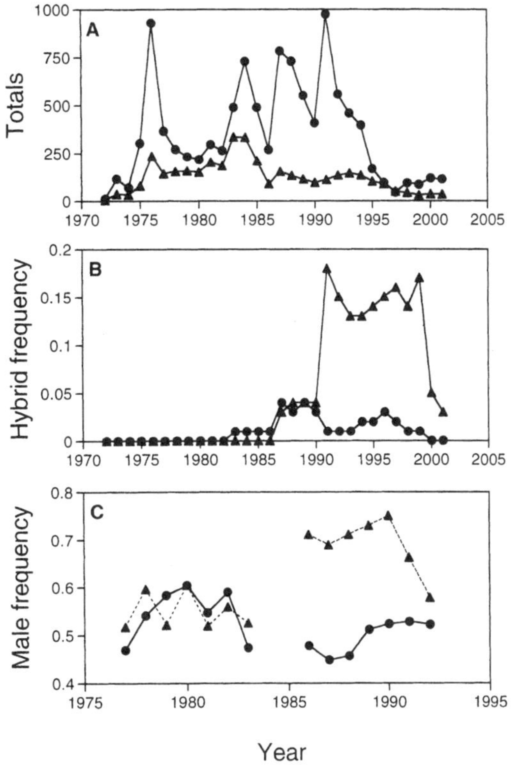

Fig. 4. Total samples of measured birds (**A**), proportional contributions of F_1 hybrids and first-generation backcrosses to the totals (**B**), and minimal frequencies of adult males (**C**) before and after habitat and demographic changes were caused by the El Niño event of 1983; *G. fortis* (●) and *G. scandens* (▲). In (B), the decline in proportion of hybrids in the *G. scandens* samples after 1999 reflects the addition of birds of unknown parents. They have been classified as *G. scandens*, but some were probably unidentified hybrids and backcrosses. In (C), birds of unknown sex in female-like plumage (0 to 10%) have been added to the female samples, and therefore proportions of males are minima. Sexes of all birds were known in some years (e.g., 1989). Data for 1984 and 1985 are not shown because of the large proportion of birds of unknown sex (>15%) produced in 1983 and 1984, comprising both males and females.

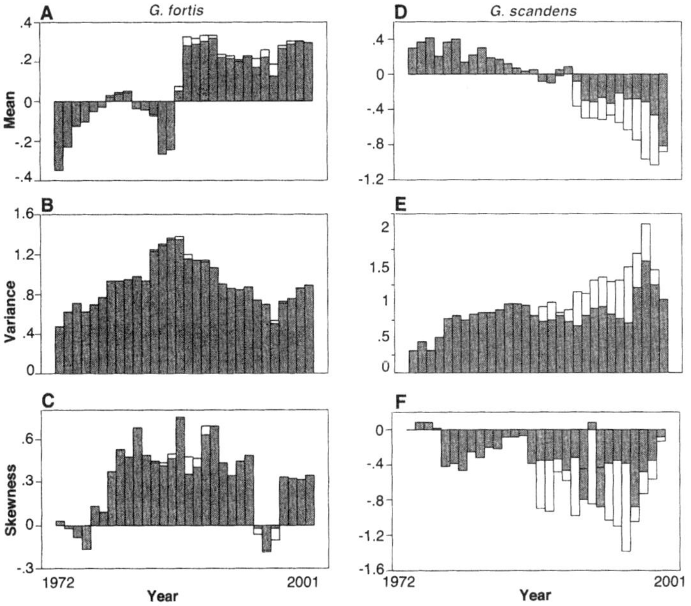

Fig. 5. Effects of hybrids and backcrosses on the mean, variance, and skewness of the beak shape frequency distributions of *G. fortis* (**A** to **C**) and *G. scandens* (**D** to **F**). Removal of F_1 hybrids and first-generation backcrosses (white bars) has a strong effect on *G. scandens* but a minor effect on *G. fortis* parameters (*45*). The proportion of known hybrids declined after 1991; not all hybrids and backcrosses could be identified with pedigree information, because some nestlings were not banded after 1991 (*16*), and therefore some may have been included as *G. fortis* and *G. scandens*. Negative skewness indicates a prolonged tail toward less pointed beaks. Additional removal of *G. fortis* × *G. fuliginosa* F_1 hybrids and backcrosses has scarcely detectable effects in these diagrams.

irregular, as well as on its demography and genetic architecture (*33*, *46*).

The study also illustrates how the value of long-term studies increases with time. Not only is regular monitoring at short intervals desirable, but sampling for many years is to be recommended, especially for long-lived organisms like vertebrates and perennial plants. Yet evolutionary studies are rarely pursued in the field for as many as 10 years (*33*). If we had stopped sampling after 10 years, our conclusions would have been different because at that time the only difference from the starting point was in beak size of *G. fortis*. By persisting beyond then, we witnessed a natural-selection event that affected beak shape in *G. fortis*, documented interbreeding and morphological effects of introgression on *G. scandens*, and gained a better quantitative estimate of the frequency of evolutionary events.

Unlike the Marsh Fritillary study (*4*), we did not witness a release of genetic variation following a population crash. The evolutionary dynamics were different. Natural selection occurred frequently in our study, occasionally strongly, unidirectionally in one species and oscillating in direction in the other as a result of their dependence on different food supplies (*16*). Introgressive hybridization, a phenomenon whose importance has been relatively underappreciated until recently, except in plants (*47*), had different effects on the two species for demographic reasons. Hybridization and selection are often connected through the selective disadvantage experienced by hybrids and backcrosses (*13*, *40*, *47*, *48*). In the present study they appear to have been connected synergistically in the sense that interbreeding may have been facilitated in part by selection for more pointed, *G. scandens*–like, beaks in the *G. fortis* population in the mid-1980s. Choice of mates is partly determined by imprinting on parental beak morphology, as well as on paternal song (*41*). The principal causes of selection have been identified as changes in food supply (*23–26*) mediated in large part by droughts. The ultimate cause of repeated natural selection and introgressive hybridization may have been a change in the seasonal movement of water masses in the eastern subtropical and tropical Pacific (*49*, *50*), triggering altered climatic patterns, including the intensification of El Niño and La Niña cycles.

Regardless of the precise chain of causality, field studies such as ours, in conjunction with multigenerational studies of microorganisms in the laboratory (*51*, *52*) and experimental studies of selection in the field (*53–55*), provide an improved basis for extrapolating from microevolution to patterns of macroevolution; in the present case, from evolutionary dynamics of populations on the scale of decades to speciation and further adaptive radiation on the scale of hundreds of thousands of years (*56*). In conclusion, the long-term unpredictability of evolutionary change that arises from unpredictable ecological change, together with the need to strengthen generalizations about the frequency and importance of selection and hybridization, are reasons for encouraging additional, continuous, long-term studies of evolution in nature.

References and Notes

1. G. Likens, Ed., *Long-term Studies in Ecology: Approaches and Alternatives* (Springer-Verlag, New York, 1989).
2. J. H. Brown, T. G. Whitham, C. K. M. Ernest, C. A. Gehring, *Science* **293**, 643 (2001).
3. P. Kareiva, J. G. Kingsolver, R. B. Huey, Eds., *Biotic Interactions and Global Change* (Sinauer, Sunderland, MA, 1993).
4. H. D. Ford, E. B. Ford, *Trans. Entomol. Soc. London* **78**, 345 (1930).
5. P. R. Ehrlich, L. G. Mason, *Evolution* **20**, 165 (1965).
6. D. A. Jones, *Trends Ecol. Evol.* **4**, 298 (1989).
7. H. L. Carson, in *Population Biology and Evolution*, R. C. Lewontin, Ed. (Syracuse Univ. Press, Syracuse, NY, 1968), pp. 123–137.
8. W. W. Anderson *et al.*, *Proc. Natl. Acad. Sci. U.S.A.* **88**, 10367 (1991).
9. R. H. Cowie, J. S. Jones, *Biol. J. Linn. Soc.* **65**, 233 (1998).
10. L. F. Keller, *Evolution* **52**, 240 (1998).
11. M. E. N. Majerus, *Melanisms: Evolution in Action.* (Oxford Univ. Press, Oxford, UK, 1998).
12. M. E. Visser, A. J. van Noordwijk, J. M. Tinbergen, C. M. Lessells, *Proc. R. Soc. London B* **265**, 1867 (1998).
13. S. E. Carney, K. A. Gardner, L. H. Rieseberg, *Evolution* **54**, 462 (2000).
14. O. Halkka, L. Halkka, K. Roukka, *Biol. J. Linn. Soc.* **74**, 571 (2001).
15. L. E. B. Kruuk, J. Merilä, B. C. Sheldon, *Am. Nat.* **158**, 557 (2001).
16. P. R. Grant, B. R. Grant, in *Long-term Studies of Vertebrate Communities*, M. L. Cody, J. A. Smallwood, Eds. (Academic Press, San Diego, CA, 1996), pp. 343–390.
17. For each species, two principal-components analyses were performed on the correlation matrix with untransformed data from all birds [males, females, and birds of unknown sex of both species; see (*18*)]. Mass (weight), wing length, and tarsus length were used in the first analysis, and beak length, depth, and width (*19*) were used in the second analysis. For *G. fortis* (n = 3204), PC1 in analysis 1 is interpreted as a body size factor (69.2% variance explained), because factor loadings were high and of the same sign: 0.856 (mass), 0.839 (wing), and 0.800 (tarsus). PC1 in analysis 2 is interpreted as a beak size factor (85.8% variance explained); factor loadings are 0.890 (length), 0.951 (depth), and 0.937 (width). PC2 is a beak shape factor (10.1% additional variance explained) with factor loadings of different sign: 0.455 (length), -0.167 (depth), and -0.262 (width). Factor loadings and their interpretation were similar in analyses of *G. scandens* (n = 1037). Percentage variance was 67.5 for PC1 in the first analysis, 69.8 for PC1 in the second analysis, and 23.2 for PC2 in the second analysis. Separate analyses of males and females gave similar results.
18. L. F. Keller, P. R. Grant, B. R. Grant, K. Petren, *Heredity* **87**, 325 (2001).
19. The six measured traits (*17*) have high repeatabilities (*20*). Nine specimens of *G. fortis* in the Charles Darwin Research Station museum, Galápagos, were measured by P.R.G. in 1975, 1976, and again in 2001 to check for possible changes in methods of linear measurement. No heterogeneity among years was found for any of the five traits [analyses of variance (ANOVAs), all $P > 0.67$], and no difference between pairs of years was found in any trait (Fisher's protected least significant difference post hoc tests, all $P > 0.38$).
20. P. R. Grant, B. R. Grant, *Evolution* **48**, 297 (1994).
21. The first samples in 1973 (221 *G. fortis* and 72 *G. scandens*) were compared by ANOVAs with the last samples in 2001 (114 *G. fortis* and 35 *G. scandens*). The data were trimmed to 2.5 SD on either side of the mean by removing one to three individuals from the samples of each species. This corrected for skewness and unequal variances. Sex was included in two-factor ANOVAs because males are generally larger than females. Mean body size was significantly smaller in *G. fortis* ($F_{1,132}$ = 7.773, P = 0.0061) and in *G. scandens* ($F_{1,50}$ = 11.272, $P < 0.0001$) in 2001 than in 1973. There was a significant effect of sex in each species ($P < 0.002$) but no sex-by-year interaction ($P > 0.1$). Mean beak size did not differ between years in either *G. fortis* ($F_{1,166}$ = 0.004, P = 0.9480) or *G. scandens* ($F_{1,50}$ = 3.108, P = 0.0840); sex effects were significant in both species ($P < 0.007$), but there were no sex-by-year interactions ($P > 0.1$). Beak shape differed between years in both species. For *G. scandens* there was a strong year effect ($F_{1,72}$ = 17.168, $P < 0.0001$), a weak sex effect ($F_{1,72}$ = 5.943, P = 0.0172), and no interaction. The *G. fortis* sexes do not differ in beak shape (P = 0.9715), and therefore a one-factor ANOVA was performed with adult males, females, and birds of unknown sex. It demonstrated a strong difference between years ($F_{1,287}$ = 30.246, $P < 0.0001$).
22. P. R. Grant, B. R. Grant, K. Petren, *Genetica* **112–113**, 359 (2001).
23. P. T. Boag, P. R. Grant, *Science* **214**, 82 (1981).
24. T. D. Price, P. R. Grant, P. T. Boag, H. L. Gibbs, *Nature* **309**, 787 (1984).
25. H. L. Gibbs, P. R. Grant, *Nature* **327**, 511 (1987).
26. P. R. Grant, B. R. Grant, *Evolution* **49**, 241 (1995).
27. ———, in *Adaptive Genetic Variation in the Wild*, T. A. Mousseau, B. Sinervo, J. A. Endler, Eds. (Oxford Univ. Press, New York, 2000), pp. 3–40.
28. Directional selection differentials were calculated as the difference in trait means before and after selection, then standardized in each case by dividing the difference by the standard deviation of the sample before selection (*32*). After checking for normality, differences between survivors and nonsurvivors were tested by two-tailed t tests. Significant skewness (*57*) was eliminated by deleting outliers (more than 2.5 SD beyond the mean), with minor effects on probability values. Equality of variances was tested with F tests. Only beak traits displayed unequal variances. Correcting for skewness reduced the number of significant variance inequalities from three to two in *G. fortis* and from nine to four in *G. scandens*. Means in the remaining six cases were tested with a median test (*57*); five were nonsignificant ($P > 0.1$). Selection was diversifying in three of the five cases. Selection differentials measure the combined direct effects of selection on a trait and indirect effects of selection on correlated traits. Direct effects on the individual measured traits have been assessed by partial regression (selection gradient analysis) in some cases (*24–26*), yielding similar interpretations; body size, beak size, and beak shape are independently selected traits.
29. Body size and beak size are strongly correlated traits in *G. fortis* (r^2 = 0.464 to 0.650; median 0.540) and *G. scandens* (r^2 = 0.393 to 0.694; median 0.553) in all 30 years at $P < 0.0001$. Beak shape is occasionally and weakly correlated with body size in *G. fortis* (r^2 = 0.001 to 0.082; median 0.015) and *G. scandens* (r^2 = 0.001 to 0.179; median 0.026).
30. *G. scandens* body size was selected in 8 out of 22 years in which a minimum sample size of 75 was met, and beak size was selected in 7 of those years. Frequency of selection in *G. fortis* was 8 out of 25 years for body size, 5 out of 26 years for beak size, and 2 out of 26 years for beak shape. These are minimal frequencies; selection below the level of statistical detectability may have occurred in other years.
31. P. R. Grant, B. R. Grant, *Ecology* **73**, 766 (1992).
32. J. A. Endler, *Natural Selection in the Wild* (Princeton Univ. Press, Princeton, NJ, 1986).
33. J. G. Kingsolver *et al.*, *Am. Nat.* **157**, 245 (2001).
34. H. E. Hoekstra *et al.*, *Proc. Natl. Acad. Sci. U.S.A.* **98**, 9157 (2001).
35. B. R. Grant, P. R. Grant, *Evolutionary Dynamics of a Natural Population. The Large Cactus Finch of the Galápagos* (Univ. of Chicago Press, Chicago, IL, 1989).
36. C. Barbraud, *J. Evol. Biol.* **13**, 81 (2000).
37. J. Merilä, L. E. B. Kruuk, B. C. Sheldon, *Nature* **412**, 76 (2001).
38. T. Price, *Evolution* **38**, 327 (1984).
39. P. R. Grant, B. R. Grant, *Proc. R. Soc. London B* **267**, 131 (2000).
40. ———, *Philos. Trans. R. Soc. London B* **340**, 127 (1993).
41. ———, *Science* **256**, 193 (1992).
42. ———, in *Endless Forms: Species and Speciation*, D. J. Howard, S. H. Berlocher, Eds. (Oxford Univ. Press, New York, 1998), pp. 404–422.
43. Eight F_1 hybrids (both sexes), all from families with a *G. scandens* father that sang a *G. scandens* song, backcrossed to *G. scandens* and produced 16 measured offspring. Six F_1 hybrids (both sexes), all from families with a *G. scandens* father that sang a *G. fortis* song, backcrossed to *G. fortis* and produced 26 measured offspring. Other F_1 hybrids and backcrosses were produced but not captured and measured.
44. The 1991 sample of *G. scandens* (n = 112), the last with complete pedigree information, included 20 hybrids (seven F_1 hybrids and 13 first-generation backcrosses). The hybrids were smaller in body size (ANOVA, $F_{1,109}$ = 38.522, $P < 0.0001$) and beak size ($F_{1,109}$ = 34.583, $P < 0.0001$) and had less pointed beaks ($F_{1,109}$ = 55.116, $P < 0.0001$) than the rest. In the same year, one F_1 hybrid and nine first-generation backcrosses in the sample of 510 *G. fortis* were larger in body size than the rest ($F_{1,507}$ = 21.817, $P < 0.0001$) and had more pointed beaks ($F_{1,507}$ = 8.323, P = 0.0041) but did not differ in beak size ($F_{1,507}$ = 1.098, P = 0.2953).
45. The fractional contribution to beak shape variance made by hybrids was on average 0.265 ± 0.140 SD for *G. scandens* (maximum 0.528; n = 15 years) and 0.024 ± 0.018 for *G. fortis* (maximum 0.064; n = 17 years).
46. H. Teotónio, M. R. Rose, *Evolution* **55**, 653 (2001).
47. M. L. Arnold, *Natural Hybridization and Evolution* (Oxford Univ. Press, New York, 1997).
48. T. Veen *et al.*, *Nature* **411**, 45 (2001).
49. T. P. Guilderson, D. P. Schrag, *Science* **281**, 240 (1998).
50. R. H. Zhang, L. M. Rothstein, A. J. Busalacchi, *Nature* **391**, 879 (1998).
51. R. E. Lenski, M. Travisiano, *Proc. Natl. Acad. Sci. U.S.A.* **91**, 6608 (1994).
52. P. B. Rainey, M. Travisiano, *Nature* **394**, 69 (1998).
53. D. N. Reznick, F. H. Shaw, F. H. Rodd, R. G. Shaw, *Science* **275**, 1934 (1997).
54. B. Sinervo, E. Svensson, T. Comendant, *Nature* **406**, 985 (2000).
55. J. R. Etterson, R. G. Shaw, *Science* **294**, 151 (2001).
56. P. R. Grant, *Ecology and Evolution of Darwin's Finches* (Princeton Univ. Press, Princeton, NJ, 1999).
57. R. R. Sokal, F. J. Rohlf, *Biometry: The Principles and Practice of Statistics in Biological Research* (Freeman, New York, ed. 3, 1996).

58. We thank I. Abbott, L. Abbott, P. T. Boag, H. L. Gibbs, L. F. Keller, K. Petren, T. D. Price, J. N. M. Smith, and many field assistants acknowledged earlier (*31*); the Galápagos National Parks Service and Charles Darwin Research Station for permits and logistical support; M. Hau, K. Petren, D. L. Stern, and M. Wikelski for comments on the manuscript; and McGill University, Natural Sciences and Engineering Research Council (Canada, 1973 to 1978), and NSF (1978 to 2002) for the long-term financial support that has made this study possible.

28 January 2002; accepted 20 March 2002

Carroll, S. B. 2003. Genetics and the making of *Homo sapiens*. *Nature* 422:849–857.

A draft of the DNA sequence of the chimpanzee genome was published on 1 September 2005. In the genome regions shared by humans and chimpanzees, the two species are 98 to 99 percent identical: One or two percent difference seems very little, but it amounts to a difference of 30–60 million DNA letters out of the nearly 3 billion in each genome. Twenty-nine percent of the enzymes and other proteins encoded by the genes are identical in both species. Out of the one hundred to several hundred amino acids that make up each protein, the 71 percent of nonidentical proteins differ between humans and chimps by only two amino acids on average.

If one takes into account DNA segments found in one species but not the other, the two genomes are about 96 percent identical, rather than 98 to 99 percent as for the sequences shared by both species. About 3 percent of the DNA, or some 90 million DNA letters, has been inserted or deleted since humans and chimps initiated their separate evolutionary ways, seven million years ago. Most of this DNA does not contain genes coding for proteins.

Comparison of the two genomes provides insights into the rate of evolution of particular genes. Genes active in the brain have changed more in the human lineage than in the chimp lineage. Also significant is that the fastest-evolving human genes are those coding for *transcription factors*, switch proteins that control the expression of other genes by determining when they are turned on and off. There are 585 genes, including genes involved in resistance to malaria and tuberculosis, that have been identified as evolving faster in humans than in chimps. (Malaria is a severe disease for humans but not for chimps.)

Genes located on the Y chromosome have been much more conserved in the human than in the chimpanzee lineage, in which several genes have incorporated disabling mutations that make the genes nonfunctional. Several regions of the human genome contain genes that have rapidly evolved within the past 250,000 years. One of these is the *FOXP2* gene, involved in speech. Other regions that show higher rates of evolution in humans than in chimpanzees and other animals include 49 segments, dubbed human accelerated regions, or ARs. The greatest observed difference occurs in *HAR1F*, an RNA gene expressed specifically in neurons of the neocortex between weeks 7 and 19 of the developing embryo (Pollard et al. 2006).

The features that distinguish us as humans begin early in development, well before birth, as the linear information encoded in the genome gradually becomes expressed into a four-dimensional individual, an individual who changes as time goes by. In an important sense, the most distinctive human features are those expressed in the brain, those that account for the human mind and for human identity.

In this review article, the developmental geneticist Sean Carroll reviews genomic findings on *Homo sapiens* and revisits Shakespeare's question, "What is a man . . . ?"

RELATED READING

Ayala, F. J. 2010. Am I a monkey? Johns Hopkins University Press, Baltimore, Maryland, USA.

Cela-Conde, C. J., and F. J. Ayala. 2007. Human evolution: trails from the past. Oxford University Press, Oxford, England.

Pollard, K. S., S. R. Salama, N. Lambert, M.-A. Lambot, S. Coppens, J. S. Pedersen, S. Katzman, B. King, C. Onodera, A. Siepel, A. D. Kern, C. Dehay, H. Igel, M. Ares Jr., P. Vanderhaeghen, and D. Haussler. 2006. An RNA gene expressed during cortical development evolved rapidly in humans. Nature 443:167–172.

Wood, B., editor. 2011. Encyclopedia of human evolution, 2 vols. Wiley-Blackwell, Oxford, England.

Genetics and the making of *Homo sapiens*

Sean B. Carroll

Howard Hughes Medical Institute and Laboratory of Molecular Biology, University of Wisconsin, 1525 Linden Drive, Madison, Wisconsin 53706, USA

Understanding the genetic basis of the physical and behavioural traits that distinguish humans from other primates presents one of the great new challenges in biology. Of the millions of base-pair differences between humans and chimpanzees, which particular changes contributed to the evolution of human features after the separation of the *Pan* and *Homo* lineages 5–7 million years ago? How can we identify the 'smoking guns' of human genetic evolution from neutral ticks of the molecular evolutionary clock? The magnitude and rate of morphological evolution in hominids suggests that many independent and incremental developmental changes have occurred that, on the basis of recent findings in model animals, are expected to be polygenic and regulatory in nature. Comparative genomics, population genetics, gene-expression analyses and medical genetics have begun to make complementary inroads into the complex genetic architecture of human evolution.

What is a man,
If his chief good and market of his time
Be but to sleep and feed? a beast, no more.
Sure, he that made us with such large discourse,
Looking before and after, gave us not
That capability and god-like reason
To fust in us unused.
W. Shakespeare, *Hamlet* IV:iv

What makes modern humans different from the great apes and earlier hominids? In what hominids and when in evolution did important physical traits and behaviours appear? Where in our larger brains do human-specific capabilities reside? These have been long-standing questions in palaeoanthropology and comparative anatomy, since the discovery of Neanderthal skulls and the first studies of great apes in the nineteenth century. Now, the mystery of human origins is expanding beyond the description and history of human traits, towards the genetic mechanisms underlying their formation and evolution. With the characterization of the human genome, and that of our chimpanzee cousin on the way, the quest to discover the genetic basis of the physical and behavioural traits that distinguish us from other apes is rapidly gaining momentum.

Genomes diverge as a function of time, and most of the sequence changes that accumulate between any two related species are selectively neutral or nearly neutral in that they do not contribute to functional or phenotypic differences. The great challenge is to elucidate the number, identity and functions of genes, and the specific changes within them, that have shaped the evolution of traits. This has been accomplished for only a few traits in model systems, so it is a difficult task for human features about which we know little, and an enormous prospect to consider the whole arc of human evolution.

In this article, I will examine both the physical and genetic scope of human evolution and the approaches being used to try to understand it. I will first review the current state of our understanding of human evolution from the viewpoint of the fossil record, comparative anatomy and development. These disciplines point to many key traits to be considered, and define the magnitude and nature of evolutionary change in the human lineage. I will preview the picture of human evolution that we might expect to emerge in view of our current knowledge of the genetic architecture of trait evolution in model systems. I will then examine the variety of methods being used on a genome-wide scale and at the level of individual loci to identify genes that may have contributed to the evolution of key traits. I will discuss some of the crucial methodological challenges in distinguishing causative from potentially large numbers of candidate loci. Finally, I will address some of the disciplines in which future advances are likely to have a central role in furthering our knowledge of the genetic and developmental basis of human evolution.

Hominid evolution

The hominid tree

To approach the origins of human traits at the genetic level, it is essential to have as a framework a history of our lineage and the characters that distinguish it. It is inadequate and misleading to consider just the comparative anatomy and development (or genomes) of extant humans, chimpanzees and other apes, and then to attempt to infer how existing differences might be encoded and realized. Each of these species has an independent lineage that reaches back as far or further than hominins ('hominins' refers to humans and our evolutionary ancestors back to the separation of the human and ape lineages; 'hominids' to humans and the African apes) (Fig. 1). The evolution of 'modern' traits was not a linear, additive process, and ideas about the tempo, pattern and magnitude of change can only be tested through fossil evidence, which is always subject to revision by new finds. The fossil record continues to shape views of three crucial issues in hominid evolution. First, what distinguishes hominins from the apes? Second, what distinguishes modern humans (*Homo sapiens*) from earlier hominins? And third, what was the nature of the last common ancestor of hominins and the *Pan* lineage?

Inferences about the chronological order and magnitude of the evolution of hominin characters depend on a model of the hominin evolutionary tree. This has always been a contentious issue and remains unsettled, in part because of the pace of exciting new fossil finds over the past two decades. An emerging view portrays hominin evolution as a series of adaptive radiations in which many different branches of the hominin lineage were formed, but died out[1,2] (Fig. 1). One prediction of this model is that various anatomical features would be found in different combinations in hominins through their independent acquisition, modification and loss in different species. For example, the recent, stunning discovery of a 6–7 million year (Myr)-old fossil cranium of *Sahelanthropus tchadensis* that had a chimpanzee-sized brain but hominin-like facial and dental features[3] is the sort of morphology that would be consistent with a radiation of ape-like animals from which the stem of the hominin lineage emerged (although the interpretation of this fossil's affinities is controversial[4]).

What makes a human?

Evolutionary trends in fossil hominins. Ideally, if every possible fossil human and ape species were identified a nd m any fairly complete specimens were available, one could reconstruct the emergence of human and chimpanzee features through time. But that is not the case for most lineages; in fact, there are no identified archaic chimpanzee fossils. We must make do with a partial and often confusing picture of human trait evolution. From a rather extensive list of qualitative and quantitative features that distinguish humans from other apes[5,6] (Box 1), our large brain, bipedalism, small canine teeth, language and advanced tool-making capabilities[7,8] have been the focus of palaeoanthropology. The major physical traits are generally not singular elements, but entail concomitant changes in skeletal features involved in locomotion (for example, in the vertebral column, pelvis and feet, and in limb proportions), grasping (hand morphology and an opposable, elongated thumb) and chewing of food (the mandible and dentition), as well as life-history traits such as lifespan. It is fortunate that most of the skeletal features lend themselves to detailed quantitative studies of the fossil record.

Some trends in the evolution of body size, brain size and dentition are evident within the hominins. More recent species are characterized by larger body mass, relatively larger brains, longer legs relative to the trunk and small teeth, whereas earlier species had, in general, smaller brains and bodies (Table 1), shorter legs relative to the trunk and large teeth[7,9,10]. I highlight these traits to focus attention on the magnitude and timescale of character evolution, and on the (increasing) number of generally recognized hominin taxa. Whatever the branching pattern in the hominin tree, sub-

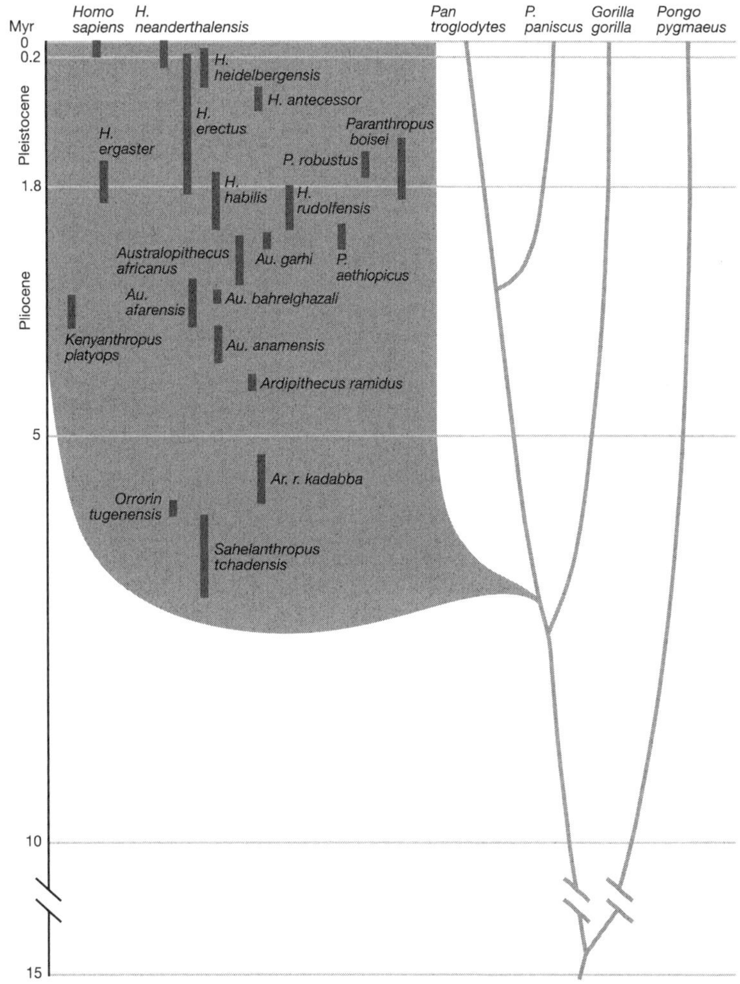

Figure 1 The timescale and phylogeny of hominids. Ape relationships are shown in grey for the chimpanzee (*Pan troglodytes*), bonobo (*P. paniscus*), gorilla and orangutan (*Pongo pygmaeus*). The approximate times of divergence are derived from molecular data (summarized in ref. 89). The phylogenetic relationships among hominins (shaded) are uncertain. The solid red bars denote the time span of the fossil species and/or the uncertainty of fossil ages. The identity of the last common ancestor of chimpanzees and humans (LCA) is not known. Note that the estimated age of *Sahelanthropus tchadensis* predates molecular estimates of the time of the chimpanzee–human divergence. This species could pre- or postdate the LCA. Also note that *Homo sapiens* represent only the last 3% of the time span of hominin evolution. Hominin distributions and nomenclature are based primarily on refs 1, 90.

stantial relative changes occurred over an extended time span and a significant number of speciation events. There was a marked increase in absolute brain size by the Early Pleistocene and again in the Middle Pleistocene, with a long interval of perhaps 1 Myr during which brain size did not change significantly[9,11] (Table 1). With regard to modern *H. sapiens*, it is interesting to note that body and brain size were even greater in *H. neanderthalensis*; there is no obvious physical explanation for the success of *H. sapiens* and the demise of *H. neanderthalensis*[11].

A beautiful mind: insights from comparative neuroanatomy. The relative increase in brain size, although marked, is only a crude index of a potential increase in cognitive abilities. Because it has long been appreciated that there are discrete areas of the brain that process various cognitive, motor and sensory functions, comparative neuroanatomists have sought to identify areas that might be central to the evolution of human capacities. There is a long-standing notion that the frontal cortex (involved in planning, organization, personality, behaviour and other 'higher' cognitive functions) is disproportionately larger in humans[12], but this now seems not to be the case (it is larger, but not disproportionately so[13]). As gross anatomical differences do not account for cognitive capabilities, relative differences in the size, cellular composition, detailed cytoarchitecture and/or connectivity of human and great ape brain areas have been sought to explain the emergence of human capabilities.

Of paramount interest is the production and understanding of speech. Two areas in particular have commanded the greatest attention. One is Broca's area in the frontal lobe of the neocortex (Fig. 2). This region is larger in the left hemisphere of the brain than in the right, an asymmetry that has been correlated with language ability. From magnetic resonance images of chimpanzees, bonobos and gorillas, a similar left–right asymmetry has been found in these great apes[14] (Fig. 2). This indicates that the neuroanatomical substrate of left-hemisphere dominance in speech production preceded the origin of hominins. The left hemisphere also usually controls right-handedness, so it is interesting to note that in captive apes, manual gestures are right-hand biased, and this bias is increased when vocalization is combined with gesturing[15], indicating a left-hemisphere-controlled communication process.

A second area of interest is Wernicke's posterior receptive language area in the temporal lobe (Fig. 2). A site within this area, the planum temporale, is implicated in human communication (both spoken and gestural) and musical talent, and also shows a left-hemisphere dominance. In most humans, the Sylvian fissure associated with the left planum temporale extends more posteriorly. Evidence for this asymmetry has been found in fossil endocasts in *H. habilis*, *H. erectus* and *H. neanderthalensis*[16]. More importantly, an asymmetrical planum temporale pattern has recently been demonstrated in chimpanzees[17,18] (Fig. 2).

Several hypotheses have been forwarded to explain the presence of these two human-communication-associated neuroanatomical landmarks in great apes[14,17]. Although it is possible that they arose and acquired their functions independently in each lineage, the most parsimonious explanation is that the common ancestor of great apes and humans had asymmetrical centres that were involved in communication, and that these structures underwent independent evolutionary modifications in chimpanzees and hominins. If this is the case, then the challenge to comparative neuroanatomy is to identify more subtle differences in suborganization (that is, 'microanatomy') that affect the interconnections of cortical regions, in local circuitry and/or cytoarchitecture[19,20] that might be unique to human brains. Recently, it has been found that the dimensions of the vertical columns of neurons in the cortex, known as 'minicolumns', differ between humans and chimpanzees in the planum temporale[21]. In addition, area 10 of the prefrontal cortex, which is involved in higher cognitive functions, has been shown to be enlarged and specialized in humans relative to apes[22]. These observations suggest that human capacities are more a product of quantitative changes in specialized areas than of neuroanatomical novelties.

Development of hominid features

The morphological differences between modern humans, earlier hominins and the great apes are, of course, the product of changes during development. Comparative studies of human and chimpanzee skull development[23], remarkably detailed examinations of Neanderthal craniofacial ontogeny[22,24–26], and early hominin dentition formation[27] have yielded crucial insights into a variety of developmental shifts that underlie modern human cranial size and morphology.

One of the long-appreciated, fundamental differences in chimpanzee and human development is the relative rate of skull growth and maturation. Human neonates have less mature skulls in terms of shape in comparison to young chimpanzees, but much larger skulls (and brains)[23,28]. These are classically described as heterochronic changes, which produce neotenic features in which maturation is retarded, size increases and shape resembles juvenile forms of ancestors[29]. Chimpanzee and human skulls eventually grow to the same size, reflecting further relative shifts in juvenile and adolescent growth periods, and marked differences in face size and cerebral volume. Importantly, all of the skeletal changes associated with bipedalism are structural innovations independent of neoteny. These observations suggest that the human brain is not a product of simple shifts in growth relationships, but of multiple, independent and superimposed modifications.

Box 1
Selected traits that distinguish humans from other apes[5–7]

- Body shape and thorax
- Cranial properties (brain case and face)
- Relative brain size
- Relative limb length
- Long ontogeny and lifespan
- Small canine teeth
- Skull balanced upright on vertebral column
- Reduced hair cover
- Elongated thumb and shortened fingers
- Dimensions of the pelvis
- Presence of a chin
- S-shaped spine
- Language
- Advanced tool making
- Brain topology

Table 1 **Evolution of brain and body size in hominids**

Species	Estimated age[1,3,7,86–88] (Myr ago)	Body size[7] (kg)	Brain size[3,7] (cm^3)
Homo sapiens	0–0.2	53	1,355
H. neanderthalensis	0.03–0.25	76	1,512
H. heidelbergensis	0.3–1	62	1,198
H. erectus	0.2–1.9	57	1,016
H. ergaster	1.5–1.9	58	854
H. rudolfensis	1.8–2.4	-	752
H. habilis	1.6–2.3	34	552
Paranthropus boisei	1.2–2.2	44	510
Australopithecus africanus	2.6–3	36	457
Au. afarensis	3–3.6	-	-
Au. anamensis	3.5–4.1	-	-
Ardipithecis ramidus kadabba	5.2–5.8	-	-
Sahelanthropus tchadensis	6–7	-	~320–380

This list does not include all recognized species.

With respect to more recent hominin species, modern human craniofacial form appears to have been shaped by changes in elements that influence the spatial position of the face, neurocranium and cranial base. Modern humans are marked by greater roundedness of the cranial vault and facial retraction (the anteroposterior position of the face relative to the cranial base and neurocranium)[25]. Comparisons with Neanderthal skull development inferred from fossils of different aged individuals suggest that the characteristic cranial differences between Neanderthal and modern human features arise early in ontogeny[24,26].

Our prolonged childhood, delayed sexual maturation and long lifespan are life-history traits that shape important aspects of human society. Insights into the evolution of these development shifts in evolution are offered by detailed comparative analysis of dental development, which is correlated with stages of primate growth and development. Rates of enamel formation in fossil hominins suggest that tooth-formation times were shorter in australopiths and early members of the genus *Homo* than they are in modern humans[27]. This indicates that the modern pattern of dental formation and correlated developmental traits appeared late in human evolution. When considered in the context of other traits, such as brain size and body proportions, a mosaic pattern of evolution emerges with different traits appearing at different times and perhaps in different combinations in hominin history[30].

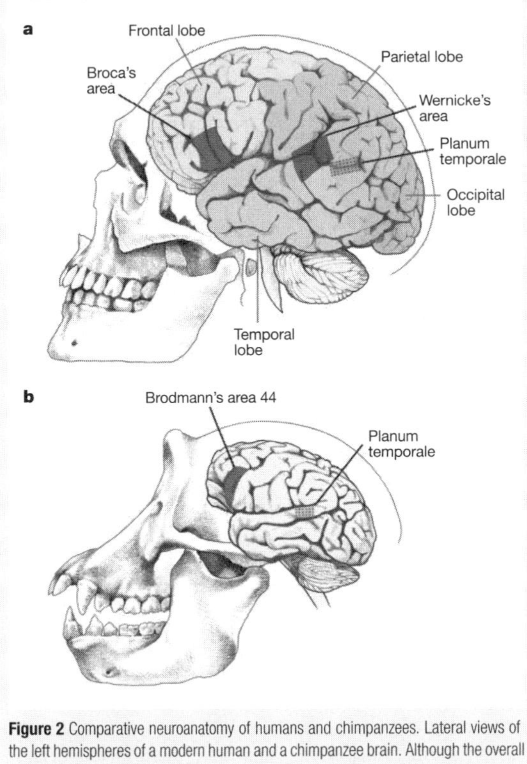

Figure 2 Comparative neuroanatomy of humans and chimpanzees. Lateral views of the left hemispheres of a modern human and a chimpanzee brain. Although the overall skull sizes are roughly comparable, the human cranial capacity and brain are much larger. **a**, Two areas of the human brain that are associated with communication are shown: Broca's area in the frontal lobe and Wernicke's area, which overlaps the posterior temporal lobe and parts of the parietal lobe. In the left hemisphere, Broca's area is larger, as is the planum temporale, which lies below the surface in Wernicke's area. **b**, These asymmetries have been found in corresponding regions of chimpanzee brains[15,17], suggesting that the areas in humans might be elaborations of a pre-existing communication centre in a common ancestor of apes and humans.*

Was human evolution special?

The magnitude, rate and pattern of change during hominin evolution, inferred from the fossil record, comparative neuroanatomy and embryology, provide the essential foundation for approaching the genetics of human evolution. From the studies discussed above, five key points emerge that have a bearing on attempts to reconstruct the genetic events that underlie the origin and modification of human traits.

First, trait evolution was nonlinear. The ~1,000-cm^3 increase in brain size over 5–7 Myr did not occur at the same relative rate in hominin phylogeny: it was static at times, faster in some intervals, and reversed slightly more recently. Second, most trait evolution can be characterized as simple quantitative changes (that is, traits are continuous). Third, evolutionary rates were not at all exceptional with respect to mammalian evolution. For example, fossil horse lineages in the late Pliocene–Pleistocene show similar rates of body-size and other character changes as do those of hominids[31]. Fourth, much evolutionary change preceded the origin of the *Homo* genus and of *H. sapiens*: the history of our species represents just the last 3% of the time span of hominin evolution (Fig. 1). And fifth, many characters are present not only in humans, but also in apes. This suggests that modification of existing structures and developmental pathways, rather than the invention of new features, underlies much of human evolution.

These observations indicate that morphological evolution in hominins was not special, but the product of genetic and developmental changes typical of other mammals and animals.

Genetics of human evolution

Genetic architecture of trait evolution

Given the dimensions of hominin evolution, inferred from the fossil record and comparative anatomy, what can we expect in terms of the genetic complexity underlying trait evolution? For example, there is a long-standing tendency for events that are perceived to be relatively 'rapid' in the fossil record to be ascribed to perhaps one or a few radical mutations[32], including recent human evolution[33,34]. Could the relative increase in brain size over 5 Myr, or its expanded cognitive function, be due to just one or a few genetic changes? The best (and, at present, the only available) guides for this question are detailed genetic studies in model organisms, which have achieved success in dissecting the genetics of complex trait formation, variation and evolution. Six essential general concepts have been established in model systems that pertain to the potential genetic architecture of human trait evolution:

(1) Variation in continuous, quantitative traits is usually polygenic. Studies of variation in model species[35–38] reveal that many genes of small effect, and sometimes one or a few genes of large effect, control trait parameters. In humans, a study of variation in 20 anthropometric variables in two different ethnic human populations suggested that more than 50% of variation was polygenic[39].

(2) The rate of trait evolution tells us nothing about the number of genes involved. Studies of artificial selection[35,38,40] and of interspecific divergence[41] indicate that the intensity of selection and heritability are more important determinants of evolutionary rate than is the genetic complexity of the traits under selection. There is considerable standing variation in traits, including characters that might be thought of as highly constrained, such as limb morphology in tetrapods[42]. In general, the observed rates of evolution under natural selection are far slower than is potentially possible[43,44]. Genetic variation or genetic complexity is not the limiting factor[32]; indeed, considerable genetic variation underlies even phenotypically invariant traits[45–48]. Because the rate at which a trait emerges in the fossil record tells us nothing about genetic architecture, the temptation to invoke macromutational models for 'rapid change'[33,49] must be resisted in the absence of genetic evidence.

(3) Morphological variation and divergence are associated with

* *This figure is reproduced in color following the index.*

genes that regulate development. Comparisons of the developmental basis of body-pattern evolution in animals suggest that morphological evolution is a product of changes in the spatiotemporal deployment of regulatory genes and the evolution of genetic regulatory networks[50–52]. Developmental changes in the human lineage are expected to be associated with genes that affect developmental parameters, such as those that encode transcription factors and members of signal-transduction pathways.

(4) Mutations responsible for trait variation are often in non-coding, regulatory regions. When it has been possible to localize variation in genes that underlie phenotypic variation or protein-level differences, insertions or substitutions in regulatory regions and non-coding regions are often responsible[53–57].

(5) Multiple nucleotide replacements often differentiate alleles. Fine-scale analysis of quantitative trait loci has often revealed that functional differences between alleles are due to multiple nucleotide differences[55,56]. It also indicates that non-additive interactions between sites within a locus may be key to the differentiation of alleles, and that the contribution of any individual site may be modest (and difficult to detect).

(6) There is some concordance between genes responsible for intraspecific variation and interspecies divergence. Genetic analyses of interspecies divergence is only possible under certain circumstances, when laboratory breeding can overcome species barriers and traits can be mapped. In some cases, it has been found that some of the same loci are involved in both within-species variation and between-species divergence[35,58]. This raises some hope that studies of intraspecific variation in humans could lead to genes that have been important in human history.

Since human trait evolution has followed a similar, incremental course as traits studied in model systems, these six concepts suggest that we should expect a highly polygenic basis for complex traits such as brain size, craniofacial morphology and development, cortical speech and language areas, hand and digit morphology, dentition and post-cranial skeletal morphology. We should also anticipate that multiple changes in non-coding regulatory regions and in regulatory genes are of great importance. But how can we find them?

The arithmetic of human sequence evolution

All genetic approaches to human origins are fundamentally comparative, and seek to identify genetic changes that occurred specifically in the human lineage and contributed to the differentiation of humans from our last common ancestor with either apes or other species of *Homo*. Our primary comparative reference is the genome of the chimpanzee (*Pan troglodytes*), our closest living relative, with whom we share a common ancestor that lived 5–7 Myr ago. The arithmetic that sets the problem for human evolutionary genetics is as follows: first, the most extensive comparison of chimpanzee and human genomic sequences indicates an average substitution level of ~1.2% in single-copy DNA[59]; second, the human genome comprises $\sim 3 \times 10^9$ base pairs; third, it is reasonable to assume that one-half of the total divergence between chimpanzees and humans occurred in the human lineage (~0.6%); and fourth, this amounts to $\sim 18 \times 10^6$ base-pair changes. In addition, there are an unknown number of gene duplications and pseudogene, transposon and repetitive element changes in each lineage. A recent small-scale survey indicated that insertions and deletions (indels) might account for another 3.4% of differences between chimpanzee and human genomes, with the bulk of that figure contributed by larger indels[60]. A good deal of genomic change might be the noise of neutral substitutions and the gain and loss of repetitive elements over long time spans (more than 46% of human DNA is composed of interspersed repeats), but some small fraction of the changes in genomic sequence is responsible for the hereditary differences between species. The crux of the challenge is how to identify specific changes that are biologically meaningful from the many that are not.

In the case of human evolution, there are three basic genetic issues that we would like to grasp. First, how many genes were directly involved in the origin of human anatomy, physiology and behaviour (a few, dozens, hundreds or thousands)? Second, which specific genes contributed to the emergence of particular human traits? And third, what types of change in these genes contributed to evolution (for example, gene duplications, amino-acid replacements or regulatory sequence evolution)? In the few pioneering studies that are directly addressing the genetic basis of human–chimpanzee divergence, different but somewhat complementary strategies are being pursued that are beginning to reveal the scope of human genetic evolution and, in some cases, specific genes that might have been under selection in the course of recent human evolution.

Comparative genomics

The most readily detected differences between animal genomes are expansions or contractions of gene families. Although the full chimpanzee genome is not yet available, a partial comparative map indicates that there are regions of the human genome that might not be represented in chimpanzees or other apes[61]. Such regions could be due to duplications or insertions that occurred in the hominin lineage or to deletions in the chimpanzee lineage. One gene family, dubbed *morpheus*, underwent expansion as part of a segmental duplication on human chromosome 16 (ref. 62). This expansion is shared by other great apes, but it seems that there were human lineage-specific duplications as well.

On the basis of comparisons with other genomes, particularly the recently reported draft mouse sequence[63], such lineage-specific duplications are expected. In the 75 Myr or more since the divergence of the common ancestor of mice and humans, several dozen clusters of mouse-specific genes arose that are generally represented by a single gene in the human genome[63]. The shorter divergence time between humans and apes suggests that the human-specific gene set will be smaller. It is interesting to note that a significant fraction of the mouse gene clusters encode proteins with roles in reproduction, immunity and olfaction. This indicates that sexual selection, pathogens and ecology can shape the main differences in coding content between mammals. It should also be noted that 80% of mouse genes have a 1:1 orthologue in the human genome, and that more than 99% have some homologue[63]. These figures and synteny data suggest that there is a gene repertoire that is qualitatively nearly identical among mammals. The presence or absence of particular gene duplicates might reflect adaptively driven change, but further evidence will be necessary to determine whether positive selection has acted on genes.

Thousands of adaptive changes in the human proteome?

The first place that adaptive genetic changes have been looked for is in the coding sequences for proteins. If the 18×10^6 substitutions in the human lineage are evenly distributed throughout the genome, only a small fraction will be expected to fall within coding regions. Assuming that the average protein is ~400 amino acids in length, and that there are ~30,000 protein-coding genes, only $\sim 3.5 \times 10^7$ base pairs (or a little more than 1.5% of the genome) consists of coding regions[64,65]. So, assuming neutrality and ignoring the selective removal of deleterious changes in protein sequences, ~1.5% of these 18×10^6 substitutions (or 270,000 sites) may contribute to protein evolution. A fraction of these (roughly one-quarter) are synonymous substitutions, so the total number of amino-acid replacements in the human lineage could be of the order of ~200,000. This figure is in good agreement with observed average rates of amino-acid replacement in mammals[66].

Various methods have been developed to detect whether amino-acid replacements could be the result of positive selection—that is, adaptive evolution[67,68]. To estimate the extent of positive selection in human protein evolution, Fay *et al.*[69] surveyed sequence-divergence data for 182 human and Old World monkey genes, and polymorphism data for a similar number of human genes. Taking into consideration the frequency of common polymorphisms (ignoring rare alleles), a greater-than-expected degree of amino-acid replacements was observed, which is evidence of selection. When extrapolated to the entire proteome, 35% of amino-acid substitutions between human and Old World monkeys were estimated to have been driven by positive selection. Applied to human–chimpanzee divergence, this would extrapolate to ~70,000 adaptive substitutions in the human lineage. This figure is substantially larger than would be expected if most mutations were neutral or nearly neutral[70]. If it is even the correct order of magnitude, it forecasts a nightmare for the identification of key genes under selection, because this figure suggests that, on average, two or more adaptive substitutions have occurred in every human protein in the last 5 Myr.

It is possible that the figure, based on the study of less than 0.5% of the human proteome, is an overestimate of the fraction or distribution of adaptive replacements. It is clear that some proteins are under strong pressure to remain constant, whereas others, especially those involved in so-called 'molecular arms races', are under pressure to change. For example, major histocompatibility complex proteins, which interact with diverse and changing foreign substances, show clear signatures of selection[71]. Proteins involved in reproduction that play a part in sperm competition or gamete recognition also appear to evolve faster and under some degree of positive selection[72]. A host of human male reproductive proteins have greater-than-average ratios of amino-acid replacements[73]. Although accelerated protein evolution can also be the consequence of relaxed constraints, the correlation of higher levels of amino-acid replacements in proteins that have a role in reproduction and immunity seems to be biologically and selectively driven.

The population genetics- and protein-sequence-based statistical estimates of adaptive evolution require three caveats regarding how much they tell us about human evolution. First, there are generally no direct functional data that either test or demonstrate whether a human protein is indeed functionally diverged from an ape orthologue. Second, the proteins for which signatures of selection have been detected generally do not affect development. And third, the proteome is just part of the whole picture of genome evolution. Non-coding sequences, including transcriptional *cis*-regulatory elements, the untranslated regions of messenger RNAs, and RNA-splicing signals, contribute considerably to evolution by affecting the time, place and level of gene expression (see above). Ever since the pioneering comparative analysis of ape and human protein-sequence divergence nearly three decades ago[74], it has generally been anticipated that changes in gene regulation are a more important force than coding-sequence evolution in the morphological and behavioural evolution of hominins.

Box 2
Selective sweeps

If a change in a gene is favoured, then selection may drive the allele bearing that change to fixation (left and centre of the figure). In the process, neutral variation at linked sites 'hitchhikes' along with the selected site; this is known as a 'selective sweep'. The physical limits of the sweep depend on the strength of the linkage between selected and adjacent sites. After a sweep, variation may again begin to build up, and initially there will be a relatively high frequency of rare polymorphisms (right of the figure). Tajima[91] proposed a statistic (*D*) that tests for selective neutrality. If the frequencies of polymorphisms are skewed, with an excess of rare types, this gives a negative value (neutral value = 0) and can be indicative of a recent selective sweep. Tajima's *D* is sensitive to other factors apart from selection that can also yield a negative value. A recent expansion in population size from a relatively small population will produce similar patterns of genetic variation and *D* values. In human populations, population history (for example, drift and expansion) and population structure (ethnicity, migration and immigration) will affect *D* values at all loci, whereas selection will affect *D* values at selected and linked loci. The mean *D* values for 437 loci range from −0.69 to −1.25, depending on sampling methods[92], indicating that population structure and history has had an effect. These negative values underscore a challenge in human evolutionary genetics to distinguish selective sweeps at loci from population-based effects. The *D* value obtained for the human *FOXP2* locus was −2.20, the second largest negative value among all human genes surveyed so far[84]. Other methods have been developed to detect positive selection: for example, by identifying areas of extended haplotype homozygosity[81]. It is important to emphasize that all of these methodologies detect signs of recent selection in the *Homo sapiens* lineage. The preceding 5–6 Myr of hominin genetic history, a period when we know from the fossil record that many human features arose, is not addressed by these methods.

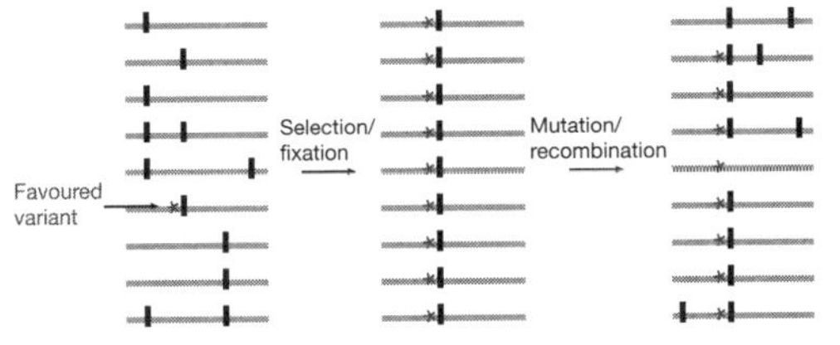

Evolution of human gene expression

How large is the functional compartment of non-coding sequences—the other 98% of the genome? A recent estimate suggests that perhaps twice as much non-coding DNA is under selection than coding DNA[63]. So, we would also expect a large number of substitutions in the human lineage, of the order of several hundred thousand, with potential functional consequences in non-coding DNA. Even if one applies a much smaller, more conservative estimate of the fraction of adaptive substitutions in non-coding DNA, such as 2% (ref. 75), one still reaches a figure of more than 10,000 adaptive substitutions in human genes and their regulatory regions. The problem is that regulatory sequences are more difficult to analyse: we have no algorithms that can infer biological function from tracts of intergenic or intronic sequence, let alone to decipher how base-pair changes affect function. It is therefore perhaps understandable why non-coding regions have received little attention at the level of population genetics. However, a growing body of work in quantitative genetics and on the evolution of development has shown that regulatory sequences are central to changes in gene expression and morphology. New methodologies have been required to detect the evolution of gene expression and regulatory sequences.

A first step towards the identification of human-specific gene-expression patterns was recently taken by Enard *et al.*[76], who used genome microarrays to analyse within- and between-species differences in primate gene expression. Analysis of RNA expression profiles from the left prefrontal lobe (Brodmann area 9, which is thought to be involved in cognitive functions) of adult male humans, chimpanzees and an orangutan, and from the neocortex of humans, chimpanzees and macaques, indicated an apparent acceleration of gene-expression differences in the human brain relative to other primates and to other tissues. Protein-expression analyses were also consistent with the idea that relative changes in protein-expression levels were accelerated in the evolution of the human brain, and could be detected for ~30% of the proteins surveyed. The concordance between RNA and protein-level data indicates that regulatory changes have occurred in a substantial fraction of genes. Indeed, a recent survey of humans heterozygous at 13 loci revealed allelic variation in gene-expression levels at 6 loci[77]. Both intraspecific variations and interspecific divergence in gene expression are probably due to substitutions in non-coding regions that influence transcript or protein abundance through transcriptional or post-transcriptional mechanisms. These data further suggest that quantitative changes in gene expression should be expected as a general feature that accompanies species divergence, and that the raw material for evolutionary changes in gene expression appears to be widely available in non-coding DNA.

The microarray experiments raise many challenges for future progress. Specifically, how can changes that contribute to human anatomy, physiology or behaviour be sorted out from those that don't? Gene-expression data are correlative, not definitive in terms of identifying cause and effect. Many developmental and genetic mechanisms could contribute to the overall pattern observed. For example, a change in the composition of a tissue (for example, the relative proportions of cell types) will be accompanied by altered expression profiles, but many of these changes will be an indirect consequence of a developmental change, not the cause. Similarly, changes in levels or activities of regulatory proteins may affect batteries of downstream genes, but again are indirect and do not necessarily involve substitutions at the loci whose expression changes. Therefore, different approaches have to be taken to identify primary changes in regulatory pathways.

Candidate genes in human evolution

The ultimate goal of microarray analyses, quantitative trait genetics, population genetics or other comparative genetic methods is the identification of genes that are candidates for being causally associated with phenotypic divergence. Although genome-wide, large-scale surveys provide an overview, rigorous tests of causality demand a gene-by-gene approach. In choosing genes to be studied in greater detail, molecular geneticists will be opportunistic, focusing on those loci for which additional information from human or model-animal biology suggests an association with a trait of greater evolutionary interest, such as craniodental development[78]. Therefore, it is unlikely that all traits will be pursued with equal vigour or success.

To implicate a gene in human evolution, two types of data need to be assessed. First, functional evidence that a gene is involved in a developmental, behavioural or physiological trait is required to formulate hypotheses about the role of an individual gene. This may come from analysis of human mutations at a locus (see below). Second, the molecular evolution and population genetics of the locus need to be analysed for evidence of natural selection. Comparison of orthologues from chimpanzees and other primates and mammals, and analyses of intraspecific variation in humans, can reveal signs of positive selection at the sequence level or of a recent 'selective sweep' through a locus (Box 2). Evidence of positive selection has been found at several human loci[73,79–81]. Although these might be physiologically important (for example, in immunity or reproduction), most genes studied so far are not expected to contribute to the divergence of morphological or behavioural traits. More recently, attention has turned to candidate genes identified from human mutations that do affect such traits.

The evolution of a gene affecting speech. Human medical genetics has made substantial progress, and sophisticated mapping techniques for polymorphisms are accelerating the characterization of genes involved in complex traits, particularly those of medical interest. One of the most provocative reports of late was the identification of the gene *FOXP2* (forkhead box P2), mutations of which are associated with a speech and language disorder[82]. The gene encodes a transcription factor and is therefore expected to control the expression of other genes. The excitement surrounding *FOXP2* stems from the observation that affected individuals appear to have not an overt impairment, but a lesion in the neural circuitry that affects language processes[62,82,83].

Is this a novel human 'language' gene? No, the gene is found in other species. In fact, the human FOXP2 protein differs from the gorilla and chimpanzee sequence at just two residues, and from the orangutan and mouse sequences at three and four residues, respectively[76]. This history is typical of human and other species' genes, in that most genes have orthologues in other mammals and animals. However, there is the possibility that the two replacements in the FOXP2 protein that evolved in the human lineage are of functional significance to the origin of language.

To examine whether the *FOXP2* gene has been the target of selection during human evolution, a detailed analysis was undertaken of nucleotide variation over a 14-kilobase (kb) subregion of the large *FOXP2* locus, of amino-acid polymorphism in a segment of the protein, and of chimpanzee and orangutan sequences[84]. An unusual excess was found of rare alleles at the human *FOXP2* locus, and of high-frequency alleles. Reduced genetic variation in neutral linked regions is a predicted consequence of a selective sweep (Box 2), so these observations are consistent with natural selection acting on the *FOXP2* locus. Estimates of the time of fixation of the two amino-acid replacements place them within the last 200,000 yr of human evolution, an intriguing correlation with the estimated age of *H. sapiens*.

However, it should be noted that there are no biological data to support the hypothesis that these amino-acid replacements are functionally important. In the 14-kb region surveyed, more than 100 fixed differences exist; the entire *FOXP2* locus is large (267 kb), and more than 2,000 differences would be expected to exist between the *FOXP2* genes of humans and chimpanzees. No assessment has been made of potential non-coding regulatory sequences that might have contributed to a divergence in the role of FOXP2 in hominids. Trait differences are often due to changes in regulatory networks that govern development, and need not be in coding regions (although that would be much more convenient, given just two changes in the human FOXP2 protein). Because FOXP2 is a transcription factor, changes in *FOXP2* expression could be of functional and evolutionary significance.

The typical genetic architecture that underlies complex traits makes it extremely unlikely that *FOXP2* was the only gene under selection in the evolution of our language capabilities. However, we have no means of assessing the relative contribution of *FOXP2* and other candidate genes. The encouraging lesson of *FOXP2* is that medical genetics has provided an interesting lead into a regulatory network that affects the development of speech ability. Further study of *FOXP2* should lead, at a minimum, to a better understanding of the neurodevelopmental biology of speech and language, and perhaps to more genes with interesting evolutionary histories.

The functions of selected genes

The three genome-scale approaches highlighted here—population genetics, comparative genomics and gene-expression profiling—have all succeeded in finding what each sought: thousands of potential adaptive coding substitutions, regulatory differences in gene expression, and gene duplications and rearrangements. Each has yielded many candidates through which to sift and, interestingly, because of the different search regimens used, there is virtually no overlap in the sets of genetic changes that have been surveyed. It is almost certain that, as in other lineages, all of these types of genetic mechanism have contributed to hominid evolution. The crucial challenge now is to obtain functional data for individual genes and to scrutinize the molecular evolution of candidates for signatures of selection.

To place any candidate gene into a functional context of human trait evolution, advances in primate and human developmental neurobiology will be essential. Non-primates are limited as models of the development and function of primate and hominid neocortex, and thus as models of the function of proteins such as FOXP2 in the development and elaboration of neural networks. Direct empirical work on developing primates, which faces serious methodological constraints as well as bona fide ethical questions, will be necessary to advance beyond associations and correlations. Testing the functional role of what may be subtle changes in human orthologues of primate genes, a daunting task in the most technically developed model species, will be even more difficult.

There are two immediate avenues to increasing the power of human evolutionary genetics. First, we would increase the value of chimpanzee–human comparative genomics by sequencing the gorilla genome, which is the next earliest branching ape to humans and chimpanzees. This would help us to determine the polarity of genetic changes by distinguishing those changes in the human lineage from those in the chimpanzee lineage. Second, 6×10^9 interbreeding humans is a very large resource for identifying rare mutations (for example, in *FOXP2*) with subtle behavioural or developmental effects, and for mapping genetic variation that underlies morphological variation, both of which could lead to genes that govern the formation of human traits and that might have played a part in hominid evolution.

The fine print below the headlines

It is easy to foresee the media headlines that will announce the completion of the chimpanzee genome. One aim of this article has been to anticipate both the excitement that accomplishment warrants and the more sobering aspects of complex trait genetics and genome-scale evolution. Despite our enhanced understanding of functional genetic architecture, there remains a tendency to associate the development, function or evolution of a trait with single genes (genes 'for' speech, cancer and so on). The ghost of 'hopeful monsters' still haunts biology and is, unfortunately, a prevalent misconception in the scientific and general press. Perhaps wishful thinking is also an intrinsic part of human nature, but it seems unlikely that the traits that interest us most—bipedalism, skeletal morphology, craniofacial morphology, brain size and speech—were the products of selection of just a few major genes. Just as palaeoanthropology now recognizes a complex pattern of hominin phylogeny and the uncertainties in identifying long-sought common ancestors, and comparative neurobiology now searches for more subtle explanations of human capabilities, the lessons of model-system genetics and comparative genomics should prepare us for the finding that the genetics of hominid trait evolution are, in fact, subtle and complicated.

I underscore this point not just for its scientific relevance, but also because of the larger issues at stake—the meaning of the pursuit of the material basis of human evolution. Evolutionary biology has always faced public resistance. It has been difficult enough to gain acceptance of fundamental ideas using humble finches or fruitflies as examples. We can anticipate even more hostile challenges to human evolutionary genetics. Opponents will be sure to exploit any instances where claims or hypotheses are founded on weak or contradictory data. Witness how the recent scrutiny of data supporting the classic paradigm of industrial melanism has been hijacked by the anti-evolution agenda[85]. The sequencing of the chimpanzee genome will reveal no more directly about the origin of human traits than the sequence of the human genome tells us about how to construct a healthy baby. Headlines may claim more, but we would be well advised to describe this as just the beginning of a large, complex and profoundly important story. □

doi:10.1038/nature01495.

1. Wood, B. Hominid revelations from Chad. *Nature* **418,** 133–135 (2002).
2. Leakey, M. *et al.* New hominin genus from eastern Africa shows diverse middle Pliocene lineages. *Nature* **410,** 433–440 (2001).
3. Brunet, M. *et al.* A new hominid from the Upper Miocene of Chad, Central Africa. *Nature* **418,** 145–151 (2002).
4. Wolpoff, M., Senut, B., Pickford, M. & Hawks, J. Palaeoanthropology: *Sahelanthropus* or '*Sahelpithecus*'? *Nature* **419,** 581–582 (2002).
5. Groves, C. P. in *Comparative Primate Biology* Vol. 1 (eds Swindler, D. R. & Erwin, J.) 187–218 (Alan R. Liss, New York, 1986).
6. Klein, J. & Takahata, N. *Where Do We Come From? The Molecular Evidence for Human Descent* (Springer, New York, 2002).
7. Wood, B. & Collard, M. The human genus. *Science* **284,** 65–71 (1999).
8. Relethford, J. H. *Genetics and the Search for Modern Human Origins* (Wiley–Liss, New York, 2001).
9. Ruff, C. B., Trinkhaus, E. & Holliday, T. W. Body mass and enchephalization in Pleistocene *Homo*. *Nature* **387,** 173–176 (1997).
10. Conroy, G. C. *et al.* Endocranial capacity in an early hominid cranium from Sterkfontein, South Africa. *Science* **280,** 1730–1731 (1998).
11. Conroy, G. C., Weber, G. W., Seidler, H., Recheis, W. & Zur Nedden, E. Endocranial capacity of the Bodo cranium determined from three-dimensional computed tomography. *Am. J. Phys. Anthropol.* **113,** 111–118 (2000).
12. Brodmann, K. Neue Ergebnisse über die vergleichende histologische Lokalisation der Grosshirnrinde mit besonderer Berücksichtigung des Stirnhirns. *Anat. Anz.* **41,** 157–216 (1912).
13. Semendeferi, K., Lu, A., Schenker, N. & Damasio, H. Humans and great apes share a large frontal cortex. *Nature Neurosci.* **5,** 272–276 (2002).
14. Cantalupo, C. & Hopkins, W. D. Asymmetric Broca's area in great apes. *Nature* **414,** 505 (2001).
15. Hopkins, W. D. & Leavens, D. A. The whole-hand point: The structure and function of pointing from a comparative perspective. *J. Comp. Psychol.* **112,** 95–99 (1998).
16. Holloway, R. L. Indonesian "Solo" (Ngandong) endocranial reconstructions: Some preliminary observations and comparisons with Neandertal and *Homo erectus* groups. *Am. J. Phys. Anthropol.* **53,** 285–295 (1980).
17. Gannon, P. J., Holloway, R. L., Broadfield, D. C. & Braun, A. R. Asymmetry of chimpanzee planum temporale: Humanlike pattern of Wernicke's brain language area homolog. *Science* **279,** 220–222 (1998).
18. Hopkins, W. D., Marino, L., Rilling, J. K. & MacGregor, L. Planum temporale asymmetries in great apes as revealed by magnetic resonance imaging (MRI). *NeuroReport* **9,** 2913–2918 (1998).
19. Hof, P. R., Nimchinsky, E. A., Perl, D. P. & Erwin, J. M. An unusual population of pyramidal neurons in the anterior cingulate cortex of hominids contains the calcium-binding protein calretinin. *Neurosci. Lett.* **307,** 139–142 (2001).
20. Nimchinsky, E. A. *et al.* A neuronal morphological type unique to humans and great apes. *Proc. Natl Acad. Sci. USA* **96,** 5268–5273 (1999).
21. Buxhoeveden, D., Switala, A., Litaker, M., Roy, E. & Casanova, M. Lateralization of minicolumns in human planum temporale is absent in nonhuman primate cortex. *Brain Behav. Evol.* **57,** 349–358 (2001).
22. Semendeferi, K., Armstrong, E., Schleicher, A., Zilles, K. & Van Hoesen, G. W. Prefrontal cortex in humans and apes: A comparative study of area 10. *Am. J. Phys. Anthropol.* **114,** 224–241 (2001).
23. Penin, X., Berge, C. & Baylac, M. Ontogenetic study of the skull in modern humans and the common chimpanzees: Neotenic hypothesis reconsidered with a tridimensional procrustes analysis. *Am. J. Phys. Anthropol.* **118,** 50–62 (2002).
24. Ponce de León, M. S. & Zollikofer, C. P. E. Neanderthal cranial ontogeny and its implications for late hominid diversity. *Nature* **412,** 534–538 (2001).
25. Lieberman, D. E., McBratney, B. M. & Krovitz, G. The evolution and development of cranial form in *Homo sapiens*. *Proc. Natl Acad. Sci. USA* **99,** 1134–1139 (2002).
26. Williams, F. L., Godfrey, L. R. & Sutherland, M. R. in *Human Evolution through Developmental Change* (eds Minugh-Purvis, N. & McNamara, K. J.) 405–441 (Johns Hopkins Univ. Press, Baltimore, 2002).
27. Dean, C. *et al.* Growth processes in teeth distinguish modern humans from *Homo erectus* and earlier hominins. *Nature* **414,** 628–631 (2001).
28. Rice, S. H. in *Human Evolution through Developmental Change* (eds Minugh-Purvis, N. & McNamara, K. J.) 154–170 (Johns Hopkins Univ. Press, Baltimore, 2002).
29. Gould, S. J. *Ontogeny and Phylogeny* (Belknap, Cambridge, Massachusetts, 1977).
30. Moggi-Cecchi, J. Questions of growth. *Nature* **414,** 595–597 (2001).
31. MacFadden, B. J. Fossil horses from "Eohippus" (*Hyracotherium*) to *Equus*: Scaling, Cope's Law, and the evolution of body size. *Paleobiology* **12,** 355–369 (1986).
32. Charlesworth, B., Lande, R. & Slatkin, M. A neo-Darwinian commentary on macroevolution. *Evolution* **36,** 474–498 (1982).
33. Schwartz, J. H. Homeobox genes, fossils, and the origin of species. *Anat. Rec.* **257,** 15–31 (1999).
34. Klein, R. G. Archeology and the evolution of human behavior. *Evol. Anthropol.* **9,** 17–36 (2000).
35. Mackay, T. F. C. Quantitative trait loci in *Drosophila*. *Nature Rev. Genet.* **2,** 11–20 (2001).
36. Atchley, W. R., Plummer, A. A. & Riska, B. Genetic analysis of size-scaling patterns in the mouse mandible. *Genetics* **111,** 579–595 (1985).
37. Atchley, W. R. & Zhu, J. Developmental quantitative genetics, conditional epigenetic variability and growth in mice. *Genetics* **147,** 765–776 (1997).
38. Doebley, J. & Stec, A. Inheritance of the morphological differences between maize and teosinte: Comparison of results for two F2 populations. *Genetics* **134,** 559–570 (1993).
39. Livshits, G., Roset, A., Yakovenko, K., Trofimov, S. & Kobyliansky, E. Genetics of human body size and shape: Body proportions and indices. *Ann. Hum. Biol.* **29,** 271–289 (2002).
40. Brakefield, P. *et al.* Development, plasticity and evolution of butterfly eyespot patterns. *Nature* **384,** 236–242 (1996).
41. True, J. R., Liu, J., Stam, L. F., Zeng, Z. B. & Laurie, C. C. Quantitative genetic analysis of divergence in male secondary sexual traits between *Drosophila simulans* and *Drosophila mauritiana*. *Evolution* **51,** 816–832 (1997).
42. Shubin, N., Wake, D. B. & Crawford, A. J. Morphological variation in the limbs of *Taricha granulosa* (Caudata: Salamandridae): Evolutionary and phylogenetic implications. *Evolution* **49,** 874–884 (1995).
43. Gingerich, P. D. Rates of evolution on the time scale of the evolutionary process. *Genetics* **112/113,** 127–144 (2001).
44. Gingerich, P. D. Rate of evolution: Effects of time and temporal scaling. *Science* **222,** 159–161 (1983).
45. Lauter, N. & Doebley, J. Genetic variation for phenotypically invariant traits detected in teosinte: Implications for the evolution of novel forms. *Genetics* **160,** 333–342 (2002).
46. Polaczyk, P. J., Gasperini, R. & Gibson, G. Naturally occurring genetic variation affects *Drosophila* photoreceptor determination. *Dev. Genes. Evol.* **207,** 462–470 (1998).

47. Rutherford, S. L. & Lindquist, S. Hsp90 as a capacitor for morphological evolution. *Nature* **396,** 336–342 (1998).
48. Gibson, G., Wemple, M. & van Helden, S. Potential variance affecting homeotic *Ultrabithorax* and *Antennapedia* phenotypes in *Drosophila melanogaster. Genetics* **151,** 1081–1091 (1999).
49. Goldschmidt, R. *The Material Basis of Evolution* (Yale Univ. Press, New Haven, Connecticut, 1940).
50. Carroll, S. B., Grenier, J. K. & Weatherbee, S. D. *From DNA to Diversity: Molecular Genetics and the Evolution of Animal Design* (Blackwell Scientific, Malden, Massachusetts, 2001).
51. Davidson, E. H. *Genomic Regulatory Systems: Development and Evolution* (Academic, San Diego, 2001).
52. Wilkins, A. S. *The Evolution of Developmental Pathways* (Sinauer Associates, Sunderland, Massachusetts, 2002).
53. Laurie, C. C. & Stam, L. F. Molecular dissection of a major gene effect on a quantitative trait: The level of alcohol dehydrogenase expression in *Drosophila melanogaster. Genetics* **144,** 1559–1564 (1996).
54. Long, A. D., Lyman, R. F., Morgan, A. H., Langley, C. H. & Mackay, T. F. C. Both naturally occurring insertions of transposable elements and intermediate frequency polymorphisms at the *achaete-scute* complex are associated with variation in bristle number in *Drosophila melanogaster. Genetics* **154,** 1255–1269 (2000).
55. Wang, R.-L., Stec, A., Hey, J., Lukens, L. & Doebley, J. The limits of selection during maize domestication. *Nature* **398,** 236–239 (1999).
56. Long, A. D., Lyman, R. F., Langley, C. H. & Mackay, T. F. C. Two sites in the *Delta* gene region contribute to naturally occurring variation in bristle number in *Drosophila melanogaster. Genetics* **149,** 999–1017 (1998).
57. Lai, C., Lyman, R. F., Long, A. D., Langley, C. H. & Mackay, T. F. C. Naturally occurring variation in bristle number and DNA polymorphisms at the *scabrous* locus of *Drosophila melanogaster. Science* **266,** 1697–1702 (1994).
58. Nuzhdin, S. V. & Reiwitch, S. G. Are the same genes responsible for intra- and interspecific variability for sex comb tooth number in *Drosophila. Heredity* **84,** 97–102 (2000).
59. Chen, F.-C. & Li, W.-H. Genomic divergences between humans and other hominoids and the effective population size of the common ancestor of humans and chimpanzees. *Am. J. Hum. Genet.* **68,** 444–456 (2001).
60. Britten, R. J. Divergence between samples of chimpanzee and human DNA sequences is 5%, counting indels. *Proc. Natl Acad. Sci. USA* **99,** 13633–13634 (2002).
61. Fujiyama, A. *et al.* Construction and analysis of a human–chimpanzee comparative clone map. *Science* **295,** 131–134 (2002).
62. Johnson, M. E. *et al.* Positive selection of a gene family during the emergence of humans and African apes. *Nature* **413,** 514–518 (2001).
63. Mouse Genome Sequencing Consortium. Initial sequencing and comparative analysis of the mouse genome. *Nature* **420,** 520–562 (2002).
64. Venter, J. C. *et al.* The sequence of the human genome. *Science* **291,** 1304–1323 (2001).
65. International Human Genome Sequencing Consortium. Initial sequencing and analysis of the human genome. *Nature* **409,** 860–921 (2001).
66. Li, W.-H. *Molecular Evolution* (Sinauer Associates, Sunderland, Massachusetts, 1997).
67. Kreitman, M. Methods to detect selection in populations with applications to the human. *Ann. Rev. Genomics Hum. Genet.* **1,** 539–559 (2000).
68. McDonald, J. H. & Kretiman, M. Adaptive protein evolution at the *Adh* locus in *Drosophila. Nature* **351,** 652–654 (1991).
69. Fay, J. C., Wyckoff, G. J. & Wu, C.-I. Positive and negative selection on the human genome. *Genetics* **158,** 1227–1234 (2001).
70. Ohta, T. Near-neutrality in evolution of genes and gene regulation. *Proc. Natl Acad. Sci. USA* **99,** 16134–16137 (2002).
71. Hughes, A. L. *Adaptive Evolution of Genes and Genomes* (Oxford Univ. Press, New York, 1999).
72. Swanson, W. J. & Vacquier, V. D. The rapid evolution of reproductive proteins. *Nature Rev. Genet.* **3,** 137–144 (2002).
73. Wyckoff, G. J., Wang, W. & Wu, C.-I. Rapid evolution of male reproductive genes in the descent of man. *Nature* **403,** 304–309 (2000).
74. King, M.-C. & Wilson, A. C. Evolution at two levels in humans and chimpanzees. *Science* **188,** 107–116 (1975).
75. Nachman, M. W. Single nucleotide polymorphisms and recombination rate in humans. *Trends Genet.* **17,** 481–485 (2001).
76. Enard, W. *et al.* Intra- and interspecific variation in primate gene expression patterns. *Science* **296,** 340–343 (2002).
77. Yan, H., Yuan, W., Velculescu, V. E., Vogelstein, B. & Kinzler, K. W. Allelic variation in human gene expression. *Science* **297,** 1143 (2002).
78. McCollum, M. A. & Sharpe, P. T. Developmental genetics and early hominid craniodental evolution. *BioEssays* **23,** 481–493 (2001).
79. Andolfatto, P. Adaptive hitchhiking effects on genome variability. *Curr. Opin. Genet. Dev.* **11,** 635–641 (2001).
80. Diller, K. C., Gilbert, W. A. & Kocher, T. D. Selective sweeps in the human genome: A starting point for identifying genetic differences between modern humans and chimpanzees. *Mol. Biol. Evol.* **19,** 2342–2345 (2002).
81. Sabetl, P. C. *et al.* Detecting recent positive selection in the human genome from haplotype structure. *Nature* **419,** 832–837 (2002).
82. Lai, C. S. L., Fisher, S. E., Hurst, J. A., Vargha-Khadem, F. & Monaco, A. P. A forkhead-domain gene is mutated in a severe speech and language disorder. *Nature* **413,** 519–523 (2001).
83. Pinker, S. Talk of genetics and vice versa. *Nature* **413,** 465–566 (2001).
84. Enard, W. *et al.* Molecular evolution of *FOXP2*, a gene involved in speech and language. *Nature* **418,** 869–872 (2002).
85. Coyne, J. A. Evolution under pressure. *Nature* **418,** 19–20 (2002).
86. Kimbel, W. H., Johanson, D. C. & Rak, Y. The first skull and other new discoveries of *Australopithecus afarensis* at Hadar, Ethiopia. *Nature* **368,** 449–451 (1994).
87. Haile-Selassie, Y. Late Miocene hominids from the Middle Awash, Ethiopia. *Nature* **412,** 178–181 (2001).
88. Asfaw, B. *et al.* Remains of *Homo erectus* from Bouri, Middle Awash, Ethiopia. *Nature* **416,** 317–320 (2002).
89. Hacia, J. G. Genome of the apes. *Trends Genet.* **17,** 637–645 (2001).
90. Richmond, B. G., Aiello, L. C. & Wood, B. A. Early hominid limb proportions. *J. Hum. Evol.* **43,** 529–548 (2002).
91. Tajima, F. Statistical method for testing the neutral mutation hypothesis by DNA polymorphism. *Genetics* **123,** 585–596 (1989).
92. Ptak, S. E. & Przeworski, M. Evidence for population growth in humans is confounded by fine-scale population structure. *Trends Genet.* **18,** 1–5 (2002).

Acknowledgements Thanks to B. Hopkins and C. Cantalupo for guidance on Fig. 2, and to L. Olds for illustrations; to B. Williams, A. Kopp, S. Paddock, A. Rokas, D. Bownds, J. Doebley, N. Shubin and J. Crow for comments on the manuscript; to P., N. and J. Carroll for inspiration, and to J. Carroll for preparation of the manuscript. S.B.C. is an Investigator of the Howard Hughes Medical Institute.

Correspondence and requests for materials should be addressed to the author (e-mail: sbcarrol@facstaff.wisc.edu).

Elena, S. F., and R. E. Lenski. 2003. Evolution experiments with microorganisms: the dynamics and genetic bases of adaptation. *Nature Reviews Genetics* 4:457–469.

One criticism of evolutionary biology is that the field is primarily descriptive rather than experimental. This shortcoming applies with equal force to historical analyses in the social sciences, where controlled experimentation likewise is not feasible. Yet both evolutionary biologists and social historians are able to draw critical conclusions about historical phenomena by using analytical methods that entail sound comparative analyses of multiple possible outcomes. Historians can learn valuable lessons from history even if they cannot control it. Moreover, evolutionists can test their inferences concerning past history by observations and experiments. As a simple example, the hypothesis that humans are more closely related to chimpanzees than to gorillas and orangutans leads to the prediction that the DNA is more similar between humans and chimps than between them and gorillas or orangutans, a prediction that has been thoroughly corroborated.

In addition, it is the case that evolution can in fact be observed in suitable situations, and nowhere is this more true than in the microbial realm. As detailed in this article by Santiago Elena and Richard Lenski, microbiotas (including viruses, bacteria, and yeast) have several biological properties that make them ideal subjects for experimental evolution in controlled laboratory settings, and many evolutionary lessons have been learned from this approach. None of this is to suggest that evolution in action can be observed only in microbiotas (macrobiotas offer some fine examples too, as described in the references below). Nevertheless, the microbes offer exceptional opportunities to address evolutionary processes in a critical experimental framework.

RELATED READING

Harvey, P. H., and M. D. Pagel. 1991. The comparative method in evolutionary biology. Oxford University Press, Oxford, England.

Palumbi, S. R. 2001. The evolution explosion: how humans cause rapid evolutionary change. Norton, New York, New York, USA.

Weiner, J. 1994. The beak of the finch: a story of evolution in our time. Alfred A. Knopf, New York, New York, USA.

EVOLUTION EXPERIMENTS WITH MICROORGANISMS: THE DYNAMICS AND GENETIC BASES OF ADAPTATION

Santiago F. Elena and Richard E. Lenski‡*

Microorganisms have been mutating and evolving on Earth for billions of years. Now, a field of research has developed around the idea of using microorganisms to study evolution in action. Controlled and replicated experiments are using viruses, bacteria and yeast to investigate how their genomes and phenotypic properties evolve over hundreds and even thousands of generations. Here, we examine the dynamics of evolutionary adaptation, the genetic bases of adaptation, tradeoffs and the environmental specificity of adaptation, the origin and evolutionary consequences of mutators, and the process of drift decay in very small populations.

MICROBIAL GENETICS

Throughout its history, evolutionary biology has relied primarily on comparative studies of living organisms, supplemented whenever possible by data from the fossil record. Such comparative studies have become increasingly powerful with the emergence of molecular data for determining phylogenies[1] and with the increased rigour with which the comparative method has been applied[2]. However, biologists have long been interested in observing the dynamics of evolutionary change more directly. Indeed, Charles Darwin remarked in 1859 that "in looking for the gradations by which an organ in any species has been perfected, we ought to look exclusively to its lineal ancestors; but this is scarcely ever possible, and we are forced in each case to look to species of the same group, that is to the collateral descendants from the same original parent-form"[3]. So, Darwin used a comparative approach by necessity, and admitted how valuable it would be to observe the actual processes of descent with modification and adaptation by natural selection.

Since Darwin's day, many examples of evolution in action have been studied in nature, ranging from the emergence of antibiotic resistance in bacteria[4] to changes in the beak morphology of Darwin's finches[5]. Beyond simply observing evolution in action, some biologists sought to carry out experiments that ran for many generations, with controls and replication, to test hypotheses about the evolutionary process. Beginning with T. H. Morgan, and for several subsequent decades, population geneticists that studied fruitflies were the main practitioners of experimental evolution. A few groups used other organisms, including bacteria[6,7], in evolution experiments but, except in the fruitfly school, this approach did not take hold. In the case of microorganisms, a rift developed as most microbiologists pursued ever more molecular approaches but largely ignored evolution.

The situation has changed in recent years, with many groups now carrying out evolution experiments on diverse organisms including plants, vertebrates and, especially, microorganisms. On one side, there was a recognition by ecologists, including those that were interested in evolutionary ecology, of the need for rigorous experiments to test hypotheses[8]. On the other side, many microbiologists realized the value of an evolutionary perspective, stimulated by the discovery of the Archaea[9] and the rapidly increasing genomic data. These new attitudes led to the realization that microbes offer powerful systems for experimental evolution (BOX 1).

There are several other reasons for which microbial evolution experiments have received increasing attention. By virtue of the control that can be exerted over many variables in a laboratory setting, and the power that is afforded by the direct observation of any

*Instituto de Biología Molecular y Celular de Plantas, Consejo Superior de Investigaciones Científicas, 46022 Valencia, Spain.
‡Department of Microbiology and Molecular Genetics, Michigan State University, East Lansing Michigan 48824, USA.
e-mails: sfelena@ibmcp.upv.es; lenski@msu.edu
doi:10.1038/nrg1088

Box 1 | **Advantages of microorganisms for evolution experiments**

Microorganisms that have been used in evolution experiments include many bacteria and viruses, as well as unicellular algae and fungi. These organisms are well suited for such experiments for many practical reasons:

- They are easy to propagate and enumerate.
- They reproduce quickly, which allows experiments to run for many generations.
- They allow large populations in small spaces, which facilitates experimental replication.
- They can be stored in suspended animation and later revived, which allows the direct comparison of ancestral and evolved types.
- Many microbes reproduce asexually and the resulting clonality enhances the precision of experimental replication.
- Asexuality also maintains linkage between a genetic marker and the genomic background into which it is placed, which facilitates fitness measurements (BOX 2).
- It is easy to manipulate environmental variables, such as resources, as well as the genetic composition of founding populations.
- There are abundant molecular and genomic data for many species, as well as techniques for their precise genetic analysis and manipulation.

dynamic process, many questions about evolution can be probed with greater rigour than would otherwise be possible. For example, the reproducibility of evolutionary outcomes can be studied in microbial populations that are founded by the same ancestor and placed in identical environments. Although a 'natural experiment' that involves the colonization of several neighbouring islands by the same insect species could provide insights into this question, it would be difficult to exclude the possibility that subtle environmental differences promoted divergence or, alternatively, that mutations that were already present in the source population contributed to parallel responses.

Also, many microbes are of great importance to humans, not only as pathogens but for numerous essential ecosystem services. Therefore, it is crucial to understand the mechanisms and dynamics of microbial evolution. The most important difference between microbial and 'macrobial' evolution is probably not organismal size, or even the speed of generations, but is the fact that most microbes can reproduce asexually whereas most plants and animals are sexual[10]. All microorganisms must not be viewed as alike. However, there are grounds for optimism that some generalizations (or at least strong tendencies) do exist because experiments on such different types of microbe as viruses, bacteria and yeast often support broadly similar conclusions. Moreover, many experiments with microbes are designed to test general hypotheses that are derived from evolutionary theory that has been developed for other organisms[11].

One final reason for the growing interest in microbial evolution experiments is the satisfaction that often comes from observing in 'real time' a process that is usually inferred indirectly. Indeed, the field of evolutionary biology has long been hounded by sceptics who question, for example, whether evolution can produce adaptation if it depends on random mutations, most of which are deleterious. Of course, such ill-founded criticisms can be dismissed by showing their logical flaws with the appropriate mathematics, but it is also nice to be able to point to experiments that confirm the basic underpinnings of a scientific field.

Most experiments in microbial evolution are conceptually simple. Populations are established (often from single clones), then propagated in a controlled and reproducible environment for many generations. A sample of the ancestral population is stored indefinitely (for example, frozen at −80 °C), as are samples from various time points in the experiment. After a population has been propagated for some time, the ancestral and derived genotypes can be compared with respect to any genetic or phenotypic properties of interest, which provides information on the dynamics of the evolutionary process and the extent of evolutionary change. Importantly, adaptation can be quantified by measuring changes in FITNESS in the experimental environment, in which fitness reflects the propensity to leave descendants[12,13]. With microorganisms, fitness can be measured using head-to-head competition between, for example, an evolutionarily derived line and its ancestor that is genetically marked (BOX 2). In brief, the population growth rates that are achieved by each type as they compete for a pool of resources are measured. Different markers can be used to distinguish competitors, such as those that produce visible reactions with dyes, resistance to antibiotics or diagnostic PCR fragments. Provided that the organisms are asexual, as in most microbial evolution experiments, the marker serves as a proxy for the entire genome. By using control assays to measure any effect of the marker, and by replicating competitions, it is possible to reliably quantify evolutionary changes in fitness.

Of course, relative fitness depends not only on the genotypes but also on the environment in which it is measured. As discussed later, it is possible to test the specificity of adaptation that occurred in an evolution experiment by measuring fitness in different environments. Unless otherwise specified, however, it should be understood that fitness is measured under conditions that are similar or identical to those that prevailed during an evolution experiment.

In this review of evolution experiments with microorganisms, we examine the dynamics of evolutionary adaptation, the genetic bases of adaptation, tradeoffs and the environmental specificity of adaptation, the origin and consequences of mutators, and the process of drift decay in small populations. Experiments that address complex interactions, including compensatory adaptation, the maintenance of genetic diversity, social conflict and cooperation, the effects of sexual recombination and host–parasite interactions, will be discussed in a second review, to be published in a future issue of *Nature Reviews Genetics*.

Even with two reviews, there are topics that we cannot cover in the available space. In organizing our subject, we chose to focus on evolution experiments that are open ended and long term in approach. We do not review selection experiments that targeted specific, often new, metabolic functions, even though this work is of great interest. Reviews of these targeted selection experiments can be found elsewhere[14,15], as can reviews of the

FITNESS
The average reproductive success of a genotype in a particular environment. Often expressed relative to another genotype, such as the ancestor in evolution experiments.

Box 2 | **Measuring fitness**

The fitness of an evolved type is generally expressed relative to its ancestor. Relative fitness is measured by allowing the ancestral and evolved types to compete with one another. Unless otherwise specified, the competition environment is the same as that used for the experimental evolution. The following description presents the protocol used in the long-term serial-transfer experiment with *Escherichia coli*[13,25,26], but similar procedures are used in experiments with many microorganisms.

The two competitors are grown separately in the competition environment to ensure that they are comparably acclimated to the test conditions. They are then mixed (usually at a 1:1 ratio) and diluted (100-fold in this case) in the competition environment. Initial densities at timepoint $t = 0$ are estimated by diluting and spreading the cells on an indicator agar that distinguishes the evolved and ancestral types by colony colour, which differs owing to an engineered marker that is selectively neutral. In this case, red and white colonies correspond to Ara^- and Ara^+ phenotypes, respectively. After one day ($t = 1$) (corresponding to the serial-transfer cycle in the evolution experiment), final densities are estimated by plating cells, as before, on the indicator agar. The growth rate of each competitor is calculated as the natural logarithm of the ratio of its final density to its initial density (adjusted for dilution during plating). Relative fitness is then defined simply as the ratio of the realized growth rates of the evolved and ancestral types. *

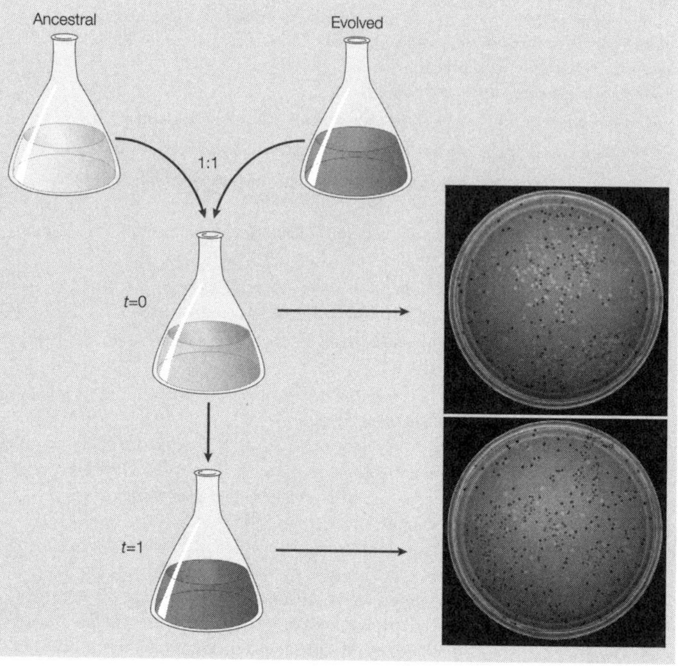

sometimes related and controversial claim that some mutations that produce new functions are either 'directed' by the organism or induced by stress[16,17]. Nor do we review experiments in which genotypes are deliberately engineered to test models of the relationship between genotype, phenotype and fitness. Reviews of these 'bottom-up' studies are also available elsewhere[18,19].

Dynamics of evolutionary adaptation

Adaptation by natural selection occurs through the spread and substitution of mutations that improve the performance of an organism and its reproductive success in its environment. An important focus of evolution experiments using microorganisms has been to investigate the dynamics of this process. Among the questions of general interest are whether genetic adaptation can continue indefinitely even in a constant environment[7,20], the magnitude of the contributions of individual mutations to fitness improvement[21,22] and the overall reproducibility of evolutionary changes[23,24].

One feature that is seen in several experiments, with both bacteria and viruses, is that fitness gains are initially rapid but tend to decelerate over time[13,25–30]. Such dynamics indicate that populations, after being placed in a new environment, are evolving from a region of low fitness towards an adaptive peak or plateau (FIG. 1). For example, in a long-term experiment with 12 *Escherichia coli* populations, the average fitness gain in the first 5,000 generations was approximately tenfold greater than that between 15,000 and 20,000 generations[26]. Even so, there was some significant improvement over the last interval, which indicated that the populations still had not reached their maximum fitness.

RANDOM DRIFT
The change in frequency of genotypes in a population that is caused by chance differences in survival and reproduction, as opposed to consistent differences in their fitness.

At first glance, it might seem surprising that the potential for genetic adaptation was not exhausted after thousands of generations in a constant environment. However, several factors contribute to continued adaptation. First, the amount of time that is required for a beneficial mutation to increase in frequency from a single individual to most of the population is inversely proportional to its advantage. Given the population size in the long-term experiment with *E. coli*, a new mutation that has a 10% advantage would take approximately 250 generations to become the majority status[13]. By comparison, a mutation with only a 0.1% advantage would require 25,000 generations to reach that frequency; so, there had not been enough time for such small beneficial mutations to have been substituted in that experiment. Second, many beneficial mutations are lost by RANDOM DRIFT while they are rare. The probability that a beneficial mutation survives extinction by drift is about twice its selective advantage[13]. A mutation with a 10% advantage requires on average five 'tries' before it is established, whereas a mutation with only a 0.1% advantage would need ~500 tries to avoid extinction by random drift. Both of these factors imply that adaptation using mutations with progressively smaller benefits can continue indefinitely, albeit ever more slowly, without depleting the supply of useful variants. Third, as discussed later, asexual populations are subject to 'clonal interference' that is caused by competition

* *This figure is reproduced in color following the index.*

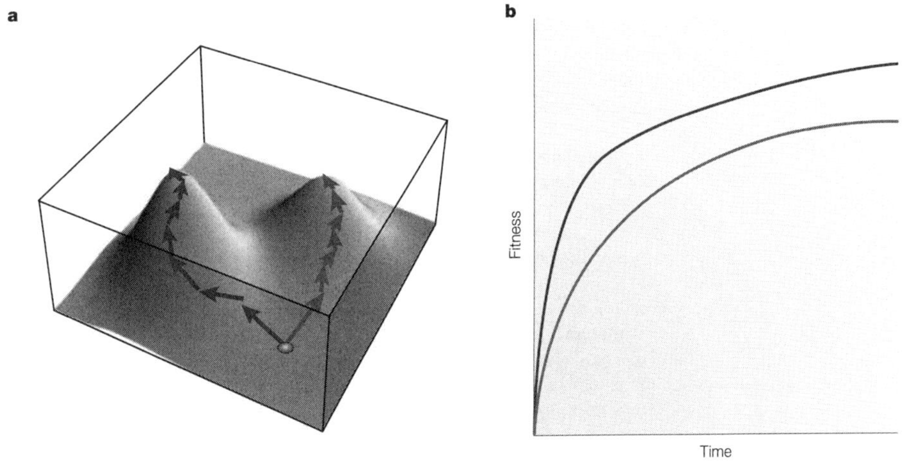

Figure 1 | **Fitness landscapes and evolutionary dynamics. a** | A hypothetical adaptive landscape with two fitness peaks. The red and green arrows show two of the possible trajectories for replicate populations that are founded from the same ancestral genotype and evolve independently in the same environment. **b** | Beneficial substitutions tend to have larger effects early in an experiment, when a population is far from an adaptive peak, than later as it approaches a local peak. Two replicate populations might reach different final fitness levels if, through the random effects of mutation and drift, they move into the domains of attraction of peaks of unequal height. *

among beneficial mutations that occur in different clones. The substitution of small beneficial mutations is especially affected by this phenomenon, further delaying their spread. Finally, it is likely that there are more mutations that confer small advantages than those that provide large benefits[21,22]. Hence, the supply of small beneficial mutations will not be exhausted as readily as the supply of large beneficial mutations (above and beyond the other dynamical effects).

Over the long term, fitness trajectories might seem smooth and continuous. However, if they are measured with sufficient temporal resolution, then fitness[13,25,31] and traits that are correlated with fitness (such as cell size[32] in *E. coli*) change with a step-like dynamic. Each step probably corresponds to the spread of a beneficial mutation[13,33]. The step-like aspect occurs because any new beneficial mutation must increase from a low initial frequency; during its ascendancy, it has little effect on mean fitness until it is present in a substantial fraction of the population. Also, whereas sexual reproduction allows two or more beneficial mutations to be substituted simultaneously, adaptation in asexual populations occurs by sequential substitutions that appear as successive steps.

The step-like dynamics of adaptive evolution provide one way of measuring the fitness effects of mutations that are substituted in evolving populations[13,25,31], and several other approaches have also provided such data[6,34,35]. These studies support some general points. Evolutionary adaptation in experimental microbial populations typically occurs through the substitution of relatively few mutations that confer large benefits, as opposed to countless mutations with small benefits. The percentage of all mutations that are beneficial is small, but is sufficient to allow adaptation given the supply of mutations even in populations that are relatively small by microbial standards[31,33–35]. It is also clear, from both theory and experiment, that the fitness gains of substituted mutations are not representative of the effects of beneficial mutations more generally. Instead, the most extreme beneficial mutations are greatly overrepresented owing to selection, and this bias is especially strong in asexual organisms as a result of clonal interference. As a consequence, it is difficult to test the assertion that the underlying distribution of beneficial mutations has many more with small than with large effects. Indirect support for this conjecture comes from the continued, but much slower, adaptation in the long-term *E. coli* populations after thousands of generations[25,26]. One study directly tested this hypothesis by varying the population size of the RNA virus $\phi6$ (REF. 31). Small populations tended to improve by more numerous, but smaller, steps than did large populations, which confirmed the preponderance of beneficial mutations with small effects in the underlying distribution.

As was briefly noted previously, an important consequence of asexuality is clonal interference. Clones that carry different beneficial mutations compete with one another and thereby interfere with each other's spread and substitution in the population[33]. In general, all but one lineage will be excluded by the clone with the most beneficial mutation or combination of mutations (unless several clones partition the environment). Clonal interference has several important implications. First, the probability of substitution of a given beneficial mutation should decline with increasing population size or mutation rate. Second, as population size or mutation rate increases, individual substitutions should entail larger fitness gains. Third, the rate of fitness improvement should show diminishing returns with

This figure is reproduced in color following the index.

an increasing supply of beneficial mutations that is caused by larger population size or higher mutation rate. Fourth, the rate of spread of a beneficial mutation should be slower than otherwise predicted from its fitness advantage. Fifth, many beneficial mutations should become transiently common but later be excluded by interfering mutations. Sixth, such transient dynamics might give rise to a 'leapfrog' event, in which the most common genotype at a given time is genealogically more distantly related to the immediately preceding dominant type than to an earlier dominant type. All of these effects have been reported in experiments with bacteria[34–38] and viruses[31,39–41].

Although founded by the same clone, and evolving in identical environments, REPLICATE POPULATIONS often diverge from one another in their relative fitness[13,25,42,43], demographic components of fitness[44], morphological features[13,42,44–46] and performance in other environments[26,46–49]. This divergence might indicate that they are approaching different local peaks in the adaptive landscape (FIG. 1), especially if fitness differences in the selective environment itself are sustained indefinitely. Experiments that start with different genotypes have also been run to examine how this affects the dynamics and extent of adaptation. In an experiment to investigate the role of historical contingencies, *E. coli* lines that had diverged greatly in their fitness on maltose as they evolved in glucose for 2,000 generations were then placed in a maltose environment and allowed to evolve for 1,000 generations[50]. Lines that started with the lowest fitness on maltose improved most rapidly, and all the lines tended to converge towards a similar fitness on that sugar. However, convergence does not always occur. Replicate populations that were founded by two different genotypes of virus ϕ6 consistently evolved to different fitness levels, which indicated that the descendants of one founder might have been 'trapped' in the domain of a lower fitness peak[51]. In other words, the mutational pathways that led from one founder to the higher fitness peak might have included maladapted intermediate genotypes that would be disfavoured by natural selection.

REPLICATE POPULATIONS
Two or more populations that started from the same ancestral genotype and were propagated under identical conditions as part of an evolution experiment. By having replicates in each of several environments, it is possible to distinguish statistically between systematic responses of the populations to a particular environmental feature (for example, temperature) and other responses that might reflect the chance effects of mutation and drift.

ISOGENIC
Genotypes that have been engineered to be identical, with the exception of one or more mutations of interest.

HITCHHIKING
The process by which a neutral, or even deleterious, mutation increases in frequency owing to its physical linkage with a beneficial mutation elsewhere in the genome.

EPISTASIS
Any non-additive interaction between two or more mutations at different loci, such that their combined effect on a phenotype deviates from the sum of their individual effects.

The genetic bases of evolutionary adaptation

Throughout the history of microbial genetics, most experiments have proceeded by disrupting organismal functions rather than improving them. This approach has been productive in terms of identifying the genes that encode the molecular components that allow various biochemical and physiological functions to occur. However, this emphasis on defective mutants does not provide much insight into how organismal function can improve. Evolution experiments, by contrast, offer opportunities to study beneficial mutations. Of course, the particular mutations that are beneficial will depend on the genomic and environmental contexts; for example, different genes and pathways might change as *E. coli* populations adapt to resource abundance versus scarcity. But several more general questions can be posed. What types of molecular event are involved in adaptation? Are point mutations more important than genomic rearrangements, or vice versa? What types of gene are affected? Are most beneficial mutations located in structural genes or in regulatory elements? Does the reproducibility of evolutionary outcomes at the phenotypic level usually indicate parallel mutations, or do similar phenotypic adaptations often arise from different underlying mutations?

Of course, the fact that one clone is more fit than another in a certain environment tells us nothing about the genetic basis of their difference. To achieve this understanding requires three steps: finding mutations that were substituted, manipulating ancestral and derived alleles to make clones that are ISOGENIC except for known mutations, and measuring the fitness consequences of those mutations in the relevant environment. The second and third steps are needed because neutral and even deleterious mutations might be fixed by random drift or by HITCHHIKING with beneficial mutations at other loci. Hitchhiking is especially important in asexual populations, in which the entire genome acts as a single linkage group. Sometimes a strong argument can be made that some mutations are beneficial without (or before) performing the last two steps. In particular, if several lines independently substitute the same or similar mutations (and if there is no reason to suspect hypermutability at those sites), then such parallel changes provide compelling evidence that they spread by selection and, hence, were adaptive.

The first challenge is to find mutations that distinguish between the evolved and ancestral genotypes. One approach is to sequence as many genes as possible; another is to focus on candidate loci that are indicated by particular phenotypic changes that have evolved, an understanding of selective factors in the environment, or both. For many viruses, it is now practical to sequence the entire genomes of several evolved lines and their ancestor. In viruses generally, and especially those with RNA genomes, mutations can accumulate quickly owing to high per-site mutation rates[52]. An interesting observation from sequencing viral genomes that have been obtained by experimental evolution is the extent of parallel changes at the nucleotide level across replicate lines that evolved in the same environment[28,41,53–55]. These parallel changes are presumably beneficial. In one study, a number of nucleotide substitutions in experimental lines of bacteriophage ϕX174 and S13 recapitulated evolution that also occurred in nature, based on the genome sequences of these closely related viruses[55]. In another experiment, several beneficial mutations were identified in ϕX174, one of which was moved into different genetic backgrounds that represented various intermediate evolutionary stages[56]. The benefit of this mutation was reduced in the later evolutionary stages, which indicated 'diminishing-returns' EPISTASIS. Such a pattern of gene interaction could contribute to the general pattern of decelerating fitness gains that was noted previously. In the RNA-encoded vesicular stomatitis virus (VSV), some parallel substitutions were synonymous and others were in non-coding regions, which indicated that selection might have been at the level of RNA folding or RNA–protein interactions[41].

For bacteria, it is not yet practical to obtain entire genome sequences for several evolved lines and their ancestor. However, it is possible to sequence many regions at random to infer the extent of genomic changes and perhaps find mutations for further study. One study, which used the 12 long-term *E. coli* lines, randomly chose 36 genes and sequenced 500-bp regions in four clones from each line and their ancestor[57]. Several mutations were found in a few lines that evolved mutator phenotypes (see later), but no mutations were found in any of the eight lines that retained functional DNA repair throughout the 20,000-generation experiment among the 18,374 bp that were sequenced from each clone. Although this study did not find any compelling mutations for further research, the data provide an important baseline against which to compare patterns of change at candidate loci. For example, using genomic arrays to look for parallel changes in gene expression in these lines, several candidate regulatory genes were identified including *spoT*, which encodes a protein that controls the level of the important effector molecule ppGpp (REF. 58). Sequencing *spoT* found point mutations that caused amino-acid replacements in 8 of the 12 lines, which indicated much more evolution than in the baseline of random genes. By moving an evolved *spoT* allele into the ancestral genome and running fitness assays, it was confirmed that the mutation was beneficial in the SERIAL-TRANSFER regime that was used in the long-term evolution experiment[58].

In CHEMOSTAT cultures, bacteria are faced with perpetual resource limitation. The transport systems for the limiting resource are probable targets of selection, and the genes that encode those systems are, therefore, candidate loci. Consistent with these expectations, *E. coli* that evolved in lactose-limited chemostats substituted mutations in *ompF* that improved permeability across the outer membrane of the nonspecific OmpF PORIN by reducing channel constriction[59]. In glucose-limited chemostats, *E. coli* evolved diverse mutations at several loci that increased glucose permeability through the LamB porin and the binding protein-dependent transport of glucose across the inner membrane into the cell[60–62]. Several clones with different alleles at these loci often increased in tandem, probably indicating clonal interference. In another study of *E. coli* in glucose-limited chemostats, several clones coexisted through a cross-feeding interaction, in which one type secreted acetate that another used as a resource; mutations in regulatory regions upstream of *acs*, which encodes acetyl-coA synthetase, were partly responsible for this interaction[63,64]. Although *E. coli* in chemostats face limiting resources, their growth rate might still be much faster than they can usually achieve in nature. By allowing *E. coli* that had been recently taken from nature to evolve for 280 generations in chemostats, it was shown that the bacteria tended to converge on the rapid-growth phenotype of strains with long histories in the laboratory[65]. Changes in the kinetic properties of the ribosomes, which increased their translational efficiency, were responsible for much of this adaptation.

Bacteria that are kept indefinitely in STATIONARY PHASE experience even more severe limitations than those in chemostats, because no new resources are provided following the initial growth in fresh medium. To survive, cells must scavenge whatever becomes available through excretion or death in an otherwise starving population. After a period of mortality, mutants emerge that can survive and grow, albeit slowly, under such conditions[66–68]. Null mutations in *rpoS*, which encodes the σ^S transcription factor, confer this advantage[66]. Other mutations, including some that enhance amino-acid catabolism under carbon starvation, also emerge in successive rounds of adaptation to the increasingly dire conditions[67,68].

From *spoT* in the serial-transfer regime[58] to *rpoS* during prolonged starvation[66], many of the evolutionarily important mutations are found in global regulatory genes, rather than in genes that might improve single enzymatic steps. This conclusion is also supported by studies that have found, after a few hundred generations, widespread changes in patterns of protein expression in a chemostat-evolved population of *E. coli*[69] and parallel changes in the transcription levels of many genes that are involved in central metabolism in three chemostat-evolved lines of *Saccharomyces cerevisiae*[70]. Beside these global regulators, many adaptive mutations involved specific regulatory changes, including the increased glucose transport[60] and acetate cross-feeding[64] in chemostat-evolved *E. coli* populations. Evolved changes in gene expression similarly implicate mutations in specific regulatory elements during adaptation by *E. coli* to high temperature[71] and by *Candida albicans* to the antifungal compound fluconazole[72,73]. Also, several studies in which bacteria have been exposed to substrates that they cannot normally use have shown that regulatory mutations that cause the increased expression of an enzyme with marginal activity on the substrate are important in the early stages of acquiring new catabolic functions[14,15]. By showing that substantial adaptation can involve a few mutations in regulatory genes, microbial evolution experiments support the famous conjecture by Mary-Claire King and Allan Wilson, based on the high degree of genetic similarity between humans and chimpanzees, that relatively few changes in regulatory genes might be responsible for important phenotypic differences[74].

Diverse mutations emerge in evolution experiments, including point mutations, small insertions and deletions that cause frame shifts, and larger rearrangements. These rearrangements usually involve transposable elements that generate insertions, as well as inversions and deletions, through recombination between homologous elements in yeast[75,76] and bacteria[37,77–80]. These elements are active in starving, as well as growing, populations. Some experiments have found sustained bursts of transposition that lead to an increase in the copy number of particular elements[37,75]. The underlying causes of these bursts are not well understood, but they might reflect a type of mutator activity (see later). Beyond their inherent interest, transposable elements are useful foci for genetic analyses of experimental lines because the mutations they cause are usually easier to find by molecular

SERIAL TRANSFER
A culture regime in which some proportion of a population is periodically diluted into fresh medium, in which the population grows until it exhausts the limiting resource and then waits until the next transfer cycle. Selection favours rapid exponential growth as well as the ability to respond quickly following transfer into fresh medium.

CHEMOSTAT
A device that allows the continuous growth of a bacterial population on a growth-rate-limiting resource. The resource flows into the chemostat at a constant rate; depleted medium and cells are washed out at the same rate. The population grows and consumes the resource until the bacteria reach an equilibrium density at which their growth rate equals the flow rate through the vessel.

PORIN
A protein channel across the outer membrane of a Gram-negative bacterium that allows the diffusion of molecules into the periplasm, which is located between the outer and inner membranes.

STATIONARY PHASE
The period in a serial-transfer regime after the limiting resource has been depleted, such that population growth ceases. A population can be kept in this phase indefinitely by never transferring it to fresh medium, and it eventually declines owing to starvation.

methods than are point mutations. In some cases, different replicate lines have substituted insertions and point mutations in the same gene, which implies that either type of mutation can produce a similar advantage[64]. In other cases, one type of mutation might prevail. For example, the 12 long-term *E. coli* lines lost the ability to catabolize ribose as a result of various deletions that were all mediated by an IS*150* element located just upstream of the *rbs* operon[79]. Genetic manipulations confirmed that such mutations provided a fitness advantage in the glucose-limited environment in which the deletions had evolved.

Tradeoffs and the specificity of adaptation

Sets of related genotypes, populations and species often show tradeoffs in their relative fitness across different environments[11,81]. Indeed, without tradeoffs, a single type would be expected to prevail across all environments, precluding any comparison among the different types. If individuals usually encounter only one of these environments, then tradeoffs will tend to promote the evolution of specialists (FIG. 2). By contrast, if most individuals encounter a mix of environments, this might favour a generalist type that has the highest average performance, even if it is suboptimal in any constant environment[11,82]. In principle, several mechanisms can produce tradeoffs. The simplest mechanism is antagonistic PLEIOTROPY (AP), in which a particular mutation that is beneficial in one environment is harmful in the other. A second mechanism is mutation accumulation (MA), in which mutations accumulate by drift in genes the products of which are not used in one environment but are useful in another. These mutations are, therefore, neutral in the environment in which they were substituted, but deleterious in the other environment. The third mechanism that can produce tradeoffs is the independent adaptation of organisms to alternative environments. If each of two populations substitutes a mutation that is beneficial in one environment and neutral in the other, then each population will be more fit in one environment than the other. A population does not suffer a decline in fitness relative to its progenitor under this third mechanism, unlike the first two. Under all three mechanisms, the net effect is a tradeoff in which different genotypes, populations or species are maximally fit in alternative environments. Although tradeoffs are widespread in nature[81], the underlying mechanisms are rarely known. Evolution experiments with microorganisms offer the opportunity to distinguish among the mechanisms.

AP or MA? There are many compelling examples of AP. Several experiments in which *E. coli* have evolved resistance to virulent phage show tradeoffs, such that the resistant bacteria are inferior competitors against their ancestors in the absence of the phage[83–85]. These cases represent AP, and not MA, because the evolving bacteria would still benefit from being efficient competitors for resources; selection for resource use was not eliminated, but selection for resistance was added. The cost of resistance (the magnitude of fitness loss in the phage-free environment) varies for different phage[84] and even among mutations that confer resistance to the same phage. Resistance to phage T4 occurs by mutations that cause defects in the lipopolysaccharide core of the cell envelope, and the cost is greater for mutations that produce defects that are more basal in this structure[85]. Another example of AP comes from *E. coli* that evolved in lactose-limited chemostats; the same mutations that enhanced the permeability of the nonspecific OmpF porin to lactose also increase the susceptibility of the cells to certain antibiotics[59]. In the virus ϕX174, several mutations that had beneficial effects at high temperatures reduced fitness at lower temperatures[56].

During a long experiment, many mutations will be substituted in an evolving population. It therefore becomes difficult to test whether the same mutations that produce adaptation to one environment cause tradeoffs in other environments (AP), or whether different mutations produce the direct fitness gain and the correlated losses (MA). Even without testing each mutation, the temporal dynamics of change can give insights into the underlying process. During 20,000 generations on glucose, 12 *E. coli* populations tended to evolve reduced catabolic function against a battery of other substrates[26]. This decay was fastest early in the experiment, which mirrored the trajectory of fitness gains on glucose, as expected for AP. Also, several populations evolved mutator phenotypes that had ~100-fold higher mutation rates, but showed only slight increases in their rates of catabolic decay, contrary to the expectation

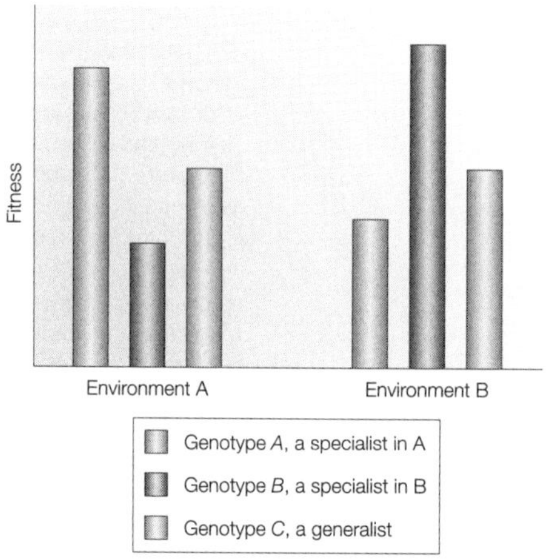

Figure 2 | **Tradeoffs and specificity of adaptation.** The fitness of three genotypes in two environments. The blue and red bars show specialists that are adapted to environments A and B, respectively. Although both genotypes perform well in their respective environments, each is poorly adapted to the other environment. One possible explanation for this negative fitness correlation, or tradeoff, is that those mutations that are beneficial in one environment have antagonistic pleiotropic effects in the other. The green bars show a generalist that performs moderately well in both environments but has lower fitness than either specialist in its preferred environment. If environments vary in space or time, the generalist might have an overall advantage.*

PLEIOTROPY
The side-effect of a mutation that affects a primary trait or function on a secondary trait or function.

*This figure is reproduced in color following the index.

under MA. Another study with evolving *E. coli* populations found very different results — the decay rate of unused catabolic functions was much faster in mutator lines than in lines with functional DNA repair, and their rate of decay did not decelerate over time[86]. At a first glance these two studies seem contradictory, but there is a crucial difference that, once understood, explains the different outcomes. In the study in which the effect of MA was dominant, the lines were propagated through single-cell BOTTLENECKS that amplify the effects of random drift and eliminate the role of selection. Indeed, these lines became much less fit even under the conditions of their evolution, which indicated that there was no adaptation but only decay.

The evolutionary effects of mutators and very small populations are discussed further in two subsequent sections.

Another evolution experiment with *E. coli*, this one carried out in germ-free mice, found evidence for MA and showed that adaptive evolution had also occurred[87]. In particular, mutator populations had a fivefold higher load of maladapted AUXOTROPHIC mutants than did populations with functional DNA repair. However, this study cannot evaluate the relative contributions of AP and MA to tradeoffs, because it did not test for any signal of AP. More generally, it should be recognized that AP and MA are not mutually exclusive, as both processes can occur in the same population[26].

An experiment with *E. coli* that evolved for 2,000 generations under several temperature regimes found a high level of thermal specificity — all the lines improved in fitness relative to their ancestor at the temperatures at which they evolved[47]. However, in many cases, these lines did not lose fitness relative to their ancestor at nearby temperatures. This pattern corresponds to independent adaptation, as defined at the beginning of this section. Across a wider range of temperatures, the situation was more complex. Most lines that evolved at 20 °C lost fitness relative to the ancestor when they competed at 40 °C and above[88], whereas most lines that evolved at 41.5 °C did not lose fitness at 20 °C and below[47], which indicated asymmetries in correlated responses to selection in different environments. Moreover, although five lines that adapted to high temperature had no fitness loss relative to the ancestor at low temperature, one line did, which shows that correlated responses can be heterogeneous among replicate lines. Complex patterns of correlated responses are also evident when *E. coli* evolve on either glucose or maltose, with all other environmental factors held constant. Lines that have evolved on glucose show no improvement, on average, if competed against their ancestor on maltose, and the glucose-adapted lines are highly variable in their fitness on maltose[89]. By contrast, lines that have evolved on maltose show consistent fitness gains on glucose as well as on maltose[90]. This asymmetry indicates that the genetic adaptations to maltose might be a subset of the adaptations to glucose[90]. Several studies have also shown that bacteria that evolved at a particular resource concentration showed greater fitness improvement when tested at that same concentration than at other concentrations[44,49,91,92]. However, despite this specificity of adaptation with respect to concentration, the overall similarity between the test environments was such that tradeoffs were usually absent and most correlated responses were positive.

Generalists and specialists. In *Chlamydomonas* that had been subjected to alternating light and dark conditions for several hundred generations, generalists evolved that were more fit than their ancestor under both conditions[93]. However, the generalists were not as fit in either constant regime as specialists that had evolved under the corresponding condition. Similar outcomes were obtained with *E. coli* lines that evolved with either alternating or constant temperatures[47]. These patterns conform to the expectation shown in FIG. 2.

Several experiments with viruses that are able to infect more than one host have found that viruses that evolved on one host became less fit (or at least did not improve) on alternative hosts[94–98]. These results imply tradeoffs and that viral adaptation is host specific. However, if viral populations evolved on two alternating hosts, they sometimes improved as much on each host as those that had evolved on a single host[96–99], in apparent contradiction to the evidence for tradeoffs and host specificity. To reconcile these results, it is possible to imagine that there are two classes of beneficial mutations. One class is beneficial only on a particular host, and has antagonistic pleiotropic effects during growth on other hosts; mutations in genes that affect interactions with host receptors or other host-specific molecules are candidates for this class. The other class produces beneficial effects on all hosts; mutations in genes that are involved in RNA processing and elongation might be candidates for this class. Even if mutations with host-specific benefits were more common than the generally beneficial mutations, the latter class would be differentially enriched in viral populations that evolved on alternating host types. Further studies to identify and characterize both the direct and correlated fitness effects of individual mutations in viruses that evolved on single and alternating hosts would allow a test of these ideas. Such work is also relevant for developing attenuated vaccines and managing parasite virulence[100].

Emergence and consequences of mutators

A 'mutator' is a genotype that has an increased mutation rate throughout its genome owing to a mutation that disrupts some aspect of DNA replication or repair[101]. This effect can be large; in bacteria that have become defective in the methyl-directed mismatch-repair pathway, for example, the genomic mutation rate is increased by the order of 100-fold (REF. 101). Genotypes with enhanced activity of a transposon might also behave like mutators, because the genome-wide rate of insertions is increased[102]. The past few years have seen growing interest in mutators from various perspectives, including theoretical modelling[33,103] and surveys of natural populations[104,105], as well as evolution experiments. The experiments aim to understand how mutators

BOTTLENECK
A severe reduction in population size that causes the loss of genetic variation. The role of random drift is increased, whereas the power of selection is reduced, by bottlenecks.

AUXOTROPHIC
A mutant that cannot synthesize a required nutrient, such as an amino acid.

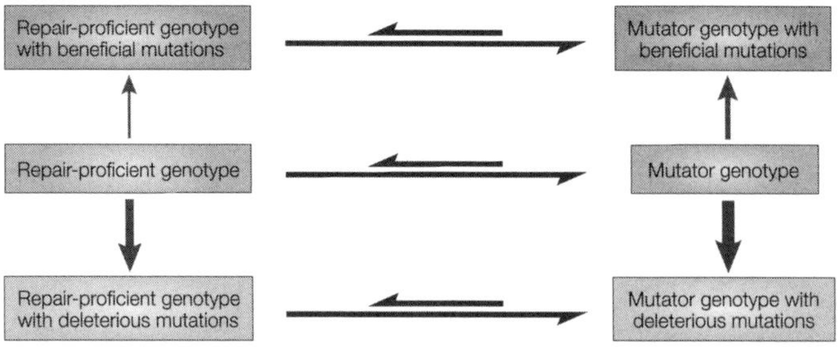

Figure 3 | **Role of mutators in generating variation.** Both mutator and normal (DNA-repair proficient) bacteria produce more deleterious than beneficial mutations. On a *per capita* basis, mutators produce more of both types. During adaptation to a new environment, mutators might promote faster adaptation by producing more beneficial mutations than do normal bacteria. However, this advantage is offset by the greater load of deleterious mutations that mutators produce. Figure modified with permission from REF. 103.

reach high frequencies in populations given their increased GENETIC LOAD, and the evolutionary consequences of their spread. Towards these goals, some experiments showed the spontaneous emergence of mutators in populations that were founded by non-mutators with functional DNA repair[38,106–108], whereas others deliberately introduced mutators and documented the effects on rates of adaptive evolution[36,87,109].

At first glance, it is tempting to directly link the emergence and consequences of mutators by assuming that they reach high frequency because they accelerate adaptive evolution. Although this view might have some merit, it also presents several difficulties that demand a closer look. First, beneficial mutations are much less common than are deleterious mutations, and mutators suffer from a higher load of deleterious mutations (FIG. 3). The cost of an elevated mutation rate in terms of a higher load, coupled with the potential benefit of increased evolvability, represents a tradeoff. Such a cost also explains why most organisms retain DNA-repair functions. Second, the potential advantage of a high mutation rate in terms of accelerating adaptive evolution is limited in large asexual populations[33,36]. In large populations, beneficial mutations occur even at low mutation rates, and asexuality gives rise to clonal interference which impedes the substitution of many beneficial mutations. Third, the view that mutators can become common by promoting adaptive evolution is teleological, or goal directed; that is, mutators might accelerate adaptive evolution once they become sufficiently common in a population to be an important contributor to the supply of beneficial mutations, but that cannot explain how they become common enough to have this effect.

This third point was shown by mixing different initial ratios of mutator and normal (repair proficient) bacteria, and propagating the mixtures until a beneficial mutation emerged in one clone that took over the population[109]. When the initial ratio of mutator to normal cells was low (for example, 1:1000), the mutators tended to decline slowly in the short term owing to their greater load of deleterious mutations, and then much more quickly after the normal clone acquired a beneficial mutation. Although each mutator cell had a higher *per capita* probability of acquiring the first beneficial mutation than a normal cell, the combined probability of the mutator clone was lower than that of the normal clone owing to the difference in their total numbers. By contrast, if the initial ratio of mutator to normal cells was increased sufficiently (for example, 1:10), then the mutator clone prevailed, following a slight initial decline, because it was more likely to acquire the first beneficial mutation. A series of such experiments showed a threshold ratio below which the normal cells generally prevailed and above which the mutators usually won. This threshold is understood by realizing that the production of beneficial mutations depends on the product $N\mu_B$, in which N is the cell number and μ_B is the beneficial mutation rate. If this product is greater for the mutator clone than for the normal clone, a mutator will probably produce the first beneficial mutation, and vice versa. This experiment shows the difficulty of understanding how a mutator can spread after it first emerges in a population.

Nonetheless, mutators can become common and take over a population. In the long-term experiment with *E. coli*, 3 of the 12 populations spontaneously evolved into mutators in 10,000 generations[107] and a fourth by 20,000 generations[26]. The mutations that produced these mutator phenotypes provided no direct competitive advantage, but spread by hitchhiking with beneficial mutations elsewhere in the genome[38]. But how did a mutator clone become common enough to produce a beneficial mutation? The answer lies in understanding that the critical frequency of mutators, explained previously, represents a stochastic and not a deterministic threshold; in other words, there is a certain probability that a mutator, although its frequency is below the threshold, will produce the next beneficial mutation. Also, although mutators are purged by selection owing to their load of deleterious mutations, new mutators are constantly generated by mutations in genes that encode DNA-repair functions, giving rise to a quasi-equilibrium frequency of mutators. Although this frequency is below the threshold, each beneficial substitution in an evolving population provides another opportunity for a mutator to produce that mutation and hitchhike with it. If 10 beneficial substitutions occurred in every population, then each population had 10 chances to be converted to a mutator. Although the odds were against such conversion in any single case, the 12 populations collectively had 120 opportunities for conversions. Calculations using rough estimates of the relevant parameters (for example, the rate of mutation to mutator status) support this model[110]. Following a conversion in which a mutator is substituted, the process is unlikely to be reversed until a population has become so well adapted to its present environment that the best mutation is one that reduces genetic load by restoring the lost repair function. However, a mutator might rise transiently to high frequency, then be eliminated, if the non-mutator type produces an even more beneficial mutation than that produced by the mutator[103,108].

GENETIC LOAD
The loss of fitness that is caused by producing offspring that carry deleterious mutations, and the resulting decrease in the rate of population growth.

In the long-term *E. coli* experiment, populations that became mutators showed fitness gains that were only slightly, if at all, greater than those that remained DNA-repair proficient[26,107]. This finding indicates that evolvability *per se* might not have increased, despite the evidence that mutators spread by hitchhiking with beneficial mutations. This paradox highlights an important distinction between the causes and consequences of mutators. Theory indicates that the extent to which a mutator will accelerate adaptive evolution depends crucially on population size[33]. When populations are moderate in size, a mutator can accelerate adaptive evolution by shortening the 'waiting time' for a beneficial mutation to emerge. However, as population size becomes very large, even a low mutation rate suffices to generate beneficial mutations. Assuming that the population is asexual, clonal interference means that only one beneficial mutation can be substituted at a time. Hence, there is a 'speed limit' on the rate of adaptation in asexual populations[33]. The predictions of this model were supported by experiments with *E. coli* and VSV in which population size and mutation rate were manipulated and the effects on the rate of adaptation were measured[36,39,40].

Let us summarize the present understanding of the emergence and consequences of mutators. Rapid adaptive evolution, as often occurs when a population encounters a new environment, causes many beneficial mutations to be substituted, and every such substitution provides another opportunity for a mutator to emerge. Whether mutators appreciably accelerate adaptive evolution depends on population size. In populations of moderate size, mutators might accelerate adaptation by reducing the waiting time for beneficial mutations. But in large populations, mutators might not accelerate adaptive evolution because the supply rate of beneficial mutations is not limiting. Two other factors also become important under certain conditions. First, as discussed in the section on tradeoffs, the accumulation of mutations in genes that experience relaxed selection will cause more rapid fitness loss by mutators if they later encounter environments in which those genes are important[26,87]. Second, as discussed in the next section, mutators are deleterious in small populations that fight a losing battle between random mutation and drift, on one side, and selection to remove deleterious mutations, on the other.

Drift and decay in very small populations

So far, we have focused on adaptive evolution, which can occur because selection finds and amplifies rare beneficial mutations by means of differential survival and reproduction. But selection loses its discriminating power in very small populations in which success becomes a matter of chance. At the extreme limit, at which a population has only a single individual, there is no variation on which selection can act. Mutations do not stop, however, and are substituted at random. Over time, the average number of mutations in the genome increases (FIG. 4) and fitness declines because many more mutations are harmful than are beneficial. So, how and why would one study this process of genetic decay?

It would be difficult to follow a single cell, and remove one of two daughter cells when it divides, for many generations. However, it is possible to achieve a similar effect by periodically plating a population of cells (or viruses), randomly choosing a single colony (or plaque) and repeating the process indefinitely (FIG. 4). Each colony contains millions of individuals, but they are all derived from a single individual. Hence, this procedure is a simple way to subject a population to extreme bottlenecks. Although the population grows and mutations occur in the intervening generations, variation is eliminated at every bottleneck, such that random mutation and drift dominate the evolutionary process[111,112]. By measuring the rate of fitness loss as well as the variation among replicate lines in the extent of their loss, it is possible to estimate the genomic rate of deleterious mutations and their average effect[113,114]. The shape of the decay trajectories might also provide information on epistatic interactions between mutations[115,116]. Besides genetic interest in these quantities, they are important for understanding the evolution of sex[111,116–118], and the survival and evolution of small populations[119–121].

Experiments with repeated single-individual bottlenecks have been carried out with several RNA viruses[122–126], a retrovirus[127], bacteria[128,129] and yeast[130,131]. In all these studies fitness declined, but the estimated mutational parameters differed greatly. The spontaneous deleterious mutation rate in the RNA-encoded VSV was estimated as ~1 per genome per generation[125], whereas for *E. coli*, the corresponding rate was only $\sim 2\times10^{-4}$ (REF. 128). Estimates from yeast are even lower[130,131], but the rate is increased in a mutator strain[121,131]. It is not surprising that RNA viruses, despite their small genomes, have much higher rates of deleterious mutation than do cellular forms that possess DNA-repair mechanisms. Some of these estimates might be too low, however, because the experiments lacked the ability to resolve mutations that are only slightly deleterious[130]. Estimates of the average fitness effects of deleterious mutations also vary, with the largest values in yeast and the lowest in VSV. However, it is unclear whether the differences in average effect are biologically meaningful or, instead, might reflect a statistical problem of estimating the mutation rate and average effect from the same data[113,114].

It is possible that RNA viruses, by virtue of their high mutation rates, have been selected to minimize the harmful effects of mutation by moving into regions of SEQUENCE SPACE in which many mutations have less impact on performance. This process has been shown experimentally with DIGITAL ORGANISMS — self-replicating computer programs that can mutate and evolve[132]. There are also other ways to reduce the harmful effects of mutations. For example, molecular CHAPERONES help mediate the proper folding of proteins, perhaps including those that might otherwise fold improperly owing to mutations. Their elevated expression has been proposed to increase the mutational robustness of endosymbiotic bacteria, which face severe bottlenecks and, therefore, the potential for decay by random drift and mutation[120].

SEQUENCE SPACE
The universe of all possible sequences or genotypes. For example, even a small viral genome of 1,000 nucleotides has 3,000 one-step neighbours, nearly 9,000,000 two-step neighbours, and more than 10^{600} variants at all possible distances of the same genome length.

DIGITAL ORGANISMS
A type of computer-based artificial life that can be used to investigate certain scientific questions. The genomes of digital organisms are computer programs and, like computer viruses, are able to self-replicate. Digital organisms can also mutate and evolve spontaneously, whereas computer viruses are deliberately modified by hackers.

CHAPERONES
A class of protein that binds to other proteins and thereby promotes their proper folding during synthesis or following damage.

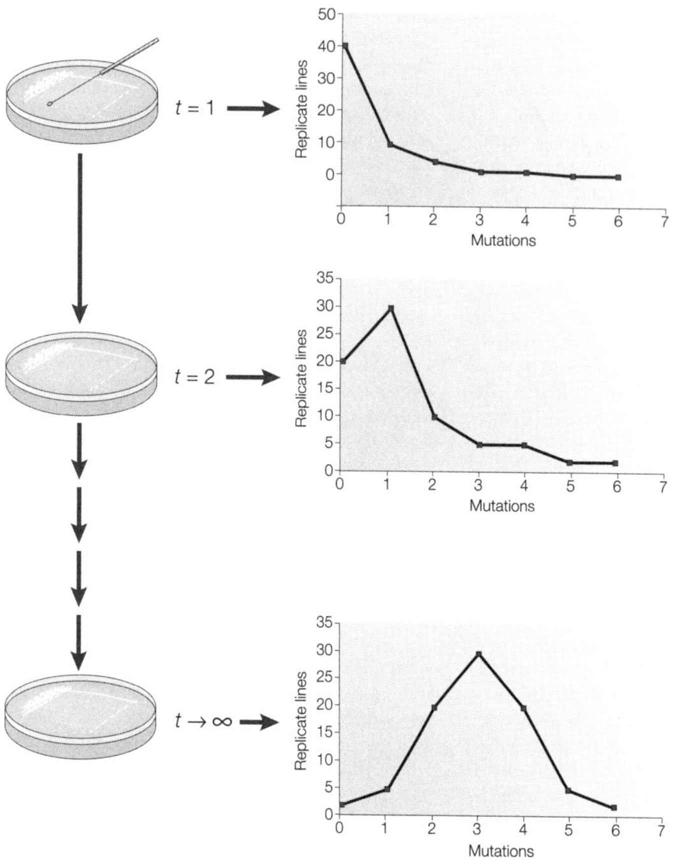

Figure 4 | **Population bottlenecks and accumulation of deleterious mutations.** The panel on the left illustrates a mutation-accumulation experiment with bacteria in which a population that is evolving over time (*t*) is repeatedly passed through single-cell bottlenecks. The bottlenecks are achieved by growing the bacteria on agar plates as discrete colonies, each of which is derived from a single cell, and then taking the cells from a single colony and spreading them over a fresh agar plate. A bottleneck size of 1 effectively eliminates natural selection and maximizes the effects of random drift. Mutations, therefore, accumulate at random, in contrast to evolution in large populations in which selection favours certain mutations and eliminates others. The graphs on the right show the changing distribution of numbers of mutations in a set of bottlenecked lines. The number of mutations tends to increase, although different lines will, by chance, accumulate fewer or more mutations, which leads to an increased variance among lines over time. Because more mutations are deleterious than are beneficial, fitness also generally declines in mutation-accumulation experiments.

To test this hypothesis, *E. coli* lines that had accumulated deleterious mutations were engineered to overexpress the GroEL chaperone protein[129]. This change reversed over half of the harmful effects of the accumulated mutations in one of two test environments. Individual deleterious mutations can often be compensated by mutations elsewhere in the genome, although this process would not enhance robustness in general[133]. Also, sexual reproduction might provide a mechanism to purge deleterious mutations from populations[116–118]. These last two processes will be examined in detail in our second review of evolution experiments with microorganisms.

Conclusions

The experimental approach to studying evolution, especially using microorganisms, has greatly expanded over the past decade, with a wealth of studies that address a wide range of issues. We have focused here on those studies that inform our understanding of the temporal dynamics of evolutionary adaptation, the genetic changes that underlie adaptation, the causes and generality of tradeoffs during adaptation, the emergence and effects of mutator genotypes, and the process of genetic decay in very small populations. With respect to these issues, microbial evolution experiments support the following generalizations. First, populations adapt rapidly when they are introduced into new environments. However, they might continue to improve indefinitely, albeit slowly, even in a constant environment because beneficial mutations with ever smaller effects become increasingly accessible to selection. Second, genetic comparisons of ancestral and evolved organisms provide striking examples of parallel molecular evolution in replicate populations, including cases of adaptive mutations in genes that encode important global regulators. In contrast to these genetic targets of selection, most genes do not change, even over thousands of generations. Third, genetic adaptation to one environment is often, but not always, associated with fitness loss in other environments. Antagonistic pleiotropy is responsible for many of these tradeoffs, although elevated mutation rates can also reduce ecological breadth owing to mutation accumulation in genes that are under relaxed selection. Fourth, rapidly evolving asexual populations provide repeated opportunities for hypermutable genotypes to spread, along with the beneficial mutations that they occasionally generate. However, the emergence of such mutators does not always substantially accelerate adaptive evolution. Finally, in very small populations, the random processes of mutation and drift overwhelm the capacity of natural selection to retain well-adapted genotypes, and fitness tends to decline; mutators hasten the genetic decay in such populations.

In the future, we look forward to the increased integration of genetic and phenotypic analyses, as well as to the improved temporal resolution of evolutionary dynamics. As an example, how is it possible to reconcile the evidence for the phenotypic divergence of replicate populations with the findings of parallel molecular changes? We see three possibilities. According to the first, different mutations in the genes that show parallel adaptation have heterogeneous pleiotropic effects on correlated traits. The second explanation proposes that although many genes undergo parallel evolution, it is unique adaptive mutations at other loci that are responsible for this phenotypic divergence. The third hypothesis is that mutations in genes that are under relaxed selection, which do not contribute to adaptation during the evolution experiment, cause the phenotypic divergence.

We also anticipate growing interest in evolution experiments that are carried out *in silico*, including simulation models that are based on microorganisms and studies that use abstract digital organisms[132,134–137]. These experiments will allow more detailed analyses of complex phenomena and, in turn, might stimulate increasingly sophisticated evolution experiments with real organisms. Already, many microbial evolution experiments have begun to explore the interesting dynamics that can emerge from complex interactions among several components of a larger system. Our second article for *Nature Reviews Genetics* will, therefore, focus on evolution experiments that address some of the fascinating interactions between multiple mutations in the same genome, between different genotypes in a population, and between interacting microbial species.

1. Zuckerkandl, E. & Pauling, L. in *Evolving Genes and Proteins* (eds Bryson, V. & Vogel, H. J.) 97–166 (Academic Press, New York, 1965).
2. Harvey, P. H. & Pagel, M. D. *The Comparative Method in Evolutionary Biology* (Oxford Univ. Press, Oxford, UK, 1991).
3. Darwin, C. *On the Origin of Species by Means of Natural Selection* (Murray, London, 1859).
4. Chadwick, D. J. & Goode, J. (eds) *Antibiotic Resistance: Origins, Evolution, Selection and Spread* (Wiley, Chichester, UK, 1997).
5. Grant, P. R. *Ecology and Evolution of Darwin's Finches* (Princeton Univ. Press, Princeton, 1999).
6. Atwood, K. C., Schneider, L. K. & Ryan, F. J. Periodic selection in *Escherichia coli*. *Proc. Natl Acad. Sci. USA* **37**, 146–155 (1951).
7. Ryan, F. J. Evolution observed. *Sci. Am.* **189**, 78–82 (1953).
8. Hairston, N. G. Evolution under interspecific competition: field experiments on terrestrial salamanders. *Evolution* **34**, 409–420 (1980).
9. Woese, C. R. & Fox, G. E. Phylogenetic structure of the prokaryotic domain: the primary kingdoms. *Proc. Natl Acad. Sci. USA* **74**, 5088–5090 (1977).
10. Levin, B. R. & Bergstrom, C. T. Bacteria are different: observations, interpretations, speculations, and opinions about the mechanisms of adaptive evolution in prokaryotes. *Proc. Natl Acad. Sci. USA* **97**, 6981–6985 (2000).
11. Bell, G. C. *Selection* (Chapman & Hall, New York, 1997).
12. Holland, J. J., de la Torre, J. C., Clarke, D. K. & Duarte, E. A. Quantitation of relative fitness and great adaptability of clonal populations of RNA viruses. *J. Virol.* **65**, 2960–2967 (1991).
13. Lenski, R. E., Rose, M. R., Simpson, S. C. & Tadler, S. C. Long-term experimental evolution in *Escherichia coli*. I. Adaptation and divergence during 2,000 generations. *Am. Nat.* **138**, 1315–1341 (1991).
14. Hall, B. G. in *Evolution of Genes and Proteins* (eds Nei, M. & Koehn, R. K.) 234–257 (Sinauer, Sunderland, Massachusetts, 1983).
15. Mortlock, R. R. (ed.) *Microorganisms as Model Systems for Studying Evolution* (Plenum, New York, 1984).
16. Sniegowski, P. D. & R. E. Lenski. Mutation and adaptation: the directed mutation controversy in evolutionary perspective. *Ann. Rev. Ecol. Syst.* **26**, 553–578 (1995).
17. Roth J. R. *et al.* Regulating general mutation rates: examination of the hypermutable state model for Cairnsian adaptive mutation. *Genetics* **163**, 1483–1496 (2003).
18. Dykhuizen, D. E. & Dean, A. M. Enzyme activity and fitness: evolution in solution. *Trends Ecol. Evol.* **5**, 257–262 (1990).
19. Dykhuizen, D. E. in *Population Genetics of Bacteria* (eds Baumberg, S., Young, J. P. W., Saunders, S. R. & Wellington, E. M. H.) 161–173 (Cambridge Univ. Press, Cambridge, UK, 1995).
20. Dykhuizen, D. E. Experimental studies of natural selection in bacteria. *Ann. Rev. Ecol. Syst.* **21**, 373–398 (1990).
21. Fisher, R. A. *The Genetical Theory of Natural Selection* (Oxford Univ. Press, Oxford, UK, 1930).
22. Orr, H. A. The population genetics of adaptation: the distribution of factors fixed during adaptive evolution. *Evolution* **52**, 935–949 (1998).
23. Gould, S. J. *Wonderful Life: the Burgess Shale and the Nature of History* (Norton, New York, 1989).
24. Conway Morris, S. *The Crucible of Creation* (Oxford Univ. Press, Oxford, UK, 1998).
25. Lenski, R. E. & Travisano, M. Dynamics of adaptation and diversification: a 10,000-generation experiment with bacterial populations. *Proc. Natl Acad. Sci. USA* **91**, 6808–6814 (1994).
26. Cooper, V. S. & Lenski, R. E. The population genetics of ecological specialization in evolving *Escherichia coli* populations. *Nature* **407**, 736–739 (2000).
This article shows the dynamics of adaptation to glucose over 20,000 generations, along with resource specialization that is caused primarily by antagonistic pleiotropy.
27. Novella, I. S. *et al.* Exponential increases of RNA virus fitness during repeated transmission. *Proc. Natl Acad. Sci. USA* **92**, 5841–5844 (1995).
28. Bull, J. J. *et al.* Exceptional convergent evolution in a virus. *Genetics* **147**, 1497–1507 (1997).
29. Elena, S. F. *et al.* Evolutionary dynamics of fitness recovery from the debilitating effects of Muller's ratchet. *Evolution* **52**, 309–314 (1998).
30. De Visser, J. A. G. M. & Lenski, R. E. Long-term experimental evolution in *Escherichia coli*. XI. Rejection of non-transitive interactions as cause of declining rate of adaptation. *BMC Evol. Biol.* **2**, 19 (2002).
31. Burch, C. L. & Chao, L. Evolution by small steps and rugged landscapes in the RNA virus ϕ6. *Genetics* **151**, 921–927 (1999).
32. Elena, S. F., Cooper, V. S. & Lenski, R. E. Punctuated evolution caused by selection of rare beneficial mutations. *Science* **272**, 1802–1804 (1996).
33. Gerrish, P. J. & Lenski, R. E. The fate of competing beneficial mutations in an asexual population. *Genetica* **102/103**, 127–144 (1998).
34. Imhof, M. & Schlötterer, C. Fitness effects of advantageous mutations in evolving *Escherichia coli* populations. *Proc. Natl Acad. Sci. USA* **98**, 1113–1117 (2001).
35. Rozen, D. E., De Visser, J. A. G. M. & Gerrish, P. J. Fitness effects of fixed beneficial mutations in microbial populations. *Curr. Biol.* **12**, 1040–1045 (2002).
36. De Visser, J. A. G. M., Zeyl, C. W., Gerrish, P. J., Blanchard, J. L. & Lenski, R. E. Diminishing returns from mutation supply rate in asexual populations. *Science* **283**, 404–406 (1999).
37. Papadopoulos, D. *et al.* Genomic evolution during a 10,000-generation experiment with bacteria. *Proc. Natl Acad. Sci. USA* **96**, 3807–3812 (1999).
38. Shaver, A. C. *et al.* Fitness evolution and the rise of mutator alleles in experimental *Escherichia coli* populations. *Genetics* **162**, 557–566 (2002).
39. Miralles, R., Gerrish, P. J., Moya, A. & Elena, S. F. Clonal interference and the evolution of RNA viruses. *Science* **285**, 1745–1747 (1999).
40. Miralles, R., Moya, A. & Elena, S. F. Diminishing returns of population size in the rate of RNA virus adaptation. *J. Virol.* **74**, 3566–3571 (2000).
41. Cuevas, J. M., Elena, S. F. & Moya, A. Molecular basis of adaptive convergence in experimental populations of RNA viruses. *Genetics* **162**, 533–542 (2002).
42. Korona, R., Nakatsu, C. H., Forney, L. J. & Lenski, R. E. Evidence for multiple adaptive peaks from populations of bacteria evolving in a structured habitat. *Proc. Natl Acad. Sci. USA* **91**, 9037–9041 (1994).
43. Goho, S. & Bell, G. The ecology and genetics of fitness in *Chlamydomonas*. IX. The rate of accumulation of variation of fitness under selection. *Evolution* **54**, 416–424 (2000).
44. Vasi, F., Travisano, M. & Lenski, R. E. Long-term experimental evolution in *Escherichia coli*. II. Changes in life-history traits during adaptation to a seasonal environment. *Am. Nat.* **144**, 432–456 (1994).
45. Lenski, R. E. & Mongold, J. A. in *Scaling in Biology* (eds Brown, J. H. & West, G. B.) 221–235 (Oxford Univ. Press, Oxford, UK, 2000).
46. Riley, M. S., Cooper, V. S., Lenski, R. E., Forney, L. J. & Marsh T. L. Rapid phenotypic change and diversification of a soil bacterium during 1000 generations of experimental evolution. *Microbiology* **147**, 995–1006 (2001).
47. Bennett, A. F. & Lenski, R. E. Evolutionary adaptation to temperature. II. Thermal niches of experimental lines of *Escherichia coli*. *Evolution* **47**, 1–12 (1993).
48. Travisano, M. & Lenski, R. E. Long-term experimental evolution in *Escherichia coli*. IV. Targets of selection and the specificity of adaptation. *Genetics* **143**, 15–26 (1996).
49. Cooper, V. S. Long-term experimental evolution in *Escherichia coli*. X. Quantifying the fundamental and realized niche. *BMC Evol. Biol.* **2**, 12 (2002).
50. Travisano, M., Mongold, J. A., Bennett, A. F. & Lenski, R. E. Experimental tests of the roles of adaptation, chance, and history in evolution. *Science* **267**, 87–90 (1995).
51. Burch, C. L. & Chao, L. Evolvability of an RNA virus is determined by its mutational neighbourhood. *Nature* **406**, 625–628 (2000).
52. Drake, J. W. & Holland, J. J. Mutation rates among RNA viruses. *Proc. Natl Acad. Sci. USA* **96**, 13910–13913 (1999).
53. Cunningham, C. W. *et al.* Parallel molecular evolution of deletions and nonsense mutations in bacteriophage T7. *Mol. Biol. Evol.* **14**, 113–116 (1997).
54. Wichman, H. A., Badgett, M. R., Scott, L. A., Boulianne, C. M. & Bull, J. J. Different trajectories of parallel evolution during viral adaptation. *Science* **285**, 422–424 (1999).
55. Wichman, H. A., Yarber, C. D., Scott, L. A. & Bull, J. J. Experimental evolution recapitulates natural evolution. *Phil. Trans. R. Soc. Lond. B* **355**, 1–8 (2000).
A remarkable study of parallel and convergent molecular evolution that is based on the whole-genome sequencing of viruses.
56. Bull, J. J., Badgett, M. R. & Wichman, H. A. Big-benefit mutations in bacteriophage inhibited with heat. *Mol. Biol. Evol.* **17**, 942–950 (2000).
57. Lenski, R. E., Winkworth, C. L. & Riley, M. A. Rates of DNA sequence evolution in experimental populations of *Escherichia coli* during 20,000 generations. *J. Mol. Evol.* **56**, 498–508 (2003).
58. Cooper, T. F., Rozen, D. E. & Lenski, R. E. Parallel changes in gene expression after 20,000 generations of evolution in *E. coli*. *Proc. Natl Acad. Sci. USA* **100**, 1072–1077 (2003).
Whole-genome arrays show parallel changes in gene expression, which led to the discovery of beneficial mutations in an important regulatory gene.
59. Zhang, E. & Ferenci, T. OmpF changes and the complexity of *Escherichia coli* adaptation to prolonged lactose limitation. *FEMS Microbiol. Lett.* **176**, 395–401 (1999).
60. Notley-McRobb, L. & Ferenci, T. Adaptive *mgl*-regulatory mutations and genetic diversity evolving in glucose-limited *Escherichia coli* populations. *Env. Microbiol.* **1**, 33–43 (1999).
61. Notley-McRobb, L. & Ferenci, T. The generation of multiple co-existing *mal*-regulatory mutations through polygenic evolution in glucose-limited populations of *Escherichia coli*. *Env. Microbiol.* **1**, 45–52 (1999).
Candidate loci that are involved in glucose transport show diverse mutational pathways to enhanced fitness in chemostat-adapted populations.
62. Notley-McRobb, L. & Ferenci, T. Experimental analysis of molecular events during mutational periodic selections in bacterial evolution. *Genetics* **156**, 1493–1501 (2000).
63. Rosenzweig, R. F., Sharp, R. R., Treves, D. S. & Adams, J. Microbial evolution in a simple unstructured environment: genetic differentiation in *Escherichia coli*. *Genetics* **137**, 903–917 (1994).
64. Treves, D. S., Manning, S. & Adams, J. Repeated evolution of an acetate-crossfeeding polymorphism in long-term populations of *Escherichia coli*. *Mol. Biol. Evol.* **15**, 789–797 (1998).
65. Mikkola, R. & Kurland, C. G. Selection of laboratory wild-type phenotype from natural isolates of *Escherichia coli* in chemostats. *Mol. Biol. Evol.* **9**, 394–402 (1992).
66. Zambrano, M. M., Siegele, D. A., Almiron, M., Tormo, A. & Kolter, R. Microbial competition: *Escherichia coli* mutants that take over stationary phase cultures. *Science* **259**, 1757–1760 (1993).
Selection under starvation conditions favours mutations in a σ-factor.
67. Finkel, S. E. & Kolter, R. Evolution of microbial diversity during prolonged starvation. *Proc. Natl Acad. Sci. USA* **96**, 4023–4027 (1999).
68. Zinser, E. R. & Kolter, R. Prolonged stationary-phase incubation selects for *lrp* mutations in *Escherichia coli* K-12. *J. Bacteriol.* **182**, 4361–4365 (2000).
69. Kurlandzka, A., Rosenzweig, R. F. & Adams, J. Identification of adaptive changes in an evolving population of *Escherichia coli*: the role of changes with regulatory and highly pleiotropic effects. *Mol. Biol. Evol.* **8**, 261–281 (1991).
70. Ferea, T. L., Botstein, D., Brown, P. O. & Rosenzweig, R. F. Systematic changes in gene expression patterns following adaptive evolution in yeast. *Proc. Natl Acad. Sci. USA* **96**, 9721–9726 (1999).
The first application of gene-expression arrays to experimental evolution, showing parallel changes in central metabolism in three yeast lines.

71. Riehle, M. R., Bennett, A. F., Lenski, R. E. & Long, A. D. Evolutionary changes in heat-inducible gene expression in lines of *Escherichia coli* adapted to high temperature. *Physiol. Genomics* (in the press).
72. Cowen, L. E. *et al.* Evolution of drug resistance in experimental populations of *Candida albicans*. *J. Bacteriol.* **182**, 1515–1522 (2000).
73. Cowen, L. E. *et al.* Population genomics of drug resistance in *Candida albicans*. *Proc. Natl Acad. Sci. USA* **99**, 9284–9289 (2002).
74. King, M.-C. & Wilson, A. C. Evolution at two levels in humans and chimpanzees. *Science* **188**, 107–116 (1975).
75. Wilke, C. M., Maimer, E. & Adams, J. The population biology and evolutionary significance of Ty elements in *Saccharomyces cerevisiae*. *Genetica* **86**, 155–173 (1992).
76. Dunham, M. J. *et al.* Characteristic genome rearrangements in experimental evolution of *Saccharomyces cerevisiae*. *Proc. Natl Acad. Sci. USA* **99**, 16144–16149 (2002).
77. Naas, T., Blot, M., Fitch, W. M. & Arber, W. Dynamics of IS-related genetic rearrangements in resting *Escherichia coli* K-12. *Mol. Biol. Evol.* **12**, 198–207 (1995).
Shows the usefulness of transposable elements for discovering cryptic genetic changes, even in cultures that are 'stored' at room temperature.
78. Schneider, D., Duperchy, E., Coursange, E., Lenski, R. E. & Blot, M. Long-term experimental evolution in *Escherichia coli*. IX. Characterization of insertion sequence-mediated mutations and rearrangements. *Genetics* **156**, 477–488 (2000).
79. Cooper, V. S., Schneider, D., Blot, M. & Lenski, R. E. Mechanisms causing rapid and parallel losses of ribose catabolism in evolving populations of *Escherichia coli* B. *J. Bacteriol.* **183**, 2834–2841 (2001).
80. Riehle, M. M., Bennett, A. F. & Long, A. D. Genetic architecture of thermal adaptation in *Escherichia coli*. *Proc. Natl Acad. Sci. USA* **98**, 525–530 (2001).
81. Schluter, D. *The Ecology of Adaptive Radiation* (Oxford Univ. Press, Oxford, UK, 2000).
82. Levins, R. *Evolution in Changing Environments* (Princeton Univ. Press, Princeton, 1968).
83. Chao, L., Levin, B. R. & Stewart, F. M. A complex community in a simple habitat: an experimental study with bacteria and phage. *Ecology* **58**, 369–378 (1977).
84. Lenski, R. E. & Levin, B. R. Constraints on the coevolution of bacteria and virulent phage: a model, some experiments, and predictions for natural communities. *Am. Nat.* **125**, 585–602 (1985).
85. Lenski, R. E. Experimental studies of pleiotropy and epistasis in *Escherichia coli*. I. Variation in competitive fitness among mutants resistant to virus T4. *Evolution* **42**, 425–432 (1988).
86. Funchain, P., Yeung, A. Stewart, J. L., Lin, R., Slupsa, M. M. & Miller, J. H. The consequences of growth of a mutator strain of *Escherichia coli* as measured by loss of function among multiple gene targets and loss of fitness. *Genetics* **154**, 959–970 (2000).
87. Giraud, A. *et al.* Costs and benefits of high mutation rates: adaptive evolution of bacteria in the mouse gut. *Science* **291**, 2606–2608 (2001).
Pioneering work that shows that rigorous evolution experiments can be carried out in animal hosts, and also indicates important effects of mutator genotypes.
88. Mongold, J. A., Bennett, A. F. & Lenski, R. E. Evolutionary adaptation to temperature. IV. Adaptation of *Escherichia coli* at a niche boundary. *Evolution* **50**, 35–43 (1996).
89. Travisano, M., Vasi, F. & Lenski, R. E. Long-term experimental evolution in *Escherichia coli*. III. Variation among replicate populations in correlated responses to novel environments. *Evolution* **49**, 189–200 (1995).
90. Travisano, M. Long-term experimental evolution in *Escherichia coli*. VI. Environmental constraints on adaptation and divergence. *Genetics* **146**, 471–479 (1997).
91. Dykhuizen, D. & Hartl, D. Evolution of competitive ability in *Escherichia coli*. *Evolution* **35**, 581–594 (1981).
92. Velicer, G. J. & Lenski, R. E. Evolutionary tradeoffs under conditions of resource abundance and scarcity: experiments with bacteria. *Ecology* **80**, 1168–1179 (1999).
93. Reboud, X. & Bell, G. Experimental evolution in *Chlamydomonas*. III. Evolution of specialist and generalist types in environments that vary in space and time. *Heredity* **78**, 507–514 (1997).
94. Novella, I. S. *et al.* Extreme fitness differences in mammalian and insect hosts after continuous replication of vesicular stomatitis virus in sandly cells. *J. Virol.* **69**, 6805–6809 (1995).
95. Crill, W. D., Wichman, H. A. & Bull, J. J. Evolutionary reversals during viral adaptation to alternating hosts. *Genetics* **154**, 27–37 (2000).
96. Turner, P. E. & Elena, S. F. Cost of host radiation in an RNA virus. *Genetics* **156**, 1465–1670 (2000).
A study, using viruses, that shows the tradeoffs on the original host during adaptation to new hosts, as well as selection for generalists in fluctuating environments.
97. Cooper, L. A. & Scott, T. W. Differential evolution of eastern equine encephalitis virus populations in response to host cell type. *Genetics* **157**, 1403–1412 (2001).
98. Weaver, W. C., Brault, A. C., Kang, W. & Holland, J. J. Genetic and fitness changes accompanying adaptation of an arbovirus to vertebrate and invertebrate cells. *J. Virol.* **73**, 4316–4326 (1999).
99. Novella, I. S., Hershey, C. L., Escarmís, C., Domingo, E. & Holland, J. J. Lack of evolutionary stasis during alternating replication of an arbovirus in insect and mammalian cells. *J. Mol. Biol.* **287**, 459–465 (1999).
100. Ebert, D. Experimental evolution of parasites. *Science* **282**, 1432–1435 (1998).
101. Friedberg, E. C., Walker, G. C. & Siede, W. *DNA Repair and Mutagenesis* (ASM, Washington, 1995).
102. Chao, L., Vargas, C., Spear, B. B. & Cox, E. C. Transposable elements as mutator genes in evolution. *Nature* **303**, 633–635 (1983).
103. Taddei, F. *et al.* Role of mutator alleles in adaptive evolution. *Nature* **387**, 700–702 (1997).
A theoretical model that shows the complex dynamics and effects of alleles that cause increased mutation rates in asexual populations.
104. LeClerc, J. E., Li, B., Payne, W. L. & Cebula, T. A. High mutation frequencies among *Escherichia coli* and *Salmonella* pathogens. *Science* **274**, 1208–1211 (1996).
105. Oliver, A., Canton, R., Campo, P., Baquero, F. & Blázquez, J. High frequency of hypermutable *Pseudomonas aeruginosa* in cystic fibrosis lung infection. *Science* **288**, 1251–1253 (2000).
106. Mao, E. F., Lane, L. Lee, J. & Miller, J. H. Proliferation of mutators in a cell population. *J. Bacteriol.* **179**, 417–422 (1997).
107. Sniegowski, P. D., Gerrish, P. J. & Lenski, R. E. Evolution of high mutation rates in experimental populations of *E. coli*. *Nature* **387**, 703–705 (1997).
This article documents the spontaneous evolution of repair-defective mutators during long-term adaptation to a new environment.
108. Notley-McRobb, L., Seeto, S. & Ferenci, T. F. Enrichment and elimination of *mutY* mutators in *Escherichia coli* populations. *Genetics* **162**, 1055–1062 (2002).
109. Chao, L. & Cox, E. C. Competition between high and low mutating strains of *Escherichia coli*. *Evolution* **37**, 125–134 (1983).
An elegant experiment showing how the fate of a mutator allele depends, in a stochastic manner, on its abundance in a population.
110. Lenski, R. E. Phenotypic and genomic evolution during a 20,000-generation experiment with the bacterium, *Escherichia coli*. *Plant Breed. Rev.* (in the press).
111. Muller, H. J. The relation of recombination to mutational advance. *Mut. Res.* **1**, 2–9 (1964).
112. Haigh, J. The accumulation of deleterious genes in a population: Muller's ratchet. *Theor. Pop. Biol.* **14**, 251–267 (1978).
113. Bateman, A. J. The viability of near-normal irradiated chromosomes. *Int. J. Radiat. Biol.* **1**, 170–180 (1959).
114. Keightley, P. D. Inference of genome-wide mutation rates and distribution of mutation effects for fitness traits: a simulation study. *Genetics* **150**, 1283–1293 (1998).
115. Mukai, T. The genetic structure of natural populations of *Drosophila melanogaster*. VII. Synergistic interaction of spontaneous mutant polygenes controlling viability. *Genetics* **61**, 749–761 (1969).
116. Chao, L. Evolution of sex in RNA viruses. *J. Theor. Biol.* **133**, 99–112 (1988).
117. Kondrashov, A. S. Deleterious mutations and the evolution of sexual reproduction. *Nature* **336**, 435–440 (1988).
118. Kondrashov, A. S. Classification of hypotheses on the advantage of amphimixis. *J. Hered.* **84**, 372–387 (1993).
119. Lynch, M. & Gabriel, W. Mutation load and the survival of small populations. *Evolution* **44**, 1725–1737 (1990).
120. Moran, N. A. Accelerated evolution and Muller's ratchet in endosymbiotic bacteria. *Proc. Natl Acad. Sci. USA* **93**, 2873–2878 (1996).
121. Zeyl, C., Mizesko, M. & De Visser, J. A. G. M. Mutational meltdown in laboratory yeast populations. *Evolution* **55**, 909–917 (2001).
122. Chao, L. Fitness of RNA virus decreased by Muller's ratchet. *Nature* **348**, 454–455 (1990).
An early study that shows the rapid decline in fitness of populations propagated through severe bottlenecks.
123. Duarte, E. A., Clarke, D. K., Moya, A., Domingo, E. & Holland, J. J. Rapid fitness losses in mammalian RNA virus clones due to Muller's ratchet. *Proc. Natl Acad. Sci. USA* **89**, 6015–6019 (1992).
124. Escarmís, C. *et al.* Genetic lesions associated with Muller's ratchet in a RNA virus. *J. Mol. Biol.* **264**, 255–267 (1996).
125. Elena, S. F. & Moya, A. Rate of deleterious mutation and the distribution of its effects on fitness in vesicular stomatitis virus. *J. Evol. Biol.* **12**, 1078–1088 (1999).
126. de la Peña, M., Elena, S. F. & Moya, A. Effect of deleterious mutation-accumulation on the fitness of RNA bacteriophage MS2. *Evolution* **54**, 686–691 (2000).
127. Yuste, E., Sánchez-Palomino, S., Casado, C., Domingo, E. & López-Galíndez, C. Drastic fitness loss in human immunodeficiency virus type 1 upon serial bottleneck events. *J. Virol.* **73**, 2745–2751 (2000).
128. Kibota, T. T. & Lynch, M. Estimate of the genomic mutation rate deleterious to overall fitness in *E. coli*. *Nature* **381**, 694–696 (1996).
129. Fares, M. A., Ruiz-González, M. X., Moya, A., Elena, S. F. & Barrio, E. Endosymbiotic bacteria: GroEL buffers against deleterious mutations. *Nature* **417**, 398 (2002).
130. Wloch, D. M., Szafraniec, K., Borts, R. H. & Korona, R. Direct estimate of the mutation rate and the distribution of fitness effects in the yeast *Saccharomyces cerevisiae*. *Genetics* **159**, 441–452 (2001).
131. Zeyl, C. & De Visser, J. A. G. M. Estimates of the rate and distribution of fitness effects of spontaneous mutation in *Saccharomyces cerevisiae*. *Genetics* **157**, 53–61 (2001).
132. Wilke, C. O, Wang, J., Ofria, C., Lenski, R. E. & Adami, C. Evolution of digital organisms at high mutation rate leads to survival of the flattest. *Nature* **412**, 331–333 (2001).
133. Moore, F. B.-G., Rozen, D. E. & Lenski, R. E. Pervasive compensatory adaptation in *Escherichia coli*. *Proc. R. Soc. Lond. B* **267**, 515–522 (2000).
134. Endy, D., You, L. C., Yin, J. & Molineux, I. J. Computation, prediction, and experimental tests of fitness for bacteriophage T7 mutants with permuted genomes. *Proc. Natl Acad. Sci. USA* **97**, 5375–5380 (2000).
135. You, L. C. & Yin, J. Dependence of epistasis on environment and mutation severity as revealed by *in silico* mutagenesis of phage T7. *Genetics* **160**, 1273–1281 (2002).
136. Ibarra, R. U., Edwards, J. S. & Palsson, B. O. *Escherichia coli* K-12 undergoes adaptive evolution to achieve *in silico* predicted optimal growth. *Nature* **420**, 186–189 (2002).
137. Lenski, R. E., Ofria, C., Pennock, R. T. & Adami, C. The evolutionary origin of complex features. *Nature* **423**, 139–144 (2003).

Acknowledgements
We thank the reviewers for comments, and N. Hajela for the photographs used in Box 2. S.F.E. is supported at present by the Spanish Consejo Superior de Investigaciones Científicas. R.E.L. is funded by the National Science Foundation and the National Aeronautics and Space Administration.

Online links

DATABASES
The following terms in this article are linked online to:
BioCyc: http://biocyc.org/ecocyc
acs | GroEL | OmpF | *ompF* | *rpoS* | *spoT*

FURTHER INFORMATION
Adam K. Chippindale's website:
http://www.achippindale.com/ee
E. coli Long-term Experimental Evolution Project Site:
http://myxo.css.msu.edu/ecoli
Access to this interactive links box is free online.

Epilogue *Science & the Public*

The overwhelming majority of scientists accept evolution. Those who know professionally the evidence for evolution cannot deny it. Scientists agree that the evolutionary origin of animals, plant, and other organisms is a scientific conclusion beyond reasonable debate. The evidence is compelling and all-encompassing, because it comes from all biological disciplines including those—such as genetics and molecular biology—that did not exist in Darwin's time. Evolutionary research nowadays seeks greater knowledge of how evolution occurs, the causes and mechanisms of the process, as the 48 seminal papers in this book attest. Evolutionary biology has been and continues to be a cornerstone of modern science. Moreover, an understanding of evolution has made major contributions to human well-being, to preventing and treating human disease, to developing new plant crops and animal breeds, and to creating new industrial products.

The scientific understanding of evolutionary biology differs dramatically from the *perception* of evolution by the general public. Opinion polls conducted during the past three decades consistently show that about one-third of adults in first world countries firmly reject the concept of biological evolution, and even more people remain skeptical. Public acceptance of evolution in the United States is lower than in Japan and in 32 of 33 surveyed countries (only Turkey has lower acceptance), apparently because of "widespread fundamentalism and the politization of science in the United States" (Miller et al. 2006). One unfortunate consequence of societal resistance to evolution was identified in a report issued jointly by the U.S. National Academy of Sciences and the Institute of Medicine: "The pressure to downplay evolution or emphasize nonscientific alternatives in public schools compromises science education" (National Academy of Sciences 2008). Evolution tends to be contentious in society, because it concerns the origins of humans and biodiversity, which are also central concerns of religion. However, theologians and many scientists have pointed out that evolution and faith need not be in conflict.

Low public esteem for the evolutionary sciences contrasts with the higher level of respect afforded other hard sciences such as physics, astronomy, and geology, where few lay people feel compelled to challenge basic discoveries such as the theory of gravity, the heliocentric theory of the solar system, or the theory of plate tectonics. Evolution is today a scientific conclusion established with the kind of certainty attributable to such concepts as the roundness of the Earth, the motions of the planets, and the molecular composition of matter. New discoveries are not likely to challenge the evolutionary origin of organisms. Evolutionary research continues, however, as scientists

Many statements from both scientists and theologians have been issued to the effect that the evolutionary sciences and religion can peacefully coexist (National Academy of Sciences 2008). Consider for example the following statement that was signed by more than 10,000 clergy members:

> Christian clergy from many different traditions believe that the timeless truths of the Bible and the discoveries of modern science may comfortably coexist. We believe that the theory of evolution is a foundational scientific truth, one that has stood up to rigorous scrutiny and upon which much of human knowledge and achievement rests. To reject this truth or to treat it as "one theory among others" is to deliberately embrace scientific ignorance and transmit such ignorance to our children. We believe that among God's good gifts are human minds capable of critical thought and that the failure to fully employ this gift is a rejection of the will of our Creator . . . We urge school board members to preserve the integrity of the science curriculum by affirming the teaching of the theory of evolution as a core component of human knowledge.

seek to ascertain the origin and relationships of particular groups of organisms, as well as to further understand how and why evolution takes place. New lines of inquiry and new technologies (such as from molecular biology) enable observations or experiments that previously were impossible.

The papers reprinted in this volume describe scientific findings and concepts that materially modified contemporary evolutionary wisdom, thus contributing substantively to the evolution of evolutionary biology. Modification in the face of new evidence is a hallmark of science and a key feature that distinguishes scientific exploration from nonscientific modes of understanding. We hope that this collection of reprinted articles has enlightened readers not merely about the history and the current knowledge in evolutionary biology but importantly also about the vibrant dynamic nature of scientific investigation in the evolutionary sciences.

SOURCES AND FURTHER READING

Miller, J. D., E. C. Scott, and S. Okamoto. 2006. Public acceptance of evolution. Science 313:765–766.

National Academy of Sciences. 2008. Science, evolution, and creationism. National Academies Press, Washington, D.C., USA.

Author Index

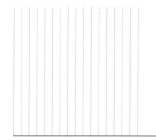

These images are reproduced in color for clarity.

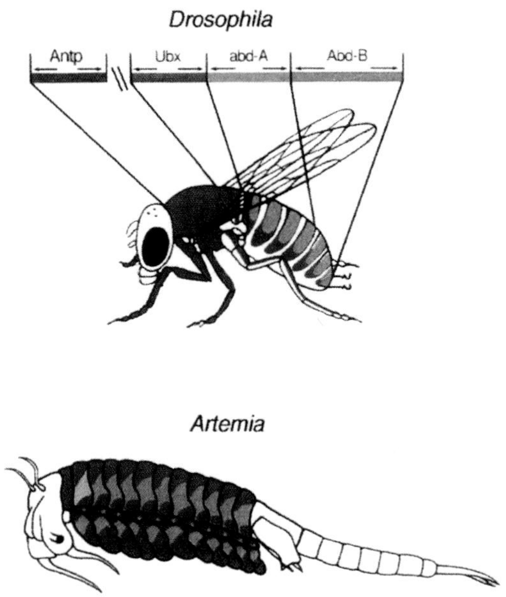

FIG. 3 Evolution of *Hox* gene regulation and the crustacean and insect body plans. The domains of BX-C gene expression have shifted with respect to each other in the thorax and abdomen of insects and crustacea. Top, the *Antp* (purple), *Ubx* (brown), *abd-A* (orange) and *Abd-B* (green) genes sculpt the morphology of the insect trunk. Bottom, *Ubx* and *abd-A* are expressed in relatively more anterior positions in crustacea[36].

Figure on page 471

Fig. 1. The classic scenario of an ecological speciation event, from beginning to end. Reproductive isolation builds in allopatry (green) as an incidental by-product of adaptation to alternative environments (by-product mechanism). Reinforcement of premating isolation, driven by reduced hybrid fitness, completes the speciation process during the sympatric phase (blue). The timing of secondary contact is flexible (indicated by arrows at the boundary between the allopatric and sympatric phases).

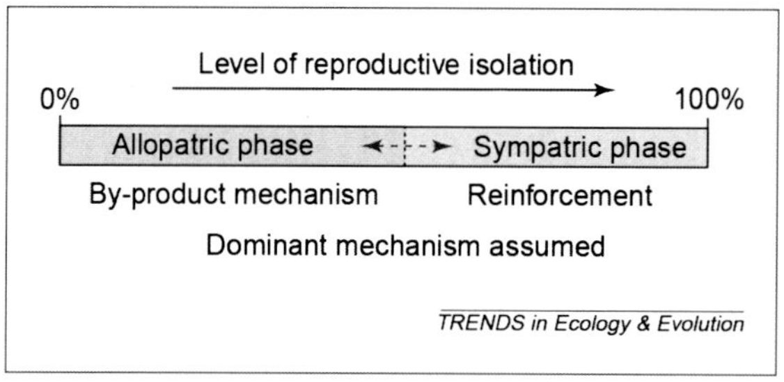

Fig. 2. Mating compatibility of independent experimental lines of *Drosophila* raised separately over multiple generations in similar or in different environments. Circles represent the proportion of mating events that occurred between individuals from different lines relative to intra-line matings. Data are from *D. pseudoobscura*[6] (green symbols) and *D. melanogaster*[5] (blue symbols). Modified from a figure to be published by Cambridge University Press (Ref. 64).

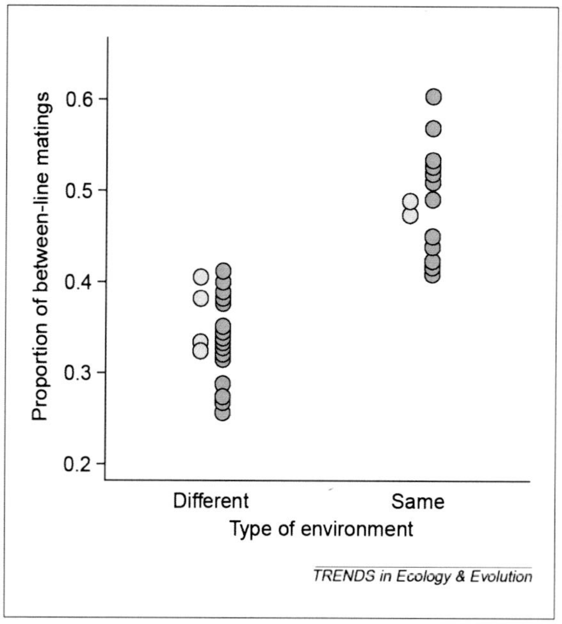

Figures on page 501

Fig. 3. A phylogenetic tree showing the possible correlation between ecological diversification and speciation rate in the Darwin's finches (*Geospiza* spp.). The high diversity of beak traits among species within the CLADE of tree and ground finches (outlined in red) contrasts with the lower diversity of beak traits among species of the three older LINEAGES. Speciation rates are also highly uneven, being significantly greater in the tree and ground finch clade than in the rest of the tree ($P = 0.011$, calculated using the Nee *et al.*[62] equal-rates test for multiple lineages). Data are taken from Ref. 63. Bird images are reproduced, with permission, from Ref. 65.

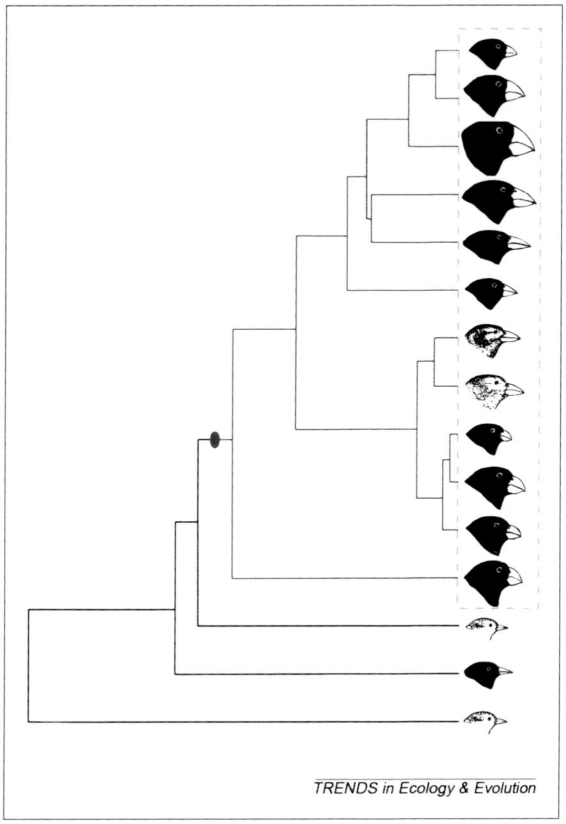

Fig. 4. Parallel evolution of premating isolation between benthic and limnetic threespine sticklebacks (*Gasterosteus* spp.) from three lakes. Each circle indicates frequency of matings between males and females from a pair of populations, measured in no-choice laboratory mating trials. Pairs of populations from the same lake are indicated in green; values in column D are mating frequencies between males and females from the same population. Pairs of populations from different lakes are indicated in blue. Pairs of populations occurring in the same type of environment (C and D) mate with higher frequency than do pairs of populations from different types of environments (A and B). Modified, with permission, from Ref. 55.

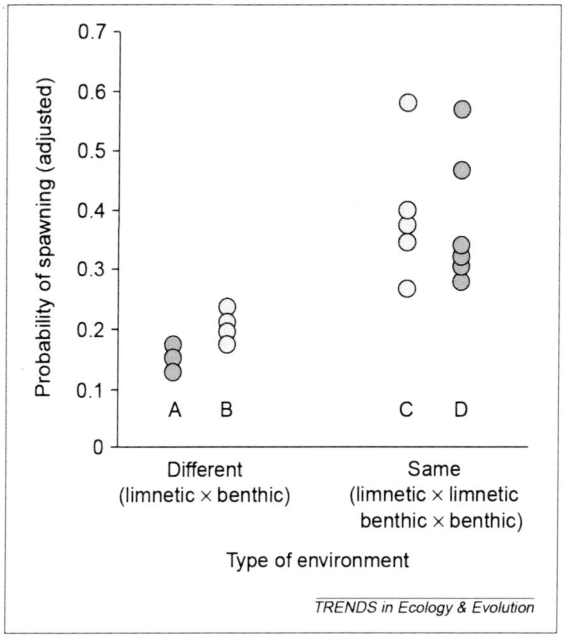

Figures on pages 503 and 504

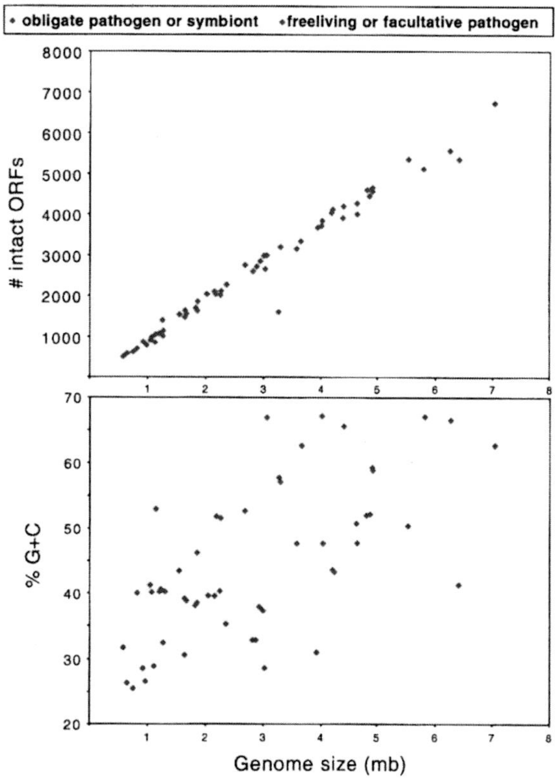

Figure 1. Size of Eubacterial Genomes in Relation to Number of intact ORFs (top) and % G+C Content in Genome (bottom)

The red spots correspond to obligate pathogens, belonging to a variety of unrelated lineages. *M. leprae* contains inactivated genes not included in the tally of intact ORFs. Included are all eubacterial genomes available January 2002 in NCBI Entrez Genomes (http://www.ncbi.nlm.nih.gov:80/PMGifs/Genomes/eub_g.html), with numbers of intact ORFS taken from this annotation.

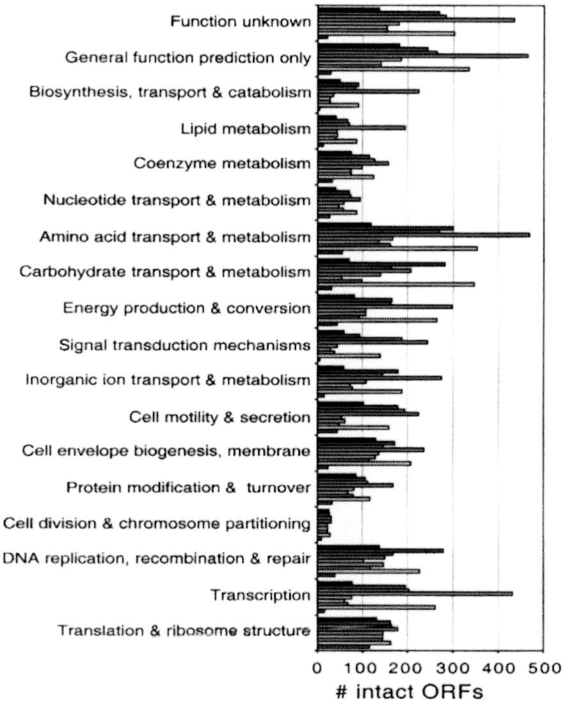

Figure 2. Numbers of ORFs Assigned to Different Functional Categories for Fully Sequenced Genomes of Members of the β- and γ-Proteobacteria

From top, they are *Xylella fastidiosa*, *Yersinia pestis*, *Vibrio cholerae*, *Pseudomonas aeruginosa*, *Pasteurella multocida*, *Neisseria meningitidus*, *Haemophilus influenzae*, *Escherichia coli* K12, and *Buchnera aphidicola*. *Buchnera*, the only organism in this group that is obligately associated with hosts and that has a highly reduced genome, shows reduced numbers of ORFs in all categories. Categories are those corresponding to the COGs database (Tatusov, R.L., Koonin, E.V., and Lipman, D.J., 1997, Science *278*, 631-637, http://www.ncbi.nlm.nih.gov/COG/xindex.html).

Figures on page 510

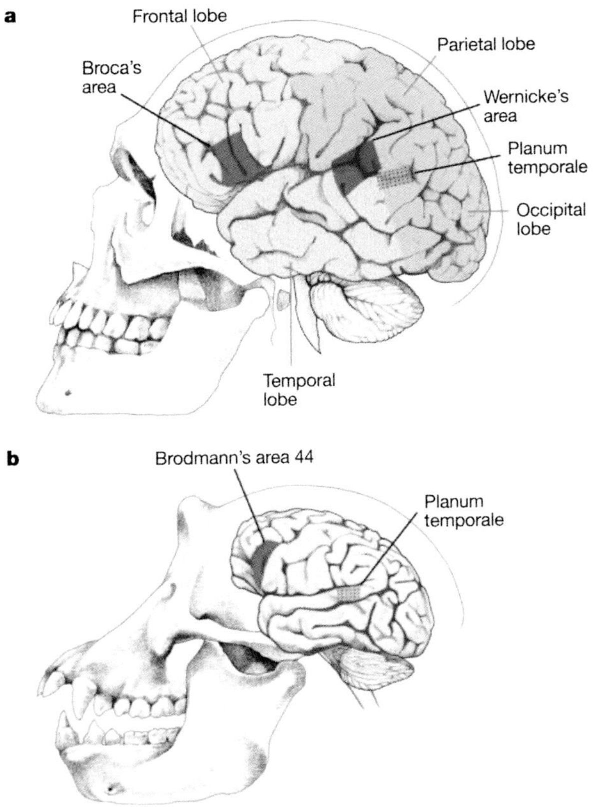

Figure 2 Comparative neuroanatomy of humans and chimpanzees. Lateral views of the left hemispheres of a modern human and a chimpanzee brain. Although the overall skull sizes are roughly comparable, the human cranial capacity and brain are much larger. **a**, Two areas of the human brain that are associated with communication are shown: Broca's area in the frontal lobe and Wernicke's area, which overlaps the posterior temporal lobe and parts of the parietal lobe. In the left hemisphere, Broca's area is larger, as is the planum temporale, which lies below the surface in Wernicke's area. **b**, These asymmetries have been found in corresponding regions of chimpanzee brains[15,17], suggesting that the areas in humans might be elaborations of a pre-existing communication centre in a common ancestor of apes and humans.

Figure on page 523

Box 2 | **Measuring fitness**

The fitness of an evolved type is generally expressed relative to its ancestor. Relative fitness is measured by allowing the ancestral and evolved types to compete with one another. Unless otherwise specified, the competition environment is the same as that used for the experimental evolution. The following description presents the protocol used in the long-term serial-transfer experiment with *Escherichia coli*[13,25,26], but similar procedures are used in experiments with many microorganisms.

The two competitors are grown separately in the competition environment to ensure that they are comparably acclimated to the test conditions. They are then mixed (usually at a 1:1 ratio) and diluted (100-fold in this case) in the competition environment. Initial densities at timepoint $t = 0$ are estimated by diluting and spreading the cells on an indicator agar that distinguishes the evolved and ancestral types by colony colour, which differs owing to an engineered marker that is selectively neutral. In this case, red and white colonies correspond to Ara^- and Ara^+ phenotypes, respectively. After one day ($t = 1$) (corresponding to the serial-transfer cycle in the evolution experiment), final densities are estimated by plating cells, as before, on the indicator agar. The growth rate of each competitor is calculated as the natural logarithm of the ratio of its final density to its initial density (adjusted for dilution during plating). Relative fitness is then defined simply as the ratio of the realized growth rates of the evolved and ancestral types.

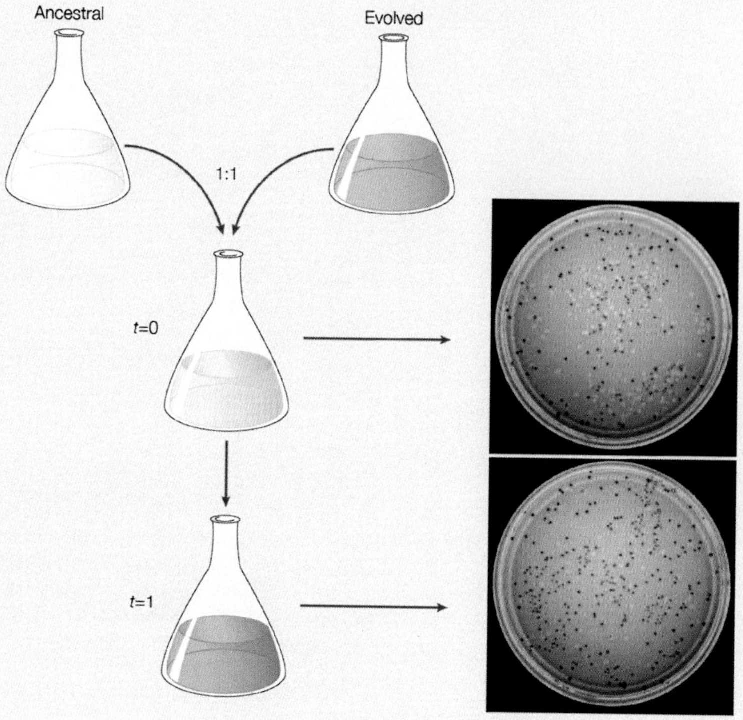

Figure on page 532

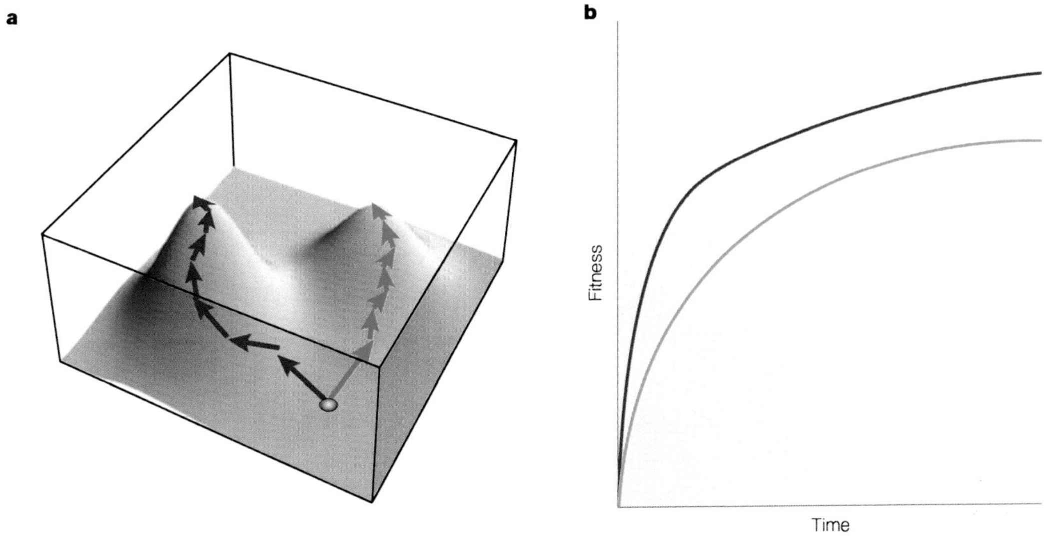

Figure 1 | **Fitness landscapes and evolutionary dynamics. a** | A hypothetical adaptive landscape with two fitness peaks. The red and green arrows show two of the possible trajectories for replicate populations that are founded from the same ancestral genotype and evolve independently in the same environment. **b** | Beneficial substitutions tend to have larger effects early in an experiment, when a population is far from an adaptive peak, than later as it approaches a local peak. Two replicate populations might reach different final fitness levels if, through the random effects of mutation and drift, they move into the domains of attraction of peaks of unequal height.

Figure on page 533

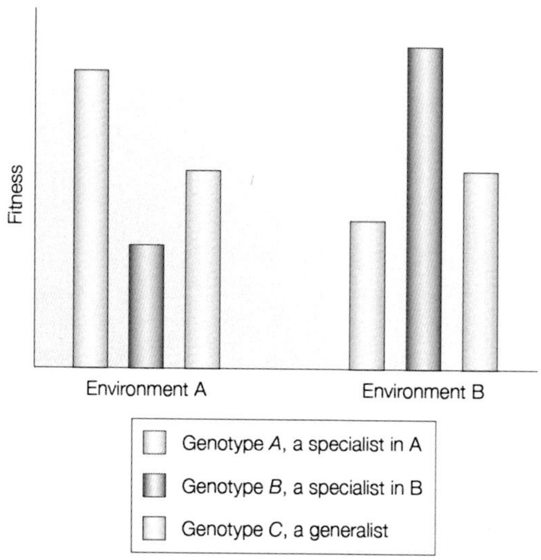

Figure 2 | **Tradeoffs and specificity of adaptation.** The fitness of three genotypes in two environments. The blue and red bars show specialists that are adapted to environments A and B, respectively. Although both genotypes perform well in their respective environments, each is poorly adapted to the other environment. One possible explanation for this negative fitness correlation, or tradeoff, is that those mutations that are beneficial in one environment have antagonistic pleiotropic effects in the other. The green bars show a generalist that performs moderately well in both environments but has lower fitness than either specialist in its preferred environment. If environments vary in space or time, the generalist might have an overall advantage.

Figure on page 536